Dioxins and Health

Dioxins and Health

Edited by

Arnold Schecter

State University of New York
Health Science Center at Syracuse–
Clinical Campus at Binghamton
Binghamton, New York

Plenum Press • New York and London

Library of Congress Cataloging-in-Publication Data

Dioxins and health / edited by Arnold Schecter.
p. cm.
Includes bibliographical references and index.
ISBN 0-306-44785-1
1. Dioxins--Toxicology. I. Schecter, Arnold.
RA1242.D55D58 1994
615.9'512--dc20 94-40007
CIP

ISBN 0-306-44785-1

A Division of Plenum Publishing Corporation
233 Spring Street, New York, N.Y. 10013

Printed in the United States of America

Contributors

Olav Axelson • Department of Occupational Medicine, University Hospital, S-581 85 Linköping, Sweden

Pier Alberto Bertazzi • Institute of Occupational Health, Epidemiology Section, University of Milan, 20122 Milan, Italy

Linda S. Birnbaum • Environmental Toxicology Division, Health Effects Research Laboratory, United States Environmental Protection Agency, Research Triangle Park, North Carolina 27711

Yung-Cheng Joseph Chen • Department of Psychiatry, National Cheng Kung University Medical College, Tainan 70428, Taiwan, Republic of China

George C. Clark • National Institute of Environmental Health Sciences, Laboratory of Biochemical Risk Analysis, Research Triangle Park, North Carolina 27709

Barry Commoner • Center for the Biology of Natural Systems, Queens College–CUNY, Flushing, New York 11367

Peter L. deFur • Environmental Defense Fund, Washington, D.C. 20009

Michael J. DeVito • Center for Environmental Medicine and Lung Biology, University of North Carolina at Chapel Hill, Chapel Hill, North Carolina 27514

Alessandro di Domenico • Laboratory of Comparative Toxicology and Ecotoxicology, Italian National Institute of Health, 00161 Rome, Italy

Mikael Eriksson • Department of Oncology, University Hospital, S-901 85 Umeå, Sweden

John P. Giesy • Department of Fisheries and Wildlife, Pesticide Research Center and Institute of Environmental Toxicology, Michigan State University, East Lansing, Michigan 48824

Yue-Liang Leon Guo • Department of Environmental and Occupational Health, National Cheng Kung University Medical College, Tainan 70428, Taiwan, Republic of China

Lennart Hardell • Department of Oncology, Örebro Medical Centre, S-701 85 Örebro, Sweden

Chen-Chin Hsu • Department of Psychiatry, National Cheng Kung University Medical College, Tainan 70428, Taiwan, Republic of China

James Huff • Environmental Carcinogenesis Program, National Institute of Environmental Health Sciences, Research Triangle Park, North Carolina 27709

Nancy I. Kerkvliet • Department of Agricultural Chemistry and Environmental Health Sciences Center, Oregon State University, Corvallis, Oregon 97331

Philip J. Landrigan • Department of Community Medicine, Mount Sinai School of Medicine, New York, New York 10029

George W. Lucier • National Institute of Environmental Health Sciences, Laboratory of Biochemical Risk Analysis, Research Triangle Park, North Carolina 27709

James P. Ludwig • The SERE Group, Ltd., Victoria, British Columbia, Canada V8P 3C8

Yoshito Masuda • Daiichi College of Pharmaceutical Sciences, Fukuoka 815, Japan

William J. Nicholson • Department of Community Medicine, Mount Sinai School of Medicine, New York, New York 10029

James R. Olson • Department of Pharmacology and Toxicology, State University of New York at Buffalo, Buffalo, New York 14214

Richard E. Peterson • School of Pharmacy and Environmental Toxicology Center, University of Wisconsin, Madison, Wisconsin 53706

Christoffer Rappe • Institute of Environmental Chemistry, University of Umeå, S-901 87 Umeå, Sweden

Walter J. Rogan • Intramural Research Program, National Institute of Environmental Health Sciences, Research Triangle Park, North Carolina 27709

Susan L. Schantz • Institute for Environmental Studies, University of Illinois at Urbana–Champaign, Urbana, Illinois 61801

Arnold Schecter • Department of Preventive Medicine, University of New York Health Science Center-Syracuse, Binghamton, New York 13903

Richard F. Seegal • New York State Department of Health, Wadsworth Center for Laboratories and Research, Albany, New York 12201, and School of Public Health, University at Albany, State University of New York, Albany, New York 12203

Ellen K. Silbergeld • Environmental Defense Fund, Washington, D.C., 20009, and Department of Epidemiology and Preventive Medicine, University of Maryland Medical School, Baltimore, Maryland 21201

James R. Startin • Ministry of Agriculture, Fisheries and Food, Food Science Laboratory, Norwich Research Park, Colney, Norwich NR4 7UQ, United Kingdom

Anne Sweeney • School of Public Health, University of Texas Health Science Center at Houston, Houston, Texas 77025

H. Michael Theobald • School of Pharmacy, University of Wisconsin, Madison, Wisconsin 53706

Donald E. Tillitt • National Fisheries Contaminant Research Center, U.S. Fish and Wildlife Service, Columbia, Missouri 65201

Angelika M. Tritscher • National Institute of Environmental Health Sciences, Laboratory of Biochemical Risk Analysis, Research Triangle Park, North Carolina 27709

Mary K. Walker • School of Pharmacy, University of Wisconsin, Madison, Wisconsin 53706

Thomas Webster • Department of Environmental Health, Boston University School of Public Health, Boston, Massachusetts 02118

Mei-Lin M. Yu • Department of Public Health, National Cheng Kung University Medical College, Tainan 70428, Taiwan, Republic of China

Sheila Hoar Zahm • Occupational Studies Section, Environmental Epidemiology Branch, Division of Cancer Etiology, National Cancer Institute, Rockville, Maryland 20892

Douglas R. Zook • Institute of Environmental Chemistry, University of Umeå, S-901 87 Umeå, Sweden. *Present address:* University of Groningen, Mass Spectrometry Center, 9713 AW Groningen, The Netherlands

Preface

This book originated in a series of cross-disciplinary conversations in the years 1984–1990 between the editor, who is a physician–researcher involved in clinical and laboratory research, and a dioxin toxicologist.

During the years in which the conversations took place, an extraordinary amount of new scientific literature was published related to dioxins, defined for purposes of this text as the chlorinated dibenzo-*p*-dioxins, dibenzofurans, polychlorinated biphenyls (PCB's) and other compounds that are structurally and toxicologically similar to 2,3,7,8-tetrachlorodibenzo-*p*-dioxin (2,3,7,8-TCDD), the most extensively studied and most toxic of this group of chemicals. Dioxins also began to interest not only chemists and toxicologists, but also specialists from diverse disciplines such as wildlife and environmental science, immunology, neuroscience, public health, epidemiology, medicine, government, law, sociology, and journalism.

Specialists from such varied disciplines, while familiar with their own literature, frequently did not have time to follow the dioxin literature outside their specialty area. In addition, each specialty had unique knowledge, methods, and perspectives. Cross-disciplinary conversation was necessary, but all too frequently, specialists from the various disciplines did not speak the same language, resulting in misunderstanding. This text was written to help facilitate cross-disciplinary discourse in the dioxin field both by providing a one volume summary of the new and cutting edge dioxin research of the past few years and by offering a perspective as to how this data may relate to real issues of human and environmental health. Most major scientific areas involving dioxins have been included; however, the coverage has been somewhat condensed deliberately in order not to overwhelm the reader.

The chapter authors are all scientists with international reputations in the dioxin scientific arena. They have been asked to address their chapters to an audience of well-educated and intelligent professionals who may not be familiar with the authors' specialties, and have included highlights of their fields with a representative, but not necessarily encyclopedic, sampling of important references. Through this approach, the book should be accessible to the broadest range of health professionals and non-health professionals interested in dioxins, as well as to policy makers and the general public.

Since the middle of this century, dioxins have been wide-spread and persistent environmental contaminants in the United States and other industrial countries. Because of their ubiquity, persistence, and known toxicity in animal laboratory studies, considerable concern arose regarding their presence in the food chain and in human tissue, and in the multiple types of potential health consequences in humans and animals, such as cancer, immune system compromise, reproductive and developmental disorders, neurological damage and endocrine system alterations. Although considerable controversy exists at the present time in many areas of dioxin research, with increasing human as well as laboratory studies, the extent of dioxin toxicity has become better characterized in many species, including humans.

In their introductory overview of the dioxin debate, Thomas Webster and Barry Commoner provide a summary of some of the major current dioxin controversies. A review of the current understandings and controversies in dioxin risk assessment is presented in considerable depth by Ellen Silbergeld and Peter deFur. Douglas Zook and Christopher Rappe, in reviewing chemical aspects of dioxins, examine the question of the environmental sources of dioxins, noting that the source of much (or most) dioxin production and discharge is unknown at this time. James Startin provides a review of what is currently known with respect to dioxins in the food chain, which is the immediate source of over 90% of the dioxin body burden in the general population in industrial countries. He describes the challenges involved in both analyzing the presence of dioxin in foods and in interpreting existing studies.

In separate articles, Michael DeVito and Linda Birnbaum; James Olson; Nancy Kerkvliet; and Angelika Tritscher, George Clark, and George Lucier; many of whom were responsible for generating the original data presented in their chapters, discuss the toxicology, pharmacokinetics, immunotoxicology, and dose-response of dioxins and related chemicals at the cellular and molecular level, based on a variety of toxicological and biochemical studies. The discussion regarding the threshold level of dioxins, below which there may be no serious health effects, is of particular importance in light of the present research and debate in this area. John P. Giesy, James Ludwig, and Donald Tillitt; and Mary Walker and Richard Peterson review the research on the effects of dioxin and related chemicals on wildlife health in general, and in biota in aquatic environments. The authors observe that many biological outcomes and mechanisms are similar between species; however, correlation between the effects of dioxins on other animal species and humans is still not completely characterized.

Specific laboratory and related findings on the developmental and reproductive toxicity of dioxins, dioxins and mammalian carcinogenesis, and the neurochemical and behavioral sequelae of exposure to dioxins and PCB's are discussed respectively in articles by H. Michael Theobald and Richard Peterson; James Huff; and Richard Seegal and Susan Schantz. The findings of Theobald and Peterson that low level maternal exposure of rats to dioxin at a sensitive time during pregnancy can lead to behavioral, reproductive, and endocrine alterations in male offspring are especially significant when viewed in conjunction with the discussion of human epidemiological studies in later chapters.

The final chapters in the text are devoted to more specific discussion of dioxins and related chemicals and human health. The difficulties in measuring dioxins and related chemicals in human tissue and the current "state of the art" are reviewed in a chapter by the editor, followed by a comprehensive overview of the literature on the human health effects of polychlorinated biphenyls provided by William J. Nicholson and Phillip J. Landrigan. The epidemiological studies of dioxins and cancer and dioxin and reproduction are described in greater depth in respective articles by Lennart Hardell, Mikael Eriksson, Olav Axelson and Sheila Hoar Zahm; and Anne Sweeney. The work concludes with reviews and analyses of three major incidents of dioxin exposure in humans by scientists who have been long involved in their study. Pier Bertazzi and Alessandro di Domenico discuss the dioxin industrial incident in Seveso, Italy in 1976. Yoshito Masuda describes the Japanese "Yusho" rice oil poisoning of 1968, and Chen-Chin Hsu, Mei-Lin M. Yu, Yung-Cheng Joseph Chen, Yue-Liang Leon Guo, and Walter Rogan review and interpret the Taiwan Yu-cheng rice oil poisoning of 1979.

While differences of opinion are reflected in the text, the authors generally agree that dioxins are highly dangerous and persistent synthetic chemicals which may cause a large number of different types of health consequences, and which can initiate harmful biological responses in a dose-dependent fashion. Dioxins appear to be causally linked to cancers, adverse reproductive and developmental outcomes, and immune system compromise. They have also been associated with endocrine, neurological, liver and skin damage, which may occur at relatively low doses in genetically sensitive individuals. The very long half lives of dioxins, especially 2,3,7,8-TCDD, the most toxic of the dioxins, in the environment and in humans are cause for concern. It is not known, however, whether all biological effects of dioxins are harmful, or whether some are attempts to adapt. Also, there are sex differences in cancer response and other dioxin toxic end points.

It is now relatively well established that dioxins interact with the cytoplasmic aryl hydrocarbon, or aH receptor, to initiate a cascade of changes in signal transduction and gene expression. Because the aH receptor may have some normal functions, action via a transcription factor, and likely via altered cell differentiation and proliferation, may well provide significant information about disease states. Dioxin may be toxic due to the inappropriate movement of the dioxin receptor complex at an inappropriate time and/or for an inappropriate period. One exciting recent development in the field of dioxin research reviewed in these chapters is the research into the mechanisms for dioxin activity and the consideration of the meaning of this knowledge for human health and the possibilities of developing molecular biomarkers of susceptibility as well as biomarkers of exposure.

The preparation of this book has taken several years. Each year has brought important new developments. Both the chapter authors and the editor realize that, although we have made every effort to make this text the most independent, comprehensive, and current available reference on dioxin and health, supplements and new editions will be necessary. Even as the text is delivered to the publisher, new discoveries are being made, and a bi-yearly update is being planned. By the next edition, new knowledge from laboratory and human research may have eliminated many of the current controversies surrounding dioxin and health. Until then, both the authors and editor hope that the reader derives as much enjoyment in reading this volume as we have derived from performing the research and preparing the text.

Acknowledgments

I wish to acknowledge and thank the chapter authors for their extraordinary effort and patience during editing. In addition, the chemists who helped generate the data of very high quality which I reviewed in my chapter on exposure assessment are gratefully acknowledged. Scientists and dioxin-exposed persons in a number of countries who participated in obtaining specimens, including those from Africa, Cambodia, China, India, Israel, Russia, Thailand, the U.S.A., Vietnam, and elsewhere, are gratefully acknowledged. Further thanks are extended to my family for the inspiration to perform scientific and scholarly work, especially in public health and medicine, and for the patience and encouragement shown by my wife and children during the many years of time spent away from them during preparation of this volume.

Arnold Schecter, MD, MPH

Binghamton, New York

Contents

Chapter 14

Exposure Assessment: Measurement of Dioxins and Related Chemicals in Human Tissues

Arnold Schecter

Chapter 15

Human Health Effects of Polychlorinated Biphenyls

William J. Nicholson and Philip J. Landrigan

Chapter 1

Overview
The Dioxin Debate

Thomas Webster and Barry Commoner

1. INTRODUCTION

To the general public, dioxin is the archetype of toxic chemicals, a substance that in minute amounts causes cancer and birth defects. Raised to a high level of visibility by the use of Agent Orange in Vietnam, it continues to generate environmental issues that capture public attention: Times Beach, Seveso, Love Canal, herbicide spraying in the United States, waste incineration, and chlorine-bleached paper.

Public fear has engendered a counterreaction. Some have claimed that dioxin causes no harm to humans other than chloracne, a disfiguring skin disease.[1,2] Others have compared the public attitude toward dioxin with witchhunts. Dioxin, they say, is a prime example of "chemophobia," the irrational fear of chemicals.[3,4] The counterreaction made front-page news in 1991. U.S. Assistant Surgeon General Vernon Houk claimed that the evacuation of Times Beach, Missouri, had been a mistake.[5,6] Administrator William Reilly of the U.S. Environmental Protection Agency ordered a reevaluation of the toxicity of dioxin. He stated that "I don't want to prejudge the issue, but we are seeing new information on dioxin that suggests a lower risk assessment for dioxin should be applied."[6]

In our opinion, the public fears are largely justified. The current scientific evidence argues not only that dioxin is a potent carcinogen, but also that the noncancer health and environmental hazards of dioxin may be more serious than previously believed. Indeed, dioxin appears to act like an extremely persistent synthetic hormone, perturbing important physiological signaling systems. Such toxic mimicry leads to a host of biological changes, especially altered cell development, differentiation, and

Thomas Webster • Department of Environmental Health, Boston University School of Public Health, Boston, Massachusetts 02118. Barry Commoner • Center for the Biology of Natural Systems, Queens College–CUNY, Flushing, New York 11367.

Dioxins and Health, edited by Arnold Schecter. Plenum Press, New York, 1994.

regulation. Perhaps the most troubling consequence is the possibility of reproductive and developmental effects at the levels of dioxin and dioxinlike compounds now present in the bodies of the average person. This phenomenon has already been observed in wildlife, especially fish-eating birds from the Great Lakes. These findings suggest that the environment is already overburdened with these dangerous compounds.

The pendulum of official opinion may be swinging back. Contradicting Houk, U.S. Assistant General Barry Johnson testified in June, 1992, that the evacuation of Times Beach was not a mistake.[7] The USEPA's reevaluation, although still incomplete at this writing, is currently indicating that the "danger from dioxin may be broader and more serious than previously thought."[8] In this overview, we will discuss the technical basis for this dramatic turnaround and its logical implication: a policy directed toward exposure reduction and pollution prevention.

2. DIOXIN AND DIOXINLIKE COMPOUNDS

Although commonly used in public discourse, the term *dioxin* is not proper chemical nomenclature. Correctly named, the polychlorinated dibenzo-*p*-dioxins are a group of 75 congeners, including the well-known 2,3,7,8-tetrachlorodibenzo-*p*-dioxin (2,3,7,8-TCDD or, as we shall refer to it, TCDD). Based on toxicity similar to that of TCDD, a wider group of halogenated aromatic compounds have been recognized as dioxinlike. These include certain polychlorinated dibenzofurans (PCDFs), biphenyls (PCBs), diphenyl ethers, naphthalenes and others. Brominated and chloro/bromo versions of these compounds may be dioxinlike as well.[9–11]

Membership in the class is defined biologically: dioxinlike compounds produce a similar spectrum of toxic effects thought to be caused by a common mechanism.[12–14] The key step in the presumed mechanism is binding of the dioxinlike compound to a cytoplasmic receptor protein, the Ah receptor. The molecule's planar shape facilitates binding to the receptor, and its relative potency depends to a large degree on how well it fits the receptor. TCDD binds the Ah receptor with a very high affinity and is extremely potent. Other planar molecules of about the same size and shape—including a number of the polyhalogenated dibenzo-*p*-dioxins and dibenzofurans—fit almost as well and are also very active. Although certain types of PCBs bind to the receptor only weakly, their relative abundance in the environment nevertheless makes them biologically important. PCBs with chlorines in positions that prevent the molecule from assuming a planar position do not bind to the Ah receptor and are not dioxinlike in their biological effects. Some of these PCBs can exert toxicity through other mechanisms, however.[14,15]

3. SOURCES

Large-scale industrial production of the dioxinlike polychlorinated naphthalenes began during World War I. Production of PCBs followed in the late 1920s (Table 1).

Table 1
Selected "History" of Dioxin and Dioxinlike Compounds

1899	Chloracne characterized
1918	Outbreaks of chloracne following exposure to chlorinated naphthalenes
1920–1940	Dramatic increase of PCDD and PCDF levels in North American lake sediments (reported 1984)
1929	U.S. commercial production of PCBs begins
1947	"X" disease described in cattle in the United States
1949	Monsanto incident (Nitro, West Virginia)[a]
1952–1954	Boehringer incident[a]
1953	BASF incident[a]
1957	TCDD identified as unwanted contaminant in the manufacture of trichlorophenols
	"Chick edema disease" outbreak in poultry in southeastern United States
1962–1970	Agent Orange used in Southeast Asia
1965–1966	Holmesburg prison experiments
Mid-1960s	Outbreaks of reproductive and developmental effects noted in Great Lakes fish-eating birds
1968	Yusho "oil disease" (Japan)
1971	TCDD found to cause birth defects in mice
	Contamination of Times Beach and other Missouri sites
1972–1976	Ah receptor hypothesis developed
1973	Polybrominated biphenyls accidentally added to cattle feed in Michigan
1974	TCDD detected in human breast milk from South Vietnam
1976	Accident in Seveso, Italy
1977	U.S. commercial production of PCBs halted
	TCDD found to cause cancer in rats
1978	Discovery of dioxin emissions from trash incinerators
	Kociba *et al.* cancer study of rats exposed to TCDD
1979	USEPA emergency suspension of some 2,4,5-T uses
	Yu-cheng "oil disease" (Taiwan)
	Association of soft tissue sarcoma with TCDD and phenoxyacetic acid herbicides
	TCDD found to modulate hormones and their receptors
1980	Evacuation of Love Canal
1981	Capacitor fire in Binghamton, NY, contaminates state office building with PCBs and PCDFs
1983–1985	General public found to be contaminated with PCDD and PCDF
1985	USEPA health assessment of TCDD
1986	Production of dioxin by chlorine-bleached paper mills discovered although proposed earlier
1988	First USEPA reassessment on dioxins
1990	Second Banbury conference on dioxins
1991	NIOSH cancer mortality study of U.S. chemical workers
	Second USEPA reassessment begins
1992	U.S.–Canadian International Joint Commission Sixth Biennial Report

[a]Three of many chemical accidents involving dioxin.

The thermal and chemical stability of PCBs, among other properties, led to their widespread use in transformers, capacitors, heat transfer and hydraulic fluids as well as carbonless copy paper, plasticizers, and numerous other applications. Health and environmental problems led to curbs on their industrial production, but not until decades later. In the meantime, about 650,000 metric tons were produced in the United States and about 1.5 million metric tons worldwide.[16] It is estimated that about 20–30% of this amount has entered the environment. Much of the remainder is still in stock or in uses such as capacitors and transformers.[17]

In contrast, the polychlorinated dibenzo-*p*-dioxins (PCDDs) and polychlorinated dibenzofurans (PCDFs) are unwanted by-products. Knowledge of their origins has increased considerably, beginning with the identification of TCDD as an unwanted by-product of the production of certain trichlorophenols and herbicides,[18,19] in particular Agent Orange, a 1:1 mixture of the *n*-butyl esters of 2,4,5-trichlorophenoxyacetic acid (2,4,5-T) and 2,4-dichlorophenoxyacetic acid (2,4-D). TCDD formation simply requires combining two molecules of 2,4,5-trichlorophenol under the right conditions.[19] More highly chlorinated PCDDs and PCDFs are formed during the production of pentachlorophenol, a compound still used in the United States and elsewhere as a wood preservative.[20,21] PCDFs also occur as low-level contaminants of PCBs. Much higher levels are generated by heating PCBs under the right conditions of heat and oxygen.[22] This phenomenon was a major contributor to the Yusho and Yu-cheng ("oil disease") tragedies, where cooking oil was contaminated with PCBs.[23,24] It also occurred in numerous incidents involving capacitor and transformer fires.[25,26]

It was thought for a while that the dioxin problem was limited to a few reactions of closely related chemicals. Unfortunately, this is not the case. PCDD and PCDF were discovered in ash from trash-burning incinerators in 1977[27] and later in their air emissions. It was not known at first whether the emissions were related to unburned PCDD and PCDF in the fuel, formation from chlorinated organic precursors, or *de novo* synthesis.[27,28] Based on the observation of fly-ash-catalyzed chlorination of organic residues,[29] it was suggested that PCDD and PCDF were synthesized as exhaust gases cooled down in the boiler and air pollution control devices.[30,31] This was soon confirmed by tests conducted at a Canadian incinerator. Little or no PCDD and PCDF were found in the gases leaving the furnace, but these compounds were detected in the cooler stack gases.[32] Laboratory studies indicate significant formation at about 300°C.[33] Various mechanisms have been postulated for these reactions, in which metals play an important role.[34,35] Both organic and inorganic sources of chlorine may contribute to the formation of PCDD and PCDF. However, the relationship between emissions of these compounds and levels of chlorine in fuel is still not completely understood.[36–38]

Such findings imply that dioxinlike compounds may be formed during virtually any combustion process when chlorine is present. This was proposed in the "trace chemistry of fire" hypothesis, which stated that PCDD formation was a natural consequence of combustion.[39,40] This led, in turn, to the claim that nonindustrial sources such as forest fires are potentially significant or even the dominant source of dioxin.[41–43] However, this claim is at odds with a number of observations. Levels of PCDD and PCDF are higher in people living in industrialized countries than in residents of less industrialized nations.[44,*] Lake sediments from North America and Europe show that PCDD and PCDF levels were very low until approximately 1920–1940.[45–50] Similarly, the levels of these compounds in ancient mummies and 100- to 400-year-old

*Levels of PCDD and PCDF—especially TCDD—are elevated in both river sediments and human tissues of southern Vietnam relative to northern Vietnam, reflecting both differences in industrialization and the millions of gallons of Agent Orange sprayed in the south.[44]

frozen Eskimos are far lower than those currently found in the average resident of an industrialized country.[51–53] Hence, while trace amounts of PCDD and PCDF may have been present in preindustrial times, the current levels represent a huge increase over these low levels.

In North America, the dramatic increase of PCDD and PCDF in lake sediments matches the beginnings of large-scale industrial chlorine chemistry—and combustion of its products—during the period 1920–1940. Concentrations of PCDD and PCDF in stored vegetation and soil samples from the United Kingdom increased at the turn of the century,[54] perhaps reflecting the advanced development of industrial processes in that country. Coal burning has been suggested as a major contributor; however, large-scale coal burning antedates the increase of PCDD/PCDF in North American sediments. PCDD and PCDF have been detected in fly ash and soot from coal combustion, but not yet in flue gas.[22,55] It is possible that the relatively high levels of sulfur in coal inhibit formation of PCDD/PCDF.[56,57] Low levels of PCDD and PCDF found in British soil and vegetation samples from the mid-1800s may reflect even earlier industrial activity. For instance, the Leblanc process for producing alkali, a forerunner of modern industrial chlorine chemistry, first found widespread application in the United Kingdom at this time.[58]

PCDD and PCDF are emitted when other chlorine-containing fuels are burned, including chemical waste, hospital waste,[59] and sewage sludge.[22] Dioxin-containing wastes have not proven as easy to destroy in hazardous waste incinerators as once claimed,[60,*] perhaps because of resynthesis. The exhausts from automobiles burning leaded gasoline contain both chlorinated and mixed halogenated dioxins and dibenzofurans, apparently arising from ethylene dichloride and ethylene dibromide used as lead scavengers. Much lower levels of PCDD and PCDF have been found in exhaust from vehicles burning unleaded gasoline, presumably reflecting the low level of chlorine in this fuel.[22,63,64]

The identification of additional PCDD and PCDF sources strengthens the connection between these compounds and industrial chlorine chemistry. PCDD and PCDF are formed in the bleaching of pulp and paper with chlorine, not surprising given the rich aromatic content of the lignin found in wood.[22,65] Suggested as early as 1974,[66] this phenomenon was not confirmed until the mid-1980s when high concentrations of TCDD were found in fish downstream of bleached pulp mills[67,68] and then in the mills themselves.[69,70] Dioxin may also be formed during dry cleaning,[22] metal degreasing and finishing,[71] and chlorine regeneration of metal catalysts used in petroleum refining.[72]

Large amounts of PCDD and PCDF are also produced by certain types of metal processing, perhaps reflecting the catalytic properties of a number of metals. PCDD and PCDF are emitted by the burning of scrap metal such as copper cable coated with PVC plastic insulation.[73,74] Other sources include aluminum smelting, magnesium and nickel production, scrap metal melting, and iron and steel production.[75–77] In these manufacturing processes, chlorine either is used or is contained in cutting oils, plastic and other contaminants.[76,77]

*The USEPA's response was that the regulation requiring an incinerator to achieve 99.9999% destruction or removal of dioxin-containing wastes does not actually apply to the dioxin itself.[61,62]

Dioxinlike compounds are formed at the heart of the chlorine industry as well. Large amounts of PCDD and PCDF have been found in the sludge from chloralkali plants that used graphite electrodes, once widely employed.* Most modern facilities now use other kinds of electrodes.[78] PCDD/PCDF have been detected in some common chlorinated hydrocarbons.[79] They are formed during the production of ethylene dichloride (EDC).[80–83] EDC is mainly used to produce vinyl chloride, the precursor to polyvinyl chloride (PVC) plastic. About 4.2 million metric tons of PVC was produced in the United States in 1991[84], making it the single largest use of chlorine in the country.

Preliminary reports indicate formation of octachlorinated dioxin from pentachlorophenol at near ambient temperatures in sewage sludge.[85] It is not known whether this is an abiotic chemical reaction or whether an enzymatic process is involved.[86]

In sum, the range of sources has expanded to the point that virtually all industrial chlorine chemistry can be suspected of generating dioxinlike compounds at some point during production, use, or disposal. The unwanted production of PCDD and PCDF may reflect the relative stability of these compounds. They can be thought of as thermodynamic sinks that are likely to accumulate in reactions involving chlorine and organic materials, and may therefore be expected to occur in a very wide range of reactions.

What are the largest sources? National emissions inventories have recently been performed in Germany, Austria, the Netherlands, Sweden, and the United Kingdom.[22,87–90] These estimate that waste incineration—municipal, hospital, hazardous, and industrial—and metal processing are the largest current sources of emissions to the atmosphere. More difficult to quantify, but potentially important, are secondary releases from stocks of pentachlorophenol and PCB.[22,90] Cases of massive dioxin contamination have been reported in Russia.[91–94] Rappe[95] has attempted to balance emissions against levels found in the environment. This is made difficult by the problems of representativeness of samples, persistence of past contamination, and long-distance transport. Nevertheless, environmental levels seem to exceed the level implied by known sources. It is possible that major sources have not yet been properly accounted for or identified.*

4. ENVIRONMENTAL FATE AND EXPOSURE

Dioxin is essentially a 20th century problem, a by-product of industrial chlorine chemistry and the combustion of chlorine-containing fuels. The growth of these processes during the last century has dramatically increased the levels of these compounds in the environment and in biota. Accumulation also depends on the environmental

*The chloralkali process manufactures chlorine and sodium hydroxide (an alkali) from sodium chloride brine via electrolysis. The graphite may provide a source of carbon for the generation of PCDD and PCDF.

*Recent tests of a Columbus, Ohio, municipal solid waste incinerator built in 1983 indicate that it could be emitting as much as 1 kg of TEQ per year.[96] It doesn't take too many sources of this magnitude to seriously affect emissions inventories.[97]

behavior of dioxinlike compounds, a consequence of their chemical and physical properties: low vapor pressure and water solubility, high lipophilicity, and relative chemical stability.[98] When the metabolic inertness of many congeners is added to the list, the profile is complete: dioxinlike compounds tend to persist and bioaccumulate.

Combustion sources emit large quantities of dioxinlike compounds into the atmosphere, where they are both dispersed and subjected to selective degradation. The more highly chlorinated compounds tend to adsorb onto airborne particulates at ambient temperature, greatly reducing their rate of degradation.[99] Larger fractions of the lower chlorinated congeners are found as vapor, making them more susceptible to photolysis and attack by hydroxyl radicals. These phenomena partly explain the shift toward a preponderance of the more highly chlorinated compounds seen in many abiotic environmental samples relative to the typical pattern found in combustion emissions.[48] Atmospheric residence times of particulate-bound congeners are determined by dry and wet deposition of the particulate.[100]

Thus, many dioxinlike compounds are sufficiently stable to travel long distances in the atmosphere, leading to their ubiquitous presence in the environment. Deposition from the air contaminates soil, water, and vegetation. Deposition of both particulate and vapor onto plants provides a significant entry into the terrestrial food chain.[101–104] Human exposure via milk and beef may be in some cases hundreds to thousands of times higher than via inhalation,[101,104] making it a major issue in the permitting of air emission sources.[105–107]

Because of their low water solubilities and vapor pressures, PCDD and PCDF tend to partition into soil and sediment. The half-life of TCDD may be on the order of a decade[22] or more in soil and probably longer in sediment. As a result, these two media can act as reservoirs, leading to recontamination of other media.

In aquatic systems, the highly lipophilic and hydrophobic dioxinlike compounds tend to bioconcentrate from water to aquatic animal and then biomagnify up the multistep food chain.[108–110] Levels of PCBs found in fish-eating birds, animals near the top of the aquatic food chain, can reach concentrations tens of millions times higher than that dissolved in water.[108] The combined effects of bioaccumulation and the action of sediment as a reservoir make direct discharge of these compounds into aquatic systems particularly problematic.

Humans are also high on the food chain, eating the meat and milk of herbivores as well as fish and plants. The average person in an industrial country is thought to be exposed to PCDD and PCDF primarily via these animal products.[111] The average daily dose is about 1–3 pg/kg of PCDD and PCDF expressed as 2,3,7,8-TCDD equivalents,[111] dioxinlike compounds considered equivalent in toxicity to 2,3,7,8-TCDD (reviewed in Refs. 9, 10, 14). General contamination of the environment and food sources may explain the relatively similar levels of PCDD and PCDF found in the average residents of industrialized countries.[44]

Dioxinlike compounds primarily accumulate in people's body lipid, especially adipose tissue. Their elimination depends on metabolic degradation—slight or nil for many congeners—and on the rate of excretion which is almost completely via the feces.[112] As a result, the half-life of 2,3,7,8-TCDD in humans is very long, on the order of a decade[113,114]; PCDD may have a half-life as long as 50 years.[114,115] This biological persistence leads to another route of exposure. After accumulating in the mother over decades, dioxinlike compounds can be passed to the developing fetus *in utero*[116]—a particularly vulnerable period—or to newborns via lactation.[44] Similarly, birds and fish accumulate these compounds and pass them to the egg.[108,109,117]

Like the increase of atmospheric chlorine caused by CFCs and certain chlorinated solvents, the dioxinlike compounds represent a perturbation of the planet's chemistry. This might have been only a curious and little-noticed sidelight of the industrial age if not for an additional factor: the extraordinarily powerful biological effects of the dioxinlike compounds.

5. BIOCHEMISTRY AND TOXICITY

5.1. Biological Persistence

As noted earlier, the central event that instigates the biological effects of TCDD and dioxinlike substances is thought to be their binding to a receptor protein. This aryl hydrocarbon or Ah receptor was first postulated during studies of PCDDs and polycyclic aromatic hydrocarbons (PAHs) such as 3-methylcholanthrene.[118–120] Once such a ligand is bound to the receptor, a series of intracellular processes ensue including shedding and binding of cofactor(s) and migration of the resulting complex into the nucleus where it influences the rate of transcription of specific messenger RNAs, thereby altering the rate of synthesis of the related proteins. Thus, the binding of an appropriate ligand to the receptor changes the cellular concentration of certain proteins by regulating the expression of genes governing their synthesis.[121,122] The apparent necessity of one or more cofactors, for instance the ARNT protein,[123] may account for differences in gene expressions by tissue.[124] Molecular mechanisms that do not require transcription are also possible.

Among the proteins induced by the Ah receptor are two closely related enzymes, cytochromes P450IA1 and P450IA2 (in recent nomenclature, CYP1A1 and CYP1A2). These so-called phase I enzymes oxidize "foreign" (xenobiotic) substances including PAHs, plant constituents such as flavones, aromatic amines, and some pharmaceutical drugs. One consequence of this metabolic conversion is that the xenobiotic substance, typically lipid- rather than water-soluble, is then subject to further enzymatic conversion. Phase II enzymes add hydrophilic groups, enhancing water solubility and excretion from the body. In this manner, the Ah-induced enzymes can reduce the biological effect of some environmental agents by facilitating their metabolic degradation.[125,126]

On the other hand, the oxidative transformation of a xenobiotic compound by the Ah-induced cytochrome P450 enzymes may greatly enhance its biological activity.[125] A classical example is 2-acetylaminofluorene (AAF), a potent liver carcinogen in rats. A number of studies have shown that the proximal carcinogen is not AAF, but an oxidized metabolite produced via cytochrome P450IA2.[125,127] Preparations containing these enzymes are used in microbial mutagenesis tests to activate otherwise inert genotoxins.

When the oxidative degradation of the xenobiotic compound facilitates its excretion so that the intracellular concentration is reduced, a negative feedback is established: Binding of the compound to the receptor is also reduced; transcription decreases; the level of the cytochrome P450 enzymes diminishes and the system returns to its initial condition.

TCDD and related halogenated compounds strongly induce cytochrome P450IA1 and P450IA2, but are not readily oxidized by these enzymes. They are apparently protected from attack by the presence of halogen atoms in certain positions of the molecule. Hence, they are excreted very slowly, resulting in a prolonged and amplified response. In effect, the feedback system that governs behavior of other Ah-binding substances (such as the PAHs) is inoperative in the case of TCDD. Thus, the extraordinary biological potency of dioxinlike substances appears to be the consequence of their unique combination of two properties: a high affinity for the Ah receptor and biological persistence. The relevance of persistence is evident from a comparison of the behavior of dioxinlike compounds and PAHs. Although some PAHs bind to the Ah receptor with an affinity almost equal to that of TCDD, their *in vivo* potency (as measured by enzyme induction) is many orders of magnitude less.[12]

5.2. Perturbation of Hormones and Growth Factors

Although the induction of cytochrome P450IA1 is the most well characterized of the biochemical effects of dioxinlike compounds, it is by no means the only one. The expression of a growing number of genes are thought to be regulated by the Ah receptor.[128] This may explain how dioxinlike compounds perturb the regulation of hormones, growth factors, and other molecular messengers that control growth and differentiation with diverse and potentially devastating impact. Some examples follow.

TCDD may alter the levels of certain hormones through its influence on the enzymes that primarily metabolize xenobiotic compounds. For instance, TCDD induces one form of UDP-glucuronyltransferase (UDPGT), a phase II enzyme. In addition to metabolizing xenobiotics, UDPGT also conjugates and enhances the excretion of thyroxine (T_4), causing reduced serum levels of this thyroid hormone in rats.[129] Among the resultant complications is a perturbation of an important biological feedback system: The pituitary responds to low T_4 with increased secretion of thyroid stimulating hormone (TSH). When prolonged, this may lead to thyroid tumors,[130] a sensitive endpoint in TCDD-exposed rats.[131–135]

TCDD does not bind to steroid hormone receptors and steroid hormones do not bind to the Ah receptor.[136] Nevertheless, TCDD affects steroid hormone regulation in more subtle ways. Thus, TCDD decreases ("downregulates") the number of estrogen receptors in certain organs of the female rodent, making tissues less responsive to this hormone.[137–139] This may decrease fertility and the incidence of tumors of these organs, as has been suggested in rats exposed to TCDD.[140]

TCDD reduces testosterone levels in adult male rats by decreasing the production of testosterone from cholesterol in the testes at a critical rate-limiting step. The pituitary (and/or hypothalamus) normally responds to the low testosterone concentration by increasing secretion of luteinizing hormone, causing increased production of testosterone. TCDD interferes with this feedback system, preventing the compensatory increase of luteinizing hormone.[141–144]

TCDD also affects growth factors. In the female rat liver, TCDD may increase migration of epidermal growth factor receptor (EGFR) internally from the cell membrane, providing a stimulus for mitosis.[136,145] This effect appears to be dependent on ovarian hormones[145]; interactions between EGFR and estrogen receptors have been noticed elsewhere.[146] TCDD may affect EGFR by increasing the levels of TGFα, a ligand for EGFR.[135] In mice, TCDD alters the differentiation of certain tissues in the developing palate. This may be caused by perturbation of growth factors and their receptors, including EGFR. The palatal shelves come into contact but fail to fuse, resulting in cleft palate.[147,148]

TCDD inhibits the proper maturation and differentiation of both T and B cells, important in cell-mediated immunity and the production of antibodies.[149–151] B-cell maturation may be disturbed by TCDD-induced phosphorylation of the amino acid tyrosine. Tyrosine kinases often play important roles in regulating growth factor receptors and cell differentiation.[152,153] A recent report indicates that TCDD leads to increased tyrosine phosphorylation of cdc2-kinase, a protein playing a key role in the cycle of cell growth and division.[154] TCDD also influences a number of other chemical messengers including the glucocorticoid hormone receptor, plasminogen activator inhibitor, protein kinase C, interleukin 1β, and other cytokines.[128]

TCDD can be considered a persistent "environmental hormone."[155] Its fundamental molecular mechanism—binding to a receptor that regulates gene expression—has certain similarities to steroid hormones[156] as well as differences.[157,158] It alters cell growth and differentiation. It affects other hormones and growth factors, including altering the levels of their receptors. Finally, like hormones, TCDD causes significant effects at very low doses. This new knowledge of dioxin's biochemistry increases our concern over its widespread occurrence in the environment.

Does the body possess some unidentified hormone that binds to the Ah receptor, serving an important but unknown function? Such a situation is not unprecedented. A number of such "orphan" receptors, i.e., receptors without known ligands, have been found.[159] Dioxin may be a case of "toxic mimicry," possessing a molecular shape similar to that of its putative natural counterpart. The long residence time of TCDD in the body may alter expression of Ah-regulated genes for an inappropriately long period of time. It is also possible that the supposed natural ligand of the Ah receptor might normally function during a specific period of development; TCDD may activate the system at the wrong time.[160] This hypothesis might be explored using a "knockout" strategy, examining whether animals without a functional receptor develop normally.[158]

5.3. Toxicity

From the foregoing account it is apparent that TCDD is capable of disrupting a wide variety of biochemical processes which are likely to lead to an equally broad spectrum of macroscopic toxic effects in animals. The latter include acute toxicity, "wasting" and death, atrophy of the thymus, liver damage, epidermal changes, immunotoxicity, birth defects, reduced fertility, and cancer.[12,160] It is generally thought that the Ah receptor mediates most if not all such effects,[161] although there may be exceptions.[149] Hence, it is assumed that other dioxinlike compounds will also cause these effects.

The relative sensitivity of various toxic endpoints appears to vary with tissue and species, implying that humans may be less sensitive than laboratory animals for some effects and more sensitive for others.[160] In this review, we will focus on two: cancer and reproductive/developmental effects. They have formed the primary basis for regulatory efforts by a number of countries and are central to the current EPA reassessment of TCDD toxicity. Immunotoxicity promises to be an increasingly important topic. TCDD affects the immune systems of laboratory animals in minute doses (reviewed in Refs. 149–151); human immune system alterations were observed in the Yusho and Yu-cheng rice-oil poisonings (reviewed in Refs. 23, 24, 150).

6. CANCER

6.1. Mechanism

There is no doubt that TCDD causes cancer in animals. This has been shown in both sexes of several species.[134,162] The contentious points surround its carcinogenic mechanism(s) and their implications for human exposure at low doses. The development of cancer is generally thought to proceed in several steps: (1) an initial permanent alteration of a cell, typically some kind of genetic damage; (2) clonal proliferation of the altered cell; (3) another permanent alteration in at least one of these cells, followed by more cell replication.[136] This last step may be repeated several times, a process called tumor progression.

So-called two-stage cancer experiments attempt to partially dissect these steps: an animal is given a dose of a DNA-damaging ("initiating") agent followed by chronic exposure to a "promoting" agent. In such an experiment, a classic promoter greatly enhances the number of tumors and precancerous lesions, but causes little or no cancer by itself. Its action is considered reversible, i.e., removal of the promoter causes the tumor to regress. Note that initiation and promotion are operationally defined by this experimental protocol and are not necessarily synonymous with DNA damage and cell proliferation.[163]

In two-stage experiments involving rat liver and mouse skin, TCDD is an extremely potent promoter and displays little or no initiating activity.[135,164–166] The latter finding is puzzling, given TCDD's ability to generate substantial numbers of tumors in the rat liver (and other organs) in long-term bioassays when given "alone," i.e., without a known initiator.[140] These divergent results explain why some researchers consider TCDD a promoter, while others consider it a "complete" carcinogen.[134,162] The discrepancy might be related to differences in the food or environment of the test animals and/or the shorter length of exposure in the two-stage experiments.

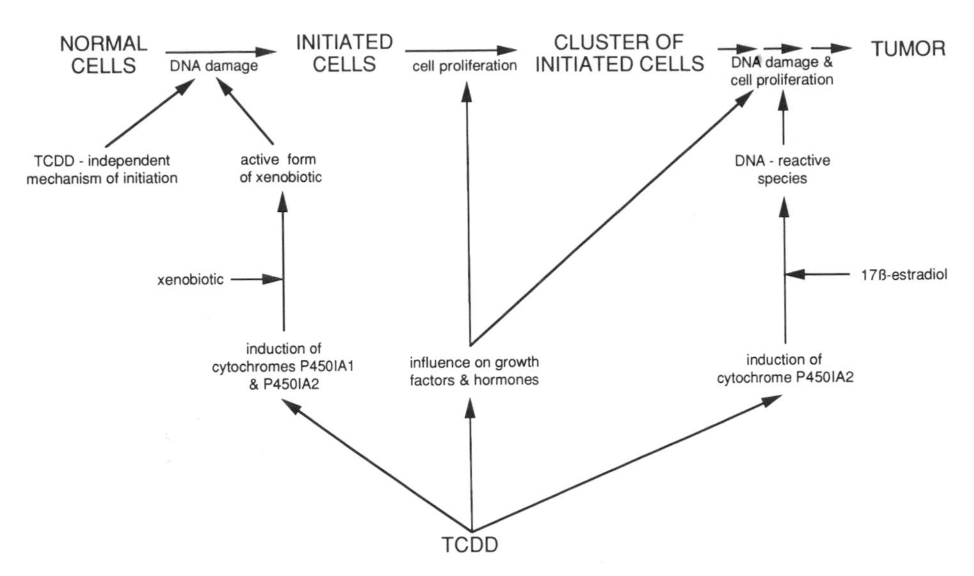

Figure 1. Some possible roles of TCDD in carcinogenesis of the female rat liver. (Adapted from Lucier.[136])

While it is possible that TCDD promotes "background" initiated cells damaged by some independent process, other evidence suggests that TCDD may act throughout the carcinogenic process (Fig. 1). TCDD does not directly cause mutations in several common assays and therefore appears to lack the direct genetic effect characteristic of an initiator. However, it is possible that TCDD indirectly contributes to DNA damage. By inducing cytochrome P450IA1 and P450IA2, dioxinlike compounds may in some cases increase the conversion of other compounds into mutagens.[167] For instance, lymphocytes from people previously exposed to dioxinlike compounds show increased frequency of sister chromatid exchange when cultured with α-naphthoflavone (ANF); ANF is metabolized to an active form by cytochrome P450IA1.[168] On the other hand, some experiments show reduced DNA damage and cancer in animals dosed with certain PAHs after exposure to TCDD compared with animals exposed to PAHs alone.[169] The balance between metabolic activation and deactivation may depend on the compound and the dosing regime.[135]

Two-stage studies of the female rat liver support the hypothesis that TCDD-induced cell proliferation is involved in promotion, although this may not be the only mechanistic step. Preneoplastic foci were greatly increased in rats exposed to diethylnitrosamine (DEN) followed by TCDD, but not in animals exposed to TCDD alone. Cell proliferation was greatly increased in animals exposed to TCDD. These elegant studies also demonstrate the involvement of estrogenic hormones. Cell proliferation and preneoplastic foci were significantly higher in intact rats compared with ovariectomized animals.* TCDD may enhance cell proliferation via the epidermal

*In these experiments, lung cancer was seen in ovariectomized females, but not in the intact animals. In long-term bioassays, TCDD significantly increases liver tumors in female but not male rats.[131,140] However, TCDD produces liver tumors in both sexes of mice.[131] The reason for this difference is unknown.

growth factor pathway. Internalization of EGFR, a mitogenic stimulus, is enhanced by TCDD in intact but not ovariectomized animals.[145,166] Note that the initiator DEN is metabolically activated by P450 enzymes other than cytochrome P450IA1 and P450IA2.[170,171] Induction of these enzymes was unaffected by removal of the ovaries.

Evidence from rat liver studies suggests that TCDD may also play a role in tumor progression. Some precancerous lesions regress when administration of TCDD ends, but others continued to grow in size.[172] One possible explanation is that additional permanent alterations occurred in these lesions making them promoter-independent. TCDD may indirectly contribute to these events by another estrogen-dependent mechanism. Metabolism of the ovarian hormone 17β-estradiol by cytochrome P450IA2 can lead to the production of DNA-reactive species.[136,145]

While most of the experimental work has been done on the liver, TCDD alters tumor incidence at numerous sites in long-term bioassays. This finding implies that a number of mechanisms may be involved. As discussed earlier, prolonged secretion of TSH in response to TCDD-induced degradation of T_4, may increase thyroid tumors in rats. TCDD decreases the number of estrogen receptors in the uterus and breast; this may be connected to the apparent reduction of tumors in these organs in the rat.

TCDD may contribute to cancer in other ways including interaction with viruses,[173] increased expression of protooncogenes,[174] decreased expression of tumor suppressor genes, and suppression of cell-mediated immunity.[175] Epstein–Barr virus, which is widespread in the human population, may cause B-cell proliferation and immortalization. Impairment of cell-mediated immunity by TCDD and other chemicals may allow continued proliferation and development into non-Hodgkin's lymphoma.[176] Finally, while TCDD may not be a direct mutagen, long-term changes in gene expression may lead to irreversible cell damage; current concepts of genotoxicity may be too narrow.[177]

In sum, TCDD may act at a number of steps in the carcinogenic process in conjunction with endogenous hormones, exogenous compounds, and viruses in an organ-specific fashion. Many aspects of its carcinogenic mechanism remain unknown.

6.2. Cancer Risk Assessment and Reassessment

Debate over the mechanism of TCDD carcinogenicity has played a central role in regulation of exposure and in the EPA reassessments of its potency. The general goal of these efforts has been the identification of a "safe" or "acceptable" daily dose of dioxin. As Table 2 shows, the values used by a number of countries and government agencies range from 0.006 to over 20 pg/kg body wt per day, a factor of several thousand. The extreme toxicity of TCDD is reflected in these levels: they deal with picograms or trillionths of a gram.

These risk assessments have generated intense controversy. The average citizen of an industrialized country is exposed to about 1–3 pg/kg per day of 2,3,7,8-TCDD equivalents (TEQ). If values from the upper end of the range are used, then the average dose is "tolerable." On the other hand, average exposure greatly exceeds the lower

Table 2
"Acceptable" Daily Doses of TCDD

Organization[a]	Level (pg/kg/day)	Methodology	Basis[b]
USEPA	0.006	LMS model[c]	Kociba
California	0.007	LMS model[c]	NTP mouse
Centers for Disease Control	0.03	LMS model[c]	Kociba
U.S. Food and Drug Administration	0.06	LMS model[c]	Kociba
National Research Council of Canada	0.07	LMS model[c]	Kociba
Germany	1–10	Safety factor	Kociba, Murray
Netherlands	4	Safety factor	Kociba
Canada and Ontario	10	Safety factor	Kociba, Murray
World Health Organization	10	Safety factor	Kociba, Murray
Washington State Dept. of Health	20–80	Safety factor	Receptor occupancy

[a]Sources: USEPA[185] except Washington State Department of Health[198] and World Health Organization, Summary Report: Consultation on Tolerable Daily Intake from Food of PCDDs and PCDFs, Regional Bureau for Europe (1991).
[b]Most of the agencies relied on the rat cancer study of Kociba *et al.*[140] and/or the rat reproduction study of Murray *et al.*[278] California used a mouse cancer study.[131]
[c]For an upper-bound lifetime cancer risk of 10^{-6}.

estimates of acceptable dose, a situation some interpret as requiring remedial action.[178] In practice, the USEPA only regulates incremental exposure from a single source or medium. Nevertheless, many sources fail to meet even these standards.

One can immediately see one reason for the political pressure placed on the USEPA position: its estimate of the "acceptable" dose is one of the lowest in Table 2. A number of dioxin-generating industries and owners of dioxin-contaminated sites have, perhaps not surprisingly, maintained that higher values are more appropriate.

Another fundamental reason for disagreement lies in scientific differences about how to construct a "tolerable" dose. Practical considerations restrict the number of animals that can be used in a cancer bioassay. This imposes a limit on the ability to detect an increased number of tumors. In order to avoid false negatives, the doses employed are typically much larger than those commonly experienced by people. Judging the safety of the latter requires extrapolation from high to low dose and from animal to human. Depending on how these extrapolations are carried out, the same data can readily yield very different conclusions.

The primary theoretical difference between the high and low estimates of "acceptable" dose is the shape of the dose–response curve at low doses, in particular the presence or absence of a threshold (Fig. 2). The high values are derived from the view that there is a dose of TCDD below which there is no effect. They are typically estimated by applying a safety factor to an experimentally defined "no observed adverse effect level" (NOAEL) or "lowest observed adverse effect level" (LOAEL). Such levels depend on both the biology of the phenomenon and the methodology and statistics of the experiment. Thus, an effect might occur at some low dose but the particular experimental design may not be powerful enough to distinguish it from the

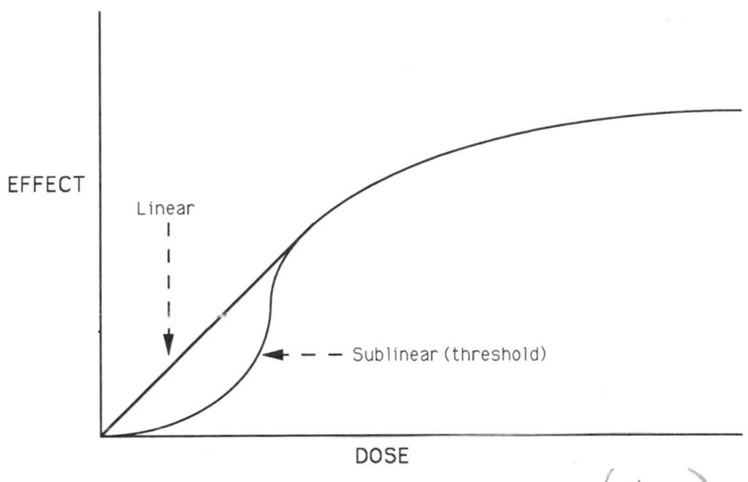

Figure 2. Two possible dose–response curves. (class)

control value. Ontario's tolerable average daily intake, 10 pg/kg, is based on a presumed no-effect level in animals of 1000 pg/kg per day for cancer and reproductive effects with a safety factor of 100.[179]

In contrast, the low "acceptable" doses are based on the theory that there is no threshold for cancer. The probability of cancer is assumed to be directly proportional to dose at low dose. Since there is no completely safe dose, the slope factor or "potency" of the chemical is used to calculate the dose resulting in a certain lifetime risk of cancer that is regarded as "acceptable." The USEPA followed this procedure in its landmark 1985 assessment of TCDD carcinogenicity, the basis of most current regulations in the United States.[133] The slope factor was estimated from rat tumor data from Kociba *et al.*[140] using the linearized multistage (LMS) model of cancer. A small additional factor adjusted for possible interspecies differences. According to this estimate, an average daily dose of 0.006 pg/kg corresponds to an upper-bound excess lifetime cancer risk of one in a million (10^{-6}). Hence, if one million people received this level of exposure over their lifetime, less than one case of cancer would be expected. The USEPA considers this level of risk "acceptable" (*de minimus*) as a matter of policy, although the agency sometimes uses or approves higher values (10^{-4} was recently approved for drinking water[180]).

Assumptions about the shape of the dose–response curve for cancer depend critically on the mechanism of carcinogenicity. The LMS model used by the USEPA is derived from the theory that cancer involves a sequence of irreversible stages.[181,182] There have been numerous criticisms of this approach as it applies to TCDD. One group has argued that TCDD is a promoter, not an initiator, and is therefore subject to a threshold. In this case, the NOAEL/safety factor approach would be more appropriate. Others argue that the LMS model is not appropriate because it does not take into account cell proliferation. However, other work indicates that even a "pure" promoter can act linearly at low dose if its effect is additive to background processes.[183]

A third source of controversy concerns the role of scientific uncertainty in regulatory policy. While some argue that pollution should be allowed until proof of harm is certain, others advocate more careful approaches. For example, the USEPA has determined that "in the absence of adequate information to the contrary, the linearized multistage procedure will be employed,"[184] assuming that this approach will be more protective of public health.

Some of these arguments were raised during the USEPA's 1988 reassessment of TCDD's slope factor. The USEPA concluded that TCDD may cause cancer through a variety of mechanisms and that the LMS model would be retained, in part because there was no adequate alternative model. Nevertheless, the agency's Dioxin Workgroup argued that the 1985 slope factor estimate "is likely to have led to an overestimate of risk".[185] Although the degree of overestimation was unspecified, they proposed raising the "acceptable" dose to 0.1 pg/kg per day, simply as a matter of policy. This proposal was rejected by the agency's Science Advisory Board because no new scientific evidence had been presented to justify the change in the risk estimate.[186] However, the board expressed concern about the applicability of the LMS model to TCDD and encouraged development of new risk models which would incorporate additional mechanistic research into risk assessment, in particular, receptor mediation of toxicity.

6.3. Reassessment II

By the late 1980s a new political factor entered the dioxin arena: the interest of the paper and allied industries. Dioxin had been discovered in effluent and products from pulp mills using chlorine as a bleaching agent. The industry was concerned about possible legal action and impending surface water quality standards. The paper industry began a campaign to "get EPA to 'rethink' dioxin risk assessment."[187,188] The chlorine-producing industry became an ally, presumably because the paper industry consumed a significant fraction of their output.

A new challenge to the USEPA's cancer slope factor came in 1989 during consideration of a water quality standard for the State of Maine. Female liver tissue samples from the 1978 Kociba experiment were reviewed and reclassified based on new criteria for the presence of tumors.[189] Tumors from other sites and bioassays were not examined, although some—male rat thyroid, male mouse liver—also produce high slope factor estimates.[131,132,134] The revised liver tumor data might reduce the TCDD cancer slope factor estimate by a factor of about two to three, an insignificant amount in view of the underlying uncertainties.[190] While some suggest that the liver tumors were a secondary response to hepatotoxicity,[191] a review by the USEPA and Food and Drug Administration disagreed.[192]

In October, 1990, the Chlorine Institute and the USEPA cosponsored a scientific meeting entitled the "Biological Basis for Risk Assessment of Dioxins and Related Compounds" at the Banbury Center. There was general agreement among the toxicologists and biochemists present at the meeting that the currently known toxic effects of dioxin are mediated by the Ah receptor. While this was not particularly radical thinking, some participants drew a controversial conclusion: receptor mediation implied a threshold for the biological effects of dioxin. They stated that a certain number of receptors must be occupied for any biological effect to occur. Rather than being linear at low doses, the dose–response curve was shaped like a hockey stock: flat or increasing very slightly at first and only becoming linear at higher doses. Furthermore, a practical threshold for toxic effects could be determined from the dose of TCDD necessary to induce cytochrome P450IA1, the presumed most sensitive endpoint. The resulting rough and rapidly calculated estimate was about 1–3 pg/kg per day, much greater than the USEPA's value of 0.006 pg/kg.[2,161] A public relations firm hired by the Chlorine Institute went further, claiming in a press packet that the attendees had formally reached consensus on the threshold concept. This was not correct, prompting vociferous protests and creating a minor scandal in the scientific press.[193]

This meeting set the stage for another dioxin reassessment by the USEPA, which was announced in April, 1991. The primary focus would be the development of a new "biologically-based" model for dioxin toxicity, developing the ideas from the Banbury Conference and the earlier comments of the Science Advisory Board.[194–196] Incorporating the latest scientific findings, the new model would provide an alternative to the safety factor and LMS approaches to risk assessment. News of the reassessment was widely reported, including the notion that TCDD was much less toxic than previously thought.

Meanwhile, the paper industry used the supposed outcome of the Banbury Conference to argue for relaxed TCDD water quality standards in a number of states.[e.g.,197] The Washington State Department of Health issued revised guidelines for fish consumption based on a tolerable daily intake in the range of 20–80 pg/kg. This was calculated by applying a safety factor to the dose estimated to give 5% occupation of Ah receptors in the rat liver, a level assumed necessary for any biological response. No references to the scientific literature were given for this crucial assumption.[198]

There are good reasons to be skeptical of the claim that involvement of a receptor requires a threshold. The classic, simple model for receptors predicts a linear relationship between low concentrations of TCDD and the amount of receptor-bound dioxin.* Biological responses would not have a threshold if they are proportional to the amount of receptor-bound TCDD. Of course, the dose–response curves of more complicated biological responses might deviate from linearity, but this is only a possibility and not a requirement of receptor theory.[177]

The simple threshold model was seriously weakened at the Eleventh International Symposium on Chlorinated Dioxins and Related Compounds, held in North Carolina in September, 1991. Studies yielded data on induction of cytochrome P450IA1 and P450IA2 in the rat liver that were consistent with both threshold and nonthreshold (low-dose linear) models. The no-threshold models provided the best mathematical fit.

*Some people may have been confused by different ways of plotting the fraction of occupied receptors as a function of ligand concentration. Plotted on a linear scale, the curve is linear at low dose; plotted on a log scale the curve looks nonlinear at low doses. The latter version was printed in the *Science* article covering the Banbury conference.[161]

Similar results were found for dioxin-induced loss of EGFR from plasma membrane.[199,200,*] These results relied on extrapolation from experimental doses, so it is possible that a deviation from linearity may exist at lower doses. However, increased messenger RNA for cytochrome P450IA1 has been detected in rats at doses corresponding to background tissue levels in humans.[200,201] Hence, if there is a threshold for this effect, human tissues may already be above it. If the dose–response behavior of P450IA1, P450IA2, and EGFR are used as surrogates of toxicity, the cancer risks posed by TCDD may be as high or higher than previously estimated by USEPA.[199,202] In sum, the model proposed by some at the Banbury meeting is incorrect: action of TCDD through the Ah receptor does not necessitate a threshold.

The next question, of course, is whether these biochemical markers are reasonable surrogates of cancer or other toxic effects. As noted earlier, P450IA1, P450IA2, and internalization of EGFR may be related to cancer in the rat liver. On the other hand, liver cell proliferation and preneoplastic foci show no detectable increase at low doses of dioxin. The dose–response curve for these higher-level biological responses may be nonlinear, but substantial variability among experimental animals clouds the issue.[136,166]

The current dioxin reassessment is producing biologically more realistic models, but these still contain substantial uncertainty, especially with respect to the gap between biochemical markers and more complicated biological responses. Although much of this work has gone into modeling cancer of the rat liver, TCDD causes cancer in other organs as well. Since their mechanistic details appear to differ, they will require additional modeling efforts. Biologically realistic models will also have to address other complications such as interactions with other compounds and viruses.

6.4. Human Effects

Is dioxin merely a powerful rodenticide? Are humans somehow exempt or less susceptible to the biological effects of dioxin? Is it true that "chloracne is [the] only adverse effect associated with human exposure"?[203] Such arguments are sometimes offered for downgrading the toxicity of dioxin. If correct, the "allowable" doses of dioxin overstate the real hazard since they are based on animal research.

The reasons for the primary reliance on animal research are well known. First, the underlying biology of animals and humans is generally similar. Second, experimentation on humans is usually considered unethical. Only at Holmesburg prison, Pennsylvania, were people (other than the scientists involved) deliberately exposed to TCDD to test toxicity, in this case the dose necessary to cause chloracne. The fate of most of these unfortunate individuals is not known.[19]

The inadvertent exposure of people to large amounts of dioxinlike compounds

*The difference depends on whether the action of TCDD is additive or independent of existing processes.[199] Heterogeneity of liver cell response suggests a further complication. Low doses of TCDD appear to induce cytochrome P450IA1 and P450IA2 maximally in some cells and little or none in others; increasing the dose "turns on" more cells.[200]

began at the turn of the century (see Table 1). Chloracne was first described in 1899 in workers handling tarry wastes from the production of chlorine using graphite electrodes.[204,205] Chloracne was observed during World War I following exposure to polychlorinated naphthalenes; these "halowaxes" were used in the production of gas masks.[206] Several of the best known occupational exposures to dioxin occurred around midcentury in facilities manufacturing TCDD-contaminated herbicides. In the Yusho and Yu-cheng incidents, people consumed rice oil contaminated with PCBs, PCDFs, and related compounds. Large populations were exposed by the use of the herbicide Agent Orange in Southeast Asia, the spreading of dioxin-contaminated wastes in Missouri, and the chemical accident at Seveso, Italy. Indeed, it was learned in the 1980s that we are all exposed: the general population of the industrialized world carries some quantity of PCDD and PCDF in their bodies (reviewed in Ref. 44).[207–212]

Information on the effects of dioxin on humans has been obtained from some of these experiences using epidemiological methods to compare the rates and patterns of disease in exposed and reference populations. Given the uncontrolled nature of human exposure, it is important to take note of the limitations of such studies. Certain effects, such as cancer, may not occur until many years, even decades, after exposure. Other effects, for instance subtle neurobehavioral abnormalities in children, may be missed unless specifically looked for. It is often difficult to exclude other factors that might contribute to changes in disease incidence. Effects may not be detected unless the exposed population is sufficiently numerous and the difference in exposure from control groups is relatively large. Estimating who was exposed and at what levels is made particularly difficult by the ubiquity and persistence of dioxinlike compounds. As a result, many supposedly "negative" studies are in reality merely inconclusive.[213] When strong, biologically plausible effects are seen in a number of carefully performed studies, the implications need to be taken very seriously.

6.4.1. Qualitative Evidence for Cancer in Humans

Much of the debate about human effects has centered on the question of whether TCDD causes cancer in people. That it causes cancer in animals should not be in doubt. Given similar biology, this strongly suggests that it will cause cancer in humans as well.

As of the late 1980s, the human epidemiological evidence was mixed, including both positive and "negative" studies. Uncertainty was increased by the difficulty of establishing exposure and of separating the possible effects of TCDD from other chemicals to which people were often coexposed, e.g., phenoxyacetic acid herbicides. Some denied any connection between increased human cancer and exposure to phenoxyacetic acid herbicides and/or their dioxin contaminants.[214–216] Others such as the Agent Orange Scientific Taskforce—a group of independent scientists on which one of us (B.C.) served—concluded that there was sufficient evidence to legally qualify Vietnam veterans for compensation for several types of cancer and disease.[217,218] The USEPA and the International Agency for Research on Cancer (IARC) rated TCDD as a probable human carcinogen, based on what they considered sufficient evidence in animals and inadequate evidence in humans.[133,219]

The position at that time can be illustrated with two sets of studies. Beginning in the late 1970s, Hardell and others found significantly* increased numbers of a very rare cancer, soft tissue sarcoma (STS), in Swedish forestry and agricultural workers exposed to phenoxyacetic acid herbicides and/or chlorophenols which are frequently, but not always, contaminated with dioxins.[220–224] On the other hand, several studies of chemical workers thought to have been exposed to high levels of TCDD during the manufacture of herbicides and chlorophenols were considered negative. In particular, no significant increase in cancer mortality was observed in workers exposed following an accident in 1949 at a Monsanto facility in Nitro, West Virginia.[225–227] According to a 1989 report by the World Health Organization, this was one of only two such incidents that "have been adequately followed up epidemiologically with matched control groups."[205] Others have called it a "major source of information about the effects of high-level dioxin exposure."[214] In retrospect, the pioneering work of Hardell *et al.* on STS has largely survived criticism, while confidence in the Monsanto studies appears to have been misplaced. The latter are at the very least flawed by exposure misclassification[228]; some have raised even more serious questions about the integrity of the research.[229,230] Similar controversy surrounds a number of human health studies.[231–233]

Establishment of exposure played a crucial role in these debates. A technical breakthrough came with the ability to measure PCDD and PCDF in human tissues, first in breast milk,[234] and later in adipose tissue and blood (reviewed in Ref. 44). Since PCDDs and PCDFs are persistent in the body, this provides a useful measure of past exposure (although the absence of elevated levels does not necessarily preclude exposure[217,235]). The U.S. National Institute of Occupational Safety and Health (NIOSH) followed this approach in its landmark retrospective cohort study of male American chemical workers thought to have been exposed to TCDD.[236] Over 5000 workers were included from 12 plants (including Monsanto's Nitro, West Virginia plant). Blood serum from a sample of workers was used to validate estimates of exposure made on the basis of work history.

NIOSH found a statistically significant 15% increase (relative risk = 1.15, 95% confidence level = 1.02–1.30) in total cancer mortality in the entire cohort. Tumors at a number of specific sites were elevated but not significantly. In a subcohort of 1520 men with more than 1 year of exposure and over 20 years of latency—a group most likely to show effects—there were significant increases of overall cancer (RR = 1.46, CI = 1.21–1.76) as well as mortality from STS (RR = 9.22, CI = 1.90–26.95) and respiratory tract cancer (RR = 1.42, CI = 1.03–1.92). NIOSH conservatively concluded that these results were "consistent with the status of TCDD as a carcinogen."[236,*]

*Because of space constraints, some results are presented here in terms of statistical significance instead of the more informative relative risks and confidence intervals. Note that an increased relative risk lends support, albeit weak, to a hypothesis of causation even if it is not statistically significant.[213]

*NIOSH could not completely exclude the possible contribution of smoking and exposure to other chemicals. Although their statistics were calculated using information from death certificates, only two of four STS cases were confirmed in pathologic review. On the other hand, hospital records showed two other STS cases that were not recorded on death certificates.

NIOSH's results were partly confirmed by two small studies of German chemical workers employed by Boehringer[237] and BASF.[238] Death from all cancers combined was significantly increased in high-exposed subgroups of both studies. Significantly increased mortality from cancers of all sites combined, lung, and the hematopoietic system were observed in men from Boehringer relative to reference gas workers.

Another new study of over 18,000 chemical workers or sprayers from ten other countries found significantly increased STS, thyroid, and lung cancer as well as combined cancer mortality in certain subgroups, but detailed exposure information was not available for all workers.[239] Although preliminary, data from Seveso show statistically significant increases in cancer morbidity in certain zones of the contaminated area. These include STS in men from zone R as well as "a remarkably increased risk" (RR = 5.23, CI = 2.1–13.2) for gallbladder and biliary tract cancer in women from zone AB.[240] These results cover only the period 1977 through 1986.

The increase in mortality from all cancers combined noted in several occupational studies is unusual, as chemical carcinogens are often associated with a particular organ. In this respect, the human results appear consistent with the animal experiments, in which TCDD causes cancer at multiple sites. There are also parallels between rodents and humans with respect to specific sites: fibrosarcoma, lung, and thyroid cancer (reviewed in Refs. 133–135). Tumors of the liver are observed in TCDD-exposed rodents, but generally not observed in TCDD-exposed people. If TCDD-related liver cancer in humans is dependent on ovarian hormones as it is in rats (but not mice), this may be related to the fact that most occupational studies have examined men.* Primary human liver cancer is very rare outside of sub-Saharan Africa and Asia.[241]

These new studies shift the weight of evidence toward considering TCDD as a human carcinogen. The agreement of several studies of occupationally exposed men with reasonable checks on exposure provides the strongest evidence thus far. On this basis, the USEPA and IARC should revise their qualitative ranking of TCDD upward, from inadequate to at least limited, and possibly sufficient human evidence. The case for sufficiency is increased by the biological plausibility of the observed types of cancer and the action of dioxinlike compounds on potentially relevant human biochemistry and cell biology.[e.g.,160,173,242]

6.4.2. Quantitative Evidence for Cancer in Humans

The publication of the long-awaited NIOSH findings in January, 1991, provided part of the scientific rationale for EPA's reassessment.[196] The study also played a curious role in the public discussion of dioxin. Although it strengthens the qualitative evidence for cancer in humans, it was often portrayed as showing reduced danger from

*Liver, respiratory, and total cancer mortality were significantly increased in PCB- and PCDF-exposed male Yusho patients from one of two prefectures. Liver cancer in females was elevated but not statistically significant.[23] The Boehringer study included 54 women, but very few were highly exposed; breast carcinoma was elevated and of borderline significance.[237] Ovariectomy increases lung cancer in two-stage experiments with TCDD in the female rat, suggesting a partial hormonal dependence for this tumor.[145]

TCDD.[243–245] Some had the story completely wrong, reporting (without qualification) that cancer mortality was not significantly elevated in the cohort.[246,247] Others reported the NIOSH results as indicating that TCDD only causes cancer in humans at very high doses. NIOSH found statistically significant increases of cancer in the entire cohort as well as in a "high" exposed subcohort. They had also analyzed a "less" exposed subcohort (less than 1 year of exposure and over 20 years of latency). While cancers of some sites were elevated in this group, none were statistically significant at a 95% level. These results may have been interpreted by some as showing a threshold, in apparent agreement with the alleged consensus of the Banbury Conference held only a few months earlier.[248] However, the epidemiological evidence is ambiguous on this point; such results could arise merely from lack of statistical power, e.g., from an insufficient number of subjects.[249]

Another argument is that much less cancer was observed in the exposed chemical workers than was expected based on rat experiments. The question of whether TCDD is less potent in humans than in animals were also discussed during EPA's 1988 reassessment.[250] The NIOSH data presented a better opportunity to test this idea. One can estimate the carcinogenic slope factor of TCDD from the increased cancer mortality and the dose these men received, projected from the current serum levels. The results are approximately the same as, or in some cases higher than, those derived by the USEPA based on rat data.[251,252,*] Similar results may emerge from the USEPA's reassessment. While the impact of coexposure to other compounds cannot be completely excluded, these results argue against the premise that TCDD is a weaker carcinogen in humans than in animals.

Although the epidemiological studies of the last few years have yielded significant new evidence for TCDD's carcinogenicity in humans, arguments continue to rage.[253–258] In the meantime, other signs of danger come from a completely different direction.

7. REPRODUCTIVE AND DEVELOPMENTAL EFFECTS

7.1. Livestock and Wildlife

Twentieth century chlorine chemistry has exposed livestock and wildlife as well as humans to dioxinlike compounds (Table 1). A mysterious cattle malady known as "X disease" marked by thickened skin (hyperkeratosis) was described in 1947[259]; its cause was later determined to be chlorinated naphthalenes.[260] Polybrominated biphenyls were inadvertently added to cattle feed in Michigan in 1973–1974[261] (some PBBs have dioxinlike activity). This episode has been called "the most costly and disastrous accidental contamination ever to occur in United States agriculture."[262] Sickness and death of horses and other animals was one of the first signs of trouble at Times

*Rough dose estimates are obtainable from current serum levels using the half-life of the TCDD, approximate first-order kinetics, the time since end of exposure, and certain assumptions about the type and length of exposure. The incremental lifetime average daily dose for the "high" exposed group was estimated to be about 80 pg/kg.[252] Slope factor estimates were obtained using additive and relative risk models.[250,252]

Beach. Chick edema disease was first described in 1957 in the southeastern United States.[263,264] Millions of chickens have since died or had to be killed as a result of consuming feed contaminated with dioxinlike compounds. Several outbreaks were caused by higher chlorinated PCDDs and PCDFs present in "toxic fat" from animal hides treated with chlorophenols.[260,265,266]

While these episodes with livestock were the result of specific contaminations, a much more ominous phenomenon appeared in wildlife during the 1960s (reviewed in Refs. 266, 267). Epidemics of reproductive and developmental problems have since been observed in fish-eating birds from the Great Lakes and elsewhere. There is good evidence that some of these episodes have been caused by dioxinlike compounds. The Great Lakes embryo mortality, edema and deformities syndrome (GLEMEDS) has very close parallels to chick edema disease. The effects are consistent with the results of laboratory studies with dioxinlike compounds.[266,267] In one of the most striking demonstrations, a strong inverse correlation was found between the egg mortality of double-crested cormorants and the levels of dioxinlike compounds found in the eggs, as measured by enzyme induction.[268] Forster's terns from Green Bay hatched fewer eggs in 1983 than those from a less contaminated area. Eggs from the contaminated area had significantly higher levels of dioxinlike compounds.[269] A cross-fostering experiment—in which eggs were switched between the two areas—showed that parental behavior played a role as well. Adults from the contaminated area took less care of their eggs.[269] The neurotoxic effects of certain nondioxinlike PCBs may have contributed to the latter problem.[15] Dioxinlike compounds may cause bill deformities and embryonic abnormalities in double-crested cormorants and Caspian terns.[270–272]

Although certain locations such as Green Bay have a particularly high incidence of GLEMEDS, the problem appears to be relatively widespread in the Great Lakes. Egg mortality and bill defects of double-crested cormorants are generally more prevalent in Great Lakes birds than in reference areas.[268,270] Persistent dioxinlike PCBs appear to be the major problem,[268,269] although outbreaks among Lake Ontario herring gulls in the 1970s were probably caused mainly by TCDD discharged from chemical manufacturing and waste dumps on the Niagara River.[266]

Reproductive and developmental problems have also been observed in lake trout, which are particularly sensitive to TCDD during early development. Failure of restocked fish to reproduce in Lake Ontario during the 1970s was probably related to TCDD and dioxinlike compounds.[109,273] The decreased number of mink and otter found around the Great Lakes may be connected to PCBs and dioxinlike compounds, although more research may be needed before causality can be inferred.[274] Great blue heron chicks living near a pulp mill in British Columbia suffered depressed growth and greater edema than birds from a less contaminated area.[275] Crossed bills have been observed in cormorants from a number of other North American locations.[270] Concentrations of dioxinlike compounds in the yolk sacs of Dutch cormorants correlated with developmental impairment.[14]

These experiences indicate that the developing animal may be particularly sensitive to dioxinlike compounds. Persistent dioxinlike compounds bioaccumulate in the adult and are passed along to the vulnerable developing organism. These studies also suggest that entire ecosystems are or have been overburdened with dioxinlike compounds. This is a profoundly disturbing finding as it implies that significant problems may be related to general environmental contamination rather than specific accidents or spills. It also serves as a warning of potential hazards to the human population. In this respect, wildlife can be thought of as the modern equivalent of the miner's canary.

7.2. Sensitive Reproductive and Developmental Effects in Mammals

The dioxinlike compounds perturb hormones and growth factors, powerful regulators of growth and development. This suggests that organisms may be quite susceptible to the effects of these compounds during certain periods of development. Considerable laboratory work has been performed in this field and in reproductive toxicology (reviewed in Refs. 148, 276, 277). We will only discuss some of the most sensitive results (Table 3).

Reproductive effects occur in rats at very low doses. Male and female rats exposed to TCDD over three generations showed decreased fertility at 10 ng/kg per day,[278] with controversy over indications of effects at a tenfold lower dose.[276,279]

Central nervous system effects have been found in rhesus monkeys exposed to TCDD *in utero* and through lactation. The infant monkeys displayed "subtle" alterations in certain learning behaviors at chronic maternal doses as low as 0.13 ng/kg per day.[280,281] Alterations of reproductive behavior have not been reported. Prenatal mortality was increased at doses of 0.64 ng/kg per day.[282]

Strong evidence for the sensitivity of transgenerational effects comes from elegant studies of the developing rat. Earlier we mentioned the TCDD-related reduction of testosterone in adult male rats. The dose causing a 50% change (ED_{50}) for this and related effects is about 15 μg/kg.[283] Recent work shows that the developing male rat is much more sensitive than the mature animal. The ED_{50} for changes in spermatogenesis, size of secondary sex organs, and sexual behavior was only 0.16 μg/kg for a single dose administered to the mother on day 15 of pregnancy. Some effects were seen at the lowest dose tested, 0.064 μg/kg. By day 15 of gestation most organs have been generated and the testis begins to secrete testosterone.[284–287] Since chronic dosing of the mother leads to accumulation in her tissue and transfer to the fetus, even lower daily doses would be expected to produce the same results. A cross-fostering experiment would determine whether transplacental or lactational exposure plays the dominant role.

Some of these effects may be caused by TCDD-induced reduction of hormone levels during critical periods of development. In the rat, testosterone and its metabolites play a crucial role in development of male accessory sex organs, spermatogenesis, and the central nervous system. It is thought that normal development of reproductive behavior in the adult male rat requires the production of 17β-estradiol from testosterone in the brain near the end of pregnancy or soon after birth. TCDD may interfere with this process by reducing testosterone levels, slowing production of estradiol or "downregulating" brain estrogen receptors.[286,288] Examination of certain brain structures in these animals, e.g., the sexually dimorphic nucleus of the preoptic

Table 3
Selected Reproductive and Developmental Effects of TCDD and Related Compounds[a]

Body burden (ng/kg)	LOAEL Dose (ng/kg/day)	Species	Effect	Comment
290[b]	10	Rat	Litter size[c]	Steady-state body burden in F_1 and F_2
113[b]	0.64	Rhesus monkey	Prenatal mortality[d]	Maternal body burden from chronic dosing
[64][e]	64 single dose	Rat	Depressed spermatogenesis and altered reproductive behavior	Single dose to dam on gestation day 15
11–≥25[f] 45–≥154[g]	?	Human	Reduced serum testosterone in adult male	Current TEQ body burden and estimate at end of high exposure
22[b]	0.13	Rhesus monkey	Object learning[h]	Maternal body burden from chronic dosing
7[i]	0.001–0.003	Human	?	Average U.S. body burden and chronic dose of PCDD/PCDF (TEQ)

[a]This table reflects in part the work of C. L. Hughes, Jr., and T. Webster while serving on the USEPA's dioxin reassessment review committee. The agency has not taken an official position yet.
[b]Body burden estimates from an addendum table provided by R. Peterson.[277]
[c]Murray *et al.*[278] Nisbet and Paxton[279] argued for a LOAEL of 1 ng/kg/per day, corresponding to a body burden of 29 ng/kg.
[d]Bowman *et al.*[282]
[e]Transient body burden arising from a single dose. This may not be comparable to chronic body burdens as a result of differences in distribution. Mably *et al.*[286,287]
[f]Reduced testosterone was more prevalent in workers with serum lipid TCDD levels of 20–75 ppt.[292,293] Assume that almost all TEQ is in body lipid which makes up roughly 20–25% of body mass. Body burden of TEQ was approximated by dividing of 4 and adding "background" body burden of other PCDD and PCDF.
[g]It is unknown if the current effect is related to the current body burden, highest body burden in the past, or some cumulative effect of long-term exposure. The body burden of TCDD at the end of exposure was estimated using first-order kinetics, a 7.1-year half-life, and 21 years since last exposure.
[h]Bowman *et al.*[280], Schantz and Bowman.[281]
[i]Adipose tissue levels from 1987 NHATS[318] were converted to TEQ and divided by 4 to approximate body burden.

area of the hypothalamus, would be interesting.[289,290] Although effects on females have been studied less, females exposed to dioxinlike PCBs *in utero* and through lactation also display altered reproductive behavior.[291]

7.3. Human Effects

Loss of libido was reported early on in chemical workers exposed to TCDD,[205] providing a possible sign of hormonal effects in humans. More definitive evidence was reported quite recently. Reduced testosterone and/or elevated luteinizing hormone was found in occupationally exposed men.[292,293] In one statistical analysis, reduced testosterone correlated with serum TCDD levels in men who handled Agent Orange in Vietnam ("Ranch Hands").[294]

The experimental studies with rats suggest that dioxinlike compounds may cause reproductive effects in developing human males at maternal doses that are much lower than those associated with testosterone reduction in men exposed as adults. Although exposure would take place transplacentally or via lactation, many of these effects would not be observed until puberty. No studies appear to have examined spermatogenesis or other reproductive effects in the offspring of women exposed to high levels of dioxinlike compounds. Such studies may now be possible as children of women exposed at Seveso, Vietnam, and the Japanese and Taiwanese rice oil incidents have reached or are nearing puberty.

Children of women exposed in the Yu-cheng incident showed a range of developmental effects including smaller size and abnormalities of the gums, nails, skin, teeth, and lungs. Psychomotor development was delayed relative to controls, an effect consistent with the neurobehavioral effects in monkeys exposed to TCDD. The wide spectrum of effects suggests a general developmental impact on tissue derived from ectoderm. Some of these effects have also been noted in the Yusho incident. PCDFs have been suggested as one of the principal toxic agents.[24,295] However, physical defects and developmental delay were weakly related. Nondioxinlike PCBs may play a role in the observed neurotoxicity, possibly via effects on dopamine and other mechanisms.[14,15,296,297]

Neurobehavioral effects and smaller birth size have also been noted in more general populations exposed to complex mixtures of dioxinlike and nondioxinlike compounds.[298–300] Reduced short-term memory was found in infants and 4-year-old children of women who consumed Lake Michigan fish. The effect was more highly correlated with levels of total PCBs in umbilical cord serum than with lactational exposure, suggesting an *in utero* effect.[298–300] This interpretation is supported by cross-fostering PCB experiments in rats.[301] A number of potential confounding factors were controlled statistically, including exposure to certain other toxics. However, a contribution by unreported xenobiotics is possible. The most disturbing aspect of the study is the relatively low exposure. The women in the study only consumed, on average, the equivalent of two to three salmon or lake trout meals per month.[298–300] Cognitive effects in children were associated with prenatal exposure to PCBs in a North Carolina cohort drawn from the general population. However, they did not persist as long as in the Michigan group.[297,302,303]

A recent Dutch study suggests hormonal changes in the general population. Levels of certain thyroid hormones in 38 healthy newborn infants depended significantly on the concentrations of TCDD equivalents in the breast milk of their mothers. The average concentration of TCDD equivalents in breast milk fat was 38 ng/kg in the "high" group and 19 ng/kg in the "low" group. These infants were exposed both *in utero* and via breast milk.[304] For comparison, average levels of about 20–30 ng/kg are found in breast milk fat from the general population of industrialized countries and roughly 3–13 ng/kg in less industrialized nations (with the exception of southern Vietnam).[44] A contribution to this effect by other (unreported) xenobiotics cannot be excluded. Levels of T_4 were elevated in the "high" group. Exposure to TCDD increases T_4 levels in hamsters and decreases the hormone in rats.[129]

In sum, the epidemiological evidence suggests that humans are susceptible to reproductive and developmental effects of dioxinlike compounds and that children exposed transplacentally are probably more sensitive than adults. Although the influence of other compounds may be present in some cases, the results appear to be generally consistent with the effects of dioxinlike compounds seen in animals.

7.4. Reproductive and Developmental Effects of the Background Human Body Burden*

Might dioxinlike compounds be causing reproductive and developmental effects in the general human population? This question can be investigated by comparing average exposure with the levels observed to cause effects. Table 3 lists some of the most sensitive effects identified from animal experiments and human studies. Several of these are transgenerational, in which the infant, exposed *in utero* and/or via lactation, is more sensitive than the adult. We assume that these effects are probably mediated by the Ah receptor and that current TCDD-equivalents factors are applicable.

A so-called tolerable daily dose of 10 pg/kg for reproductive and developmental effects is used by some governmental bodies (Table 2). This value is based on the multigenerational rat fertility experiment of Murray *et al.*[278] with a presumed NOAEL of 1 ng/kg per day and a safety factor of 100. By this logic, the average human daily background dose of 1–3 pg/kg TEQ is often considered "tolerable."

This conclusion may not be sufficiently protective of public health.[305] The monkey data shown in Table 3 may support a tolerable daily intake of 1 pg/kg or less. A reanalysis of the rat experiment indicated effects at 1 ng/kg per day,[279] although this conclusion is controversial.[276]

There is even less room for optimism when the data are examined in terms of tissue levels. Since most dioxinlike compounds are biologically persistent, they accumulate in the body. Such measures of internal exposure may be more relevant for many toxic effects than daily dose, a measure of the external rate of exposure.[280–282,306,307] In addition, direct comparison of doses ignores interspecies differences in the half-life of TCDD: on the order of 7–10 years in humans, months in monkeys, and several weeks in rats.[114,282] Given the same daily dose, the overall concentration in people at steady state will be considerably higher than that in rats.[115] Ideally, one would want to compare concentrations in target tissues over critical time periods, taking into account differences in distribution, sensitivity, and the biology of the endpoint. Human cells and tissues may be at least as sensitive as rat or mice cells and tissues to several biochemical endpoints (including induction of P450IA1) by dioxinlike compounds.[242,308] Since knowledge of the pharmacokinetics of these compounds is still fairly rudimentary (especially in humans), we will use a less ideal measure: body burden or overall concentration.

The simplest comparisons are between the average body burden found in people

*This section reflects in part the work of Dr. C. L. Hughes, Jr., and one of the authors (T.W.) while serving on the USEPA's dioxin peer review and risk characterization panel in September, 1992. The agency has not taken an official position yet.

and the steady-state body burdens that cause effects in animals. As Table 3 shows, effects occur in the monkey at body burdens that are uncomfortably close to the present average levels in the population of industrialized countries.

Abnormally low testosterone was two to three times more prevalent in workers with TCDD serum levels of 20–75 ppt than among controls with mean serum TCDD levels of 7 ppt.[293] Adding in background levels of other PCDDs and PCDFs, the workers' body burdens were about 11–25 ng/kg TEQ at the time of measurement. It is not known if the effect is related to the current levels of dioxinlike compounds, the highest body burden in the past, or some cumulative effect of long-term exposure. Hence, Table 3 also estimates their body burden at the end of high occupational exposure. Even these numbers are only 6–22 times higher than background.

These findings in adult men increase our concern about the possibility of even more sensitive effects in the developing male. The single maternal dose in rats that depresses spermatogenesis and alters reproductive behavior in male offspring is less than a factor of 10 above the average human body burden. This comparison may not be valid since acute and chronic doses differ in their distribution, at least initially. This uncertainty could be reduced with a chronic dosing experiment and better knowledge of the biology of the phenomenon, especially critical time periods. Reduction of human sperm counts has potentially serious consequences. Although the decreased spermatogenesis seen in the rats did not affect their fertility, rats produce a large excess of sperm. This is not true of humans.[287,288] There are some indications of a decline in human sperm counts since the 1930s.[309] Environmental factors, including the dioxinlike compounds, may have contributed to this trend.[310]

Unfortunately, reproductive and developmental effects may occur at body burdens that are within a factor of 10 of the average U.S. resident. For a number of reasons, this provides little or no margin of safety: (1) effects may be seen at lower body burdens (the figures in Table 3 are low effect levels); (2) body burden varies between individuals, with some members of the general population being higher than average; (3) some individuals may be more sensitive than others (shown for certain other Ah-mediated effects[308]); (4) the relative sensitivity of humans and animals to these effects is unknown; (5) inclusion of dioxinlike PCBs may significantly increase total body burden.[311–313]

8. TRENDS OF DIOXINLIKE COMPOUNDS IN THE ENVIRONMENT

Concentrations of TCDD in Lake Ontario sediment peaked in early 1960s,[273] shortly before the first observations of high embryo mortality in herring gulls.[266] Levels in sediment and herring gull eggs decreased dramatically in the early 1970s, probably as a result of the halting of 2,4,5-trichlorophenol production along the Niagara River.[314] This may have permitted the recovery of the lake's herring gull population from GLEMEDS.[266] However, TCDD levels in herring gull eggs and lake trout changed much less in the following decade. Chemical waste sites such as Love Canal may still be a major source.[314] PCDD and PCDF concentrations reached a maximum in the sediments of other Great Lakes in the 1970s and declined somewhat afterwards.[48] One likely cause was the passage in the early 1970s of the U.S. Clean Air Act and other environmental legislation.[48]

As noted above, the dioxinlike PCBs are thought to be biologically more significant than PCDDs and PCDFs in several Great Lakes wildlife incidents. Concentrations of PCBs in Lake Ontario sediments rose until manufacturing was banned in the 1970s and then began to fall.[315] Concentrations in Great Lakes biota initially declined after the ban, but there were some indications of a leveling off in the 1980s. PCBs may be reaching a steady state in this ecosystem with input from long-distance transport, emissions from remaining uses, and other sources roughly matching output.[268,316] This phenomenon may explain the continuing effects noted in Great Lakes double-crested cormorants.[268] Similar PCB trends have been seen in parts of the Baltic.[17]

If contaminant levels fall further, more subtle effects may be noticed. The susceptibility of experimental animals to neurobehavioral and hormonal disturbances suggests an examination of these effects in wildlife.[317] "Masking" of subtle effects by more overt ones has occurred earlier. High levels of DDE (a persistent metabolite of DDT) and associated egg shell thinning probably caused populations of double-crested cormorants to drop in the 1960s and 1970s. Yet, as DDE levels fell, GLEMEDS emerged.[268]

There may be a downward trend over time in the body burdens of PCDD and PCDF in the average U.S. resident. This is suggested by the National Human Adipose Tissue Survey (NHATS) results of 1982 and 1987 and a Veterans Administration/USEPA analysis of stored NHATS samples from 1971 to 1982 for males aged 17–46.[318] According to the authors, this trend may reflect several factors: real decline in body burdens, advances in analytical methods, and loss of integrity of tissue during storage.[318] While it is likely that at least some of the observed decline is real, this does not mean that the problem is solved. Relatively small declines in body burden may not provide a sufficient margin of safety for reproductive and developmental effects. We may be observing the delayed consequence of a drop in environmental levels of PCDD and PCDF in the late 1970s and early 1980s. The experience with PCBs in Great Lakes biota indicates that rapid initial declines may not continue. Although PCB levels in human adipose tissue were still decreasing slowly in the United States in 1985,[319] they may have reached steady state in Japan.[320]

9. CONCLUSIONS

The dioxin debate has largely centered around two issues. First, are humans less sensitive to TCDD than laboratory animals? The weight of evidence indicates that people experience many of the toxic responses observed in animals. In particular, we believe that there is now sufficient evidence that TCDD causes cancer in humans. The data suggest a quantitative similarity between the animal and human responses for some effects as well, including certain biochemical endpoints and carcinogenicity.

The second primary issue has been the choice of model for the low-dose effects of TCDD, in particular the appropriateness of a low-dose linear model for cancer. The possibility of an alternative threshold model for biological responses based on receptor mediation formed one of the motivations for the USEPA's most recent reassessment. Since some receptor-mediated biochemical responses appear to be linear at low dose, this simple model can now be discounted. Given the "background" dose of dioxinlike compounds experienced by residents of industrialized countries, questions about thresholds at lower doses—e.g., whether one molecule can cause effects—are essentially moot. Furthermore, if the mechanism by which TCDD causes cancer is additive to some ongoing process, then the cancer dose–response curve for dioxinlike compounds should be regarded as probably linear at low doses.[321]

Gaps clearly remain between our limited biochemical understanding of TCDD and the biology of cancer. A truly realistic model would have to take into account the complexity of the observed phenomenon that includes: Ah receptor mediation; multiple tumor sites; multiple biochemical mechanisms; indirect genotoxicity; anticarcinogenic effects; hormonal and growth factor involvement; stimulation of cell proliferation; interactions with other chemicals as well as viruses; and interindividual variation.

Interactions between dioxin and other compounds may pose a particularly difficult modeling problem. TCDD's enzyme-inducing ability may significantly affect the toxicity of other exogenous compounds. The result may depend on the specific compound and the dosing regime. Risk assessors typically treat the cancer risks of exposure to multiple carcinogens as being additive. While it is generally argued that this is a reasonable approximation for low-dose exposure, it is unclear that this applies when a "promoter" is concerned, especially one that may not have a threshold.[322] Nonadditive effects may also be seen in multifactor, noncancer endpoints. Recent work suggests that mixtures of dioxinlike and nondioxinlike PCBs have a synergistic effect on fetal mortality in the mink.[323]

Research on the biological mechanism of dioxin-induced effects is an endlessly fascinating scientific undertaking that will no doubt provide enough questions for decades of work. However, the construction of "biologically more realistic models"—a current goal of EPA—should not be used as an excuse for delay and inaction. We may aspire to ever closer approximations of reality, but the outcome of the dioxin reassessment so far reinforces something that we have—or should have—known for a long time: production of dioxin should be avoided. The experience with dioxin does not bode well for regulating other compounds. Recall that TCDD is one of the most studied of all toxic compounds. There are thousands of commercial chemicals produced in large quantities for which little or no toxicological information is available.[324]

The development of improved cancer models alone will not compensate for an inadequate regulatory structure. One premise of the focus on cancer is the notion that protecting for cancer will also guard against other effects. Yet, there is a growing concern that immunotoxicity and reproductive/developmental effects of dioxinlike compounds are more sensitive or important than cancer.[155,266,317,325] At current environmental loads, reproductive and developmental effects in Great Lakes wildlife appear to be a great problem than cancer.[266,325] But there are subtleties to the sensitivity question. For instance, the relative position of apparent "no-effect levels" depends not only on the biological phenomenon, but also on the measurement techniques and the statistical power of the experiment. The shapes of the dose–response curves for cancer and other effects remain unclear. The relative importance of various effects at current body burdens and environmental levels is an even more difficult question, akin to asking whether a certain increased cancer risk is more or less important than a certain deficit in cognitive functioning.

However, the premise that standards based on cancer will prevent other effects certainly fails in another respect. Current USEPA regulations of dioxin are based on so-called "acceptable" risks of cancer from *incremental* exposure, i.e., exposure to one source or through one medium. For instance, the water quality standards ignore other routes of exposure and disregard dioxinlike compounds other than TCDD. Even using traditional methodology, this procedure fails to ensure that total exposure of dioxinlike compounds—incremental plus background—falls below tolerable doses for noncarcinogenic effects.

Indeed, the current background levels of dioxinlike compounds in humans may pose a reproductive and developmental hazard. The reproductive and developmental problems observed in certain species of wildlife indicate that some ecosystems are overburdened with these compounds. Instead of spending another decade arguing about the shape of the dose–response curve, we believe that the focus of policy needs to shift toward simply reducing exposure. The most sensible approach is pollution prevention: elimination of sources.

The International Joint Commission, the U.S.–Canadian agency with environmental responsibilities regarding the Great Lakes, has already made a path-breaking recommendation in this direction.[326,327] The Commission calls for "zero discharge" of persistent toxic substances, including PCBs and dioxin, into the Great Lakes ecosystem. These and other persistent toxic compounds—some undoubtably still unidentified—are products or by-products of industrial chlorine chemistry and the combustion of chlorine-containing fuels. Arguing that chemical-by-chemical regulation has largely failed in this arena, the Commission advocates a precautionary approach: "the Parties, in consultation with industry and other affected interests, develop timetables to sunset the use of chlorine and chlorine-containing compounds as industrial feedstocks and that the means of reducing or eliminating other uses be examined."[327] Sunsetting means restriction, phase out, and eventual banning of the substance. After studying air quality in the Detroit–Windsor/Port Huron–Sarnia region, the Commission also recommended that "incineration facilities in the region be phased out of use or required to eliminate the production and emission of dioxins, furans, PCBs. . . ."[328]

Because of its notoriety, dioxin remains a target of those who wish to convince the public that environmental contamination is not a problem.[329,330] Much of the media coverage of the dioxin debate has consisted of trying to convince the public that their common sense is wrong and that experts know best. In this case, the public's view has been largely correct. Dioxin is a dangerous and unwanted environmental pollutant. Dioxin policy should strive to eliminate the sources and thereby prevent pollution. Rather than telling the public that certain dioxin risks are "acceptable"—and regarding the resultant opposition as merely hysterical and uninformed—regulators ought to listen to their concerns. The public does not respond according to a one-dimensional measure of risk, but in keeping with a much richer set of criteria. These include, not surprisingly, fairness, democratic choice, and an examination of alternatives.[331]

Note added in proof. The case for potential hazards at current body burdens of dioxin-like compounds is strengthened by the recent observation of increased prevalence and severity of endometriosis in rhesus monkeys chronically exposed to TCDD.[332] The implications of this finding are discussed by Webster.[333]

REFERENCES

1. Editorial, Dioxin destroys ships, *Wall Street Journal* January 25 (1993).
2. The Chlorine Institute, Conference Summary Report: Biological Basis for Risk Assessment of Dioxins and Related Compounds, Banbury Center, NY, October 21–24, 1990, and Washington, DC, November 1 (1990).
3. Symposium on chemophobia, *Chemosphere* **15,** N1–N45 (1986).
4. Thc nanogram mafia, *Wall Street Journal* June 29 (1993).
5. T. Uhlenbrock, Dioxin scare now called mistake, *St. Louis Post-Dispatch* May 23 (1991).
6. K. Schneider, U.S. officials say dangers of dioxin were exaggerated, *New York Times* August 12 (1991).
7. Testimony of Dr. Barry Johnson, Subcommittee on Human Resources and Intergovernmental Relations, Committee on Government Operations, U.S. House of Representatives, June 10 (1992).
8. R. Gutfeld, Dioxin's health risks may be greater than believed, EPA memo indicates, *Wall Street Journal* October 16 (1992).
9. S. Safe, Polychlorinated biphenyls (PCBs), dibenzo-p-dioxins (PCDDs), dibenzofurans (PCDFs), and related compounds: Environmental and mechanistic considerations which support the development of toxic equivalency factors (TEFs), *CRC Crit. Rev. Toxicol.* **21,** 51–88 (1990).
10. S. Safe, Development, validation and limitations of toxic equivalency factors, *Chemosphere* **25,** 61–64 (1992).
11. A. Hanberg, F. Waern, L. Asplund, E. Haglund, and S. Safe, Swedish Dioxin Survey: Determination of 2,3,7,8-TCDD toxic equivalent factors for some polychlorinated biphenyls and naphthalenes using biological tests, *Chemosphere* **20,** 1161–1164 (1990).
12. A. Poland and J. Knutson, 2,3,7,8-Tetrachlorodibenzo-p-dioxin and related halogenated aromatic hydrocarbons: Examination of the mechanism of toxicity, *Annu. Rev. Pharmacol. Toxicol.* **22,** 517–554 (1982).
13. J. Goldstein and S. Safe, Mechanism of action and structure–activity relationships for the chlorinated dibenzo-p-dioxins and related compounds, in: *Halogenated Biphenyls, Terphenyls, Naphthalenes, Dibenzodioxins and Related Products* (R. D. Kimbrough and A. A. Jensen, eds.), pp. 239–293, Elsevier, Amsterdam (1989).
14. U. Ahlborg, A. Brouwer, M. Fingerhut, J. Jacobson, S. Jacobson, S. Kennedy, A. Kettrup, J. Koeman, H. Poiger, C. Rappe, S. Safe, R. Seegal, J. Tuomisto, and M. van den Berg, Impact of polychlorinated dibenzo-p-dioxins, dibenzofurans, and biphenyls on human and environmental health, with special emphasis on application of the toxic equivalency factor concept, *Eur. J. Pharmacol. Environ. Toxicol. Pharmacol. Sect.* **228,** 179–199 (1992).
15. R. Seegal, B. Bush, and W. Shain, Neurotoxicity of ortho-substituted polychlorinated biphenyls, *Chemosphere* **23,** 1941–1949 (1991).

16. P. De Voogt and U. Brinkman, Production, properties and usage of polychlorinated biphenyls, in: *Halogenated Biphenyls, Terphenyls, Naphthalenes, Dibenzodioxins and Related Products* (R. D. Kimbrough and A. A. Jensen, eds.), pp. 3–45, Elsevier, Amsterdam (1989).
17. P. Reijnders and S. Brasseur, Xenobiotic induced hormonal and associated developmental disorders in marine organisms and related effects in humans: An overview, in: *Chemically-Induced Alterations in Sexual and Functional Development: The Wildlife/Human Connection* (T. Colborn and C. Clement, eds.), pp. 159–174, Princeton Scientific Publishing Co., Princeton, NJ (1992).
18. J. Kimmig and K. H. Schultz, Chlorierte aromatische zyklische. Ather also Ursache der Sogenannten Chlorakne, *Naturwissenschaften* **44,** 337–338 (1957).
19. A. Hay, *The Chemical Scythe: Lessons of 2,4,5-T and Dioxin,* Plenum Press, New York (1982).
20. E. McConnell, J. Huff, M. Hejtmancik, A. Peters, and R. Persing, Toxicology and carcinogenesis studies of two grades of pentachlorophenol in B6C3F1 mice, *Fundam. Appl. Toxicol.* **17,** 519–532 (1991).
21. C. Van Strum and P. Merrell, *The Politics of Penta,* Greenpeace USA (1989).
22. H. Fiedler, O. Hutzinger, and C. Timms, Dioxins: Sources of environmental load and human exposure, *Toxicol. Environ. Chem.* **29,** 157–234 (1990).
23. M. Kuratsune, Yusho, with reference to Yu-Cheng, in: *Halogenated Biphenyls, Terphenyls, Naphthalenes, Dibenzodioxins and Related Products* (R. D. Kimbrough and A. A. Jensen, eds.), pp. 381–400, Elsevier, Amsterdam (1989).
24. W. Rogan, Yu-Cheng, in: *Halogenated Biphenyls, Terphenyls, Naphthalenes, Dibenzodioxins and Related Products* (R. D. Kimbrough and A. A. Jensen, eds.), pp. 401–415, Elsevier, Amsterdam (1989).
25. A. Schecter and K. Charles, The Binghamton State Office Building transformer incident after one decade, *Chemosphere* **23,** 1307–1321 (1991).
26. P. O'Keefe and R. Smith, PCB capacitor/transformer accidents, in: *Halogenated Biphenyls, Terphenyls, Naphthalenes, Dibenzodioxins and Related Products* (R. D. Kimbrough and A. A. Jensen, eds.), pp. 417–444, Elsevier, Amsterdam (1989).
27. K. Olie, P. Vermeulen, and O. Hutzinger, Chlorodibenzo-p-dioxins and chlorodibenzofurans are trace components of fly ash and flue gas of some municipal incinerators in the Netherlands, *Chemosphere* **8,** 455–459 (1977).
28. J. Lustenhouwer, K. Olie, and O. Hutzinger, Chlorinated dibenzo-p-dioxins and related compounds in incinerator effluents: A review of measurements and mechanisms of formation, *Chemosphere* **9,** 501–522 (1980).
29. G. Eiceman and H. Rghei, Chlorination reactions of 1,2,3,4-tetrachlorodibenzo-p-dioxin on fly ash with HCl in air, *Chemosphere* **11,** 833–839 (1982).
30. B. Commoner, M. McNamara, K. Shapiro, and T. Webster, Environmental and Economical Analysis of Alternative Municipal Waste Disposal Technologies. II. The Origins of Chlorinated Dioxins and Dibenzofurans Emitted by Incinerators that Burn Unseparated Municipal Solid Waste, and an Assessment of Methods for Controlling Them, CBNS, Queens College, Flushing, NY (1984).
31. B. Commoner, K. Shapiro, and T. Webster, The origin and health risks of PCDD and PCDF, *Waste Manage. Res.* **5,** 327–346 (1987).
32. Environment Canada, *The National Incinerator Testing and Evaluation Program: Two-stage Combustion (Prince Edward Island),* EPS 3/UP/1 (1985).
33. H. Vogg and L. Stieglitz, Thermal behavior of PCDD/PCDF in fly ash from municipal incinerators, *Chemosphere* **15,** 1373–1378 (1986).
34. R. Hoffman, G. Eiceman, Y.-T. Long, M. Collins, and M.-Q. Lu, Mechanism of chlorina-

tion of aromatic compounds adsorbed on the surface of fly ash from municipal incinerators, *Environ. Sci. Technol.* **24,** 1635–1641 (1990).
35. L. Stieglitz, H. Vogg, G. Zwick, J. Beck, and H. Bautz, On formation conditions of organohalogen compounds from particulate carbon of fly ash, *Chemosphere* **23,** 1255–1264 (1991).
36. D. Lenoir, A. Kaune, O. Hutzinger, G. Muetzenich, and K. Horch, Influence of operating parameters and fuel type on PCDD/F emissions from a fluidized bed incinerator, *Chemosphere* **23,** 1491–1500 (1991).
37. J. Vikelsoe, P. Nielsen, P. Blinksbjerg, H. Madsen, and O. Manscher, Significance of chlorine sources for the generation of dioxins during incineration of MSW, Presented at Dioxin '90, Bayreuth, *Organohalogen Compounds,* Vol. 3, pp. 193–196 (1990).
38. I. Fangmark, S. Marklund, C. Rappe, B. Stromberg, and N. Berge, Use of a synthetic refuse in a pilot combustion system for optimizing dioxin emission, Part II. *Chemosphere* **23,** 1233–1243 (1991).
39. R. Rawls, Dow finds support, doubt for dioxin ideas, *Chem. Eng. News* February 12, pp. 23–29 (1979).
40. R. Bumb, W. Crummett, S. Artie, J. Gledhill, R. Hummel, R. Kagel, L. Lamparski, E. Luoma, D. Miller, T. Nestrick, L. Shadoff, R. Stehl, and J. Woods, Trace chemistries of fire: A source of chlorinated dioxins, *Science* **210,** 385–390 (1980).
41. T. Nestrick and L. Lamparski, Isomer-specific determination of chlorinated dioxins for assessment of formation and potential environmental emission from wood combustion, *Anal. Chem.* **54,** 2292–2299 (1982).
42. A. Sheffield, Sources and releases of PCDD's and PCDF's to the Canadian environment, *Chemosphere* **14,** 811–814 (1985).
43. G. Gribble, Naturally occurring organohalogen compounds—A survey, *J. Nat. Products (Lloydia)* **55,** 1353–1395 (1992).
44. A. Schecter, Dioxins and related chemicals in humans and the environment, in: *Biological Basis for Risk Assessment of Dioxins and Related Compounds* (M. A. Gallo, R. J. Scheuplein, and K. A. van der Heijden, eds.), pp. 169–212, Banbury Report 35, Cold Spring Harbor Laboratory Press, Cold Spring Harbor, NY (1991).
45. J. Czuczwa and R. Hites, Environmental fate of combustion-generated polychlorinated dioxins and furans, *Environ. Sci. Technol.* **18,** 444–450 (1984).
46. J. Czuczwa, B. McVeety, and R. Hites, Polychlorinated dibenzo-p-dioxins and dibenzofurans in sediments from Siskiwit Lake, Isle Royale, *Science* **226,** 568–569 (1984).
47. J. Czuczwa, F. Niessen, and R. Hites, Historical record of polychlorinated dibenzo-p-dioxins and dibenzofurans in Swiss lake sediments, *Chemosphere* **14,** 1175–1179 (1985).
48. J. Czuczwa and R. Hites, Airborne dioxins and dibenzofurans: Sources and fates, *Environ. Sci. Technol.* **20,** 195–200 (1986).
49. H. Hagenmaier, H. Brunner, R. Haag, and A. Berchtold, PCDDs and PCDFs in sewage sludge, river and lake sediments from south west Germany, *Chemosphere* **15,** 1421–1428 (1986).
50. R. Smith, P. O'Keefe, K. Aldous, R. Briggs, D. Hilker, and S. Connor, Measurements of PCDFs and PCDDs in air samples and lake sediments at several locations in upstate New York, *Chemosphere* **25,** 95–98 (1992).
51. A. Schecter, A. Dekin, N. Weerasinghe, S. Arghestani, and M. Gross, Sources of dioxins in the environment: A study of PCDDs and PCDFs in ancient, frozen Eskimo tissue, *Chemosphere* **17,** 627–631 (1988).
52. H. Tong, M. Gross, A. Schecter, S. Monson, and A. Dekin, Sources of dioxins in the environment. Second stage study of PCDD/Fs in ancient human tissue and environmental samples, *Chemosphere* **20,** 987–992 (1990).

53. W. Ligon, Jr., S. Dorn, R. May, and M. Allison, Chlorodibenzofuran and chlorodibenzo-p-dioxin levels in Chilean mummies dated to about 2800 years before the present, *Environ. Sci. Technol.* **23,** 1286–1290 (1989).
54. L.-O. Kjeller, K. Jones, A. Johnston, and C. Rappe, Increases in the polychlorinated dibenzo-p-dioxin and -furan content of soils and vegetation since the 1840s, *Environ. Sci. Technol.* **25,** 1619–1627 (1991).
55. S. Harrad, A. Fernandes, C. Creaser, and E. Cox, Domestic coal combustion as a source of PCDDs and PCDFs in the British environment, *Chemosphere* **23,** 255–261 (1991).
56. R. Griffin, A new theory of dioxin formation in municipal solid waste combustion, *Chemosphere* **15,** 1987–1990 (1986).
57. B. Gullett, K. Bruce, and L. Beach, Effect of sulfur dioxide on the formation mechanism of polychlorinated dibenzodioxin and dibenzofuran in municipal waste combustors, *Environ. Sci. Technol.* **26,** 1938–1943 (1992).
58. L. Niedringhaus Davis, *The Corporate Alchemists,* Morrow, New York (1984).
59. G. Lindner, A. Jenkins, J. McCormack, and R. Adrian, Dioxins and furans in emissions from medical waste incinerators, *Chemosphere* **20,** 1793–1800 (1990).
60. P. Costner, The Incineration of Dioxin in Jacksonville, Arkansas: A Review of Trial Burns and Related Air Monitoring at Vertac Site Contractors Incinerator, Jacksonville, AR, Greenpeace Toxics Campaign, Washington, DC (1992).
61. S. Lowrance, Director, Office of Solid Waste, USEPA, Memorandum: Assuring Protective Operation of Incinerators Burning Dioxin-listed Wastes, September 22 (1992).
62. Environmental Research Foundation, New memo says all hazardous waste incinerators fail to meet regulations, *Rachel's Hazardous Waste News #312,* P.O. Box 5036, Annapolis, MD 21403-7036, November 18 (1992).
63. S. Marklund, R. Andersson, M. Tysklind, C. Rappe, K.-E. Egeback, E. Bjorkman, and V. Grigoriadis, Emissions of PCDDs and PCDFs in gasoline and diesel fueled cars, *Chemosphere* **20,** 553–561 (1990).
64. H. Hagenmaier, N. Dawidowsky, U. Weberruss, O. Hutzinger, K. Schwind, H. Thoma, U. Essers, U. Buehler, and R. Greiner, Emission of polyhalogenated dibenzodioxins and dibenzofurans from combustion-engines, Presented at Dioxin '90, Bayreuth, *Organohalogen Compounds,* Vol. 2, pp. 329–334 (1990).
65. B. Hrutfiord and A. Negri, Dioxin sources and mechanisms during pulp bleaching, *Chemosphere* **25,** 53–56 (1992).
66. W. Sandermann, Polychlorierte aromatische Verbindungen als Umweltgifte, *Naturwissenschaften* **61,** 207–213 (1974).
67. USEPA, Office of Water Regulations and Standards, *The National Dioxin Study, Tiers 3,5,6 and 7,* EPA 440/4-87-003, February (1987).
68. C. Van Strum and P. Merrell, *No Margin of Safety: A Preliminary Report on Dioxin Pollution and the Need for Emergency Action in the Pulp and Paper Industry,* Greenpeace USA (1987).
69. G. Amendola, D. Barna, R. Blosser, L. LaFleur, A. McBride, F. Thomas, T. Tiernan, and R. Whittemore, The occurrence and fate of PCDDs and PCDFs in five bleached Kraft pulp and paper mills, *Chemosphere* **18,** 1181–1188 (1989).
70. USEPA, Office of Water Regulations and Standards, *USEPA/Paper Industry Cooperative Dioxin Study "The 104 Mill Study" Statistical Findings and Analysis,* July 13 (1990).
71. W. Drechsler, The formation of PCDDs and PCDFs from the industrial use of chlorinated compounds, Presented at Dioxin '92, Tampere, Finland, *Organohalogen Compounds,* Vol. 8, p. 231 (1991).
72. T. Thompson, R. Clement, N. Thornton, and J. Luyt, Formation and emission of PCDDs/PCCFs in the petroleum refining industry, *Chemosphere* **20,** 1525–1532 (1990).

73. W. Christmann, K. Kloeppel, H. Partscht, and W. Rotard, Determination of PCDD/PCDF in ambient air, *Chemosphere* **19,** 521–526 (1989).
74. A. Riss, H. Hagenmaier, U. Weberruss, C. Schlatter, and R. Wacker, Comparison of PCDD/PCDF levels in soil, grass, cow's milk, human blood and spruce needles in an area of PCDD/PCDF contamination through emissions from a metal reclamation plant, *Chemosphere* **21,** 1451–1456 (1990).
75. J. Aittola, J. Paasivirta, and A. Vattulainen, Measurements of organochlorine compounds at a metal reclamation plant, Presented at Dioxin '92, Tampere, Finland, *Organohalogen Compounds,* Vol. 9, pp. 9–12 (1992).
76. M. Oehme, S. Mano, and B. Bjerke, Formation of polychlorinated dibenzofurans and dibenzo-p-dioxins by production processes for magnesium and refined nickel, *Chemosphere* **18,** 1379–1389 (1989).
77. M. Tysklind, G. Soderstrom, C. Rappe, L.-E. Hagerstedt, and E. Burstrom, PCDD and PCDF emissions from scrap metal melting processes at a steel mill, *Chemosphere* **19,** 705–710 (1989).
78. C. Rappe, L.-O. Kjeller, and S.-E. Kulp, Levels, profile and pattern of PCDDs and PCDFs in samples related to the production and use of chlorine, *Chemosphere* **23,** 1629–1636 (1991).
79. A. Heindl and O. Hutzinger, Search for industrial sources of PCDD/PCDF. III. Short-chain chlorinated hydrocarbons, *Chemosphere* **16,** 1949–1957 (1987).
80. E. Evers, J. von Berghem, and K. Olie, Exploratory data analysis of PCDD and PCDF measurements in sediments from industrialized areas, *Chemosphere* **19,** 459–466 (1989).
81. E. Evers, M. Buring, K. Olie, and H. Govers, Catalytic oxychlorination processes of aliphatic hydrocarbons as new industrial sources of PCDDs and PCDFs, Presented at Dioxin '89, Toronto, Ontario, Abstract SOU 14 (1989).
82. Greenpeace International, Dioxin Factories: A Study of the Creation and Discharge of Dioxins and Other Organochlorines from the Production of PVC, Amsterdam (1993).
83. A. Miller, Dioxin emissions from EDC/VCM plants, *Environ. Sci. Technol.* **27,** 1014–1015 (1993).
84. *Chem. Eng. News,* Facts and figures for the chemical industry, June 29, p. 38 (1992).
85. L. Oberg, Presented at Dioxin '92, Tampere, Finland.
86. L. Oberg and C. Rappe, Biochemical formation of PCDD/Fs from chlorophenols, *Chemosphere* **25,** 49–52 (1992).
87. A. Riss and K. Scheidl, Statutory and research action in Austria, Presented at Dioxin '90, Bayreuth, *Organohalogen Compounds,* Vol. 4, pp. 51–56 (1990).
88. H. Bremmer and A. Sein, Sources of dioxins in the Netherlands, Presented at Dioxin '91, Research Triangle Park, NC (1991).
89. C. Rappe, Sources of PCDDs and PCDFs. Introduction. Reactions, levels, patterns, profiles and trends. *Chemosphere* **25,** 41–44 (1992).
90. S. Harrad and K. Jones, A source inventory and budget for chlorinated dioxins and furans in the United Kingdom environment, *Sci. Total Environ.* **126,** 89–107 (1992).
91. L. Fedorov and B. Myasoedov, Dioxins: Analytical chemical aspects of the problem, *Russ. Chem. Rev.* **59**(11)**,** 1063–1092 (1990). Translated from *Usp. Khim.* **59,** 1818–1866 (1990).
92. L. A. Fedorov, Ecological problems in Russia caused with dioxin emissions of chemical industry, Presented at Dioxin '92, Tampere, Finland, *Organohalogen Compounds,* Vol. 9, pp. 75–78 (1992).
93. L. A. Fedorov, Dioxin problems in Russian chemical industry, Presented at Dioxin '92, Tampere, Finland, *Organohalogen Compounds,* Vol. 10, pp. 313–316 (1992).

94. E. Green, Poisoned legacy: Environmental quality in the newly independent states, *Environ. Sci. Technol.* **27,** 590–595 (1993).
95. C. Rappe, Sources of and human exposure to PCDDs and PCDFs, in: *Biological Basis for Risk Assessment of Dioxins and Related Compounds* (M. A. Gallo, R. J. Scheuplein, and K. A. van der Heijden, eds.), pp. 121–129, Banbury Report 35, Cold Spring Harbor Laboratory Press, Cold Spring Harbor, NY (1991).
96. Energy and Environmental Research Corporation, Permit Compliance Test Report, Columbus Solid Waste Reduction Facility No. 6, Submitted to Ohio EPA, September 11 (1992). [Unit 6 (the only one tested) emitted at an average rate of about 8.8 μg/sec.]
97. Work On Waste USA, *Waste Not #122* (quoting the Sud Deutsche Zeitung of 9/24/90), 82 Judson St., Canton, NY, October 25 (1990).
98. D. Mackay, W.-Y. Shiu, and K. Ma, *Illustrated Handbook of Physical–Chemical Properties and Environmental Fate of Organic Chemicals Volume II,* Lewis Publishers, Chelsea, MI (1992).
99. C. Koester and R. Hites, Photodegradation of polychlorinated dioxins and dibenzofurans adsorbed to fly ash, *Environ. Sci. Technol.* **26**(3)**,** 502–507 (1992).
100. R. Atkinson, Atmospheric lifetimes of dibenzo-p-dioxins and dibenzofurans, *Sci. Total Environ.* **104,** 17–33 (1991).
101. P. Connett and T. Webster, An estimation of the relative human exposure to 2,3,7,8-TCDD emissions via inhalation and ingestion of cow's milk, *Chemosphere* **16,** 2079–2084 (1987).
102. T. Webster and P. Connett, The use of bioconcentration factors in estimating the 2,3,7,8-TCDD content of cow's milk, *Chemosphere* **20,** 779–786 (1990).
103. M. McLachlan and O. Hutzinger, Accumulation of organochlorine compounds in agricultural food chains, Presented at Dioxin '90, Bayreuth, *Organohalogen Compounds,* Vol. 1, pp. 479–484 (1990).
104. USEPA, *Estimating Exposure to Dioxin-Like Compounds: Draft,* EPA/600/6-88/005B (1992).
105. R. Guimond, Acting Assistant Administrator, USEPA Office of Solid Waste and Emergency Response, Memo to Carol Browner, Administrator: WTI Incinerator Issues, January 22 (1993).
106. Work On Waste USA, *Waste Not #226,* 82 Judson St., Canton, NY, January 30 (1993).
107. J. Aldrich, United States District Court, Northern District of Ohio, Eastern Division, Order: Greenpeace, Inc., et al. vs. Waste Technologies Industries, et al. March 5 (1993).
108. T. Clark, K. Clark, S. Paterson, D. Mackay, and R. Norstrom, Wildlife monitoring, modeling, and fugacity, *Environ. Sci. Technol.* **22,** 120–127 (1988).
109. P. Cook, D. Kuehl, M. Walker, and R. Peterson, Bioaccumulation and toxicity of TCDD and related compounds in aquatic ecosystems, in: *Biological Basis for Risk Assessment of Dioxins and Related Compounds* (M. A. Gallo, R. J. Scheuplein, and K. A. van der Heijden, eds.), pp. 143–165, Banbury Report 35, Cold Spring Harbor Laboratory Press, Cold Spring Harbor, NY (1991).
110. P. Jones, G. Ankley, D. Best, R. Crawford, N. DeGalan, J. Giesy, T. Kubiak, J. Ludwig, J. Newsted, D. Tillitt, and D. Verbrugge, Biomagnification of bioassay derived 2,3,7,8-tetrachlorodibenzo-p-dioxin equivalents, *Chemosphere* **26,** 1203–1212 (1993).
111. P. Fuerst, C. Fuerst, and K. Wilmers, Body burden with PCDD and PCDF from food, in: *Biological Basis for Risk Assessment of Dioxins and Related Compounds* (M. A. Gallo, R. J. Scheuplein, and K. A. van der Heijden, eds.), pp. 133–142, Banbury Report 35, Cold Spring Harbor Laboratory Press, Cold Spring Harbor, NY (1991).
112. T. Webster and P. Connett, Estimating bioconcentration factors and half-lives in humans

using physiologically based pharmacokinetic modelling: 2,3,7,8-TCDD, *Chemosphere* **23,** 1763–1768 (1991).

113. J. Pirkle, W. Wolfe, D. Patterson, L. Needham, J. Michalek, J. Miner, M. Peterson, and D. Phillips, Estimates of the half-life of 2,3,7,8-tetrachlorodibenzo-p-dioxin in Vietnam veterans of Operation Ranch Hand, *J. Toxicol. Environ. Health* **27,** 165–171 (1989).
114. C. Schlatter, Data on kinetics of PCDDs and PCDFs as a prerequisite for human risk assessment, in: *Biological Basis for Risk Assessment of Dioxins and Related Compounds* (M. A. Gallo, R. J. Scheuplein, and K. A. van der Heijden, eds.), pp. 215–226, Banbury Report 35, Cold Spring Harbor Laboratory Press, Cold Spring Harbor, NY (1991).
115. T. Webster, Physiologically based pharmacokinetic modelling of dioxin: Structural analysis and extension to other congeners, Presented at Dioxin '92, Tampere, Finland, *Organohalogen Compounds,* Vol. 8, p. 389 (1992).
116. A. Schecter, O. Paepke, and M. Ball, Evidence for transplacental transfer of dioxin from mother to fetus: Chlorinated dioxin and dibenzofuran levels in the livers of stillborn infants, *Chemosphere* **21,** 1017–1022 (1990).
117. J. Nosek, S. Craven, J. Sullivan, J. Olson, and R. Peterson, Metabolism and disposition of 2,3,7,8-tetrachlorodibenzo-p-dioxin in ring-necked pheasant hens, chicks, and eggs, *J. Toxicol. Environ. Health* **35,** 153–164 (1992).
118. D. W. Nebert, F. M. Goujon, and J. E. Gielen, Aryl hydrocarbon hydroxylase by polycyclic hydrocarbons: Simple autosomal dominant trait in the mouse, *Nature New Biol.* **236,** 107–110 (1972).
119. A. Poland and E. Glover, Chlorinated dibenzo-p-dioxins: Potent inducers of delta-aminolevulinic acid synthetase and aryl hydrocarbon hydroxylase. II. A study of the structure–activity relationship, *Mol. Pharmacol.* **9,** 736–747 (1973).
120. A. Poland, E. Glover, and A. S. Kende, Stereo-specific, high affinity binding of 2,3,7,8-tetrachlorodibenzo-p-dioxin by hepatic cytosol. Evidence that the binding species is a receptor for induction of aryl hydrocarbon hydroxylase, *J. Biol. Chem.* **251,** 4936–4946 (1976).
121. J. Landers and N. Bunce, The Ah receptor and the mechanism of dioxin toxicity, *Biochem. J.* **276,** 273–287 (1991).
122. J. Whitlock, Jr., Genetic and molecular aspects of 2,3,7,8-tetrachlorodibenzo-p-dioxin action, *Annu. Rev. Pharmacol. Toxicol.* **30,** 251–277 (1990).
123. E. Hoffman, H. Reyes, F.-F. Chu, F. Sander, L. Conley, B. Brooks, and O. Hankinson, Cloning of a factor required for activity of the Ah (dioxin) receptor, *Science* **252,** 954–958 (1991).
124. T. Gasiewicz and E. Henry, Different forms of the Ah receptor: Possible role in species- and tissue-specific responses to TCDD, in: *Biological Basis for Risk Assessment of Dioxins and Related Compounds* (M. A. Gallo, R. J. Scheuplein, and K. A. van der Heijden, eds.), pp. 321–333, Banbury Report 35, Cold Spring Harbor Laboratory Press, Cold Spring Harbor, NY (1991).
125. D. Nebert and N. Jensen, The Ah locus: Genetic regulation of the metabolism of carcinogens, drugs, and other environmental chemicals by cytochrome P-450-mediated monooxygenases, *CRC Crit. Rev. Biochem.* **6,** 401–437 (1979).
126. D. Nebert and F. Gonzalez, P450 genes: Structure, evolution and regulation, *Annu. Rev. Biochem.* **56,** 945–993 (1987).
127. C. Ioannides and D. Parke, The cytochromes P-448—A unique family of enzymes involved in chemical toxicity and carcinogenesis, *Biochem. Pharmacol.* **36,** 4197–4207 (1987).
128. T. Sutter and W. Greenlee, Classification of members of the Ah gene battery, *Chemosphere* **25,** 223–226 (1992).

129. E. Henry and T. Gasiewicz, Changes in thyroid hormones and thyroxine glucuronidation in hamsters compared with rats following treatment with 2,3,7,8-tetrachlorodibenzo-p-dioxin, *Toxicol. Appl. Pharamcol.* **89,** 165–174 (1987).
130. R. Hill, L. Erdreich, O. Paynter, P. Roberts, S. Rosenthal, and C. Wilkinson, Thyroid follicular cell carcinogenesis, *Fundam. Appl. Toxicol.* **12,** 629–697 (1989).
131. National Toxicology Program (NTP), *Carcinogenesis Bioassay of 2,3,7,8-Tetrachlorodibenzo-p-Dioxin in Osborne–Mendel Rats and B63CF1 Mice (Gavage Study),* U.S. National Toxicology Program, Box 12233, Research Triangle Park, NC (1982).
132. C. Portier, D. Hoel, and J. Van Ryzin, Statistical analysis of the carcinogenesis bioassay data relating to the risks from exposure to 2,3,7,8-tetrachlorodibenzo-p-dioxin, in: *Public Health Risks of the Dioxins* (W. W. Lowrance, ed.), pp. 99–120, William Kaufmann, Los Altos, CA (1984).
133. USEPA, *Health Assessment Document for Polychlorinated Dibenzo-p-Dioxins,* EPA/600/8-84/014F (1985).
134. L. Zeise, J. Huff, A. Salmon, and N. Hopper, Human risks from 2,3,7,8-tetrachlorodibenzo-p-dioxin and hexachlorodibenzo-p-dioxins, *Adv. Mod. Environ. Toxicol.* **17,** 293–342 (1990).
135. USEPA, *Chapter 6: Carcinogenicity in Animals,* Review Draft, EPA/600/AP-92/001f (1992).
136. G. W. Lucier, Receptor-mediated carcinogenesis, in: *Mechanisms of Carcinogenesis in Risk Identification* (H. Vainio, P. Magee, D. McGregor, and A. McMichael, eds.), pp. 87–112, International Agency for Research on Cancer, Lyon, France (1992).
137. M. DeVito, T. Umbreit, T. Thomas, and M. Gallo, An analogy between the actions of the Ah receptor and the estrogen receptor for use in the biological basis for risk assessment of dioxin, in: *Biological Basis for Risk Assessment of Dioxins and Related Compounds* (M. A. Gallo, R. J. Scheuplein, and K. A. van der Heijden, eds.), pp. 427–440, Banbury Report 35, Cold Spring Harbor Laboratory Press, Cold Spring Harbor, NY (1991).
138. S. Safe, B. Astroff, M. Harris, T. Zacharewski, R. Dickerson, M. Romkes, and L. Biegel, 2,3,7,8-Tetrachlorodibenzo-p-dioxin (TCDD) and related compounds as antioestrogens: Characterization and mechanism of action, *Pharmacol. Toxicol.* **69,** 400–409.
139. M. DeVito, T. Thomas, E. Martin, T. Umbreit, and M. Gallo, Antiestrogenic action of 2,3,7,8-tetrachlorodibenzo-p-dioxin: Tissue-specific regulation of estrogen receptor in CD1 mice, *Toxicol. Appl. Pharmacol.* **113,** 284–292 (1992).
140. R. J. Kociba, D. G. Keyes, J. E. Beyer, R. M. Carreon, C. E. Wade, D. A. Dittenber, R. P. Kalnins, L. E. Frauson, C. N. Park, S. D. Barnard, R. A. Hummel, and C. G. Humiston, Results of a two-year chronic toxicity and oncogenicity study of 2,3,7,8-tetrachlorodibenzo-p-dioxin in rats, *Toxicol. Appl. Pharmacol.* **46,** 279–303 (1978).
141. R. Bookstaff, F. Kamel, R. Moore, D. Bjerke, and R. Peterson, Altered regulation of pituitary gonadotropin-releasing hormone (GnRH) receptor number and pituitary responsiveness to GnRH in 2,3,7,8-tetrachlorodibeno-p-dioxin-treated male rats, *Toxicol. Appl. Pharmacol.* **105,** 78–92 (1990).
142. R. Bookstaff, R. Moore, and R. Peterson, 2,3,7,8-tetrachlorodibenzo-p-dioxin increases the potency of androgens and estrogens as feedback inhibitors of luteinizing hormone secretion in male rats, *Toxicol. Appl. Pharmacol.* **104,** 212–224 (1990).
143. R. Moore, C. Jefcoate, and R. Peterson, 2,3,7,8-Tetrachlorodibenzo-p-dioxin inhibits stereoidogenesis in the rat testis by inhibiting the mobilization of cholesterol to cytochrome P450scc, *Toxicol. Appl. Pharmacol.* **109,** 85–97 (1991).
144. R. Moore, R. Bookstaff, T. Mably, and R. Peterson, Differential effects of 2,3,7,8-tetrachlorodibenzo-p-dioxin on responsiveness of male rats to androgens, 17beta-estradiol,

luteinizing hormone, gonadotropin-releasing hormone, and progesterone, *Chemosphere* **25,** 91–94 (1992).
145. G. Clark, A. Tritscher, R. Maronpot, J. Foley, and G. Lucier, Tumor promotion by TCDD in female rats, in: *Biological Basis for Risk Assessment of Dioxins and Related Compounds* (M. A. Gallo, R. J. Scheuplein, and K. A. van der Heijden, eds.), pp. 389–400, Banbury Report 35, Cold Spring Harbor Laboratory Press, Cold Spring Harbor, NY (1991).
146. D. Ignar-Trowbridge, K. Nelson, M. Bidwell, S. Curtis, T. Washburn, J. McLachlan, and K. Korach, Coupling of dual signaling pathways: Epidermal growth factor action involves the estrogen receptor, *Proc. Natl. Acad. Sci. USA* **89,** 4658–4662 (1992).
147. L. A. Couture, B. D. Abbott, and L. S. Birnbaum, A critical review of the developmental toxicity and teratogenicity of 2,3,7,8-tetrachlorodibenzo-p-dioxin: Recent advances toward understanding the mechanism, *Teratology* **42,** 619–627 (1990).
148. L. Birnbaum, Developmental toxicity of TCDD and related compounds: Species sensitivities and differences, in: *Biological Basis for Risk Assessment of Dioxins and Related Compounds* (M. A. Gallo, R. J. Scheuplein, and K. A. van der Heijden, eds.), pp. 51–64, Banbury Report 35, Cold Spring Harbor Laboratory Press, Cold Spring Harbor, NY (1991).
149. M. P. Holsapple, D. L. Morris, S. C. Wood, and N. K. Snyder, 2,3,7,8-Tetrachlorodibenzo-p-dioxin-induced changes in immunocompetence: Possible mechanisms, *Annu. Rev. Pharmacol. Toxicol.* **31,** 73–100 (1991).
150. J. Vos, H. van Loveren, and H.-J. Schuurman, Immunotoxicity of dioxin: Immune function and host resistance in laboratory animals and humans, in: *Biological Basis for Risk Assessment of Dioxins and Related Compounds* (M. A. Gallo, R. J. Scheuplein, and K. A. van der Heijden, eds.), pp. 79–88, Banbury Report 35, Cold Spring Harbor Laboratory Press, Cold Spring Harbor, NY (1991).
151. J. G. Vos and M. I. Luster, Immune alterations, in: *Halogenated Biphenyls, Terphenyls, Naphthalenes, Dibenzodioxins and Related Products* (R. D. Kimbrough and A. A. Jensen, eds.), pp. 295–322, Elsevier, Amsterdam (1989).
152. G. Clark, J. Blank, D. Germolec, and M. Luster, 2,3,7,8-Tetrachlorodibenzo-p-dioxin stimulation of tyrosine phosphorylation in B lymphocytes: Potential role in immunosuppression, *Mol. Pharmacol.* **39,** 495–501 (1991).
153. M. Luster, S. Holladay, B. Blaylock, D. Germolec, G. Clark, C. Comment, J. Heindel, and G. Rosenthal, TCDD inhibits murine thymocyte and B lymphocyte maturation, *Chemosphere* **25,** 115–118 (1992).
154. M. Devito, J. Diliberto, and L. Birnbaum, Comparative ability of TCDD to induce hepatic and skin cytochrome P-450IA1 activity following 13 weeks of treatment, Presented at Dioxin '92, Tampere, Finland, *Organohalogen Compounds,* Vol. 10, pp. 41–44 (1992).
155. K. Schmidt, Dioxin's other face: Portrait of an "environmental hormone," *Science News* **141,** 24–27 (1992).
156. R. Evans, The steroid and thyroid hormone receptor superfamily, *Science* **240,** 889–895 (1988).
157. K. Burbach, A. Poland, C. Bradfield, Cloning of the Ah-receptor cDNA reveals a distinctive ligand-activated transcription factor, *Proc. Natl. Acad. Sci. USA* **89,** 8185–8189 (1992).
158. O. Hankinson, Research on the aryl hydrocarbon (dioxin) receptor is primed to take off, *Arch. Biochem. Biophys.* **300,** 1–5 (1993).
159. P. Fuller, The steroid receptor superfamily: Mechanisms of diversity, *FASEB J.* **5,** 3092–3099 (1991).

160. E. Silbergeld and T. Gasiewicz, Dioxins and the Ah receptor, *Am. J. Ind. Med.* **16,** 455–474 (1989).
161. L. Roberts, Dioxin risks revisited, *Science* **251,** 624–626 (1991).
162. J. Huff, 2,3,7,8-TCDD: A potent & complete carcinogen in experimental animals, *Chemosphere* **25,** 173–176 (1992).
163. F. Perera, Perspectives on the risk assessment for nongenotoxic carcinogens and tumor promoters, *Environ. Health Perspect.* **94,** 231–235 (1991).
164. H. Pitot, T. Goldsworthy, H. A. Campbell, and A. Poland, Quantitative evaluation of the promotion by 2,3,7,8-tetrachlorodibenzo-p-dioxin of hepatocarcinogenesis from diethylnitrosamine, *Cancer Res.* **40,** 3616–3620 (1980).
165. A. Poland, D. Palen, and E. Glover, Tumour promotion by TCDD in skin of HRS/J hairless mice, *Nature* **300,** 271–273 (1982).
166. G. Lucier, A. Tritscher, T. Goldsworthy, J. Foley, G. Clark, J. Goldstein, and R. Maronpot, Ovarian hormones enhance 2,3,7,8-tetrachlorodibenzo-p-dioxin-mediated increases in cell proliferation and preneoplastic foci in a two-stage model for rat hepatocarcinogenesis, *Cancer Res.* **51,** 1391–1397 (1991).
167. R. E. Kouri, T. H. Rude, R. Joglekar, P. M. Dansette, D. M. Jerina, S. A. Atlas, I. S. Owens, and D. W. Nebert, 2,3,7,8-Tetrachlorodibenzo-p-dioxin as co-carcinogen causing 3-methylcholanthrene-initiated subcutaneous tumors in mice genetically "nonresponsive" at Ah locus, *Cancer Res.* **38,** 2777–2783 (1978).
168. C. Thompson, M. Andries, K. Lundgren, J. Goldstein, G. Collman, and G. Lucier, Humans exposed to polychlorinated biphenyls (PCBs) and polychlorinated dibenzofurans (PCDFs) exhibit increased SCE frequency in lymphocytes when incubated with alpha-naphthoflavone: Involvement of metabolic activation by P-450 isozymes, *Chemosphere* **18,** 687–694 (1989).
169. G. Cohen, W. Bracken, R. Iyer, D. Berry, J. Selkirk, and T. Slaga, Anticarcinogenic effects of 2,3,7,8-tetrachlorodibenzo-p-dioxin on benzo(a)pyrene and 7,12-dimethylbenz(a)anthracene tumor initiation and its relationship to DNA binding, *Cancer Res.* **39,** 4027–4033 (1979).
170. H. Yamazaki, Y. Inui, C. Yun, F. Guengerich, and T. Shimada, Cytochrome P450 2E1 and 2A6 enzymes as major catalysts for metabolic-activation of N-nitrosodialkylamines and tobacco-related nitrosamines in human liver-microsomes, *Carcinogenesis* **13,** 1789–1794 (1992).
171. H. Yamazaki, Y. Oda, Y. Funae, S. Imaoka, Y. Inui, F. Guengerich, and T. Shimada, Participation of rat-liver cytochrome P450-2E1 in the activation of N-nitrosodimethylamine and N-nitrosodiethylamine to products genotoxic in an acetyltransferase-overexpressing Salmonella-typhimurium strain (NM2009), *Carcinogenesis* **13,** 979–985 (1992).
172. Y. Dragan, X. Xu, T. L. Goldsworthy, H. A. Campbell, R. R. Maronpot, and H. C. Pitot, Characterization of the promotion of altered hepatic foci by 2,3,7,8-tetrachlorodibenzo-p-dioxin in the female rat, *Carcinogenesis* **13,** 1389–1395 (1992).
173. J. Yang, P. Thraves, A. Dritschilo, and J. S. Rhim, Neoplastic transformation of immortalized human keratinocytes by 2,3,7,8-tetrachlorodibenzo-p-dioxin, *Cancer Res.* **52,** 3478–3482 (1992).
174. K. Tullis, H. Olsen, D. W. Bombick, F. Matsumura, and J. Jankun, TCDD causes stimulation of c-ras expression in the hepatic plasma membranes in vivo and in vitro, *J. Biochem. Toxicol.* **7**(2), 107–116 (1992).
175. E. Silbergeld, Carcinogenicity of dioxins, *J. Natl. Cancer Inst.* **83,** 1198–1199 (1991).

176. P. Vineis, F. D'Amore, and Working Group on the Epidemiology of Hematolymphopoietic Malignancies in Italy, The role of occupational exposure and immunodeficiency in B-cell malignancies, *Epidemiology* **3,** 266–270 (1992).
177. E. Silbergeld, Dioxin: Receptor-based approaches to risk assessment, in: *Biological Basis for Risk Assessment of Dioxins and Related Compounds* (M. A. Gallo, R. J. Scheuplein, and K. A. van der Heijden, eds.), pp. 441–445, Banbury Report 35, Cold Spring Harbor Laboratory Press, Cold Spring Harbor, NY (1991).
178. B. Commoner, T. Webster, and K. Shapiro, Environmental levels and health effects of chlorinated dioxins and furans, Presented at AAAS meeting, Philadelphia, May 28 (1986).
179. Ontario Ministry of the Environment, Scientific Criteria Document for Standard Development No. 4-84: Polychlorinated dibenzo-p-dioxins (PCDDs) and polychlorinated dibenzofurans (PCDFs) (1985).
180. USEPA, *Fed. Regist.* **57,** 31816, July 17 (1992).
181. A. Whittemore and J. Keller, Quantitative theories of carcinogenesis, *SIAM Rev.* **20,** 1–30 (1978).
182. E. Anderson and the US Environmental Protection Agency Carcinogen Assessment Group, Quantitative approaches in use to assess cancer risk, *Risk Anal.* **3,** 277–295 (1983).
183. C. Portier, Statistical properties of a two-stage model of carcinogenesis, *Environ. Health Perspect.* **76,** 125–131 (1987).
184. USEPA, Guidelines for carcinogen risk assessment, *Fed. Regist.* **51,** 33992–34003 (1986).
185. USEPA, *A Cancer Risk-Specific Dose Estimate for 2,3,7,8-TCDD: Draft,* EPA/60/6-88/007Aa (1988).
186. USEPA Science Advisory Board Ad Hoc Dioxin Panel, Review of Draft Documents "A Cancer Risk-Specific Dose Estimate for 2,3,7,8-TCDD and Estimating Exposure to 2,3,7,8-TCDD," November 28 (1989).
187. Dioxin Public Affairs Plan, April (1987), Discussed in P. von Stackleberg, Whitewash: The Dioxin Cover-up, *Greenpeace* **14**(2)**,** 7–11 (1989).
188. J. Bailey, Dueling studies: How two industries created a fresh spin on the dioxin debate, *Wall Street Journal* February 20 (1992).
189. D. Goodman and R. Sauer, Hepatotoxicity and carcinogenicity in female Sprague–Dawley rats treated with 2,3,7,8-tetrachlorodibenzo-p-dioxin (TCDD): A pathology working group reevaluation, *Regul. Toxicol. Pharmacol.* **15,** 245–252 (1992).
190. T. Webster, Downgrading dioxin's cancer risk: Where's the science? *J. Pestic. Reform* **11**(1)**,** 11–14 (1991).
191. W. Brown, Implications of the reexamination of the liver sections from the TCDD chronic rat bioassay, in: *Biological Basis for Risk Assessment of Dioxins and Related Compounds* (M. A. Gallo, R. J. Scheuplein, and K. A. van der Heijden, eds.), pp. 13–18, Banbury Report 35, Cold Spring Harbor Laboratory Press, Cold Spring Harbor, NY (1991).
192. W. Farland and R. Scheuplein, EPA and FDA Position on the Pathology Working Group (PWG) Report on Slide Review of Liver in Female Sprague–Dawley Rats in the Kociba, et al., Study of 2,3,7,8-Tetrachlorodibenzo-p-dioxin (TCDD), January 9 (1991).
193. L. Roberts, Flap erupts over dioxin meeting, *Science* **251,** 866–867 (1991).
194. W. Reilly, Administrator USEPA, Memorandum. Dioxin: Follow-up to Briefing on Scientific Developments, April 8 (1991).
195. D. Hanson, EPA to take another hard look at dioxin health risk, *Chem. Eng. News* April 29, pp. 13–14 (1991).
196. USEPA, EPA's Scientific Reassessment of Dioxin: Background Document for Public Meeting on November 15, 1991 (1991).

197. K. Rhyne, King & Spalding, Letter to Charles Baker, Chairman North Carolina Environmental Management Commission, January 16 (1991).
198. K. Marien, G. Patrick, and H. Ammann, *Health Implications of TCDD and TCDF Concentrations Reported from Lake Roosevelt Fish,* Washington State Department of Health, April (1991).
199. C. Portier, A. Tritscher, M. Kohn, C. Sewall, G. Clark, L. Edler, D. Hoel, and G. Lucier, Ligand/receptor binding for 2,3,7,8-TCDD: Implications for risk assessment, *Fundam. Appl. Toxicol.* **20,** 48–56 (1993).
200. A Tritscher, J. Goldstein, C. Portier, Z. McCoy, G. Clark, and G. Lucier, Dose–response relationships for chronic exposure to 2,3,7,8-tetrachlorodibenzo-p-dioxin in a rat tumor promotion model: Quantification and immunolocalization of CYP1A1 and CYP1A2 in the liver, *Cancer Res.* **52,** 3436–3442 (1992).
201. J. Vanden Heuvel, G. Lucier, G. Clark, A. Tritscher, W. Greenlee, and D. Bell, Use of reverse-transcription polymerase chain reaction to quantitate mRNA for dioxin-responsive genes in the low-dose region in rat liver, Presented at Dioxin '92, Tampere, Finland, *Organohalogen Compounds,* Vol. 10, pp. 377–379 (1992).
202. G. Lucier, G. Clark, A. Tritscher, J. Foley, and R. Maronpot, Mechanisms of dioxin tumor promotion: Implications for risk assessment, *Chemosphere* **25,** 177–180 (1992).
203. R. E. Keenan, R. J. Wenning, A. H. Parsons, and D. J. Paustenbach, A Re-evaluation of the Tumor Histopathology of Kociba et al. (1978) Using 1990 Criteria: Implications for Risk Assessment Using the Linearized Multistage Model. Overheads presented to the Minnesota Pollution Control Agency and Minnesota Department of Health, January 10 (1991).
204. K. Herxheimer, *Muench. Med. Wochenschr.* **46,** 278 (1899).
205. World Health Organization, International Programme on Chemical Safety, *Polychlorinated Dibenzo-para-dioxins and Dibenzofurans,* Environmental Health Criteria 88, Geneva (1989).
206. R. Kimbrough and P. Grandjean, Occupational exposure, in: *Halogenated Biphenyls, Terphenyls, Naphthalenes, Dibenzodioxins and Related Products* (R. D. Kimbrough and A. A. Jensen, eds.), pp. 485–507, Elsevier, Amsterdam (1989).
207. M. Gross, J. Lay, P. Lyon, D. Lippstreu, N. Kangas, R. Harless, S. Taylor, and A. Dupuy, 2,3,7,8-Tetrachlorodibenzo-p-dioxin levels in adipose tissue of Vietnam veterans, *Environ. Res.* **33,** 261–268 (1984).
208. J. J. Ryan, D. T. Williams, and B. Lau, Analysis of human fat tissue for 2,3,7,8-tetrachlorodibenzo-p-dioxin and chlorinated dibenzofuran residues, in: *Chlorinated Dioxins & Dibenzofurans in the Total Environment II* (L. H. Keith, C. Rappe, and G. Choudhary, eds.), pp. 205–214, Butterworths, Boston (1985).
209. A. Schecter, T. O. Tiernan, M. L. Taylor, G. F. Van Ness, J. H. Garrett, D. J. Wagel, G. Gitlitz, and M. Bogdasarian, Biological markers after exposure to polychlorinated dibenzo-p-dioxins, dibenzofurans, biphenyls, and biphenylenes. Part I: Findings using fat biopsies to estimate exposure, in: *Chlorinated Dioxins & Dibenzofurans in the Total Environment II* (L. H. Keith, C. Rappe, and G. Choudhary, eds.), pp. 215–245, Butterworths, Boston (1985).
210. M. Graham, F. Hileman, D. Kirk, J. Wendling, and J. Wilson, Background human exposure to 2,3,7,8-TCDD, *Chemosphere* **14,** 925–928 (1985).
211. M. Ono, T. Wakimoto, R. Tatsukawa, and Y. Masuda, Polychlorinated dibenzo-p-dioxins and dibenzofurans in human adipose tissue of Japan, *Chemosphere* **15,** 1629–1634 (1985).
212. C. Rappe, M. Nygren, G. Lindstrom, and M. Hansson, Dioxins and dibenzofurans in biological samples of European origin, *Chemosphere* **15,** 1635–1639 (1986).

213. A. Ahlbom, O. Axelson, E. S. Hansen, C. Hogstedt, U. J. Jensen, and J. Olsen, Interpretation of "negative" studies in occupational epidemiology, *Scand. J. Work Environ. Health* **16,** 153–157 (1990).
214. M. Gough, *Dioxin, Agent Orange: The Facts,* Plenum Press, New York (1986).
215. M. Gough, Human health effects: What the data indicate, *Sci. Total Environ.* **104,** 129–158 (1991).
216. Dioxin Toxicity, *Am. Fam. Physician* **47,** 855–861 (1993). [Reprinted from an Agency for Toxic Substances and Disease Registry monograph.]
217. R. Clapp, B. Commoner, J. Constable, S. Epstein, P. Kahn, J. Olson, and D. Ozonoff, *Report of the Task Force on Human Health Effects Associated with Exposure to Herbicides and/or Their Associated Contaminants (chlorinated dioxins),* National Veteran's Legal Services Project, Washington, DC (1990).
218. R. Clapp and J. Olson, A new review of the dioxin literature in the context of compensation for Vietnam veterans, *New Solutions* Spring, pp. 31–37 (1991).
219. International Agency for Research on Cancer, World Health Organization, *IARC Monographs on the Evaluation of Carcinogenic Risks to Humans, Overall Evaluations of Carcinogenicity: An Updating of IARC Monographs Volumes 1 to 42,* Supplement 7, Lyon, France (1987).
220. L. Hardell, Soft-tissue sarcomas and exposure to phenoxy acids: A clinical observation, *Lakartidningen* **74,** 2753–2754 (1977).
221. L. Hardell and A. Sandstrom, A case–control study: Soft-tissue sarcomas and exposure to phenoxyacetic acids or chlorophenols, *Br. J. Cancer* **39,** 711–717 (1979).
222. M. Eriksson, L. Hardell, N. Berg, T. Moller, and O. Axelson, Soft-tissue sarcomas and exposure to chemical substances: A case-referent study, *Br. J. Ind. Med.* **38,** 27–33 (1981).
223. L. Hardell and M. Eriksson, The association between soft tissue sarcomas and exposure to phenoxyacetic acids. A new case-referent study, *Cancer* **62,** 652–656 (1988).
224. M. Eriksson, L. Hardell, and H.-O. Adami, Exposure to dioxins as a risk factor for soft tissue sarcoma: A population-based case–control study, *J. Natl. Cancer Inst.* **82,** 486–490 (1990).
225. J. Zack and R. Suskind, The mortality experience of workers exposed to tetrachlorodibenzodioxin in a trichlorophenol process accident, *J. Occup. Med.* **22,** 11–14 (1980).
226. J. Zack and W. Gaffey, A mortality study of workers employed at the Monsanto Company plant in Nitro, West Virginia, *Environ. Sci. Res.* **26,** 575–591 (1983).
227. R. Suskind and V. Hertzberg, Human health effects of 2,4,5-T and its toxic contaminants, *J. Am. Med. Assoc.* **251,** 2372–2380 (1984).
228. A. Hay and E. Silbergeld, Assessing the risk of dioxin exposure, *Nature* **315,** 102–103(1985).
229. C. van Strum and P. Merrell, Dioxin human health damage studies: Damaged data? *J. Pestic. Reform* **10**(1), 8–12 (1990).
230. L. Roberts, Monsanto studies under fire, *Science* **251,** 626 (1991).
231. F. Rohleder, Dioxins and Cancer Mortality—Reanalysis of the BASF Cohort, Presented at the Ninth International Symposium on Chlorinated Dioxins and Related Compounds, Toronto, Ontario (1989).
232. L. Hardell and M. Eriksson, The association between cancer mortality and dioxin exposure: A comment on the hazard of repetition of epidemiological misinterpretation, *Am. J. Ind. Med.* **19,** 547–549 (1991).
233. P. Connett and B. Elmore (eds.), *The First Citizen's Conference on Dioxin: Proceedings,* NC WARN, 5301 Rolling Hill Road, Sanford, NC 27330 (1992).

234. R. W. Baughman, Tetrachlorodibenzo-p-dioxins in the environment: High resolution mass spectrometry at the picogram level, Ph.D. thesis, Harvard University, Cambridge, MA (1974). [Referenced in No. 44.]
235. R. A. Albanese, The chemical 2,3,7,8-tetrachlorodibenzo-p-dioxin and U.S. Army Vietnam-era veterans, *Chemosphere* **22,** 597–603 (1991).
236. M. A. Fingerhut, W. E. Halperin, D. A. Marlow, L. A. Piacitelli, P. A. Honchar, M. H. Sweeney, A. L. Greife, P. A. Dill, K. Steenland, and A. J. Suruda, Cancer mortality in workers exposed to 2,3,7,8-tetrachlorodibenzo-p-dioxin (TCDD), *N. Engl. J. Med.* **324,** 212–218 (1991).
237. A. Manz, J. Berger, J. H. Dwyer, D. Flesch-Janys, S. Nagel, and H. Waltsgott, Cancer mortality among workers in chemical plant contaminated with dioxin, *Lancet* **338,** 959–964 (1991).
238. A. Zober, P. Messerer, and P. Huber, Thirty-four year mortality follow-up of BASF employees exposed to 2,3,7,8-TCDD after the 1953 accident, *Int. Arch. Occup. Environ. Health* **62,** 139–157 (1990).
239. R. Sarraci, M. Kogevinas, P.-A. Bertazzi, B. H. Bueno De Mesquita, D. Coggon, L. M. Green, T. Kauppinen, K. A. L'Abbe, M. Littorin, E. Lynge, J. D. Mathews, M. Neuberger, J. Osman, N. Pearce, and R. Winkelmann, Cancer mortality in workers exposed to chlorophenoxy herbicides and chlorophenols, *Lancet* **338,** 1027–1032 (1991).
240. A. Pesatori, D. Consonni, A. Tironi, M. Landi, C. Zocchetti, and P. Bertazzi, Cancer morbidity in the Seveso area, 1976–1986, *Chemosphere* **25,** 209–212 (1992).
241. L. Tomatis (Editor-in-Chief), *Cancer: Causes, Occurrence and Control,* International Agency for Research on Cancer, World Health Organization, Lyon, France (1990).
242. G. W. Lucier, Humans are a sensitive species to some of the biochemical effects of structural analogs of dioxin, *Environ. Toxicol. Chem.* **10,** 727–735 (1991).
243. L. Roberts, High dioxin dose linked to cancer, *Science* **251,** 625 (1991).
244. M. Gladwell, Extensive study finds reduced dioxin danger, *Washington Post* January 24 (1991).
245. Anon, Dioxin re-examined: A dose of dissent, *The Economist* March 16, p. 87 (1991).
246. T. Wray, Dioxin. Studies examine toxicity of dioxin, *Hazmat World* March, p. 80 (1992).
247. C. R. Dempsey and E. T. Oppelt, Incineration of hazardous waste: A critical review update, *Air Waste* **43,** 25–73 (1993).
248. V. N. Houk, Dioxin Risk Assessment for Human Health: Scientifically Defensible or Fantasy? Presented at the 25th Annual Conference on Trace Substances in Environmental Health, University of Missouri, Columbus, May (1991). Available from the Centers for Disease Control, Atlanta, GA.
249. L. Tollefson, Use of epidemiology data to assess the cancer risk of 2,3,7,8-tetrachlorodibenzo-p-dioxin, *Regul. Toxicol. Pharmacol.* **13,** 150–169 (1991).
250. S. Bayard, Quantitative implications of the use of different extrapolation procedures for low-dose cancer risk estimates from exposure to 2,3,7,8-TCDD, in: USEPA, *A Cancer Risk-Specific Dose for 2,3,7,8-TCDD: Appendices A Through F,* USEPA/600/6-88/007Ab (1988).
251. L. Goldman, D. Hayward, D. Siegel, and R. Stephens, Dioxin and mortality from cancer, *N. Engl. J. Med.* **324,** 1811 (1991).
252. T. Webster, Estimation of the cancer risk posed to humans by 2,3,7,8-TCDD, Presented at Dioxin '91, Research Triangle Park, NC (1991).
253. J. J. Collins, J. F. Acquavella, and B. R. Friedlander, Reconciling old and new findings on dioxin, *Epidemiology* **3,** 65–69 (1992).

254. M. Fingerhut, K. Steenland, M. Haring Sweeney, W. E. Halperin, L. A. Piacitelli, and D. A. Marlow, Old and new reflections on dioxin, *Epidemiology* **3,** 69–72 (1992).
255. T. Sinks, Misclassified sarcomas and confounded dioxin exposure, *Epidemiology* **4,** 3–6 (1992).
256. J. J. Collins, M. E. Strauss, G. J. Levinskas, and P. R. Conner, The mortality experience of workers exposed to 2,3,7,8-tetrachlorodibenzo-p-dioxin in a trichlorophenol process accident, *Epidemiology* **4,** 7–13 (1993).
257. J. J. Collins, Letter to the Editor, *Epidemiology* **4,** 187 (1993).
258. T. Sinks, Reply to Collins, *Epidemiology* **4,** 187–188 (1993).
259. P. Olafson, *Cornell Vet.* **37,** 279–291 (1947). [Referenced in No. 260.]
260. E. McConnell, Acute and chronic toxicity and carcinogenesis in animals, in: *Halogenated Biphenyls, Terphenyls, Naphthalenes, Dibenzodioxins and Related Products* (R. D. Kimbrough and A. A. Jensen, eds.), pp. 161–193, Elsevier, Amsterdam (1989).
261. M. Reich, *Toxic Politics: Responding to Chemical Disasters,* Cornell University Press, Ithaca NY (1991). [This book also examines the Seveso and Yusho incidents.]
262. H. Anderson, General population exposure to environmental concentrations of halogenated biphenyls, in: *Halogenated Biphenyls, Terphenyls, Naphthalenes, Dibenzodioxins and Related Products* (R. D. Kimbrough and A. A. Jensen, eds.), pp. 325–344, Elsevier, Amsterdam (1989).
263. V. Sanger, L. Scott, A. Handy, C. Gale, and W. Pounden, *J. Am. Vet. Med. Assoc.* **133,** 172–176 (1958). [Referenced in No. 260.]
264. C. Simpson, W. Pritchard, and R. Harms, *J. Am. Vet. Med. Assoc.* **134,** 410–416 (1959). [Referenced in No. 260.]
265. D. Firestone, Etiology of chick edema disease, *Environ. Health Perspect.* **5,** 59–66 (1973).
266. M. Gilbertson, T. Kubiak, J. Ludwig, and G. Fox, Great Lakes embryo mortality, edema, and deformities syndrome (GLEMEDS) in colonial fish-eating birds: Similarity to chick-edema disease, *J. Toxicol. Environ. Health* **33,** 455–520 (1991).
267. M. Gilbertson, Effects on fish and wildlife populations, in: *Halogenated Biphenyls, Terphenyls, Naphthalenes, Dibenzodioxins and Related Products* (R. D. Kimbrough and A. A. Jensen, eds.), pp. 103–127, Elsevier, Amsterdam (1989).
268. D. E. Tillitt, G. T. Ankley, J. P. Giesy, J. P. Ludwig, H. Kurita-Matsuba, D. V. Weseloh, P. S. Ross, C. A. Bishop, L. Sileo, K. L. Stromborg, J. Larson, and T. J. Kubiak, Polychlorinated biphenyl residues and egg mortality in double-crested cormorants from the Great Lakes, *Environ. Toxicol. Chem.* **11,** 1281–1288 (1992).
269. T. J. Kubiak, H. J. Harris, L. M. Smith, T. R. Schwartz, D. L. Stalling, J. A. Trick, L. Sileo, D. E. Docherty, and T. C. Erdman, Microcontaminants and reproductive impairment of the Forster's tern on Green Bay, Lake Michigan—1983, *Arch. Environ. Contam. Toxicol.* **18,** 706–727 (1989).
270. G. A. Fox, B. Collins, E. Hayakawa, D. V. Weseloh, J. P. Ludwig, T. J. Kubiak, and T. C. Erdman, Reproductive outcomes in colonial fish-eating birds: A biomarker for developmental toxicants in Great Lakes food chain. II. Spatial variation in the occurrence and prevalence of bill defects in young double-crested cormorants in the Great Lakes, 1979–1987, *J. Great Lakes Res.* **17,** 158–167 (1991).
271. N. Yamashita, S. Tanabe, J. P. Ludwig, H. Kurita, M. E. Ludwig, and R. Tatsukawa, Embryonic abnormalities and organochlorine contamination in double-crested cormorants (Phalacrocorax auritus) and Caspian terns (Hydroprogne caspia) from the upper Great Lakes in 1988, *Environ. Pollut.* **79,** 163–173 (1993).

272. J. P. Ludwig, H. Kurita, H. J. Auman, M. E. Ludwig, N. Yamashita, and S. Tanabe, Embryonic abnormalities and organochlorine contamination in double-crested cormorants (Phalacrocorax auritus) and Caspian terns (Hydroprogne caspia) from the upper Great Lakes. I. Description and rates of abnormalities in dead eggs, 1987 and 1988, submitted for publication (1992).
273. P. Cook, Characterization of Toxicity and Risks of 2,3,7,8-TCDD and Related Chemicals in Aquatic Environments, USEPA (1992).
274. C. Wren, Cause–effect linkages between chemicals and populations of mink (Mustela vison) and otter (Lutra canadensis) in the Great Lakes Basin, *J. Toxicol. Environ. Health* **33,** 549–585 (1991).
275. L. E. Hart, K. M. Cheng, P. E. Whitehead, R. M. Shah, R. J. Lewis, S. R. Ruschkowski, R. W. Blair, D. C. Bennett, S. M. Bandiera, R. J. Norstrom, and G. D. Bellward, Dioxin contamination and growth and development in great blue heron embryos, *J. Toxicol. Environ. Health* **32,** 331–344 (1991).
276. G. L. Kimmel, Reproductive and developmental toxicity of 2,3,7,8-TCDD, in: USEPA, *A Cancer Risk-Specific Dose Estimate for 2,3,7,8-TCDD: Appendices A Through F. Draft,* EPA/600/6-88/007Ab (1988).
277. USEPA, *Chapter 5: Reproductive and Developmental Toxicity,* Review Draft, EPA/600/AP-92/001e (1992).
278. F. Murray, F. Smith, K. Nitschke, C. Humiston, R. Kociba, and B. Schwetz, Three-generation reproduction study of rats given 2,3,7,8-tetrachlorodibenzo-p-dioxin (TCDD) in the diet, *Toxicol. Appl. Pharmacol.* **50,** 241–252 (1979).
279. I. Nisbet and M. Paxton, Statistical aspects of three-generation studies of the reproductive toxicity of 2,3,7,8-TCDD and 2,4,5-T, *Am. Stat.* **36,** 290–298 (1982).
280. R. Bowman, S. Schantz, M. Gross, and S. Ferguson, Behavioral effects in monkeys exposed to 2,3,7,8-TCDD transmitted maternally during gestation and for four months of nursing, *Chemosphere* **18,** 235–242 (1989).
281. S. Schantz and R. E. Bowman, Learning in monkeys exposed perinatally to 2,3,7,8-tetrachlorodibenzo-p-dioxin (TCDD), *Neurotoxicol. Teratol.* **11,** 13–19 (1989).
282. R. Bowman, S. Schantz, N. Weerasinghe, M. Gross, and D. Barsotti, Chronic dietary intake of 2,3,7,8-tetrachlorodibenzo-p-dioxin (TCDD) at 5 or 35 parts per trillion in the monkey: TCDD kinetics and does–effect estimate of reproductive toxicity, *Chemosphere* **18,** 243–252 (1989).
283. R. Moore, C. Potter, H. Theobald, J. Robinson, and R. Peterson, Androgenic deficiency in male rats treated with 2,3,7,8-tetrachlorodibenzo-p-dioxin, *Toxicol. Appl. Pharmacol.* **79,** 99–111 (1985).
284. T. Mably, R. Moore, D. Bjerke, and R. Peterson, The male reproductive system is highly sensitive to in utero and lactational TCDD exposure, in: *Biological Basis for Risk Assessment of Dioxins and Related Compounds* (M. A. Gallo, R. J. Scheuplein, and K. A. van der Heijden, eds.), pp. 69–78, Banbury Report 35, Cold Spring Harbor Laboratory Press, Cold Spring Harbor, NY (1991).
285. T. Mably, R. Moore, and R. Peterson, In utero and lactational exposure of male rats to 2,3,7,8-tetrachlorodibenzo-p-dioxin. 1. Effects on androgenic status, *Toxicol. Appl. Pharmacol.* **114,** 97–107 (1992).
286. T. Mably, R. Moore, R. Goy, and R. Peterson, In utero and lactational exposure of male rats to 2,3,7,8-tetrachlorodibenzo-p-dioxin. 2. Effects on sexual behavior and the regulation of luteinizing hormone secretion in adulthood, *Toxicol. Appl. Pharmacol.* **114,** 108–117 (1992).

287. T. Mably, D. Bjerke, R. Moore, A. Gendron-Fitzpatrick, and R. Peterson, In utero and lactational exposure of male rats to 2,3,7,8-tetrachlorodibenzo-p-dioxin. 3. Effects on spermatogenesis and reproductive capability, *Toxicol. Appl. Pharmacol.* **114,** 118–126 (1992).
288. R. Peterson, R. Moore, T. Mably, D. Bjerke, and R. Goy, Male reproductive system ontogeny: Effects of perinatal exposure to 2,3,7,8-tetrachlorodibenzo-p-dioxin, in: *Chemically-Induced Alterations in Sexual and Functional Development: The Wildlife/ Human Connection* (T. Colborn and C. Clement, eds.), pp. 175–193, Princeton Scientific Publishing Co., Princeton, NJ (1992).
289. K. A. Faber and C. L. Hughes, Jr., The effect of neonatal exposure to diethylstilbesterol, genistein, and zearalenone on pituitary responsiveness and sexually dimorphic nucleus volume in the castrated adult rat, *Biol. Reprod.* **45,** 649–653 (1991).
290. K. A. Faber and C. L. Hughes, Jr., Anogenital distance at birth as a predictor of volume of the sexually dimorphic nucleus of the preoptic area of the hypothalamus and pituitary responsiveness in castrated adult rats, *Biol. Reprod.* **46,** 101–104 (1992).
291. A. Smits-van Prooije, J. Lammers, D. Waalkens-Berendsen, and B. Kulig, Effects of 2,3,5,3′,4′,5′-hexachlorobiphenyl alone and in combination with 3,4,3′,4′-tetrachlorobiphenyl on the reproduction capacity of rats, Presented at Dioxin '92, Tampere, Finland, *Organohalogen Compounds,* Vol. 10, pp. 217–219. (1992).
292. C. Egeland, M. Sweeney, M. Fingerhut, W. Halperin, K. Willie, and T. Schnorr, Serum dioxin (2,3,7,8-TCDD) and total serum testosterone, and gonadotropins in occupationally exposed men, Abstract for presentation at the Society for Epidemiologic Research Annual Meeting, Minneapolis, MN, June, National Institute for Occupational Safety and Health, Cincinnati, OH 45226 (1992).
293. USEPA, *Chapter 7: Epidemiology/Human Data,* Review Draft, EPA/600/AP-92/001g (1992).
294. W. H. Wolfe, J. E. Michalek, J. C. Miner, R. H. Roegner, W. D. Grubbs, M. B. Lustik, A. S. Brockman, S. C. Henderson, and D. E. Williams, The Air Force Health Study: An epidemiologic investigation of health effects in Air Force personnel following exposure to herbicides, serum dioxin analysis of 1987 examination results, *Chemosphere* **25,** 213–216 (1992).
295. W. Rogan, B. Gladen, K.-L. Hung, S.-L. Koong, L.-Y. Shih, J. Taylor, Y.-C. Wu, D. Yang, N. Ragan, and C.-C. Hsu, Congenital poisoning by polychlorinated biphenyls and their contaminants in Taiwan, *Science* **241,** 334–336 (1988).
296. M.-L. Yu, C.-C. Hsu, B. Gladen, and W. Rogan, In utero PCB/PCDF exposure: Relation of developmental delay to dysmorphology and dose, *Neurotoxicol. Teratol.* **13,** 195–202 (1991).
297. W. Rogan and B. Gladen, Neurotoxicology of PCBs and related compounds, *Neurotoxicology* **13,** 27–36 (1992).
298. G. Fein, J. Jacobson, S. Jacobson, P. Schwartz, and J. Dowler, Prenatal exposure to polychlorinated biphenyls: Effects on birth size and gestational age, *J. Pediatr.* **105,** 315–320 (1984).
299. J. L. Jacobson, S. W. Jacobson, and H. E. B. Humphrey, Effects of in utero exposure to polychlorinated biphenyls and related contaminants on cognitive functioning in young children, *J. Pediatr.* **116,** 38–45 (1990).
300. J. L. Jacobson, S. W. Jacobson, R. J. Padgett, G. A. Brumitt, and R. L. Billings, Effects of prenatal PCB exposure on cognitive processing efficiency and sustained attention, *Dev. Psychol.* **28,** 297–306 (1992).
301. H. Lilienthal and G. Winneke, Sensitive periods for behavioral toxicity of polychlorinated

biphenyls determination by cross-fostering in rats, *Fundam. Appl. Toxicol.* **17,** 368–375 (1991).

302. B. Gladen, W. Rogan, P. Hardy, J. Thullen, J. Tingelstad, and M. Tully, Development after exposure to polychlorinated biphenyls and dichlorodiphenyl dichloroethene transplacentally and through human milk, *J. Pediatr.* **113,** 991–995 (1988).
303. B. Gladen and W. Rogan, Effects of perinatal polychlorinated biphenyls and dichlorodiphenyl dichloroethene on later development, *J. Pediatr.* **119,** 58–63 (1991).
304. H. J. Pluim, J. G. Koppe, K. Olie, J. W. vd Slikke, J. H. Kok, T. Vulsma, D. van Tijn, and J. J. M. de Vijlder, Effects of dioxins on thyroid function in newborn babies, *Lancet* **339,** 1303 (1992).
305. R. Frakes, Health-Based Water Quality Criteria for 2,3,7,8-Tetrachlorodibenzo-p-Dioxin (TCDD), Final Document, Bureau of Health, Maine Department of Human Services (1990).
306. Universities Associated for Research and Education in Pathology, Inc., *Human Health Aspects of Environmental Exposure to Polychlorinated Dibenzo-p-Dioxins and Polychlorinated Dibenzofurans,* Bethesda (1988).
307. J. Ryan, T. Gasiewicz, and J. Brown, Human body burden of polychlorinated dibenzofurans associated with toxicity based on the Yusho and Yucheng incidents, *Fundam. Appl. Toxicol.* **15,** 722–731 (1990).
308. G. Clark, A. Tritscher, D. Bell, and G. Lucier, Integrated approach for evaluating species and interindividual differences in responsiveness to dioxins and structural analogs, *Environ. Health Perspect.* **98,** 125–132 (1992).
309. E. Carlsen, A. Giwercman, N. Keiding, and N. Shakkebaek, Evidence for decreasing quality of semen during past 50 years, *Br. Med. J.* **305,** 609–613 (1992).
310. R. M. Sharpe and N. E. Shakkebaek, Are oestrogens involved in falling sperm counts and disorders of the male reproductive tract? *Lancet* **341,** 1392–1395 (1993). [Note that TCDD appears to be antiestrogenic and to alter testosterone levels and sperm counts by other mechanisms outlined above.]
311. A. Schecter, H. McGee, J. Stanley, and K. Boggess, Dioxin, dibenzofuran, and PCB, including coplanar PCB levels in the blood of Vietnam veterans in the Michigan Agent Orange Study, *Chemosphere* **25,** 205–208 (1992).
312. A. Schecter, H. McGee, J. Stanley, and K. Boggess, Chlorinated dioxin, dibenzofuran and coplanar PCB levels in the blood and semen of Michigan Vietnam veterans, Presented at Dioxin '92, Tampere, Finland, *Organohalogen Compounds,* Vol. 9, pp. 231–234 (1992).
313. A. Schecter, Agent Orange, dioxins and Vietnam, Presented at the Scientific Workshop on Exposure Assessment, Committee to Review the Health Effects in Vietnam Veterans of Exposure to Herbicides, Institute of Medicine, National Academy of Sciences, December 8 (1992).
314. Environment Canada, *Toxic Chemicals in the Great Lakes and Associated Effects, Volume I: Contaminant Levels and Trends* (1991).
315. S. Eisenreich, P. Capel, J. Robbins, and R. Bourbonniere, Accumulation and diagenesis of chlorinated hydrocarbons in lacustrine sediments, *Environ. Sci. Technol.* **23,** 1116–1126 (1989).
316. P. C. Baumann and D. M. Whittle, The status of selected organics in the Laurentian Great Lakes: An overview of DDT, PCBs, dioxins, furans, and aromatic hydrocarbons, *Aquat. Toxicol.* **11,** 241–257 (1988).
317. *Chemically-Induced Alterations in Sexual and Functional Development: The Wildlife/Human Connection* (T. Colborn and C. Clement, eds.) Princeton Scientific Publishing Co., Princeton, NJ (1992).

318. J. S. Stanley and J. Orban, *Chlorinated Dioxins and Furans in the General U.S. Population: NHATS FY87 Results,* USEPA 560/5-91-003 (1991).
319. P. E. Robinson, G. A. Mack, J. Remmers, R. Levy, and L. Mohadjer, Trends of PCB, hexachlorobenzene and beta-benzene hexachloride levels in the adipose tissue of the US population, *Environ. Res.* **53,** 175–192 (1990).
320. B. G. Loganathan, S. Tanabe, Y. Hidaka, M. Kawano, H. Hidaka, and R. Tatsukawa, Temporal trends of persistent organochlorine residues in human adipose tissue from Japan, 1928–1985, *Environ. Pollut.* **79,** 31–39 (1993).
321. K. S. Crump, D. G. Hoel, C. H. Langley, and R. Peto, Fundamental carcinogenic processes and their implications for low dose risk assessment, *Cancer Res.* **36,** 2973–2979 (1976).
322. R. Kodell, D. Krewski, and J. Zielinski, Additive and multiplicative relative risk in the two-stage clonal expansion model of carcinogenesis, *Risk Anal.* **11,** 483–490 (1991).
323. J. E. Kihlstrom, M. Olsson, S. Jensen, A. Johansson, J. Ahlbom, and A. Bergman, Effects of PCB and different fractions of PCB on the reproduction of the mink (Mustela vison), *Ambio* **21,** 563–569 (1992).
324. National Research Council, *Toxicity Testing: Strategies to Determine Needs and Priorities,* National Academy Press, Washington, DC (1984).
325. T. Colborn, A. Davidson, S. Green, R. Hodge, C. Jackson, and R. Liroff, *Great Lakes: Great Legacy?* The Conservation Foundation and the Institute for Research on Public Policy, Baltimore (1990).
326. B. Hileman, Concerns broaden over chlorine and chlorinated hydrocarbons, *Chem. Eng. News* April 19, pp. 11–20 (1993).
327. International Joint Commission, *Sixth Biennial Report,* Ottawa, Washington, DC (1992).
328. International Joint Commission, *Air Quality in the Detroit–Windsor/Port Huron–Sarnia Region,* Ottawa, Washington, DC (1992).
329. Environmental Research Foundation, Detoxifying dioxin and everything else, *Rachel's Hazardous Waste News #346,* P.O. Box 5036, Annapolis, MD 21403-7036, July 15 (1993).
330. V. Monks, See no evil, *American Journalism Review* June, pp. 18–25 (1993). [Critique of coverage by The New York Times.]
331. B. Commoner, Pollution Prevention: Putting Comparative Risk Assessment in its Place, Presented at Resources for the Future Conference on Setting National Environmental Priorities: The EPA Risk-Based Paradigm and its Alternatives, November (1992).
332. S. E. Rier, D. C. Martin, R. E. Bowman, W. P. Dmowski and J. L. Becker, Endometriosis in rhesus monkeys (Macaca mulatta) following chronic exposure to 2,3,7,8-tetrachlorodibenzo-p-dioxin, *Fundam. Applied Toxicol.* **21,** 433–441 (1993).
333. T. Webster, *Dioxin and Human Health: A Public Health Assessment of Dioxin Exposure in Canada,* Department of Environmental Health, Boston University of Public Health, Boston (1994).

Chapter 2

Risk Assessments of Dioxinlike Compounds

Ellen K. Silbergeld and Peter L. deFur

1. INTRODUCTION

The dioxins and related compounds, including halogenated dibenzo-*p*-dioxins, dibenzofurans, and polyhalogenated biphenyls, have been of major environmental concern for over three decades. Prior to that, primarily in occupational settings, workers were exposed to these compounds in chemical production and waste disposal for nearly 100 years, from the earliest years of the synthetic organic chemical industry in Germany, England, and the United States (see Silbergeld *et al.*[1] for a brief history of occupational health and the dioxins). Even before the dioxins and furans were definitively identified by Kimmig and Schulz in 1957 as contaminants and by-products in chemical synthesis,[2] a number of reports had appeared in the clinical literature describing serious adverse health effects, including death, in exposed workers and in some cases in their families, who were exposed by the transport of material from the workplace to the home.[3] Starting in the 1930s, animal experiments funded by industry were conducted with materials associated with worker intoxication. Explosions and other industrial accidents resulted in substantial occupational exposures, which were also studied by industry. However, the results of these studies were not, in most cases, communicated to the affected workers, the public, or authorities regulating occupational health and safety.[1,2,4]

Three events precipitated more general public concern, and the consequent involvement of environmentalists, regulatory agencies, and industry around the world in the risk assessment and risk management of these compounds. First, in the early

---

Ellen K. Silbergeld • Environmental Defense Fund, Washington, D.C. 20009, and Department of Epidemiology and Preventive Medicine, University of Maryland Medical School, Baltimore, Maryland 21201. Peter L. deFur • Environmental Defense Fund, Washington, D.C. 20009.

Dioxins and Health, edited by Arnold Schecter. Plenum Press, New York, 1994.

1970s, analytic chemists reported finding polychlorinated biphenyls and other halogenated hydrocarbons widely dispersed in the environment; these findings were associated with observed declines in reproductive rates of wildlife, particularly in fish-eating birds and marine mammals.[5–8] Second, in three separate incidents of nonoccupational exposures—in 1968 in Yusho, Japan; in 1976 in Seveso, Italy; and in 1976 in Taiwan—significant numbers of persons were massively exposed to PCBs, dioxins, and/or furans under conditions that induced observable acute and chronic health effects. Third, the intensive use of dioxin-contaminated herbicides by the U.S. armed forces in Vietnam in the 1960s provoked first the concern of ecologists, such as Arthur Westing, and then that of environmental health experts in many countries around the world.[2] The publication of data demonstrating endemic contamination by dioxins of all ecosystems and human populations[9]—including human breast milk[10]—intensified public concern in the 1980s.

By the mid-1980s public opinion in the United States was strongly in favor of controls on these chemicals. But of all these chemicals, only PCBs had been regulated. Congress, in response to public pressure, explicitly banned the production and use of PCBs in the Toxic Substances Control Act of 1976.[5,11] Control of dioxins and furans has been less expeditious for many reasons. This chapter will focus mostly on the scientific controversies related to the risk assessment of the dioxins, but it should be understood that much of the continuing delay by government in implementing comprehensive management of that risk arises not only from scientific uncertainty but also from the politics and economics of controlling specific dioxin sources. Although the dioxins and furans are generated as unintended by-products of chemical manufacture, paper pulp production, and waste disposal, their regulation necessarily involves regulating these diverse primary sources. The involvement of dioxins in the Agent Orange controversy, and the larger dimensions of national division over the Vietnam War, has contributed to hesitancy by government in resolving many outstanding issues, including appropriate compensation of veterans in both Australia and the United States.[12]

The interactions of source, product, and waste management contribute to the complicated regulatory aspects of dioxin and PCB risk management. This complexity of regulatory interaction has been a source of continuing controversy, within government, on the risks of the dioxins. The EPA has been challenged to act using most of the existing environmental statutes—the Clean Water Act, the Clean Air Act, Toxic Substances Control Act (TSCA), Federal Insecticide Fungicide and Rodenticide Act (FIFRA), Resource Conservation and Recovery Act (RCRA), and Superfund. Because dioxins are found in food and paper products that contain food, as well as potentially in some medical devices and other products, the FDA has also been under pressure to act. The Centers for Disease Control have been involved in setting advisories and guidance levels for human exposures to dioxins in the workplace and general environment. At the state level, health departments, environment and natural resource agencies have had to grapple with the formulation and implementation of policies to control dioxins.

In other countries, the dioxins have also been matters of public concern, and government response, for both ecological and human health in the environment and workplace. Most of the countries in the Organization for Economic Cooperation and Development (OECD) have now banned or restricted the uses of PCBs in commerce and industry; many OECD countries have regulated sources of dioxins and furans, including hazardous waste management (the Seveso doctrine of the European Community), incineration, and paper pulp production. Contamination of the food supply has been a major concern in several European countries and in Japan: measurements reported to the World Health Organization (WHO) of dioxin and furan contamination of breast milk in samples collected in Europe and elsewhere generated a strong public reaction (see Schecter[10]). While this chapter will focus on risk assessments for the dioxins conducted in the United States, we will discuss similar activities within international organizations, such as WHO, the European Community, and NATO, and regulatory activities within certain other countries. Our focus on the United States is not meant to be chauvinistic, but reflects our relatively more extensive knowledge of recent history and science-based public policy-making in our own country. Since our analysis of much recent public policy-making on the dioxins in the United States is rather critical, this focus should not be interpreted as a recommendation for emulation by other countries.

Concern over human health impacts of these chemicals has largely dominated the policy debate in Europe and the United States. Ecological impacts have received considerably less attention from the public, government agencies, or the regulated community, reflecting the general diminution of concern for ecosystem integrity related to toxic chemicals. In some cases, usually at the local level in the context of specific sources, such as paper mills, there has been some public response to ecosystem risks, but, as discussed below, the relative lack of sophistication in environmental risk assessment methods has limited the debate in this arena. Ecological and nonhuman species protection in U.S. environmental policy is largely mandated by FIFRA, the pesticides law that has been used to cancel registration of pesticides toxic to endangered species, such as bald eagles (in the case of dicofol). As an example of the relative lack of emphasis for ecological issues, it took specific federal regulation to restrict uses of the biocide tributyltin in boat paints. Nevertheless, dioxin is an important example of a chemical whose ecological impacts have been extensive, and this chapter will discuss the critical need to increase vigilance over such risks by improving more generally the methods for ecological risk assessment.

The authors of this chapter have been involved at several instances in the risk assessments prepared by U.S. government agencies for the dioxins; we shall draw on that experience in this chapter, but the opinions and views in this chapter are our own and do not necessarily reflect those of our colleagues, inside and out of government, with whom we have participated in these undertakings.

2. RISK ASSESSMENT—THEORY AND PRACTICE

Understanding the development of human health risk assessments for the dioxins requires some background on the general context of risk assessments by regulatory and health agencies. "Risk assessment," as a term of art, refers to a process of hazard identification, quantitative dose–response evaluation, extrapolation of experimental data to human health, and finally exposure assessment in order to generate a quantitative estimate of the risks, or probability of disease or death, associated with specified exposures of individuals and populations (see NRC[13] for a standard description of this process and its terminology). In the United States, risk assessment in practice frequently refers to a very specific methodology developed for generating point estimates of increased risk of cancer associated with lifetime exposures to chemical carcinogens.[14,15] This methodology was originally based on inferences drawn from epidemiological and experimental data on radiation- and chemical-induced carcinogenesis.[15] It has been used by regulatory agencies in the United States since the early 1980s to provide a scientific basis for regulation, remediation, and prioritization of chemical hazards in the environment, workplace, and food supply. The risks are calculated in the models for individuals, and population risks are generally calculated by simple multiplication of individual or unit risks, without regard for the distribution of ages and other factors within a specific population.

In the hazard identification phase, risk assessment utilizes data from studies of humans—rarely available—and from experimental studies on laboratory rodents, such as the so-called cancer bioassay developed by the National Toxicology Program. Both the U.S. EPA and the International Agency for Research on Cancer have developed largely similar guidelines for the use of animal data in hazard identification, that is, the criteria for defining a chemical, or other exposure, as carcinogenic (see IARC[16]).

In addition, the likelihood of human carcinogenicity is inferred on the basis of short-term *in vitro* tests of genotoxicity, including bacterial mutation and chromosomal aberration assays.[17] It is important to note that these means of defining carcinogens are primarily operational, that is, they depend on the responses elicited in these types of tests. Since we do not yet understand all of the molecular events involved in carcinogenesis, we cannot be sure that our methods of hazard identification are complete, that we are correctly identifying chemicals as carcinogenic or noncarcinogenic. Our test methods are relatively limited: on one extreme we have the "gold standard" of the lifetime animal bioassay, in which tumors and precarcinogenic lesions are observed after prolonged *in vivo* exposure of rodents; on the other extreme, we have a few tests of genotoxicity in defined cell systems (such as the Ames bacterial mutation assay), which may represent some but not all of the critical events in carcinogenesis. Both types of tests have been extensively criticized by scientists and policy analysts,[e.g.,18] but several expert panels have been unable to recommend substantial alteration of test protocol or interpretation guidelines [for instance, in 1992 the Committee on Risk Assessment Methodologies of the U.S. National Academy of Sciences recommended continued use of relatively high doses in the bioassay design, despite concerns over possible "false positives" produced by this method (see Ref. 19)].

Methods for identifying risks to ecosystems and nonhuman species, i.e., ecological risk assessment, are not as well developed as those for humans. This places ecological targets at considerable risk because of our failure to develop and validate sensitive methods to detect effects more subtle than species depletion. EPA only recently[20] finalized general guidelines for conducting ecological risk (ecorisk) assessments, although EPA is attempting to employ these in several current proceedings (see below).[21–23] The basic steps in ecorisk assessments are similar to those described above for human risk assessments, but the terminology differs somewhat: the problem of identification is the same, although "hazard" is replaced by "effects assessment," and "exposure" is replaced with "stressor," usually incorporating some measure of bioaccumulation in the initial steps. The goals of the dose response and risk characterization steps are essentially the same, the limitations being the dearth of information on dose-response for wildlife and aquatic life. The goal of an ecorisk assessment, however, is often fundamentally different from risk assessments for human health.[24]

Unlike human assessments, ecorisk effects (hazard) assessments are not founded on a risk of individual mortality, such as cancer. Rather, the emphasis is theoretically placed on protection of a population, species, or group of species, rather than on the individual member of the species.[25] Unless the organism is a rare or endangered one, there is no federal law requiring explicit or implicit protection of *individual* plants and animals. Most of the population-based endpoints are based on drastic population reductions caused by death, as in the loss of trout from Lake Ontario, or diminished reproductive rates, as caused by DDT in osprey.[5,8] Few sublethal endpoints, such as immune system impairment or birth deformities, have been incorporated into ecorisk assessments. Furthermore, EPA has no quantitative risk management guidelines for protecting populations of nonhuman species, similar to the 10^{-6} increase in individual risk of cancer that is often used as a "rule of thumb" or "bright line" by EPA for human risk assessments.

Estimates of exposure of aquatic life, wildlife, or biological communities used in ecorisk assessments are complicated by the need to know how chemicals behave in the environment and how ecological systems function. In the case of dioxin, this requires knowledge of the physical and chemical properties of the dioxins (and related compounds) under a wide range of conditions. In addition, information is required on the critical components of an ecosystem and how they interact, quantitatively and qualitatively in space and time. These requirements cannot be met at present, even for 2,3,7,8,-TCDD, the best known of the dioxins, and for ecosystems that have been extensively studied.

The behavior of dioxins in the environment is not well understood from a mass balance perspective. It is clear that physical and chemical movement of dioxinlike chemicals in the environment (fugacity and solubility) are low, and that ingestion of contaminated materials and subsequent transfer to predator species are the most significant avenues of exposure. But there are few reliable studies available from which to estimate the biotic parameters. Dioxins have such a low solubility in water and extreme lipophilicity that most dioxin discharged into the water is associated with particulate matter in the water or sediment (or soil).

Exposure assessments for ecological risks are also complicated. For the dioxins, their hydrophobicity and very low water solubility cause them to be adsorbed strongly to organic particles, to partition or dissolve almost exclusively in lipids, and to remain associated with soils and sediments. Natural processes of chemical and biological degradation of dioxin are so slow as to be immeasurable. For instance, dioxin concentrations in the sediments of Newark Bay have remained unchanged for 30 years.[26] The most important exposure route for dioxin in aquatic systems, where such chemicals accumulate, is from contaminated sediments into the aquatic trophic system. Eventually, this exposure route leads to humans, terrestrial birds, and mammals consuming fish contaminated by eating contaminated material. Thus, efforts to quantify exposures of natural habitats or entire ecosystems involve estimating the extent to which dioxin concentrates in both ecological and physiological compartments in order to model the movement and accumulation of dioxin in a realistic food web. In theory, these two approaches (ecological and physiological) should yield similar results for a generalized ecosystem of sufficient size such that the errors inherent in assumptions are small relative to the size of the data base. Small ecosystems with more modest data sets will be less predictable from a general model, and they require data, models, and equations modified for site-specific conditions.

In human risk assessment, epidemiological data can and have been the source of exposure information on carcinogenic hazards in this century, particularly before the collection of experimental data. Both cohort and case–control data have been collected on populations exposed to chemical carcinogens, primarily in the workplace, and on persons with specific cancers and other diseases (see Ref. 27). In the case of the dioxins, there is now on balance a substantial set of well-conducted epidemiological studies that demonstrate statistically significant increases in mortality from cancer in groups exposed to dioxins in the workplace[28,29] and the general environment, through accidental contamination episodes (Seveso,[30,31] Yusho[32]). Most scientists in the field now conclude that TCDD is a human carcinogen.[33,34]

Epidemiological data are rarely useful to develop quantitative estimates of risk, because of uncertainties in exposure assessment. Because of this, extrapolation and inference are required to derive quantitative dose–response estimates, particularly at low dose/response. Extrapolation from species to species is required when experimental animals are the source of data on response and humans are the species of concern, as is often the case. Because of statistical limitations on the certainty of data and practical limits on the size of experiments, inference rules are required to develop risk estimates in the range of concern, where no data are available. In studies of ecological risks, the target species may be used in the laboratory to identify a hazard, but the problems of extrapolating beyond the experimental dose–response range still remain.

An important inference rule, used in quantitative risk assessment of carcinogens by the U.S. EPA since 1983, is that a chemical carcinogen (for any species, including humans) is assumed to have no threshold unless information on mechanism of action clearly supports another assumption.[15] The no-threshold assumption was derived from studies of radiation and chemical mutagens, and is based on the observation that many mutagens—DNA-damaging agents—are carcinogenic. A single point mutation can result from the interaction of one reactive molecule with one base pair of DNA.[35] This mutation may be sufficient to set in motion the molecular events that culminate in cancer. Thus, for purposes of risk assessment, one such event increases the probability of cancer by some finite amount > 0. Obviously, many other events, including DNA repair, can intervene between the interaction of a chemical and DNA and the induction of cancer, but that interaction by itself increases the likelihood of cancer. If the chemical must be metabolically transformed into a DNA-reactive compound, then there may be a threshold for the activation of converting enzymes, but the genotoxic event is still considered to have no threshold, once the genotoxic agent has reached its nuclear target within a cell.

In determining the relationship between dose and response or stressor and effect for wildlife and aquatic life, ecorisk assessments apply a form of the safety factor approach. Under ideal circumstances, the reference dose would be one that has been shown to have no effect on the target species. Safety factors are applied to account for uncertainty, variation among species, and differences between acute and chronic exposure. In reality, toxicity data for dioxin are available for only a few species and these must be used to set protective standards for dozens of other animals. The justification for the reference dose approach is the same as for humans—that there is some low dose that will have no effect.

These assumptions underlie most of the risk assessments proposed by EPA and OSHA over the past decade. Computerized models, the so-called linearized multistage (LMS) models, have been developed for the quantitative estimation of risk.[14] One major area of controversy related to risk assessments for the dioxins concerns the applicability of these models, and these inference assumptions, specifically to the dioxins.[36] Most of the controversy has centered on the appropriateness of assuming a no-threshold dose–response model in the low dose range for dioxin. It has been argued that since there is little evidence that the dioxins are mutagenic, as operationally defined by tests, it is inappropriate to apply a model that is based on the mutation model of cancer.[36,37]

Another approach to human cancer risk assessment has been applied by FDA and by many European government agencies. This approach makes no assumption of low dose linearity or lack of threshold for carcinogens, even mutagens. A safety factor, or uncertainty factor, is applied to a defined dose on the basis of judgment or decision factors related to the type of evidence available. This factor, usually a multiple of 10, reduces the defined dose by several orders of magnitude to generate an "acceptable" or "virtually safe dose." Much depends on the defined dose that determines the starting point for these calculations: it may be the lowest dose at which a statistically significant increase in the incidence of tumors is observed in an experimental study, or the highest dose at which no statistically significant increase is detected. Determining this dose is therefore substantially limited by the statistical power of the data sets available. As has been extensively discussed, it is not practically possible to conduct an experiment that yields reliable data unless the increase in incidence is at least a doubling of the background rate of cancer. Yet as a matter of public health policy, most people are reluctant to accept, or impose, risks of that magnitude. Courts, regulators, and others have generally accepted as reasonable increased risks in the range of 10^{-3} to 10^{-6} (i.e., an increased incidence or risk of cancer of 1 per 1000 or 1 per 1,000,000 persons exposed).[11] It will be noted that the safety factor approach assumes a linearity between the dose defined in the study and the acceptable dose, since the latter is calculated by simple division of the former. However, the results of the two methods of quantitative risk assessment—the LMS and the safety factor approach—are usually quite different. This is because the LMS model is strongly influenced by the inclusion of the zero added dose point, while the safety factor model is strongly influenced by data in the experimental range.

The biological basis for the safety factor model is an assumption that carcinogenicity is not different in kind from other types of cellular toxicity, that there may be thresholds and nonlinearities of dose and response for carcinogens as there may be for noncarcinogens.[38] Recently, several investigators have argued that for many chemicals, carcinogenicity is a consequence of other types of cellular toxicity, such as cell death, reactive oxygen-induced damage, and lipid peroxidation.[18,39] Because of the lack of mechanistic data for almost all chemical carcinogens, it is not usually possible to make data-based choices among these assumptions. However, recent data on the mechanisms of action of TCDD (see below) have encouraged some scientists to develop other approaches.[40–43]

3. RISK ASSESSMENT AT THE EPA, CDC, AND FDA OVER THE PAST DECADE

While these fundamental scientific issues remained unresolved, regulatory agencies have had to formulate risk assessments in making policy decisions over the past decade. As noted above, EPA has formalized a set of methodological rules for the quantitative assessment of chemical carcinogens, largely based on the recommendations of two federal advisory panels, the Interagency Regulatory Liaison Group of the late 1970s and the White House Office of Science and Technology Policy in the early 1980s (see OTA[15] for a history). EPA accepts both animal and human evidence for hazard identification, and EPA will utilize animal data alone for the generation of quantitative risk assessments. Both extrapolation and inference rules—related to such issues as size conversion among species—have been stipulated by the agency.

FDA has not adopted the formal rules of quantitative risk assessment published by EPA in 1986. FDA had published, earlier, a set of guidelines for cancer risk assessment, but these remain less than clear on such critical issues as threshold and linearity. In practice, FDA appears to apply an inference rule of the safety factor approach unless the chemical is clearly genotoxic, in which case the LMS model is applied.[37]

EPA only recently completed and published general guidelines for conducting ecological risk assessments[20] and the framework is being applied in three proceedings related to dioxin.[21–23] Perhaps the most contentious of these three is the development of water quality criteria for the Great Lakes states, known as the Great Lakes Initiative (GLI).[21] The GLI was mandated by Congress in order to set uniform water quality standards for all of the states in the Great Lakes watershed, including standards for the protection of aquatic life and wildlife. EPA had to pull together a large set of disparate data in order to prepare a coherent proposal that would adequately address such a diverse biological group. However, the widespread application of a single set of numeric criteria to numerous species in all of the Great Lakes states has prompted opposition at the state level.

The other two cases in which ecorisk assessments treat dioxin have received much less attention. The substantial database on wildlife and aquatic life for the GLI has been used to develop the initial methods for setting water quality criteria for the protection of wildlife nationally.[22] This preliminary effort is the first attempt by EPA to protect nonhuman species in a national effort via numeric criteria as set forth in the Clean Water Act. The same database was incorporated into the risk characterization chapter of the recent dioxin "reassessment" begun in mid-1991.

Lastly, the ecorisk characterization in the EPA dioxin reassessment[23] is admittedly more preliminary than the other nine EPA documents, and identifies numerous data gaps and uncertainties. This reflects the lower priority given ecorisk assessments generally, and dioxin specifically. EPA did not have sufficient data to attempt addressing the PCBs, furans, and other dioxins via a toxic equivalency factor (TEF) analysis (see below). This lower priority for nonhuman endpoints may well prove to be the undoing of some interest groups if wildlife require more stringent standards than do humans, as proposed in the GLI.[22]

CDC, which is not a regulatory agency, has no formal risk assessment guidelines. In its only published risk assessment on dioxin[44], CDC scientists used an LMS model to generate guidance for responding to situations with dioxin contamination of soil. Other statements from CDC officials suggest a reluctance toward adopting EPA guidelines, but no clear policy alternates have been proposed by this agency.

Until recently, there was no formal federal interagency coordination on risk assessment, despite recommendations from the National Academy of Science and suggestions from Congress. Through the early 1990's, an interagency council [Federal Committee to Coordinate Science and Technology (FCCST)] was established, under the leadership of Office of Science and Technology Policy (OSTP) of the White House, to examine the science and policy of risk assessment in theory and practice throughout the federal government. The exact charge of this group is not clear, nor is its position with respect to regulatory decisions by the regulatory agencies, which are under varying statutory obligations to solicit scientific and public review of their risk assessments. Nevertheless, it has recently entered the arena of dioxin risk assessments, at least in general terms.

Another important, but even less open, coordinating body for federal policy on dioxins was the Agent Orange Working Group, which was active through the Reagan Administrations largely in response to tort litigation against the government and industry for damages alleged to have been incurred by Vietnam veterans exposed to dioxin in herbicides.[12] This group was coordinated for a time in the 1980's by Attorney General Edwin Meese through the Domestic Policy Council of the White House.

The National Academy of Sciences has also been active in risk assessment methodology, as noted above, and in the specific issue of dioxin risk assessments. In the late 1960s, several members of the NAS, notably Matthew Meselson of Harvard, attempted to engage NAS resources in the study of ecological and other impacts of herbicide use in Vietnam.[2] NAS has published two studies on dioxins and one on PCBs, and the Institute of Medicine (a sister organization of NAS) recently completed a review of epidemiologic studies on the health effects of herbicides in Vietnam veterans.[46]

Because Agent Orange, a phenoxyacetic acid herbicide contaminated in manufacture with 2,3,7,8-TCDD, was used by the U.S. armed forces in Vietnam between 1962 and 1970, the Veterans Administration (VA, now Department of Veterans Affairs) and the Defense Department have also been involved in research and policy analysis on the health risks of the dioxins. These activities have been highly controversial, because of perceptions of conflict of interest and exclusion of affected parties from close access to study design and data interpretation.[42] Prodded by Congress, notably Senator Thomas Daschle, the VA has slowly admitted a compensable association between

Agent Orange exposure and several diseases, including neuropathies, certain cancers, chloracne, and toxic (hepatic) porphyria.[46]

The lack of coherent federal policy on dioxin risk assessment, and the failure of agencies to act on available scientific and technical knowledge, has caused a paralysis of environmental permit and enforcement actions over the past decade. Notwithstanding EPA's claims to proceed with "business as usual" during its most recent reassessment (1988–1994), during this time, permits, criteria, standards, regulations, guidelines, and site-specific actions (such as CERCLA cleanups of Newark Bay) have all been held up pending the outcome of EPA's process. Even with the completion of the latest health risk assessment in September 1994, it is likely to take over a year for regulations to be proposed. In the absence of a consistent federal presence the states have been in disarray. States have issued water quality standards that vary by at least two orders of magnitude between states with shared rivers, under the guise of "state flexibility."[47] More fundamentally, the nominal cancer risk attributed to chemicals has been altered and the acceptable level of risk has been raised by agencies in several southeastern states, such as Virginia and Maryland, in order to accommodate higher allowable inputs of dioxin, specifically by pulp mills.

4. CONTROVERSIES IN DIOXIN RISK ASSESSMENTS

Risk assessments for the dioxins remain controversial, and public policy for their management is still unresolved. In this section, we will deal with the scientific origins of these controversies related to dioxin risk assessment, but no one should be oblivious to the political and economic origins of these controversies. If dioxin were less controversial in these other contexts, risk assessments might have been accepted and risk management might have proceeded more quickly than has been the case.

While we do not know everything about the mechanistic biology of the dioxins, or the range of human response, we know more about these topics with regard to dioxin than with regard to almost any other significant environmental chemical. Controversies continue over the nature and interpretation of experimental evidence for dioxin carcinogenicity (see, for instance, Ref. 48). No significant new experimental data similar to a cancer bioassay have been developed since the Dow 1978 study and the NTP 1982 bioassay.[49] These data sets continue to be the source of information utilized in a range of statistically based risk assessments (see Ref. 50 for a review) that are distinguished primarily by assumptions as to mechanism of action rather than actual dose-response.

4.1. Mechanisms of Action

A major source of controversy in the risk assessments of dioxin relates to the mechanisms of action proposed for dioxin-induced cancer. As discussed above, mechanistic assumptions underlie the selection of risk assessment models. The LMS model, used by EPA for most chemical carcinogens, assumes that the chemical is (or acts like) a mutagen or genotoxic agent, inducing cancer through direct action on DNA in a manner analogous to radiation or other directly DNA-damaging agent (such as an electrophilic adduct that disrupts base pair association). The definition of genotoxicity has been largely derived from studies of such mutagens, and the test systems used to define genotoxicity are systems that can detect these types of DNA or chromosomal events.

Cancer is presently considered a multistage process that starts with an alteration in DNA (initiation), followed by a proliferation of cells in which the altered DNA is present, and then followed by the progression of these proliferating cells into an irreversible, rapidly growing tumor.[38] Thus, the first step is the initiating event that damages DNA through breakage, deletion, or adduct formation. If not repaired by excision or selective killing of the altered cell, the damaged DNA is "fixed" in the tissue through cell division that replicates the genetic injury into a clone of identical mutant cells. The next step is the promotion and progression of the growth of these altered cells into a preneoplastic state, which can be recognized histologically in certain organs. This may involve another mutation, or alteration in gene expression, associated with exposure to another carcinogen (a promoter) or repeated exposure to the same carcinogen that initiated the process. Promotion may also result from the action of endogenous hormones, such as estrogen, and growth factors.[51] The promoting ability of hormones has become an important issue in risk assessment, and the similarities between dioxin and endogenous steroid hormones a major issue in new risk assessment approaches.[40,43,52] Finally, the preneoplastic lesion progresses to an end stage of cancer, in which the growth of the neoplasm now threatens the functional integrity of the organ and organism.

TCDD, and other dioxins, are not genotoxic by the test systems we currently use to define genotoxicity.[49] There is a small amount of direct DNA binding of dioxin,[2,53] but dioxin is not genotoxic in *in vitro* test systems.

This lack of genotoxicity data has convinced some that dioxin should not be evaluated by the LMS model approach described above, which was derived for application to mutagens and similarly acting carcinogens.[36] If dioxin is not genotoxic, it was argued, then it must be a late-stage carcinogen, or a promoter.[38,54] The definition of a promoter is its dependence on *prior* exposure to an initiator, and its failure to increase tumor rates in animals when administered by itself. In fact, while in the experimental systems developed to detect promotion, such as two-stage skin painting, TCDD was shown to have promoterlike activity,[54] dioxin is also a complete carcinogen, i.e., capable of increasing tumor rates when administered by itself.[49] This observation has, in the past, been sufficient to support application of the LMS model because it has been assumed that complete carcinogens must have the properties of both initiators and promoters. Moreover, dioxin produces cancer in several target organs of rodents, including lung and thyroid, and not solely in the liver of rodents, so that the controversies over the relevance of rodent liver carcinogens[18] cannot be employed to discredit the data on dioxin as a complete carcinogen.

The alternative suggested is to apply other models of quantitative risk assessment. Multiple-stage models of carcinogenesis have generated methods for estimating risks of agents acting at later stages; the most prominent is the socalled Moolgavkar–Knudson model.[55] The important quantitative feature of these models is that they can include a low dose threshold, and they assume nonlinearities within the low dose range. Thus, in practice, these models, using threshold and nonlinearities in dose-response, tend to generate quantitative risk estimates that are considerably lower (in terms of risk per unit dose) than those generated by the LMS model using the same data.[50,55]

The apparent inconsistency between lack of genotoxicity and ability to induce cancer may point to limitations on our current assumptions related to the mechanisms of cancer and the ways in which our risk assessment models reflect these assumptions. Specifically, our definition of "genotoxicity," which is operationally driven, may be incomplete. A broader definition of genotoxicity, not limited by mutagenicity data, might include alterations in gene *expression* as well as alterations in gene *structure*. As discussed below, there is considerable evidence that the fundamental cellular action of dioxin is to alter gene expression.[56]

Does amplification of our concepts of genotoxicity to include gene expression imply modification of the statistical models for risk assessment? The LMS model assumes that the initial, gene-level event can be triggered by one hit, or the interaction of one molecule with one site on DNA. Can gene expression be altered by one hit, or by one molecule, other than through the modification of one base in DNA (mutation)? These are the critical issues involved in new approaches to dioxin risk assessment.[57]

4.2. Molecular Theories of Dioxin Action and Integration into Risk Assessments

Over the past 3 years, considerable research has been focused on increasing knowledge of the mechanisms of action of dioxin. This research challenges not only some current risk assessments of dioxin, but also the general definition of "genotoxicity" in risk assessment theory. Most of the known biological effects of dioxin, including carcinogenicity, appear to involve the ability of dioxin to act, through a cytosolic receptor (an endogenous cell protein), to regulate transcription of specific genes. We assume, although our knowledge is not complete, that most of the critical effects of TCDD follow on the binding of TCDD to this receptor, which has been called the dioxin or Ah (aryl hydrocarbon) receptor.

The role of this initial binding event, to later cellular events, is shown schematically in Fig. 1.[57] It can be seen that several important steps occur subsequent to the binding of TCDD to its receptor; the receptor–TCDD complex is associated with another cytosolic protein, the so-called Ah receptor nuclear transport (ARNT) protein, undergoes a conformation change and stabilization in the bound form, and is translocated to the nucleus.[58] In the nucleus, the complex binds to specific *d*ioxin (or *x*enobiotic) *r*ecognition *e*lements (DREs or XREs) in the genome that appear to have a consensus base pair sequence.[59] Once bound to DNA, the receptor complex acts (possibly with other molecules) to affect cellular function through altering the expression of target genes downstream from the recognition or DNA binding sites.[60]

The identification of all genes regulated in this manner by TCDD, and the role of these gene products in cell response, remains largely unknown.[61–63] Our understand-

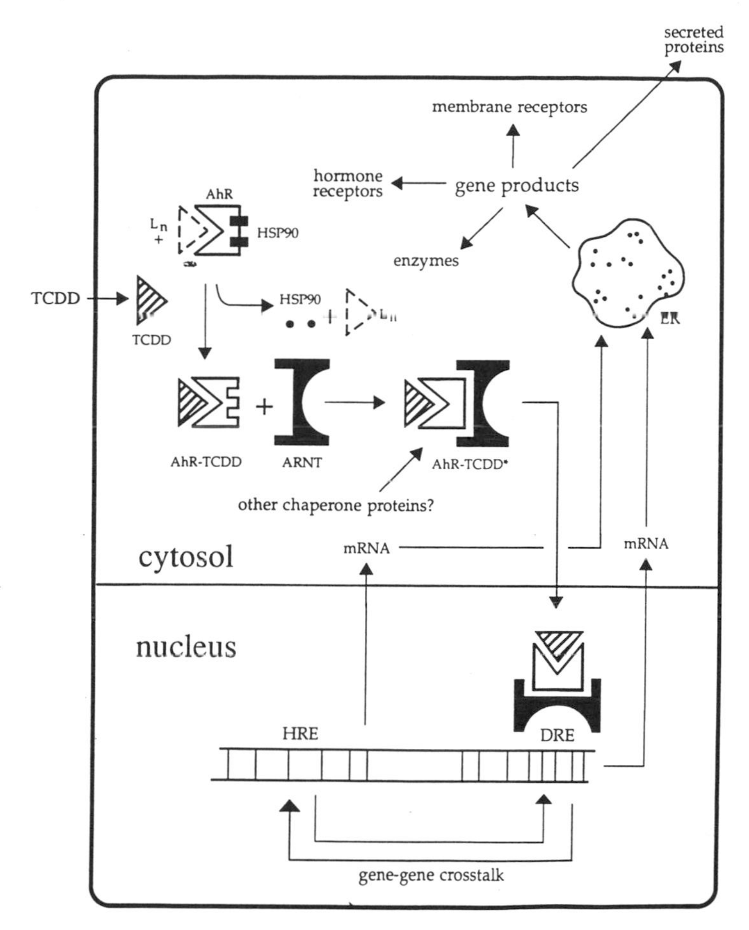

Figure 1. Model for the cellular models of dioxin (from Silbergeld[57]). AhR, Ah receptor; HSP90, heat shock protein 90 kDa; ARNT, Ah receptor nuclear translocator protein; L_n, natural ligand; ER, endoplasmic reticulum; mRNA, messenger RNA; HRE, hormone-responsive element; DRE, dioxin-responsive element.

ing of mechanism at this time is largely based on intensive research on one such event, for which most of the steps have been elucidated: the action of TCDD to induce the P450IA1 enzymes through increasing transcription of the CYP450IA1 gene.[56,60] Although enzyme induction may parallel other events,[64] it does not appear to be directly involved in cell- or organ-level toxicity of the dioxins.[41] Thus, while this information may be a useful guide to our risk assessments, it remains an assumption that other molecular events parallel those that have been studied for enzyme induction.[60,65]

This information has formed the basis for incorporating receptor theory into current approaches to mechanism-based risk assessment for the dioxins, as first suggested by Silbergeld and Gasiewicz in 1989.[52] Assuming a critical role for the dioxin receptor, risk assessments can draw on models of steroid hormone action, at the genetic level.[66–68] Quantitative models can be based on the kinetics of receptor–ligand interactions (binding or affinity constants, and number of binding sites). Models can also incorporate the relationships between receptor binding and the earliest cellular responses to receptor occupancy, increases or decreases in specific gene transcripts. For this approach, it is critical to identify the gene-level events that are clearly associated with dioxin–receptor binding and that are mechanistically relevant to toxicity. Unfortunately, the best studied early cellular response to dioxin, the increase in mRNA for CYP450IA1, is not likely to be related to carcinogenesis. We can assume that an as yet unknown gene product (or set of products) that is involved in carcinogenesis behaves similarly, at the molecular level, as does CYP450IA1. This approach has been utilized by Portier *et al.*[42] It is encouraging to note that the effects of TCDD on the synthesis of tumor necrosis factor (TNF) and the receptor for epidermal growth factor (EGF) appear to be kinetically similar to those found for CYP450IA1.[69–71] These similarities suggest that models built on our knowledge of enzyme induction may not be grossly inaccurate quantitatively.

However, more complex considerations may have to be introduced into a receptor-based model for dioxin risk assessment. First, dioxin probably regulates the synthesis and/or availability of its own receptor. Second, there are extensive interactions among dioxin and other steroid hormone receptors mediated through the binding of dioxin to its receptor. These interactions could either amplify or dampen the molecular signals induced by dioxin in the cell. Moreover, these interactions suggest that dioxin will interact with endogenous steroids, and it may be the coordinated expression of many hormonelike actions that determines the functional effect of dioxin in a cell. The interactions between dioxin and estrogen[72] and between dioxin and glucocorticoids[73,74] indicate that age, sex, and hormonal status may determine the toxicity of dioxins at the organ and organism level.

There is nothing inherent in a receptor-based model that supports the inference of a threshold in the dose–response curve for dioxin.[57,69] In receptor theory, at low concentrations of ligand, the binding of a ligand to its receptor is linearly related to the concentration of ligand (since [ligand] $<$ [receptor]). Thus, the first event in the molecular signaling pathway is likely to be linearly related to concentration of dioxin within the cell. However, it has been argued[65] that receptor-based mechanisms have thresholds because some finite number of receptors must be occupied before significant biological transduction occurs. While it is unlikely that the occurrence of one receptor binding event does not increase the probability of the next step at all, there may be significant nonlinearities in the relationship between receptor occupancy and the next step. This would not be the case if dioxin were additive to an endogenous ligand for the Ah receptor; in this case, the presence of dioxin would be added onto the concentration of the endogenous ligand to calculate the kinetics of the next reaction. It is difficult to believe that the Ah receptor, and the highly specific genetic recognition sites for the receptor in its liganded form, are present in cells of most organisms for no purpose other than responding to xenobiotics. Most dioxins and dioxinlike compounds are of relatively recent origin, although there may be compounds in plants that bind to the Ah receptor.[75]

Even assuming that there is no endogenous ligand for the receptor, the argument for nonlinearity between molecular dose and response is not confirmed by any experimental data so far obtained. As shown by Tukey *et al.*[76] several years ago, there is a linear relationship between the concentration of liganded Ah receptor and the presence of the receptor–ligand complex in the nucleus, and a linear relationship between nuclear translocation and transcription of CYP450 mRNA. More recent studies by Lucier's group have confirmed a linear relationship between dioxin dose, tissue concentration, and appearance of mRNA for the EGF receptor.[42] Experimental studies cannot resolve this issue at the molecular level. Further understanding of the mechanistic steps between receptor binding and gene transcription will enable us to infer on a biological basis the most reasonable shape of the molecular dose–response curve at ultralow concentrations.

A receptor-based model for carcinogenesis is not necessarily different in concept from the old mutation-based model of the LMS risk assessment method. DNA can be considered a receptor for adducts, chemicals that form covalent bonds with the base pairs of genetic material. But the formation of a covalent bond is thought to be irreversible, unlike the binding of a ligand to its receptor. Of course, covalent adduction of chemical carcinogens to DNA can be "reversed" by excision and repair of the adducted base pair; this is known to happen by the appearance of DNA adducts in urine of persons exposed to mutagens. The binding of dioxin to its receptor, while reversible, is stabilized by the subsequent binding of additional proteins.[58] The binding of these proteins to the dioxin–receptor complex reduces the likelihood that dioxin will be dissociated from its receptor and favors the translocation of the liganded receptor to nuclear recognition sites on DNA.[59,77] All of these factors suggest that simple Michaelis–Menten kinetics, which assume a steady-state relationship between the bound and unbound state of a receptor, may not model the interactions between dioxin and its receptor in actual cells.

Multiple binding sites for the receptor occupied or liganded by dioxin have been identified on DNA.[63,78] The existence of more than one target gene for dioxin may explain its ability to cause cancer by itself; one gene-level event may result in an "initiation"-like event, and a subsequent event may result in a "promotion"-like event. Alternatively, the actions of dioxin on gene transcription may resemble initiation in the old model, and its interactions with steroid hormones and their receptors may produce the events associated with promotion and progression—induction of cell division and further alterations in cell physiology that commit the clonal set of transformed cells toward tumorigenesis. Assuming a simple model, whereby dioxin acts through its receptor to alter (sequentially) the expression of three different genes (allowing the expression of an oncogene, increasing expression of a growth factor gene, and down-regulating the expression of a tumor suppressor gene, for example), this model would permit a molecular threshold of three binding events before cellular response occurs, but after these three events have occurred, the response of the cell to this battery of effects would be linear.

5. OTHER TOXIC ENDPOINTS

The dioxins are multipotent agents that produce a range of toxic effects involving several organ systems in animal models. Relatively few of these have been investigated in human populations. Aside from cancer, among the most controversial endpoints are the reproductive and developmental deficits associated with dioxin exposure in rodents and primates. In humans, the most controversial association is between exposures of men and adverse reproductive outcomes in their children.[79] There are more data on reproductive and developmental effects in humans mediated via maternal exposure to dioxins, dibenzofurans, and PCBs before and/or during pregnancy.[80] At higher exposures, these chemicals have been associated with increased rates of miscarriage and early neonatal death. At lower doses, and in pregnancies after cessation of exposure, fetotoxic and developmental deficits have been reported in children born to exposed mothers.[81]

The toxic effects of dioxin on reproduction and development in animals include teratogenesis, fetal death, miscarriage, and developmental deficits.[82] While there are some species differences in response, particularly teratogenesis,[80,82] the induction of neurodevelopmental deficits has been reported to occur after single-dose *in utero* exposures to extremely low doses of TCDD.[83]

Risk assessments based on reproductive toxicity of dioxin generally employ the safety factor approach rather than a no-threshold linearized model. Nevertheless, very low "acceptable" or "reference" doses are generated by these assessments, on the order of 0.1 to 1 pg/kg per day.[9] We know very little about biologically based models for assessing the risks of reproductive or developmental toxicants and so these numbers must be considered highly provisional. We do not have methods for assessing risks that are highly sensitive to the period in development in which they occur. Like DES,[84] many of the effects of TCDD are greatest early in organ system development. Recent studies on the neuroendocrine and neurobehavioral effects of dioxin in male rats have shown dysfunctions subsequent to single low-dose exposures in late gestation.[85]

Dioxin is also highly immunotoxic (see Kerkvliet[86]). The consequences of immunotoxicity include reduced resistance to infection, reduced ability to respond to humoral or cell-mediated challenge, altered ratios of lymphocyte cell subtypes, and thymic dysplasia.[87,88] In addition, the immunotoxicity of dioxin may be functionally related to its toxic effects on reproduction and its carcinogenicity. Successful reproduction involves a suppression of the normal graft versus host response, to override the maternal rejection of the embryo at implantation; the progression of carcinogenesis depends on the ability of transformed cells to evade surveillance and death mediated by T lymphocytes (NK cells). In addition, virally induced cancers (which include certain sarcomas and lymphomas) may be expressed more often when the immune system is suppressed, as in AIDS.

Assessing the risks of dioxin on reproduction and immune system function introduces new challenges to risk assessment. The cancer model has been based on lifetime exposures, and allows long latencies between exposure and effect. In contrast, both epidemiological and experimental data on the developmental toxicity of dioxin indicate that very limited prenatal exposures may induce significant and long-lasting effects on neuroendocrine development and immune system function (see below). We do not as yet have adequate models to develop risk assessments for these types of effects, or these types of exposures. The finding of such effects in young animals after exposure of the mother during pregnancy also raises concerns for exposures prior to pregnancy because of the long half-life of dioxin. Models for reproductive risks of dioxin may be relevant to cancer as well, if dioxin acts similarly *in utero* to DES, which is a powerful transplacental carcinogen.[84]

In addition, effects of dioxin on the immune system may play a role in carcinogenesis, as discussed above. We have no models for risk assessment of immunotoxins,[87] which is recognized as a high priority for dioxin health risk assessment. Again, the great developmental sensitivity of the immune system to dioxin[88] suggests a need for very cautious assumptions concerning limited exposures early in life. The immunotoxic effects of TCDD can be produced at very low exposures in animal models. However, we remain uncertain as to the connection between some of the low-dose changes in specific cell number or intracellular signal molecules, and the functionality of the immune system. If we were to base risk assessments on the *earliest* (or lowest dose) indicators of altered immunology, then immunotoxic effects of dioxin would drive our risk assessments.

6. THE TOXIC EQUIVALENCY FACTOR (TEF) CONCEPT

The most toxic dioxins and dibenzofurans are a group of closely related structures that can be characterized by molecular conformation as roughly planar molecules. Their molecular similarity confers similarity of toxic action, which is thought to result from relative goodness-of-fit to a specific cellular receptor (see above). Correlations between receptor binding and a number of short- and long-term responses have been catalogued: these include enzyme induction, teratogenicity, hepatic toxicity, and certain types of immunotoxicity (reviewed in Refs. 89, 90).

Based on this, the Environmental Defense Fund in 1985 first proposed that these compounds should be assessed and regulated as a class, based on a biologically based scheme of relative toxicity. EDF advanced this hypothesis in a petition to the EPA under TSCA; a major motivation for this proposal was that since dioxins and furans are usually found in mixtures in chemical products, wastes, and emissions, considerable efficiencies could be achieved by regulating these substances as a group. EDF's proposal stimulated considerable analysis by EPA and other agencies, which culminated in the policy statement of the so-called toxic equivalency factors (TEFs). This concept, of integrated risk assessment for a number of dioxins and dibenzofurans, has been adopted by many countries and international organizations.[90,91]

What remains controversial is the appropriate ranking scale to use for deriving TEF's relative to TCDD. At one point, octachlorodibenzodioxin (OCDD) was given a factor of zero, because of its extremely low affinity for the dioxin receptor in *in vitro* binding assays. However, Birnbaum and colleagues produced data demonstrating significant *in vivo* chronic toxicity of OCDD.[92] Other discrepancies have been observed between predicted TEFs, based on receptor binding or *in vitro* enzyme induction, and actual relative toxicity potencies, based on semichronic *in vivo* exposures.[90] The largest source of discrepancies seems to be the relative *in vivo* half lives and potencies of various isomers. Moreover, the basis for *in vivo* potency estimates is limited, since there are no comprehensive data on many isomers for such major endpoints as teratogenicity, immunotoxicity, developmental toxicity, or cancer. There also appear to be significant differences in predicted and actual TEFs between mammals and fish.[93,94]

The accurate valuation of the TEFs is important because 2,3,7,8-TCDD—the standard against which TEFs are derived—is often a small component of real-world mixtures of halogenated dibenzodioxins and furans in environmental media and human tissues. For instance, in incinerator emissions, OCDD predominates[95] whereas in pulp mill effluents the chlorinated dibenzofurans predominate. Changing the TEF for OCDD from zero to 0.001, as suggested by Ahlborg *et al.*,[90] greatly increases the sum total risk of incinerator emissions.

Even more controversial is the issue of whether the PCBs should be included in a comprehensive TEF scheme. Clearly, the coplanar PCBs interact with the Ah receptor, and possess many of the same *in vitro* and *in vivo* biological properties as the dioxins and furans. Yet some PCBs may, as weak agonists for the dioxin receptor, function as antagonists to the highly potent dioxins.[89] As with all agonist–antagonist interactions, such pharmacologic interactions are highly dependent on dose. If PCBs are treated as weak dioxins, and their risks assessed on the same scale and with the same methods as are used for the dioxins, then the significance of very small exposures to dioxins is greatly enlarged in terms of marginal contribution to overall risk.

7. "BACKGROUND" EXPOSURES TO DIOXIN AND RELATED COMPOUNDS

The evaluation of the significance of specific exposures (and sources and routes of exposure) to dioxin is properly done in the context of knowledge of all other sources of exposure. That is, the importance of *incrementally* increasing exposures, and risks, can only be understood with information on exposures, and risks, in addition to the specific exposure being evaluated. This holistic approach to exposure and risk assessment is particularly important for the dioxins because of their high degree of environmental stability and long biological persistence within target organisms. Regardless of whether or not one assumes a linear or nonlinear dose–risk curve, as shown in Fig. 2, it is critical to estimate where on the dose curve an individual or a population already falls. If the population or individual is already exposed to doses that exceed the level of acceptable risk (in a linear model) or the "threshold" level (in a nonlinear model), then different assumptions as to the existence of a threshold at very low doses no longer matter for purposes of making public policy, as shown in Fig. 2.

Much of the population of industrialized societies, where dioxins and related compounds have been released through industrial discharges, incinerator emissions, and dispersive uses of contaminated chemicals and herbicides, is already above the low dose range of exposures. Levels of dioxins in human breast milk are substantial, relative to the toxic effects that can be extrapolated for developing organisms. Recent data on human body burdens in the United States are shown in Table 1 (see also

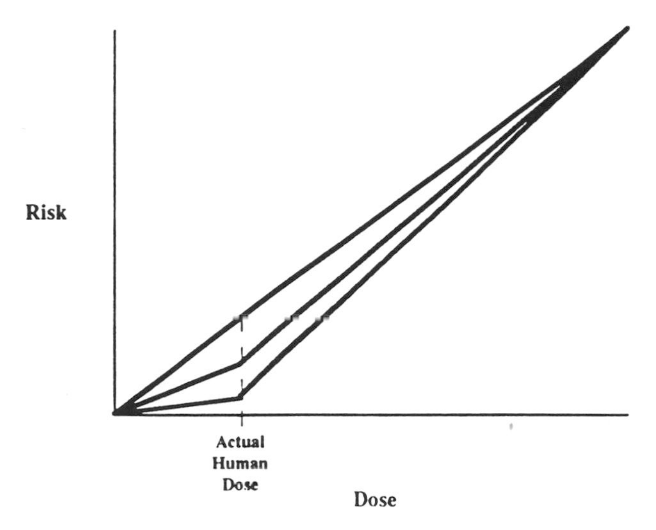

Figure 2. Theoretical dose–response curves for (from top to bottom) a linear, nonlinear, and hockey-stick model. If actual human dose is at the point indicated, then all models predict a linear dose–response relationship for increased exposures.

Schecter[10]). Under such conditions, the only prudent public policy is to take all feasible actions to reduce ongoing exposures and environmental inputs.

8. ECOLOGICAL AND WILDLIFE RISK ASSESSMENT

The realization that dioxins, furans, and PCBs are found throughout the environment and in living tissues was accompanied by the discovery that aquatic and terrestrial

Table 1
Body Burden of Dioxins and Furans in U.S. Population 1982[a]

Compound	Census regions			
	Northeast	South	North central	West
2,3,7,8-TCDD	6.6 ± 1.6[b]	6.1 ± 1.2	7.1 ± 1.6	4.1 ± 1.2
HxCDD	160 ± 50	100 ± 26	110 ± 30	120 ± 47
1,2,3,7,8-PeCDD	120 ± 145	60 ± 18	62 ± 21	73 ± 30
2,3,4,7,8-PeCDF	49 ± 15	30 ± 7.1	38 ± 9.8	52 ± 18
OCDD	750 ± 180	780 ± 140	920 ± 190	850 ± 250

[a]Data on specimens collected in FY 1982 from 763 individuals (composited into 46 samples). [EPA, *NHATS Broad Scan Analysis: Population Estimates from Fiscal Year 1982 Specimens*, EPA 560-5-90-001, EPA, Washington, DC (1989).]
[b]Lipid-adjusted weighted estimates of the average concentration ± standard errors, pg/g.

wildlife species were at least as sensitive to dioxin as lab animals (see Refs. 22, 23, 93). Mink[96] and Ontario lake trout[8] turned out to be two of the most sensitive species yet examined, showing reproductive and/or developmental effects at current ambient dioxin concentrations. Applying these data using current ecorisk procedures results in a water quality standard of 0.0085 ppq, lower than the value of 0.013 ppq for protecting humans against cancer.[97,98] Not surprisingly, the regulated community and some states have objected vociferously.

One of the critical components of ecorisk assessments is the quantification of the extent to which dioxin accumulates and biomagnifies in the trophic system. In the mid-1980s, EPA used a single number to correct for this phenomenon [bioconcentration factor (BCF)], calculated as the simple ratio of tissue to water concentrations at equilibrium (see Ref. 99 for review). Further research on the process, however, revealed that the original estimate of 5000 as a BCF grossly underestimated accumulation, perhaps by 20- to 30-fold.[100] The uptake and concentration of dioxin and related compounds turned out to be much more complex, however, than reflected in a simple BCF ratio. Much, if not all, of the uptake actually occurs via food and sediment (especially the organic carbon fraction) and, once ingested, is concentrated in lipids. Thus, total accumulation from all sources expressed on a lipid basis is considered more accurate.[93] The term bioaccumulation factor (BAF) has come to replace BCF, although EPA continues to regulate on the basis of BCF.

Although BAF is an improvement over BCF, accurate determination of its value requires extensive laboratory and field measurements. Models have to account for ecosystem properties such as number of trophic levels, as well as physiological changes such as seasonal lipid variations in fish.[101] The difficulty in making the required measurements has prompted the development of computer models to estimate bioaccumulation from trophic structures and physicochemical properties of dioxin.[25,99,102] But the models are not a panacea and have their own limitations and weaknesses (see deFur[103] for discussion).

There are ample data associating current and past environmental exposures to dioxins and related compounds with actual effects on wildlife populations (see Refs. 8, 23, 93, 104). While these studies correlating laboratory with field data provide the type of "smoking gun" that is often missing in ecotoxicocology, the presence of multiple contaminants (the many active dioxin and furan isomers and PCBs) confounds the problem of regulating sources of specific individual chemicals. The presence of multiple contaminants contributing to the toxic effects in Great Lakes wildlife[104] demonstrates the need for regulatory agencies such as EPA to treat these chemicals as a class, rather than on a chemical-by-chemical basis. The diversity of effects caused by this class of compounds in wildlife (Table 2) underscores the need for integrated, comprehensive treatment.

The total input of PCBs, dioxins, and furans into the Great Lakes ecosystem has apparently declined over the past decade, and some wildlife populations have begun to recover.[8,104] However, other species have not responded as expected, related in part to the detection of additional toxic effects at lower doses that may have been obscured by more acute effects associated with higher exposures in the 1970s.

Table 2
Toxic Effects of Dioxin and Related Compounds on Selected Wildlife Species

Species	Concentration or dose	Effect	Ref.
Lake trout	40 ppt[a] TCDD[b] in eggs	Death of fry	Spitsbergen et al. 1991
Rainbow trout	0.0011 ppt TCDD in water[c]	Death of fry	101
Mink	5–10 ppt in food	Reproductive abnormalities	24, 97 Hochstein et al. 1988
	1000 ppt body burden	Death	
Herring gulls[d]	125 ppt in embryo[c] chicks	Feminize male	8
Domestic chickens	5.8 ppt in eggs	Developmental deformities	Cheung, 1981
Cormorants	50 ppt in eggs total	Egg mortality	Tillett et al. 1992

[a]ppt, part per trillion.
[b]TCDD, tetrachlorodibenzo-*p*-dioxin.
[c]Concentrations estimated from loading or other data.
[d]The herring gulls were also exposed to DDE, PCBs, mirex, and heavy metals.

9. IMPLICATIONS OF DIOXIN RISK ASSESSMENTS FOR POLICY AND CORPORATE BEHAVIOR

When former EPA administrator William Reilly announced another "reassessment" of the health risks of dioxin in 1991, he also noted that EPA would continue "business as usual" in the regulatory programs. That statement has turned out to justify a policy that allowed state agencies to set whatever water quality standards for dioxin they chose. The policy was articulated earlier in January, 1990, when the Assistant Administrator for Water issued a guidance memo to the EPA regional administrators informing them of the states' flexibility to set standards based on local conditions.[47] In spite of the failure of state agencies to provide specific data to actually document lower standards, EPA approved dioxin standards that were 100 times higher (less protective) than EPA guidance. And all but one of the southeastern states followed this lead set by Maryland and Virginia. In each case, EPA approved the use of older data, out-of-date fish consumption values, and bioaccumulation values that could not be supported by data. The reevaluation thus turned out to be an interim period for industry to obtain approval of lax permits and standards.

The failure of public policy during the reassessment period (1991–1994) has had broader implications for the control of toxic chemicals by federal agencies. The Clean Water Act is intended to prevent the use of environmental regulations and standards for economic benefit by state governments. One state is not supposed to be able to lure industries by setting lax and unprotective environmental standards. But economic competition is exactly what is happening as industry argues for relaxation of water quality standards (Maine Department of Environmental Protction, November 5–6, 1992). Even before the reassessment, this 100-fold variation among states was recognized as a policy problem, and legislation was proposed to address the gap (H.R. 2084[105]).

The public frequently depends on news reports for information on environmental policy and scientific developments. Shifts in policy may not be manifested in actions that affect the "average" citizen. Citizen groups, including environmental organizations, have therefore become increasingly important sources of information and representatives of the "public."

10. CONCLUSIONS

There is an urgent need for government agencies to expeditiously regulate ongoing sources of dioxins in the environment, and to come to closure on acceptable standards for concentrations of dioxins in environmental media. Delays in this process result in continued releases and widespread propagation of these stable contaminants in the biosphere.

Despite a very large body of clinical, epidemiological, and experimental studies on the hazards of dioxin, we have been uanble to resolve continuing political controversies over risk assessments of the dioxins. Most of these controversies have focused on the identification of dioxin as a carcinogenic hazard for humans; now, after the publication of several well-conducted epidemiological studies, it is generally conceded that dioxin is carcinogenic for humans. However, other important toxic endpoints have been neglected in this debate: notably, the developmental and immunotoxic effects of dioxin. It is sometimes suggested that these effects may be even more significant, at low dose, than any cancer risk. In fact, all of these effects may be fundamentally interrelated.

In addition, although the TEF concept has provided partial remedy for the chemical-by-chemical approach to regulation, it is now clear that several other halogenated organics, such as PCBs, must also be considered in comprehensive risk assessment and regulation. Moreover, the cumulative impacts of all of these chemicals must be incorporated.

The focus of controversy on human health effects of dioxin has caused us to neglect the substantial ecological damage these compounds have already inflicted on wildlife. Major ecosystems, such as the Great Lakes, are perhaps irreversibly contaminated by dioxins and related compounds. All members of the food web (including humans) are at risk of exposure in these ecosystems. Damage to reproduction, immune function, and cancer have been described in fish and birds.[104] However, until more sensitive methods of ecological risk assessment are applied to risks such as dioxin, the importance of ecological impacts will continue to be neglected as we develop and implement a coherent, science-based policy of reducing risks associated with these unnecessary chemicals before the legacy of their mismanagement is practically irreversible.

ACKNOWLEDGMENTS. The work of the Environmental Defense Fund on dioxin-related issues has been supported by the CS Mott Foundation, the Public Welfare Foundation, and the Noble Foundation. Research in the laboratory of Dr. E. Silbergeld at the University of Maryland has been supported by the Veterans Administration and by a cooperative agreement with the Environmental Protection Agency. The authors also wish to thank colleagues in other organizations, particularly Joseph Guth, formerly with the Natural Resources Defence Council, Thomas Webster of the Center for the Biology of Natural Systems, Richard Christian of the American Legion, and Peter Washburn of the Natural Resources Council of Maine.

REFERENCES

1. E. K. Silbergeld, M. Kelly, and M. Gordon, Dioxin at Diamond Shamrock, in: *Toxic Circles* (R. Wedeen and H. Sheehan, eds.), pp. 55–80, Rutgers University Press, New Brunswick, NJ (1993).
2. A. Hay, *The Chemical Scythe,* Plenum Press, New York, (1982).
3. A. Thelwell Jones, The etiology of acne with special reference to acne of occupational origin, *J. Ind. Toxicol.* **23,** 290–312 (1941).
4. E. K. Silbergeld, Dioxin: A case study in chloracne, in: *Dermatotoxicology* (F. Marzulli and H. Maibach, eds.), pp. 667–686, Hemisphere Publishing, Washington, DC (1991).
5. Environmental Defense Fund, *Malignant Neglect,* Knopf, New York (1979).
6. R. J. Norstrom, D. C. G. Muir, C. A. P. Ford, M. Simon, C. R. MacDonald, and P. Beland, Polychlorinated-p-dioxins and furans in marine mammals of the Canadian North, *Environ. Pollut.* **66,** 1–19 (1990).
7. E. M. Silberhorn, H. P. Glauert, and L. W. Robertson, Carcinogenesis of polyhalogenated biphenyls: PCBs and PBBs, *CRC Crit. Rev. Toxicol.* **20,** 439–464 (1990).
8. G. A. Fox, Epidemiologic and pathobiological evidence of contaminant-induced alteration in sexual development of free living wildlife, in: *Chemically Induced Alterations in Sexual and Functional Development: The Wildlife/Human Connection* (T. Colburn and C. Clement, eds.), pp. 147–158, Princeton Science Publishers, Princeton, NJ (1992).
9. U.S. EPA, *Health Effects Assessment Document for Polychlorinated Dibenzo-p-Dioxins,* U.S. EPA ECAO, Cincinnati (EPA 600/8-64/0146) (1985).
10. See Schecter, this volume.
11. R. Percival, A. Miller, and H. Leap, *Environmental Law,* Little, Brown, Boston (1992).
12. P. Schuck, *Agent Orange on Trial,* Belknap/Harvard, Cambridge, MA (1986).
13. National Research Council, *Risk Assessment in the Federal Government: Managing the Process,* NAS Press, Washington, DC (1983).
14. E. L. Anderson, Quantitative approaches in use to assess cancer risk, *Risk. Anal.* **3,** 277–295 (1983).
15. Office of Technology Assessment, *Identifying and Regulating Carcinogens in the Federal Government,* GPO, Washington, DC (1987).
16. IARC (International Agency for Research on Cancer), *Mechanisms of Carcinogenesis in Risk Identification,* WHO, Lyon, France (1992).
17. R. Tennant and E. Zieger, Genetic toxicology: The current status of methods in carcinogen identification, *Environ. Health Perspect.* **100,** 307–315 (1993).
18. B. N. Ames and L. S. Gold, Chemical carcinogenesis: Too many rodent carcinogens, *Proc. Natl. Acad. Sci. USA* **87,** 7772–776 (1990).
19. National Research Council, *Issues in Risk Assessment,* NAS Press, Washington, DC (1992).

20. U.S. EPA, *A Framework for Ecological Risk Assessment,* EPA/630/R-92-001, Risk Assessment Forum, Washington, DC (1992).
21. U.S. EPA, *Great Lakes Water Quality Initiative Technical Support Document for the Procedures for Deriving Criteria for the Protection of Wildlife,* U.S. EPA draft document, Region V, Chicago (1992).
22. U.S. EPA, *Proc. National Wildlife Criteria Methodologies Mtg.,* draft report, Charlottesville, VA, EPA Office of Water, Criteria and Standards, Washington, DC, (1992).
23. U.S. EPA, *Interim Report on Assessment of Data and Methods for of 2,3,7,8-Tetrachlorodibenzo-p-dioxin to Aquatic Life and Associated Wildlife,* EPA/600 (R-93/055), ORD, ERL–Duluth, Duluth, MN (1993).
24. S. B. Norton, D. J. Rodier, W. H. Gentile, W. P. van der Schalie, W. Wood, and W. M. Slimak, A framework for ecological risk asessment at the EPA, *Environ. Toxicol. Chem.* **11,** 1663–1672 (1992).
25. S. M. Bartell, R. H. Gardner, and R. V. O'Neill, *Ecological Risk Estimation,* Lewis Publishers, Chelsea, MI (1992).
26. R. F. Bopp, M. L. Gross, H. Tong, H. J. Simpson, S. J. Monson, B. L. Deck, and F. S. Moser, A major incident of dioxin contamination: Sediments of New Jersey estuaries, *Environ. Sci. Technol.* **25,** 951–956 (1991).
27. See Hardell *et al.,* this volume.
28. M. A. Fingerhut, W. E. Halprin, D. A. Marlow, L. A. Piacitelli, P. A. Honchar, M. H. Sweeney, A. L. Greife, P. A. Dill, K. Steenland, and A. J. Suruda, Cancer mortality in workers exposed to 2,3,7,8-tetrachlorodibenzo-p-dioxin, *N. Engl. J. Med.* **324,** 212–218 (1991).
29. A. Manz, J. Berger, J. H. Dwyer, D. Flesch-Janys, B. Nagel, and H. Waltsgott, Cancer mortality among workers in chemical plant contaminated with dioxin, *Lancet* **338,** 959–962 (1991).
30. P. A. Bertazzi, C. Zocchetti, A. C. Pesatori, S. Guercilena, M. Sanarico, and L. Radice, Ten-year mortality study of the population involved in the Seveso incident in 1976, *Am. J. Epidemiol.* **129,** 1187–1200 (1989).
31. P. A. Bertazzi, A. C. Pesatori, D. Consonni, A. Tironi, M. T. Landi, and C. Zocchetti, Cancer incidence in a population accidentally exposed to 2,3,7,8-TCDD, *Epidemiology* **4,** 398–406 (1993).
32. M. Kuratsune, Yusho, in: *Halogenated Biphenyls, Terphenyls, Naphthalenes, Dibenzodioxins and Related Products* (R. D. Kimbrough, ed.), pp. 287–302, Elsevier/North-Holland, Amsterdam (1980).
33. A. H. Smith and M. L. Warner, Biologically measured human exposure to 2,3,7,8-tetrachlorodibenzo-p-dioxin and human cancer, *Chemosphere* **25,** 219–222 (1992).
34. E. K. Silbergeld, Carcinogenicity of the dioxins, *J. Natl. Cancer Inst.* **85,** 1198–1199 (1991).
35. F. Williams and J. H. Weisburger, Chemical carcinogens, in: *Casarett and Doull's Toxicology,* 3rd ed. (C. D. Klaassen, M. O. Amdur, and J. Doull, eds.), pp. 90–173, Macmillan Co., New York (1986).
36. H. P. Shu, D. J. Paustenbach, and F. J. Murray, A critical evaluation of the use of mutagenesis, carcinogenesis, and tumor promotion data in a cancer risk assessment of 2,3,7,8-tetrachlorodibenzo-p-dioxin, *Regul. Toxicol. Pharmacol.* **7,** 57–68 (1987).
37. R. L. Sielken, Quantitative cancer risk assessments for TCDD, *Food Chem. Toxicol.* **25,** 257–267 (1987).
38. H. C. Pitot and Y. P. Dragan, Facts and theories concerning the mechanisms of carcinogenesis, *FASEB J.* **5,** 2280–2286.

39. S. M. Cohen and L. B. Ellwein, Cell proliferation and carcinogenesis, *Science* **249,** 1007–1011 (1990).
40. G. W. Lucier, C. Portier, and M. A. Gallo, Receptor mechanisms and dose–response models for the effects of dioxins, *Environ. Health Perspect.* **101,** 36–44 (1993).
41. A. M. Tritscher, J. A. Goldstein, C. J. Portier, Z. McCoy, G. C. Clark, and G. W. Lucier, Dose–response relationships for chronic exposure to 2,3,7,8-tetrachlorodibenzo-p-dioxin in a rat tumor model: Quantification and immunolocalization of CYP450IA1 and CYP450IA2 in the liver, *Cancer Res.* **52,** 3436–3442 (1992).
42. C. Portier, A. Tritscher, M. Kohn, C. Sewall, G. Clark, L. Edler, D. Hoel, and G. Lucier, Ligand/receptor binding for 2,3,7,8-TCDD: Implications for risk assessment, *Fundam. Appl. Toxicol.* **20,** 48–56 (1993).
43. E. K. Silbergeld, Dioxin: Receptor-based approaches to risk assessment, in: *Biological Basis for Risk Assessment of Dioxins and Related Compounds* (M. A. Gallo, R. Scheuplein, and K. A. van der Heijden, eds.), pp. 441–455, Cold Spring Harbor Laboratory Press, Cold Spring Harbor, NY (1991).
44. R. D. Kimbrough, H. Falk, P. Stehr, and G. F. Fries, Health implications of TCDD contamination of residential soil, *J. Toxicol. Environ. Health* **14,** 47–53 (1984).
45. E. Zumwalt, Report to the Secretary, Department of Veteran Affairs (1990).
46. Institute of Medicine, National Academy of Sciences, *Veterans and Agent Orange: A Review of the Evidence,* NAS Press, Washington, DC (1993).
47. L. Wilcher, Assist. Admin., Office of Water, U.S. EPA Headquarters, Limitations related to 2,3,7,8-TCDD in surface water, January 5 (1990).
48. W. R. Brown, Implications of the reexamination of the liver sections from the TCDD chronic rat bioassay, in: *Biological Basis for Risk Assessment of Dioxins and Related Compounds* (M. Gallo, R. Scheuplein, and K. A. van der Heijden, eds.), pp. 13–18, Cold Spring Harbor Laboratory Press, Cold Spring Harbor, NY (1991).
49. J. E. Huff, 2,3,7,8-TCDD: A potent and complete carcinogen in experimental animals, *Chemosphere* **25,** 173–176 (1992).
50. L. Zeise, J. E. Huff, A. G. Salmon, and N. K. Hooper, Human risks from 2,3,7,8-tetrachlorodibenzo-p-dioxin and hexachlorodibenzo-p-dioxins, *Adv. Mod. Environ. Toxicol.* **17,** 293–342 (1990).
51. B. E. Henderson, R. K. Ross, M. C. Pike, and J. T. Casagrande, Endogenous hormones as a major factor in human cancer, *Cancer Res.* **42,** 3232–3239 (1982).
52. E. K. Silbergeld and T. A. Gasiewicz, Dioxins and the Ah receptor, *Am. J. Ind. Med.* **16,** 455–474 (1989).
53. A. Poland and E. Glover, An estimate of the maximum in vivo covalent binding of 2,3,7,8-tetrachlorodibenzo-p-dioxin to rat liver protein, ribosomal RNA, and DNA, *Cancer Res.* **39,** 3341–3344 (1970).
54. H. C. Pitot, T. L. Goldsworthy, H. A. Campbell, and A. Poland, Quantitative evaluation of the promotion by 2,3,7,8-tetrachlorodibenzo-p-dioxin of hepatocarcinogenesis from diethylnitrosamine, *Cancer Res.* **40,** 3616–3621 (1980).
55. U.S. EPA, *A Cancer Risk Specific Dose Estimate for 2,3,7,8-TCDD,* ORD, Washington, DC (1988).
56. J. M. Fisher, K. W. Jones, and J. P. Whitlock, Activation of transcription as a general mechanism of 2,3,7,8-tetrachlorodibenzo-p-dioxin, *Mol. Carcinogen.* **1,** 216–221 (1989).
57. E. K. Silbergeld, Biologically based approaches to dioxin risk assessment: Implications for theories of chemical carcinogenesis, in: *Hormone-Dependent Mechanisms of Carcinogenesis* (T. Slaga, ed.), pp. 273–293, Plenum Press, New York, (1994).
58. O. Hankinson, H. Reyes, E. C. Hoffman, B. A. Brooks, B. Johnson, J. Nanthur, A. J.

Watson, and K. Weir-Brown, A genetic analysis of Ah receptor action, *Chemosphere* **25,** 37–40 (1992).
59. T. A. Gasiewicz, C. J. Elferink, and E. C. Henry, Characterization of multiple forms of the Ah receptor: Recognition of a dioxin-responsive enhancer involves heteromer formation, *Biochemistry* **30,** 2909–2916.
60. J. R. Whitlock, Mechanism of dioxin action: Relevance to risk assessment, in: *Biological Basis for Risk Assessment of Dioxins and Related Compounds* (M. A. Gallo, R. Scheuplein, and K. A. van der Heijden, eds.), pp. 351–359, Cold Spring Harbor Laboratory Press, Cold Spring Harbor, NY (1991).
61. T. Sutter and W. F. Greenlee, Classification of members of the Ah gene battery, *Chemosphere* **25,** 223–226 (1992).
62. Y. Aoki, B. A. Fowler, S. R. Max, and E. K. Silbergeld, Effects of TCDD in vivo and in vitro on gene expression, *Biochem. Pharmacol.* **42,** 1195–1201 (1991).
63. T. R. Sutter, K. Guzman, K. M. Dold, and W. F. Greenlee, Targets for dioxin: Genes for plasminogen activator inhibitor-2 and interleukin-1 beta, *Science* **318,** 415–418 (1991).
64. S. Flodstrom and U. G. Ahlborg, Promotion of hepatocarcinogenesis in rats by PCDDs and PCDFs, in: *Biological Basis for Risk Assessment of Dioxins and Related Compounds* (M. A. Gallo, R. J. Scheuplein, and K. A. van der Heijden, eds.), pp. 405–414, Cold Spring Harbor Laboratory Press, Cold Spring Harbor, NY (1991).
65. A. Poland, Receptor-mediated toxicity: Reflections on a quantitative model for risk assessment, in: *Biological Basis for Risk Assessment of Dioxins and Related Compounds* (M. A. Gallo, R. Scheuplein, and K. A. van der Heijden, eds.), pp. 417–424, Cold Spring Harbor Laboratory Press, Cold Spring Harbor, NY (1991).
66. S. Cuthill, A. Wilhelmsson, G. Mason, M. Gillner, L. Poellinger, and J. A. Gustafsson, The dioxin receptor: A comparison with the glucocorticoid receptor, *J. Steroid Biochem.* **30,** 277–280 (1988).
67. K. R. Yamamoto, Steroid receptor regulated transcription of specific genes and gene networks, *Annu. Rev. Genet.* **19,** 209–252 (1985).
68. J. H. Clark, W. T. Schrader, and B. W. O'Malley, Mechanisms of steroid action, in: *Mechanisms of Action of Steroid Hormones and Peptides* (J. Roth, ed.), pp. 33–75, Liss, New York (1990).
69. G. W. Lucier, Receptor mediated carcinogenesis, in: *Mechanisms of Carcinogenesis in Risk Identification,* pp. 87–112, IARC, Lyon, France (1992).
70. G. C. Clark, M. J. Taylor, A. M. Tritscher, and G. W. Lucier, Tumor necrosis factor involvement in 2,3,7,8-tetrachlorodibenzo-p-dioxin-mediated endotoxin hypersensitivity in C57BL/6J mice congenic at the Ah locus, *Toxicol. Appl. Pharmacol.* **111,** 422–431 (1991).
71. G. A. Clark, A. Tritscher, D. Bell, and G. Lucier, Integrative approach for evaluating species and interindividual differences in responsiveness to dioxins and structural analogs, *Environ. Health Perspect.* **98,** 125–132 (1992).
72. T. Devito, T. Thomas, T. Umbreit, and M. Gallo, Antiestrogenicity of TCDD involves the downregulation of estrogen receptor mRNA and protein, *Toxicology* **10,** 981 (1991).
73. S. R. Max and E. K. Silbergeld, Skeletal muscle glucocorticoid receptor and glutamine synthetase activity in the wasting syndrome in rats treated with 2,3,7,8-TCDD, *Toxicol. Appl. Pharmacol.* **87,** 523–527 (1987).
74. R. P. Ryan, G. I. Sunahara, G. W. Lucier, L. S. Birnbaum, and K. G. Nelson, Decreased ligand binding to the hepatic glucocorticoid and epidermal growth factor receptors after 2,3,4,7,8-pentachlorodibenzofuran treatment of pregnant mice, *Toxicol. Appl. Pharmacol.* **98,** 454–464 (1991).

75. D. W. Nebert, The Ah locus: Genetic differences in toxicity, cancer, mutation, and birth defects, *CRC Crit. Rev. Toxicol.* **20,** 153–179 (1989).
76. R. H. Tukey, R. R. Hannah, M. Negishi, D. W. Nebert, and H. J. Elsen, The Ah locus: Correlation of intranuclear appearance of inducer–receptor complex with induction of cytochrome P1-450, *Cell* **31,** 275–289 (1982).
77. J. P. Landers and N. J. Bunce, The Ah receptor and the mechanism of dioxin toxicity, *Biochem. J.* **276,** 273–287 (1991).
78. M. S. Denison, P. A. Bank, and E. F. Yao, DNA-sequence-specific binding of transformed Ah receptor to a dioxin responsive enhancer: Looks aren't everything, *Chemosphere* **25,** 33–36 (1992).
79. E. K. Silbergeld and D. R. Mattison, The effects of dioxin on reproduction: Experimental and clinical studies, *Am. J. In. Med.* **11,** 131–144 (1987).
80. L. S. Birnbaum, Developmental toxicity of TCDD and related compounds: Species sensitivities and differences, in: *Biological Basis for Risk Assessment of Dioxins and Related Compounds* (M. A. Gallo, R. Scheuplein, and K. A. van der Heijden, eds.), pp. 51–68, Cold Spring Harbor Laboratory Press, Cold Spring Harbor, NY (1991).
81. S. T. Hsu, C. L. Ma, S. H. Hsu, S. S. Wu, N. H. M. Hsu, and C. C. Yeh, Discovery and epidemiology of PCB poisoning in Taiwan, *Am. J. Ind. Med.* **5,** 71–80 (1984).
82. L. A. Couture, D. B. Abbott, and L. S. Birnbaum, A critical review of the developmental toxicity and teratogenicity of 2,3,7,8-TCDD: Recent advances toward understanding the mechanism, *Teratology* **42,** 627 (1990).
83. T. A. Mably, R. W. Moore, and R. E. Peterson, In utero and lactational exposure of male rats to 2,3,7,8-tetrachlorodibenzo-p-dioxin, *Toxicol. Appl. Pharmacol.* **114,** 108–117 (1992).
84. J. A. McLachlan, K. S. Korach, R. R. Newbold, and G. H. Degen, Diethylstilbestrol and other estrogens in the environment, *Fundam. Appl. Toxicol.* **4,** 686–691 (1984).
85. R. Peterson, T. A. Mably, R. W. Moore, and R. W. Goy, In utero and lactational exposure of male rats to 2,3,7,8-TCDD: Effects on sexual behavior and the regulation of luteinizing hormone secretion in adulthood, *Chemosphere* **25,** 157–160 (1992).
86. See Kerkvliet, this volume.
87. J. H. Dean and L. D. Lauer, Immunological effects following exposure to 2,3,7,8-tetrachlorodibenzo-p-dioxin: A review, in: *Public Health Risks of the Dioxins* (W. W. Lowrance, ed.), pp. 275–294, William Kaufman, Los Altos, CA (1984).
88. M. P. Holsapple, D. L. Morris, S. C. Wood, and N. K. Snyder, 2,3,7,8-Tetrachlorodibenzo-p-dioxin induced changes in immunocompetence, *Annu. Rev. Pharmacol. Toxicol.* **31,** 73–97 (1991).
89. S. H. Safe, Comparative toxicology and mechanism of action of polychlorinated dibenzo-p-dioxins and dibenzofurans, *Annu. Rev. Pharmacol. Toxicol.* **26,** 371–399 (1986).
90. U. G. Ahlborg, M. A. Brouwer, J. L. Jacobson, S. W. Jaconson, S. W. Kennedy, A. A. F. Kettrup, J. H. Coeman, H. Poiger, C. Rappe, S. H. Safe, R. F. Seegal, J. Puomisto, and M. van den Berg, Impact of polychlorinated dibenzo-p-dioxins, dibenzofurans, and biphenyls on human and environmental health with special emphasis on application of the toxic equivalency factor concept, *Eur. J. Pharmacol.* **228,** 179–199 (1992).
91. S. Safe, Development, validation and limitations of toxic equivalent factors, *Chemosphere* **25,** 61–64 (1992).
92. L. A. Couture, M. R. Elwell, and L. S. Birnbaum, Dioxin-like effects observed in male rats following exposure to octachlorodibenzo-p-dioxin during a 13-week study, *Toxicol. Appl. Pharmacol.* **93,** 31–46 (1988).
93. P. M. Cook, M. K. Walker, D. W. Kuehl, and R. E. Peterson, Bioaccumulation and

toxicity of TCDD and related compounds in aquatic ecosystems, in: *Biological Basis for Risk Assessment of Dioxins and Related Compounds* (M. A. Gallo, R. Scheuplein, and K. A. van der Heijden, eds.), pp. 143–168, Cold Spring Harbor Laboratory Press, Cold Spring Harbor, NY (1991).
94. P. H. Peterman and D. L. Stalling, Toxicity and bioaccumulation of 2,3,7,8-tetrachlorodibenzo-p-dioxin and 2,3,7,8-tetrachlorodibenzofuran in rainbow trout, *Environ. Toxicol. Chem.* **7,** 47 (1988).
95. C. Rappe, Sources of PCDDs, and PCDFs, *Chemosphere* **25,** 41–44 (1992).
96. R. J. Aulerich, S. J. Burstein, and A. C. Napolitano, Biological effects of epidermal growth factor and 2,3,7,8-tetrachlorodibenzo-p-dioxin on developmental parameters of neonatal mink, *Arch. Environ. Contam. Toxicol.* **17,** 27–31 (1988).
97. U.S. EPA, *Water Quality Criteria Document for 2,3,7,8-tetrachlorodibenzo-p-dioxin,* EPA 440/5-84-007, Washington, DC (1984).
98. U.S. EPA, *Interim Report on Data and Methods for Assessments of 2,3,7,8-tetrachlorodibenzo-p-dioxin Risks to Aquatic Life and Associated Wildlife,* EPA/600/R-93/055, Washington, DC (1993).
99. P. F. Landrum, H. Lee, and M. J. Lydy, Toxicokinetics in aquatic systems: Model comparisons and use in hazard assessment, *Environ. Toxicol. Chem.* **11,** 1709–1725 (1992).
100. P. M. Mehrle, D. R. Buckler, E. E. Little, L. M. Smith, J. D. Petty, P. H. Peterman, and D. L. Stalling, Toxicity and bioaccumulation of 2,3,7,8-tetrachlorodibenzo-p-dioxin and 2,3,7,8-tetrachlorodibenzo-p-furan in rainbow trout, *Environ. Toxicol. Chem.* **7,** 47 (1988).
101. J. Kent, C. L. Prosser, and G. Graham, Alterations in liver compositions of channel catfish (Ictalurus punctatus) during seasonal acclimatization, *Physiol. Zool.* **65,** 867–884 (1992).
102. R. V. Thomann, Bioaccumulation model of organic chemical contamination in aquatic food chains, *Environ. Sci. Technol.* **23,** 699–707 (1989).
103. P. L. deFur, Bioaccumulation modeling: Regulation and policy, *Health Environ. Dig.* **6,** 4–5.
104. T. Colburn and C. Clement (eds.), *Chemically Induced Alterations in Sexual and Functional Development: The Wildlife/Human Connection,* Princeton Science Publishers, Princeton, NJ (1992).
105. H.R. 2084, Bill to Establish a Minimum Requirement for Dioxin, 102nd Congress, 2nd Session (Introduced by Rep. R. Tallon *et al.*).

Chapter 3

Environmental Sources, Distribution, and Fate of Polychlorinated Dibenzodioxins, Dibenzofurans, and Related Organochlorines

Douglas R. Zook and Christoffer Rappe

1. INTRODUCTION AND CHAPTER OVERVIEW

The intent of this chapter is to bring to the reader an appreciation of polychlorinated dioxin (PCDD) and polychlorinated dibenzofuran (PCDF) sources and to present general information regarding their subsequent environmental distribution, levels, and fate. This brief overview is intended to serve as an informative and up-to-date summary for the interested nonspecialist. Selected citations will be found to allow for further study of pertinent or representative literature up to early 1993. Factors that influence the passage of PCDD/F through the environment will be noted in an effort toward understanding how these substances are transported from their places of origin and transformed during deposition in the environment. Following deposition PCDD/F can come into contact with biological systems, including humans.

In the absence of potential occupational or accidental contact, the major route of human exposure to PCDD/F is through the diet. The World Health Organization currently recommends limiting dietary intake to 10 pg (ten trillionths of a gram) of

Douglas R. Zook and Christoffer Rappe • Institute of Environmental Chemistry, University of Umeå, S-901 87 Umeå, Sweden. Present address of D.R.Z.: Institute of Physical Chemistry, Academy of Sciences, ul. Kasprzaka 44/52, 01-224 Warsaw, Poland.

Dioxins and Health, edited by Arnold Schecter. Plenum Press, New York, 1994.

2,3,7,8-tetraCDD (TCDD) or an equal quantity of structurally similar (so-called 2,3,7,8-substituted) PCDD/F counted in terms of international dioxin toxic equivalency factors (I-TEFs) to arrive at 10 pg dioxin toxic equivalents (TEQs) per kg body wt per day.[1] The United States Environmental Protection Agency (USEPA) feels that a smaller dioxin daily uptake is even more protective of human health. The I-TEQ expresses the total quantity of combinations of PCDD/F isomers in terms of 2,3,7,8-TCDD through current consensus estimations of relative toxicity to 2,3,7,8-TCDD.[1,2] For example, octaCDD (OCDD; the most abundant of PCDD/F found in the environment, as well as the most abundant found in human tissues) is thought to be approximately 1000 times less toxic than 2,3,7,8-TCDD.[1,2]

Following a summary of established primary PCDD/F sources, information regarding their distribution, levels, and fate will be presented using selected examples from North American and European locales for which most information is presently available. Unknown environmental sources of PCDD/F are briefly considered in light of apparent discrepancies between estimated source emissions and observed environmental deposition levels. This chapter concludes with a brief summary of compounds closely related to PCDD/F that merit attention. The reader should keep in mind that at present many questions have yet to be answered with respect to the source, distribution, level, and fate overview we are seeking to describe.

Dioxins and Society

Polychlorinated dibenzo-*para*-dioxins and dibenzofurans (PCDDs and PCDFs) are two distinct but related classes of tricyclic aromatic chemicals comprised of the elements carbon, hydrogen, oxygen, and chlorine. Members in each class vary only in the placement of hydrogen and chlorine nuclei about well-defined frameworks of carbon and oxygen, as depicted in Fig. 1, for a combined set of 210 (= 135 PCDD + 75 PCDF) unique chemical compounds (more formally referred to as isomers).

PCDDs and PCDFs (PCDD/Fs) have in recent decades generated wide interest in both scientific and public settings as a result of the pronounced toxicity and persistence of members within these compound classes. The 2,3,7,8-TCDD isomer, frequently referred to as "TCDD" or "dioxin," has received the most attention since it is believed to be among the most toxic of all known organochlorines, and is the most carcinogenic chemical yet tested with laboratory animals by the USEPA. It has been estimated that the U.S. government has spent billions of dollars studying PCDD/Fs.[3] Globally the

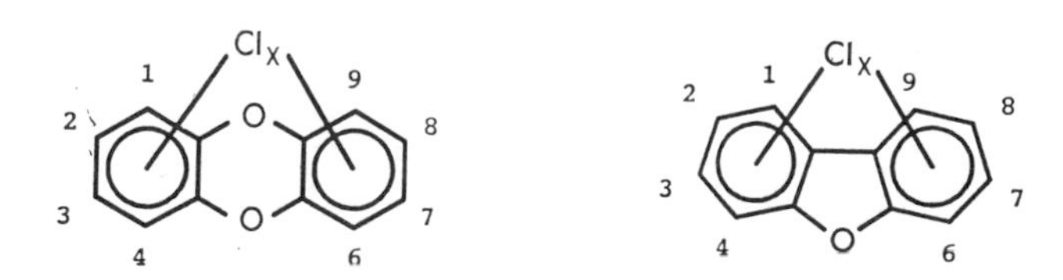

Figure 1. Structures of (left) polychlorinated-para-dioxins and (right) dibenzofurans.

costs associated with study, abatement, and remediation (as well as litigation in the United States surrounding human exposure to PCDD/Fs) must certainly exceed many billions more. An international "Dioxin" research conference meets annually to discuss developments regarding these environmental contaminants. PCDD/Fs have thus become a virtual academic cornucopia pouring forth a diverse assortment of scientific publications, news reports, and books, placing them among the most studied of environmental contaminants in these recent years of popular environmental consciousness.

Many informed citizens and medical professionals are acquainted with issues stemming from PCDD/F contamination. Major topics have concerned the presence of PCDD/F in the environment, biological effects deduced or inferred from laboratory studies, and of foremost importance to many, possible human health risks following ingestion or dermal contact. Now classic references to direct human exposure include use of the defoliant Agent Orange (a herbicide contaminated by 2,3,7,8-TCDD) by the U.S. military in Vietnam, use of 2,3,7,8-TCDD-contaminated waste oils for dust abatement in Missouri, mismanagement of industrial wastes contaminated with 2,3,7,8-TCDD resulting in the evacuation of Love Canal, New York, an explosion at a chlorophenol plant releasing PCDD/F near Seveso, Italy, polychlorinated biphenyl (PCB) (transformer oil) fires, such as the one in Binghamton, New York, and the accidental PCB and PCDF contamination of rice oil in southwestern Japan.[1] Major litigation has followed health claims by Vietnam veterans and others over the past two decades to the present. Citizens have been alarmed by evacuations from areas contaminated by PCDD/F, and in particular by the U.S. government buy-out of Times Beach, Missouri. Concerns over PCDD/F contamination have perhaps been unrivaled by any other class of environmental contaminant. PCDD/Fs have thus become a controversial focal point in discussions concerning the health consequences of chemical-technological activity on both nature and humans.

The discovery of chlorine gas production in the past century, along with the facile reactivity of chlorine with organic compounds, provided society with the necessary tools to usher in the so-called chlorine age. Chlorine chemistry forms the basis for the large-scale manufacture of polyvinyl chloride (PVC) plastics, numerous organo chlorine chemicals and solvents, organochlorine pesticides, bleached paper products, and municipal water treatment. Additionally, chlorine compounds are used as chemical intermediates during pharmaceutical production. Globally 40 billion kg (about 90 billion pounds) of chlorine is produced each year from sodium chloride (ordinary table salt) for chlorine-based commodities.[4] Over one-fourth of the total, 12 billion kg, is consumed in North America alone.[5] Of the North American total, 16% of the chlorine is used directly for pulp bleaching, 5% for water treatment, and the remaining 79% is used for chlorine-containing products or products derived in some manner from chlorine or chlorine-containing precursors.[5] Initially the many benefits derived from a range of organochlorine products (such as crop protection, lightweight materials, and water disinfection) were regarded as purely positive technological advances. Aromatic organochlorines, however, are in general quite chemically stable and are thus largely retained and recycled in the environment. Today many organochlorine pesticides such as dichloro-diphenyl-trichloro-ethane (DDT) and technical organochlorine products such as PCBs can be readily detected in the lipids of virtually all terrestrial and marine organisms, including human tissues and breast milk,[1,6] even in regions where they have been long since banned from use. Many more commercial organochlorine products have also come to be recognized as extremely harmful in broad ecological and human health contexts and have been ultimately restricted or banned from use, but unfortunately only after extensive environmental application and release. PCDD/Fs, never intentionally produced (except for research purposes), are present in much lower environmental and biological concentrations than the majority of well-known organochlorine contaminants but behave in a like bioaccumulative manner.

2. ENVIRONMENTAL SOURCES OF PCDD/F

In this section the known environmental sources of PCDD/F will be reviewed. Prior to doing so, some aspects regarding the analysis of PCDD/F, now measurable in minute quantities in biotic and abiotic matrices the world over, will be given along with a summary of their global temporal occurrence.

2.1. Analysis, Ubiquity, and Time Trends

Interest in measuring PCDD/F, often present below the detection limit of the analytical instruments initially used for their study, has led to major improvements in the methods used for the identification and quantification of these and other ultratrace organic contaminants. The current state of the art in PCDD/F analyses allows for identification of individual isomers and their routine quantification down to part-per-trillion (ppt, not to be confused with part-per-thousand) concentrations in a given sample of interest. For some samples in which sample quantity is unrestricted (e.g., water), detection limits have been lowered even further, down to part-per-quadrillion (ppq) concentrations. The synthesis and purification of PCDD/F reference materials or chemical standards, the development of methods for sample preparation, the refinement of capillary gas chromatography used to separate individual PCDD/F isomers, and the optimization of high-resolution mass spectrometric techniques used for their detection, were arduous but necessary steps leading to today's sophisticated state of routine analysis. The remarkable sensitivity of current techniques is testimony of the extreme interest in measuring PCDD/F throughout the environment. The skill and perseverance shown by successful workers cannot be understated. Because of high instrumentation costs for gas chromatogram–mass spectrometers (U.S. $500,000–1,000,000) and meticulous sample preparation demands, relatively few facilities worldwide have the resources to reliably perform PCDD/F analysis.

As in all chemical analytical work, facilities must utilize established methodologies and be subject to evaluation as participants in interlaboratory calibration studies to ensure competence and reliability in routinely reporting part-per-trillion concentrations of PCDD/F isomers.[7,8] Further analytical difficulties are encountered in obtaining representative samples, as in studies of combustion emissions from a smokestack, or in taking samples that are not fixed to one locale, such as with ambient air or fish. Without stringent controls on the analysis of PCDD/F, which require detection limits orders of magnitude below other major environmental contaminants, results obtained may be of highly dubious value. It is for these many reasons that PCDD/F analyses are so costly to perform. The interested reader may refer to several up-to-date references regarding PCDD/F analytical methods.[1,9–12]

Interest in human exposure to PCDD/F initially began in the 1970s following recognition of the extreme toxicity of 2,3,7,8-TCDD to certain laboratory animals, and awareness of its presence at trace levels in 2,4,5-trichlorophenoxyacetic acid (2,4,5-T), a herbicide and important component of the defoliant Agent Orange.[1] Subsequent examination of exposed Vietnam military personnel, exposed Vietnamese, and chemical workers documented human exposures to PCDD/F. Other studies revealed the trace presence of PCDD/F in chemical products including chlorophenols, PCBs, polychlorinated naphthalenes, and polychlorodiphenyl ethers. Eventual examination of a range of environmental samples demonstrated the global ubiquity of PCDD/F. Following these many developments, interest in the identification of environmental sources expanded.

In an effort to investigate the temporal occurrence of PCDD/F, various datable matrices representative of preindustrial times have been studied by several workers.[13–20] Motivation for such study hinges on source apportionment, and further to determine if releases from anthropogenic activity are superimposed on possible natural sources. Hashimoto and co-workers investigated sediment samples to examine the chronology of PCDD/F deposition in Japanese coastal areas.[13] After consideration of possible experimental errors they concluded that a relatively constant PCDD/F background is observable in studied Japanese sediments over eight millennia to the present. A substantial surge in the level of PCDD/F was indicated at a sediment horizon representing a period 500 years before the present. Reference was made to runoff from burnt forestland as a potential PCDD/F source to explain the noted increases. This report thus implies that natural sources of PCDD/F exist (e.g., forest fires or volcanic activity), a conclusion made previously by others in connection with natural combustion processes.[14] Czuczwa and Hites studied radioisotopically dated lake sediment cores from both North America and Europe which revealed, in stark contrast, very definite temporal dependencies.[15] Essentially no measurable quantities of PCDD/F were detected in any examined core sediments deposited prior to 1940, but from 1940 sediment PCDD/F concentrations increased abruptly, reaching a maximum sometime in the 1970s followed by moderate decreases to the present. Since sediment PCDD/F profiles so closely resemble those observed for atmospheric particulates, it was assumed that no significant degradation occurs following deposition and movement of PCDD/F through the water column, or after final deposition in a sediment layer.[15] If postdeposition degradation were important, it is expected that an alteration in the PCDD/F profile would occur. U.S. chloroaromatic production increased in 1940 from below 22 million kg to a maximum output of nearly 360 million kg in 1960, with subsequent, moderate decreases to the present.[15] Hence, comparisons between the temporal profile of PCDD/F in lake sediments and figures for the U.S. temporal production profile of chloroaromatic substances over the same period provide compelling evidence for a relationship between environmental PCDD/F levels and the expansion of the chlorine industry.[15] Presumably, decreases observed from the 1970s to the present are related in part to abatement practices which are discussed further in Section 2.3.

Macdonald and co-workers have studied dated sediment cores from Howe Sound, British Columbia, Canada.[16] They reported that 2,3,7,8-TCDF first appeared in Howe Sound sediments upon initiation of chlorine bleaching of pulp in the region in the 1960s. They also reported that OCDD, now widely distributed throughout the region, first appeared around 1940 and reached a maximum sediment concentration around 1970, a result very similar to that obtained by Czuczwa and Hites.[15] Kjeller and co-workers have examined archived topsoil and herbage samples from the United Kingdom for which low but detectable PCDD/F concentrations were reported as early as 1846, followed by increases into the first half of this century.[17] Herbage samples from 1979–1988 contained slightly lower concentrations of PCDD/F than those from 1960–1970, indicating a time trend similar to those reported by Czuczwa and Hites [15] and Macdonald *et al.*[16] In addition to these temporal variation studies, other workers have investigated ancient human samples for the presence of PCDD/F in which virtually none were detected.[18–20] OCDD was reported to be present in mummified Chilean Indians dating from 2800 years before the present, but only at a small fraction of the level commonly found today in humans from industrialized countries.[19] Frozen Eskimo remains dated to 400 years ago were also reported to contain essentially no PCDD/F compared with levels typically observed in human tissues today.[20] Numerous studies have thus correlated the environmental loading of PCDD/Fs with chloroaromatic production this century, indications that global PCDD/F contamination is primarily a contemporary development associated with anthropogenic activities.

2.2. Primary Environmental Sources

Primary sources release PCDD/F directly into the aquatic, atmospheric, or terrestrial environment. PCDD/Fs are unique among the myriad of mainstream organochlorines of environmental interest in that they were never intentionally produced as desired commercial products. Instead, PCDD/F have entered the environment as products of various uncontrolled chemical reactions. Primary environmental sources of PCDD/F include chemical side reactions associated with the manufacture of a variety of chloroaromatic substances, in the delignification of cellulose (i.e., the bleaching of brown pulp for white paper production),[1] and the production of chlorine,[21] all described in Section 2.2.1. Other primary sources include numerous forms of combustion which are presented in Section 2.2.2. Combustion-related sources are today widely believed to be the major known sources of global PCDD/F and include all forms of waste incineration, many forms of metal production, and fossil fuel and wood combustion.[9,22] Recent studies seem to indicate that biogenic processes involving microbes occurring in soils, sludges, or composts may unexpectedly result in PCDD/F formation from chloroaromatic precursors (such as chlorophenols) present as environmental contaminants.[23] Secondary PCDD/F sources take the form of deposition reservoirs which can reintroduce PCDD/F into the environment and include PCDD/F-contaminated soils, sediments, landfills, sludges, composts, chemical wastes, and vegetation.[1,21,23–28] Secondary sources are hence associated with the recycling of PCDD/F previously formed from primary sources.

2.2.1. Uncontrolled Solution Chemistry

The production of many organochlorines, as well as the delignification of wood fibers, inadvertently give rise to PCDD/F. They can result from uncontrolled chemical reactions occurring under various solution phase conditions as described below.

2.2.1a. Organochlorine Products. Uncontrolled solution phase chemistry is known to produce PCDD/Fs in small yields during production of many organochlorines of commerce including chlorophenols, PCBs, chlorodiphenylether herbicides, hexachlorophene, and hexachlorobenzene, each at one time important commercial products.[1] PCDD/F release with organochlorine products coincides with their intentional or accidental dispersion into the environment, for example through use of contaminated herbicides or in mismanagement of industrial wastes. Chlorophenols, under certain conditions, are important precursors of PCDD/Fs. Depending on the desired isomer, chlorophenols are synthesized either by the direct chlorination of phenol or by the alkaline hydrolysis of chlorobenzene isomers.[1] Chlorophenols were first introduced in the 1930s. In 1978 the annual world production of chlorophenols was estimated at 200 million kg.[1] At a total PCDD/F concentration of 130 ppm,[29] this would correspond to roughly 26 tons of chlorophenol-related PCDD/F produced annually, the majority being OCDD. The wood products industry has consumed a substantial portion of the total chlorophenol output for wood preservation.[1] The presence of PCDD/Fs in chlorophenoxy herbicides such as 2,4,5-T results from contamination relating to their presence in 2,4,5-trichlorophenol (formed by the alkaline hydrolysis of 1,2,4,5-tetrachlorobenzene) which is a chemical precursor to 2,4,5-T. The concentrations of 2,3,7,8-TCDD in 2,4,5-T and related formulations have been reported to vary from parts per billion up to nearly 50 ppm.[1,30]

The number of PCDD/F isomers present in a given sample may approach the theoretical limit of 210. Hence, data reduction and interpretation can be a formidable task. Homologue profiles of the PCDD/F show only the sum total of PCDD/F at each level of chlorination (i.e., the sum of all TCDD isomers, the sum of all pentaCDD (PeCDD) isomers, and so forth). Despite excluding important isomer-specific information, homologue profiles simply illustrate how PCDD/F vary between differing samples of interest. As chemical precursors and reaction conditions differ in production of particular organochlorine products, such as for chlorophenols or PCBs, so too do the possible PCDD/F isomers vary. The homologue profile of PCDD/Fs in two different chlorophenols[29] and one common PCB formulation[31] are shown in Fig. 2. Among the obvious differences, it is seen that OCDD is typically a dominant isomer formed in chlorophenol but is only a minor component in common PCB mixtures.

Mechanistic details are complicated to describe but the combination of appropriate precursors (such as aromatic substrates and free chlorine) and solution conditions

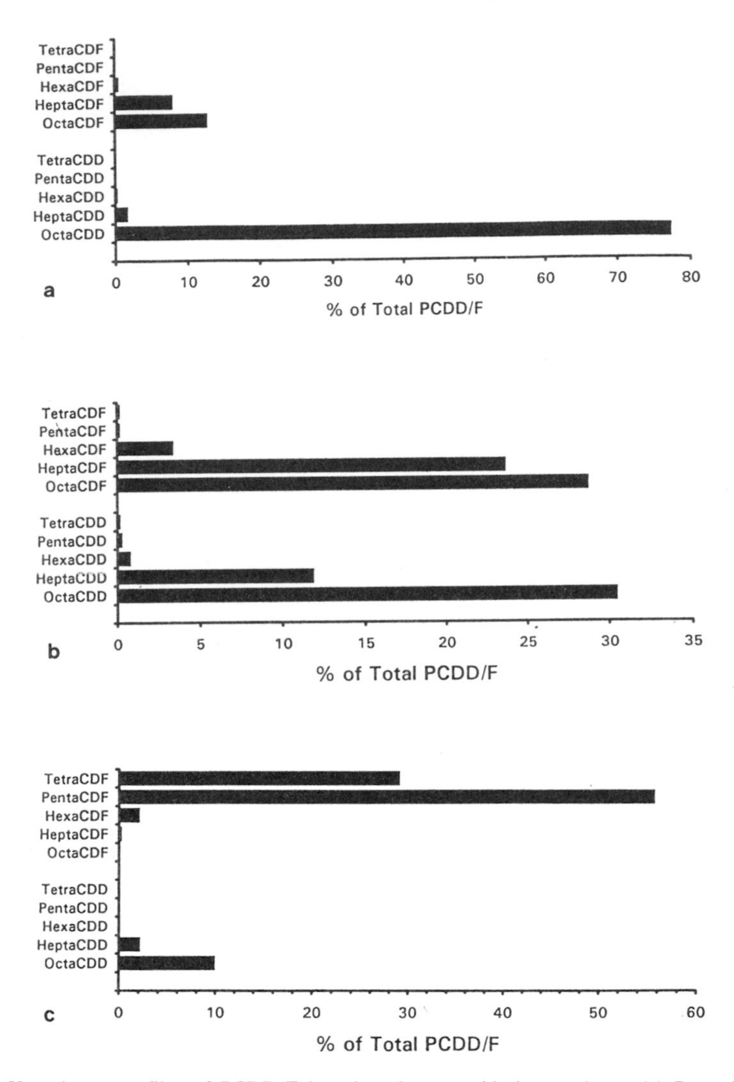

Figure 2. Homologue profiles of PCDD/F in selected organochlorine products. (a) Pentachlorophenol, (b) sodium pentachlorophenate, and (c) Chlophen 30A (PCB) formulation.

favoring radical reactions (including moderately high temperatures and alkaline media) are thought to promote PCDD/F formation.[9] For example, PCDDs are believed to arise from thermal dimerization and conversion of chlorinated phenoxyphenols (so-called predioxins), contaminants in chlorophenols which have been estimated to comprise as much as 1–5% of typical formulations.[32] PCDFs, often present at ppm levels,[1] are more abundant than PCDDs in PCB formulations, as seen for Chlophen 30A (Fig. 2, data from Refs. 29 and 31). Additional solution phase work has shown the existence of photoinduced formation mechanisms which have been experimentally demonstrated for chloroaromatics in various solvents. Examples include PCBs in phenol, PCDPEs in ketone, and PCDPEs in alcohol.[33]

2.2.1b. Wood Products Industry. Uncontrolled solution phase processes during the chlorine bleaching of lignin for paper products are also well known to result in the release of numerous chlorinated compounds, including chlorolignins, chlorinated phenolics, and PCDD/Fs.[34] PCDD/Fs are formed from the bleaching of pulp, especially if chlorine gas is used as the oxidizing agent. In terms of I-TEQs (see Section 1), both 2,3,7,8-TCDD and 2,3,7,8-TCDF have received attention as principal isomers found in pulps, waste effluents, and waste sludges.[9] These two isomers have also been reported as important by-products from studied Kraft pulp mills.[35] Much variability, however, has been noted between Kraft mills and between specific samples obtained at a given mill.[35] PCDD/Fs found in sediment samples have been identified as likely products from a large pulp mill in Sweden[36] which is described further in Section 3.2. Typical bleached paper products such as coffee filters, sanitary napkins, paper plates, and cups may contain ppt levels of PCDD/F.[37] The preservation of wood by treatment with chlorophenols has been of interest[38–40] since chlorophenols can contain ppm levels of PCDD/F. Pine needles have recently been studied as indicators of such activity, with PCDD/F homologue profiles reported to resemble those from chlorophenol formulations and with amounts of PCDD/F on pine needles decreasing with distance away from the wood-preserving site.[38] A Finnish study of abandoned sawmills concluded that site soils were contaminated with low ppm levels of PCDD/F, higher levels of chlorophenoxyphenols (predioxins), and still higher levels of chlorophenols.[40] The PCDD/F were reported to be immobile in soils, whereas chlorophenols could migrate. Chlorophenols have been proposed as secondary environmental sources of PCDD/F[28] as discussed in Section 5. Comments regarding PCDD/F abatement are found in Section 2.3.

2.2.1c. Chloralkali Industry. Graphite electrodes used in the chloralkali industry are receiving increasing interest as a result of the very high associated levels of PCDF contamination. PCDDs, however, are not formed in appreciable quantities. The cathodes used by the chloralkali industry were originally prepared from wood pitch, material providing an array of aromatic precursors from which to form PCDFs during electrolysis.[21] Because of a lack of appropriate precursors in the pitch, however, PCDDs are not present in significant concentrations in such electrode sludges. Hence, localized terrestrial contamination by spent cathodes has given rise to a so-called "chlorine pattern" which is dominated by PCDFs as seen in Fig. 3.[21] Of studied cases the total PCDD/F concentrations in electrode sludges have been found to be as high as 650 ng/g, nearly half of which is represented by 2,3,7,8-substituted isomers.[21]

2.2.2. Incomplete Combustion and Pyrolysis

PCDD/Fs are also known to result from heating various carbon and chlorine precursors to a sufficient temperature. The most significant known anthropogenic

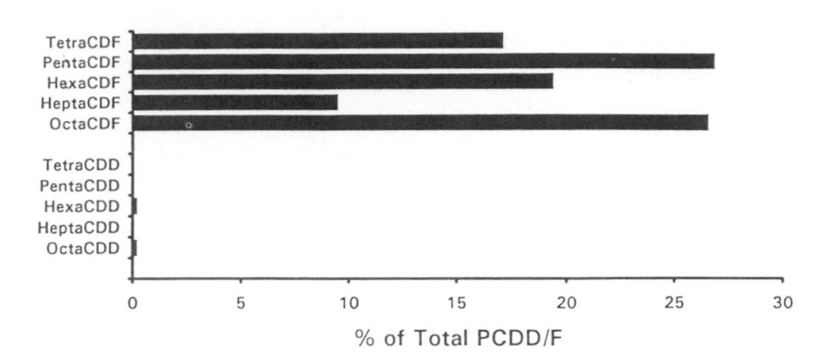

Figure 3. Homologue profile of PCDD/F from chloralkali electrodes.

environmental sources of PCDD/F include waste incineration, metals production, petroleum refinement, and fossil fuel (petroleum and coal) combustion.[9] If combusted with complete efficiency, compounds containing carbon and hydrogen are oxidized to carbon dioxide and water with energy released as heat. If chlorine is present in some form, hydrogen chloride gas is also produced. The combustion of carbon, hydrogen, and chlorine together in real-world incinerators is typically incomplete, however. Consequently, many other compounds may result, among them numerous organochlorines. In the following sections, waste incineration, metals production, fossil fuel and petroleum refinement, the pyrolysis of organochlorine products, and natural combustion will be briefly discussed with reference to PCDD/F formation.

2.2.2a. Waste Incineration. Waste incineration has increasingly become a mode of waste management in favor of land filling and for conversion of waste to energy. Because of incomplete combustion, hundreds to thousands of chemical compounds may be emitted during normal operation of a waste incinerator. The identification of PCDD/Fs in soot and ash residues taken from combustion emissions associated with incineration plants was a major advance in understanding PCDD/F sources.[41–43] Theoretical[44] and experimental[45,46] work has addressed the formation of PCDD/F via waste incineration, and the parameters of importance have been summarized.[47,48] Emissions from automobiles burning leaded fuels with halogenated scavengers[49,50] have also been shown to contain PCDD/Fs. Processes leading to PCDD/Fs during municipal and hazardous waste incineration involve reactions in the gas phase as well as reactions on particulate surfaces within and following the combustion zone.[47] It is believed that the presence of various metals bound to particulates acts to catalyze (i.e., speed up) reactions leading to PCDD/F. PCDD/F are thought to escape primarily with fly ash rather than with particulates or vapors. Processes leading to PCDD/F formation appear to be most favored at temperatures around 300 to 450°C. Collectively the factors of importance which influence PCDD/F formation during combustion include the composition of the fuel material, the availability of oxygen and chlorine, temperature, combustion residence time, the presence of metal catalysts, and the availability of particulate surfaces on which reactions may take place. Related events occurring in the postcombustor cooling zones are also thought to be of importance as shown in model incineration studies.[51]

PCDD/F contaminants in organochlorine products (mentioned in Section 2.2.1) are relatively limited in the total number of PCDD/F isomers present. In contrast, combustion processes can give rise to most of the 210 possible PCDD/F isomers. Figure 4a (from Ref. 49) is a gas chromatograph–mass spectrometric trace showing the individual TCDF isomers found in typical municipal waste incineration (MWI) samples. Figure 4b is a typical MWI homologue profile (used for simplification as discussed in Section 2.2.1a). Figure 4a gives an impression of the number of individual isomers present in a given incineration sample; numerous other isomers are present at each level of chlorination, the sums at each level of chlorination seen in Fig. 4b (from Ref. 3). Very often trace impurities which are extremely difficult to isolate from the PCDD/F may be detected as such as shown in Fig. 4a. Most often only TCDD/F

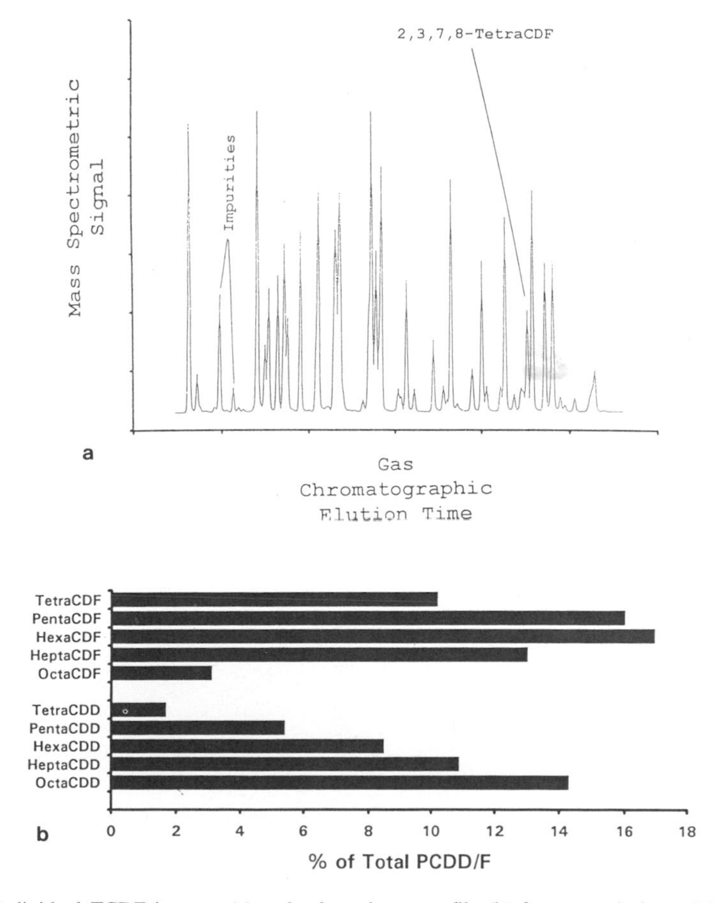

Figure 4. Individual TCDF isomers (a) and a homologue profile (b) from a typical municipal waste incinerator.

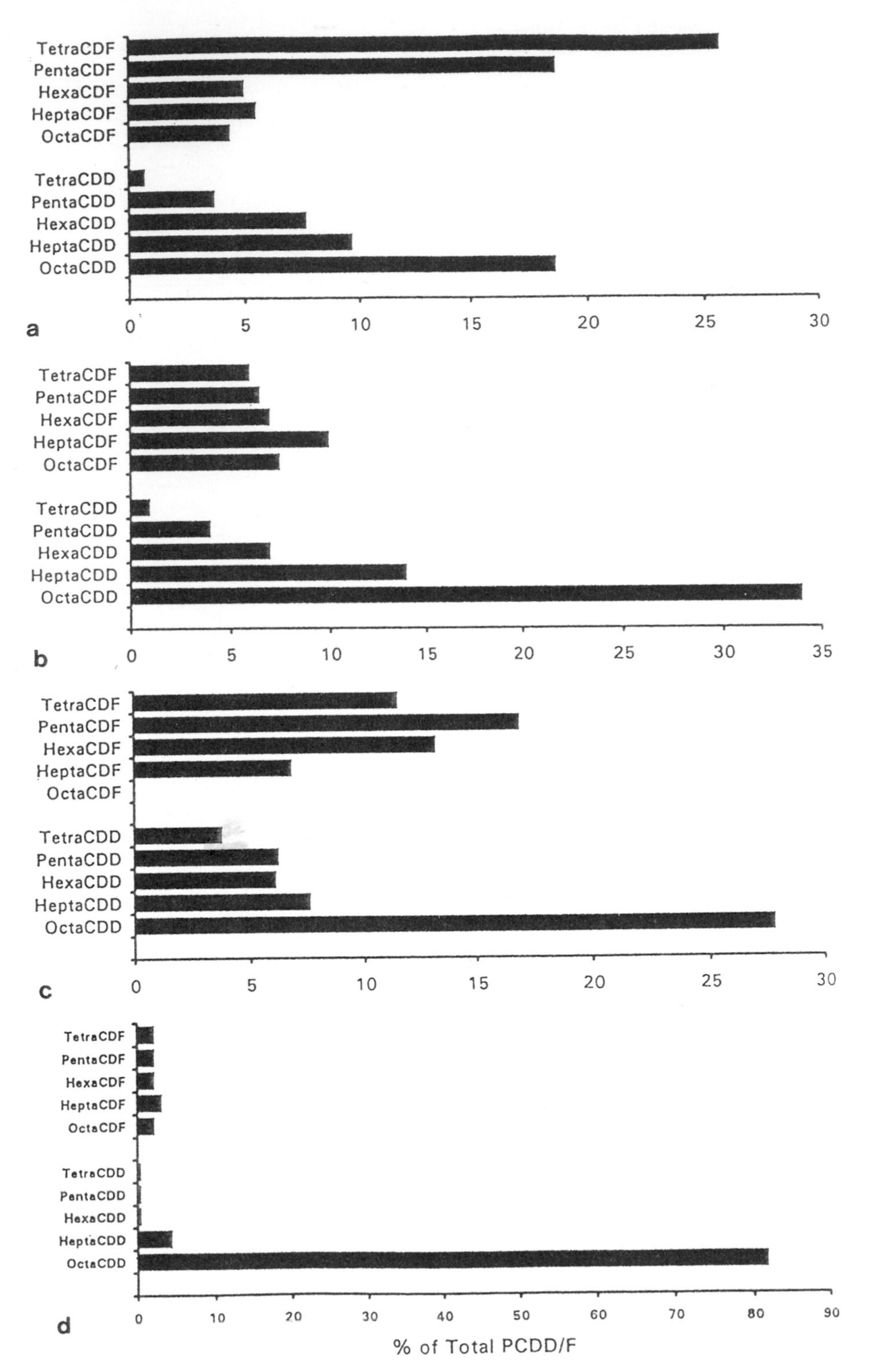

Figure 5. Homologue profiles of PCDD/F in selected atmospheric and terrestrial samples. (a) Ambient air, (b) snow, (c) soil, (d) sediment.

through OCDD/F are monitored for a given sample since PCDD/F containing four or more chlorines (in the 2,3,7,8 positions, see Fig. 1) have been of most interest because of their toxicity. It can be seen in Fig. 4b that the OCDD isomer dominates the PCDDs, with each lower chlorination level diminishing in abundance. The hexaCDF (HxCDF) isomers are seen to dominate the PCDFs, progressively diminishing in the direction of lower or higher chlorination. Figures 5a and b show typical PCDD/F homologue profiles observed in two background atmospheric samples, air and snow (from Ref. 52 and 53, respectively). Figures 4a, 5a and b might seem to suggest homologue profile alterations occurring during the atmospheric dispersion and deposition of PCDD/F from waste incineration. The predominance of OCDD in many abiotic environmental samples is thought to reflect an enrichment process (described previously by Hites in Ref. 3 as discussed in Sections 3.2 and 4.5) since OCDD is the least environmentally degraded of all PCDDs and is often the most abundant isomer present in snow, soil, or sediment as seen in Fig. 5b, c, and d (Fig. 5c is from Ref. 54). It is noteworthy that many abiotic samples such as sediment (Fig. 5d from Ref. 3) can resemble chlorophenol homologue profiles (Fig. 2a), and hence the relative importance of these two PCDD/F sources may be unclear for environmental samples from certain locales.

The overall importance of chemical products such as chlorophenols and PCBs has been estimated to be a minor contribution to the total global PCDD/F burden in comparison with incomplete combustion-related sources in the United States.[55] Chlorophenols have, however, been proposed as significant contributors to the total PCDD/F flux in the United Kingdom[28] and Canada.[56]

2.2.2b. Metals Production. The production or recycling of iron, steel, magnesium, nickel, lead, and aluminum are known to result in PCDD/F.[9] Iron and steel production are presently considered to be the most significant sources of PCDD/Fs in the Swedish environment.[57] Magnesium production is recognized as an important source of PCDD/Fs in Norway.[58] The combustion of chlorine-based plastics, such as PVC, has been implicated in the formation of PCDD/F. The recovery of scrap metals in the presence of PVC, as for plastic-coated copper wire, is one such example.[59] The presence of copper oxide may serve to enhance such formation processes. Chloroparaffins used as cutting oils have also been implicated in PCDD/F formation.[59]

2.2.2c. Fossil Fuels and Petroleum Refinement. Coal has been considered to be a potentially important environmental source of PCDD/F in the United Kingdom,[28] but in North America coal-related PCDD/F appears to be dwarfed by output attributed to chlorine chemistry and subsequent environmental circulation of organochlorine compounds in various forms through incineration.[3] Variations in the natural chlorine content of coal might influence PCDD/F production.[9] The refinement of petroleum has also been reported to result in formation of PCDD/F.[9,60] The regeneration of spent catalyst material used during refinement has been identified as the cause.[60]

Automotive traffic is known to be a major source of various types of roadside environmental pollution. Levels of lead and PCDD/F have both been shown to decrease fourfold over a distance of 20 m (62 ft) from a traffic lane.[61] Chlorinated hydrocarbons such as dichloroethane added to leaded fuels are thought to be the chlorine source for automotively formed PCDD/F.[48,49] Marklund and co-workers have estimated a total PCDD/F emission of about 40 pg/km driven from a single automobile operated on unleaded fuel with a catalytic converter exhaust system.[49] Vehicles operated on leaded fuels can produce 20-fold more PCDD/F.[49] It may be assumed that the phaseout of leaded fuels has lowered emissions attributable to automobiles. Travis and Hattemer-Frey estimate that automobile emissions amount to roughly 5% of the known 2,3,7,8-TCDD sources in the United States.[62]

2.2.2d. Organochlorine Product Pyrolysis. Pyrolysis is the heating or incomplete combustion of matter resulting in a chemical change. Buser and co-workers have studied the formation of PCDD/F by pyrolysis of several important organochlorine compounds.[63–66] The pyrolysis of PCBs[63] results in PCDD/F formation as illustrated by cases involving local contamination caused by accidental fires in which PCBs were used as the heat transfer medium in electrical capacitors and transformers.[67] The pyrolysis of polychlorinated diphenyl ethers (PCDPEs) as for PCBs leads primarily to PCDFs[64] but the pyrolysis of chlorobenzenes leads to both PCDDs and PCDFs.[65] The pyrolysis of chlorophenates results in formation of PCDDs.[66] Polybrominated diphenyl ethers have been widely used as flame retardants and have also been shown to result in polybrominated dioxins and furans.[68] These compounds are briefly described further in Section 6.1.

2.2.2e. Natural Combustion. Speculation regarding possible natural combustion sources for PCDD/F includes forest fires and volcanic activity in connection with the so-called "trace chemistry of fires hypothesis."[14] The hypothesis states that natural incomplete combustion of organic material in the presence of inorganic chlorine, both abundant in nature, can result in at least slight formation of PCDD/F. The argument was put forth years ago by workers employed by a chemical firm. Though of academic interest it initially seemed motivated toward defending the chemical industries encountering litigation over the burden of responsibility for environmental PCDD/F contamination. Forest fires have not been clearly demonstrated to be important global sources of PCDD/F.[69] Domestic wood combustion is, however, now well known to result in PCDD/F formation.[50] A recent study of the mummified remains of Chilean Indians who in their time were constantly exposed to domestic cooking fires concluded that the combustion of preindustrial, untreated wood is not a significant environmental PCDD/F source.[19] Hence, the controversy continues. One remote study of volcanic activity involved ash from the Mount St. Helens, Washington, eruption of 1980. It appeared that ash particulates actually absorbed background PCDD/F in the vicinity of the volcano[70] thereby precluding evidence for volcanism as a PCDD/F source. Numerous temporal studies (Ref. 13–20, described in Section 2.1) appear to convincingly correlate the increase of synthetic organochlorines in this century to increases in global PCDD/F.

2.3. PCDD/F Source Abatement

Since incomplete combustion is presently regarded to be among the most important of PCDD/F sources, much attention has been directed toward reducing municipal and chemical waste incineration emissions. As described in Section 2.2.2a, numerous parameters influence PCDD/F production. A sufficient supply of oxygen, maintenance of high combustion temperatures (> 850°C), long combustion gas residence times, and thorough gas mixing assist in reducing PCDD/F emissions.[9] The degree of combustion efficiency must be very high (> 99.9999%) in order to reduce PCDD/F emissions to meet many Western regulatory standards.[9] Aromatic compounds are known to be among the most difficult to combust. Since incineration is a major mode of waste management and remediation of aromatic compounds, it is important that pyrolytic conditions (as mentioned in Section 2.2.2d) be avoided. Available technology for reduction of PCDD/F from combustion sources can lower emissions to below 100 pg of I-TEQs (defined in Section 1) per cubic meter of released stack gas. This limit is a current regulatory guideline in several European countries.[9] Earlier recognized contamination problems involving dairy production arose from emissions which were perhaps 100- to 1000-fold higher. One form of incineration source abatement is based on the dry scrubbing of stack gases with lime and activated charcoal followed by fabric filters. Another gas scrubbing technique involves spray dryer adsorption which removes PCDD/F with activated carbon particulates.[9] Decreases in sediment PCDD/F concentrations reported to begin around 1970 are thought to reflect the initiation of emission controls[15] and perhaps further were the result of restrictions applied to the production of organochlorines.

As mentioned in Section 2.2.2c, automobiles (particularly those operated with leaded fuels containing organochlorine additives) have been relatively minor environmental sources of PCDD/F.[49,50] Vehicles with emission control systems (catalytic converters) and operated with unleaded fuels almost invariably release less PCDD/F in the exhaust. Older vehicles without emission control devices might release lower PCDD/F than similar newer models since extensive carbon deposition within the exhaust system might act to adsorb PCDD/F to some degree.[71] Diesel-fueled engines appear to generally have lower associated PCDD/F emissions than gasoline-fueled engines.[49]

In recent years, the wood products industries in the Western world have lowered the total amount of chlorinated by-products formed in the delignification (bleaching) of pulp in response to concerns over environmental contamination. Chlorine gas and sodium hypochlorite are inexpensive delignifiers but result in levels of PCDD/F exceeding alternative nonchlorine methods. It is known that the quantity of chlorine gas used has a major effect on PCDD/F formation, decreased chlorine expectedly resulting in lower levels of PCDD/F. The Kraft process (currently the most widely used method for pulp bleaching) employs chlorine dioxide in the first stage of delignification. Use of chlorine dioxide as an oxidizing agent results in less efficient chlorination of PCDD/F precursors as obtained through direct chlorine bleaching. It is currently possible to reduce emissions of PCDD/F during pulp production to 1 pg I-TEQs/g (1 ppt) in pulp intended for manufacture of paper products intended for food packaging. Hydrogen peroxide–oxygen-bleached pulp has been reported to have one-third less PCDD/F than chlorine-bleached pulp.[72]

Increasing the regulation of effluents and emissions from the wood products and combustion-related industries has been useful in lowering the output of chlorine contaminants including PCDD/F. Regulations have encouraged development of delignification methods which utilize less chlorine or employ alternative bleaching agents. Regulations have also hastened the optimization of combustion parameters and stack scrubbing practices. Lowered public demand for bleached white consumer paper products has lessened chlorine use as well. Several environmental groups including the Sierra Club, the National Wildlife Federation, the U.S. Public Interest Group, the International (American–Canadian) Joint Commission on the Great Lakes, Greenpeace, and the Natural Resources Defense Council support a phase-out of chlorine and organochlorine production. Studies from the United Kingdom suggest that the widespread use of chlorophenols by the wood products industry may significantly contribute to the background PCDD/F flux in Britain.[28] Hence, other obvious forms of PCDD/F abatement include the restrictions on use or banning further production of organochlorine products known to be contaminated with PCDD/F such as chlorophenols and PCB formulations.

3. ENVIRONMENTAL DISTRIBUTION AND LEVELS OF PCDD/F

PCDD/F are present in all compartments of the environment—atmospheric, aquatic, and terrestrial. PCDD/F are distributed from point sources as combustion-related emissions dispersed on air masses or as effluents from liquid waste discharges released into aquatic systems. Though revolatilization may occur, PCDD/F are primarily associated with particulate matter under normal ambient conditions. The resedimentation and resuspension of initially deposited particulate matter are forms of PCDD/F recirculation. Details regarding the environmental distribution of PCDD/F are summarized below.

3.1. Atmospheric Distribution

Today numerous environmental pollutants can be measured in the Arctic.[73] It is clear that they have originated from lower (industrial) latitudes having been transported and deposited out from cooled, northerly flowing, air masses.[74] Combustion processes (as stated in Section 2.2.2) are presently regarded to be the most important environmental sources of PCDD/F. The following two sections present details concerning the distribution of PCDD/F introduced into the atmosphere. The long-range transport of particulates will first be described followed by information regarding the depositional removal of these particulates from the atmosphere.

3.1.1. Long-Range Atmospheric Transport

The physical properties of any material dictate how it may be distributed through the atmosphere. PCDD/F are solids at room temperature and have rather low volatility.[1] Early evidence of PCDD/F long-range atmospheric transport came from the detection of PCDD/F in remote locales free of inputs from industrial releases. Stalling and co-workers first reported the presence of PCDF in fish sampled from Siskiwit Lake, a remote aquatic system on Isle Royal, Lake Superior, USA.[75] It was immediately apparent that their occurrence in this pristine area, free of local point source discharges, was the result of atmospheric transport. Long-range transport of PCDD/F has been repeatedly confirmed by similar studies involving environmental samples well isolated from anthropogenic activities.[55,76] Subsequently, aquatic depositional fluxes for PCDD/F were estimated by Hites and co-workers from samples collected from Siskiwit Lake as well as Lake Erie, renowned for its proximity to industrial activity.[55,77] The depositional flux from the atmosphere into each aquatic system was calculated from sediment PCDD/F concentrations, sedimentation rates, and sediment densities. The importance of distance from known combustion sources was seen to be of importance in determining depositional flux, but other factors such as sediment focusing during settling through the water column were considered to potentially influence flux estimates. More recent investigations have examined ambient air concentrations of PCDD/F in relation to the bulk atmospheric movement of air masses.[52] It is now apparent that concentrations of PCDD/F follow those of many inorganic pollutants, and further, tend to increase if air movement trajectories originate from industrial–urban centers. Depending on location and atmospheric conditions, a "typical" air parcel may contain a total of 1 to 10 pg PCDD/F per cubic meter, and up to 100 pg PCDD/F per cubic meter in more extreme situations. Rural locales, as might be expected, are typically on the low end of this scale, below 1 pg per cubic meter of ambient air.[52]

3.1.2. Modes of Atmospheric Deposition

Wet deposition, referring to removal of PCDD/F through any precipitation event, was shown to be more efficient for particle-bound PCDD/F than so-called rain scavenging of vapor-phase PCDD/F.[78] The importance of dry atmospheric deposition for PCDD/F, earlier thought to have been underestimated as indicated by lower wet depositions to both semirural and urban settings in Indiana, when compared with aquatic fluxes to the Great Lakes, was also examined. Indeed, dry depositional mechanisms were found to be not only significant but of equal or greater importance for PCDD/F, particularly for deposition occurring closer to atmospheric sources. Furthermore, dry deposition was shown to increase in efficiency with the degree of chlorination, as for wet deposition, and both processes were shown to be temperature dependent.[78] Dry deposition has earlier been shown to be up to five times more efficient than wet deposition for various studied PCBs and chlorinated pesticides.[79] Broman has reported that dry deposition dominates over wet deposition by a factor of 6 for PCDD/F in the Baltic environment.[80] An equilibrium exists between the vapor and particulate-bound PCDD/F present in the atmosphere, with only a small fraction present on particulates under typical atmospheric conditions.[81] During the summer the fraction found on particulates increases with chlorination level. At lower temperatures there is a shift in the particulate-bound–vapor-phase PCDD/F equilibrium further favoring particle adsorption onto particulates. Hence, PCDD/F depositional processes increase in efficiency at lower ambient temperatures.

Comparisons between the homologue profiles for incineration, atmospheric, and sediment samples (such as shown in Figs. 4b, 5a–d) would appear to reveal selection forces occurring during transport such that released PCDD/F possessing the greatest affinity for particulate adsorption (i.e., the most chlorinated, least volatile) are most efficiently transported and deposited to aquatic or terrestrial environments. Such an enrichment mechanism potentially explains why OCDD is the most abundant PCDD/F isomer found in many abiotic environmental samples, such as sediment or soil, as seen when comparing typical MWI homologue profiles (Fig. 4b) with those thought to be representative of background soil or sediment (Fig. 5c and d). Though homologue profiles are very similar between wet and dry deposition (not shown), both differ from ambient air since only particulates are removed by either depositional mode, the vapor component (which is inefficiently removed by rain scavenging) being left behind in the atmosphere.[78] Vapor phase PCDD/F are perhaps more readily photodegraded than the particulate-bound PCDD/F. Deposition samples are thus enriched in the less volatile, highly chlorinated PCDD/F as the atmospheric PCDD/F load is at least temporarily enriched in the more volatile, lower chlorinated isomers. Reported variations in atmospheric and aquatic depositional flux measurements have ranged from a total of 50 ng to 9μg PCDD/F per square meter per year, with atmospheric concentrations reflecting proximity to anthropogenic PCDD/F point sources.[78,79,82] The air homologue profile (shown in Fig. 5a) corresponds to 35 pg per cubic meter of air.[24] The soil PCDD/F homologue profile shown in Fig. 5c corresponds to about 60 pg PCDD/F per gram of soil. The sediment PCDD/F homologue profile shown in Fig. 5d corresponds to a total of about 2 ng PCDD/F per gram of sediment.[3]

3.2. Aquatic Distribution

Since PCDD/F are nonpolar substances, they exhibit very low solubility in water.[1] On mixing with water they behave much like oil, and tend to associate with materials which exhibit similar hydrophobic properties, as well as with particulate matter which they may adsorb to or be absorbed by. In this manner PCDD/F released from aquatic point sources, such as wood products industry effluents, are dispersed into the aquatic environment. Furthermore, particulate-bound PCDD/F can be carried with runoff into drainages and distributed onward into drainage basins where they may settle as sediments.

Sediment cores have been examined in the area surrounding a large wood products mill located in Sweden which annually produces 280,000 tons of fully bleached Kraft pulp.[36] The study showed that decreasing environmental levels of PCDD/F are correlated with increasing distance from the mill. There were exceptions in the observed spatial distribution, however. 1,2,3,6,7,8-HxCDD and OCDD were found at maximum concentrations in sediment at a distance of 8 km from the mill. These two isomers are major contaminants in pentachlorophenol and perhaps may have originated from resuspension of chlorophenol-contaminated sediments, though no certain explanation has yet been found. A further point of interest was that 2,3,4,7,8-PeCDF, a major PCDD/F contaminant in Baltic herring, showed no correlation to distance from the mill. Macdonald and co-workers have linked sediment samples from Howe Sound, British Columbia, to a pulp mill located 30 km away.[16] The low metabolic capacity of crabs has allowed Oehme and co-workers to distinguish PCDD/F source patterns arising from a magnesium plant from pulp mill or combustion sources along the Norwegian coast.[58]

Investigations of municipal water supplies have demonstrated very low concentrations of PCDD/F, often less than 1 pg per liter of treated water, primarily due to OCDD.[83,84] These low water concentrations have been attributed to filtration stages during treatment which remove lipid and particulate matter with which PCDD/F can be associated.

3.3. Terrestrial Distribution

PCDD/F adsorb tightly to particulates and as a consequence undergo slow migration through soils. Particulate resuspension may occur through the weathering of soils and sediments which serve as dominant PCDD/F repositories. As mentioned above, terrestrially deposited PCDD/F adsorbed to particulates may be transferred into aquatic systems as runoff. Depending on locale, soil samples have been reported to contain up to several micrograms of PCDD/F per kilogram of soil (see Table 1). By comparison, rural soil samples have generally been reported to contain less, on the order of a few

Table 1
Approximate Ranges of PCDD/F Concentrations Reported in Each Environmental Compartment

Environmental compartment	Sample type	Concentration range	References
Atmospheric	Urban air	1–50 pg/m^3	24, 25, 86, 87, 89
	Rural air	<1 pg/m^3	
Aquatic	Raw water	1–1000 pg/liter	1, 84
	Treated water	0.01–100 pg/liter	
	Sediment	1–10 ng/g	
Terrestrial	Soil	1–2500 pg/g	24, 25, 86, 90

hundred nanograms of PCDD/F per kilogram of soil. Such spatial trends reflect the importance of proximity to point sources.

3.3.1. Reemission from Secondary Sources

Secondary sources are reservoirs of PCDD/F arising from deposition related to primary sources. An inventory of secondary PCDD/F sources in the United Kingdom concludes that soil and sediments comprise the bulk (> 99%) of PCDD/F reservoirs in Britain.[28] Their thermodynamic stability and resistance to degradation allow for possible environmental recirculation after emission from any of the primary sources mentioned in Section 2.2. Secondary sources (i.e., deposition reservoirs) are of importance since they are associated with the most common human exposure routes, i.e., through ingestion of food products.

3.4. Environmental Levels

Levels, or more correctly environmental concentrations, of PCDD/F have been reviewed for many environmental sample types.[1,24,25,85,86] In principle, results should be comparable but intercalibration studies have shown that this is not always the case.[78] Differences in sampling techniques and analytical methods may significantly influence the final results obtained. In this section, selected results will be given in an effort to convey ranges of typical environmental levels in each environmental compartment though "typical" is difficult to define since many features such as proximity to local sources, source strengths, and ambient conditions are so variable. Table 1 gives a simplified overview of selected results obtained for some environmental samples which seem to be fairly representative of those found in the literature. It is noteworthy that the observed PCDD/F isomer patterns in many abiotic samples resemble those characteristic of combustion sources.[85]

4. ENVIRONMENTAL FATE

Following release into the environment, PCDD/F may undergo transformation and degradation en route to deposition. Transformational or degradational processes may potentially occur in each environmental compartment in which PCDD/F are found. PCDD/F as a group have high melting and boiling points, have low volatilities, and are very insoluble in water.[1] PCDD/F are lipophilic, meaning that they are soluble in nonpolar (lipidlike) media and thereby accumulate readily in lipid-containing matrices. The near-planar configuration of PCDD/Fs impart a high adsorptivity to particulate matter. Both the degrcc of involatility and the relative solubility with lipid increase with the level of PCDD/F chlorination. In the following sections, transformation and degradation processes will be considered with respect to atmospheric, aquatic, and terrestrial environments.

4.1. Atmospheric Transformation–Degradation

Because of long-range transport, observed background ambient air homologue profiles may slightly differ depending on bulk air movement from various industrial–urban centers to other locales.[52] Generally air homologue profiles (Fig. 5a) sharply increase from TCDD to OCDD (which dominates the homologue profile) while PCDF profiles are characterized by a progressive decrease from TCDF to OCDF, with OCDF the least prevalent of all. In addition to the wet and dry depositional mechanisms (described in Section 3.2) which remove PCDD/F from the atmosphere, chemical or photochemical transformations may occur during transport from sources to sinks, as well as after deposition to either aquatic or terrestrial environments. The dominant atmospheric losses of vapor-phase PCDD/F are thought to be photolysis and reactions with hydroxyl radical which is naturally present in the troposphere.[91] Hydroxyl radical is known to atmospheric chemists as the "tropospheric janitor." It is the primary oxidant of atmospheric pollutants despite being a million times lower in concentration than ozone. In solution, hydroxyl radical can readily break down certain organochlorines by means of oxidation reactions. In the case of PCBs suspended in solution, oxidation is hindered if the carbon atom is protected from attack by a chlorine atom[92] which perhaps helps to explain why OCDD (the most fully chlorinated isomer) is the most persistent and most abundant PCDD/F isomer in the environment.

As stated above, PCDD/F can be photodegraded on exposure to UV radiation. Early laboratory studies revealed that OCDF can be transformed to lower chlorinated PCDF.[93] Most OCDF was reported to be degraded after 24 hr of exposure to a UV source with HxCDFs as the major photoproducts.[93] It has since become apparent that adsorption onto particulate surfaces may provide an effective shield against photodegradation.[94] Evidence for profile altering mechanisms is seen by comparing air and sediment samples, Fig. 5a and d, respectively. The sediment homologue profile (Fig. 5d) is dominated by OCDD whereas several PCDD/F isomers can be sampled from air (Fig. 5a). Current evidence suggests that atmospheric photodegradation is occurring for the lower chlorinated PCDD/F to a much greater degree, as they have a higher tendency to desorb from suspended atmospheric particulates. The end result of presently established environmental transformations appears to be enrichment in higher chlorinated PCDD/F as stated in Section 3.1.2 Photodegradative mechanisms are less efficient for higher chlorinated congeners related in part to their lower degradability to UV light, but also to their lower volatility thus binding them more tightly to atmospheric particulates which seems to reduce[95] or even preclude[94] photodegadation. Estimated tropospheric lifetimes for atmospheric degradation of vapor-phase PCDD/F range from a half-day for monoCDD (MCDD) to 10 days for OCDD, and from roughly 1 day for MCDF to 39 days for OCDF.[91] Lifetimes from photodegradation are thought to increase tenfold when adsorbed onto a particulate.[92] Thus, when bound to particulates, PCDD/F atmospheric lifetimes clearly allow for long-range transport as discussed in Section 3.1.2.

4.2. Aquatic Transformation–Degradation

Reported similarities between PCDD/F homologue profiles for atmospheric particulates and sediments suggest that degradation or transformation events are negligible after a particulate is deposited onto surface waters and moves through the water column.[15]

4.3. Terrestrial Transformation–Degradation

When adsorbed to a surface (as for the case of a particulate suspended in the atmosphere discussed above), soil-bound PCDD/F tend to be resistant to photodegradation.[94,95] Photosensitization by components able to transfer energy to chemical bonds and result in their rupture may be necessary for photodegradation to occur on surfaces.[77] 2,3,7,8-TCDD has been reported to be persistent in soil with minimal losses from photolysis or volatilization, even after a 1-year period.[96] The soil half-life of 2,3,7,8-TCDD has been estimated to be between 10 and 12 years.[97] It is apparent that photolysis can be a major degradative force but shielding from sunlight by surface soil particulates (as seen by examining soil profiles) extends the lifetime of subsurface PCDD/F.[96,98] In the presence of organic solvents, however, the photolysis of PCDD/F in soils is greatly enhanced.[99] It has been reported that OCDD is photodegraded to lower chlorinated congeners on soil particulates, with preference toward formation of toxic 2,3,7,8-substituted CDD isomers.[100,101] This observation may be of interest since OCDD is the most widely distributed PCDD/F isomer present in the environment.

4.3.1. Biological Transformation–Degradation

Correlations between primary sources (such as combustion processes or PCB formulations) or secondary sources (such as polluted river sediments) and biological samples are complicated since many organisms exhibit a preferential retention or accumulation of 2,3,7,8-substituted PCDD/F. The propensity to accumulate organochlorines by living organisms is variable; lower marine organisms such as crustaceans seem to accumulate PCDD/F indiscriminantly and the crab hepatopancreas is the most PCDD/F-contaminated food product known.[102] Crustaceans have been used as marker organisms for monitoring local and regional contamination.[58] In contrast, seals, higher marine organisms, appear to possess a biodegradative–elimination mechanism which limits the amount of PCDD/F they accumulate in comparison with their expected dietary intake of contaminated fish.[103]

Since soil and sediments are often heavily populated by microorganisms, it is not easy to separate purely physical from biological degradation–transformation mechanisms. Specific soil features are expected to influence the survival of PCDD/F, and microbial and natural degradation of PCDD/F are dependent on soil conditions.[96] Microbial breakdown of chloroaromatic compounds such as PCDD/F is thought generally to be slow. Lower chlorinated PCDD/F appear to be the most readily biodegraded.[104] Absorption of PCDD/F onto sediment particulates appears to significantly prolong possible biodegradation routes.[105]

5. PCDD/F FORMATION DISCREPANCIES

Estimates of various local, well-publicized, PCDD/F contamination problems include 4 to 8 kg of 2,3,7,8-TCDD in Newark Bay, NJ, 6 kg of 2,3,7,8-TCDD in Times Beach, MO, and 0.25 to 3 kg of 2,3,7,8-TCDD in Seveso, Italy.[108] Travis and co-workers estimate that the total yearly emission of 2,3,7,8-TCDD in the United States is between 25 and 120 kg.[106] Eitzer and Hites have estimated that the total deposition of PCDD/F over the United States is roughly 9500 kg per year.[109] In terms of I-TEQs the estimate by Eitzer and Hites for the United States might be 100-fold less, and thus be comparable to the estimates by Travis and co-workers. The estimated total PCDD emission to air, water, soil, and bound to particulates or other matter has been estimated to be roughly 10 kg expressed as I-TEQs per year for Germany, Britain, and the Netherlands combined.[9] Extrapolated over the whole of Europe, these rough estimates seem comparable to estimates for the United States.

Though the relative importance of individual known PCDD/F sources for the total global environment are uncertain (as well as the totals which are emitted), it is apparent from several temporal studies (Section 2.1) that the increased presence of chlorine in the environment has increased its potential for recycling in various forms, such as through formation of PCDD/F during municipal and chemical waste incineration. The major known PCDD/F sources as described in Section 2.2 include waste incineration, metals production, fossil fuel combustion, petroleum refinement, pulp bleaching, pyrolysis of organochlorines, and contamination originating from organochlorine products such as chlorophenols, for example in connection with wood preservation. Analysis of environmental samples indicates that OCDD is the most abundant globally distributed isomer, apparently related in part to its resistance to degradational forces which were discussed in Section 4. As mentioned in Section 1, OCDD is currently considered to be 1000 times less toxic than 2,3,7,8-TCDD.

Quantitative information regarding emissions of PCDD/F is uncertain and to date only rudimentary estimates as presented above have been put forward for discussion. The global relative order of importance of known PCDD/F sources has thus not been firmly established. Combustion-related sources are considered to be the most important sources in the global PCDD/F budget.[9] Travis and Hattemer-Frey estimate that incineration of municipal and hospital wastes combined accounts for less than 5% of the total 2,3,7,8-TCDD released in the United States.[62] Automobile emissions are estimated to comprise another 5% and residential wood burning is estimated to contribute

1% of the total 2,3,7,8-TCDD released in the United States. The remaining nearly 90% is unaccounted for. Other crude estimates concur with Travis, that roughly 10% of PCDD/F distributed through the environment may be accounted for by presently known sources.[28,107] The PCDD/F budget difficulties are manyfold. Source strengths, i.e., the quantities of PCDD/F emitted from a given primary source, are not accurately known[9] and atmospheric flux estimates of primary sources may be complicated by reemission from secondary deposition sources.[28] PCDD/F detected in the environment are not uniform, and additionally natural transformation–degradation processes (introduced in Section 4) may preclude reliable deposition estimates. The PCDD/F mass balance discrepancy may possibly be attributed to underestimation from established sources, unrecognized contributions from long-range transport, unidentified secondary sources such as emissions from chlorophenol-treated substrates, and/or resuspension of PCDD/F bound to particulates.[28] Information is limited but unknown formation routes yet to be revealed may help explain the apparent discrepancy between sources and sinks.

Unknown Formation Routes

As discussed above, changes in the distribution of PCDD/F during atmospheric transport are primarily thought of in terms of degradative mechanisms that remove individual isomers with varying degrees of efficacy. It is generally believed that higher chlorinated PCDD/F are more resistant to biodegradation than lower chlorinated ones. Transformation reactions are generally thought of as steps toward complete degradation. It has recently been proposed that transformations occurring in limnic environments such as the natural chlorination of humus may actually contribute to the environmental level of organically bound halogen compounds.[110] The enzymatically mediated halogenation of organic matter is a subject gaining interest since it may result in the natural formation of organohalogens.[110] Recent studies suggest the potential importance of "natural" PCDD/F formation mechanisms operative in composts and sludges from unidentified chlorinated precursors.[23] In other studies, polychlorinated phenoxyphenols (which are present in chlorophenols) have been proposed as PCDD/F precursors, or so-called predioxins.[32]

6. RELATED ORGANOCHLORINES

In this section a brief introduction to compounds currently found in the environment which resemble PCDD/F, and hence merit attention, is given. The working definition of related molecules are those which are lipophilic, of similar molecular dimension, possess a rigid planar configuration (exist in one geometric plane) or can be easily deformed to near-planarity, and which have halogen atoms in 2,3,7,8-TCDD/F-like positions. These features are thought to be of utmost importance in determining PCDD/F biological activity and toxicity. Additionally, the binding to certain enzymes, in particular the AHH receptor, is widely believed to be a fundamentally important criterion for harmful biological effects caused by 2,3,7,8-TCDD.[111]

Because of the high specificity associated with the techniques used for analysis of PCDD/F, information regarding closely related environmental pollutants is typically not sought for in a given sample of interest. There are a number of PCDD/F relatives which have been only recently identified as environmental contaminants, and as such have received only limited attention since methods and reference materials to assist in their identification are presently only in developmental stages. Related compounds include polybrominated dibenzo-*para*-dioxins and dibenzofurans (PBDD/Fs), polychlorinated dibenzothiophenes (PCDTs) and polychlorinated thianthrenes (PCTAs), and alkyl-substituted dibenzofurans (R-PCDFs). Polychlorinated biphenyls (PCBs) which can assume a planar orientation and polychlorinated naphthalenes (PCNs) are two additional well-recognized PCDD/F relatives. These PCDD/F relatives are briefly introduced below.

6.1. Polybrominated Dibenzodioxins and Dibenzofurans

Polybrominated dibenzodioxins and furans (PBDD/F) are the closest relatives to PCDD/F, bromine rather than chlorine atoms situated about the basic dibenzo-*p*-dioxin or dibenzofuran structures, as shown in Fig. 6a. As mentioned in Section 2.2.2d, PBDD/F have been shown to arise from the combustion of brominated diphenyl ethers which have been widely used as flame retardants.[68] Television sets also have been investigated as potential sources. Other work has shown the trace formation of PBDD/F from fly ash and automotive exhausts.[50,112]

6.2. Polychlorinated Dibenzothiophenes and Thianthrenes

Sulfur rather than oxygen comprises the heteroatom(s) in the central ring for the polychlorinated dibenzothiophenes (PCDTs) and thianthrenes (PCTAs), as shown in Fig. 6b and c, respectively.[113,114] PCDT sources include incineration and are thought to be specific indicators of activities involving the production of iron and steel.[115] PCDTs have also recently been reported to occur from pulp mill discharges.[116] Octachlorothianthrene has been reported to occur widely, but as for PCDD/F the environmental sources have not yet been clearly determined.[117] 2,4,6,8-TCDTs have been identified in aquatic organisms from Newark Bay, NJ.[118]

6.3. Alkylated Polychlorinated Dibenzofurans

One or more methyl (CH_3-) or ethyl (CH_3CH_2-) substituents may appear on the PCDF framework as shown in Fig. 6d thereby forming an alkylated (R-) PCDF.

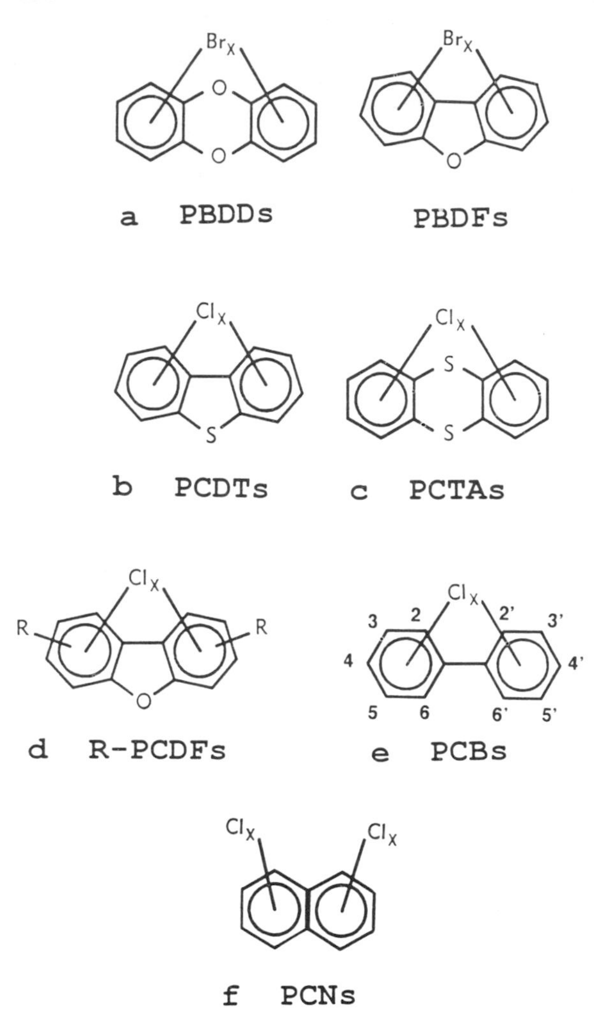

Figure 6. Structures of various PCDD/F-related organochlorines. (a) Polybrominated dibenzo-*p*-dioxins and dibenzofurans; (b) polychlorinated dibenzothiophenes; (c) polychlorinated thianthrenes; (d) alkylated polychlorinated dibenzofurans; (e) planar polychlorinated biphenyls; (f) polychlorinated naphthalenes.

R-PCDFs have primarily been identified in samples related to pulp mills at levels 10–100 times higher than PCDD/Fs.[119] Interestingly, they do not appear to accumulate in higher marine organisms though they have been identified in the hepatopancreas of crab. The toxicological significance of R-PCDFs is presently unknown but they appear to serve as antagonists to 2,3,7,8-TCDD, competing for access to the AHH receptor.

6.4. "Planar" Polychlorinated Biphenyls

Less structurally similar environmental pollutants include the so-called planar (i.e., mono-*ortho* and non-*ortho* chloro-substituted) chlorobiphenyls (PCBs). The basic PCB framework is shown in Fig. 6e. Environmental contamination by PCBs has been a subject of concern over the past nearly three decades. They have been released into the environment for much of this century from a number of sources utilizing PCB as dielectric fluids, nonflammable oils, and plasticizers, and are now banned or restricted in many countries. A few particular PCB isomers are of interest because of the positions of chlorine atoms allowing them to assume a planar configuration and hence making them structurally similar to PCDD/F. These include PCB 77 (3,3′,4,4′), PCB 105 (2,3,3′,4,4′), PCB 118 (2,3′,4,4′,5), PCB 126 (3,3′,4,4′,5) PCB 129 (2,2′,3,3′,4,5), PCB 156 (2,3,3′,4,4′,5), PCB 157 (2,3,3′,4,4′,5′), PCB 167 (2,3′,4,4′,5,5′), and PCB 169 (3,3′,4,4′,5,5′). These individual PCB isomers are often present in environmental samples at low levels requiring special analytical techniques analogous to those for PCDD/F. A recent study suggests that these PCBs may be more important environmental contaminants than the 2,3,7,8-TCDD/F when considered in terms of I-TEQs as defined in Section 1.[121] Some PCBs also have nondioxinlike toxicity.[2]

6.5. Polychlorinated Naphthalenes

PCNs, depicted in Fig. 6f, have been used as electronic component materials and furthermore, like PCDD/Fs, occur in PCB formulations as contaminants.[121] PCNs exhibit similar biological properties to PCDD/F and as a result of their wide environmental occurrence are of interest to the environmental scientist, though environmental levels may be decreasing as for the so-called planar PCBs noted above.[120]

7. CHAPTER SUMMARY

The objective of this chapter has been to review several essential details regarding the formation, environmental distribution, levels, and fate of PCDD/F. The relative importance of known PCDD/F sources is not firmly established, but incomplete combustion of organic matter in the presence of chlorine is among the most discussed. Implicated combustion sources include all types of waste incineration, metal processing of many forms, and fossil fuel combustion. In addition, natural combustion mechanisms (such as forest fires) have been thought to result in PCDD/F formation though anthropogenic sources connected to chlorine chemistry appear to account for the majority of the global PCDD/F burden. Several temporal studies indicate that the advent of the chlorine chemistry industry during this century, and not the advent of fire, is correlated to the environmental occurrence of most PCDD/F. Details of combustion-related formation mechanisms are largely speculative but are known to involve reactions in the gas phase as well as reactions on particulate surfaces, both at elevated temperatures with the optimum range between 300 and 450°C. In solution phase, uncontrolled condensation side reactions of chlorophenol precursors, as well as other chlorinated aromatics, are known to yield PCDD/Fs. Delignification (pulp bleaching) by the wood products industry produces a range of chlorinated substances including

PCDD/Fs, the quantities released dependent on the delignification method in use. Atmospheric flux and deposition estimates indicate that the environmental release of technical products such as chlorophenol and PCB, although important environmental problems in themselves, seems secondary in comparison with combustion sources as contributors to the global PCDD/F burden, though this is unclear. In the United Kingdom the secondary release of PCDD/F from chlorophenols has been proposed as a potential secondary source since pentachlorophenol is often contaminated with the higher chlorinated congeners. These compounds tend to be resistant to degradation when adsorbed onto particulates on which they can be transported through the atmosphere over long distances. Discrepancies between emission and depositional mass balance estimates may suggest that presently unknown PCDD/F sources exist which remain to be identified. Speculations include formation mechanisms involving appropriate PCDD/F precursors, or predioxins. It has been proposed that unknown formation processes may include unexpected natural transformations such as enzymatically mediated halogenation or transformation of organic matter or existing organochlorine precursors.

Physical properties of chlorinated pollutants such as PCDD/F, including low vapor pressures and high affinity for adsorption onto particulate matter, control modes by which they are dispersed from combustion sources throughout the environment. The apparent enhanced resistance of higher chlorinated PCDD/F against photodegradation in the vapor phase, as well as their greater tendency to exist adsorbed onto particulate surfaces where photodegradation is thought to be minimal, provide explanations for the homologue patterns observed in atmospheric particulates and sediments which favor higher chlorinated PCDD/F.

The prevalence of chlorine in waste products and the intensive use of incineration for waste management require that waste incineration (and related) industries optimize combustion conditions and continue application of available abatement technologies to minimize releases of PCDD/Fs. Progress continues to be made toward lowering environmental releases of PCDD/Fs by development and application of appropriate abatement technologies to combustion sources as well as wood product and chemical industries. Related pollutants such as the PCDTs and R-PCDFs, which have only recently received attention, point to a continual need for increasing our knowledge of related toxic synthetic or natural substances presently being emitted into the environment. These related compounds are also potential indicators of specific industrial activities. Other related contaminants, such as the so-called planar PCBs and PCNs with properties which resemble PCDD/Fs, also merit attention as environmental contaminants. Finally, recycling of PCDD/Fs from wide-scale dispersion and deposition into soil, sediments, solid wastes, and other reservoirs, will continue to be important since PCDD/F are thermally stable and accumulate in abiotic and biotic matrices with potentially long lifetimes against degradation.

REFERENCES

1. World Health Organization, Polychlorinated Dibenzo-para-dioxins and Dibenzofurans, *Environmental Health Criteria '88*, Geneva (1989).

2. U. L. Ahlborg, A. Brouwer, M. A. Fingerhut, J. L. Jacobson, S. W. Jacobson, S. W. Kennedy, A. A. F. Kettrup, J. H. Koeman, H. Poiger, C. Rappe, S. H. Safe, R. F. Seegal, J. Tuomisto, and M. Van den Berg, Impact of polychlorinated dibenzo-p-dioxins, -dibenzofurans, and -biphenyls on human health and environmental health, with special emphasis on application to the toxic equivalence factor concept, *Eur. J. Pharmacol.* **228,** 179–199 (1992).
3. R. A. Hites, Environmental behavior of chlorinated dioxins and furans, *Acc. Chem. Res.* **23,** 194–201 (1991).
4. M. Rossberg, Chlorinated hydrocarbons, in: *Ullman's Encyclopedia of Industrial Chemistry,* 5th ed. (W. Gerhartz, ed.), VCH Publishers, New York (1986).
5. *Chemical and Engineering News,* pp. 11–12, May 10 (1993).
6. A. A. Jensen and S. A. Slorach, *Chemical Contaminants in Human Milk,* CRC Press, Boca Raton, FL (1990).
7. R. D. Stephens, C. Rappe, D. G. Hayward, M. Nygren, J. Startin, A. Esbøll, J. Carlé, and E. J. Yrjänheikki, World Health Organization International Intercalibration Study on Dioxins and Furans in Human Milk and Blood, *Anal. Chem.* **64,** 3109–3117 (1992).
8. World Health Organization, Regional Office for Europe, Levels of PCBs, PCDDs, and PCDFs in Breast Milk: Results of WHO coordinated interlaboratory quality control studies and analytical field studies, in: *Environmental Health Series, No. 34,* Copenhagen (1989).
9. Society for Clean Air in the Netherlands, *Expertise on the Measurement and Control of Dioxins,* Vereniging Lucht, Delft, The Netherlands (1991).
10. H.-R. Buser, Review of methods of analysis for polychlorinated dibenzo-p-dioxins and dibenzofurans, in: *Environmental Carcinogens: Methods of Analysis and Exposure Measurement, Vol. 11, PCDD/Fs* (C. Rappe, H.-R. Buser, B. Dodet, and I. K. O'Neill, eds.), pp. 105–147, IARC Scientific Publishers, Lyon, France (1989).
11. C. Rappe and H.-R. Buser, Chemical and physical properties, analytical methods, sources, and environmental levels of halogenated dibenzodioxins and dibenzofurans, in: *Halogenated Biphenyls, Terphenyls, Naphthalenes, Dibenzodioxins and Related Products* (R. D. Kimbrough, ed.), pp. 71–102, Elsevier, Amsterdam (1989).
12. R. E. Clement, Ultratrace dioxin and dibenzofuran analysis: 30 years of advances, *Anal. Chem.* **63,** 1130A–1137A (1991).
13. S. Hashimoto, T. Wakimoto, and R. Tatsukawa, PCDDs in (Japanese) sediments accumulated about 8120 years ago from coastal areas, *Chemosphere* **21,** 825–835 (1990).
14. R. R. Bumb, W. B. Crummett, S. S. Cutie, J. R. Glendhill, R. H. Hummel, R. O. Kagel, L. L. Lamparski, E. V. Luoma, D. L. Miller, T. J. Nestick, L. A. Shadoff, R. H. Stehl, and J. S. Woods, Trace chemistries of fire: A source of chlorinated dioxins, *Science* **210,** 385–390 (1980).
15. J. M. Czuczwa and R. A. Hites, Airborne dioxins and dibenzofurans, sources and fates, *Environ. Sci. Technol.* **20,** 95–200 (1986).
16. R. W. Macdonald, W. J. Cretney, N. Crewe, and D. Paton, A history of octachlorodibenzo-p-dioxin, 2,3,7,8-tetrachlorodibenzofuran, and 3,3′,4,4′-tetrachlorobiphenyl contaminants in Howe Sound, British Columbia, *Environ. Sci. Technol.* **26,** 1544–1550 (1992).
17. L.-O. Kjeller, K. C. Jones, A. E. Johnson, and C. Rappe, Increases in polychlorinated dibenzo-p-dioxin and furan content of soil and vegetation since the 1840's, *Environ. Sci. Technol.* **25,** 1619–1627 (1991).
18. H. Y. Tong, M. L. Gross, A. Schecter, S. J. Monson, and A. Dekin, Sources of dioxins in the environment: Second stage study of PCDD/Fs in ancient human tissue and environmental samples, *Chemosphere* **20,** 987–992 (1990).
19. W. V. Ligon, S. B. Dorn, R. J. May, M. J. A Chlorodibenzofurans and chlorodibenzo-

p-dioxins in Chilean mummies dated to about 2800 years before the present, *Environ. Sci. Technol.* **23,** 1286–1290 (1989).
20. A. Schecter, A. Dekin, N. C. A. Weerasinghe, S. Arghestani, and M. L. Gross, Sources of dioxins in the environment: A study of PCDDs and PCDFs in ancient, frozen Eskimo tissue, *Chemosphere* **17,** 627–631 (1988).
21. C. Rappe, L.-O. Kjeller, S.-E. Kulp, C. de Wit, I. Hasselsten, and O. Palm, Levels, profiles, and patterns of PCDDs and PCDFs in samples related to the production and use of chlorine, *Chemosphere* **23,** 1629–1636 (1991).
22. R. Bacher, M. Swerev, and K. Ballschmiter, Profiles and patterns of monochloro- through octachlorodibenzodioxins and dibenzofurans in chimney deposits from wood burning, *Environ. Sci. Technol.* **26,** 1649–1655 (1992).
23. L. Öberg, N. Wågman, R. Andersson, and C. Rappe, *De novo Formation of PCDD/Fs in Compost and Sewage Sludge—A Status Report,* Dioxin '93, 13th International Symposium on Chlorinated Dioxins and Related Compounds, Vienna, *Organohalogen Compounds,* Vol. 2, pp. 297–302 (1993).
24. C. Rappe and L.-O. Kjeller, PCDDs and PCDFs in environmental samples: Air, particulates, sediments, and soil, *Chemosphere* **16,** 1775–1780 (1987).
25. C. Rappe, R. Anderson, P.-A. Bergqvist, C. Brohende, M. Hansson, L.-O. Kjeller, G. Lindström, S. Marklund, S. Nygren, S. E. Swanson, M. Tysklind, and K. Wiberg, Overview of environmental fate of chlorinated dioxins and dibenzofurans. Sources, levels, and isomeric patterns in various matrices, *Chemosphere* **16,** 1603–1618 (1987).
26. C. Rappe, P.-A. Bergqvist, and L.-O. Kjeller, Levels, Trends and patterns of PCDDs and PCDFs in Scandinavian environmental samples, *Chemosphere* **18,** 651–658 (1989).
27. S. J. Harrad, T. A. Malloy, M. A. Khan, and T. D. Goldfarb, Levels and sources of PCDDs, PCDFs, chlorophenols (CPs) and chlorobenzenes (CBzs) in composts from a municipal yard waste composting facility, *Chemosphere* **23,** 181–191 (1991).
28. S. Harrad, A. P. Sewart, and K. C. Jones, PCDD/Fs in the British environment—Sinks, sources and temporal trends, in: *Proceedings of the 12th International Symposium on Dioxins and Related Compounds,* Tampere, Finland, *Organohalogen Compounds,* Vol. 9, pp. 89–92 (1992).
29. H.-R. Buser and H. P. Bosshardt, Determination of polychlorinated dibenzo-p-dioxins and dibenzofurans in commercial pentachlorophenols by combined gas chromatography–mass spectrometry, *J. Assoc. Off. Anal. Chem.* **59,** 562–569 (1976).
30. H.-R. Buser and H.-P. Bosshardt, Determination of 2,3,7,8-tetrachlorinated dibenzo-1,4-dioxins at part-per-billion levels in technical grade 2,4,5-trichlorophenoxyacetic acid, in 2,4,5-T alkyl ester, and 2,4,5-T amine salt herbicide formulations by quadrupole mass fragmentography, *J. Chromatogr.* **90,** 71–77 (1974).
31. H. Brunner, Untersuchungen zu Herkunft und Vorkommen Polychlorierter Dibenzodioxine und Dibenzofurane in der Umwelt, Dissertation Universität Tubingen, Germany (1990).
32. C.-A. Nilsson, Å. Norström, K. Andersson, and C. Rappe, Impurities in commercial products related to chlorophenols, in: *Pentachlorophenol: Chemistry, Pharmacology, and Environmental Toxicology* (R. Rao, ed.), pp. 313–324, Plenum Press, New York.
33. G. C. Chouhury and G. R. B. Webster, Environmental photochemistry of polychlorinated dibenzofurans and dibenzo-p-dioxins: A review, *Toxicol. Environ. Chem.* **14,** 43–61 (1987).
34. S. E. Swanson, C. Rappe, J. Malmström, and K. P. Kringstad, Emissions of PCDDs and PCDFs from the pulp industry, *Chemosphere* **17,** 681–691 (1988).
35. G. Amendola, D. Barna, R. Blossom, L. LaFleur, A. McBride, F. Thomas, T. Tiernan, and R. Whitemore, The occurrence and fate of PCDDs and PCDFs in fine bleached Kraft pulp and paper mills, *Chemosphere* **18,** 1181–1188 (1989).

36. P. Jonsson, C. Rappe, L.-O. Kjeller, A. Kierkegaard, L. Håkanson, and B. Jonsson, Pulp-mill related polychlorinated organic compounds in Baltic Sea sediments, *Ambio* **22,** 37–43 (1993).
37. G. L. Lebel, D. T. Williams, and F. M. Benoit, Chlorinated dibenzodioxins and dibenzofurans in consumer paper products, *Chemosphere* **25,** 1683–1690 (1992).
38. S. Safe, K. W. Brown, K. C. Donnelly, C. S. Anderson, K. V. Markiewicz, M. S. McLachlan, A. Reischl, and O. Hutzinger, Polychlorinated dibenzo-p-dioxins and dibenzofurans associated with wood-preserving chemical sites: Biomonitoring with pine needles, *Environ. Sci. Technol.* **26,** 394–396 (1992).
39. P. Mckee, A. Burt, D. McCurvin, D. Hollinger, R. Clement, D. Sutherland, and W. Neaves, Levels of dioxins, furans, and other organic contaminants in harbour sediments near a wood preserving plant using pentachlorophenol and creosote, *Chemosphere* **20,** 1679–1686 (1990).
40. V. H. Kitunen and M. S. S. Salkinoja-Salonen, Soil contamination of abandoned sawmill areas, *Chemosphere* **20,** 1671–1677 (1990).
41. K. Olie, P. L. Vermeulen, and O. Hutzinger, Chlorodibenzo-p-dioxins and chlorodibenzofurans are trace components of fly ash and flue gas of some municipal incinerators in the Netherlands, *Chemosphere* **8,** 455–459 (1977).
42. H.-R. Buser and H.-P. Bosshardt, Polychlorinated dibenzo-p-dioxins, dibenzofurans, and benzenes in the ashes from municipal and industrial incinerators, *Mitt. Geb. Lebensmittelunter. Hyg.* **69,** 191–199 (1978) [in German].
43. H.-R. Buser, H.-P. Bosshardt, and C. Rappe, Identification of polychlorinated dibenzo-p-dioxin isomers in fly ash, *Chemosphere* **2,** 165–172 (1978).
44. W. M. Schaub and W. Tsang, Dioxin formation in incinerators, *Environ. Sci. Technol.* **17,** 721–730 (1983).
45. L. Stieglitz, G. Zwick, J. Beck, W. Roth, and H. Vogg, On the de novo synthesis of PCDD/PCDF on fly ash of municipal waste incinerators, *Chemosphere* **18,** 1219–1226 (1989).
46. F. W. Karasek and L. C. Dickson, Model studies of polychlorinated dibenzo-p-dioxin formation during municipal refuse incineration, *Science* **237,** 754–756 (1987).
47. P. T. Williams, The formation and control of dioxins and furans from the incineration of municipal waste, *Trans. Inst. Chem. Eng.* **68(B),** 122–126 (1990).
48. P. Acharya, S. G. DeCicco, and R. G. Novak, Factors that can influence and control the emissions of dioxins and furans from hazardous waste incinerators, *J. Air Waste Manage. Assoc.* **41,** 1605–1615 (1991).
49. S. Marklund, R. Andersson, M. Tysklind, C. Rappe, K. E. Egebäck, E. Björkman, and V. Grigoriadis, Emissions of PCDDs and PCDFs in gasoline and diesel fueled cars, *Chemosphere* **20,** 553–561 (1990).
50. R. Bacher, U. Riehle, M. Swerev, and K. Ballschmiter, Patterns and levels of polychlorinated (Br^-, Cl^-) dibenzodioxins and dibenzofurans in automobile traffic related samples, *Chemosphere* **23,** 1151–1171 (1991).
51. I. Fängmark, B. Strömberg, N. Berge, and C. Rappe, The influence of post-combustion temperature profiles on the formation of PCDD, PCDF, PCBz, and PCB in a pilot incinerator, submitted to *Environ. Sci. Technol.* **27,** 1602–1610 (1993).
52. M. Tysklind, I. Fängmark, S. Marklund, A. Lindskog, L. Thaning, and C. Rappe, Atmospheric transport and transformation of polychlorinated dibenzo-p-dioxins and dibenzofurans, *Environ. Sci. Technol.* **27,** 2190–2197 (1993).
53. P. Andersson, S. Marklund, and C. Rappe, Levels and profiles of PCDDs and PCDFs in environmental samples as determined in snow deposited in northern Sweden, in: *Proceed-*

ings from the 12th International Symposium on Dioxins and Related Compounds, Tampere, Finland, *Organohalogen Compounds,* Vol. 8, pp. 307–310 (1992).
54. I. Zebühr, Trace analysis of polychlorinated dibenzo-p-dioxins, dibenzofurans and related compounds in environmental matrices, Stockholm Universitet, Sweden (1992).
55. J. M. Czuczwa, B. D. McVeety, and R. A. Hites, Polychlorinated dibenzodioxins and dibenzofurans in sediments from Siskiwit Lake, Isle Royale, *Science* **226,** 568–569 (1984).
56. A. Sheffield, Sources and releases of PCDDs and PCDFs in the Canadian environment, *Chemosphere* **14,** 811–814 (1985).
57. Swedish Environmental Protection Board (SNV), *Persistant Organic Substances in the Environment* (in Swedish, SNV Report 4136), SNV, Solna, Sweden (1993).
58. M. Oehme, A. Bartonova, and J. Knutzen, Estimation of polychlorinated dibenzofuran and dibenzo-p-dioxin contamination of a coastal region using isomer profiles in crabs, *Environ. Sci. Technol.* **24,** 1836–1841 (1990).
59. M. Tysklind, G. Söderström, C. Rappe, L.-E. Hägerstedt, and E. Burström, PCDD and PCDF emissions from scrap metal melting processes at a steel mill, *Chemosphere* **19,** 705–710 (1989).
60. T. S. Thompson, R. E. Clement, N. Thorton, and J. Luyt, Formation and emission of PCDDs/PCDFs in the petroleum refining industry, *Chemosphere* **20,** 1525–1532 (1990).
61. E. Benfenati, S. Valzacchi, G. Mariani, L. Airoldi, and R. Fanelli, PCDD, PCDF, PCB, PAH, cadmium, and lead in roadside soil: Relationship between road distance and concentration, *Chemosphere* **24,** 1077–1083 (1992).
62. C. C. Travis and H. A. Hattemer-Frey, Dioxin—Research needs for risk assessment, *Chemosphere* **20,** 729–742 (1990).
63. H.-R. Buser, H.-P. Bosshardt, and C. Rappe, Formation of polychlorinated dibenzofurans (PCDFs) from the pyrolysis of PCBs, *Chemosphere* **7,** 109–119 (1978).
64. R. Lindahl, C. Rappe, and H.-R. Buser, Formation of polychlorinated dibenzofurans and polychlorinated dibenzo-p-dioxins from the pyrolysis of polychlorinated diphenyl ethers, *Chemosphere* **9,** 351–361 (1980).
65. H.-R. Buser, Formation of polychlorinated dibenzofurans and dibenzo-p-dioxins from the pyrolysis of chlorobenzenes, *Chemosphere* **8,** 415–424 (1979).
66. C. Rappe, S. Marklund, H.-R. Buser, and H.-P. Bosshardt, Formation of polychlorinated-p-dibenzodioxins (PCDDs) and dibenzofurans (PCDFs) by burning or heating chlorophenates, *Chemosphere* **7,** 269–281 (1978).
67. World Health Organization, PCBs, PCDDs and PCDFs: Prevention and control of accidental and environmental exposures, *Environmental Health Series,* Vol. 23, Copenhagen (1987).
68. H.-R. Buser, Polybrominated dibenzofurans and dibenzo-p-dioxins: Thermal reaction products of polybrominated diphenyl ether flame retardants, *Environ. Sci. Technol.* **20,** 404–408 (1986).
69. C. Tashiro, R. E. Clement, B. J. Stockes, L. Radke, W. R. Cofer, and P. Wood, Preliminary report: Dioxin and furans in prescribed burns, *Chemosphere* **20,** 1533–1536 (1990).
70. L. Lamparski, personal communication.
71. S. Marklund, personal communication.
72. J. Towara, R. Fuchs, and O. Hutzinger, PCDD/F in the pulp and paper production, in: *Proceedings of the 12th International Symposium on Dioxins and Related Compounds,* Tampere, Finland, *Organohalogen Compounds,* Vol. 9, p. 287 (1992).
73. B. T. Hargrave, G. C. Harding, W. P. Vass, P. E. Erickson, B. R. Fowler, and V. Scott,

Organochlorine pesticides and polychlorinated biphenyls in the Arctic Ocean food web, *Arch. Environ. Contam. Toxicol.* **22,** 41–54 (1992).
74. F. Wania and D. Mackay, Global fractionation and cold condensation of low volatility organochlorine compounds in polar regions, *Ambio* **22,** 10–18 (1993).
75. D. L. Stalling, L. M. Smith, J. D. Petty, J. W. Hogan, J. L. Johnson, C. Rappe, and H.-R. Buser, Residues of polychlorinated dibenzo-p-dioxins and dibenzofurans in Laurentian Great Lakes fish, *Environ. Sci. Res.* **26,** 221–240 (1983).
76. C. Rappe, S. Marklund, L.-O. Kjeller, and A. Lindskog, Long range transport of PCDDs and PCDFs on airborne particles, *Chemosphere* **18**(1–6), 1283–1290 (1989).
77. B. D. Eitzer and R. A. Hites, Atmospheric transport and deposition of polychlorinated dibenzo-p-dioxins and dibenzofurans, *Environ. Sci. Technol.* **23,** 1396–1401 (1989).
78. C. J. Koester and R. A. Hites, Wet and dry deposition of chlorinated dioxins and furans, *Environ. Sci. Technol.* **26,** 1375–1382 (1992).
79. S. J. Eisenreich, B. B. Looney, and J. D. Thorton, Airborne organic contaminants in the Great Lakes ecosystem, *Environ. Sci. Technol.* **15,** 30–38 (1981).
80. D. Broman, Transport and fate of hydrophobic organic compounds in the Baltic environment—Polycyclic aromatic hydrocarbons, polychlorinated dibenzodioxins and dibenzofurans, Dissertation, Stockholm Universitet, Sweden (1990).
81. T. Nakano, M. Tsuji, and T. Okuno, Distribution of PCDDs, PCDFs, and PCBs in the atmosphere, *Atmos. Environ.* **24A,** 1361–1368 (1990).
82. C. Näf, D. Broman, H. Pettersen, C. Rolff, and Y. Zebühr, Flux estimates and pattern recognition of particulate polycyclic aromatic hydrocarbons, polychlorinated dibenzo-p-dioxins, and dibenzofurans in the waters outside various emission sources on the Swedish Baltic coast, *Environ. Sci. Technol.* **26,** 1444–1457 (1992).
83. B. Jobb, M. Uza, R. Hunsinger, K. Roberts, H. Tosine, R. Clement, B. Bobbie, G. LeBel, D. Williams, and B. Lau, A survey of drinking water supplies in the Province of Ontario for dioxins and furans, *Chemosphere* **20,** 1553–1558 (1990).
84. C. Rappe, L.-O. Kjeller, and R. Andersson, Analysis of PCDDs and PCDFs in sludge and water samples, *Chemosphere* **19,** 13–20 (1989).
85. C. Rappe, Sources of exposure, environmental concentrations, and exposure assessment of PCDDs and PCDFs, *Chemosphere* **27,** 211–225 (1993).
86. C. Rappe, Environmental concentrations and ecotoxicological effects of PCDDs, PCDFs, and related compounds, *Dioxin '93, International Symposium on Dioxins and Related Compounds,* Vienna, *Organohalogen Compounds* Vol. 12, 163–170 (1993).
87. T. O. Tiernan, D. J. Wagel, G. F. Vanness, J. H. Garrett, J. G. Solch, and L. A. Harden, PCDD/PCDF in the ambient air of a metropolitan area in the U.S., *Chemosphere* **19**(1–6), 541–546 (1989).
88. D. Broman, C. Näf, C. Rolff, and Y. Zebühr, Occurrence and dynamics of polychlorinated dibenzo-p-dioxins and dibenzofurans and polycyclic aromatic hydrocarbons in the mixed surface layer of remote coastal and offshore waters in the Baltic, *Environ. Sci. Technol.* **25,** 1850–1864 (1991).
89. B. D. Eitzer and R. A. Hites, Polychlorinated-p-dioxins and dibenzofurans in the ambient atmosphere of Bloomington, Indiana, *Environ. Sci Technol.* **23,** 1389–1395 (1989).
90. W. Rotard, W. Christman, and W. Knoth, Background levels of PCDD/PCDF in soils of western Germany, Paper presented at the 11th International Symposium on Chlorinated Dioxins and Related Compounds, Abstract 116, Research Triangle Park, NC (1991).
91. R. Atkinson, Atmospheric lifetimes of dibenzo-p-dioxins and dibenzofurans, *Sci. Total Environ.* **104,** 17–33 (1991).

92. D. L. Sediak and A. W. Andren, Aqueous-phase oxidation of polychlorinated biphenyls by hydroxyl radicals, *Environ. Sci. Technol.* **25,** 1419–1427 (1991).
93. H.-R. Buser, Preparation of qualitative standard mixtures of polychlorinated-p-dioxins and dibenzofurans by ultraviolet and gamma-irradiation of the octachloro compounds, *J. Chromatogr.* **129,** 303–307 (1976).
94. C. J. Koester and R. A. Hites, Photodegradation of polychlorinated dioxins and dibenzofurans adsorbed to fly ash, *Environ. Sci. Technol.* **26,** 502–507 (1992).
95. M. Tysklind, A. E. Carey, C. Rappe, and G. C. Miller, Photolysis of OCDF and OCDD on soil, in: *Proceedings of the 12th International Symposium on Dioxins and Related Compounds,* Tampere, Finland, *Organohalogen Compounds,* Vol. 9, pp. 293–296 (1992).
96. S. Kapila, A. F. Yanders, C. E. Orazio, J. E. Meadows, S. Cerlesi, and T. E. Clevenger, Field and laboratory studies on the movement and fate of tetrachlorodibenzo-p-dioxin in soil, *Chemosphere* **18,** 1297–1304 (1989).
97. A. L. Young, H. K. Kang, and B. M. Shepard, Chlorinated dioxins as herbicide contaminants, *Environ. Sci. Technol.* **17,** 5320A–532A (1983).
98. C. deWit, B. Jansson, M. Strandell, P. Jonsson, P.-A. Bergqvist, S. Bergek, L.-O. Kjeller, C. Rappe, M. Olsson, and S. Slovach, Results from the first year of the Swedish Dioxin Survey, *Chemosphere* **20,** 1473–1480 (1990).
99. S. Kieatiwong, L. V. Nguyen, V. R. Hebert, M. Hackett, M. Miller, J. Miille, and R. Mitzel, Photolysis of chlorinated dioxins in organic solvents and on soils, *Environ. Sci. Technol.* **24,** 1575–1580 (1990).
100. G. C. Miller, V. R. Hebert, M. J. Miille, R. Mitzel, and R. G. Zepp, Photolysis of octachlorodibenzo-p-dioxin on soils: Production of 2,3,7,8-TCDD, *Chemosphere* **18,** 1265–1274 (1989).
101. R. G. Orth, C. Ritchie, and F. Hileman, Measurement of the photoinduced loss of vapor phase TCDD, *Chemosphere* **18,** 1275–1282 (1989).
102. C. Rappe, P.-A. Bergqvist, L.-O. Kjeller, S. Swanson, T. Belton, B. Ruppel, K. Lockwood, and P. C. Kahn, Levels and patterns of PCDD and PCDF contamination in fish, crabs, and lobsters from Newark Bay and the New York Bight, *Chemosphere* **22,** 239–266 (1991).
103. S. Bergek, P.-A. Bergqvist, M. Hjelt, M. Olsson, C. Rappe, A. Roos, and D. Zook, Concentrations of PCDDs and PCDFs in seal from Swedish waters, *Ambio* **21,** 553–556 (1993).
104. J. R. Parsons and M. C. M. Storms, Biodegradation of chlorinated dibenzo-p-dioxins in batch and continuous cultures of strain JB1, *Chemosphere* **19,** 1297–1308 (1989).
105. J. R. Parsons, Influence of suspended sediment on the biodegradation of chlorinated dibenzo-p-dioxins, *Chemosphere* **25,** 1973–1980 (1992).
106. C. C. Travis, H. A. Hattemer-Frey, and E. Silbergeld, Dioxin, dioxin, everywhere, *Environ. Sci. Technol.* **23,** 1061–1063 (1989).
107. C. Rappe, Sources of and human exposure to PCDDs and PCDFs, in: *Biological Basis for Risk Assessment of Dioxins and Related Compounds* (M. A. Gallo, R. J. Scheuplein, and K. A. van der Heijden, eds.), pp. 121–131, Banbury Report 35, Cold Spring Harbor Laboratory Press, Cold Spring Harbor, NY (1991).
108. R. F. Boop, M. L. Gross, H. Yong, H. J. Simpson, S. J. Monson, B. L. Deck, and F. C. Moser, A major incident of dioxin contamination: Sediments of New Jersey estuaries, *Environ. Sci. Technol.* **25,** 951–956 (1991).
109. B. D. Eitzer and R. A. Hites, Comment on "Airborne Dioxins and Dibenzofurans: Sources and Fates," *Environ. Sci. Technol.* **21,** 922–924 (1987).

110. G. Asplund and A. Grimvall, Organochlorines in nature: More widespread than previously assumed, *Environ. Sci. Technol.* **25,** 1347–1350 (1991).
111. S. Safe, T. Sawyer, S. Bandiera, L. Safe, B. Zmudzka, G. Mason, M. Romkes, M. A. Denomme, and T. Funita, Binding of the 2,3,7,8-TCDD receptor and AHH/EROD induction: In vitro QSAR, in: *Biological Mechanism of Dioxin Action* (A. Poland and R. D. Kimbrough, eds.), pp. 135–149, Banbury Report 18, Cold Spring Harbor Laboratory Press, Cold Spring Harbor, NY (1984).
112. P. Haglund, K. E. Egebäck, and B. Jansson, Analysis of polybrominated dioxins and furans in vehicle exhaust, *Chemosphere* **17,** 2129–2140 (1988).
113. H.-R. Buser, S. Dolezar, M. Wolfensberger, and C. Rappe, Polychlorodibenzothiophenes, the sulphur analogues of the polychlorodibenzofurans identified in incineration samples, *Environ. Sci. Technol.* **25,** 1637–1643 (1991).
114. H.-R. Buser, Identification and sources of dioxin-like compounds: I. Polychlorodibenzothiophenes and polychlorothianthrenes, the sulfur-analogues of the polychlorodibenzofurans and polychlorodibenzodioxins, *Chemosphere* **25,** 45–48 (1992).
115. C. Rappe, Dioxin. Patterns and source identification, *Fresenius J. Anal. Chem.* **348,** 63–75 (1994).
116. S. Sinkkonen, J. Paasivirta, J. Koistinen, M. Lahtiperä, J. Tarhanen, and R. Lammi, Determination of polychlorinated dibenzothiophenes in stack gases and in bleached pulp mill effluents, in: *Proceedings of the 12th International Symposium on Dioxins and Related Compounds,* Tampere, Finland, *Organohalogen Compounds,* Vol. 9, pp. 271–274 (1992).
117. T. Benz, H. Hagenmaier, C. Lindig, and J. She, Occurrence of the sulphur analogue of octachlorodibenzo-p-dioxin in the environment and investigations on its potential source, *Fresenius J. Anal. Chem.* **344,** 286–291 (1992).
118. H.-R. Buser and C. Rappe, Determination of polychlorinated thiophenes, the sulfur analogues of polychlorinated dibenzofurans, using various gas chromatographic–mass spectrometric techniques, *Anal. Chem.* **63,** 1210–1217 (1991).
119. H. R. Buser, L.-O. Kjeller, S. E. Swanson, and C. Rappe, Methyl-, polymethyl-, and alkylpolychlorinated-dibenzofurans identified in pulp mill sludge and sediments, *Environ. Sci. Technol.* **23,** 1130–1137 (1989).
120. U. A. T. Brinkman and H. G. M. Reymer, Polychlorinated naphthalenes, *J. Chromatogr.* **127,** 203–243 (1976).
121. U. Järnberg, L. Asplund, C. de Wit, A.-K. Grafström, P. Haglund, B. Janson, K. Lexén, M. Olsson, and B. Jonsson, Polychlorinated biphenyls and polychlorinated naphthalenes in Swedish sediment and biota: Levels, patterns, and time trends, *Environ. Sci. Technol.* **27,** 1364–1374 (1993).

Chapter 4

Dioxins in Food

James R. Startin

1. INTRODUCTION

The purpose of this chapter is to summarize the information that is currently available on the concentrations of PCDDs and PCDFs in foods and the resulting dietary intake. As is made clear in other chapters of this book, PCDDs and PCDFs are present at measurable concentrations in human tissues even when the subjects have no history of occupational or accidental exposure. It is accepted that this is related very largely to the presence of PCDDs and PCDFs in foods, and that the proportion of the body burden of background subjects that stems from inhalation or from direct contact with PCDDs and PCDFs in the environment is probably no more than a few percent.

PCDDs and PCDFs are highly lipophilic and tend to accumulate in fatty tissues. It is well established that the 17 congeners that contain chlorine at the 2, 3, 7, and 8 positions are persistent and accumulate in many organisms and tend to dominate the congener pattern in higher animals and fish, even though these congeners only form a small proportion of the total output from many sources and of the environmental load. As discussed in Chapter 5 it is also these 2,3,7,8-substituted congeners that are regarded as significantly toxic and that have become the main focus of interest.

To allow the available data to be summarized in a reasonably concise form, considerable use is made in this chapter of 2,3,7,8-TCDD equivalent (TEQ) values for the total (summed) concentration of the 2,3,7,8-substituted PCDDs and PCDFs. Although various weighting schemes have been used in the literature, wherever possible the data given in this chapter have been recalculated using the widely accepted International Toxic Equivalency Factors (I-TEF).[1] It is worth noting that the use of TEQs in assessing exposures has recently been criticized since the TEFs are based on relative toxicities of the compounds when once they have been absorbed by the organism, but

---

James R. Startin • CSL Food Science Laboratory, Norwich Research Park, Colney, Norwich NR4 7UQ, United Kingdom.

Dioxins and Health, edited by Arnold Schecter. Plenum Press, New York, 1994.

do not take account of the very different absorption characteristics of different congeners.[2]

In considering the results that have been reported from the different studies and assessing their usefulness, it is necessary to give careful consideration to the quality of the analysis (whether the results properly reflect the sample that was analyzed and are congener-specific and of adequate sensitivity), to the sampling (whether the samples that were analyzed were fully representative of the food supply in a particular location), and to the location represented (whether this was "background" or influenced by localized contamination). This chapter includes references to most of the available literature, but greatest emphasis is given to studies that appear to meet these analytical criteria and to be of the most general significance.

The analysis of PCDDs and PCDFs in foods is particularly challenging because of the very low concentrations that are involved and the consequent need to maintain high sensitivity coupled with low analytical blanks. The former requires extensive and time-consuming sample cleanup as well as sophisticated analytical instruments, and the latter scrupulous attention to technical detail and quality assurance.

Over the years continuing improvements in the sensitivity of analytical instruments and refinements in the detailed implementation of laboratory methods have allowed a progressive improvement in sensitivity so that congener-specific measurements of PCDDs and PCDFs can now be made at very small concentrations. Many of the earlier studies were conducted without the benefit of these advances and were not congener-specific or did not achieve detection limits that matched the levels now known to be present in foods, so that the results are of limited value. Even in some more recent studies the levels that are reported of some individual congeners often appear to be close to the analytical detection limits and may be subject to considerable imprecision in measurement.

There are a number of other pitfalls in comparing results from different studies and in drawing general conclusions. The analytical standards, on which the accuracy of quantification depends, are normally available only as dilute solutions and, despite the efforts of commercial suppliers and standards organizations, there are good reasons to suspect that the initial accuracy and long-term consistency of the standards used in some laboratories is poor.

In addition, some of the studies that have been reported included steps in the analytical method that are now known to introduce inaccuracies. In particular, the use of vigorous alkaline saponification (for example, by refluxing in ethanolic potassium hydroxide) as a step in the treatment of fatty samples has been shown to result in loss of some of the more highly chlorinated congeners and formation of other congeners with fewer chlorine atoms.[3]

2. OVERVIEW OF FOOD STUDIES

Some of the data on foods have been derived from studies of one particular commodity, sometimes incidentally in the pursuit of work with a different focus, and these are dealt with below in the sections on specific foods. There have also been a number of studies, conducted in different countries or regions, that have included many different foods with the aim of establishing, in conjunction with food consumption statistics, average dietary intakes. Although much of the data will be discussed in sections on individual foods, it is useful to consider the coverage and general features of some of these studies separately.

Some surveys of foods have been triggered by the association between pentachlorophenol (PCP) and the more highly chlorinated (Cl_6–Cl_8) PCDDs. The results from Canadian studies of meat collected in 1980 have been published by Ryan *et al.*[4] but the lower chlorinated congeners were not detected. Firestone *et al.*[5] have published the results of analyses for the higher chlorinated PCDDs in various foods collected in the United States by the Food and Drug Administration in a 5-year period beginning in 1979. PCDFs and the lower chlorinated PCDDs including TCDD were not studied. In view of the changes in PCP usage over the last decade, and since these studies do not provide information on the concentrations of the lower chlorinated compounds, they are of limited relevance to a more general consideration of PCDDs and PCDFs in foods.

Stanley and Bauer have reported a study of composites of selected foods collected from San Francisco and Los Angeles and giving emphasis to foodstuffs of Californian origin.[6] The foodstuffs included saltwater fish, freshwater fish, beef, chicken, pork, cow's milk, and eggs. Between five and eight composite samples were analyzed for each commodity, and each composite was formed from between six and ten individual samples. The analytical methods and quality control were rigorous but the limits of detection were somewhat higher than those obtained in some other recent studies and few measurable levels were found.

The studies cited above of foods in the United States give little information about the concentrations of the toxicologically more significant Cl_4 and Cl_5 compounds. This is partly redressed by studies of particular food commodities (listed in later sections), some of which have dealt exclusively with 2,3,7,8-TCDD or with TCDD and TCDF isomers. More recently, results from the analysis, with high sensitivity, of a number of individual retail samples of various commodities including dairy, meat, and fish products purchased from a supermarket in New York have been reported by Schecter *et al.*[7]

A study of the food supply in Ontario, Canada, has been reported by Birmingham *et al.*[8] Samples of beef, pork, chicken, eggs, milk, apples, peaches, potatoes, tomatoes, and wheat from Ontario sources and from other locations were analyzed with an average detection limit of 1–4 ng/kg (whole product) and pooled extracts from the Ontario samples were reanalyzed with detection limits a factor of 10 lower. Some samples were found to contain low levels of the higher chlorinated congeners but no Cl_4 or Cl_5 PCDDs or PCDFs were detected. Although the results for market-basket samples from Ontario that had been published previously by Davies[9] had shown high levels of PCDDs and PCDFs in a composite fruit sample (consisting mainly of apples) and also measurable levels in vegetables, meat, milk, and eggs, these results were challenged and discounted by Birmingham *et al.*[8]

Wider geographical coverage of foods available in Canada has been addressed by Ryan *et al.*[10,11] who analyzed fatty food composites from six major cities, obtained through the Canadian Total Diet Programme and collected between 1985 and 1988. The major compounds found, with respect to both frequency and concentration, were the 2,3,7,8-substituted HxCDD isomers, 1,2,3,4,6,7,8-HpCDF, 1,2,3,4,6,7,8-HpCDD, and OCDD. In addition, 2,3,7,8-TCDF and, to a lesser extent, 2,3,7,8-TCDD were above the detection limit in cow's milk composites and this was attributed to migration from bleached paperboard cartons used as milk containers.

Broadly based studies of foods in Germany have been published by Beck *et al.*[12] and by Fürst *et al.*[13] The study of Beck *et al.*[12] was based on analysis of 12 random purchases of different foods (chicken, eggs, butter, pork, red fish, cod, herring, vegetable oil, cauliflower, lettuce, cherries, and apples) and of 8 samples of cow's milk that had been reported earlier.[14] The analysis achieved very good sensitivity so that most of the congeners with a 2,3,7,8-substitution pattern were reported as positive values in most of the samples. Although the use of single individual samples to represent the main types of food other than milk must create doubts as to extent to which the results can represent the whole food supply, this study has been widely cited. Fürst *et al.*[13] concentrated on fatty foods and foods of animal origin in a study involving over 100 individual samples, again conducted with high sensitivity. The results of both of these studies are included below in sections dealing with specific food types.

A comprehensive assessment of Dutch foodstuffs has been reported by Liem *et al.*[15] The sampling scheme was designed after considering the individual consumption data of approximately 6000 individuals from 2200 families over a 2-day period. Animal fat and liver from six different types of animals were collected from slaughterhouses and cereal products, cow's milk, dairy products, meat products, nuts, eggs, fish, and game from retail sources. In each of these groups, duplicate collection of samples was carried out in each of four Dutch regions and proportional pooling performed to provide a pair of duplicate samples for each group and region. Further combination of four regional pooled samples into one national composite was also made before analysis. This study is of particular interest because of the inclusion of non-*ortho* PCBs in the analysis, but the results, other than for 2,3,7,8-TCDD, are so far only available in the form of total TEQs.

Results of the analysis of various foods in the United Kingdom have been reported.[16] Although considerable emphasis was placed on cow's milk, results for other types of food, based on composite samples from the U.K. Total Diet Survey, are included.

A survey of PCDD and PCDF levels in Norwegian food has also been in progress and some results are available[17] covering fish, cod liver and cod liver oil, crabs, margarine (based on fish oils), butter, sheep tallow, and eggs. Further work is said to be in progress on milk and dairy products and on beef and pork.

Several studies have been conducted of foods in the Japanese diet, but some doubt must be attached to the results. Ono *et al.* [18] reported the analysis of vegetables, cooking oils, cereals, fish, pork, beef, poultry, and eggs collected in Matsuyame in 1986. In contrast to other studies, the congeners that dominate in the fish and animal samples are generally not those with 2,3,7,8-substitution, suggesting that the results do not reflect biologically incurred residues but some other form of contamination. OCDD (which is normally the congener of highest concentration) was not consistently found and it is possible that the alkaline saponification procedure used for some samples resulted in dechlorination of OCDD and distortion of the congener profile. Other studies by Takizawa and Muto[19] and by Ogaki *et al.*[20] also employed alkaline saponification and the reports do not include congener-specific results.

Reliable data on PCDD and PCDF concentrations in foods consumed in other parts of the world are sparse. Results from the analysis of samples from North and South Vietnam have been provided by Schecter and co-workers[21–23] and some data on samples from the Soviet Union have also been reported.[23]

An alternative to studies of separate food commodities (whether as individual samples or composites) is the use of duplicate diet or food basket samples. The analysis of the solid phase of food basket samples representing the Swedish diet has been reported by de Wit *et al.*[24] Because of the very low levels in such composites and the contribution of the laboratory blank, it was found to be impossible to obtain data of sufficient quality to calculate an average intake of PCDDs and PCDFs.

3. LEVELS IN SPECIFIC FOODS

3.1. Vegetables

Growing plants may be exposed to PCDDs and PCDFs via soil, groundwater, and by direct aerial deposition. Studies on uptake from soil, which have been reviewed,[25] are not entirely consistent but it is generally accepted that systemic uptake through the roots and translocation within plants is not detectable.

The outer surface of root crops can obviously become contaminated by soil contact; low concentrations of PCDDs and PCDFs have been found in root crops such as carrots and potatoes grown in contaminated soils but were largely removed by peeling.[26,27]

PCDD and PCDF contamination of foliage and fruits is also surfaceborne and may include contributions from direct deposition of airborne particulates and from absorption of vapor-phase contaminants from the air, including those that are attributable to evaporation from the soil.[28] Using apples and pears grown in a highly contaminated area, Müller *et al.*[29] have shown that peeling removes most of the PCDD and PCDF contamination although washing had less effect.

In studies of field-grown vegetables, measurable levels have generally been found only in areas where a specific PCDD and PCDF contamination problem exists. Concentrations of 2,3,7,8-TCDD of 100 ng/kg were detected in the peel of fruits grown on soils contaminated in the Seveso incident, but not in the flesh.[30] Recent investigations in the vicinity of a wire reclamation incinerator showed contamination of leaf vegetables at levels of 5 to 10 ng/kg TEQ and rather less of fruits.[31]

Apart from this localized contamination, PCDD and PCDF concentrations in fruits and vegetables are usually immeasurably small. In the survey of foodstuffs available in West Berlin reported by Beck *et al.*,[12] five vegetable samples were analyzed (including cauliflower, lettuce, cherries, and apples) and PCDDs and PCDFs were not found, subject to a detection limit of 0.01 ng/kg for each isomer. Similarly, PCDDs and PCDFs were not detected to any significant extent in vegetable samples from the Total Diet Survey schemes in the United Kingdom[16] and in Canada.[10]

A number of reports have also dealt with vegetable oils. Beck *et al.*,[12] Fürst *et al.*,[13] and Liem *et al.*[15] all report that the concentrations of PCDDs and PCDFs were below the limit of detection, apart from some low levels of the hepta- and octa-chlorinated congeners.

3.2. Animal Products

The main routes that can be envisaged for the exposure of terrestrial animals to PCDDs and PCDFs are ingestion of fodder (which will contain traces of PCDDs and PCDFs as a result of deposition from the atmosphere or transfer from soil), ingestion of contaminated soil itself, and contact with PCP-treated wood (which has in the past been used in the construction of agricultural buildings). As a result of selectivity in uptake and elimination, animal tissues usually contain only or mainly the 2,3,7,8-substituted congeners.

3.2.1. Milk

The detection of PCDDs and PCDFs in milk was first reported by Rappe *et al.*[32] who analyzed milk from cows from six locations in Switzerland and found levels in the range 0.03 to 0.27 ng TEQ/kg on a whole milk basis. The levels of 2,3,7,8-TCDD varied from < 0.01 ng/kg to 0.05 ng/kg. It is noteworthy that 2,3,4,7,8-PeCDF was found at levels in the range 0.07 to 0.43 ng/kg; the relatively high concentration of this congener in European samples has been noted frequently and, with an I-TEF value of 0.5, it is often the largest single contributor to the total TEQ.

Subsequently a number of reports have dealt with background levels of PCDDs and PCDFs in cow's milk from various European countries. In the United Kingdom a survey of commercial milk from different areas showed an overall mean concentration of 0.08 ng TEQ/kg whole milk, with a range of 0.05 to 0.13 ng TEQ/kg.[16] Results from individual farms in the United Kingdom selected as being remote from both major roads and urban/industrial areas gave a slightly lower mean of 0.05 ng TEQ/kg with a very narrow range of 0.04 to 0.06 ng TEQ/kg.[33]

The results reported by Beck *et al.*[14] for eight samples of milk from Germany had a mean of 1.53 ng TEQ/kg fat (equivalent to 0.06 ng TEQ/kg on a whole milk basis assuming a 4% fat content). Fürst *et al.*[13] found the mean level in ten samples to be 3.8 ng TEQ/kg fat (0.15 ng TEQ/kg whole milk). These samples were restricted to farms from a small area and also had slightly elevated PCB levels; some specific local contamination may be implicated. More recently, Fürst *et al.*[34] have reported on a more extensive sampling of milk, cream, butter, and cheese from commercial dairies in North Rhine–Westphalia showing levels in the range 0.76 to 2.62 ng TEQ/kg fat with a mean of 1.35 ng TEQ/kg.

In Holland the level in composite samples was 1.5 ng TEQ/kg fat.[15] The results on samples from nine commercial dairies in Switzerland have recently been reported.[35] Although 2,3,7,8-TCDD was not determined, the concentrations of other congeners were very similar to the other results from western Europe. By inserting an average concentration for this congener taken from the literature, Schmid and Schatter[35] arrived at an average level of 1.31 ng TEQ/kg fat with a relative standard deviation for the nine different sources of 19%.

Rappe *et al.*[36] have provided data on commercial milk from five locations in Sweden with results in the range 0.46 to 1.74 ng TEQ/kg milk fat (approximately 0.02 to 0.07 ng TEQ/kg based on whole milk). It is interesting that the upper end of this range, representing southern Sweden, is fairly similar to central European data but that levels appear to be slightly lower in northern Sweden.

In contrast, a considerably lower background (0.006 ng TEQ/kg) has been reported in milk from New Zealand where the population is sparse and where muncipal waste incineration is not practiced.[37]

Birmingham *et al.*[8] reported that the mean level in fresh whole milk from Ontario, Canada, was 0.11 ng TEQ/kg, and Ryan *et al.*[10,11] found between 0.008 and 0.27 ng TEQ/kg (mean 0.15) in partially defatted milk (about 1.5% fat) collected from six Canadian cities, reflecting in part migration of 2,3,7,8-TCDF and 2,3,7,8-TCDD from the paperboard milk cartons.

There are very little good data on milk from the United States. Although Stanley and Bauer[6] have reported the analysis of eight composite samples, most PCDDs and PCDFs were not detected and the detection limits were slightly higher than typical background concentrations found in Europe. LaFleur *et al.*[38] have reported a level of 0.002 ng/kg of 2,3,7,8-TCDD in milk used in a study of migration from milk cartons but did not analyze higher congeners. Glidden *et al.*[39] also analyzed for these congeners and reported that they could not be detected in milk that had not been packaged in paperboard cartons. Some information on levels in dairy products is discussed below, and this tends to indicate that levels in milk from the United States are in fact quite similar to levels in Europe.

A number of studies have demonstrated the localized influence of incinerators and other sources. In surveillance around incinerators in Holland,[40] levels of up to 13.5 ng TEQ/kg fat were found. The highest dioxin concentrations were usually found within about 2 km of the source. The Dutch authorities adopted an action level of 6 ng TEQ/kg fat, and milk from the area found to exceed this level was withdrawn from the public supply. Riss *et al.*[41] have investigated the contamination caused by a metal reclamation plant at Brixlegg in Austria and found levels in cow's milk in the range 13.5 to 37.0 ng TEQ/kg fat (using the German scheme of toxic equivalent values, which tends to give a somewhat lower total than the I-TEF scheme) while a control sample gave 3.6 ng TEQ/kg. In the United Kingdom,[16] milk from a number of farms in different areas close to urban or industrial centers had concentrations in the range 0.07 to 0.38 ng TEQ/kg milk, with a mean value of 0.12 ng TEQ/kg (the equivalent range on a fat basis is approximately 1.75 to 9.5 ng TEQ/kg and mean 3.0 ng TEQ/kg). In addition, two farms were found in one location producing milk with concentrations around 1.9 ng TEQ/kg milk (~ 47 ng TEQ/kg fat). Subsequently, a

Table 1

Selected Congener-Specific Data for PCDDs and PCDFs in Milk (pg/g fat)

	a	*b*	*c*	*d*	*e*	*f*	*g*	*h*	*i*	*j*
2,3,7,8-TCDD	0.20	0.40	<0.40	<0.10	—	—	<0.25	1.25	0.50	0.69
1,2,3,7,8-PeCDD	0.20	1.20	0.49	<0.20	0.46	1.23	<0.50	1.25	0.50	0.98
1,2,3,4,7,8-HxCDD	0.30	0.80	0.30	<0.20	0.21	0.48	0.75	1.75	1.25	1.00
1,2,3,4,7,8-HxCDD	1.10	4.00	1.50	0.30	0.49	1.13				
1,2,3,7,8,9-HxCDD	0.40	0.80	<0.30	<0.20	0.27	0.36	<0.25	<0.75	<0.50	1.01
1,2,3,4,6,7,8-HpCDD	<2.00	6.20	3.10	1.00	0.98	1.65	<1.25	2.25	<4.50	2.86
OCDD	<10.00	11.00	3.50	1.40	2.50	2.34	7.25	9.75	13.00	3.78
2,3,7,8-TCDF	0.70	4.10	0.60	0.40	0.24	0.28	<0.25	<0.50	<0.25	2.74
1,2,3,7,8-PeCDF	0.20	0.30	0.30	0.10	0.09	0.18	<0.25	<0.25	<0.25	0.39
2,3,4,7,8-PeCDF	1.40	2.70	1.20	0.20	1.13	4.78	0.75	4.50	0.75	0.38
1,2,3,4,7,8-HxCDF	0.90	1.70	0.84	<0.10	0.56	2.93	0.50	1.50	0.25	0.70
1,2,3,6,7,8-HxCDF	0.80	1.10	0.60	<0.10	0.51	2.43	0.25	1.00	0.25	0.68
1,2,3,7,8,9-HxCDF	—	—	<0.20	<0.10	0.03	0.08	<0.25	<0.50	<0.25	0.88
2,3,4,6,7,8-HxCDF	0.70	1.30	0.30	<0.10	0.61	2.35	0.25	1.00	<0.25	0.82
1,2,3,4,6,7,8-HpCDF	0.50	1.50	0.50	0.20	0.36	1.01	<0.50	<0.75	0.75	0.87
1,2,3,4,7,8,9-HpCDF	—	—	<0.20	<0.20	0.05	0.16	<0.50	<0.50	<0.25	1.25
OCDF	<1.00	1.20	<1.50	<0.30	0.18	0.22	<1.00	<1.75	<1.75	2.67

[a] Commercial milk from Germany, mean of 8 samples, data from Ref. 14.
[b] Milk from specific area of Germany, mean of 10 samples, data from Ref. 13.
[c] Commercial milk from Malmö, Sweden, data from Ref. 36.
[d] Commercial milk from Umeå, Sweden, data from Ref. 36.
[e] Commercial milk from Switzerland, mean of 9 samples, data from Ref. 35.
[f] Milk from an area of metal recycling in Switzerland, data from Ref. 35.
[g] Milk from rural farms in the United Kingdom, mean of 9 samples, data from Ref. 16.
[h] Milk from farms in the United Kingdom close to urban or industrial areas, mean of 9 samples, data from Ref. 16.
[i] Commercial milk from the United Kingdom collected in summer, mean of 7 pooled samples, data from Ref. 16.
[j] Commercial milk from the United States (California), contributions from 49 samples, data from Ref. 6.

third farm was found where milk from suckler cows contained an average of 3.4 ng TEQ/kg (56 ng TEQ/kg fat).[42] A "maximum tolerable concentration" of 0.7 ng TEQ/kg whole milk (17.5 ng TEQ/kg fat) was set and as a result milk from farms found to exceed this value was withdrawn from the public supply.

Some results were also obtained by Schmid and Schatter[35] for milk from sites near waste incineration, metal recycling, and other industrial facilities in Switzerland, where the levels were two- to fourfold greater than in commercial samples.

Selected examples of the congener-specific results underlying these total concentrations are given in Table 1.

3.2.2. Dairy Products

The concentrations of PCDDs and PCDFs in dairy products are expected to relate closely to those in milk when expressed on a fat basis. Thus, as mentioned above, the data of Fürst *et al.*[34] for milk, cream, butter, and cheese from commercial dairies showed virtually no difference in levels. In Holland, Liem *et al.*[15] found an average of 1.6 ng TEQ/kg fat in composite samples of Dutch butter and cheese, also very similar to the milk level, and in the Norwegian survey[17] the mean of three samples of butter was 1.2 ng TEQ/kg fat. However, both Beck *et al.*[12] and Fürst *et al.* in an earlier report[13] found levels in dairy products that were about 50% of those in milk, possibly reflecting importation of these products from areas with a lower background.

As mentioned above, the levels in a number of dairy products purchased from a New York supermarket have been reported and were in the range 0.8 to 1.5 ng TEQ/kg fat which is very similar to the European values.[7]

Levels of 0.28, 0.67, 2.1, and 4.7 ng TEQ/kg fat were found in dairy products from widely separated locations in the Soviet Union.[23]

3.2.3. Meat

Beck *et al.*[12] included a number of samples of meat in their survey and found concentrations in the range 1.8 to 2.4 ng TEQ/kg on a fat basis for beef, lamb, and chicken but a much lower concentration, 0.22 ng TEQ/kg fat, in pork. The results reported by Fürst *et al.*[13] are reasonably consistent; levels in beef, lamb, chicken, and in canned meat were in the range 2.1 to 3.45 ng TEQ/kg fat while PCDDs and PCDFs other than OCDD were not detected in pork. Fürst *et al.* found a considerably higher average concentration of 7.4 ng TEQ/kg fat in veal, presumably because of the high early-life input from milk.

Similar levels were found in beef, mutton, and chicken from Holland[15] where the range was 1.3 to 2.4 ng TEQ/kg fat. Again a considerably lower level of 0.43 ng TEQ/kg fat was found in pork. This study also included horse and goat fat in which higher concentrations of 14 and 4.2 ng TEQ/kg fat, respectively, were found. Liver from the same types of animals was also analyzed and the levels, on a fat basis, were between two- and tenfold higher than in the corresponding animal fat.

The data obtained in the United Kingdom were based on two composite samples each of carcass meat, offal, poultry, and meat products. Meat from different types of animals were represented within the composites in proportion to national consumption data. The levels of PCDDs and PCDFs found were mostly around 0.2 to 0.4 ng TEQ/kg (whole sample basis) although a single sample gave a rather high result of 1.1 ng TEQ/kg. The fat contents of the samples are not available, which hinders full comparison with the results cited above, but on a lipid-adjusted basis the concentrations would almost certainly be very similar.

LaFleur *et al.*[38] have examined various examples of canned corned beef hash, ground beef, beef hot dogs, and ground pork available in the United States for 2,3,7,8-TCDD and 2,3,7,8-TCDF. The former was found in all of the beef samples at levels between 0.03 and 0.35 ng/kg fat but was not detected in pork. 2,3,7,8-TCDF was also found in most beef samples at levels between 0.03 and 0.10 ng/kg fat, and in pork at around 0.05 ng/kg fat. These individual congener results are not very different from the corresponding levels found in the European studies and it is likely that levels of other congeners are also similar.

Schecter *et al.*[7] have more recently reported on a number of individual samples of retail meat products from the United States. The levels found in four different samples of beef and beef products spanned a wide range from 0.04 to 1.5 ng TEQ/kg on a whole sample basis. Cooked ham was found to contain 0.03 ng TEQ/kg while a single pork chop contained 0.26 ng TEQ/kg and a sample of lamb sirloin 0.41 ng TEQ/kg. A very low level of 0.04 ng TEQ/kg was found in a sample of chicken.

Levels in samples from widely separated locations in the Soviet Union were 0.3, 0.33, and 3.3 ng TEQ/kg fat in beef, 0.56 ng TEQ/kg fat in pork, and 0.18 ng TEQ/kg fat in sausage.

In various animal tissue samples from Vietnam, levels ranging from 0.26 to as high as 31.5 ng TEQ/kg have been measured. It should be noted that many of the results were reported on a whole sample basis without the fat content but that some of the samples were described as being fat from particular species, and also that the higher figure was exceptional and only one other sample was found to exceed 5 ng TEQ/kg.[21–23]

Selected examples of some of the congener-specific results underlying these total concentrations are given in Table 2.

3.2.4. Eggs

Although eggs from various types of birds have been examined in studies of levels in and effects on wildlife, the information relevant to human dietary exposure remains rather sparse. In the countries for which data are available, chicken's eggs dominate consumption and are the only type for which information on PCDD and PCDF levels is available; there appears to be no information on levels in duck's eggs (apart from a single example from Vietnam) even though these might be a significant part of the diet of a small group of consumers. Although the 2,3,7,8-substituted congeners predominate in eggs, other congeners can be observed to a greater extent than in other animal-derived foods.

Table 2

Selected Congener-Specific Data for PCDDs and PCDFs in Meat (pg/g fat)

	Beef			Pork			Chicken			Sheep		Veal
	a	*b*	*c*	*d*	*e*	*f*	*g*	*h*	*i*	*j*	*k*	*l*
2,3,7,8-TCDD	0.26	0.60	<0.50	0.33	0.03	<0.50	0.50	0.30	<0.50	0.01	<0.50	<0.50
1,2,3,7,8-PeCDD	2.78	0.80	1.70	2.29	0.12	<0.50	1.64	0.70	1.00	0.50	<0.50	3.10
1,2,3,4,7,8-HxCDD	2.07	0.60	1.90	1.66	0.21	<0.50	1.51	0.50	0.60	0.30	0.80	1.90
1,2,3,6,7,8-HxCDD		1.90	3.20		0.29	<0.50		2.80	1.80	1.50	3.00	5.30
1,2,3,7,8,9-HxCDD	2.08	0.60	2.00	1.72	0.06	<0.50	1.78	0.60	0.60	0.40	0.70	1.80
1,2,3,4,6,7,8-HpCDD	5.38	18.00	3.90	13.05	2.10	0.70	8.17	6.00	4.50	15.00	11.40	14.40
OCDD	9.63	25.00	5.40	76.18	19.00	8.20	31.03	52.00	16.50	68.00	19.30	22.30
2,3,7,8-TCDF	0.55	0.30	<0.30	0.35	0.11	<0.30	0.28	0.01	0.70	0.01	<0.30	0.30
1,2,3,7,8-PeCDF	0.60	0.01	<0.30	0.68	0.01	<0.30	0.42	2.10	2.60	0.60	<0.30	0.20
2,3,4,7,8-PeCDF	0.57	1.50	2.70	0.62	0.08	<0.30	0.26	1.50	1.30	0.90	1.70	7.40
1,2,3,4,7,8-HxCDF	0.67	0.80	0.70	1.58	0.15	<0.30	0.51	0.60	1.00	0.90	0.80	2.90
1,2,3,6,7,8-HxCDF	0.63	0.60	0.50	0.65	0.07	<0.30	0.45	0.40	0.50	1.20	0.60	2.40
1,2,3,7,8,9-HxCDF	0.82	—	—	0.84	—	—	0.59	—	—	—	—	—
2,3,4,6,7,8-HxCDF	0.75	1.30	0.90	0.77	0.05	<0.30	0.54	0.30	0.70	1.50	0.50	2.80
1,2,3,4,6,7,8-HpCDF	1.37	2.20	2.00	4.04	1.10	<0.30	4.86	0.80	0.50	8.10	1.00	1.70
1,2,3,4,7,8,9-HpCDF	1.72	—	—	2.99	—	—	1.34	—	—	—	—	—
OCDF	1.73	0.20	<0.30	2.51	0.41	<0.30	1.73	0.60	1.00	0.30	<0.30	1.40

[a]Beef from the United States (California), contributions from 50 samples, data from Ref. 6.

[b]Beef from Germany, data from Ref. 14.

[c]Beef from Germany, mean of 3 samples, data from Ref. 13.

[d]Pork from the United States (California), contributions from 51 samples, data from Ref. 6.

[e]Pork from Germany, data from Ref. 14.

[f]Pork from Germany, mean of 3 samples, data from Ref. 13.

[g]Chicken from the United States (California), contributions from 51 samples, data from Ref. 6.

[h]Chicken from Germany, data from Ref. 14.

[i]Chicken from Germany, mean of 2 samples, data from Ref. 13.

[j]Mutton from Germany, data from Ref. 14.

[k]Mutton from Germany, mean of 2 samples, data from Ref. 13.

[l]Veal from Germany, mean of 4 samples, data from Ref. 13.

Levels of 0.22 and 0.16 ng TEQ/kg whole egg were found in two egg composites from the U.K. Total Diet Survey,[16] and an average of 0.2 ng TEQ/kg whole egg was also found in Norway.[17] Beck *et al.*[12] have reported results, on a fat basis, for a single sample, of 1.61 ng TEQ/kg fat, and Liem *et al.*[15] found 2.0 ng TEQ/kg fat in two composites. Assuming an average of 10% lipid in eggs, these results from Europe are in very good agreement.

Apart from the work reported by Stanley and Bauer,[6] in which no PCDDs and PCDFs were detected, the only data available from the United States are from the work of Stephens *et al.*[43–45] Although these investigations were concerned primarily with the effects of localized soil contamination, results for control samples point to a background level for chickens fed on commercial formulations that is similar to that in Europe. Considerably higher concentrations were found in eggs from free-range birds with access to moderately contaminated soils, with levels as much as 100-fold higher than commercial eggs.

In the Canadian data, only higher chlorinated congeners were detected in eggs but an average concentration of 0.59 ng TEQ/kg was used for dietary intake calculations.[8]

Results from two examples of chicken's eggs from Vietnam revealed levels of 0.55 and 1.62 ng TEQ/kg whole egg while PCDDs and PCDFs were not detected in a single sample of duck's eggs.[22]

3.3. Fish

Although PCDDs and PCDFs are usually present in aquatic systems only at very low levels, bioaccumulation of the 2,3,7,8-substituted congeners can result in significant levels in fish. As with animals, the 2,3,7,8-substituted congeners dominate the congener pattern found in fish although this is not true of crustaceans and shellfish.[46] Since different species occupy quite different trophic positions, large differences in PCDD and PCDF concentrations are to be expected.

A number of studies have provided information on the levels in fish as consumed in western Europe. In the United Kingdom, eight retail samples, including plaice, mackerel, herring, cod, skate, and coley, yielded a mean of 0.74 ng/kg and a range of 0.15–1.84 TEQ/kg on a wet weight basis.[16,33] In Germany, Beck *et al.*[12] found between 30 and 43 ng TEQ/kg fat in samples of herring, cod, and redfish, and Fürst found a mean of 34 ng TEQ/kg fat for 15 samples of saltwater fish. Somewhat higher concentrations were found in Holland for the less fatty types of sea fish where the results for two composites were 48.1 and 49.2 ng TEQ/kg fat (the fat contents of both were 0.9% and 2,3,7,8-TCDD levels 16.2 and 16.5 ng/kg fat). However, lower concentrations, on a lipid-adjusted basis, were found for saltwater fish with a high fat content where concentrations of 6.0 and 7.6 ng TEQ/kg fat were found in two composites (the fat contents were 17.8 and 14.5% and 2,3,7,8-TCDD levels 0.6 and 0.7 ng/kg fat). Other European data, such as those of Biseth *et al.*[47] and Færden,[17] generally show a rather similar range.

A study of cod and herring from the seas around Sweden has been reported.[48]

Herring from the Baltic were found to have levels in the range 6.7 to 9.0 ng TEQ/kg wet weight while considerably lower levels of 1.8 to 3.4 ng TEQ/kg were found in fish from west coast to Sweden. De Wit *et al.*[49] have also provided results demonstrating the higher levels present in fish from the Baltic Sea.

In Europe there has been relatively little attention paid to freshwater fish, but in the results from Holland[15] levels of 1.7 and 3.1 ng TEQ/kg fat were found in two composites (the corresponding fat contents were 7.9 and 7.3% and the 2,3,7,8-TCDD concentration in both cases was 0.2 ng/kg fat). Fürst *et al.*[13] have also reported some results but did not analyze for 2,3,7,8-TCDD so that it is difficult to make comparisons based on TEQs.

Some studies have included separate analysis of fish oils. Fürst *et al.*[13] found an average of 4.4 ng TEQ/kg in salmon oil capsules (which is about one-fifth of the concentration reported in saltwater fish on a fat basis), but in cod liver oil found an average concentration very similar to that in fish when expressed on a fat basis. Liem has also reported results for fish oils and found concentrations between 0.24 and 4.4 ng TEQ/kg in different samples.

Takayama *et al.*[50] have provided data on coastal and marketing fishes from Japan. The means for these two groups were 0.87 and 0.33 ng TEQ/kg, respectively, on a wet weight basis.

A number of studies have dealt with contamination of fish in the North American Great Lakes and rivers in the Great Lakes Basin,[51–57] which are among the most severely contaminated in the United States. An extensive survey of 2,3,7,8-TCDD in fish from inland waters in the United States has also been conducted[58] and over 25% of all samples were found to be contaminated at or above the detection limit which varied between 0.5 and 2.0 ng/kg. Concentrations in excess of 5.0 ng/kg were found in 10% of samples and the highest level found was 85 ng/kg. Samples collected near sites of discharge from pulp and paper manufacture had a higher frequency of TCDD contamination than other samples.

In contrast to the extensive North American attention to fish caught in specific locations, there is little information to represent the levels in the general food supply apart from the results reported by Schecter *et al.*[7] who found 0.03 and 0.13 ng TEQ/kg in haddock fillets, 0.03 ng TEQ/kg in a cod fillet, and 0.24 ng TEQ/kg in a perch fillet. These levels are considerably lower than those typical in Europe but relate to a very restricted sampling base.

Selected examples of the congener-specific results underlying these total concentrations are given in Tables 3 and 4.

3.4. Food Processing and Cooking

Körner and Hagenmaier[59] analyzed several samples of smoked meat, sausage, and fish, charcoal-grilled meat and fish, and fat that had been used for deep frying. In cold smoked pork, non-2,3,7,8-substituted congeners were found in addition to the expected 2,3,7,8-substituted compounds and the concentrations of the latter were about

Table 3
Selected Congener-Specific Data for PCDDs and PCDFs in Fish (pg/g fat)

	a	*b*	*c*	*d*	*e*	*f*
2,3,7,8-TCDD	1.21	5.02	23.00	4.70	—	6.50
1,2,3,7,8-PeCDD	1.81	10.28	1.30	12.00	9.70	7.50
1,2,3,4,7,8-HxCDD	2.64	25.68	0.01	1.20	2.40	1.00
1,2,3,6,7,8-HxCDD			17.00	5.80	14.90	7.30
1,2,3,7,8,9-HxCDD	2.36	13.37	5.20	1.00	1.80	2.80
1,2,3,4,6,7,8-HpCDD	2.08	64.70	2.10	3.60	9.90	8.80
OCDD	13.82	510.10	83.00	19.00	19.30	10.50
2,3,7,8-TCDF	21.96	3.19	98.00	57.00	52.00	70.70
1,2,3,7,8-PeCDF	1.62	1.04	48.00	16.00	8.40	12.60
2,3,4,7,8-PeCDF	1.28	1.16	3.10	29.00	32.20	27.30
1,2,3,4,7,8-HxCDF		4.66	6.90	3.00	4.20	3.80
1,2,3,6,7,8-HxCDF	1.51	1.14	13.00	4.20	2.90	3.60
1,2,3,7,8,9-HxCDF	1.96	1.47	—	—	—	—
2,3,4,6,7,8-HxCDF	2.91	1.35	8.20	3.60	3.10	4.20
1,2,3,4,6,7,8-HpCDF	2.05	1.33	10.00	1.60	2.20	3.10
1,2,3,4,7,8,9-HpCDF	2.69	26.47	—	—	—	—
OCDF	5.71	2.48	2.10	1.40	0.70	3.00

[a]Saltwater fish from the United States, contributions from 48 samples, data from Ref. 6.
[b]Freshwater fish from the United States, contributions from 48 samples, data from Ref. 6.
[c]Cod from Germany, data from Ref. 14.
[d]Herring from Germany, data from Ref. 14.
[e]Freshwater fish from Germany, mean of 18, data from Ref. 13.
[f]Saltwater fish from Germany, mean of 15, data from Ref. 13.

twice those in raw pork samples. In the case of charcoal-grilled pork, increased levels of PCDDs and PCDFs were found only in a single sample that had been salted before cooking. Pan-frying, even after salting, had no effect on concentrations. Pork fat that was used for cooking salted fish was found to have increased concentrations of PCDDs and PCDFs and the authors claim that *de novo* formation of TCDF and PeCDF congeners was evident.

4. ESTIMATES OF DIETARY INTAKE

The most usual method of assessing average dietary intake is to multiply the average consumption of each type of food by the average concentrations found in corresponding food samples, and then add together the contributions from various components of the diet. Using this approach, Birmingham *et al.*[8] calculated a daily intake from food for Canadian adults of 92 pg TEQ, the main contributors being milk, dairy products, and beef (Table 5). Beck *et al.*[12] (Table 6) and Fürst *et al.*[13] (Table 7), both using the same food consumption statistics, estimated the average West German intake as 93.5 and 85 pg, respectively, expressed in German Federal Health Office

Table 4
Selected Congener-Specific Data for PCDDs and PCDFs in Fish (pg/g wet weight)

	a	*b*	*c*	*d*
2,3,7,8-TCDD	0.36	0.04	0.07	<0.13
1,2,3,7,8-PeCDD	0.19	0.07	0.12	<0.20
1,2,3,4,7,8-HxCDD	<0.03	<0.03	<0.03	} 0.12
1,2,3,6,7,8-HxCDD	0.81	<0.30	<0.03	
1,2,3,7,8,9-HxCDD	<0.03	<0.30	<0.03	<0.04
1,2,3,4,6,7,8-HpCDD	0.89	0.04	0.52	<0.32
OCDD	2.00	0.14	0.79	2.60
2,3,7,8-TCDF	4.00	0.35	0.62	1.20
1,2,3,7,8-PeCDF	0.85	0.14	0.27	0.16
2,3,4,7,8-PeCDF	0.50	0.30	0.38	<0.66
1,2,3,4,7,8-HxCDF	<0.02	<0.02	<0.02	<0.08
1,2,3,6,7,8-HxCDF	0.79	<0.02	<0.02	<0.07
1,2,3,7,8,9-HxCDF	<0.02	<0.02	<0.02	<0.04
2,3,4,6,7,8-HxCDF	0.86	<0.20	<0.02	<0.08
1,2,3,4,6,7,8-HpCDF	0.42	0.06	1.50	<0.09
1,2,3,4,7,8,9-HpCDF	<0.06	<0.06	0.04	<0.09
OCDF	0.07	0.03	0.88	<0.20

[a]Pilchard from Japan, data from Ref. 50.
[b]Natural red sea bream from Japan, data from Ref. 50.
[c]Cultured red sea bream from Japan, data from Ref. 50.
[d]Saltwater fish from the United Kingdom, mean of 8 individual samples, data from Ref. 16.

Table 5
Estimated Dietary Intake of PCDDs and PCDFs for Canadian Adults as Calculated by Birmingham *et al.*[8,a]

Food			Daily dioxin intake	
Group	Consumption (g whole wt/ person/day)	Dioxin concentration (pg TEQ/g whole wt)	pg TEQ/person	pg TEQ/ kg body wt
Beef	55.8	0.29	16.2	0.27
Pork	19	0.03	0.6	0.01
Poultry	20	0.39	7.8	0.13
Eggs	29	0.59	17.0	0.28
Milk products	444	0.11	48.8	0.81
Fruits	190	0.004	0.76	0.06
Vegetables	220	0.002	0.44	0.007
Wheat-based products	125	0.0007	0.09	0.001
Total			91.7	1.52

[a]Daily consumption data from Ref. 65; I-TEF values were used.

Table 6
Estimated Dietary Intake of PCDDs and PCDFs in Germany as Calculated by Beck *et al.*[12,a]

Food		
Group	Consumption (g/day)	Dioxin daily intake (pg TEQ)
Meat and meat products	38 (fat)	23.5
Milk and dairy products	33 (fat)	28.5
Eggs	3.9 (fat)	4.2
Fish and fish products	1.8 (fat)	33.3
Vegetable oil	26 (fat)	0.3
Vegetables	244	2.4
Fruits	130	1.3
Total		93.5

[a]Daily consumption data from Refs. 66–68; TEQ values calculated based on factors from the German Federal Health Office.[69]

toxic equivalents (FHO-TEQ). The FHO-TEQ values are typically around 50% of the I-TEQ values calculated from the same congener-specific results. Both studies concluded that the intake was derived about equally from milk, meat, and fish.

In the United Kingdom the daily intake has been estimated as 125 pg I-TEQ derived mainly from milk, meat, and oils and fats (Table 8). In Norway a provisional

Table 7
Estimated Dietary Intake of PCDDs and PCDFs in Germany as Calculated by Fürst *et al.*[13,a]

Food			
Type	Consumption (g/day) whole	Consumption (g/day) fat	Daily intake (pg TEQ)
Cow's milk	170	6.0	13.9
Cheese	33	5.2	5.1
Butter	14	12	7.9
Beef	46	10	16.9
Veal	2	0.1	0.3
Pork	39	14	5.6
Chicken	15	1	1.4
Canned meat	8	2	2.6
Lard	1.5	1.5	0.7
Salad oil	5	5	1
Margarine	18	14	2.8
Fish and fish products	15	1.8	27
Total			85

[a]Daily consumption data and TEF factors as for Table 6.

Table 8
Estimated Dietary Intake of PCDDs and PCDFs in the United Kingdom Taken from Ref. 16[a]

Food			
Group	Consumption (kg/day)	Dioxin concentration (ng TEQ/kg)	Daily dioxin intake (pg TEQ)
Miscellaneous cereals	0.105	0.05	5.3
Carcass meat	0.032	0.68	22
Offal	0.002	0.46	0.92
Poultry	0.017	0.33	5.6
Meat products	0.048	0.21	9.8
Fish	0.016	0.48	7.7
Fats and oils	0.03	0.65	19
Eggs	0.024	0.19	4.6
Milk products	0.055	0.21	12
Green vegetables	0.043	0.02	0.65
Potatoes	0.151	0.04	5.3
Other vegetables	0.069	0.09	6.2
Fresh fruit	0.055	0.05	2.8
Milk	0.303	0.08	23
Total			125

[a]Daily consumption data from Ref. 70; I-TEF values were used.

estimate of 625 pg Nordic-TEQ/week has been made[17] (corresponding to a daily intake of 89 pg) and in a Dutch study[60] using this procedure the estimated daily intake was 115 pg I-TEQ.

The estimates above are remarkably similar. For the Italian diet, Di Domenico[61] estimated a higher average daily intake of between 260 and 480 pg I-TEQ. However, the assessment included a significant intake from vegetables, based on the assumption that the ratio of plant concentration to soil concentration was 0.1 for below-ground vegetables and 0.01 for above-ground vegetables, as suggested by Pocchiari and co-workers on the basis of observations at Seveso,[62] and Di Domenico also assumed that the fatty fraction of bread, pasta, and other cereal products contained similar concentrations to cow's milk.

It is worth noting that assessments of any possible contribution from fruits and vegetables are problematic. Although most of the results show immeasurably low concentrations of the relevant congeners, it is customary to include either 50 to 100% of the corresponding detection limit in the summation leading to a total TEQ value. Because of this and the relatively high average consumption, contributions from fruits and vegetables that may be considerable overestimates are included in several assessments.

There are some difficulties associated with the use of overall food consumption statistics to assess the intake of consumers, especially for extreme consumers of specific foods since high intakes of one commodity by an individual may be balanced by below-average consumption of other foods.

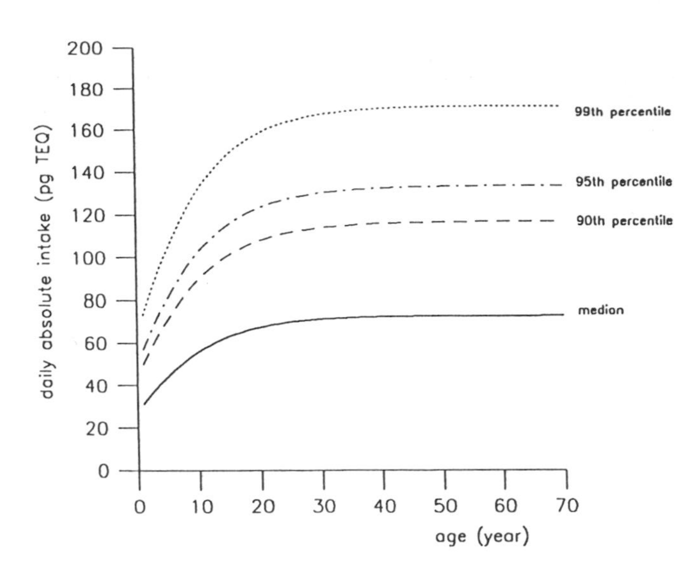

Figure 1. Median and higher percentile intakes (pg I-TEQ per day) of PCDDs and PCDFs from food in the Netherlands as a function of age. (Reprinted from Theelen *et al.*[63] with permission from Pergamon Press.)

An alternative method of assessment is to calculate the dietary intake for a large number of individuals, based on individual consumption statistics and average PCDD and PCDF levels. This approach was used to assess intakes in Holland[63] using a data base of food consumption information covering 5898 individuals on two consecutive days, and allowed the median and extreme intakes to be related to the consumers' age as shown in Fig. 1. This shows a median intake in adulthood of about 70 pg TEQ/day and a 99th percentile of about 170 pg TEQ/day. The 1st percentile curve (not shown) reached about 30 pg/day. It should be noted that the combination of this approach to calculation with average levels in foods derived from a large-scale and wide-area sampling scheme gave a considerably lower estimate of average intake than that previously made in Holland.

The assessments discussed above have used "national" data sets to calculate dietary intakes. Although the data on the global distribution of PCDDs and PCDFs in foods are far from complete, the information that is available on levels in the foods available to normal consumers in industrialized countries from several continents suggests a remarkable uniformity, despite the influences of differences in industrial history, regulatory aspects, and population density and there could be some justification for using single consensus concentrations. In particular the confounding influences of small sample statistics and analytical variability that affect some matrices in some data sets could be offset if all of the data were taken into account.

For animal products such as milk and dairy products, eggs, mutton or lamb, beef, and poultry, a remarkably consistent general level of 1 to 2 ng TEQ/kg emerges when the results are expressed on a lipid basis. A considerably lower concentration is associated with pork and a higher level of around 5 ng TEQ/kg fat seems to be a reasonable estimate for liver. It seems probable that in many areas, including much of

Europe and North America, differences in dietary intake of PCDDs and PCDFs from animal products will result more from differences in consumption patterns than from differences in the concentrations of PCDDs and PCDFs in the foodstuffs. However, levels in fish are clearly more variable, being markedly species and location dependent. The quantities and types of fish consumed also vary between different countries and ethnic and occupational groups so that very large variations in dietary intake attributable to fish might be expected.

This leads to a final matter of some importance, namely the extent to which contamination of foods produced in the vicinity of localized sources leads to above-average intake by individuals. As discussed above, considerably elevated levels in foods (particularly in cow's milk) have been measured in a number of locations. There are some examples of a direct influence. Raised blood serum levels of PCDDs and PCDFs have been measured in individuals from a Swedish fishing community with a very high intake of fish from the Baltic.[64] Riss *et al.*[41] analyzed blood from a farmer in a contaminated location (see Section 3.2.1) who consumed milk produced on his farm, and found levels well above the expected background.

In industrialized countries, however, these are both likely to be atypical circumstances. The patterns of distribution and marketing should result in most individuals receiving food from diverse sources; occasional consumption of foods representing extreme examples of contamination has not been found to present any acute risk to health and will have a minor effect on lifetime intake.

REFERENCES

1. F. W. Kutz, D. G. Barnes, D. P. Bottimore, H. Greim, and E. W. Bretthauer, The international toxicity equivalency factor (I-TEF) method of risk assessment of complex mixtures of dioxins and related compounds, *Chemosphere* **20,** 751–757 (1990).
2. M. McLachlan, Exposure toxicity equivalents (ETEs): A plea for more environmental chemistry in dioxin risk assessment, *Chemosphere* **27,** 484–490 (1993).
3. J. J. Ryan, R. Lizotte, L. G. Panopio, and B. P.-Y. Lau, The effects of strong alkali on the determination of polychlorinated dibenzofurans (PCDFs) and polychlorinated dibenzo-*p*-dioxins (PCDDs), *Chemosphere* **18,** 149–154 (1989).
4. J. J. Ryan, R. Lizotte, T. Sakuma, and B. Mori, Chlorinated dibenzo-p-dioxins, chlorinated dibenzofurans, and pentachlorophenol in Canadian chicken and pork samples, *J. Agric. Food Chem.* **33,** 1021–1026 (1985).
5. D. Firestone, R. A. Niemann, L. F. Schneider, J. R. Grindley, and D. E. Brown, Dioxin residues in fish and other foods, in: *Chlorinated Dioxins and Dibenzofurans in Perspective* (C. Rappe, G. Choudhary, and L. H. Keith, eds.), pp. 355–365, Lewis Publishers, Chelsea, MI (1986).
6. J. S. Stanley and K. M. Bauer, *Chlorinated Dibenzo-p-dioxin and Dibenzofuran Residue Levels in Food,* Report No. 8922-A, Midwest Research Institute, Missouri (1989).
7. A. Schecter, J. Startin, C. Wright, M. Kelly, O. Päpke, A. Lis, M. Ball, and J. Olson, Congener-specific levels of dioxins and dibenzofurans in U.S. food and estimated daily dioxin toxic equivalent intake for infants, children, and adults. *Environ. Health Perspec.* (in press).

8. B. Birmingham, B. Thorpe, R. Frank, R. Clement, H. Tosine, G. Fleming, J. Ashman, J. Wheeler, B. D. Ripley, and J. J. Ryan, Dietary intake of PCDD and PCDF from food in Ontario, Canada, *Chemosphere* **19,** 507–512 (1989).
9. K. Davies, Concentrations and dietary intake of selected organochlorines, including PCBs, PCDDs and PCDFs, in fresh food composites grown in Ontario, Canada, *Chemosphere* **17,** 263–276 (1988).
10. J. J. Ryan, L. G. Panopio, D. A. Lewis, D. F. Weber, and H. B. S. Conacher, PCDDs/PCDFs in 22 categories of food collected from six Canadian cities between 1985 and 1988, Paper presented at the 10th International Conference on Organohalogen Compounds, Bayreuth (1990).
11. J. J. Ryan, L. G. Panopio, D. A. Lewis, and D. F. Weber, Polychlorinated dibenzo-*p*-dioxins and polychlorinated dibenzofurans in cows' milk packaged in plastic-coated bleached paperboard containers, *J. Agric. Food Chem.* **39,** 218–223 (1991).
12. H. Beck, K. Eckart, W. Mathar, and R. Wittkowski, PCDD and PCDF body burden from food intake in the Federal Republic of Germany, *Chemosphere* **18,** 417–424 (1989).
13. P. Fürst, C. Fürst, and W. Groedel, Levels of PCDDs and PCDFs in foodstuffs from the Federal Republic of Germany, *Chemosphere* **20,** 787–792 (1990).
14. H. Beck, K. Eckart, M. Kellert, W. Mathar, C.-S. Rühl, and R. Wittkowski, Levels of PCDFs and PCDDs in samples of human origin and food in the Federal Republic of Germany, *Chemosphere* **16,** 1977–1982 (1987).
15. A. K. D. Liem, A. P. J. M. de Jong, R. M. C. Theelen, J. H. van Wijnen, P. C. Beijen, H. A. van der Schee, H. A. M. G. Vaessen, G. Kleter, and J. A. van Zorge, Occurrence of dioxins and related compounds in Dutch foodstuffs—Part 1: Sampling strategy and analytical results, Paper presented at the 11th International Symposium on Chlorinated Dioxins and Related Compounds, Research Triangle Park, NC, September 23–27 (1991).
16. Ministry of Agriculture Fisheries and Food, *Dioxins in food,* Food Surveillance Paper No. 31, HMSO, London (1992).
17. K. Færden, *Dioksiner i næringsmidler,* SNT-report 4, 1991, Norwegian Food Control Authority (Statens næringsmiddeltilsyn), Postboks 8187 Dep., 0034 Oslo, Norway (1991).
18. M. Ono, Y. Kashima, T. Wakimoto, and R. Tatsukama, Daily intake of PCDDs and PCDFs by Japanese through food, *Chemosphere* **16,** 1823–1828 (1987).
19. Y. Takizawa and H. Muto, PCDDs and PCDFs carried to the human body from the diet, *Chemosphere* **16,** 1971–1976 (1987).
20. J. Ogaki, K. Takayama, H. Miyata, and T. Kashimoto, Levels of PCDDs and PCDFs in human tissues and various foodstuffs in Japan, *Chemosphere* **16,** 2047–2056 (1987).
21. A. Schecter, R. Kooke, P. Serne, K. Olie, Do 'Quang Huy, Nguyen Hue, and J. Constable, Chlorinated dioxin and dibenzofuran levels in food samples collected between 1985–87 in the north and south of Vietnam, *Chemosphere* **18,** 627–634 (1989).
22. K. Olie, A. Schecter, J. Constable, R. M. Kooke, P. Serne, P. C. Slot, and P. de Vries, Chlorinated dioxin and dibenzofuran levels in food and wildlife samples in the north and south of Vietnam, *Chemosphere* **19,** 493–496 (1989).
23. A. Schecter, P. Fürst, C. Fürst, W. Groebel, J. D. Constable, S. Kolesnikov, A. Biem, A. Boldonov, E. Trubitsun, B. Vaslov, Hoang Dinh Cau, Le Cau Dai, and Hoang Tri Quynh, Levels of chlorinated dioxins, dibenzofurans and other chlorinated xenobiotics in food from the Soviet Union and the south of Vietnam, *Chemosphere* **20,** 799–806 (1990).
24. C. de Wit, B. Jansson, M. Strandell, P. Jonsson, P.-A. Bergqvist, S. Bergek, L.-O. Kjeller, C. Rappe, M. Olsson, and S. Slorach, Results from the first year of the Swedish Dioxin Survey, *Chemosphere* **20,** 1473–1480 (1990).
25. G. A. Kew, J. L. Schaum, P. White, and T. T. Evans, Review of plant uptake of 2,3,7,8-

TCDD from soil and potential influences of bioavailability, *Chemosphere* **18,** 1313–1318 (1989).
26. S. Facchetti, A. Balasso, C. Fichtner, G. Frare, A. Leoni, C. Mauri, and M. Vasconi, Studies on the absorption of TCDD by plant species, in: *Chlorinated Dioxins and Dibenzofurans in Perspective* (C. Rappe, G. Choudhary, and L. H. Keith, eds.), pp. 225–235, Lewis Publishers, Chelsea, MI (1986).
27. A. Hülster and H. Marschener, Transfer of PCDD/PCDF from contaminated soils into two vegetable crops, Paper presented at the 11th International symposium on Chlorinated Dioxins and Related Compounds, Research Triangle Park, NC, September 23–27 (1991).
28. A. Reischl, M. Reissinger, H. Thoma, and O. Hutzinger, Uptake and accumulation of PCDD/F in terrestial plants: Basic considerations, *Chemosphere* **19,** 467–474 (1989).
29. J. F. Müller, A. Hülster, P. Päpke, M. Ball, and H. Marschner, Transfer pathways of PCDD/PCDF to fruits, *Chemosphere* **27,** 195–201 (1993).
30. H.-K. Wipf, E. Homberger, N. Neuner, U. B. Ranalder, W. Vetter, and J. P. Vuilleumier, TCDD levels in soil and plant samples from the seveso area, in: *Chlorinated Dioxins and Related Compounds—Impact on the Environment* (O. Hutzinger, R. W. Frei, E. Merian, and F. Pocchiari, eds.), p. 115, Pergamon Press, Elmsford, NY (1982).
31. B. Prinz, G. H. M. Krause, and L. Radermacher, Criteria for the evaluation of dioxins in vegetable plants and soils, Paper presented at the 10th International Conference on Organohalogen Compounds, Bayreuth (1990).
32. C. Rappe, M. Nygren, G. Lindström, H. R. Buser, O. Blaser, and C. Wüthrich, Polychlorinated dibenzo-*p*-dioxins and other chlorinated contaminants in cow milk from various locations in Switzerland, *Environ. Sci. Technol.* **21,** 964–970 (1987).
33. J. R. Startin, M. Rose, C. Wright, I. Parker, and J. Gilbert, Surveillance of British foods for PCDDs and PCDFs, *Chemosphere* **20,** 793–798 (1990).
34. P. Fürst, C. Fürst, and K. Wilmers, Survey of dairy products for PCDDs, PCDFs, PCBs and HCB, *Chemosphere* **25,** 1039–1048 (1992).
35. P. Schmid and C. Schatter, Polychlorinated dibenzo-*p*-dioxins and polychlorinated dibenzofurans in cow's milk from Switzerland, *Chemosphere* **24,** 1013–1030 (1992).
36. C. Rappe, G. Lindström, B. Glas, and K. Lundström, Levels of PCDDs and PCDFs in milk cartons and in commercial milk, *Chemosphere* **20,** 1649–1656 (1990).
37. S. J. Buckland, D. J. Hannah, J. A. Taucher, and R. J. Weston, The migration of polychlorinated dibenzo-p-dioxins and polychlorinated dibenzofurans into milks and cream from bleached paperboard packaging, Paper presented at the 10th International Conference on Organohalogen Compounds, Bayreuth (1990).
38. L. LaFleur, T. Bousquet, K. Ramage, B. Brunck, T. Davis, W. Luksemburg, and B. Peterson, Analysis of TCDD and TCDF on the ppq-level in milk and food sources, *Chemosphere* **20,** 1657–1662 (1990).
39. R. M. Glidden, P. J. Brown, R. A. Sittig, C. N. Syvertson, and P. V. Smith, Determination of 2,3,7,8-tetrachlorodibenzo-*p*-dioxin and 2,3,7,8-tetrachlorodibenzofuran in cow's milk, *Chemosphere* **20,** 1619–1624 (1990).
40. A. K. D. Liem, R. Hoogerbrugge, P. R. Koostra, E. G. van der Velde, and A. P. J. M. de Jong, Occurrence of dioxins in cow's milk in the vicinity of municipal waste incinerators and a metal reclamation plant in the Netherlands, *Chemosphere* **23,** 1675–1684 (1991).
41. A. Riss, H. Hagenmaier, U. Weberruss, C. Schlatter, and R. Wacker, Comparison of PCDD/PCDF levels in soil, grass, cow's milk, human blood and spruce needles in an area of PCDD/PCFF contamination through emissions from a metal reclamation plant, *Chemosphere* **21,** 1451–1456 (1990).
42. Ministry of Agriculture, Fisheries and Food, Studies on dioxins in Derbyshire carried out by

the Ministry of Agriculture, Fisheries and Food: Full congener-specific results, London, UK (20 August 1992).
43. R. Chang, D. Hayward, L. Goldman, M. Harnly, J. Flattery, and R. Stephens, Foraging farm animals as biomonitors for dioxin contamination, *Chemosphere* **19,** 481–486 (1989).
44. R. D. Stephens, M. Harnly, D. G. Hayward, R. R. Chang, J. Flattery, M. X. Petreas, and L. Goldman, Bioaccumulation of dioxins in food animals. II: Controlled exposure studies, *Chemosphere* **20,** 1091–1096 (1990).
45. M. X. Petreas, L. R. Goldman, D. G. Hayward, R. R. Chang, J. J. Flattery, T. Wiesmüller, and R. D. Stephens, Biotransfer and bioaccumulation of PCDD/PCDFs from soil: Controlled exposure studies of chickens, *Chemosphere* **23,** 1731–1741 (1991).
46. M. Oehme, S. Mano, E. M. Brevik, and J. Knutzen, Determination of polychlorinated dibenzofuran (PCDF) and dibenzo-*p*-dioxin (PCDD) levels and isomer patterns in fish, crustacea, mussel and sediment samples from a fjord region polluted by Mg-production, *Fresenius Z. Anal. Chem.* **335,** 987–997 (1989).
47. A. Biseth, M. Oehme, and K. Færden, Levels of polychlorinated dibenzo-p-dioxins and polychlorinated dibenzofurans in selected Norwegian foods, Paper presented at the 10th International Conference on Organohalogen Compounds, Bayreuth (1990).
48. P.-A. Bergqvist, S. Bergek, H. Hallbäck, C. Rappe, and S. A. Slorach, Dioxins in cod and herring from the seas around Sweden, *Chemosphere* **19,** 513–516 (1989).
49. C. de Wit, B. Jansson, M. Strandell, M. Ohlsson, S. Bergek, M. Boström, P.-A. Bergqvist, C. Rappe, and Ö. Andersson, Polychlorinated dibenzo-p-dioxin and dibenzofuran levels and patterns in fish samples analyzed within the Swedish Dioxin Survey, Paper presented at the 10th International Conference on Organohalogen Compounds, Bayreuth (1990).
50. K. Takayama, H. Miyata, M. Mimura, and T. Kashimoto, PCDDs, PCDFs and coplanar PCBs in coastal and marketing fishes in Japan, *Jpn. J. Toxicol. Environ. Health (Eisei Kaqaku)* **37,** 125–131 (1991).
51. R. L. Harless, E. O. Oswald, R. G. Lewis, A. E. Dupuy, D. D. McDaniel, and H. Tai, *Chemosphere* **11,** 193–198 (1982).
52. D. L. Stalling, L. M. Smith, J. D. Petty, J. W. Hogan, J. L. Johnson, C. Rappe, and H. R. Buser, Residues of polychlorinated dibenzo-*p*-dioxins and dibenzofurans in Laurentian Great Lakes, in: *Human and Environmental Risks of Chlorinated Dioxins and Related Compounds* (R. E. Tucker, A. L. Young, and A. P. Grsy, eds.), pp. 221–240, Plenum Press, New York (1983).
53. P. O'Keefe, C. Meyer, D. Hilker, K. Aldous, B. Jelus-Tyror, K. Dillon, R. Donnelly, E. Horn, and R. Sloan, Analysis of 2,3,7,8-tetrachlorodibenzo-*p*-dioxin in Great Lakes fish, *Chemosphere* **12,** 325–332 (1983).
54. J. J. Ryan, P.-Y. Lau, J. C. Piulon, D. Lewis, H. A. McLeod, and A. Gervals, Incidence and levels of 2,3,7,8-tetrachloro-*p*-dibenzodioxin in Lake Ontario commercial fish, *Environ. Sci. Technol.* **18,** 719–721 (1984).
55. N. V. Fehringer, S. M. Walthers, R. J. Kozara, and L. F. Schneider, Survey of 2,3,7,8-tetrachlorodibenzo-*p*-dioxin in fish from the Great Lakes and selected Michigan rivers, *J. Agric. Food Chem.* **33,** 626–630 (1985).
56. D. De Vault, W. Dunn, P.-A. Bergqvist, K. Wiberg, and C. Rappe, Polychlorinated dibenzofurans and polychlorinated dibenzo-*p*-dioxins in Great lakes fish: A baseline and interlake comparison, *Environ. Toxicol. Chem.* **8,** 1013–1022 (1989).
57. A. J. Niimi and B. G. Oliver, Assessment of relative toxicity of chlorinated dibenzo-p-dioxins, dibenzofurans, and biphenyls in Lake Ontario salmonids to mammalian systems using toxic equivalent factors (TEF), *Chemosphere* **18,** 1413–1423 (1989).
58. D. W. Kuehl, B. C. Butterworth, A. McBride, S. Kroner, and D. Bahnick, Contamination

of fish by 2,3,7,8-tetrachlorodibenzo-*p*-dioxin: A survey of fish from major watersheds in the United States, *Chemosphere* **18,** 1997–2014 (1989).
59. W. Körner and H. Hagenmaier, PCDD/PCDF formation in smoked, fried and broiled meat and fish, Paper presented at the 10th International Conference on Organohalogen Compounds, Bayreuth (1990).
60. R. M. C. Theelen, Modeling of human exposure to TCDD and i-TEQ in the Netherlands: Background and occupational, in: *Banbury Report 35: Biological Basis for Risk Assessment of Dioxins and Related Compounds* (M. A. Gallo, R. J. Scheuplein, and K. A. van der Heijden, eds.), pp. 277–290, Cold Spring Harbor Laboratory Press, Cold Spring Harbor, NY (1991).
61. A. Di Domenico, Guidelines for the definition on environmental action alert thresholds for PCDDs and PCDFs, *Regul. Toxicol. Pharmacol.* **11,** 8–23 (1990).
62. F. Pocchiari, F. Caffabeni, G. Della Porta, U. Fortunati, V. Silano, and G. Zapponi, Assessment of exposure to 2,3,7,8-tetrachlorodibenzo-p-dioxin (TCDD) in the Seveso area, *Chemosphere* **15,** 1851–1865 (1986).
63. R. M. C. Theelen, A. K. D. Liem, W. Slob, and J. H. van Wijnen, Intake of 2,3,7,8 chlorine substituted dioxins, furans, and planar PCBs from food in the Netherlands: Median and distribution, *Chemosphere* **27,** 1625–1635 (1993).
64. B.-G. Svensson, A. Nilsson, M. Hansson, C. Rappe, B. Akesson, and S. Skerfving, Fish consumption and human exposure to dioxins and dibenzofurans, Paper presented at the 10th International Conference on Organohalogen Compounds, Bayreuth (1990).
65. Nutrition Canada, Food Consumption Patterns Report, Health and Welfare Canada, Health Protection Branch (1977).
66. Ernährungsbericht 1984, Deutsche Gesellschaft für Ernährung e.V., Frankfurt (1984).
67. Material zum Ernährungsbericht 1984, Deutsche Gesellschaft für Ernährung e.V., Frankfurt (1984).
68. Fischwirtschaft—Daten und Fakten 1987, Fischwirtschaftliches Marketing Institut, Schriftenreihe (1987).
69. Umweltbundesamt and Bundesgesundheitsamt, Sachstand Dioxine, Stand November 1984, Erich Schmidt Verlag, Berlin (1985).
70. Ministry of Agriculture, Fisheries and Food, Household food consumption and expenditure 1987, HMSO (1989).

Chapter 5

Toxicology of Dioxins and Related Chemicals

Michael J. DeVito and Linda S. Birnbaum

1. INTRODUCTION

Chlorinated dibenzo-*p*-dioxins are environmental contaminants present in a variety of environmental media. This class of compounds has caused great concern in the general public as well as generated intense interest in the scientific community. The public concerns stem from the characterization of dioxin as one of the most potent "man-made" toxicants ever studied. Indeed, these compounds are extremely potent at producing a variety of toxic effects in experimental animals. The most potent of these chemicals is 2,3,7,8-tetrachlorodibenzo-*p*-dioxin (2,3,7,8-TCDD). 2,3,7,8-TCDD and its congeners produce a wide spectrum of toxic effects including immunotoxicity, teratogenicity, carcinogenicity, and lethality. The interest in dioxins and its congeners in the scientific community arises because many of their toxic effects are associated with alterations in growth and development of a variety of different cell types present in different organs. Thus, dioxins can be used as tools to study mechanisms of growth and development. Recent laboratory studies have provided new insights into the mechanisms involved in the toxicity of dioxins. These findings, in conjunction with epidemiology studies, indicate that at least some portions of the population are exposed to concentrations of dioxins and related compounds which may be causing subtle effects.

This manuscript has been reviewed in accordance with the policy of the Health Effects Research Laboratory, U.S. Environmental Protection Agency, and approved for publication. Approval does not signify that the contents necessarily reflect the view and policies of the Agency, nor does mention of trade names or commercial products constitute endorsement or recommendation for use.

Michael J. DeVito • Center for Environmental Medicine and Lung Biology, University of North Carolina at Chapel Hill, Chapel Hill, North Carolina 27514. Linda S. Birnbaum • Environmental Toxicology Division, Health Effects Research Laboratory, United States Environmental Protection Agency, Research Triangle Park, North Carolina 27711.

Dioxins and Health, edited by Arnold Schecter. Plenum Press, New York, 1994.

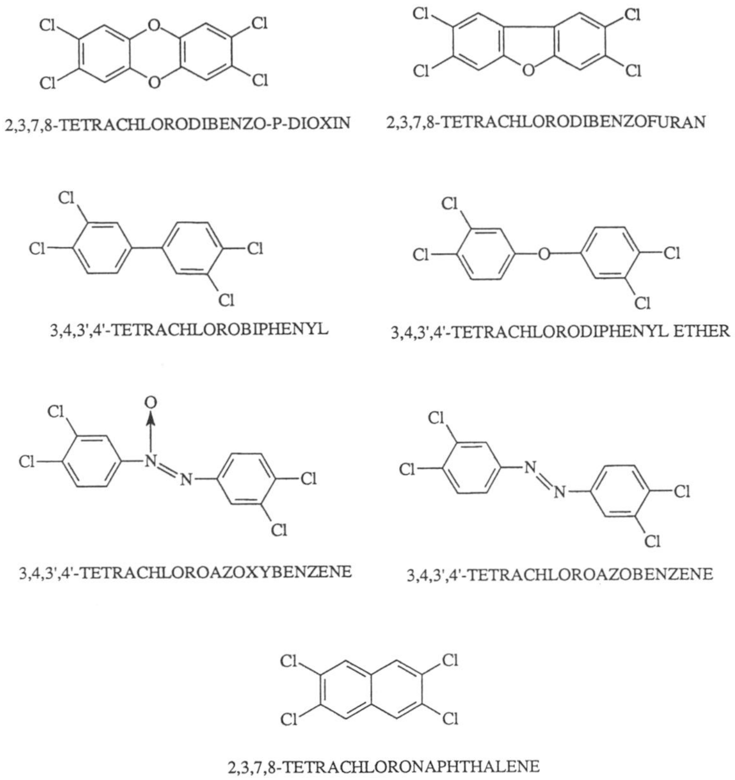

2,3,7,8-TETRACHLORODIBENZO-P-DIOXIN

2,3,7,8-TETRACHLORODIBENZOFURAN

3,4,3',4'-TETRACHLOROBIPHENYL

3,4,3',4'-TETRACHLORODIPHENYL ETHER

3,4,3',4'-TETRACHLOROAZOXYBENZENE

3,4,3',4'-TETRACHLOROAZOBENZENE

2,3,7,8-TETRACHLORONAPHTHALENE

Figure 1. 2,3,7,8-Tetrachlorodibenzo-*p*-dioxin and its structural analogues.

Chlorinated dioxins are part of a family of chemicals designated *p*oly*h*alogenated *a*romatic *h*ydrocarbons (PHAHs). The PHAHs are structurally similar to dioxins and in some instances produce similar toxicities. Examples of this class of chemicals are chlorinated and brominated dioxins, dibenzofurans, biphenyls, azo- and azoxybenzenes, diphenyl ethers, and naphthalenes (Fig. 1). The structure of these chemicals allows for multiple halogenation sites and the position and number of chlorines and bromines can be varied. The end result of these combinations is that there are 75 chlorinated dioxins, 135 chlorinated dibenzofurans, 209 chlorinated biphenyls, 209 chlorinated azobenzenes, 209 chlorinated azoxybenzenes, 209 chlorinated diphenyl ethers, and 75 chlorinated naphthalenes. An equal number of brominated compounds are possible as well as over 5000 different combinations of mixed brominated and chlorinated compounds. Fortunately, the majority of these compounds do not induce dioxinlike toxicities. However, there are a significant number of these compounds with dioxinlike activity. In the following discussion, the term *dioxins* will refer to any of the above chemicals that produce the same toxicities through the same mechanisms as 2,3,7,8-TCDD.

2. TOXICITY

The toxic effects of a chemical are often categorized by those effects that occur following acute administration versus those that occur following chronic administration. For many chemicals the acute effects are often quite different than those following chronic treatment. For example, an acute high dose of volatile organic hydrocarbons will produce lethality as a result of its narcotic effects while chronic exposure to lower doses will produce a completely different lethal pathology. For dioxins, categorizing the effects as either acute or chronic is problematic. In animals, the toxic effects of dioxins appear to be independent of the route (i.e., oral or intraperitoneal administration or dietary exposure) and rate of exposure (a single acute exposure versus repeated daily exposures). Instead, the toxicity of these chemicals is dependent on body burdens. Many dioxins have long half lives in animals. 2,3,7,8-TCDD has a half-life of approximately 10–15 days in mice[1,2] and 12–31 days in rats.[3,4] In humans, the half-life for 2,3,7,8-TCDD has been estimated to be 5–10 years.[5–7] While there are many dioxinlike compounds with shorter half lives, these chemicals still have estimated half lives on the order of months in humans. Because of the long half-life, following a single exposure animals as well as humans will be exposed to dioxins, especially 2,3,7,8-TCDD, for relatively long periods of time. Furthermore, much lower daily doses are required to produce the equivalent toxic effects in chronic exposures relative to the doses needed in acute exposures. Thus, in the following sections, the discussion of the effects of dioxins focuses on those effects that occur in both acute and chronic exposures provided sufficient body burdens of dioxins are achieved.

2.1. The Wasting Syndrome

While there are species differences in the toxic effects of dioxins, all animals studied have several toxic responses in common. Dioxin congeners are lethal when administered at a sufficient dose. Unlike most acute toxicants in which animals die within hours to days, dioxin-induced lethality takes weeks to manifest. The time to death is characteristic of the species examined. The exact cause of death is not known but is preceded by severe weight loss that is usually obvious within a few days of exposure. This wasting syndrome is accompanied by dramatic reduction of muscle mass and adipose tissue.[8,9] Sublethal doses of dioxins can also produce dose-dependent decreases in body weight.

Although the exact mechanism of the wasting syndrome is undetermined, there is a growing body of evidence indicating that dioxins alter the "set point" for body weight (reviewed in Ref. 9). Body weight and food intake are regulated by the hypothalamus. At a given age, the animal tries to maintain a constant body weight by adjusting its food intake. Dioxins decrease this "set weight" producing hypophagia or decreased food intake. This model was proposed by Peterson and co-workers in the mid-1980s based on several lines of evidence. 2,3,7,8-TCDD decreases food intake, accounting for almost all of the weight loss in rats, mice, and guinea pigs. Gastrointestinal absorption is not involved because daily fecal energy loss (kcal) as a percentage of daily feed energy intake is similar in control and 2,3,7,8-TCDD-treated rats. Body weight loss follows the same time course in both 2,3,7,8-TCDD-treated rats and pair-fed controls. These studies indicate that the wasting syndrome is related to decreased food intake which is mediated by effects on the central nervous system, in particular the hypothalamus.

There are no reported cases in humans of a wasting syndrome leading to death following acute or chronic exposure to dioxins. In the Yusho patients there are reports of appetite suppression accompanied with nausea and vomiting.[10] The lack of a wasting syndrome in humans in no way demonstrates that humans are a species insensitive to the toxic effects of dioxins. Instead, a more plausible explanation is that human exposure has not approached acutely lethal levels.

2.2. Thymic Atrophy

The thymus is a glandular organ mostly composed of lymphoid tissue and functions in cell-mediated immunity. The thymus is the site where T cells mature. Non-lethal doses of dioxins cause thymic atrophy in all experimental animals studied. Thymic atrophy is characterized by lymphocyte depletion in the thymic cortex.[11] In adult animals treated with dioxins, thymic atrophy does not correlate with alterations in immune function.[12,13] In mice, removal of the thymus does not alter the 2,3,7,8-TCDD-induced suppression of the immune system.[12] The thymus does not have a significant role in adult immunity.[14] In contrast, the thymus is critical for the developing immune system and, in fact, neonatal thymectomy produces significant reductions in the number and function of T lymphocytes.[14] Rodents exposed to dioxins either perinatally or neonatally are more sensitive to dioxin-induced thymic atrophy than adults. Developing animals exposed to dioxins at doses that produce thymic atrophy are immune compromised.[15,16] Thus, in adults, thymic atrophy occurs but is not associated with altered immune response. In the developing animal, the thymus is an extremely sensitive target organ for the toxic effects of dioxins and thymic atrophy is associated with immune suppression.

2.3. Chloracne

One of the hallmarks of dioxin toxicity is chloracne. Chloracne is acne following exposure to chlorinated organics. It is often difficult to clinically distinguish chloracne from juvenile acne. Diagnosis of chloracne is based on exposure history, time of onset in relation to age, and similar factors.[17] The only chemicals known to induce chloracne are the PHAHs. This disease involves both hyperplastic and hyperkeratotic changes in the skin and alterations in pigmentation resulting in a form of cystic acne.[17] Chloracne has been seen in humans, monkeys, hairless mice, and on rabbit ears and can occur following either dermal exposure or systemic exposure.[18] Chloracne is a high-dose response in animals and humans. In animals this response is accompanied by thymic atrophy and the wasting syndrome. Chloracne in humans is proof of dioxin exposure; however, lack of chloracne does not indicate that exposure has not taken place. It has been a rare but useful indicator of such exposure especially in chemical workers.

2.4. Hepatotoxicity

Hepatotoxicity is seen is several species exposed to dioxins, although the severity of the lesion is dependent on the species examined. Hepatic lesions may contribute to the lethal effects of dioxins in rats and rabbits. In contrast, both the guinea pig and the hamster exhibit few signs of hepatotoxicity. Hyperplasia and hypertrophy of parenchymal cells producing hepatomegaly are common findings in animals treated with dioxins. In rats, hepatic lesions are characterized by degenerative and necrotic lesions with the appearance of mononuclear cell infiltration, multinucleated giant hepatocytes, and increased numbers of mitotic figures and intracytoplasmic lipid droplets. There is also a marked increase in the smooth endoplasmic reticulum, associated with microsomal enzyme induction. In animals that exhibit hepatotoxicity, markers of hepatic damage such as serum glutamic-oxaloacetic transaminase (SGOT) and serum glutamic-pyruvic transaminase (SGPT) activities are increased. Other markers of liver function are not consistently altered by dioxins. Increases in serum cholesterol are seen in rats while serum triglycerides are unchanged.[19] In contrast, serum cholesterol is unchanged while serum triglycerides are increased in monkeys treated with dioxins.[20,21] In humans exposed to high levels of dioxins, effects on liver function have sometimes been reported immediately after exposure.[22]

2.5. Immunotoxicity

Alterations in immune function can have devastating effects on the survival of an organism. Immune suppression can result in increased susceptibility, incidence, and severity of infectious diseases. Immune suppression may also be associated with increased incidence of cancer. Conversely, loss of regulation within the immune system can lead to autoimmune diseases. The role of the immune system in homeostasis is critical, and alterations, either enhancement or suppression, must be considered toxic effects of a chemical.

The immunotoxicity of dioxins has been studied for many years, yet it is difficult to characterize a dioxinlike immunotoxic syndrome (see Kerkvliet, this volume). Dioxins suppress both humoral and cell-mediated immunity in mice. The resulting immunosuppression in mice leads to increased susceptibility to a variety of infectious agents including endotoxin,[13] *Listeria monocytogenes*,[16] *Salmonella*,[23] *Streptococcus pneumoniae*,[24] and *Trichinella*.[25] A single dose of 2,3,7,8-TCDD as low as 10 ng/kg significantly increases the mortality rate in adult mice exposed to influenza virus.[26]

These studies indicate that in mice immunosuppression occurs at extremely low doses of dioxins.

An excellent initial screen for immunotoxicants is the plaque-forming cell (PFC) assay in response to sheep red blood cells (SRBC).[27] In the PFC assay, animals are injected with SRBC. The animal responds to this challenge by producing antibodies against SRBC. Antibody production against SRBC requires an integrated response between macrophages, T cells, and B cells. Thus, the PFC assay using SRBC as the antigen is an excellent initial screen for immunotoxicants because it requires a functional response as well as cell–cell communication between macrophages, T cells, and B cells. In rats, 2,3,7,8-TCDD enhances the immune response to SRBC in the PFC assay.[28] However, similar doses suppressed host resistance to *Trichinella* infection[29] and influenza virus.[30] In mice, 2,3,7,8-TCDD suppresses the immune response to SRBC in the PFC assay.[31]

Dioxins are clearly immunotoxic in animals. However, the immune response to dioxins depends on the species studied, the dose of dioxin as well as the antigen and exposure paradigm studied. The current lack of data does not provide a strong enough foundation for determining whether dioxins are immunosuppressors or immunoenhancers in humans.

2.6. Reproductive Toxicity

2,3,7,8-TCDD reduces fertility, litter size, and uterine weights in mice, rats, and primates.[32–34] In several species, 2,3,7,8-TCDD alters ovarian function as indicated by anovulation and suppression of the estrous cycle.[32–34]

Dioxins appear to inhibit the actions of estrogens in several species. Inhibition of estrogen actions is known as an antiestrogenic effect. One possible mechanism to block estrogen actions is to decrease the concentration of circulating estrogens. Decreases in serum estradiol concentrations resulting in anovulatory menstrual cycles, and infertility occurred in female rhesus monkeys receiving high doses of dioxins.[33] These same animals also suffered from a wasting syndrome. In MCF7 cells, which is a cell line derived from a human breast adenocarcinoma and responsive to estrogens, dioxins increase the metabolism of estradiol by 100-fold.[35] Another mechanism by which dioxins can act as antiestrogens is to decrease the levels of estrogen receptors. Estrogen receptors are intracellular proteins that bind estrogens and mediate their effects. In rats and mice, the antiestrogenic actions of dioxins occur at doses that do not alter body weights and are not associated with the wasting syndrome.[34,36,37] In these species, decreases in uterine weights, an indicator of an antiestrogenic effect, are associated with decreases in uterine estrogen receptors.[36,37] Sublethal doses of 2,3,7,8-TCDD do not alter serum estrogens in rats or mice.[37,38]

Detailed mechanistic studies on the reproductive toxicity of dioxins in female animals have not been reported. In particular, very little information on developing female reproductive organs is available. Preliminary studies by Gray and co-workers at

USEPA indicate that perinatal exposure to low levels of 2,3,7,8-TCDD delays vaginal opening and induces cleft phallus/clitoris in female rats.[39] In severe cases of clefting, animals were also hypospadiac, which is an abnormal development of the urethra.[39] The mechanism of these effects is undetermined and the role of estrogens and their interactions with dioxins require further studies.

Alterations in testicular morphology have been reported in several species and are characterized as loss of germ cells, the appearance of degenerating spermatocytes and mature spermatozoa within the lumen of the seminiferous tubules, and a reduction in the number of tubules containing mature spermatozoa.[18] These effects have been reported in several species including rhesus monkeys, rats, mice, guinea pigs, and chickens.[21,32,40,41] These are high-dose effects in animals in that they are accompanied by signs of overt toxicity such as decreases in food intake and body weight. In sexually mature rats, 2,3,7,8-TCDD impairs male reproductive capabilities which are associated with reduction in testis and accessory sex organ weight, abnormal testicular morphology, and decreased spermatogenesis.[32,42] These effects are associated with decreases in serum androgen concentrations. Testicular steroidogenesis is decreased in 2,3,7,8-TCDD-treated rats and is responsible for the decreased circulating androgens in this species.[43,44] The ED_{50} for this response is 15 μg/kg and does not occur until there are signs of overt toxicity as indicated by decreases in food intake and body weights.[43] The experimental data indicate that in adult male animals, dioxin and its congeners act as antiandrogens in adult animals but require overtly toxic doses to produce these effects.

Recent epidemiological studies show an association between dioxin exposure and decreased serum testosterone levels and increased serum follicle-stimulating hormone (FSH) and luteinizing hormone (LH) in male workers exposed to 2,3,7,8-TCDD during the manufacturing of 2,4,5-trichlorophenol.[45] There are significant positive correlations between serum 2,3,7,8-TCDD levels and serum FSH and LH. Serum 2,3,7,8-TCDD levels were inversely correlated with serum testosterone. That is, as 2,3,7,8-TCDD levels increase in blood, the serum testosterone levels go down and serum FSH and LH levels go up. These findings demonstrate that the animal data are an excellent predictor of the toxic effects of 2,3,7,8-TCDD in humans. Furthermore, these data suggest that humans may be more sensitive to the antiandrogenic effects of dioxins than are rodents. The antiandrogenic effects of dioxins in experimental animals are accompanied by overt toxicity.[43] In humans this does not appear to be the case. Future studies to determine if other adverse health effects are present in this cohort are continuing.

2.7. Developmental Toxicity and Teratogenicity

Dioxins and its congeners are teratogenic in several species and are also developmental toxicants in rodents.[46] The species most sensitive to the overt teratogenic effects of dioxins are mice.[47] Low doses of dioxins, which do not produce any maternal or fetal toxicity, produce cleft palate and hydronephrosis in mice. In other species, major malformations are not produced until significant maternal toxicity and fetal lethality are evident.

Development of the fetus is regulated by the interactions of a variety of hormones and growth factors that act to control growth, differentiation, and programmed cell death. Exposure to steroidlike chemicals *in utero* has produced deleterious effects in humans. For example, male and female offspring of women taking the synthetic estrogen diethylstilbestrol (DES) have increased rates of cancers of the reproductive organs.[48] The mechanism of action of dioxins is similar to steroid hormones. Because of these similarities, the developing fetus may be at greater risk than the adult to the deleterious effects of dioxins. In fact, there are several reports that the developing fetus is more sensitive to the effects of dioxins. One example is that fetal lethality following dioxin exposure occurs at doses that do not produce maternal mortality.

Male rats are exquisitely sensitive to the reproductive toxicity of dioxin during development. Furthermore, the actions of 2,3,7,8-TCDD *in utero* appear to be irreversible. A single dose as low as 0.064 μg/kg to a pregnant rat on day 15 of gestation produces significant decreases in daily sperm production, cauda epididymal sperm number, and decreases in epididymis and cauda epididymis weight in male offspring 120 days after birth.[49] In addition, sexual behavior in male offspring was altered by perinatal exposure to 2,3,7,8-TCDD. Much of this work has been replicated and extended to another strain of rat by Gray and co-workers at USEPA.[39] The extreme sensitivity of these effects warrants concern for human exposure and the potential for alterations in male reproductive capabilities. These effects in animals are even more striking in light of the recent evidence that sperm counts have dropped approximately 50% over the past 30 years.[50]

Teratogenic and developmental effects caused by dioxins in humans have been difficult to characterize because the exposures were to mixtures of polychlorinated dibenzofurans and biphenyls.[51] These mixtures contained significant amounts of dioxinlike congeners and produced chloracne in adults. However, these mixtures also contained large amounts of nondioxinlike PCBs. Many of the children born to women who were exposed to these mixtures developed hyperpigmentation of the skin, mucous membranes, fingernails, and toenails, presence of erupted teeth in neonates, and hypersecretion of the meibomian gland. This syndrome has been called ectodermal dysplasia.[51] In addition to these malformations, *in utero* exposure to this mixture resulted in neurobehavioral abnormalities and delays in developmental milestones[52] (see Hsu *et al.*, this volume). Although these mixtures contained significant amounts of dioxinlike chemicals such as polychlorinated dibenzofurans and biphenyls, nondioxinlike congeners were also present and may have contributed to the adverse effects. The relative contribution of the dioxinlike congeners and the nondioxinlike congeners in the toxic effects of these mixtures is currently unknown. Evidence to support that some of the effects were related to the dioxinlike congeners is that many of these effects have been reported in animals receiving dioxins alone. The chloracnegenic response has been demonstrated in several species and is a clear indicator of dioxin exposure. Hypersecretion of the meibomian gland occurs in monkeys receiving dioxins.[53] These data strongly indicate that at least these effects are the result of the presence of the dioxinlike chemicals in the mixtures.

2.8. Carcinogenicity

2,3,7,8-TCDD is a potent animal carcinogen and has tested positive in 19 different studies in four different species.[54,55] In rodents, the repeated administration of 2,3,7,8-TCDD produces tumors in multiple sites in both sexes.[56–58] Recent epidemiological studies have demonstrated increased cancer risks associated with exposure to 2,3,7,8-TCDD and its congeners.[59–61] It should be stressed that the binding of dioxins and its congeners to the Ah receptor and their subsequent binding to DNA are all reversible reactions. Dioxins are not directly mutagenic[62] and no DNA adducts have been detected using methods that can detect 1 adduct in 10^{11} normal nucleotides.[63] In tumor promotion models, 2,3,7,8-TCDD is a potent promoter, but has very weak or no initiationlike activity.[64] In light of this evidence, 2,3,7,8-TCDD is considered a nongenotoxic carcinogen that acts as a promoter. However, the term *promoter* is an operational definition whose only true meaning is in the context of experimental systems. In fact, dioxins are whole and complete carcinogens in all four species tested: mice,[57] rats,[56,57] hamster,[58] and fish.[55] Given the weight of evidence from the experimental and epidemiological studies, the focus should be on what dose causes a carcinogenic response in humans. Continuing efforts to address this question using biologically based dose–response models are under development by several groups.[65,66]

3. MOLECULAR MECHANISMS

Most if not all of the effects of dioxins are mediated through their interaction with the Ah receptor.[67,68] The Ah receptor is an intracellular protein that acts as a signal transducer and transcription factor in a manner analogous to the steroid hormones. A signal transducer is a receptor protein that responds to signals, such as steroid hormones, and allows the cell to respond to the signals. In the case of the Ah receptor, the signal is 2,3,7,8-TCDD or other dioxins. *Transcription factor* is a term used to describe any protein that is involved in gene transcription. Gene transcription is the process of synthesizing a messenger RNA (mRNA) molecule from a specific sequence of DNA known as a gene. Once the mRNA is transcribed or synthesized, the message is then translated into a protein.

Originally, it was proposed that the Ah receptor was a member of the steroid hormone receptor family.[69] The recent cloning and sequencing of the Ah receptor gene demonstrates that there are significant structural differences between the Ah receptor and the steroid family of receptors and makes it unlikely that it is a member of this family.[70] For example, the protein sequence of the Ah receptor contains a basic/helix–loop–helix region; this region is involved in protein–protein interactions and allows the activated Ah receptor to bind to a second protein, ARNT, and it is this complex that binds to DNA. In contrast, the steroid family of receptors contains zinc fingers, which when bound with zinc allow the receptor to bind to DNA. Despite the structural differences between the two types of receptors, there are still many functional similarities. Both the Ah receptor and the steroid receptors bind lipophilic chemicals which are known as *ligands*. Each receptor binds structurally distinct ligands. That is, ligands that bind steroid receptors do not bind to the Ah receptor and ligands that bind the Ah receptor do not bind to steroid receptors. On ligand binding, both receptor types reversibly bind to specific regions of DNA called *response elements*. Response elements are regions of DNA that bind to specific proteins. The response elements for steroid hormones are distinct from the response elements for the Ah receptor. These regions are located in front of the site where transcription starts for a particular gene. On binding to the response element, both the steroid hormone and the Ah receptors bend the DNA. When the DNA bends, a conformational change occurs and allows greater access for other transcription factors to bind to DNA, resulting in an increase in transcription of genes which contain the response elements. The cellular responses to steroids are mediated through changes in gene transcription which subsequently produce alterations in cellular proliferation and differentiation. Alterations in gene transcription also appear to underlie the toxicity and carcinogenicity of dioxins.

One of the best characterized effects of dioxin is the increased synthesis of cytochrome P-450 1A1 (CYP1A1) (reviewed in Ref. 71). CYP1A1 is one of a family of proteins which function in the detoxification or activation of endogenous and exogenous chemicals. The induction of CYP1A1 by dioxins is mediated through the Ah receptor (Fig. 2). The Ah receptor is part of a complex which contains several proteins. Two of these proteins have been tentatively identified as heat shock proteins which have molecular masses of 90 kDa.[72] A third protein associated with the Ah receptor complex has not been fully characterized but has been designated p50.[72] On dioxin binding, the Ah receptor dissociates from the complex and interacts with a protein designated *a*ryl hydrocarbon *r*eceptor *n*uclear *t*ransferase (ARNT), forming a complex containing one Ah receptor and one ARNT protein.[73] Once bound to ARNT, the activated receptor complex binds to specific response elements on the DNA,[74] designated *dioxin responsive elements* (DRE) or *xenobiotic responsive elements* (XRE). These response elements are located in the CYP1A1 gene.[75,76] On binding to these response elements, there is an increase in transcription of the CYP1A1 gene.[71] The mechanism described above demonstrates that although the Ah receptor is necessary for the induction of CYP1A1, other factors are required. The regulation of the Ah receptor and ARNT proteins may be tissue-specific, resulting in differential tissue sensitivities.

The Ah receptor was originally found in cytosolic fractions from liver tissue.[77] It had been hypothesized that once bound to TCDD or other ligands the Ah receptor was translocated into the nucleus. In fact, the ARNT protein was thought to mediate the nuclear translocation of the Ah receptor. Recent studies have demonstrated that in some cell types the Ah receptor is a nuclear protein while in others it is a cytosolic protein.[78] The significance of this cell-specific localization of the Ah receptor is unknown.

While the mechanism for the induction of CYP1A1 is a well-characterized phe-

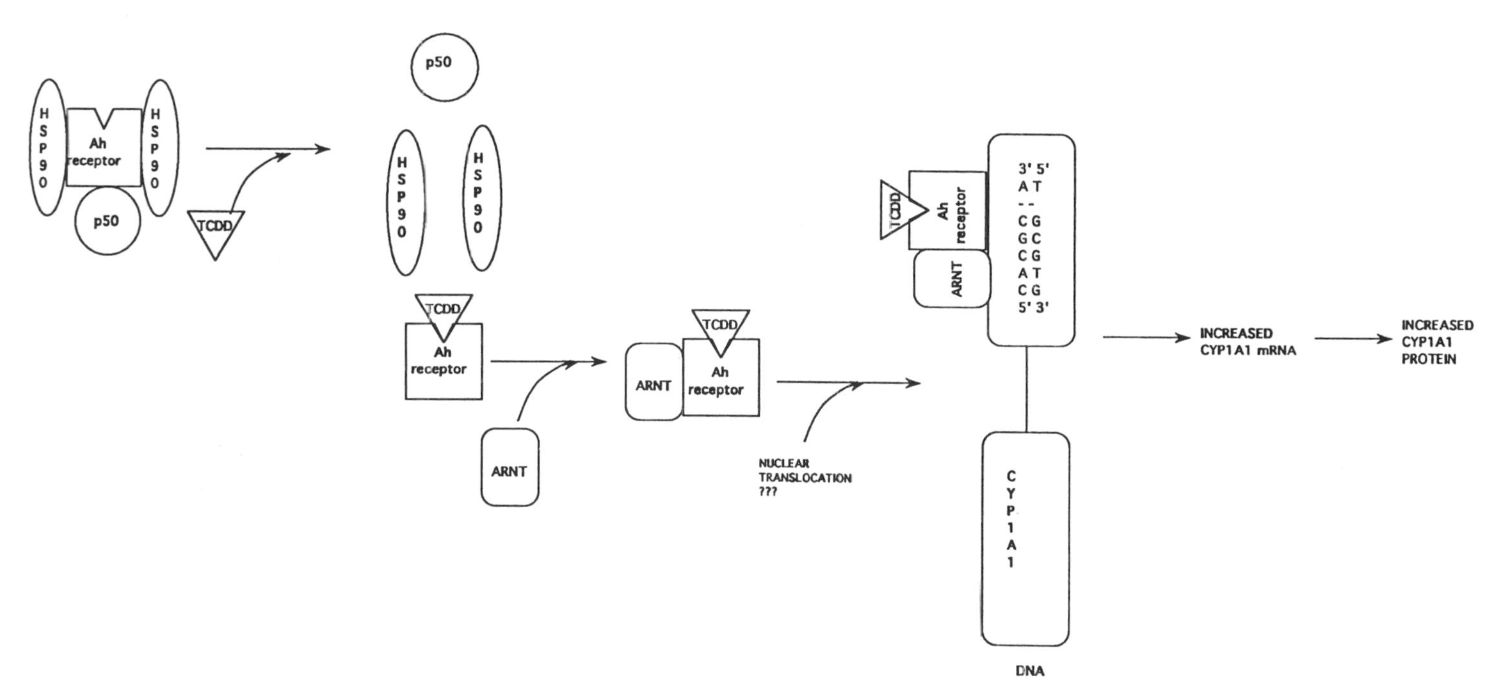

Figure 2. Mechanism of action of dioxins. Dioxins enter the cell and bind to the Ah receptor. The binding of dioxin to the receptor allows the receptor to dissociate from the heat shock proteins (HSP90) and p50. The bound receptor then binds to the ARNT protein. This complex then interacts with the specific regions of DNA designated dioxin responsive elements. When the Ah receptor binds to DNA, there is an increase in transcription of the genes downstream of the response element. The increase in transcription results in increased protein concentrations of the CYP1A1 gene. (See text for detailed description.)

nomenon, much less is known about how dioxins alter the regulation of other genes. Although there is substantial evidence indicating that the Ah receptor mediates all of the known effects of dioxins (reviewed in Ref. 67), few of the steps involved beyond binding to the Ah receptor are understood. Cytochrome P-450 1A2 (CYP1A2) is another member of the cytochrome P-450 family and is induced in hepatic tissue by dioxins. Unlike CYP1A1, CYP1A2 is normally produced in hepatic tissue and its induction by dioxins occurs through tissue-specific mechanisms.[79,80] CYP1A2 is only found in hepatic tissue and nasal mucosa, and is only inducible in hepatic tissue. While there is significant evidence that the induction of CYP1A2 is mediated by the Ah receptor, so far no DREs homologous to the CYP1A1 gene have been found upstream of the CYP1A2 gene despite intensive efforts to locate these response elements.[80]

Similarly, other biochemical alterations by dioxins such as decreases in estrogen receptor concentration are mediated by the Ah receptor. However, whether these effects involve changes in gene transcription similar to those for the induction of CYP1A1 remains undetermined. In addition, dioxins increase the activity of several proteins through mechanisms that do not involve direct effects on gene transcription.[81,82] Thus, the mechanism by which dioxins produce their effects is mediated by the Ah receptor, although there may be different pathways subsequent to Ah receptor activation which are specific for each effect.

Evidence implicating the Ah receptor in the actions of dioxins is based on two lines of research. First, 2,3,7,8-TCDD, the most potent dioxin, binds with high affinity to the Ah receptor and there are few if any chemicals that bind with equal affinity. Chemicals that bind to the Ah receptor produce the same effects as 2,3,7,8-TCDD but require higher doses because they do not bind as well to the Ah receptor. Structure–activity relationships demonstrate a direct relationship between binding affinity to the Ah receptor and the potency of the chemical to induce CYP1A1, thymic atrophy, or weight loss in animals.[83] The greater the binding affinity of a chemical for the Ah receptor, the more potent it is at producing these effects.

Second, in mice, sensitivity to dioxin and its congeners segregates with the *Ah* locus. The Ah receptor gene is encoded by several different alleles of the same locus in mice.[84] A locus is a position in a chromosome of a particular gene. A particular gene may have several different variants, called *alleles*. The C57BL/6 and DBA/2 are the prototypic sensitive and resistant strains of mice, respectively. The C57BL/6 mice possess an Ah receptor that has a high binding affinity to dioxins, while the DBA/2 mice possess a receptor with low dioxin binding affinity.[77] The C57BL/6 mice require lower doses of dioxins to produce the same effects compared with the DBA/2 mice. The sensitivity of these two strains of mice corresponds to the ability of their Ah receptor to bind dioxins. Because these strains may have differences in many other genes involved in dioxin toxicity, these studies do not prove that the Ah receptor is involved in the toxicity. To demonstrate the role of the Ah receptor, congenic mouse strains have been bred.[85,86] Congenic strains are genetically identical except at a single allele for a particular genetic locus. In mice congenic at the *Ah* locus, one strain has an Ah receptor that has a high binding affinity for dioxins while the other strain has a receptor with a low affinity for dioxins. Animals with a high-affinity receptor require lower doses of dioxins to produce a toxic effect compared with animals with low-affinity receptors which require higher doses to produce an effect. Studies using these congenic mice demonstrate that the sensitivity to the biochemical and toxic effects of dioxin congeners segregates with the *Ah* locus.[86–89]

4. BIOCHEMICAL EFFECTS

Hyperplasia, hypoplasia, metaplasia, and dysplasia are generalized responses to dioxin exposure in animals. Hyperplasia, an increase in cell number, occurs in the gastric mucosa and bile duct in monkeys in response to dioxins.[18] Guinea pigs develop urinary bladder hyperplasia following exposure to dioxins.[18] Hyperplasia of hepatic and dermal tissue occurs in several species. Hypoplasia, a decrease in number of cells, occurs in the lymphoid tissues in all species exposed to dioxins. Metaplasia is the transformation of one cell type to another. Squamous metaplasia occurs in the meibomian glands of the eyelid and the ceruminous glands of the ears, producing waxy exudates, of monkeys exposed to dioxins.[18] Dysplasia, the abnormal growth or development of an organ, occurs in ectodermal tissue in humans and primates resulting in alterations in teeth and nails following exposure to dioxins. Thus, many of the toxic actions of dioxins are related to their ability to disrupt normal growth processes.

The toxic effects of dioxins are mediated through alterations in normal homeostatic processes. These processes are regulated through the interaction of growth factors, steroid hormones, and the enzymes involved in the synthesis and degradation of these factors. The actions of dioxins can be explained, in part, as modulations in growth factors and their receptors, modulation of hormones and their receptors, and induction of enzymes involved in metabolism.

4.1. Alterations in Growth Factors and Their Receptors

Cellular processes such as proliferation and differentiation are normally held under tight control for obvious reasons. Inappropriate proliferation and alterations in differentiation can lead to deleterious effects such as developmental abnormalities and cancer. Regulation of these processes are mediated in part by a variety of growth factors. There are both stimulatory growth factors such as epidermal growth factor (EGF) and transforming growth factor-α (TGFα) as well as inhibitory growth factors. TGFβ is both an inhibitory and stimulatory growth factor depending on the cell type. Growth factors are proteins that regulate cellular proliferation, differentiation, and apoptosis (a mechanism of normal cell death) by interacting with specific membrane receptors which act as signal transducers. Dioxins can alter the levels of both growth factors and their receptors.

One example of how alterations in these factors result in a toxic effect occurs in the production of cleft palate by dioxins. In the developing palate, the medial epithelial cells cease to express EGFR,[90] stop proliferating, and ultimately transdifferentiate to mesenchymal cells allowing the left and right palate to fuse. These events are regulated by temporal changes in the levels of EGFR, EGF, TGFα, and TGFβ.[91] Alterations in the production of the proteins interfere with the fusion of the left and right palate. In the presence of dioxin, the medial epithelial cells continue to express EGFR, proliferate, and differentiate into stratified squamous orallike epithelium which prevents the left and right palatal shelves from fusing.[90] The dioxin-induced changes in medial epithelial cell differentiation are associated with increases in TGFβ and EGFR, and decreases in TGFα.[91]

Alterations in growth factors and their receptors have been characterized in many different systems. Depending on the organ or tissue, age, sex, and species studied, dioxins can increase or decrease these factors. For example, 2,3,7,8-TCDD induces a proliferative response in rat liver,[92] hairless mouse skin,[93] and the developing ureteric epithelia in mice.[94,95] EGF receptor concentration is decreased in rat liver, unchanged in hairless mouse skin, and increased in the developing ureteric and palatal epithelium. While all three tissues respond to dioxins by proliferating, the actions of dioxins on a particular growth factor are cell type specific.

4.2. Alterations in Hormones and Their Receptors

There are a variety of different types of hormones that are involved in homeostatic control at the cellular, tissue, organ, and organism levels. Steroid hormones participate in the regulation of growth and development, sexual differentiation, reproduction, and cellular metabolism. Other hormones such as insulin regulate carbohydrate and fat metabolism. Dioxins alter many of these hormones and their receptors. Decreases in estrogen receptor,[36,37] glucocorticoid,[88] and insulin receptor concentrations[96] have been reported following dioxin treatment. Dioxins also alter the concentration of several hormones such as estrogen,[33] progesterone,[33] testosterone,[43] vitamin A,[97] and thyroid hormone[98] in several tissues and in several species.

In humans, there is evidence suggesting that dioxins also produce alterations in hormone concentrations. As mentioned above, dioxins decrease testosterone and increase FSH and LH concentrations in workers exposed during the manufacturing of 2,4,5-trichlorophenol. There are also reports of an increased incidence of diabetes in workers exposed to dioxins[99] as well as those Vietnam veterans associated with Operation Ranch Hand.[100] In rats, high doses of dioxins decrease hepatic insulin receptors which may result in alterations in glucose metabolism.[96] In guinea pigs, glucose uptake is decreased in adipose tissue, pancreas, and brain.[101] While these changes are not equivalent to diabetes, these studies indicate that dioxins can alter glucose metabolism. These data suggest that dioxins can alter hormonal status in humans resulting in pathologies similar to those seen in experimental animals.

4.3. Alterations in Enzyme Activity

The best characterized action of dioxins is their effects on CYP1A1 and CYP1A2. Both of these proteins are involved in the metabolism of endogenous and exogenous chemicals. Dioxins increase the level and activity of a number of enzymes involved in cellular metabolism. For example, in rats dioxins increase UDP-glucuronyltransferases, which metabolizes thyroxine.[98,102] The increased hepatic metabolism of thyroxine may lead to decreases in circulating thyroxine levels. The pituitary gland then increases thyroid stimulating hormone (TSH) secretion in response to the lowered thyroxine levels. The increased TSH levels are accompanied by hyperplasia and hypertrophy of the thyroid. Prolonged stimulation may lead to thyroid tumors and in fact 2,3,7,8-TCDD increases thyroid tumors in both rats and mice.[57] Thus, changes in enzymes involved in the metabolism of endogenous substances can result in effects as severe as cancer.

5. SPECIES DIFFERENCES

There are many reports in the literature demonstrating large species differences in the doses needed to produce lethal effects of dioxins. For example, the most sensitive species to the acute lethal effects of 2,3,7,8-TCDD is the guinea pig, where the LD_{50} (LD_{50} is the dose of a chemical required to kill 50% of the animals treated) is approximately 0.6 μg/kg.[103] The hamster is approximately 1000–10,000 times less sensitive than the guinea pig to the lethal effects of dioxins ($LD_{50} > 3000$ μg/kg).[104] However, in many other species such as monkeys, rabbits, rats, mice, and dogs, the LD_{50} is between 100 and 300 μg/kg.[67] In contrast to the lethal effects on adult animals, doses of 2,3,7,8-TCDD that are fetotoxic are similar across several species and are lower than those of the adult animals.[105] The incidence of cleft palate in mice can reach 100% before there is an increase in fetal mortality.[47] However, in other species cleft palate is only produced at doses that are fetotoxic. A more accurate description of the species differences in susceptibility to dioxins is that for every toxic endpoint there are either extremely sensitive or resistant species; however, most species respond to the toxic effects of dioxins at similar doses.

The species differences in toxic effects do not appear to be causally related to the amount of Ah receptor present in a particular tissue.[106] The species differences may be explained, in part, by differences in the size and binding affinity of the Ah receptor. In addition, there may be inherent differences in the actions of this receptor between species. Pharmacokinetic differences between species may also play a role in the differences in sensitivity. Although the Ah receptor may be critical in the actions of dioxins, there are other factors that contribute to the toxic effects.

One of the difficulties in extrapolating the animal data to humans is determining whether humans are a sensitive or resistant species. In the early years of dioxin research, chloracne was the only documented response in humans and it occurs only after exposure to very high doses of dioxins. There is now a significant amount of evidence indicating that humans are sensitive to the effects of dioxins. Functional Ah receptors have been found in many human tissues including lymphocytes, liver, lung, and placenta.[107] Human CYP1A1 is inducible in lung and placenta.[108] Decreased activity of the EGE receptor occurs in human placenta isolated from women exposed to high levels of dioxinlike compounds[108] and this same response occurs in rodent tissue such as liver[89,109] and thymus.[110] Decreases in testosterone and increases in LH and FSH have been reported in men with blood levels of 2,3,7,8-TCDD as low as 70 ppt, which is approximately 10 times higher than "unexposed" controls.[45] In rats, 2,3,7,8-TCDD decreases testosterone and increases LH levels in a dose-dependent fashion.[111,112] Embryonic palates from mice, rats, and humans respond similarly in organ culture.[113–115] The cultured human and rat tissues respond at similar 2,3,7,8-TCDD concentrations which are 100-fold greater than that required to elicit responses in the mouse palates.[90,114,115] These and other studies demonstrate that humans are a sensitive species. We are now left with the question of which effects are likely to occur in humans and at what doses these effects will occur.

6. TOXIC EQUIVALENCY FACTORS

Dioxins and other PHAHs are often found in the environment as part of a complex mixture of chemicals. Estimating the potential health effects following exposure to these mixtures is problematic. For many of these chemicals there are very limited data available on toxicity. Detailed carinogenicity, teratogenicity, and reproductive toxicity studies are available for only a few of these compounds (reviewed in Ref. 83). Toxic equivalency factors (TEF) were developed and validated in animals[116] in order to estimate the potential health effects of these complex mixtures such as soot found in the Binghamton, New York, PCB transformer fire of 1981. A TEF is a measure of the relative potency of a chemical for which there are little toxicity data available, versus a chemical for which significant toxicity data are available as well as an understanding of the mechanism of toxicity.[116,117] For the dioxins, the prototype chemical to which all others are compared is 2,3,7,8-TCDD. The TEF approach has been successfully used to estimate the potency of mixtures containing chlorinated dioxins and dibenzofurans. TEFs for PCBs have been proposed by Safe[83] and are currently under evaluation by the USEPA.[118]

The fact that two chemicals produce the same effect is not enough evidence for development of a TEF. For example, both 2,3,7,8-TCDD and diethylnitrosamine are hepatocarcinogens in rodents. Yet a TEF cannot be developed for these chemicals since they act through completely different mechanisms and their interactive effects are synergistic and not additive. The validity of the TEF methodology is based on several criteria.[117] The class of chemicals studied should be well defined and should produce a similar spectrum of toxicity. The relative toxicity of a chemical should be independent of the endpoint studied. More importantly, the mechanism by which these chemicals act *must* be identical and the effects of mixtures *must* be additive. These criteria hold for all dioxinlike PHAHs that bind to the Ah receptor.

While the TEF approach has been invaluable in assessing potential human health and environmental risks, there are several shortcomings of this approach. The TEFs can only estimate the risk associated with dioxinlike compounds. More often than not, such compounds are minor components of an environmental sample. The remaining chemicals may be toxic through mechanisms that are independent of the Ah receptor.

In addition, there is evidence that the nondioxinlike PCBs may interact synergistically (the interaction is greater than additive) with the dioxinlike PCBs.[119] The TEF approach ignores nondioxinlike chemicals and metals and does not incorporate potential interactions among the different classes of compounds.

7. SUMMARY

Dioxins are members of a class of compounds that share several features. They are chlorinated aromatic hydrocarbons that are persistent in both environmental and biological samples. They produce a similar spectrum of toxicity which is mediated by interaction with the Ah receptor. The development of TEFs has permitted the estimation of human health risks associated with these chemicals.

The toxic effects of these chemicals can best be described by their actions as growth dysregulators. Dioxins disrupt normal homeostatic processes that tightly regulate cellular growth and differentiation. Disruption in these processes produces a variety of toxicities and pathologies. The available data indicate that humans are sensitive to the toxic effects of these chemicals. Clearer definition of human responses and the body burdens associated with such effects remains to be determined.

ACKNOWLEDGMENT. This work was supported by funds provided by the Collaborative Clinical Research on Health Effects of Exposure to Air Pollutants (U.S. EPA CR817643) through the Center for Environmental Medicine and Lung Biology, University of North Carolina, Chapel Hill.

REFERENCES

1. L. Birnbaum, Distribution and excretion of 2,3,7,8-tetrachlorodibenzo-p-dioxin in congenic strains of mice which differ at the Ah locus, *Drug Metab. Dispos.* **14,** 34–40 (1986).
2. T. A. Gasiewicz and G. Rucci, Cytosolic receptor for 2,3,7,8-tetrachlorodibenzo-p-dioxin: Evidence for a homologous nature among mammalian species, *Mol. Pharmacol.* **26,** 90–98 (1984).
3. R. Pohjanvirta, T. Vartiainen, A. Uusi-rauva, J. Monkkonen, and J. Tuomisto, Tissue distribution, metabolism, and excretion of [14C]-TCDD in a TCDD-susceptible and a TCDD-resistant rat strain, *Pharmacol. Toxicol.* **66,** 93–100 (1990).
4. J. Q. Rose, J. C. Ramsey, T. H. Wentzler, R. A. Hummel, and P. J. Gehring, The fate of 2,3,7,8-tetrachlorodibenzo-p-dioxin following single and repeated oral doses to the rat, *Toxicol. Appl. Pharmacol.* **36,** 209–226 (1976).
5. H. Poiger and C. Schlatter, Pharmacokinetics of 2,3,7,8-TCDD in man, *Chemosphere* **15,** 1489–1494 (1986).
6. J. L. Pirkle, W. F. Wolfe, and D. G. Patterson, Estimate of the half-life of 2,3,7,8-tetrachlorodibenzo-p-dioxin in Vietnam veterans of Operation Ranch Hand, *J. Toxicol. Environ. Health* **27,** 165–171 (1989).

7. M. Van den Berg and H. Poiger, Selective retention of PCDDs and PCDFs in mammals: A multiple cause problem, *Chemosphere* **18,** 677–680 (1989).
8. S. R. Max and E. K. Silbergeld, Skeletal muscle glucocorticoid receptor and glutamine synthetase activity in the wasting syndrome in rats treated with 2,3,7,8-tetrachlorodibenzo-p-dioxin, *Toxicol. Appl. Pharmacol.* **87,** 523–527 (1987).
9. R. E. Peterson, M. D. Seefeld, B. J. Christian, C. L. Potter, C. K. Kellin, and R. E. Keesey, The wasting syndrome in 2,3,7,8-tetrachlorodibenzo-p-dioxin toxicity: Basic features and their interpretation, *Banbury Report* **18,** 291–308 (1984).
10. H. Urabe, H. Koda, and M. Asahi, Present state of Yusho patients, *Ann. N.Y. Acad. Sci.* **320,** 273–275 (1979).
11. J. G. Vos, H. Van Loveren, and H.-J. Schuurman, Immunotoxicity of dioxin: Immune function and host resistance in laboratory animals and humans, *Banbury Report* **35,** 79–93 (1991).
12. A. N. Tucker, S. J. Vore, and M. I. Luster, Suppression of B cell differentiation by 2,3,7,8-tetrachlorodibenzo-p-dioxin, *Mol. Pharmacol.* **29,** 372–377 (1986).
13. J. G. Vos, J. G. Kreftenberg, H. W. B. Engel, A. Minderhoud, and L. M. van Noorle Jansen, Studies on 2,3,7,8-tetrachlorodibenzo-p-dioxin induced immune suppression and decreased resistance to infection: Endotoxin hypersensitivity, serum zinc concentration and effect of thymosin treatment, *Toxicology* **9,** 75–84 (1978).
14. E. Benjamini and S. Leskowitz, *Immunology: A Short Course,* 2nd ed., p. 26, Wiley–Liss, New York (1991).
15. J. G. Vos and J. A. Moore, Suppression of cellular immunity in rats and mice by maternal treatment with 2,3,7,8-tetrachlorodibenzo-p-dioxins, *Int. Arch. Allergy Appl. Immunol.* **47,** 777–791 (1974).
16. M. I. Luster, D. R. Boorman, J. H. Dean, M. H. Harris, R. W. Luebke, M. L. Padarathigh, and J. A. Moore, Examination of bone marrow, immunologic parameters and host susceptibility following pre- and postnatal exposure to 2,3,7,8-tetrachlorodibenzo-p-dioxin (TCDD), *Int. J. Immunopharmacol.* **2,** 310–320 (1980).
17. R. Kimbrough, Skin lesions in animals and humans: A brief overview, *Banbury Report* **18,** 357–364 (1984).
18. E. E. McConnell and J. A. Moore, Toxicopathology characteristics of the halogenated aromatics, *Ann. N.Y. Acad. Sci.* **320,** 138–150 (1979).
19. J. G. Zinkl, J. G. Vos, J. A. Moore, and B. N. Gupta, Hematologic and clinical chemistry effects of 2,3,7,8-tetrachlorodibenzo-p-dioxin in laboratory animals, *Environ. Health Perspect.* **5,** 111–118 (1973).
20. J. R. Allen, L. A. Carstens, and D. A. Barsotti, Residual effects of short-term low-level exposure of nonhuman primates to polychlorinated biphenyls, *Toxicol. Appl. Pharmacol.* **30,** 440–451 (1974).
21. E. E. McConnell, J. A. Moore, and D. W. Dalgard, Toxicity of 2,3,7,8-tetrachlorodibenzo-p-dioxin in rhesus monkeys (Macaca mulatta) following a single oral dose, *Toxicol. Appl. Pharmacol.* **43,** 175–187 (1978).
22. P. Mocarelli, A. Marocchi, P. Brambilla, P. M. Gerthoux, L. Colombo, A. Mondonico, and L. Meazza, Effects of dioxin exposure in humans at Seveso, Italy, *Banbury Report* **35,** 95–100 (1991).
23. R. D. Hinsdill, D. L. Couch, and R. S. Speirs, Immunosuppression in mice induced by dioxin (TCDD) in feed, *J. Environ. Pathol. Toxicol.* **4,** 401–412 (1980).
24. K. L. White, H. H. Lysy, J. A. McCay, and A. C. Anderson, Modulation of serum complement levels following exposure to polychlorinated dibenzo-p-dioxins, *Toxicol. Appl. Pharmacol.* **84,** 209–221 (1986).
25. R. W. Luebke, C. Copeland, D. L. Andrews, J. J. Diliberto, P. I. Acubue, and L. S.

Birnbaum, Effects of TCDD on resistance to *T. spiralis* infection in mice, *Toxicologist* **13,** 306 (1993).
26. G. R. Burelson, H. Lebrec, Y. G. Yang, J. D. Ibanes, and K. N. Pennington, Effect of 2,3,7,8-tetrachlorodibenzo-p-dioxin (TCDD) on influenza virus host resistance in mice, submitted for publication (1993).
27. M. I. Luster, J. H. Dean, and J. A. Moore, Evaluation of immune functions in toxicology, in: *Principles and Methods of Toxicology* (A. Wallace Hayes, ed.), pp. 561–586, Raven Press, New York (1984).
28. R. A. Smialowicz, M. Riddle, W. Williams, J. Diliberto, D. Andrews, C. Copeland, and D. Ross, Comparison of the immunotoxicity of 2,3,7,8-tetrachlorodibenzo-p-dioxin (TCDD) in mice and rats, *Toxicologist* **13,** 1352 (1993).
29. C. Copeland, R. W. Luebke, and D. L. Andrews, Resistance to *T. spiralis* infection in rats exposed to TCDD, *Toxicologist* **13,** 307 (1993).
30. Y. G. Yang and G. R. Burleson, Effects of 2,3,7,8-tetrachlorodibenzo-p-dioxin (TCDD) on the influenza virus augmented pulmonary natural killer activity of Fisher 344 rats, *Toxicologist* **13,** 308 (1993).
31. A. Vecchi, A. Mantovani, M. Sironi, M. Luini, W. Cairo, and S. Garattini, Effect of acute exposure to 2,3,7,8-tetrachlorodibenzo-p-dioxin on humoral antibody production in mice, *Chem. Biol. Interact.* **30,** 337–341 (1980).
32. R. J. Kociba, P. A. Keller, C. N. Park, and P. J. Gehring, 2,3,7,8-Tetrachlorodibenzo-p-dioxin (TCDD): Results of a 13-week oral toxicity study in rats, *Toxicol. Appl. Pharmacol.* **35,** 553–574 (1976).
33. D. A. Barsotti, L. J. Abrahamson, and J. R. Allen, Hormonal alterations in female rhesus monkeys fed a diet containing 2,3,7,8-tetrachlorodibenzo-p-dioxin, *Bull Environ. Contam. Toxicol.* **211,** 463–469 (1979).
34. T. H. Umbreit, E. J. Hesse, G. J. Macdonald, and M. A. Gallo, Effects of TCDD–estradiol interactions in three strains of mice, *Toxicol. Lett.* **40,** 1–9 (1987).
35. D. C. Spink, D. W. Lincoln, II, H. W. Dickerman, and G. F. Gierthy, 2,3,7,8-Tetrachlorodibenzo-p-dioxin causes an extensive alteration of 17-beta-estradiol metabolism in MCF-7 breast tumor cells, *Proc. Natl. Acad. Sci. USA* **87,** 6917–6921 (1990).
36. M. Romkes and S. Safe, Comparative activities of 2,3,7,8-tetrachlorodibenzo-p-dioxin and progesterone as antiestrogens in female rat uterus, *Toxicol. Appl. Pharmacol.* **92,** 368–380 (1988).
37. M. J. DeVito, T. Thomas, E. Martin, T. H. Umbreit, and M. A. Gallo, Antiestrogenic action of 2,3,7,8-tetrachlorodibenzo-p-dioxin: Tissue-specific regulation of estrogen receptor in CD1 mice, *Toxicol. Appl. Pharmacol.* **113,** 284–292 (1992).
38. K. T. Shiverick and T. F. Muther, 2,3,7,8-Tetrachlorodibenzo-p-dioxin (TCDD) effects on hepatic microsomal steroid metabolism and serum estradiol of pregnant rats, *Biochem. Pharmacol.* **32,** 991–995 (1983).
39. L. E. Gray, J. S. Ostby, W. Kelce, R. Marshall, J. J. Diliberto, and L. S. Birnbaum, Perinatal TCDD exposure alters sex differentiation in both female and male LE hooded rats, Paper presented at Dioxin '93, Vienna, September 20–24 (1993).
40. J. R. Allen and J. J. Lalich, Response of chickens to prolonged feeding of crude "toxic fat," *Proc. Soc. Exp. Biol. Med.* **109,** 48–51 (1962).
41. J. R. Allen and L. A. Carstens, Light and electron microscopic observations in Macaca mulatta monkeys fed toxic fat, *Am. J. Vet. Res.* **28,** 1513–1526 (1967).
42. I. Chahoud, R. Krowke, A. Schimmel, H. Merker, and D. Neubert, Reproductive toxicity and pharmacokinetics of 2,3,7,8-tetrachlorodibenzo-p-dioxin. I. Effects of high doses on the fertility of male rats, *Arch. Toxicol.* **63,** 432–439 (1989).
43. R. W. Moore, J. A. Parsons, R. C. Bookstaff, and R. E. Peterson, Androgenic deficiency

in male rats treated with 2,3,7,8-tetrachlorodibenzo-p-dioxin, *Toxicol. Appl. Pharmacol.* **79,** 99–111 (1985).

44. R. W. Moore, C. R. Jefcoate, and R. E. Peterson, 2,3,7,8-Tetrachlorodibenzo-p-dioxin inhibits steroidogenesis in the rat testis by inhibiting the mobilization of cholesterol to cytochrome $P450_{ssc}$, *Toxicol. Appl. Pharmacol.* **109,** 85–97 (1991).
45. G. M. Egeland, M. H. Sweeney, M. A. Fingerhut, W. E. Halperin, K. K. Wille, and T. M. Schnorr, Serum 2,3,7,8-tetrachlorodibenzo-p-dioxin's (TCDD) effect on total serum testosterone and gonadotropins in occupationally exposed men, Society of Epidemiological Research, Minneapolis (1992).
46. L. Birnbaum, Developmental toxicity of TCDD and related compounds: Species sensitivities and differences, *Banbury Report* **35,** 51–68 (1991).
47. L. A. Couture, B. D. Abbott, and L. S. Birnbaum, A critical review of the developmental toxicity and teratogenicity of 2,3,7,8-tetrachlorodibenzo-p-dioxin: Recent advances toward understanding the mechanism, *Teratology* **42,** 619–632 (1990).
48. R. J. Stillman, In utero exposure to diethylstilbestrol: Adverse effects on the reproductive tract and reproductive performance in male and female offspring, *Am. J. Ostet. Gynecol.* **142,** 905–921 (1982).
49. T. A. Mably, D. L. Bjerke, R. W. Moore, A. Gendron-Fitzpatrick, and R. E. Peterson, In utero and lactational exposure of male rats to 2,3,7,8-tetrachlorodibenzo-p-dioxin: 3. Effects on spermatogenesis and reproductive capability, *Toxicol. Appl. Pharmacol.* **114,** 118–126 (1992).
50. R. M. Sharpe and N. E. Skakkebaek, Are oestrogens involved in falling sperm counts and disorders of the male reproductive tract? *Lancet* **341,** 1392–1395 (1993).
51. F. Yamashita and M. Hayashi, Fetal PCB syndrome: Clinical features, intrauterine growth retardation and possible alterations in calcium metabolism, *Environ. Health Perspect.* **59,** 41–45 (1985).
52. W. J. Rogan, B. C. Gladen, and K.-L. Hung, Congenital poisoning by polychlorinated biphenyls and their contaminants in Taiwan, *Science* **241,** 334–338 (1988).
53. J. A. Moore, E. E. McConnell, D. W. Dalgard, and M. W. Harris, Comparative toxicity of three halogenated dibenzofurans in guinea pigs, mice and rhesus monkeys, *Ann. N.Y. Acad. Sci.* **320,** 151–163 (1979).
54. J. Huff, 2,3,7,8-TCDD: A potent and complete carcinogen in experimental animals, *Chemosphere* **25,** 173–176 (1992).
55. R. Johnson, J. Tietge, and S. Botts, Carcinogenicity of 2,3,7,8-TCDD to medaka, *Toxicologist* **12**(1)**,** 476 (1922).
56. R. J. Kociba, D. G. Keyes, J. E. Beyer, R. M. Carreaon, E. E. Wade, D. A. Dittenber, R. P. Kalnins, L. E. Fauson, C. N. Parks, S. D. Barnard, R. A. Hummel, and C. G. Humitson, Results of a two-year chronic toxicity and oncogenicity study of 2,3,7,8-tetrachlorodibenzo-p-dioxin in rats, *Toxicol. Appl. Pharmacol.* **46,** 279–290 (1978).
57. National Toxicology Program (NTP), Bioassay of 2,3,7,8-tetrachlorodibenzo-p-dioxin for possible carcinogenicity (gavage study), Technical Report Series No. 102, National Toxicology Program, Research Triangle Park, NC (1982).
58. M. S. Rao, V. Subbarao, J. D. Prasad, and D. G. Scarpelli, Carcinogenicity of 2,3,7,8-tetrachlorodibenzo-p-dioxin in the Syrian golden hamster, *Carcinogenesis* **9,** 1677–1679 (1988).
59. M. A. Fingerhut, W. E. Halperin, D. A. Marlow, L. A. Piacitelli, P. A. Honchar, M. H. Sweeney, A. L. Griefe, P. A. Dill, K. Steenland, and A. H. Suruda, Cancer mortality in workers exposed to 2,3,7,8-tetrachlorodibenzo-p-dioxin, *N. Engl. J. Med.* **324,** 212–218 (1991).

60. A. Manz, J. Barger, and J. H. Dwyer, Cancer mortality among workers in chemical plant contaminated with dioxin, *Lancet* **338,** 959–964 (1991).
61. A. Zober, P. Messerer, and P. Huber, Thirty-four year mortality follow up of BASF employees exposed to 2,3,7,8-TCDD after the 1953 accident, *Int. Arch. Occup. Environ. Health* **62,** 139–157 (1990).
62. J. S. Wassom, J. E. Huff, and N. A. Lotriano, A review of the genetic toxicology of chlorinated dibenzo-p-dioxins, *Mutat. Res.* **47,** 141–160 (1977).
63. K. W. Turteltaub, J. S. Felton, B. L. Gledhill, J. S. Vogel, J. R. Southon, M. W. Caffee, R. C. Finkel, D. E. Nelson, I. D. Procter, and J. C. David, Accelerator mass spectrometry in biomedical dosimetry: Relationship between low-level exposure and covalent binding of heterocyclic amine carcinogens to DNA, *Proc. Natl. Acad. Sci. USA* **87,** 5288–5292 (1990).
64. H. C. Pitot, T. Goldsworthy, H. A. Campbell, and A. Poland, Quantitative evaluation of the promotion by 2,3,7,8-tetrachlorodibenzo-p-dioxin of hepatocarcinogenesis from diethylnitrosamine, *Cancer Res.* **40,** 3616–3620 (1980).
65. M. E. Andersen, J. J. Mills, M. L. Gargas, L. Kedderis, L. S. Birnbaum, D. Neubert, and W. F. Greenlee, Modeling receptor-mediated processes with dioxins: Implications for pharmacokinetics and risk assessment, *Risk Anal.* **13,** 25–36 (1993).
66. C. Portier, A. Tritscher, M. Kohn, C. Sewall, G. Clark, L. Edler, D. Hoel, and G. Lucier, Ligand/receptor binding for 2,3,7,8-TCDD: Implications for risk assessment, *Fundam. Appl. Toxicol.* **20,** 48–56 (1993).
67. L. Birnbaum, The mechanism of dioxin toxicity: Relationship to risk assessment, *Environ. Health Perspect.* in press.
68. R. J. Scheuplein, M. A. Gallo, and D. A. van der Heijden, Epilogue, *Banbury Report* **35,** 489–490 (1991).
69. R. Evans, The steroid and thyroid hormone superfamily, *Science* **240,** 889–892 (1988).
70. K. M. Burbach, A. Poland, and C. A. Bradfield, Cloning of the Ah-receptor cDNA reveals a distinctive ligand-activated transcription factor, *Proc. Natl. Acad. Sci. USA* **89,** 8185–8189 (1992).
71. J. P. Whitlock, Genetic and molecular aspects of 2,3,7,8-tetrachlorodibenzo-p-dioxin action, *Annu. Rev. Pharmacol. Toxicol.* **30,** 251–277 (1990).
72. G. Perdew, Chemical cross-linking of the cytosolic and nuclear forms of the Ah receptor in hepatoma cell line 1c1c7, *Biochem. Biophys. Res. Commun.* **182,** 55–62 (1992).
73. E. C. Hoffman, H. Reues, F. F. Chu, F. Sander, L. H. Conley, B. A. Brooks, and O. Hankinson, Cloning of a factor required for activity of the Ah (dioxin) receptor, *Science* **252,** 954–958 (1991).
74. H. Reyes, S. Reisz-Porszasz, and O. Hankinson, Identification of the Ah receptor nuclear translocator protein (Arnt) as a component of the DNA binding form of the Ah receptor, *Science* **256,** 1193–1195 (1992).
75. M. S. Denison, J. M. Fisher, and J. P. Whitlock, Jr., The recognition site for the dioxin–Ah receptor complex: Nucleotide sequence and functional analysis, *J. Biol. Chem.* **263,** 17221–17224 (1988).
76. A. Fujisawa-Sehara, K. Sogawa, M. Uamane, and Y. Fujii-Kuriyama, A DNA-binding factor specific for xenobiotic responsive elements of P-450c gene exists as a cryptic form in cytoplasm: Its possible translocation to nucleus, *Proc. Natl. Acad. Sci. USA* **85,** 5859–5863 (1988).
77. A. Poland, E. Glover, and A. S. Kende, Stereospecific high affinity binding of 2,3,7,8-tetrachlorodibenzo-p-dioxin by hepatic cytosol, *J. Biol. Chem.* **251,** 4936–4942 (1975).

78. B. D. Abbott, J. J. Diliberto, and L. S. Birnbaum, TCDD alters expression of glucocorticoid receptor in embryonic mouse palate, *Toxicologist* **13,** 316 (1993).
79. J. A. Goldstein and P. Linko, Differential induction of two 2,3,7,8-tetrachlorodibenzo-p-dioxin inducible forms of cytochrome P-450 in extrahepatic vs. hepatic tissues, *Mol. Pharmacol.* **25,** 185–191 (1983).
80. L. C. Quattrochi and R. H. Tukey, The human cytochrome cyp1A2 gene contains regulatory elements responsive to 3-methylcholanthrene, *Mol. Pharmacol.* **36,** 66–71 (1989).
81. K. W. Gaido, S. C. Maness, L. S. Leonard, and W. F. Greenlee, 2,3,7,8-Tetrachlorodibenzo-p-dioxin-dependent regulation of transforming growth factors-alpha and beta2 expression in a human keratinocyte cell line involves both transcriptional and post transcriptional control, *J. Biol. Chem.* **267,** 24591–24595 (1992).
82. T. M. Sutter and W. F. Greenlee, Classification of members of the Ah gene battery, *Chemosphere* **25,** 223–225 (1992).
83. S. Safe, Polychlorinated biphenyls (PCBs), dibenzo-p-dioxins (PCDDs), dibenzofurans (PCDFs), and related compounds: Environmental and mechanistic considerations which support the development of toxic equivalency factors (TEFs), *Crit. Rev. Toxicol.* **21,** 219–238 (1990).
84. A. Poland and E. Glover, Characterization and strain distribution pattern of the murine Ah receptor specified by the Ah^{d} and Ah^{b-3} alleles, *Mol. Pharmacol.* **38,** 306–312 (1990).
85. A. Poland and E. Glover, 2,3,7,8-Tetrachlorodibenzo-p-dioxin: Segregation of toxicity with the Ah locus, *Mol. Pharmacol.* **17,** 86–94 (1980).
86. L. S. Birnbaum, M. M. Mcdonald, P. C. Blair, A. M. Clark, and M. W. Harris, Differential toxicity of 2,3,7,8-tetrachlorodibenzo-p-dioxin (TCDD) in C57BL/6J mice congenic at the Ah locus, *Fundam. Appl. Toxicol.* **15,** 186–197 (1990).
87. N. I. Kerkvliet, L. B. Stepan, J. A. Brauner, J. A. Deyo, M. C. Henderson, R. S. Tomar, and D. R. Buhler, Influence of the Ah locus on the humoral immunotoxicity of 2,3,7,8-tetrachlorodibenzo-p-dioxin (TCDD) immunotoxicity: Evidence for Ah receptor-dependent and Ah receptor-independent mechanisms of immunosuppression, *Toxicol. Appl. Pharmacol.* **105,** 26–36 (1990).
88. F. H. Lin, S. J. Stohs, L. S. Birnbaum, G. Clark, G. W. Lucier, and J. A. Goldstein, The effects of 2,3,7,8-tetrachlorodibenzo-p-dioxin on the hepatic estrogen and glucocorticoid receptors in congenic strains of Ah responsive and Ah non-responsive C57BL/6J, *Toxicol. Appl. Pharmacol.* **108,** 129–139 (1991).
89. F. H. Lin, G. Clark, L. S. Birnbaum, G. W. Lucier, and J. A. Goldstein, Influence of the Ah locus on the effects of 2,3,7,8-tetrachlorodibenzo-p-dioxin on the hepatic epidermal growth factor receptor, *Mol. Pharmacol.* **39,** 307–313 (1991).
90. B. D. Abbott and L. S. Birnbaum, TCDD alters medial epithelial cell differentiation during palatogenesis, *Toxicol. Appl. Pharmacol.* **99,** 287–301 (1989).
91. B. D. Abbott, M. W. Harris, and L. S. Birnbaum, Comparisons of the effects of TCDD and hydrocortisone on growth factor expression provide insight into the synergistic interaction occurring in embryonic palates, *Teratology* **45,** 35–53 (1992).
92. G. W. Lucier, A. Tritscher, T. Goldsworthy, J. Foley, G. Clark, J. Goldstein, and R. Maronpot, Ovarian hormones enhance 2,3,7,8-tetrachlorodibenzo-p-dioxin-mediated increases in cell proliferation and preneoplastic foci in a two-stage model for rat hepatocarcinogenesis, *Cancer Res.* **51,** 1391–1397 (1991).
93. S. J. Stohs, B. D. Abbott, F. H. Lin, and L. S. Birnbaum, Induction of ethoxyresorufin-o-deethylase and inhibition of glucocorticoid receptor binding in skin and liver of haired and hairless HRS/J mice by topically applied 2,3,7,8-tetrachlorodibenzo-p-dioxin, *Toxicology* **65,** 123–136 (1990).

94. B. D. Abbott, L. S. Birnbaum, and R. M. Pratt, 2,3,7,8-TCDD-induced hyperplasia of the ureteral epithelium produces hydronephrosis in murine fetuses, *Teratology* **35,** 329–334 (1987).
95. B. D. Abbott and L. S. Birnbaum, Effects of TCDD on embryonic ureteric epithelial EGF receptor expression and cell proliferation, *Teratology* **41,** 71–84 (1990).
96. B. V. Madhukar, D. W. Brewster, and F. Matsumura, Effects of in vivo-administered 2,3,7,8-tetrachlorodibenzo-p-dioxin on receptor binding of epidermal growth factor in the hepatic plasma membrane of rat, guinea pig, mouse and hamster, *Proc. Natl. Acad. Sci. USA* **81,** 7407–7411 (1984).
97. T. Thunberg, Effect of TCDD on vitamin A and its relation to TCDD-toxicity, *Banbury Report* **18,** 333–344 (1984).
98. C. H. Bastomsky, Enhanced thyroxine metabolism and high uptake goiter in rats after single dose of 2,3,7,8-tetrachlorodibenzo-p-dioxin, *Endocrinology* **101,** 292–297 (1977).
99. M. H. Sweeney, R. W. Hornung, D. K. Wall, M. A. Fingerhut, and W. E. Halperin, Prevalence of diabetes and elevated serum glucose levels in workers exposed to 2,3,7,8-tetrachlorodibenzo-p-dioxin (TCDD), presented at Dioxin '92, Tampere, Finland, August 24–28 (1992).
100. W. Wolfe, J. Michalek, J. Miner, L. Needham, and D. Patterson, Jr., Diabetes versus dioxin body burdens in veterans of Operation Ranch Hand, presented at Dioxin '92, Tampere, Finland, August 24–28 (1992).
101. E. Enan, P. C. C. Liu, and F. Matsumura, 2,3,7,8-TCDD causes reduction of glucose transporting activities in the plasma membranes of adipose tissue and pancreas from the guinea pig, *J. Biol. Chem.* **267,** 19785–19791 (1992).
102. G. W. Lucier, C. H. Sewall, J. Vanden Heuvel, J. Kanno, A. M. Tritscher, and G. C. Clark, 2,3,7,8-TCDD alterations of thyroid function in a rodent tumor promoter model, *Toxicologist* **13,** 313 (1993).
103. B. A. Schwetz, J. M. Morris, G. L. Sparschu, V. K. Rowe, P. J. Gerhring, L. L. Emerson, and G. C. Gerbig, Toxicology of chlorinated dibenzo-p-dioxins, *Environ. Health Perspect.* **5,** 87–100 (1973).
104. J. R. Olson, M. A. Holscher, and R. A. Neal, Toxicity of 2,3,7,8-tetrachlorodibenzo-p-dioxin in the golden Syrian hamster, *Toxicol. Appl. Pharmacol.* **55,** 67–78 (1980).
105. J. R. Olson and B. P. McGarrigle, Comparative developmental toxicity of 2,3,7,8-tetrachlorodibenzo-p-dioxin (TCDD), *Chemosphere* **25,** 71–74 (1991).
106. T. A. Gasiewicz, L. E. Geiger, G. Rucci, and R. A. Neal, Distribution, excretion and metabolism of 2,3,7,8-tetrachlorodibenzo-p-dioxin in C57BL/6J, DBA/2J and B6D2F$_1$/J mice, *Drug Metab. Dispos.* **11,** 397–403 (1983).
107. K. M. Burbach-Dolwick, J. V. Schmidt, L. A. Carver, and C. A. Bradfield, Cloning of the human Ah-receptor cDNA, *Toxicologist* **13,** 25 (1993).
108. G. W. Lucier, Humans are a sensitive species to some of the biochemical effects of structural analogs of dioxin, *Environ. Toxicol. Chem.* **10,** 727–735 (1991).
109. D. W. Bombick, B. V. Madhukar, D. W. Brewster, and F. Matsumura, TCDD (2,3,7,8-tetrachlorodibenzo-p-dioxin) causes increases in protein kinases particularly protein kinase C in the hepatic plasma membrane of the rat and guinea pig, *Biochem. Biophys. Res. Commun.* **127,** 296–302 (1984).
110. D. W. Bombick, J. Lankin, K. Tullis, and F. Matsumura, 2,3,7,8-Tetrachlorodibenzo-p-dioxin causes increases in expression of c-erb-A and levels of protein-tyrosine kinases in selected tissues of responsive mouse strains, *Proc. Natl. Acad. Sci. USA* **85,** 4128–4132 (1988).
111. R. W. Moore, J. A. Parsons, R. C. Bookstaff, and R. E. Peterson, Plasma concentrations

of pituitary hormones in 2,3,7,8-tetrachlorodibenzo-p-dioxin-treated male rats, *J. Biol. Toxicol.* **4,** 165–172 (1989).

112. R. C. Bookstaff, R. W. Moore, and R. E. Peterson, 2,3,7,8-Tetrachlorodibenzo-p-dioxin increases the potency of androgens and estrogens as feedback inhibitors of luteinizing hormone secretion in male rats, *Toxicol. Appl. Pharmacol.* **104,** 212–224 (1990).
113. B. D. Abbott and L. S. Birnbaum, TCDD alters embryonic palatal medial epithelial cell differentiation in vitro, *Toxicol. Appl. Pharmacol.* **100,** 119–131 (1989).
114. B. D. Abbott and L. S. Birnbaum, Rat embryonic palatal shelves respond to TCDD in organ culture, *Toxicol. Appl. Pharmacol.* **103,** 441–451 (1990).
115. B. D. Abbott and L. S. Birnbaum, TCDD exposure of human embryonic palatal shelves in organ culture alters the differentiation of medial epithelial cells, *Teratology* **43,** 119–132 (1991).
116. G. L. Eadon, J. Kaminisky, K. Silkworth, D. Aldous, P. Hilker, R. O'Keefe, J. Smith, J. Gierthy, K. Hawley, N. Kim, and A. DeCaprio, Calculation of 2,3,7,8-TCDD equivalent concentrations of complex environmental contaminant mixtures, *Environ. Health Perspect.* **70,** 221–227 (1986).
117. D. Barnes, A. Alford-Stevens, L. Birnbaum, F. W. Kutz, W. Wood, and D. Patton, Toxicity equivalency factors for PCBs? *Qual. Assur. Good Pract. Regul. Law.* **1,** 70–81 (1991).
118. M. J. DeVito, W. E. Maier, J. J. Diliberto, and L. S. Birnbaum, Comparative ability of various PCBs, PCDFs and TCDD to induce cytochrome P-450 1A1 and 1A2 activity following 4 weeks of treatment, *Fundam. Appl. Toxicol.* **20,** 125–130 (1993).
119. L. M. Sargent, G. L. Sattler, B. Roloff, Y. Xu, C. A. Sattler, L. Meisner, and H. C. Pitot, Ploidy and specific karyotypic changes during promotion with phenobarbital, 2,5,2′,5′-tetrachlorobiphenyl and/or 3,4,3′,4′-tetrachlorobiphenyl in rat liver, *Cancer Res.* **52,** 955–962 (1992).

Chapter 6

Pharmacokinetics of Dioxins and Related Chemicals

James R. Olson

1. INTRODUCTION

Polychlorinated dibenzo-*p*-dioxins (PCDDs) and polychlorinated dibenzofurans (PCDFs) are two classes of structurally and toxicologically similar persistent environmental contaminants. These compounds are formed as by-products in various chemical and combustion processes and are now global environmental contaminants. The environmental persistence and lipophilic properties of these compounds have led to effective transport of these chemicals into the food chain with pronounced accumulation at higher trophic levels, including humans. As a result, there is concern regarding possible adverse effects of these contaminants on human health and the environment.

PCDDs and PCDFs elicit a broad spectrum of congener/isomer-, species-, and tissue-specific biological and toxicological responses, with the induction of hepatic cytochrome P-450 1A1 and 1A2 and extrahepatic cytochrome P-450 1A1 representing two of the most sensitive responses associated with exposure to these compounds. Chlorine substitution at the 2, 3, 7, and 8 positions is generally considered necessary for dioxin-like activity, with 2,3,7,8-tetrachlorodibenzo-*p*-dioxin (2,3,7,8-TCDD) representing the most potent and extensively studied of the PCDDs and PCDFs. The congener-dependent binding of 2,3,7,8-substituted PCDDs and PCDFs to the cytosolic Ah receptor (aryl hydrocarbon or dioxin receptor) is generally considered the initial event necessary for the expression of the dioxin-like activity of these compounds. Congener-specific affinity of PCDDs and PCDFs for the Ah receptor and congener-specific pharmacokinetics are two factors that contribute to the relative *in vivo* potency of a given PCDD or PCDF in a given species.

James R. Olson • Department of Pharmacology and Toxicology, State University of New York at Buffalo, Buffalo, New York 14214.

Dioxins and Health, edited by Arnold Schecter. Plenum Press, New York, 1994.

The pharmacokinetics of PCDDs and PCDFs is congener-, dose-, and species-specific, with urinary and biliary excretion being dependent on the metabolism of these compounds. For PCDDs and PCDFs it is considered that the parent compounds are the causal agents, with metabolism and subsequent elimination of these compounds representing a detoxification process. In this respect, pharmacokinetics plays a significant role in determining the overall toxicity of these compounds. The disposition and pharmacokinetics of 2,3,7,8-TCDD and related compounds have been investigated in several species and under various exposure conditions. There are several reviews on this subject that focus on 2,3,7,8-TCDD and related halogenated aromatic hydrocarbons.[1–5] This chapter reviews the disposition and pharmacokinetics of these agents and identifies congener- and species-specific factors that may have an impact on the dose-related biological responses of these compounds.

2. ABSORPTION/BIOAVAILABILITY FOLLOWING EXPOSURE

The gastrointestinal, dermal, and transpulmonary absorption of these compounds are discussed herein because they represent potential routes for human exposure to this class of persistent environmental contaminants.

2.1. Oral

2.1.1. Gastrointestinal Absorption in Animals

A major source of human exposure to 2,3,7,8-TCDD and related compounds is thought to be through the diet.[6,7] Experimentally, these compounds are commonly administered in the diet or by gavage in an oil vehicle. Gastrointestinal absorption is usually estimated as the difference between the administered dose (100%) and the percentage of the dose that was not absorbed. The unabsorbed fraction is estimated as the recovery of parent compound in feces within 24–48 hr of a single oral exposure by gavage.

In Sprague–Dawley rats given a single oral dose of 1.0 μg [^{14}C]-2,3,7,8-TCDD/kg bw in acetone–corn oil (1:25, v/v), the fraction absorbed ranged from 66 to 93% with a mean of 84%.[8] With repeated oral dosing of rats at 0.1 or 1.0 μg/kg per day (5 days/week for 7 weeks), gastrointestinal absorption of 2,3,7,8-TCDD was observed to be approximately that observed for the single oral exposure.[8] Oral exposure of Sprague–Dawley rats to a larger dose of 2,3,7,8-TCDD in acetone–corn oil (50 μg/kg) resulted in an average absorption of 70% of the administered dose.[9]

Absorption was also investigated in the guinea pig and hamster, the species most sensitive and most resistant to the acute lethality of TCDD, respectively. TCDD in corn oil was generally well absorbed from both species, with 50 and 75% of the dose absorbed from the guinea pig and hamster, respectively.[9,10] Poiger and Schlatter[11] investigated the absorption of 2,3,7,8-TCDD in a 42-year-old man after ingestion of

105 ng [^{3}H]-2,3,7,8-TCDD (1.14 ng/kg bw) in 6 ml corn oil and found that > 87% of the oral dose was absorbed from the gastrointestinal tract. The results from several species suggest that TCDD is effectively absorbed following oral exposure in an oil vehicle.

The relative absorbed dose or bioavailability of 2,3,7,8-TBDD after a single oral exposure was estimated in the rat at 78, 82, 60, and 47% at dose levels of 0.001, 0.01, 0.1, and 0.5 μmole/kg, respectively. These results suggest nonlinear absorption at the higher doses, with maximal oral absorption at an exposure of ≤0.01 μmole/kg (5 μg/kg).[12]

The absorption of 2,3,7,8-TCDF has been investigated after oral exposure by gavage. Approximately 90% of the administered dose (0.1 and 1.0 μmole/kg) of 2,3,7,8-TCDF in Emulphor–ethanol (1:1) was absorbed in male Fischer 344 rats.[13] [Emulphor EL-620 is a polyethoxylated vegetable oil preparation (GAF Corp., New York, NY)]. Similarly, >90% of the administered dose (0.2 μmole/kg, 6 μg/kg, and 1–15 μg/kg) of 2,3,7,8-TCDF in Emulphor–ethanol–water (1:1:8) was absorbed in male Hartley guinea pigs.[14,15] Thus, 2,3,7,8-TCDF appears to be almost completely absorbed from the gastrointestinal tract. This may be related to the greater relative solubility of 2,3,7,8-TCDF compared with that of 2,3,7,8-TCDD or 2,3,7,8-TBDD.

The oral bioavailability of 2,3,4,7,8-PeCDF in corn oil is similar to that of 2,3,7,8-TCDD.[16] Furthermore, 2,3,4,7,8-PeCDF absorption was independent of the dose (0.1, 0.5, or 1.0 μmole/kg). Incomplete and variable absorption of 1,2,3,7,8-PeCDD in corn oil was reported in rats, with 19–71% of the dose absorbed within the first 2 days after oral exposure.[17] Birnbaum and Couture[18] found that the gastrointestinal absorption of OCDD in rats was very limited, ranging from 2 to 15% of the administered dose. Lower doses (50 μg/kg) in an *o*-dichlorobenzene–corn oil (1:1) vehicle were found to give the best oral bioavailability for this extremely insoluble compound.

The above data indicate that gastrointestinal absorption of 2,3,7,8-TCDD and related compounds is variable, incomplete, and congener specific. More soluble congeners, such as 2,3,7,8-TCDF, are almost completely absorbed, while the extremely insoluble OCDD is very poorly absorbed. In some cases, absorption has been found to be dose dependent, with increased absorption occurring at lower doses (2,3,7,8-TCDD, OCDD). The limited data base also suggests that there are no major interspecies differences in the gastrointestinal absorption of these compounds.

2.1.2. Bioavailability following Oral Exposure

Oral exposure of humans to 2,3,7,8-TCDD and related compounds usually occurs as a complex mixture of these contaminants in food, soil, dust, water, or other mixtures that would be expected to alter absorption.

The influence of dose and vehicle or adsorbent on gastrointestinal absorption has been investigated in rats by Poiger and Schlatter,[19] using hepatic concentrations 24 hr after dosing as an indicator of the amount absorbed. Administration of 2,3,7,8-TCDD in an aqueous suspension of soil resulted in a decrease in the hepatic levels of 2,3,7,8-TCDD as compared with hepatic levels resulting from administration of 2,3,7,8-TCDD in 50% ethanol. The extent of the decrease was directly proportional to the length of time the 2,3,7,8-TCDD had been in contact with the soil. When 2,3,7,8-TCDD was mixed in an aqueous suspension of activated carbon, absorption was almost totally eliminated (<0.07% of the dose in hepatic tissues).

Since 2,3,7,8-TCDD in the environment is likely to be adsorbed to soil,[20] Lucier *et al.*[21] compared the oral bioavailability of 2,3,7,8-TCDD from contaminated soil with that of 2,3,7,8-TCDD administered in corn oil in rats and guinea pigs, respectively. As indicated by biological effects and the amount of 2,3,7,8-TCDD in the liver, the intestinal absorption from Times Beach and Minker Stout, Missouri, soil was ~50% less than from corn oil. Shu *et al.*[22] reported an oral bioavailability of ~43% in the rat dosed with three environmentally contaminated soil samples from Times Beach, Missouri. This figure did not change significantly over a 500-fold dose range of 2–1450 ng 2,3,7,8-TCDD/kg bw for soil contaminated with ~2, 30, or 600 ppb of 2,3,7,8-TCDD. In studies of other soil types, Umbreit *et al.*[23,24] estimated an oral bioavailability in the rat of 0.5% for soil at a New Jersey manufacturing site and 21% for a Newark salvage yard. These results indicate that bioavailability of 2,3,7,8-TCDD from soil varies between sites and that 2,3,7,8-TCDD content alone may not be indicative of potential human hazard from contaminated environmental materials. Although these data indicate that substantial absorption occurs from contaminated soil, soil type and duration of contact may substantially affect the absorption of 2,3,7,8-TCDD from soils obtained from different contaminated sites.

2.2. Dermal Absorption

Brewster *et al.*[25] examined the dermal absorption of 2,3,7,8-TCDD and three PCDFs in male Fischer 344 rats (10 weeks old; 200–250 g) at 3 days after a single exposure using acetone as a vehicle. At an exposure of 0.1 μmole/kg, the absorption of 2,3,7,8-TCDF (49% of administered dose) was greater than that of 2,3,4,7,8-PeCDF (34%), 1,2,3,7,8-PeCDF (25%), and 2,3,7,8-TCDD (18%). For each compound, the relative absorption (percentage of administered dose) decreased with increasing dose while the absolute absorption (μg/kg) increased nonlinearly with dose. Results also suggest that the majority of the compound remaining at the skin exposure site was associated with the epidermis and did not penetrate through to the dermis. In a subsequent study, Banks and Birnbaum[26] examined the rate of absorption of 2,3,7,8-TCDD over 120 hr after the dermal application of 200 pmol (1 nmole/kg) to male Fischer 344 rats. The absorption kinetics appeared to be first-order, with an absorption rate constant of 0.005 hr^{-1}. First-order kinetics indicates that there is a constant rate of absorption (% dose absorbed/time). A first-order kinetic model can be described by the equation $S = S_0 e^{-kt}$, where S is the amount of compound at any time (t), S_0 is the initial amount administered, and k is the rate constant. This equation can be rearranged, solving for the half-time ($t^1/_2$) or the time it takes one-half of the compound to be absorbed, $t^1/_2 = \ln 2/k = 0.693/k$. With a rate constant of 0.0005 hr^{-1}, one-half of the dose of TCDD is absorbed in 139 hr. Together, these results on dermal absorption indicate that at lower doses (≤0.1 μmole/kg), a greater percentage of this administered dose of 2,3,7,8-TCDD and three PCDFs was absorbed. Nonetheless, the rate of absorption of 2,3,7,8-

TCDD is still very slow (rate constant of 0.005 hr^{-1}) even following dermal application in acetone at a dose of 200 pmol (1 nmole/kg).

Rahman *et al.*[27] and Gallo *et al.*[28] compared the *in vitro* permeation of 2,3,7,8-TCDD through hairless mouse and human skin. In both species, the amount of 2,3,7,8-TCDD permeated increased with the dose, but the percentage of the dose permeated decreased with increasing dose. The permeability coefficient of 2,3,7,8-TCDD in human skin was about one order of magnitude lower than that in mouse skin. The hairless mouse skin does not appear to be a suitable model for the permeation of 2,3,7,8-TCDD through human skin since the viable tissues were the major barrier to 2,3,7,8-TCDD permeation in hairless mouse skin, while the stratum corneum layer provided the greater resistance in human skin. A significant increase in 2,3,7,8-TCDD permeation through human skin was observed when the skin was damaged by tape-stripping. Gallo *et al.*[28] suggested that washing and/or tape-stripping of the exposed area might remove most of the 2,3,7,8-TCDD and reduce the potential for systemic exposure and toxicity since most of the 2,3,7,8-TCDD remained within the horny layer of human skin even at 24 hr following exposure.

Weber *et al.*[29] also investigated the penetration of 2,3,7,8-TCDD into human cadaver skin at concentrations of 65–6.5 ng/cm^2. This study also found that the stratum corneum acted as a protective barrier, as its removal increased the amount of 2,3,7,8-TCDD absorbed into layers of the skin. With intact skin and acetone as the vehicle, the rate of penetration into the dermis and epidermis ranged from 6 to 170 pg/hr per cm^2, while penetration into the dermis and epidermis ranged from 100 to 800 pg/hr per cm^2. With mineral oil as the vehicle, there was an approximately five- to tenfold reduction in the rate of penetration of 2,3,7,8-TCDD into the intact skin.

Bioavailability following Dermal Exposure

Dermal exposure of humans to 2,3,7,8-TCDD and related compounds usually occurs as a complex mixture of these contaminants in soil, oils, or other mixtures that would be expected to alter absorption. Poiger and Schlatter[19] presented evidence that the presence of soil or lipophilic agents dramatically reduces dermal absorption of 2,3,7,8-TCDD compared with absorption of pure compound dissolved in solvents. In a control experiment, 26 ng of 2,3,7,8-TCDD in 50 μl methanol was administered to the skin of rats, and 24 hr later the liver contained 14.8 ± 2.6% of the dose. Dermal application of 2,3,7,8-TCDD to rats in Vaseline (a lipophilic ointment) or polyethylene glycol (hydrophilic) reduced the percentage of the dose in hepatic tissue to 1.4 and 9.3%, respectively, but had no observable effect on the dose of 2,3,7,8-TCDD required to induce skin lesions (~1 μg/ear) in the rabbit ear assay. Application of 2,3,7,8-TCDD in a soil/water paste decreased hepatic 2,3,7,8-TCDD to ~2% of the administered dose and increased the amount required to produce skin lesions to 2–3 μg in rats and rabbits, respectively. Application in an activated carbon/water paste essentially eliminated absorption, as measured by percentage of dose in the liver, and increased the amount of 2,3,7,8-TCDD required to produce skin lesions to ~160 μg. These results suggest that the dermal absorption and acnegenic potency of 2,3,7,8-TCDD depend on the formulation (vehicle or adsorbent) containing the toxin.

Shu *et al.*[30] investigated the dermal absorption of soil-bound 2,3,7,8-TCDD in rats. The authors observed that the degree of uptake does not appear to be influenced significantly by the concentration of 2,3,7,8-TCDD in soil, the presence of crankcase oil has co-contaminants, or by environmentally versus laboratory-contaminated soil.

A major limitation of the above studies is the uncertainty regarding the extrapolation of dermal absorption data on these compounds from the rat to the human. The *in vitro* uptake of 2,3,7,8-TCDD has been investigated in hairless mouse and human skin.[27,28] *In vitro* dermal uptake of 2,3,7,8-TCDD from laboratory-contaminated soil found that aging of soils (up to 4 weeks) and the presence of additives (2,3,5-trichlorophenol and motor oil) in the soil did not have any significant effect on dermal uptake.[28] Since most of the 2,3,7,8-TCDD remained in the stratum corneum layer of human skin, the permeation of 2,3,7,8-TCDD was significantly lower in human than in hairless mouse skin.

2.3. Transpulmonary Absorption

The use of incineration as a means of solid and hazardous waste management results in the emission of contaminated particles that may contain TCDD and related compounds into the environment. Thus, significant exposure to TCDD and related compounds may result from inhalation of contaminated fly ash, dust, and soil. In an attempt to address the bioavailability and potential health implications of inhaling contaminated particles, Nessel *et al.*[31] examined the potential for transpulmonary absorption of TCDD after intratracheal instillation of the compound administered to female Sprague–Dawley rats either in a corn oil vehicle or as a laboratory-prepared contaminant of gallium oxide particles. Several biomarkers of systemic absorption were measured, including the dose-dependent effects of TCDD on hepatic microsomal cytochrome P-450 content, AHH activity, and liver histopathology. Significant dose-related effects were observed at an exposure of ≥ 0.55 μg TCDD/kg. The authors found that induction was slightly higher when animals received TCDD in corn oil than when animals received TCDD-contaminated particles and was comparable to induction after oral exposure. Similar bioavailability was also observed following inhalation exposure to TCDD when bound to gallium oxide particles and when bound to soil.[32]

3. DISTRIBUTION

3.1. Distribution in Blood and Lymph

Once a compound is absorbed, its distribution is regulated initially by its binding to components in blood and its ability to diffuse through blood vessels and tissue membranes. Lakshmanan *et al.*[33] investigated the absorption and distribution of 2,3,7,8-TCDD in thoracic duct-cannulated rats. Their results suggest that following gastrointestinal absorption, 2,3,7,8-TCDD is absorbed primarily by the lymphatic route and is transported predominantly by chylomicrons. Ninety percent of the 2,3,7,8-

TCDD in lymph was associated with the chylomicron fraction. The plasma disappearance of 2,3,7,8-TCDD-labeled chylomicrons followed first-order decay kinetics, with 67% of the compound leaving the blood compartment very rapidly ($t_{1/2}$ = 0.81 min), whereas the remainder of the 2,3,7,8-TCDD had a $t_{1/2}$ of 30 min. 2,3,7,8-TCDD was then found to distribute primarily to the adipose tissue and the liver.

In human blood, less than 10% of 2,3,7,8-TCDD was associated with red blood cells, indicating that most of this compound is bound to serum lipids and lipoproteins.[34] *In vitro* studies of 2,3,7,8-TCDD in human whole blood found ~80% of the compound associated with the lipoprotein fraction, 15% associated with protein (primarily human serum albumin), and 5% associated with cellular components.[35] Theoretical and limited experimental data also suggest that 2,3,7,8-TCDD and related compounds may be associated with plasma prealbumin.[36,37] The distribution of [^{3}H]-2,3,7,8-TCDD among lipoprotein fractions from three fasting, normolipemic donors indicated a greater percentage associated with LDL (55.3 ± 9.03% S.D.) than with VLDL (17.4 ± 9.07% S.D.) or HDL (27.3 ± 10.08% S.D.). The distribution of 2,3,7,8-TCDD among the lipoprotein fractions was similar to that reported earlier by Marinovich *et al.*[38] When the binding of 2,3,7,8-TCDD was calculated per mole of lipoprotein, maximal binding capacity was exerted by VLDL, followed by LDL and HDL.[38] The results also suggest that variations in the amounts of each lipoprotein class may alter the distribution of 2,3,7,8-TCDD among lipoproteins in a given subject. Significant species differences also exist; in the case of the rat, which has markedly lower plasma lipids relative to humans, 2,3,7,8-TCDD was distributed almost equally among the lipoprotein fractions.[38]

Congener-specific differences have also been observed for *in vivo* binding of the 2,3,7,8-substituted PCDDs and PCDFs to different serum fractions in the blood.[34] Binding to the lipoproteins gradually decreased with increasing chlorine content with about 75% of 2,3,7,8-TCDD bound to lipoproteins, while approximately 45% of OCDD was bound to this fraction. In contrast, binding to other proteins increased with chlorine content from approximately 20% for 2,3,7,8-TCDD to 50% for OCDD. Considerably less PCDDs and PCDFs were bound to the chylomicrons in serum, with less than 10% bound to this serum fraction.[34] In general, these *in vivo* results indicate that in serum, the higher chlorinated congeners do not partition according to the lipid content of the fractions. Thus, on absorption, 2,3,7,8-TCDD and probably related compounds are bound to chylomicrons, lipoproteins, and other serum proteins that assist in distributing these uncharged, lipophilic compounds throughout the vascular system. These compounds then partition from blood components into cellular membranes and tissues, probably largely by passive diffusion. In addition, cellular uptake may be facilitated partly through the cell membrane LDL receptor,[39] the hepatic receptor for albumin,[40] and/or other systems.

3.2. Tissue Distribution in Lab Animals

Once absorbed into blood, 2,3,7,8-TCDD and related compounds readily distribute to all organs. Tissue distribution within the first hour after exposure parallels blood levels and reflects physiological parameters such as blood flow to a given tissue and relative tissue size. For example, high initial concentrations of 2,3,7,8-TCDD and 1,2,3,7,8-PeCDF were observed in highly perfused tissue such as the adrenal glands during the 24-hr period after a single exposure.[10,41,42] A high percentage of the dose of 2,3,7,8-TCDF and 1,2,3,7,8-PeCDF was also found in muscle within the first hour after intravenous exposure, as a result of the large volume of this tissue.[4,13,42] Nevertheless, within several hours the liver, adipose tissue, and skin become the primary sites of disposition, when expressed as percentage of administered dose per gram tissue and as percentage of dose per organ. Liver, adipose tissue, skin, and thyroid were the only tissues to show an increase in the concentration of 2,3,7,8-TCDD during the initial 4 days after a single intraperitoneal exposure of rats.[41] In this study, a similar general pattern of disposition was observed in Han/Wistar and Long–Evans rats, which are respectively most resistant and susceptible to the acute toxicity of 2,3,7,8-TCDD.[41]

While the liver and adipose tissue contain the highest concentrations of 2,3,7,8-TCDD and 2,3,7,8-TCDF, there are some congener-specific differences in the relative tissue distribution of related compounds. Figure 1 illustrates the liver/adipose tissue concentration ratios for 2,3,7,8-substituted PCDDs and PCDFs on a wet weight basis in marmoset monkeys and rats.[43,44] Monkey and rat data were obtained 7 days following s.c. administration of a complex mixture of PCDDs and PCDFs in toluene–DMSO (1:2). Monkeys received a total dose of 27,800 ng/kg (464 ng I-TE/kg) while rats received a total dose of 23,222 ng/kg (388 ng I-TE/kg). For most of the 2,3,7,8-substituted congeners, the highest concentrations on a wet weight basis were detected in hepatic and adipose tissue, with correspondingly lower values detected in kidney, brain, lung, heart, thymus, or testes. The hepatic and adipose tissue concentrations were similar for 2,3,7,8-TCDD and 2,3,7,8-TCDF. However, with increasing chlo-

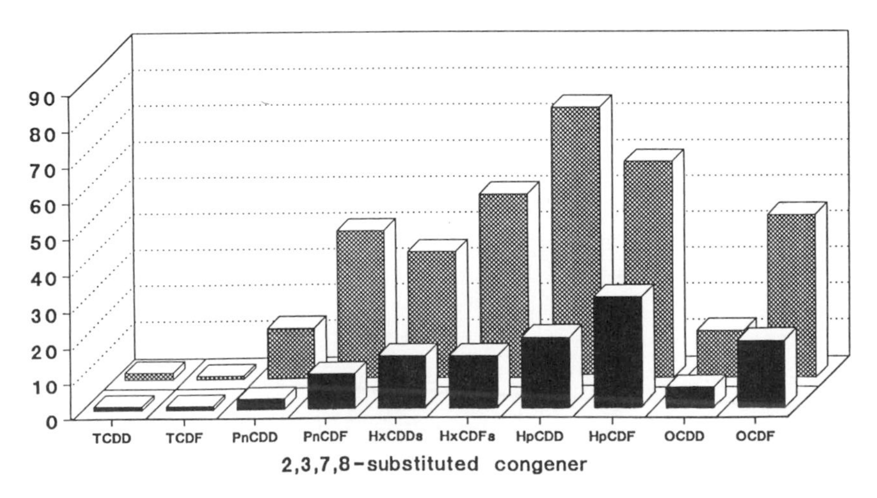

Figure 1. Liver/adipose tissue ratios for 2,3,7,8-substituted PCDDs and PCDFs in marmoset monkeys (■) and rats (▩). The ratios were obtained from tissue concentrations on a wet weight basis. Data from Abraham *et al.*[44] and Neubert *et al.*[43]

rination the relative disposition to the liver markedly increases. The preferential hepatic accumulation of 2,3,7,8-substituted congeners was more pronounced in the rat with the hepatic concentration of HpCDD being approximately 80-fold greater than that in adipose tissue. Thus, there are differences in the disposition of PCDDs and PCDFs to liver and adipose tissue. Therefore, adipose tissue and/or serum concentrations of PCDDs and PCDFs may not reflect the concentrations of specific congeners in target tissues, such as the liver.

Tissue Distribution in Humans

Facchetti *et al.*[45] reported tissue concentrations of 2,3,7,8-TCDD at levels of 1–2 ng/g in adipose tissue and pancreas, 0.1–0.2 ng/g in the liver, and $\leq$0.1 ng/g in thyroid, brain, lung, kidney, and blood in a woman who died 7 months after possible exposure to 2,3,7,8-TCDD from the Seveso accident. This pattern of 2,3,7,8-TCDD distribution, however, may not be representative for humans since the woman at the time of death had an adenocarcinoma (which was not considered related to the accident) involving the pancreas, liver, and lung.

Poiger and Schlatter[11] estimated that ~90% of the body burden of 2,3,7,8-TCDD was sequestered in the fat after a volunteer ingested [^{3}H]-2,3,7,8-TCDD in corn oil at a dose of 1.14 ng/kg. During this 135-day study, elevated radioactivity was detected in the blood only during the first 2 days after treatment. The data would be consistent with the high bioconcentration potential of 2,3,7,8-TCDD in humans, as calculated by Geyer *et al.*[46] from daily intake assumptions, levels in human adipose tissue, and pharmacokinetic models. Geyer *et al.*[46] estimated a bioconcentration factor (BCF) of between 104 and 206 for 2,3,7,8-TCDD in human adipose tissue.

Patterson *et al.*[47] developed a high-resolution gas chromatographic/high-resolution mass spectrometric analysis for 2,3,7,8-TCDD in human serum. The arithmetic mean of the individual human serum samples was 47.9 ppq on a whole-weight basis and 7.6 ppt on a lipid-weight basis. Paired human serum and adipose tissue levels of 2,3,7,8-TCDD have been compared by Patterson *et al.*,[48] Kahn *et al.*,[49] and Schecter *et al.*[50] All three groups reported a high correlation between adipose tissue and serum 2,3,7,8-TCDD levels when the samples were adjusted for total lipid content. Furthermore, their correlation was observed over a concentration range of almost three orders of magnitude.[48] This correlation indicates that serum 2,3,7,8-TCDD is a valid estimate of the 2,3,7,8-TCDD concentration in adipose tissue.

Schecter *et al.*[50] investigated the partitioning of 2,3,7,8-substituted PCDDs and PCDFs between adipose tissue and plasma lipid content in 20 Massachusetts Vietnam veterans. The distribution ratio between plasma lipid and adipose tissue increased with chlorine substitution on the PCDDs and PCDFs. While 2,3,7,8-substituted TCDD, TCDF, PeCDD, PeCDF, HxCDD, and HxCDF had a plasma lipid-to-adipose tissue ratio of about 1.0, OCDD had a ratio of about 2.0. On the other hand, whole blood PCDDs and PCDFs seem to be found at the same concentrations as in adipose tissue, on a lipid basis.[51] Schecter *et al.*[54,55] also reported the mean PCDD and PCDF levels on a wet weight basis in human autopsy tissue samples from two patients from the

Table 1
Mean PCDD and PCDF Levels on a Wet Weight Basis in Human Autopsy Tissue Samples from Two Patients from the United States[a]

Analyte	Abdominal	Subcutaneous	Adrenal	Liver	Muscle	Spleen	Kidney
2378-TCDD	6.6	4.9	3.8	2.5*	ND	1.3*	ND[b]
12378-PnCDD	9.0	7.9	4.0	3.7*	1.4	6.1	ND
123678-HxCDD	62.5	60.5	37	27.3	6.9	1.7	3.3
1234678-HpCDD	110	106.5	45.5	33.0	9.5	8.9	9.1
OCDD	555	645	405	285	123	33.0	35
23478-PnCDF	15.5	16.5	5.2	5.4*	1.1*	ND	ND
Total HxCDF	40.5	30	9.8	10.6	2	ND	2.6
1234678-HpCDF	16	18	4	4.9	ND	ND	1.7*
Total PCDD	743	825	495	352	141	51	47
Total PCDF	72	65	19	21	3	ND	4
Total PCDD/F	815	890	514	373	144	51	51
Total PCDD TEQ	29	17	10	8	2	5	0.5
Total PCDF TEQ	12	11	4	4	1	ND	0.3
Total PCDD/F TEQ	41	28	14	12	3	5	0.8
% lipid	73	71	27	14	8	2	3

[a]Data from Schecter *et al.*[54,55] Totals are rounded.
[b]ND, not detected.
*Data value from one patient.

United States. Table 1 summarizes these data, which show no apparent relationship or ratio in the levels of PCDDs and PCDFs. In general, tissues contain higher levels of the higher chlorinated congeners and tissues with a greater lipid content contain higher levels of PCDDs and PCDFs.

The disposition of 2,3,7,8-substituted PCDDs and PCDFs in human liver and adipose tissue was assessed in a study of 28 people from the Munich area.[52,53] Table 2 summarizes these results, expressed both on a lipid basis and on a wet weight basis. The concentrations of PCDDs and PCDFs in adipose tissue and liver are not the same when calculated on a lipid basis. This is in contrast to the high correlation reported between adipose tissue and serum TCDD levels when expressed on a lipid weight basis.[48–50] Furthermore, the liver/adipose tissue ratio increased with the higher chlorinated PCDDs and PCDFs. The congener-specific hepatic deposition is also similar to that observed in rats and marmoset monkeys exposed to a complex mixture of PCDDs and PCDFs (Fig. 1). Therefore, it is important to consider congener- and tissue-specific differences in disposition of PCDDs and PCDFs when blood levels are used to estimate tissue levels or body burdens.

In a study of potentially heavily exposed Vietnam veterans, the Centers for Disease Control reviewed an Air Force study of 147 out of a total of 1200 Ranch Hand veterans who were either herbicide loaders or herbicide specialists in Vietnam.[56] At 15 to 20 years following exposure, the mean serum 2,3,7,8-TCDD level of 147 Ranch

Table 2
2,3,7,8-Substituted PCDDs and PCDFs in Human Liver and Adipose Tissue[a]

	Tissue concentrations on a lipid basis (ppt)			Tissue concentrations on a wet weight basis (ppt)	
	Fat	Liver	Liver/fat	Liver[b]	Liver/fat
TCDD	8.0	16.4	2.05	1.1	0.14
PeCDD	16.4	20.1	1.22	1.4	0.09
HxCDD	94.7	166.8	1.76	11.7	0.12
HpCDD	106.7	1002.4	9.39	70.2	0.66
OCDD	373.2	4416.2	11.83	309.1	0.83
TCDF	2.5	5.5	2.20	0.4	0.15
PeCDF	35.2	173.7	4.93	12.2	0.35
HxCDF	41.5	389.5	9.38	27.3	0.66
HpCDF	14.2	218.9	15.42	15.3	1.08
OCDF	4.0	29.7	7.43	2.1	0.52

[a]Data from Thoma *et al.*[53] Values are the mean of 28 people from the Munich area.
[b]Estimated from the % fat in the liver (7.02 ± 5.33%, mean ± S.D.).

Hand personnel was 49 ppt, based on total lipid weight, while the mean serum level of the 49 controls was 5 ppt. In addition, 79% of the Ranch Hand personnel and 2% of the controls had 2,3,7,8-TCDD levels ≥ 10 ppt. The distribution of 2,3,7,8-TCDD levels in this phase of the Air Force health study indicates that some Ranch Hand personnel had unusually heavy 2,3,7,8-TCDD exposure (5 individuals with levels >200 ppt). Similar results were obtained by Kahn *et al.*[49] and Schecter *et al.*[50,57] who compared 2,3,7,8-TCDD levels in blood and adipose tissue of Agent Orange-exposed Vietnam veterans and matched controls.[49] Although these studies demonstrated a methodology that can distinguish heavily exposed men from others, the data do not address the question of identifying persons whose 2,3,7,8-TCDD exposures are small to moderate and who constitute the bulk of the population, both military and civilian, who may have been exposed to greater than background levels of 2,3,7,8-TCDD. However, Schecter[51] has reported that background tissue levels are lower in individuals from less industrialized countries, which in turn have fewer background environmental sources for exposure to PCDDs and PCDFs.

3.3. Time-Dependent Tissue Distribution

2,3,7,8-TCDD and related compounds exhibit congener-specific disposition, which depends on tissue, species, and time after a given exposure. In general, these compounds are cleared rapidly from the blood and distributed to liver, muscle, skin,

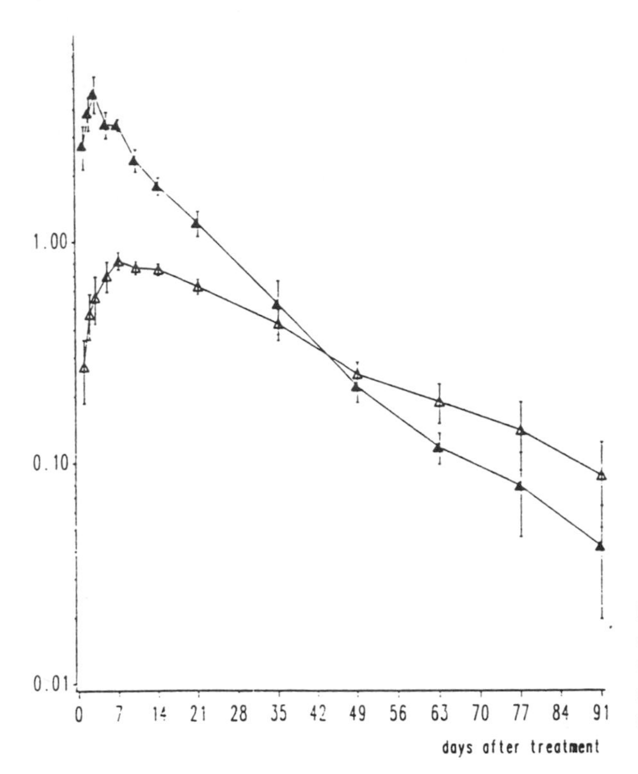

Figure 2. Time course of the concentration of [^{14}C]-TCDD in rat liver (▲) and adipose tissue (△) after a single subcutaneous injection of 300 ng TCDD/kg bw to female rats (mean ± S.D.). (From Abraham *et al.*[58])

adipose tissue, and other tissues within the first hour(s) after exposure. This is followed by redistribution primarily to the liver and adipose tissue, which exhibit increasing tissue concentrations over several days after exposure. Elimination from tissues then occurs at rates that are congener-, tissue-, and species-specific. Thus, the ratio of the concentration of 2,3,7,8-TCDD and related compounds in different tissues (i.e., liver/adipose) may not remain constant over an extended period after a single exposure. Abraham *et al.*[58] examined the concentrations of 2,3,7,8-TCDD in liver and adipose tissue of female Wistar rats over a 91-day period after a single subcutaneous exposure at a dose of 300 ng/kg bw (Fig. 2). The maximum concentration of 2,3,7,8-TCDD in the liver and adipose tissue was reached at 3 and 7 days after exposure, respectively, with 26.5% of the administered dose in the whole liver at day 7. The liver/adipose tissue concentration ratio does not remain constant over time since the concentration of 2,3,7,8-TCDD decreases more rapidly in the liver than in the adipose tissue. For example, the liver/adipose tissue concentration ratio (for 2,3,7,8-TCDD) on a wet weight basis was 10.3 at 1 day after exposure and 0.5 at 91 days after exposure. The decrease in the 2,3,7,8-TCDD concentration in adipose tissue is a linear function in the semilogarithmic plot of log concentration versus time which indicates apparent first-order elimination kinetics with a half-life of 24.5 days in rats. First-order kinetics indicates that there is a constant rate of elimination (% dose excreted/time). As with absorption, the first-order kinetic model can be described by the equation $S = S_0\mathrm{e}^{\ kt}$, where S is the amount of TCDD at any time (t), S_0 is the initial amount administered, and k is the rate constant for elimination. Liver tissue exhibits a biphasic (two-component) exponential decay pattern with a half-life of 11.5 days for the first component (days 10–49) and a half-life of 16.9 days for the second component (days 49–91). 2,3,7,8-TCDD is more persistent in the adipose tissue than in the liver. This is in contrast to the mouse, where liver and adipose tissue have similar half lives.[59] 2,3,7,8-TCDD is exceptionally persistent in the adipose tissue of the rhesus monkey, with a half-life approximately 10- to 40-fold greater than that observed in the rat and mouse.[60] Thus, the relative persistence of 2,3,7,8-TCDD is tissue-specific and exhibits marked interspecies variability.

Most experimental tissue distribution and elimination data are obtained after exposure to a single congener, while real-world exposure to 2,3,7,8-TCDD and related compounds occurs as a complex mixture of congeners. Recently, Neubert *et al.*[43] examined the persistence of various PCDDs and PCDFs in hepatic and adipose tissue of male and female marmoset monkeys. Animals received a single subcutaneous exposure to a defined PCDD/PCDF mixture (total dose of 27,800 ng/kg bw), which contained 120 ng 2,3,7,8-TCDD/kg bw. Using the I-TE (international TCDD toxic equivalence) factors,[61,62] the total administered dose corresponded to 464 ng I-TE/kg bw. The concentrations of specific congeners in liver and adipose tissue were measured at 1, 6, 16, or 28 weeks after exposure, and elimination constants and half lives were estimated assuming first-order kinetics (Table 3). Data in Table 3 were determined from pregnant and nonpregnant female and male marmosets since no obvious differences in tissue concentrations were observed among these groups. All 2,3,7,8-substituted PCDDs and PCDFs were consistently more persistent in the adipose tissue of marmoset monkeys. In general, the persistence in adipose tissue was from ~1.3- to 2.0-fold greater than that in liver, with the exception of 1,2,3,4,7,8-/1,2,3,4,7,9-HxCDF, HpCDFs, and OCDF which were >3-fold more persistent in adipose tissue. For the latter congeners and OCDD, there was marked variance in half-life values, which may be the result of delayed and incomplete absorption of the exceptionally persistent congeners and the relatively short (28 weeks) period of investigation.

The exposure of marmoset monkeys to a complex mixture of PCDDs and PCDFs included exposure to both 2,3,7,8- and non-2,3,7,8-substituted congeners.[43] One week after exposure to this complex mixture, the non-2,3,7,8-substituted PCDDs and PCDFs were present in liver and adipose tissue in relatively minor quantities when compared with 2,3,7,8-substituted congeners; however, non-2,3,7,8-substituted compounds represented a considerable percentage of the exposure mixture. In this study, none of the non-2,3,7,8-substituted TCDDs, PeCDDs, TCDFs, or PeCDFs could be detected in the liver by gas chromatography/mass spectroscopy. Some of the hexa and hepta congeners were detected in adipose tissue and liver, but after 1 week, the total amount in the liver was > 5% of the administered dose only in the case of 1,2,4,6,8,9-HxCDF. Similar results were obtained in rats after exposure to a defined, complex mixture of PCDDs and PCDFs.[44] Additional short-term studies in rats provide evidence that the low tissue concentration of non-2,3,7,8-substituted congeners, measured 1 week after exposure, were the result of rapid elimination, since these congeners were detected at higher levels in the liver 13–14 hr after exposure.[44] These results in

Table 3
Elimination Constants and Half Lives of Various 2,3,7,8-Substituted CDDs and CDFs in Hepatic and Adipose Tissue of Marmoset Monkeys[a,b]

Congener	Hepatic tissue			Adipose tissue		
	K_e (week^{-1})	Half life (weeks)	95% confidence interval (weeks)	K_e (week^{-1})	Half life (weeks)	95% confidence interval (weeks)
2,3,7,8-TCDD[c]	0.0841 ± 0.0109	8.3	6.6–11.1	0.0658 ± 0.0072	10.5	8.7–13.4
1,2,3,7,8-PeCDD[c]	0.0649 ± 0.0101	10.7	8.2–15.4	0.0490 ± 0.0057	14.2	11.5–18.3
1,2,3,4,7,8-HxCDD	0.0702 ± 0.0059	9.9	8.4–11.8	0.0411 ± 0.0083	16.9	12.1–27.9
1,2,3,6,7,8-HxCDD	0.0558 ± 0.0046	12.4	10.7–14.9	0.0373 ± 0.0073	18.6	13.4–30.2
1,2,3,7,8,9-HxCDD	0.0767 ± 0.0078	9.0	7.5–11.3	0.0525 ± 0.0089	13.2	9.9–19.7
1,2,3,4,6,7,8-HpCDD	0.0518 ± 0.0081	13.4	10.2–19.3	0.0372 ± 0.0060	18.6	14.2–27.2
OCDD	0.0089 ± 0.0084	78	27–∞[d]	0.0122 ± 0.0093	101	20–∞[d]
2,3,7,8-TCDF	0.8012 ± 0.0549	<0.87[e]	<1.00	0.4986 ± 0.0829	1.39	1.05–2.06
1,2,3,7,8-/1,2,3,4,8-PeCDF	0.7476 ± 0.0294	0.93	0.86–1.00	0.4735 ± 0.0408	1.46	1.25–1.76
2,3,4,7,8-PeCDF	0.0786 ± 0.0048	8.8	7.9–10.0	0.0563 ± 0.0059	12.3	10.2–15.5
1,2,3,4,7,8-/1,2,3,4,7,9-HxCDF	0.0307 ± 0.0039	23	18–30	0.0103 ± 0.0074	68	28–∞[d]
1,2,3,6,7,8-HxCDF	0.0486 ± 0.0037	14.3	12.4–16.7	0.0290 ± 0.0091	24	15–62
1,2,3,7,8,9-HxCDF	0.0848 ± 0.0057	8.2	7.2–9.4	not analyzed[f]	NA[g]	NA
2,3,4,6,7,8-HxCDF	0.0373 ± 0.0057	18.6	14.3–26.5	0.0182 ± 0.0082	38	20–327
1,2,3,4,6,7,8-HpCDF	0.0186 ± 0.0072	37	21–152	−0.0140 ± 0.0137	∞[d]	54–∞[d]
1,2,3,4,7,8,9-HpCDF	0.0088 ± 0.0127	79	20–∞[d]	0.0011 ± 0.0112	660	30–∞[d]
OCDF	0.0040 ± 0.0096	174	30–∞[d]	−0.0042 ± 0.0148	∞[d]	28–∞[d]

[a]Source: Neubert *et al.*[43]
[b]Animals were treated subcutaneously with a single dose of a defined CDD–CDF mixture, and the tissues were analyzed at different times following treatment. Half lives were calculated from tissue concentrations of the 2,3,7,8-substituted congeners in hepatic and adipose tissue. Values are given as elimination rate constant K_e including estimated S.D. and half-life including 95% confidence intervals.
[c]Calculated from the time period: >6 weeks after injection.
[d]Calculated half-life is apparently infinite. Data for OCDD and OCDF are unreliable because of delayed absorption.
[e]Not detected in hepatic tissue 6 weeks after treatment; limits of detection used for calculation.
[f]Due to interference.
[g]NA, not applicable.

monkeys and rats are compatible with data from analysis of human tissue samples and milk in which the non-2,3,7,8-substituted congeners have also not been shown to be present in significant concentrations, when compared with the 2,3,7,8-substituted congeners.[52,63–67]

3.4. Dose-Dependent Tissue Distribution

Recent evidence suggests that the tissue distribution of 2,3,7,8-TCDD and possibly related compounds is dose dependent. Abraham *et al.*[58] investigated the distribution of 2,3,7,8-TCDD in liver and adipose tissue of rats 7 days after a single subcutaneous exposure to 2,3,7,8-TCDD at doses of 1–3000 ng/kg bw. Greater than 97% of the administered 2,3,7,8-TCDD was absorbed at all doses, with the exception of the 3000 ng/kg group where 84% of the dose was absorbed. Figure 3 illustrates the dose-dependent disposition of 2,3,7,8-TCDD in liver and adipose tissue (% dose/g) 7 days after exposure. A sharp increase in 2,3,7,8-TCDD concentration in liver was observed at exposure levels >10 ng/kg bw. Disposition in the liver increased from ~11% of the administered dose at an exposure level of 1–10 ng/kg bw to ~37% of the dose at an exposure level of 300 ng/kg bw. The increase in distribution to the liver was accompanied by a dose-related decrease in the concentration of 2,3,7,8-TCDD in the adipose

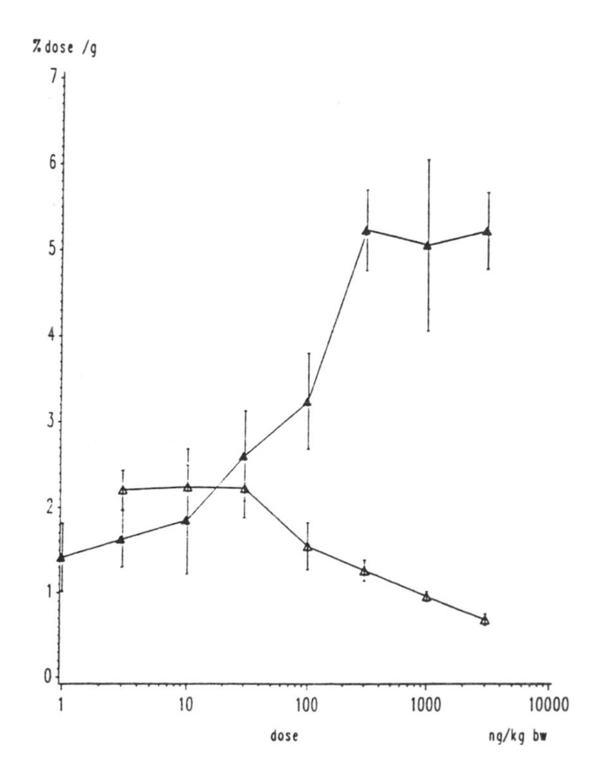

Figure 3. Dose dependency of the percentage of the administered dose of [^{14}C]-TCDD/g of tissue recovered in liver (▲) and adipose tissue (△) after a single subcutaneous exposure (values from animals treated with 3000 ng TCDD/kg bw were corrected for 84% absorption). Concentrations were measured 7 days after the injection. (From Abraham *et al.*[58])

tissue. As a result, the liver/adipose tissue concentration ratio for 2,3,7,8-TCDD increased with increasing doses, starting at an exposure level of 30 ng/kg bw. Thus, the tissue-specific disposition of 2,3,7,8-TCDD is regulated by a complex relationship, which includes species, time after a given exposure, and dose.

Other studies on the tissue disposition of 2,3,7,8-TCDD and related compounds report similar dose-dependent behavior with disproportionally greater concentrations in the liver at high doses compared with low doses.[68–73]

Chronic studies also support dose-dependent alterations in the tissue distribution of these compounds. Kociba *et al.*[74,75] found that female rats maintained on a daily dietary 2,3,7,8-TCDD intake of 100 ng/kg for 2 years had an average 2,3,7,8-TCDD content of 8100 ppt in fat and 24,000 ppt in the liver on a wet weight basis. Rats given 10 ng/kg per day had an average of 1700 ppt 2,3,7,8-TCDD in the fat and 5100 ppt in the liver. For both of these exposures the liver/adipose tissue concentration ratio of 2,3,7,8-TCDD was ~3. At the lowest dose level of 1 ng/kg per day, both fat and liver contained an average of 540 ppt 2,3,7,8-TCDD. Kociba *et al.*[76] presented evidence that steady state had been reached after about 13 weeks of feeding of 2,3,7,8-TCDD.

Other studies do not support the dose-dependent tissue distribution of 2,3,7,8-TCDD and related compounds described above.[8] Recently, Tritscher *et al.*[77] also reported a lack of a dose-dependent hepatic disposition of TCDD in female Sprague–Dawley rats exposed biweekly to TCDD for 30 weeks at doses equivalent to 3.5, 10.7, 35.7, and 125 ng/kg per day. A linear relationship between administered dose and the concentration in the liver was observed over the dose range used in this chronic exposure study.

The dose-dependent tissue distribution of 2,3,7,8-TCDD and related compounds is a critical factor that must be considered in estimating the concentration of these compounds in human tissues after chronic low-level exposure. This is particularly important since the general human population is exposed to much smaller daily doses (possibly 0.3 pg 2,3,7,8-TCDD/kg) than those used in experimental disposition studies.[7] Related at least partly to the long half-life of 2,3,7,8-TCDD in humans, however, this exposure results in concentrations of 3–6 pg/g in human adipose tissue.[51] Similar levels of 2,3,7,8-TCDD in adipose tissue (14 pg/g) were observed in rats 7 days after subcutaneous exposure to 3 ng/kg bw.[58] Under these experimental conditions, the liver/adipose tissue 2,3,7,8-TCDD concentration ratio was 0.74. Nonetheless, steady state was probably not reached under these conditions, and, with increasing time after exposure, this ratio may decrease, based on the observation that 2,3,7,8-TCDD was more persistent in adipose tissue than in liver in rats exposed to 300 ng/kg bw.[58] Human data on the liver/adipose tissue concentration ratio of 2,3,7,8-TCDD and related compounds are limited but suggest that the ratio may vary by at least an order of magnitude between individuals. Leung *et al.*[78] observed a geometric mean adipose tissue 2,3,7,8-TCDD concentration of 7.78 ppt in 26 individuals and a concentration in liver about one-tenth of that in adipose tissue on a whole weight basis. When measured on a total lipid basis, the concentrations of 2,3,7,8-TCDD in both tissues were approximately the same. In a related study of 28 people from the Munich area, Thoma *et al.*[53] reported a liver/fat ratio for TCDD of 2.05 when concentration was expressed on a lipid weight basis (Table 2). Considerable variability between individuals was observed in this study, with TCDD concentrations ranging from 2.6 to 18 ppt in adipose tissue and 1.0 to 88.9 ppt in liver on a lipid weight basis. Considerable variability in PCDD and PCDF concentrations in liver and adipose tissues was also observed between individual marmoset monkeys,[43] suggesting that individual variability may also contribute to the difficulty in assigning a constant liver/adipose tissue ratio for PCDDs and PCDFs in humans and nonhuman primates.

3.5. Potential Mechanisms for the Dose-Dependent Tissue Distribution

The observation that exposure to higher doses of 2,3,7,8-TCDD and related compounds results in a disproportionally greater hepatic concentration of these compounds may be explained by a hepatic binding protein that is induced by 2,3,7,8-TCDD and other agonists for the Ah receptor. The studies of Voorman and Aust[79,80] and Poland *et al.*[71,81] provide evidence that this binding species is cytochrome P-450IA2 (CYP1A2).

4. METABOLISM AND EXCRETION

There is evidence that a wide range of mammalian and aquatic species are capable of biotransforming 2,3,7,8-TCDD to polar metabolites.[10,82–87] Although metabolites of 2,3,7,8-TCDD have not been directly identified in humans, recent data regarding feces samples from humans in a self-dosing experiment suggest that humans can metabolize 2,3,7,8-TCDD.[88]

Investigations of 2,3,7,8-TCDD in rats, mice, guinea pigs, and hamsters found that $>$ 90% of the radiolabeled material excreted in urine and bile represented polar metabolites. Similar results were also observed for other congeners with the exception of OCDD although studies were often limited to the rat. OCDD is apparently not metabolized by the rat or metabolized to a very minimal extent.[18] For all of the congeners, essentially all of the PCDD-, and PCDF-derived radioactivity in liver, adipose tissue, and other tissues represented parent compound, suggesting that the metabolites of these compounds were readily excreted. Thus, with the exception of OCDD, the metabolism of 2,3,7,8-TCDD and related compounds is required for urinary and biliary elimination and therefore plays a major role in regulating the rate of excretion of these compounds.

4.1. Structure of Metabolites

Several metabolites of 2,3,7,8-TCDD have been identified. Sawahata *et al.*[89] investigated the *in vitro* metabolism of 2,3,7,8-TCDD in isolated rat hepatocytes. The major product was deconjugated with β-glucuronidase, derivatized with diazomethane, and separated into two compounds by HPLC. These metabolites were subsequently identified as 1-hydroxy-2,3,7,8-TCDD and 8-hydroxy-2,3,7-trichlorodibenzo-*p*-dioxin. Poiger *et al.*[86] identified six metabolites in the bile of dogs that were given a lethal dose of [^{3}H]-2,3,7,8-TCDD. The major metabolite was 1,3,7,8-tetrachloro-2-hydroxydibenzo-*p*-dioxin; however, 3,7,8-trichloro-3-hydroxydibenzo-*p*-dioxin and 1,2-dichloro-4,5-hydroxybenzene were identified as minor metabolites. The structures of the three remaining metabolites were not determined; however, two appeared to be trichlorohydroxydibenzo-*p*-dioxins and the third was apparently a chlorinated 2-hydroxydiphenyl ether. Poiger and Buser[90] reported differences in the relative amounts of various 2,3,7,8-TCDD metabolites in dog and rat bile. Trichlorodihydroxydibenzo-*p*-dioxin and tetrachlorodihydroxydiphenyl ether appear to be major metabolites in rat bile. Furthermore, conjugates, presumably glucuronides, were formed in the rat but not in the dog. The investigators also observed a generally higher metabolism rate of 2,3,7,8-TCDD in the dog.

4.2. Toxicity of Metabolites

The above discussion indicates that the metabolism of 2,3,7,8-TCDD and related compounds is required for urinary and biliary elimination and thus plays a major role in regulating the rate of excretion of these compounds. At present, metabolism is also generally considered a detoxification process.[91,92]

Data on the metabolism of 2,3,7,8-TCDD suggest that reactive epoxide intermediates may be formed. Poland and Glover[93] have investigated the *in vivo* binding of [1,6-^{3}H]-2,3,7,8-TCDD-derived radioactivity to rat hepatic macromolecules, and found maximum levels equivalent to 60 pmol nucleotide in RNA, and 6 pmol 2,3,7,8-TCDD/mol nucleotide in DNA. This corresponds to one 2,3,7,8-TCDD–DNA adduct/35 cells. These investigators suggest that it is unlikely that 2,3,7,8-TCDD-induced oncogenesis is through a mechanism of covalent binding to DNA and somatic mutation. Further studies of 2,3,7,8-TCDD and related compounds are needed to confirm these results and assess the relationship between covalent binding and the short- and long-term toxicity of these compounds.

It is possible that low levels of unextractable and/or unidentified metabolites may contribute to one or more of the toxic responses of 2,3,7,8-TCDD and related compounds. Further studies on the nature of the biotransformation products of these compounds will help to address this uncertainty.

4.3. Autoinduction of Metabolism

Accurate rate constants for metabolism are important in developing pharmacokinetic models that describe the disposition of 2,3,7,8-TCDD and related compounds. Metabolism plays a major role in regulating the excretion and relative persistence of these compounds, since metabolism is required for urinary and biliary excretion. Although the relative rate of metabolism of 2,3,7,8-TCDD and related compounds can be estimated from tissue and excretion half-life data, other factors such as relative body composition, hepatic and extrahepatic binding proteins, and direct intestinal elimination of the parent compound can also regulate the excretion of 2,3,7,8-TCDD and related compounds. Therefore, *in vivo* disposition data provide only a limited approximation of the relative rate of metabolism of a specific congener in a given species. *In vivo* disposition data were also obtained at exposures that were associated with induction of cytochromes P450IA1 and IA2 and other potentially adverse responses that could alter metabolism and disposition. Therefore, it may not be appropriate to directly extrapolate these data to predict the pharmacokinetics at low levels of exposure. Low-dose extrapolations can be assisted by assessments of the potential for autoinduction of metabolism which may occur at exposures that are associated with enzyme induction. Characterization of the dose-dependent disposition of 2,3,7,8,-TCDD and related compounds is particularly important in the extrapolation of high-exposure animal data to low-exposure human data.

The excretion of metabolites of 2,3,7,8-TCDD and related compounds into bile represents a direct means for estimating the rate of metabolism, since biliary elimination depends on metabolism and is the major route for excretion of these compounds. The small increase in metabolism and biliary excretion of 2,3,7,8-TCDD in the rat observed by Poiger and Buser[90] and the negative results of Kedderis *et al.*[69] and Curtis *et al.*[70] suggest that autoinduction of 2,3,7,8-TCDD metabolism and biliary excretion in the rat may not occur, or occurs to an extent that is not biologically relevant.

Limited data suggest that autoinduction of metabolism and biliary excretion does occur for PCDFs in contrast to PCDDs. Pretreatment of rats with 2,3,7,8-TCDF (1.0 μmole/kg, 3 days earlier) significantly increased the biliary excretion of a subsequent dose of [^{14}C]-2,3,7,8-TCDF.[94] The naive rats excreted 5.69 ± 2.35% of the dose over the initial 8 hr, while the pretreated rats excreted 13.18 ± 3.15% of the [^{14}C]-2,3,7,8-TCDF. Similarly, pretreatment of rats with 2,3,4,7,8-PeCDF (500 μg/kg, *per os*, 3 days earlier) resulted in a twofold increase in the biliary elimination of a subsequent dose of [^{14}C]-2,3,4,7,8-PeCDF.[16] These results suggest that pretreatment with 2,3,7,8-TCDF and 2,3,4,7,8-PeCDF induces the metabolism of these congeners.

Isolated hepatocytes in suspension culture have been used as an *in vitro* system for studying the autoinduction of metabolism of 2,3,7,8-TCDD and related compounds. *In vitro* results at a high substrate concentration (2.2 μM) indicate that 2,3,7,8-TCDD can induce its own rate of metabolism in the rat and hamster.[95] In contrast, 2,3,7,8-TCDD was not able to induce its own rate of metabolism in guinea pig and mouse hepatocytes.[73,96] More recently, the kinetics of 2,3,7,8-TCDD metabolism was investigated in isolated rat hepatocytes incubated with [^{3}H]-2,3,7,8-TCDD at concentrations of 0.01, 0.1, and 1.0 μM.[97] Lower 2,3,7,8-TCDD concentrations in the media result in concentrations in hepatocytes that are more similar to the levels in the liver after *in vivo* exposure. For example, the concentrations of 2,3,7,8-TCDD in hepatocytes incubated at 0.01 μM are similar to hepatic levels after *in vivo* exposure of rats at a dose of ~10 μg/kg. At 0.01 and 0.1 μM, the rate of metabolism of [^{3}H]-2,3,7,8-TCDD was similar in hepatocytes isolated from control and 2,3,7,8-TCDD-pretreated rats, while at 1.0 μM, [^{3}H]-2,3,7,8-TCDD metabolism was greater in hepatocytes isolated from

2,3,7,8-TCDD-pretreated rats. The results indicate that 2,3,7,8-TCDD can induce its own rate of metabolism in the rat, but only at high hepatic concentrations, which are generally not attained after *in vivo* exposure. Therefore, *in vitro* studies of the hepatic metabolism of TCDD (at 0.01 and 0.1 μm) are consistent with the lack of autoinduction of 2,3,7,8-TCDD metabolism and biliary excretion observed *in vivo* in the rat.[69,70]

The metabolism of [^{3}H]-2,3,7,8-TCDF was also investigated in isolated rat hepatocytes incubated at concentrations of 0.01, 0.1, and 1.0 μM.[97] At all concentrations, hepatocytes from 2,3,7,8-TCDD-pretreated rats metabolized 2,3,7,8-TCDF at a rate from 4- to 25-fold greater than that observed in hepatocytes from control rats. Results indicate that 2,3,7,8-TCDF is metabolized in rat liver by the 2,3,7,8-TCDD inducible enzyme, cytochrome P450IA1.[98] These *in vitro* results support the *in vivo* autoinduction of 2,3,7,8-TCDF metabolism and biliary elimination observed in the rat.[94] The results also suggest that 2,3,7,8-TCDF will be far more persistent following exposures at low doses which do not significantly induce CYP1A1.

4.4. Excretion in Animals

Data regarding the excretion of 2,3,7,8-TCDD and related compounds after exposure to a single radiolabeled congener support the assumption of a first-order elimination process consisting of one or more components. 2,3,7,8-TCDD was excreted slowly from all species tested, with half lives ranging from 11 days in the hamster to 2120 days in humans. 2,3,7,8-TCDD is exceptionally persistent in humans relative to other animal models.

Studies in the rat, guinea pig, hamster, and mouse have found that essentially all of the 2,3,7,8-TCDD-derived radioactivity excreted in the urine and bile corresponds to metabolites of 2,3,7,8-TCDD. The apparent absence of 2,3,7,8-TCDD metabolites in liver and fat suggests that once formed, the metabolites of 2,3,7,8-TCDD are excreted readily. Thus, urinary and biliary elimination of 2,3,7,8-TCDD depends on metabolism of the toxin. The more limited data for other compounds also suggest that this relationship may be true for 1,2,3,7,8-PeCDD,2,3,7,8-TBDD,2,3,7,8-TCDF, 1,2,3,7,8-PeCDF,2,3,4,7,8-PeCDF, and 3,3′,4,4′-TCB.

Although urine and bile appear to be free of unmetabolized 2,3,7,8-TCDD, 2,3,7,8-TCDD and its metabolites are excreted in the feces of guinea pigs, rats, mice, and hamsters treated with [^{3}H]- and/or [^{14}C]-2,3,7,8-TCDD.[1,10,84,85] The daily presence of unchanged 2,3,7,8-TCDD in feces and its absence in bile suggest that direct intestinal elimination may be the source for the fecal excretion of 2,3,7,8-TCDD. Lactation, direct intestinal elimination, and perhaps sebum may serve as routes for excretion of 2,3,7,8-TCDD, which do not depend on metabolism of the toxin. These data suggest that the *in vivo* half-life for elimination of 2,3,7,8-TCDD and related compounds only provides an approximation of the rate of metabolism of these compounds in a given animal.

The rate of excretion of 2,3,7,8-TCDD and related compounds is species- and congener-specific. 2,3,7,8-TCDD is most persistent in human and nonhuman primates. In the hamster, the least sensitive species to the acute toxicity of 2,3,7,8-

TCDD, the mean $t_{1/2}$ was 10.8 days[10,99] and in the guinea pig, the most sensitive species to the acute toxicity of 2,3,7,8-TCDD, the mean $t_{1/2}$ was 94 days.[84] 2,3,7,8-TCDF was also most persistent in the guinea pig, with a $t_{1/2}$ of 20–40 days.[14,15] Furthermore, results indicate that the relatively limited ability of the guinea pig to metabolize 2,3,7,8-TCDD and -TCDF may contribute to the greater persistence and greater acute toxicity of these congeners in the guinea pig.

The tissue distribution, metabolism, and excretion of 2,3,7,8-TCDD were also investigated in Han/Wistar and Long–Evans rats, which were, respectively, most resistant ($LD_{50} > 3000$ μg/kg) and most susceptible ($LD_{50} \sim 10$ μg/kg) to the acute toxicity of 2,3,7,8-TCDD.[41] The results suggest that the metabolism and disposition of 2,3,7,8-TCDD do not have a major role in explaining the strain differences in toxicity.

4.5. Excretion in Humans

Poiger and Schlatter[11] investigated the excretion of 2,3,7,8-TCDD in a 42-year-old man (92 kg) after ingesting 105 ng [^{3}H]-2,3,7,8-TCDD in 6 ml corn oil. The half-life for elimination was estimated to be 2120 days based on fecal excretion over a 125 day period following the single exposure (Table 4). The concentration of ^{3}H-TCDD-derived radioactivity was also measured in adipose tissue in the same individual over a 6 year period following exposure. A more accurate estimate of 2,3,7,8-TCDD half-life of 9.7 years was calculated based on adipose tissue concentrations over a 6 year period.[123] Table 4 summarizes additional half-life estimates for 2,3,7,8-TCDD and related compounds in humans, based on serum and/or adipose tissue concentrations at two or more time points. In another study, the half-life of 2,3,7,8-TCDD in humans was estimated to be ~7 years on the basis of 2,3,7,8-TCDD levels in serum samples taken in 1982 and 1987 from 36 of the Ranch Hand personnel who had 2,3,7,8-TCDD levels > 10 ppt in 1987.[100] Wolfe *et al.*[133] recently investigated the half-life of 2,3,7,8-TCDD in an expanded cohort of 337 Air Force veterans of Operation Ranch Hand that also included the 36 subjects of the earlier half-life study by Pirkle *et al.*[100] Based on paired 2,3,7,8-TCDD measurements from serum collected in 1982 and in 1987, the authors reported a mean predicted half-life of 11.6 years and a median observed half-life of 11.3 years with a nonparametric 95% confidence interval of 10.0–14.1 years. The authors also investigated how the 2,3,7,8-TCDD half-life varied with percent body fat (PBF), relative changes in PBF from 1982 to 1987, and age. They found that the 2,3,7,8-TCDD half-life increased significantly with a high PBF, suggesting that persons with more body fat tend to eliminate 2,3,7,8-TCDD more slowly. In contrast, increasing age was associated with a shorter half-life. The redistribution of fat stores from subcutaneous to abdominal areas with aging, resulting in greater mobilization of 2,3,7,8-TCDD, could in part explain the shorter half-life observed in older veterans. An increase in PBF from 1982 to 1987 was also associated with a decrease in half-life, which can be explained by a dilution of the existing body burden of 2,3,7,8-TCDD into the increasing adipose tissue mass. Future studies will examine these relationships further with data from a third serum sample collected in 1992 from these veterans. Additional data will also help address the question of the potential

Table 4
Half-Life Estimates for 2,3,7,8-TCDD and Related Compounds in Humans

Chemical	Exposure incident	Number of individuals	Sample	Time period between first and last analysis	Number of time points	Half-life (years)	Reference
CDDs							
2,3,7,8-TCDD	Male volunteer	1	Fecal excretion	125 days	28	5.8	Poiger and Schlatter[11]
2,3,7,8-TCDD	Male volunteer	1	Adipose tissue	6 years	5	9.7	Schlatter[123]
2,3,7,8-TCDD	Ranch Hand Vietnam Veterans	337	Serum	5 years	2	11.3[b]	Wolfe *et al.*[133]
2,3,7,8-TCDD	Ranch Hand Vietnam Veterans	36	Serum	5 years	2	7.1[a]	Pirkle *et al.*[100]
1,2,3,6,7,8-HxCDD	Technical pentachlorophenol in wood of home	1	Adipose tissue	28 months	2	3.5	Gorski *et al.*[132]
1,2,3,4,6,7,8-HpCDD	Technical pentachlorophenol in wood of home	1	Adipose tissue	28 months	2	3.2	Gorski *et al.*[132]
OCDD	Technical pentachlorophenol in wood of home	1	Adipose tissue	28 months	2	5.7	Gorski *et al.*[132]
CDFs							
2,3,4,7,8-PeCDF	Binghamton, New York, state office building	1	Adipose tissue	Initial 43 months	4	4.7	Schecter *et al.*[102]
			Blood	Final 29 months	4	7.2	
			Combined	Total 6 years	7	4.5	
1,2,3,4,7,8-HxCDF	Binghamton, New York, state office building	1	Adipose tissue	Initial 43 months	4	2.9	Schecter *et al.*[102]
			Blood	Final 29 months	4	4.4	
			Combined	Total 6 years	7	4.0	
1,2,3,6,7,8-HxCDF	Binghamton, New York, state office building	1	Adipose tissue	Initial 43 months	4	3.5	Schecter *et al.*[102]
			Blood	Final 29 months	4	4.3	
			Combined	Total 6 years	7	4.9	

1,2,3,4,6,7,8-HpCDF	Binghamton, New York, state office building	1	Adipose tissue	Initial 43 months	4	6.5	Schecter *et al.*[102]
			Blood	Final 29 months	4	4.1	
			Combined	Total 6 years	7	6.8	
2,3,4,7,8-PeCDF	Yu-cheng	4	Blood	Initial 2.9 years	2	1.3	Ryan and Masuda[103]
		3		Final 2.7 years	2	2.9	
		2		Total 5.6 years	3	1.7	
1,2,3,4,7,8-HxCDF	Yu-cheng	4	Blood	Initial 2.9 years	2	2.1	Ryan and Masuda[103]
		3		Final 2.7 years	2	5.1	
		2		Total 5.6 years	3	2.4	
1,2,3,4,6,7,8-HpCDF	Yu-cheng	4	Blood	Initial 2.9 years	2	1.6	Ryan and Masuda[103]
		3		Final 2.7 years	2	6.1	
		2		Total 5.6 years	3	2.4	
2,3,4,7,8-PeCDF 1,2,3,4,7,8-HxCDF 1,2,3,4,6,7,8-HpCDF	Yu-cheng	3	Blood	9 years	5–6	2–3	Ryan and Masuda[101]
2,3,4,7,8-PeCDF 1,2,3,4,7,8-HxCDF	Yusho	9	Blood	7 years	3–5	>5	Ryan and Masuda[101]
1,2,3,4,6,7,8-HpCDF	Technical pentachlorophenol in wood of home	1	Adipose tissue	28 months	2	<1.7	Gorski *et al.*[132]
OCDF	Technical pentachlorophenol in wood of home	1	Adipose tissue	28 months	2	1.8	Gorski *et al.*[132]
PCBs							
3,3′,4,4′,5-PeCB	Yu-cheng	NA[c]	Blood	NA	NA	<1	Ryan and Masuda[101]
3,3′,4,4′,5,5′-HxCB	Yu-cheng	NA	Blood	NA	NA	10	Ryan and Masuda[101]

[a]95% confidence interval about the median of 5.8–9.6 years.
[b]95% confidence interval about the median of 10.0–14.1 years.
[c]NA, not applicable.

biphasic elimination of 2,3,7,8-TCDD. These studies indicate that 2,3,7,8-TCDD is exceedingly persistent in humans. Estimated half lives for other congeners in Table 4 range from 0.8 to 10 years. The half-life values in Table 4 are rough estimates based on a small number of individuals and based on analysis at as few as two time points. Estimates also assume a simple, single-compartment, first-order elimination process.

Ryan and Masuda[101] reported on their continuing investigation into the elimination of PCDFs in humans from the Yusho and Yu-cheng rice oil poisonings. Yu-cheng individuals had PCDF blood levels on a lipid basis of 1–50 μg/kg, while Yusho patients had levels of 0.1–5 μg/kg. In the Yu-cheng individuals, half lives for three PCDFs were 2–3 years, while elimination from Yusho individuals was more variable and slower, with half lives >5 years (see Table 4) and, in several cases, no measurable elimination occurred during the 7 years in which samples were available. The limited results suggest that clearance of these PCDFs in the human is biphasic, with faster elimination at higher exposure. Schecter *et al.*[102] and Ryan and Masuda[103] also reported longer half-life values for PCDFs in humans at later time points after exposure, when concentrations are closer to the background levels of individuals with no unusual exposure.

Because of the lipophilic nature of milk, lactation can provide a relatively efficient mechanism for decreasing the body burden of 2,3,7,8-TCDD and related PCDDs and PCDFs in females. As discussed by Schecter and Gasiewicz[104,105] and Graham *et al.*,[106] this elimination of 2,3,7,8-TCDD and related compounds through mother's milk can result in high exposure levels in the infant. Since both milk and the fatty tissues of fish are essentially providing an oily vehicle, it is likely that these sources would provide 2,3,7,8-TCDD and related compounds in a form that is readily bioavailable.

Several investigators have quantified the levels of 2,3,7,8-TCDD in human milk samples. Many of the milk samples were pooled.[107] Rappe[108] reported levels of 1–3 ppt 2,3,7,8-TCDD in milk fat from five volunteers in West Germany, and in a later report, Rappe *et al.*[109] reported an average level of 0.6 ppt 2,3,7,8-TCDD in milk fat from four volunteers in northern Sweden. Furst *et al.*[110] reported an average level of 9.7 ppt 2,3,7,8-TCDD in milk fat from three individuals in the Netherlands and <1.0 ppt 2,3,7,8-TCDD in milk fat from two individuals in Yugoslavia. Nygren *et al.*[111] reported average levels of 2,3,7,8-TCDD in human milk samples from four subjects in Sweden to be 0.6 pg/g in milk fat, and in five subjects from West Germany to be 1.9 pg/g in milk fat. Schecter[51] compared PCDD and PCDF levels in human milk, in terms of milk lipid dioxin TEQs, in a number of countries characterized by varying degrees of industrialization. The United States, Japan, Canada, and Germany led with values of 20, 27, 26, and 27 ppt, respectively, while Thailand, Cambodia, and Siberia had values of 3, 3, and 12, respectively.

High levels of 2,3,7,8-TCDD have been detected in the milk of mothers exposed to high levels of 2,3,7,8-TCDD in the environment. Reggiani[112] reported levels between 2.3 and 28.0 ppt 2,3,7,8-TCDD in whole milk from mothers in Seveso. Baughman[114] reported levels between 400 and 1450 ppt 2,3,7,8-TCDD in milk lipid from mothers in South Vietnam. Reanalysis of these South Vietnamese samples, originally collected in 1973, found 77 to 230 ppt 2,3,7,8-TCDD in milk lipid.[113] Human milk samples collected in South Vietnam in 1985–1988 had 2.9 to 11.0 ppt 2,3,7,8-TCDD in milk lipid,[115] while North Vietnamese samples contained 2.1 ppt 2,3,7,8-TCDD in milk lipid.

5. PHYSIOLOGICALLY BASED PHARMACOKINETIC MODELS

Physiologically based pharmacokinetic (PB-PK) models have been developed for 2,3,7,8-TCDD in C57BL/6J and DBA/2J mice,[116] rats,[117] and humans.[118] PB-PK models incorporate known or estimated anatomical, physiological, and physicochemical parameters to describe quantitatively the disposition of a chemical in a given species. PB-PK models can assist in the extrapolation of high to low-dose kinetics within a species, estimating exposures by different routes of administration, calculating effective doses, and extrapolating these values across species.[119]

Andersen *et al.*[120] recently described a receptor-mediated PB-PK model for the tissue distribution and enzyme inducing properties of 2,3,7,8-TCDD. The data used for this analysis were from two previously published studies with Wistar rats.[58,121] The model was used to examine the tissue disposition of 2,3,7,8-TCDD and the induction of both a dioxin-binding protein (presumably cytochrome P4501A2) and cytochrome P4501A1.

Kohn *et al.*[122] recently developed a mechanistic model of the effects of dioxin on gene expression in the rat liver (referred to as the NIEHS model). The model includes the tissue distribution of 2,3,7,8-TCDD in the rat and its effect on the concentrations of CYP1A1 and CYP1A2, and the effects of 2,3,7,8-TCDD on the Ah, estrogen, and EGF receptors over a wide 2,3,7,8-TCDD dose range. Experimental data from Tritscher *et al.*[77] were incorporated into the NIEHS model. Female Sprague–Dawley rats were injected with an initiating dose of dimethylnitrosamine, and after 10 days, the rats were exposed biweekly to 2,3,7,8-TCDD in corn oil by gavage at doses equivalent to 3.5–125 ng/kg per day for 30 weeks. The NIEHS model predicts a linear relationship between administered dose and the concentration in the liver over this dose range, which is in agreement with the data of Tritscher *et al.*[77] The biochemical response curves for all these proteins were hyperbolic, indicating a proportional relationship between target tissue dose and protein concentration at low administered doses of 2,3,7,8-TCDD.

A fugacity-based PB-PK model for the elimination of 2,3,7,8-TCDD from humans was developed by Kissel and Robarge.[118] Transport within the body was assumed to be perfusion-limited. 2,3,7,8-TCDD was assumed to be uniformly distributed within each tissue or fluid phase, and tissue levels were considered to be in equilibrium with exiting fluids (blood, bile, urine). 2,3,7,8-TCDD appears to be poorly metabolized in humans, thus reducing the necessity of modeling the fate of metabolites. 2,3,7,8-TCDD also seems to exhibit fugacity-based partitioning behavior in humans as evidenced by relatively constant lipid-based tissue distribution,[78,124] although this is not the case in rodents.[72,116,117] With a human background intake of 2,3,7,8-TCDD in North America of ~50 pg/day,[125] the steady-state adipose tissue concentration predicted by the model, assuming no metabolism, was 7.7 ppt. This is similar to the lipid-based blood tissue levels reported in the general population with no known unusual exposure. The model was also used to predict the elimination of 2,3,7,8-TCDD from Ranch Hand Vietnam veterans. The model simulation assumed a background exposure of 50 pg/day and no metabolism. Under these conditions, apparent half lives of 4.4, 5.2, 5.9, 7.2, 9.1, and 20 years were estimated for individuals with adipose tissue concentrations of 100, 50, 30, 20, 15, and 10 ppt, respectively. The model-predicted half lives are similar to the experimental value of 7.1 years, based on analysis of 2,3,7,8-TCDD in blood lipids of veterans with adipose burdens greater than 10 ppt[100] (see Table 4). The apparent half lives derived from the model increased as the adipose tissue concentrations approached the steady-state level associated with background exposure. Ryan and Masuda[101] also reported a similar relationship for PCDFs, with experimentally derived half lives increasing in individuals with lower body burdens of the compounds. Finally, the model was also found to approximate the elimination of 2,3,7,8-TCDD from one volunteer as reported by Poiger and Schlatter.[11] Taken together, the comparisons described above suggest that a fugacity-based PB-PK model for 2,3,7,8-TCDD in humans can provide one method for describing the elimination of 2,3,7,8-TCDD from humans.

PB-PK models are primarily limited by the availability of congener- and species-specific data that accurately describe the dose- and time-dependent disposition of 2,3,7,8-TCDD and related compounds. As additional data become available, particularly on the dose-dependent disposition of these compounds, more accurate models can be developed. In developing a suitable model in the human, it is also important to consider that the half-life estimate of 7.1 years for 2,3,7,8-TCDD was based on two serum values taken 5 years apart, with the assumption of a single-compartment, first-order elimination process.[100] It is likely that the excretion of 2,3,7,8-TCDD in humans is more complex, involving several-compartment, tissue-specific binding proteins and a continuous daily background exposure. Furthermore, changes in body weight and body composition should also be considered in developing PB-PK models for 2,3,7,8-TCDD and related compounds in humans.

6. PHARMACOKINETICS IN SPECIAL POPULATIONS

Pregnancy and Lactation (Prenatal and Postnatal Exposure of Offspring)

Placental transfer of 2,3,7,8-TCDD in rats and mice is relatively limited, with a single oral maternal exposure on gestation day 11 of mice resulting in about 0.03% of the dose delivered to each embryo.[126] In contrast, excretion into milk represents a major pathway for maternal elimination of 2,3,7,8-TCDD and for subsequent exposure of pups. During the first 2 postnatal weeks, mouse pups were given doses of 2,3,7,8-TCDD via the milk that were, on a body weight basis, similar to those that had been administered prenatally to their mothers.[127]

The transfer of PCDDs and PCDFs through the placenta and via the milk was investigated in rats and a marmoset monkey.[128,129] All of the congeners in rat fetal and neonatal tissues were 2,3,7,8-substituted with the exception of 2,3,4,6,7-PeCDF. 2,3,7,8-TCDD and 1,2,3,7,8-PeCDD were found at the highest concentration in the liver of the newborn marmoset (about 0.15% of dose/g). For all other congeners, the concentrations in the liver of the newborn were <10% of the corresponding concentrations in adults. In contrast to liver, concentrations of 2,3,7,8-substituted congeners in the adipose tissue of the newborn marmoset were at least 33% of the levels in adults, and in the case of OCDD and OCDF, levels were threefold higher in the newborn than in the adult. The concentrations of the PCDDs and PCDFs in the newborn marmoset were highest in the adipose tissue, followed by the skin and liver. Thus, the hepatic concentrations in the marmoset fetus may not be representative of the rate of placental transfer of PCDDs and PCDFs. As expected from rodent studies, the transfer of PCDDs and PCDFs via mother's milk was considerable, resulting in hepatic concentrations of 2,3,7,8-TCDD, 1,2,3,7,8-PeCDD, and 1,2,3,6,7,8-HxCDD in the suckled infant marmoset (postnatal day 33) higher than those in the dam. Transfer of hepta- and octa-PCDDs and PCDFs to the suckled infant was rather low, only about 10% of the levels in the dam. The pre- and postnatal transfer of 2,3,7,8-TCDD to the offspring of rhesus monkeys was investigated by Bowman *et al.*[60] At weaning (4 months), the offspring had a 2,3,7,8-TCDD concentration in adipose tissue about fourfold greater than that in the mothers. The mothers excreted from 17 to 44% of their 2,3,7,8-TCDD body burden by lactation. Following weaning, the decrease in 2,3,7,8-TCDD levels in adipose tissue of young monkeys apparently followed first-order, single-component kinetics, with a half-life of about 181 days.[130] The corresponding half-life in adult rhesus monkeys was reported to range from 180 to 550 days.[60]

Furst *et al.*[131] examined the levels of PCDDs and PCDFs in human milk and the dependence of those levels on the period of lactation. The mean concentrations of PCDDs in human milk (on a fat basis) ranged from 195 ppt for OCDD to 2.9 ppt for 2,3,7,8-TCDD, with levels of the other congeners decreasing with decreasing chlorination. This is in contrast to the generally lower levels of PCDFs in human milk, which range from 25.1 ppt for 2,3,4,7,8-PeCDF to 0.7 ppt for 1,2,3,7,8-PeCDF. After breast feeding for 1 year, one mother had PCDD and PCDF levels that were 30–50% of the starting concentration, suggesting a more rapid mobilization of PCDDs and PCDFs and excretion into human milk during the first few weeks postpartum. In addition, the PCDD and PCDF levels in milk from mothers nursing their second child are on average 20–30% lower than those for their first child. Although further studies are necessary, the limited data suggest that there are time-dependent and isomer-specific differences in the excretion of PCDDs and PCDFs in human milk.

REFERENCES

1. R. A. Neal, J. R. Olson, T. A. Gasiewicz, and L. E. Geiger, The toxicokinetics of 2,3,7,8-tetrachlorodibenzo-p-dioxin in mammalian systems, *Drug Metab. Rev.* **13**(3), 355–385 (1982).
2. T. A. Gasiewicz, J. R. Olson, L. E. Geiger, and R. A. Neal, Absorption, distribution and metabolism of 2,3,7,8-tetrachlorodibenzo-p-dioxin (TCDD) in experimental animals, in: *Human and Environmental Risks of Chlorinated Dioxins and Related Compounds* (R. E. Tucker, A. L. Young, and A. P. Gray, eds.), pp. 495–525, Plenum Press, New York (1983).
3. J. R. Olson, T. A. Gasiewicz, L. E. Geiger, and R. A. Neal, The metabolism of 2,3,7,8-tetrachlorodibenzo-p-dioxin in mammalian systems, in: *Accidental Exposure to Dioxins: Human Health Aspects* (R. Coulston and F. Pocchiari, eds.), pp. 81–100, Academic Press, New York (1983).

4. L. S. Birnbaum, The role of structure in the disposition of halogenated aromatic xenobiotics, *Environ. Health Perspect.* **61,** 11–20 (1985).
5. M. Van den Berg, J. deJongh, H. Poiger, and J. R. Olson, The toxicokinetics and metabolism of PCDDs and PCDFs and their relevance for toxicity, *CRC Crit. Rev. Toxicol.* **24**(1), 1–74 (1994).
6. H. Beck, K. Eckart, W. Mathar, and R. Wittkowski, PCDD and PCDF body burden from food intake in the Federal Republic of Germany, *Chemosphere* **18,** 417–424 (1989).
7. P. Furst, C. Furst, and K. Wilmers, Body burden with PCDD and PCDF from food, in: *Biological Basis for Risk Assessment of Dioxins and Related Compounds* (M. A. Gallo, R. J. Scheuplein, and K. A. van der Heijden, eds.), pp. 133–142, Banbury Report 35, Cold Spring Harbor Laboratory Press, Cold Spring Harbor, NY (1991).
8. J. Q. Rose, J. C. Ramsey, T. H. Wentzler, R. A. Hummel, and P. J. Gehring, The fate of 2,3,7,8-tetrachlorodibenzo-p-dioxin following single and repeated oral doses to the rat *Toxicol. Appl. Pharmacol.* **36,** 209–226 (1976).
9. W. N. Piper, J. Q. Rose, and P. J. Gehring, Excretion and tissue distribution of 2,3,7,8-tetrachlorodibenzo-p-dioxin in the rat, *Environ. Health Perspect.* **5,** 241–244 (1973).
10. J. R. Olson, T. A. Gasiewicz, and R. A. Neal, Tissue distribution, excretion, and metabolism of 2,3,7,8-tetrachlorodibenzo-p-dioxin (TCDD) in the golden Syrian hamster, *Toxicol. Appl. Pharmacol.* **56**(1), 78–85 (1980).
11. H. Poiger and C. Schlatter, Pharmacokinetics of 2,3,7,8-TCDD in man, *Chemosphere* **15**(9), 1489–1494 (1986).
12. J. J. Diliberto, L. B. Kedderis, and L. S. Birnbaum, Absorption of 2,3,7,8-tetrabromodibenzo-p-dioxin (TBDD) in male rats, *Toxicologist* **10,** 54 (1990).
13. L. S. Birnbaum, G. M. Decad, and H. B. Matthews, Disposition and excretion of 2,3,7,8-tetrachlorodibenzofuran in the rat, *Toxicol. Appl. Pharmacol.* **55,** 342–352 (1980).
14. G. M. Decad, L. S. Birnbaum, and H. B. Matthews, 2,3,7,8-Tetrachlorodibenzofuran tissue distribution and excretion in guinea pigs, *Toxicol. Appl. Pharmacol.* **57,** 231–240 (1981).
15. Y. M. Ioannou, L. S. Birnbaum, and H. B. Matthews, Toxicity and distribution of 2,3,7,8-tetrachlorodibenzofuran in male guinea pigs, *J. Toxicol, Environ, Health* **12,** 541–553 (1983).
16. D. W. Brewster and L. S. Birnbaum, Disposition and excretion of 2,3,7,8-pentachlorodibenzofuran in the rat, *Toxicol. Appl. Pharmacol.* **90,** 243–252 (1987).
17. R. H. Wacker, H. Poiger, and C. Schlatter, Pharmacokinetics and metabolism of 1,2,3,7,8-pentachlorodibenzo-p-dioxin in the rat, *Chemosphere* **15**(9–12), 1473–1476 (1986).
18. L. S. Birnbaum and L. A. Couture, Disposition of octachlorodibenzo-p-dioxin (OCDD) in male rats, *Toxicol. Appl. Pharmacol.* **93,** 22–30 (1988).
19. H. Poiger and C. H. Schlatter, Influence of solvents and absorbants on dermal and intestinal absorption of TCDD, *Food Cosmet. Toxicol.* **18,** 477–481 (1980).
20. E. E. McConnell, G. W. Lucier, R. C. Rumbaugh, *et al.,* Dioxin in soil; Bioavailability after ingestion by rats and guinea pigs, *Science* **223,** 1077–1079 (1984).
21. G. W. Lucier, R. C. Rumbaugh, Z. McCoy, R. Hass, D. Harvan, and P. Albro, Ingestion of soil contaminated with 2,3,7,8-tetrachlorodibenzo-p-dioxin (TCDD) alters hepatic enzyme activities in rats, *Fundam. Appl. Toxicol.* **6,** 364–371 (1986).
22. H. Shu, D. Paustenbach, F. J. Murray, *et al.* Bioavailability of soil-bound TCDD: Oral bioavailability in the rat, *Fundam. Appl. Toxicol.* **10,** 648–654 (1988).
23. T. H. Umbreit, E. J. Hesse, and M. A. Gallo, Bioavailability of dioxin in soil from a 2,4,5-T manufacturing site, *Science* **232,** 497–499 (1986).

24. T. H. Umbreit, E. J. Hesse, and M. A. Gallo, Comparative toxicity of TCDD contaminated soil from Times Beach, Missouri and Newark, New Jersey, *Chemosphere* **15**(9–12), 2121–2124 (1986).
25. D. W. Brewster, Y. B. Banks, A. M. Clark, and L. S. Birnbaum, Comparative dermal absorption of 2,3,7,8-tetrachlorodibenzo-p-dioxin and three polychlorinated dibenzofurans, *Toxicol. Appl. Pharmacol.* **97,** 156–166 (1989).
26. Y. B. Banks and L. S. Birnbaum, Absorption of 2,3,7,8-tetrachlorodibenzo-p-dioxin (TCDD) after low dose dermal exposure, *Toxicol. Appl. Pharmacol.* **107,** 302–310 (1991).
27. M. S. Rahman, J. L. Zatz, T. H. Umbreit, and M. A. Gallo, Comparative in vitro permeation of 2,3,7,8-TCDD through hairless mouse and human skin, *Toxicologist* **12,** 80 (1992).
28. M. A. Gallo, M. S. Rahman, J. L. Zatz, and R. J. Meeker, In vitro dermal uptake of 2,3,7,8-TCDD in hairless mouse and human skin from laboratory-contaminated soils, *Toxicologist* **12,** 80 (1992).
29. L. W. D. Weber, A. Zesch, and K. Rozman, Penetration, distribution and kinetics of 2,3,7,8-TCDD in human skin in vitro, *Arch. Toxicol.* **65,** 421–428 (1991).
30. H. Shu, D. P. Tcitelbaum, A. S. Webb, *et al.*, Bioavailability of soil-bound TCDD: Dermal bioavailability in the rat, *Fundam. Appl. Toxicol.* **10,** 335–343 (1988).
31. C. S. Nessel, M. A. Amoruso, T. H. Umbreit, and M. A. Gallo, Hepatic aryl hydrocarbon hydroxylase and cytochrome P450 induction following the transpulmonary adsorption of TCDD from intratracheally instilled particles, *Fundam. Appl. Toxicol.* **15,** 500–509 (1990).
32. C. S. Nessel, M. A. Amoruso, T. H. Umbreit, R. J. Meeker, and M. A. Gallo, Pulmonary bioavailability and fine particle enrichment of 2,3,7,8-TCDD in respirable soil particles, *Fundam. Appl. Toxicol.* **19,** 279–285 (1992).
33. M. R. Lakshmanan, B. S. Campbell, S. J. Chirtel, N. Ekarohita, and M. Ezekiel, Studies on the mechanism of absorption and distribution of 2,3,7,8-tetrachlorodibenzo-p-dioxin in the rat, *J. Pharmacol. Exp. Ther.* **239**(3), 673–677 (1986).
34. D. G. Patterson, P. Furst, L. O. Henderson, D. G. Issacs, L. R. Alexander, W. E. Turner, L. L. Needham, and H. Hannon, Partitioning of in vivo bound PCDD/PCDFs among various compartments in whole blood, *Chemosphere* **19,** 135 (1989).
35. L. O. Henderson and D. G. Patterson, Jr., Distribution of 2,3,7,8-tetrachlorodibenzo-p-dioxin in human whole blood and its association with, and extractability from lipoproteins, *Bull. Environ. Contam. Toxicol.* **40,** 604–611 (1988).
36. J. D. McKinney, K. Chae, S. J. Oatley, and C. C. F. Blake, Molecular interactions of toxic chlorinated dibenzo-p-dioxins and dibenzofurans with thyroxine binding pre-albumin, *J. Med. Chem.* **28,** 375–381 (1985).
37. L. G. Pedersen, T. A. Darden, S. J. Oatley, and J. D. McKinney, A theoretical study of the binding of polychlorinated biphenyls (PCBs) dibenzodioxins and dibenzofuran to human plasma prealbumin, *J. Med. Chem.* **29,** 2451–2457 (1986).
38. M. Marinovich, C. R. Sirtori, C. L. Galli, and R. Paoletti, The binding of 2,3,7,8-tetrachlorodibenzodioxin to plasma lipoproteins may delay toxicity in experimental hyperlipidemia, *Chem. Biol. Interact,* **45,** 393–399 (1983).
39. R. B. Shireman and C. Wei, Uptake of 2,3,7,8-tetrachlorodibenzo-p-dioxin from plasma lipoproteins by cultured human fibroblasts, *Chem. Biol. Interact.* **58,** 1–12 (1986).
40. R. Weisiger, J. Gollan, and R. Ockner, Receptor for albumin on the liver cell surface may mediate uptake of fatty acids and other albumin-bound substances, *Science* **211,** 1048–1050 (1981).
41. R. Pohjanvirta, T. Vartiainen, A. Usi-Rauva, J. Monkkonen, and T. Tuomisto, Tissue

distribution, metabolism, and excretion of [^{14}C]-TCDD in a TCDD-susceptible and TCDD-resistant rat strain, *Pharmacol. Toxicol.* **66,** 93–100 (1990).
42. D. W. Brewster and L. S. Birnbaum, Disposition of 1,2,3,7,8-pentachlorodibenzofuran in the rat, *Toxicol. Appl. Pharmacol.* **95,** 490–498 (1988).
43. D. Neubert, T. Wiesmuller, K. Abraham, R. Krowke, and H. Hagenmaier, Persistence of various polychlorinated dibenzo-p-dioxin and dibenzofurans (PCDDs and PCDFs) in hepatic and adipose tissue of marmoset monkeys, *Arch. Toxicol.* **64,** 431–442 (1990).
44. K. Abraham, T. Wiesmuller, H. Brunner, R. Krowke, H. Hagenmaier, and D. Neubert, Absorption and tissue distribution of various polychlorinated dibenzo-p-dioxins and dibenzofurans (PCDDs and PCDFs) in the rat, *Arch. Toxicol.* **63,** 193–202 (1989).
45. A. Facchetti, A. Fornari, and M. Montagna, Distribution of 2,3,7,8-tetrachlorodibenzo-p-dioxin in the tissues of a person exposed to the toxic cloud at Seveso (Italy), *Adv. Mass Spectrom.* **8B,** 1405–1414 (1980).
46. H. J. Geyer, I. Scheunert, J. G. Filser, and F. Korte, Bioconcentration potential (BCP) of 2,3,7,8-tetrachlorodibenzo-p-dioxin (2,3,7,8-TCDD) in terrestrial organisms including humans, *Chemosphere* **15**(9–12), 1495–1502 (1986).
47. D. G. Patterson, Jr., L. Hampton, C. R. LaPeza, Jr., *et al.*, High resolution gas chromatographic/high resolution mass spectrometric analysis of human serum on a whole-weight and lipid basis for 2,3,7,8-tetrachlorodibenzo-p-dioxin, *Anal. Chem.* **59,** 2000–2005 (1987).
48. D. G. Patterson, Jr., L. L. Needham, J. L. Pirkle, *et al.*, Correlation between serum and adipose tissue levels of 2,3,7,8-tetrachlorodibenzo-p-dioxin in 50 persons from Missouri, *Arch. Environ. Contam. Toxicol.* **17,** 139–143 (1988).
49. P. C. Kahn, M. Gochfeld, M. Nygren, *et al.*, Dioxins and dibenzofurans in blood and adipose tissue of Agent Orange-exposed Vietnam veterans and matched controls, *J. Am. Med. Assoc.* **259**(11), 1661–1667 (1988).
50. A. Schecter, J. J. Ryan, J. D. Constable, R. Baughman, J. Bangert, P. Furst, K. Wilmers, and R. D. Oates, Partitioning of 2,3,7,8-chlorinated dibenzo-p-dioxins and dibenzofurans between adipose tissue and plasma lipid of 20 Massachusetts Vietnam veterans, *Chemosphere* **20,** 951 (1990).
51. A. Schecter, Dioxins and related chemicals in humans and the environment, in: *Biological Basis for Risk Assessment of Dioxins and Related Compounds* (M. A. Gallo, R. J. Scheuplein, and K. A. van der Heijnen, eds.), pp. 169–212, Banbury Report 35, Cold Spring Harbor Laboratory Press, Cold Spring Harbor, NY (1991).
52. H. Thoma, W. Mucke, and E. Kretschmer, Concentrations of PCDD and PCDF in human fat and liver samples, *Chemosphere* **18**(1–6), 491–498 (1989).
53. H. Thoma, W. Mucke, and G. Kauert, Comparison of the polychlorinated dibenzo-p-dioxin and dibenzofuran in human tissue and human liver, *Chemosphere* **20,** 433–442 (1990).
54. A. J. Schecter, J. J. Ryan, M. Gross, N. C. A. Weerasinghe, and J. Constable, Chlorinated dioxins and dibenzofurans in human tissues from Vietnam, 1983–1984, in: *Chlorinated Dioxins and Dibenzofurans in Perspective* (C. Rappe, G. Choudhary, and L. H. Keith, eds.), pp. 3–16, Lewis Publishers, Chelsea, MI (1986).
55. A. Schecter, J. Mes, and D. Davies, Polychlorinated biphenyl (PCB), DDT, DDE and hexachlorobenzene (HCB) and PCDD/F isomer levels in various organs in autopsy tissue from North American patients, *Chemosphere* **18,** 811–818 (1989).
56. Centers for Disease Control, Serum 2,3,7,8-tetrachlorodibenzo-p-dioxin levels in Air Force health study participants—Preliminary report, *Morbidity and Mortality Weekly Report* **3**(20), 309–311 (1988).
57. A. Schecter, J. D. Constable, J. V. Bangert, H. Tong, S. Arghestani, S. Monson, and

M. Gross, Elevated body burdens of 2,3,7,8-tetrachlorodibenzodioxin in adipose tissue of United States Vietnam veterans, *Chemosphere* **18,** 431 (1989).
58. K. Abraham, R. Krowke, and D. Neubert, Pharmacokinetics and biological activity of 2,3,7,8-tetrachlorodibenzo-p-dioxin. 1. Dose-dependent tissue distribution and induction of hepatic ethoxyresorufin O-deethylase in rats following a single injection, *Arch. Toxicol.* **62,** 359–368 (1988).
59. L. S. Birnbaum, Distribution and excretion of 2,3,7,8-tetrachlorodibenzo-p-dioxin in congenic strains of mice which differ at the Ah locus, *Drug Metab. Dispos.* **14**(1), 34–40 (1986).
60. R. E. Bowman, S. L. Schantz, N. C. A. Weerasinghe, M. L. Gross, and D. A. Barsotti, Chronic dietary intake of 2,3,7,8-tetrachlorodibenzo p dioxin (TCDD) at 5 or 25 parts per trillion in the monkey: TCDD kinetics and dose–effect estimate of reproductive toxicity, *Chemosphere* **18**(1–6), 243–252 (1989).
61. NATO/CCMS (North Atlantic Treaty Organization, Committee on the Challenges of Modern Society), *International Toxicity Equivalency Factor (I-TEF) Method of Risk Assessment for Complex Mixtures of Dioxins and Related Compounds,* Report No. 176 (1988).
62. U.S. EPA, *Interim Procedures for Estimating Risks Associated with Exposures to Mixtures of Chlorinated Dibenzo-p-dioxins and Dibenzofurans (CDDs and CDFs) and 1989 Update,* Risk Assessment Forum, Washington, DC (1989).
63. A. Schecter, T. Tiernan, F. Schaffner,, *et al.*, Patient fat biopsies for chemical analysis and liver biopsies for ultrastructural characterization after exposure to polychlorinated dioxins, furans and PCBs, *Environ. Health Perspect.* **60,** 241–254 (1985).
64. A. J. Schecter, J. J. Ryan, and J. D. Constable, Chlorinated dibenzo-p-dioxin and dibenzofuran levels in human adipose tissue and milk samples from the north and south of Vietnam, *Chemosphere* **15,** 1613–1620 (1986).
65. J. J. Ryan, Variation of dioxins and furans in human tissues, *Chemosphere* **15,** 1585–1593 (1986).
66. C. Rappe, M. Nygren, G. Linstrom, and M. Hansson, Dioxins and dibenzofurans in biological samples of European origin, *Chemosphere* **15,** 1635–1639 (1986).
67. H. Beck, K. Eckart, W. Mathar, and R. Wittkowski, Isomerenspeczifische Bestimmung von PCDD und PCDF in Human- und Lebensmittelproben, *VDI Ber.* **634,** 359–382 (1987).
68. H. Poiger, N. Pluess, and C. Schlatter, Subchronic toxicity of some chlorinated dibenzofurans in rats, *Chemosphere* **18**(1–6), 265–275 (1989).
69. L. B. Kedderis, J. J. Diliberto, P. Linko, J. A. Goldstein, and L. S. Birnbaum, Disposition of TBDD and TCDD in the rat: Biliary excretion and induction of cytochromes P450IA1 and P450IA2, *Toxicol. Appl. Pharmacol.* **111,** 163–172 (1991).
70. L. R. Curtis, N. I. Kerkvliet, L. Baecher-Steppan, and H. M. Carpenter, 2,3,7,8-Tetrachlorodibenzo-p-dioxin pretreatment of female mice altered tissue distribution but not hepatic metabolism of a subsequent dose, *Fundam. Appl. Toxicol.* **14,** 523–531 (1990).
71. A. Poland, P. Teitelbaum, E. Glover, and A. Kende, Stimulation of *in vivo* hepatic uptake and *in vitro* hepatic binding of [125I]2-iodo-3,7,8-trichlorodibenzo-p-dioxin by the administration of agonists for the Ah receptor, *Mol. Pharmacol.* **36,** 121–127 (1989).
72. H.- W. Leung, A. Poland, D. J. Paustenbach, F. J. Murray, and M. E. Andersen, Pharmacokinetics of [125I]-2-iodo-3,7,8-trichlorodibenzo-p-dioxin in mice: Analysis with a physiological modeling approach, *Toxicol. Appl. Pharmacol.* **103,** 411–419 (1990).
73. E. S. Shen and J. R. Olson, Relationship between the murine Ah phenotype and the hepatic uptake and metabolism of 2,3,7,8-tetrachlorodibenzo-p-dioxin, *Drug Metab. Dispos.* **15**(5), 653–660 (1987).

74. R. J. Kociba, D. G. Keyes, J. E. Beyer, *et al.*, Results of a two-year chronic toxicity and oncogenicity study of 2,3,7,8-tetrachlorodibenzo-p-dioxin in rats, *Toxicol. Appl. Pharmacol.* **46**(2), 279–303 (1978).
75. R. J. Kociba, D. G. Keyes, J. E. Beyer, and R. M. Carreon, Toxicologic studies of 2,3,7,8-tetrachlorodibenzo-p-dioxin (TCDD) in rats, *Toxicol. Occup. Med.* **4,** 281–287 (1978).
76. R. J. Kociba, P. A. Keeler, C. N. Park, and P. J. Gehring, 2,3,7,8-Tetrachlorodibenzo-p-dioxin results of a 13-week oral toxicity study in rats, *Toxicol. Appl. Pharmacol.* **35,** 553–574 (1976).
77. A. M. Tritscher, J. A. Goldstein, C. J. Portier, Z. McCoy, G. C. Clark, and G. W. Lucier, Dose–response relationships for chronic exposure to 2,3,7,8-tetrachlorodibenzo-p-dioxin in a rat tumor promotion model: Quantification and immunolocalization of CYP1A1 and CYP1A2 in the liver, *Cancer Res.* **52,** 3436–3442 (1992).
78. H.- W. Leung, J. M. Wendling, R. Orth, F. Hileman, and D. J. Paustenbach, Relative distribution of 2,3,7,8-tetrachlorodibenzo-p-dioxin in human hepatic and adipose tissues, *Toxicol. Lett.* **50,** 275–282 (1990).
79. R. Voorman and S. D. Aust, Specific binding of polyhalogenated aromatic hydrocarbon inducers of cytochrome P-450d to the cytochrome and inhibition of its estradiol 2-hydroxylase activity, *Toxicol. Appl. Pharmacol.* **90,** 69–78 (1987).
80. R. Voorman and S. D. Aust, TCDD (2,3,7,8-tetrachlorodibenzo-p-dioxin) is a tight binding inhibitor of cytochrome P-450d, *J. Biochem. Toxicol.* **4,** 105–109 (1989).
81. A. P. Poland, E. Teitelbaum, and E. Glover, [125I]-iodo-3,7,8-trichlorodibenzo-p-dioxin binding species in mouse liver induced by agonists for the Ah receptor: Characterization and identification, *Mol. Pharmacol.* **36,** 113–120 (1989).
82. J. C. Ramsey, J. G. Hefner, R. J. Karbowski, W. H. Braun, and P. J. Gehring, The *in vivo* biotransformation of 2,3,7,8-tetrachlorodibenzo-p-dioxin (TCDD) in the rat, *Toxicol. Appl. Pharmacol.* **65,** 180–184 (1982).
83. H. Poiger and C. Schlatter, Biological degradation of TCDD in rats, *Nature* **281,** 706–707 (1979).
84. J. R. Olson, Metabolism and disposition of 2,3,7,8-tetrachlorodibenzo-p-dioxin in guinea pigs, *Toxicol. Appl. Pharmacol.* **85,** 263–273 (1986).
85. T. A. Gasiewicz, L. E. Geider, G. Rucci, and R. A. Neal, Distribution, excretion and metabolism of 2,3,7,8-tetrachlorodibenzo-p-dioxin in C57BL/6J, and B6D2F1/J mice, *Drug Metab. Dispos.* **11**(5), 397–403 (1983).
86. H. Poiger, H. R. Buser, H. Weber, U. Zweifel, and C. Schlatter, Structure elucidation of mammalian TCDD-metabolites, *Experientia* **38,** 484–486 (1982).
87. J. M. Kleeman, J. R. Olson, and R. E. Peterson, Species differences in 2,3,7,8-tetrachlorodibenzo-p-dioxin toxicity and biotransformation in fish, *Fundam. Appl. Toxicol.* **10,** 206–213 (1988).
88. J. M. Wendling and R. G. Orth, Determination of [3H]-2,3,7,8-tetrachlorodibenzo-p-dioxin in human feces to ascertain its relative metabolism in man, *Anal. Chem.* **62,** 796–800 (1990).
89. T. Sawahata, J. R. Olson, and R. A. Neal, Identification of metabolites of 2,3,7,8-tetrachlorodibenzo-p-dioxin (TCDD) formed on incubation with isolated rat hepatocytes, *Biochem. Biophys. Res. Commun.* **105**(1), 341–346 (1982).
90. H. Poiger and H. R. Buser, The metabolism of TCDD in the dog and rat, in: *Biological Mechanisms of Dioxin Action* (A. Poland and R. D. Kimbrough, eds.), pp. 39–47, Banbury Report 18, Cold Spring Harbor Laboratory Press, Cold Spring Harbor, NY (1984).
91. H. Weber, H. Poiger, and C. Schlatter, Acute oral toxicity of TCDD-metabolites in male guinea pigs, *Toxicol. Lett.* **14,** 117–122 (1982).

92. G. Mason and S. Safe, Synthesis, biologic and toxic effects of the major 2,3,7,8-tetrachlorodibenzo-p-dioxin metabolites in the rat, *Toxicology* **41,** 153–159 (1986).
93. A. Poland and E. Glover, An estimate of the maximum *in vivo* covalent binding of 2,3,7,8-tetrachlorodibenzo-p-dioxin to rat liver protein, ribosomal RNA and DNA, *Cancer Res.* **39**(9), 3341–3344 (1979).
94. M. K. McKinley, L. B. Kedderis, and L. S. Birnbaum, The effect of pretreatment on the biliary excretion of 2,3,7,8-tetrachlorodibenzo-p-dioxin, 2,3,7,8-tetrachlorodibenzofuran, and 3,3′,4,4′-tetrachlorobiphenyl in the rat, *Fund. Appl. Toxicol.* **21,** 425–432 (1993).
95. V. J. Wroblewski and J. R. Olson, Effect of monooxygenase inducers and inhibitors on the hepatic metabolism of 2,3,7,8-tetrachlorodibenzo-p-dioxin in the rat and hamster, *Drug Metab. Dispos.* **16**(1), 43–51 (1988).
96. V. J. Wroblewski and J. R. Olson, Hepatic metabolism of 2,3,7,8-tetrachlorodibenzo-p-dioxin (TCDD) in the rat and guinea pig, *Toxicol. Appl. Pharmacol.* **81,** 231–240 (1985).
97. J. R. Olson, B. P. McGarrigle, P. J. Gigliotti, S. Kumar, and J. H. McReynolds, Hepatic uptake and metabolism of 2,3,7,8-TCDD and 2,3,7,8-TCDF, *Fund. Appl. Toxicol.* **22,** 631–640 (1994).
98. H. L. Tai, J. H. McReynolds, J. A. Goldstein, H. P. Eugster, C. Sengstag, W. L. Alworth, and J. R. Olson, Cytochrome P-4501A1 mediates the metabolism of 2,3,7,8-tetrachlorodibenzofuran (TCDF) in the rat and human, *Toxicol. Appl. Pharmacol.* **123,** 34–42 (1993).
99. J. R. Olson, M. A. Holscher, and R. A. Neal, Toxicity of 2,3,7,8-tetrachlorodibenzo-p-dioxin (TCDD) in the golden Syrian hamster, *Toxicol. Appl. Pharmacol.* **55,** 67–78 (1980).
100. J. L. Pirkle, W. H. Wolfe, D. G. Patterson, *et al.*, Estimates of the half-life of 2,3,7,8-tetrachlorodibenzo-p-dioxin in Vietnam veterans of Operation Ranch Hand, *J. Toxicol. Environ. Health* **27,** 165–171 (1989).
101. J. J. Ryan and Y. Masuda, Elimination of polychlorinated dibenzofurans (PCDFs) in humans from the Usho and Ucheng rice oil poisonings, in: *Proc. 11th Int. Symp. on Chlorinated Dioxins and Related Compounds, Dioxin '91, September 23–27, 1991,* Research Triangle Park, NC p. 70 (1991).
102. A. Schecter, J. J. Ryan, and P. J. Kostyniak, Decrease over a six year period of dioxin and dibenzofuran tissue levels in a single patient following exposure, *Chemosphere* **20**(7–9), 911–917 (1990).
103. J. J. Ryan and Y. Masuda, Half-lives for elimination of polychlorinated dibenzofurans (PCDFs) and PCBs in humans from the Yusho and Yucheng rice oil poisonings, in: *Proc. 9th Int. Symp. on Chlorinated Dioxins and Related Compounds, Dioxin '89, September 17–22, 1989,* Toronto, p. 70 (1991).
104. A. Schecter and T. Gasiewicz, Health hazard assessment of chlorinated dioxins and dibenzofurans contained in human milk, *Chemosphere* **16,** 2147 (1987).
105. A. Schecter and T. Gasiewicz, Human breast milk levels of dioxins and dibenzofurans and their significance with respect to current risk assessments, in: *Solving Hazardous Waste Problems: Learning from Dioxins, No. 191* (J. H. Exner, ed.), p. 162, American Chemical Society, Washington, DC (1987).
106. M. Graham, F. D. Hileman, R. G. Orth, J. M. Wendling, and J. W. Wilson, Chlorocarbons in adipose tissue from Missouri population, *Chemosphere* **15,** 1595–1600 (1986).
107. A. A. Jensen, Polychlorobiphenyls (PCBs), polychlorodibenzo-p-dioxin (PCDDs) and polychlorodibenzofurans (PPPCDFs) in human milk, blood and adipose tissue, *Sci. Total Environ.* **64,** 259–293 (1987).
108. C. Rappe, Analysis of polychlorinated dioxins and furans: All 75 PCDDs and 135 PPPCDFs can be identified by isomer-specific techniques, *Environ. Sci. Technol.* **18**(3), 78A–90A (1984).

109. C. Rappe, M. Nyhgren, S. Marklund, *et al.*, Assessment of human exposure to polychlorinated dibenzofurans and dioxins, *Environ. Health Perspect.* **60,** 303–304 (1985).
110. P. Furst, H.- A. Meemken, and W. Groebel, Determination of polychlorinated dibenzodioxins and dibenzofurans in human milk, *Chemosphere* **15,** 1977–1980 (1986).
111. M. Nygren, C. Rappe, G. Linstrom, *et al.*, Identification of 2,3,7,8-substituted polychlorinated dioxins and dibenzofurans in environmental and human samples, in: *Chlorinated Dioxins and Dibenzofurans in Perspective* (C. Rappe, G. Choudhary, and L. H. Keith, eds.), pp. 17–34, Lewis Publishers, Chelsea, MI (1986).
112. G. Reggiani, Acute human exposure to TCDD in Sevesco, Italy, *J. Toxicol. Environ. Health* **6**(1), 27–43 (1980).
113. A. Schecter, J. J. Ryan, and J. D. Constable, Chlorinated dibenzo-p-dioxin and dibenzofuran levels in human adipose tissue and milk samples from the north and south of Vietnam, *Chemosphere* **15,** 1613 (1986).
114. R. W. Baughman, Tetrachlorodibenzo-p-dioxins in the environment. High resolution mass spectrometry at the picogram level, Harvard University, NTIS PB75-22939 (1975).
115. A. Schecter, P. Furst, J. J. Ryan, C. Furst, H.- A. Meemken, W. Grobel, J. Constable, and D. Vu, Polychlorinated dioxin and dibenzofuran levels from human milk from several locations in the United States, Germany and Vietnam, *Chemosphere* **19,** 979 (1989).
116. H.-W. Leung, R. H. Ku, D. J. Paustenbach, and M. E. Andersen, A physiologically based pharmacokinetic model for 2,3,7,8-tetrachlorodibenzo-p-dioxin in C57BL/6J and DBA/2J mice, *Toxicol. Lett.* **42,** 15–28 (1988).
117. H.-W. Leung, D. J. Paustenbach, F. J. Murray, and M. E. Andersen, A physiological pharmacokinetic description of the tissue distribution and enzyme-inducing properties of 2,3,7,8-tetrachlorodibenzo-p-dioxin in the rat, *Toxicol. Appl. Pharmacol.* **103,** 399–410 (1990).
118. J. C. Kissel and G. M. Robarge, Assessing the elimination of 2,3,7,8-TCDD from humans with a physiologically based pharmacokinetic model, *Chemosphere* **17**(10), 2017–2027 (1988).
119. R. J. Scheuplein, S. E. Shoaf, and R. N. Brown, Role of pharmacokinetics in safety evaluation and regulatory considerations, *Annu. Rev. Pharmacol. Toxicol.* **30,** 197–218 (1990).
120. M. E. Andersen, J. J. Mills, M. L. Gargas, *et al.*, Modeling receptor-mediated processes with dioxin: Implications for pharmacokinetics and risk assessment, *Risk Anal.* **13,** 25–36 (1992).
121. R. Krowke, I. Chahoud, I. Baumann-Wilschke, and D. Neubert, Pharmacokinetics and biological activity of 2,3,7,8-tetrachlorodibenzo-p-dioxin. 2. Pharmacokinetics in rats using a loading-dose/maintenance-dose regime with high doses, *Arch. Toxicol.* **63,** 356–360 (1989).
122. M. C. Kohn, G. W. Lucier, G. W. Clark, G. C. Sewall, A. M. Tritscher, and C. J. Portier, A mechanistic model of effects of dioxin on gene expression in the rat liver, *Toxicol. Appl. Pharmacol.* **120,** 138–154 (1993).
123. C. Schlatter, Data on kinetics of PCDDs and PCDFs as a prerequisite for human risk assessment, in: *Biological Basis for Risk Assessment of Dioxins and Related Compounds* (M. A. Gallo, R. J. Scheuplein, and K. A. van der Heijnen, eds.), pp. 215–226, Banbury Report 35, Cold Spring Harbor Laboratory Press, Cold Spring Harbor, NY (1991).
124. J. J. Ryan, R. Lizotte, and D. Lewis, Human tissue levels of PCDDs and PPPCDFs from a fatal pentachlorophenol poisoning, *Chemosphere* **16**(8–9), 1989–1996 (1987).
125. C. C. Travis and H. A. Hattemer-Frey, Human exposure to 2,3,7,8-TCDD, *Chemosphere* **16**(10/12), 2331–2342 (1987).

126. H. Weber and L. S. Birnbaum, 2,3,7,8-Tetrachlorodibenzo-p-dioxin (TCDD) and 2,3,7,8-tetrachlorodibenzofuran (TCDF) in pregnant C57BL/6N mice: Distribution to the embryo and excretion, *Arch. Toxicol.* **57,** 157–162 (1985).
127. H. Nau, R. Bab, and D. Neubert, Transfer of 2,3,7,8-tetrachlorodibenzo-p-dioxin (TCDD) via placenta and milk, and postnatal toxicity in the mouse, *Arch. Toxicol.* **59,** 36–40 (1986).
128. M. Vanden Berg, C. Heeremans, E. Veenhoven, and K. Olie, Transfer of polychlorinated dibenzo-p-dioxins and dibenzofurans to fetal and neonatal rats, *Fundam. Appl. Toxicol.* **9,** 635–644 (1987).
129. H. Hagenmaier, T. Wiesmuller, G. Golor, R. Krowke, H. Helge, and D. Neubert, Transfer of various polychlorinated dibenzo-p-dioxins and dibenzofurans (PCDDs and PCDFs) via placenta and through milk in a marmoset monkey, *Arch. Toxicol.* **64,** 601–615 (1990).
130. R. E. Bowman, H. Y. Tong, M. L. Gross, S. J. Monson, and N. C. A. Weerasinghe, Controlled exposure of female rhesus monkeys to 2,3,7,8-TCDD: Concentrations of TCDD in fat of offspring, and its decline over time, *Chemosphere* **20**(7–9), 1199–1202 (1990).
131. P. Furst, C. Kruger, H.-A. Meemken, and W. Groebel, PPCDD and PPCDF levels in human milk-dependence on the period of lactation, *Chemosphere* **18**(1–6), 439–444 (1989).
132. T. Gorski, L. Konopka, and M. Brodzki, Persistence of some polychlorinated dibenzo-p-dioxins and polychlorinated dibenzofurans of pentachlorophenol in human adipose tissue, *Roczn. Pzh. T.* **35**(4), 297–301 (1984).
133. W. H. Wolfe, J. E. Michalek, J. C. Miner, J. L. Pirkle, S. P. Caudill, D. G. Patterson; and L. L. Needham, Determinants of TCDD half-life in veterans of Operation Ranch Hand, *J. Toxicol. Environ. Health* **41,** 481–488 (1994).

Chapter 7

Immunotoxicology of Dioxins and Related Chemicals

Nancy I. Kerkvliet

1. THE IMMUNE SYSTEM AS A TARGET FOR TOXIC CHEMICALS

Concern over the potential toxic effects of chemicals on the immune system arises from the critical role that the immune system plays in maintaining health. It is widely recognized that suppressed immunological function can result in increased incidence and severity of infectious diseases as well as some types of cancer. This is poignantly illustrated by the opportunistic infections and Kaposi's sarcoma that occur with great frequency in AIDS patients. Life-threatening infections and secondary cancers are also associated with immunosuppressive drug therapies used in transplant recipients and cancer patients. On the other hand, the inappropriate induction or enhancement of immune function can precipitate or exacerbate the development of allergic and autoimmune diseases. Chemical-induced allergies or autoimmunity are recognized side effects of some drugs such as penicillin, sulfa, penicillamine, and hydralazine. Thus, suppression as well as enhancement of immune function are considered to represent potential immunotoxic effects of chemicals.

2. APPROACHES TO IMMUNOTOXICITY ASSESSMENT

The immune system consists of a complex network of cells (B and T lymphocytes and macrophages) as well as soluble mediators (cytokines, interleukins) that interact in a highly regulated manner to generate immune responses of appropriate magnitude and duration. Consequently, evaluation of the immunotoxicity of chemicals must include

Nancy I. Kerkvliet • Department of Agricultural Chemistry and Environmental Health Sciences Center, Oregon State University, Corvallis, Oregon 97331.

Dioxins and Health, edited by Arnold Schecter. Plenum Press, New York, 1994.

multiple functional parameters at multiple time points. In addition, because an immune response develops in a time-dependent manner relative to antigen exposure, the immunotoxicity of a chemical may be profoundly influenced by the timing of chemical exposure relative to antigen challenge. Therefore, chronic exposure to the chemical both prior to and during the development of the immune response may be necessary to elucidate all possible immune effects. Exposures limited to discrete phases of the immune response may produce different effects on the overall immune response. Consideration of these unique complexities involved in immunotoxicology assessment are critical for the appropriate design of experiments and interpretation of the effects of chemical exposure on immune function.

Methods to assess the potential of chemicals to induce allergic contact sensitization have been used for a number of years in safety assessment.[1–3] More recently, methods for evaluating the immunosuppressive potential of chemicals have been described.[4–8] A tiered testing scheme using inbred mice has been recommended by The National Toxicology Program for evaluating the immunosuppressive effects of chemicals.[9] The first tier of assays represents a screening approach to assess immunopathology associated with chemical exposure, as well as evaluation of specific immune function. Antibody- and T-cell-mediated immune responses are measured as well as aspects of nonspecific immunity (e.g., macrophage phagocytosis, natural killer cell activity). Tier 1 assays are generally used in conjunction with a subchronic exposure regimen. If chemical exposure significantly alters Tier 1 assays, a more comprehensive evaluation of immunotoxicity is recommended in Tier 2 testing. Included in Tier 2 testing are pathogen challenge models to assess the effects of chemical exposure on the ability of the animal to fight infections or tumor growth. Based on the immunotoxic activity revealed in Tier 1 and 2 assessment, additional testing using other exposure protocols and other species may or may not be warranted. Studies to address the mechanisms of immunotoxicity of a chemical are important for extrapolation of results of animal tests to humans.

3. OVERVIEW OF TCDD IMMUNOTOXICITY

Extensive evidence has been published over the past 20 years to demonstrate that the immune system is a target for toxicity of TCDD and structurally related halogenated aromatic hydrocarbons (HAH), including other polychlorinated dioxins (PCDD), furans (PCDF), and biphenyls (PCB). This evidence has been derived from numerous studies in various animal species, primarily rodents, but also guinea pigs, rabbits, monkeys, marmosets, and cattle. Unfortunately, because of widely differing experimental designs, exposure protocols, and immunologic assays used, it has been very difficult to define a "TCDD-induced immunotoxic syndrome" in a single species, let alone across species. For example, there is only one report that directly compared the effects of TCDD on T-cell-mediated immunity of rats, mice, and guinea pigs, and, even then, different immunologic endpoints were assessed and different antigens were used in the different species.[10] In that study, the delayed-type hypersensitivity (DTH) response to tuberculin was evaluated in guinea pigs and rats for assessment of T-cell-mediated immunity, while the graft versus host (GVH) response was used to measure T cell responsiveness in mice. A decreased DTH response to tuberculin was observed in guinea pigs following 8 weekly doses of 40 ng/kg TCDD (total dose, 320 ng/kg), while the DTH response of rats to tuberculin was unaffected by 6 weekly doses of 5 μg/kg TCDD (total dose, 30,000 ng/kg TCDD). The GVH response in mice was suppressed by 4 weekly doses of 5 μg/kg TCDD (total dose, 20,000 ng/kg TCDD). The greater sensitivity of guinea pigs compared with rats to the immunosuppressive effects of TCDD is consistent with the greater sensitivity of guinea pigs to other toxic effects of TCDD.[11,12] Although these results appear to suggest that cell-mediated immunity in mice is more sensitive to TCDD than in rats, the use of different antigens complicates direct comparison. Specifically, in another study in mice, the DTH response to oxazolone was suppressed by 4 weekly doses of 4 μg/kg TCDD (total dose, 16,000 ng/kg), while the DTH response to sheep red blood cells (SRBC) was unaffected by a 10-fold higher dose of TCDD,[13] illustrating that DTH responses to different antigens are not equally sensitive to TCDD-induced suppression, even in the same species. The underlying basis for the interstudy and interspecies variability is not known.

It is generally accepted that TCDD immunotoxicity does not result from a lymphocytotoxic effect since alterations in specific immune effector functions and increased susceptibility to infectious disease have been identified at doses of TCDD well below those that cause lymphoid tissue depletion. It is also generally accepted that there are multiple cellular targets within the immune system that are altered by TCDD since both T-cell-mediated and antibody-mediated immune responses are suppressed following TCDD exposure. Limited evidence also suggests that the immune system may be indirectly targeted by TCDD-induced changes in nonlymphoid tissues (e.g., via the endocrine system). This chapter will not attempt to catalogue all of the descriptive immunotoxic effects that have been reported in animals exposed to TCDD and related HAH. Several comprehensive reviews have been published on the immunotoxic effects of HAH in general,[14,15] and TCDD in particular.[16,17] This chapter will emphasize more recent developments in the field as well as identify important gaps in our knowledge that require further research.

4. INFLUENCE OF TCDD ON HOST RESISTANCE TO DISEASE

The ability of an animal to resist and/or control viral, bacterial, parasitic, and neoplastic diseases is determined to a large extent by the state of its immune system. Decreased functional activity in any immunological compartment may result in increased susceptibility to infectious and neoplastic diseases. However, because of redundancy in effector mechanisms as well as immunological reserve, not all measurable changes in specific immune functions will translate into detectable changes in disease susceptibility. Thus, even though changes in host resistance would seem to be the "most relevant" immunotoxic endpoints, such changes serve as only one indicator of chemical-induced immune modulation.

Animal models that mimic human disease are available and have been used to assess the effects of TCDD on host resistance to disease. For example, several studies have shown that TCDD exposure increased the mortality of mice and rats challenged with the gram-negative bacterium *Salmonella.*[18–20] TCDD exposure also resulted in increased mortality following injection of *E. coli* endotoxin[20,21] suggesting that the increased susceptibility to *Salmonella* caused by TCDD may be related to the endotoxin associated with the gram-negative bacterium. However, endotoxin is not prerequisite for TCDD effects since TCDD treatment of mice also resulted in increased mortality when challenged with *Streptococcus pneumoniae,* a gram-positive bacterium.[22]

Under certain exposure conditions, TCDD has also been shown to increase susceptibility to *Listeria monocytogenes* infection. Hinsdill *et al.*[19] reported that 50 ppb TCDD in the diet increased bacteremia and mortality of Swiss Webster mice that had been injected intravenously with *Listeria.* Similarly, exposure of pregnant B6C3F1 mice to 5.0 μg/kg TCDD at day 14 of gestation and again on days 1, 7, and 14 following birth resulted in an increased mortality in pups challenged with *Listeria* when they were 8 weeks of age.[23] However, Vos *et al.*[20] reported that oral administration of 50 μg TCDD/kg once a week for 4 weeks to young Swiss mice did not alter the number of viable *Listeria* organisms per spleen, suggesting that TCDD did not affect nonspecific phagocytosis and killing of *Listeria.* Also, House *et al.*[24] reported that a single oral dose of TCDD (10, 1.0, and 0.1 μg/kg) did not enhance mortality to a *Listeria* challenge given 7 days later. Differences in study design such as TCDD dose, route, single versus multiple administrations, or mouse strain, age, and sex may contribute to these disparate effects.

Enhanced susceptibility to viral disease has been reported after TCDD administration. Clark *et al.*[25] reported that TCDD treatment significantly enhanced mortality to herpes simplex type II virus. In contrast, TCDD exposure did not alter the time to death or the incidence of mortality following *Herpesvirus suis* infection.[18] House *et al.*[24] reported that TCDD treatment enhanced mortality of mice challenged with Influenza/A/Taiwan/1/64 (H2N2).

TCDD exposure has also been shown to alter parasitic disease. Tucker *et al.*[26] studied the effects of TCDD administration on *Plasmodium yoelii* 17 XNL, a nonlethal strain of malaria, in 6- to 8-week-old B6C3F1 mice. A single dose of TCDD at 5 or 10 μg/kg *per os* resulted in increased susceptibility to *P. yoelii* as evidenced by the magnitude and duration of the peak parasitemia.

Luster *et al.*[23] demonstrated enhanced growth of transplanted polyoma virus-induced tumors in mice treated with TCDD at doses of 1.0 or 5.0 μg/kg. Mothers were given TCDD by gavage at day 14 of gestation and again on days 1, 7, and 14 following birth; host resistance studies were performed 6–8 weeks after weaning. This exposure protocol resulted in an increased incidence of tumors in pups from dams receiving repeated doses of 5.0 but not 1.0 μg TCDD/kg.

In summary, results from host resistance studies provide evidence that exposure to TCDD results in increased susceptibility to bacterial, viral, parasitic, and neoplastic disease. These effects are observed at low doses and may result from TCDD-induced suppression of immunological function. However, the specific immunological functions targeted by TCDD in most of the host resistance models remain to be fully defined. Also, nonimmunological mechanisms for altered host resistance should not be discounted without investigation. For example, Rosenthal *et al.*[27] reported that endotoxin hypersensitivity in B6C3F1 mice was associated with hepatotoxicity and decreased clearance of the endotoxin providing a nonimmunological explanation for enhanced susceptibility to *Salmonella* or other gram-negative bacteria.

5. EFFECTS OF TCDD ON THE THYMUS GLAND: ROLE IN IMMUNOTOXICITY

Thymic involution is one of the hallmarks of exposure to TCDD and related HAH in all species examined.[11,12] Because the thymus plays a critical role in the prenatal development of T lymphocytes, this observation appears to have precipitated the earliest investigations into the immunotoxicity of TCDD and PCBs. Even today, TCDD-induced thymic atrophy is often referred to as an immunotoxic effect. However, while an intact thymus is crucial to the development of the T-cell receptor repertoire during the prenatal and early postnatal period in rodents as well as humans, the physiological role played by the thymus in adult life has not been established.[28,29] In animal models, adult thymectomy has little effect on the quantity or quality of T lymphocytes, which have already matured and populated the secondary lymphoid organs.[30] Likewise, in humans, childhood and adult thymectomy produces no clearly documentable adverse consequences in terms of altered immune function, although some might argue that definitive studies have not been done. In light of these facts, it is not surprising that no direct relationship between the effects of TCDD on the thymus and immune suppression has been demonstrated in adult animals. Specifically, adult thymectomy prior to TCDD exposure did not modify the suppression of the antibody response to SRBC.[26,31] Furthermore, suppression of immune responses occurs at dose levels of TCDD or PCBs significantly lower than those required to induce thymic atrophy.[20,26,32–34] Thus, it is clear that thymic involution does not represent a surrogate marker for TCDD immunotoxicity in adult animals.

In contrast to adult animals, congenital thymic aplasia or neonatal thymectomy results in severe reduction in the number and function of T lymphocytes, and produces a potentially lethal wasting disease.[30] Accordingly, there is experimental evidence that immune suppression in rodents occurs at lower doses of TCDD or PCBs when exposure occurs during the pre/neonatal period as compared with rodents exposed as adults, and that the prenatal effects are selective for T-cell-mediated immunity.[23,35,36] Several studies have examined immune function in mice, rats, and/or guinea pigs following pre- and early postnatal exposure to TCDD or PCB.[10,21,23,35] The most sensitive indicator of immunotoxicity in these studies was an increase in the growth of transplanted tumor cells in the offspring of inbred B6C3F1 mice treated with 1 μg/kg TCDD at 4 weekly intervals. (Total TCDD dose to dam was 4 μg/kg; dose to offspring was not determined.) The offspring of Swiss mice fed a diet containing 1 ppb TCDD for 7 weeks showed enhanced mortality following endotoxin challenge, while the antibody- and T-cell-mediated immune responses were suppressed in offspring of mice fed 5.0-ppb TCDD diets. (Estimated daily dose for a 20-g dam consuming 5 g of a 5-ppb TCDD diet is equivalent to 1.25 μg/kg TCDD per day.) Rats appeared to be more resistant to the immunotoxic effects of pre/neonatal exposure to TCDD based on the finding that 5 but not 1 μg/kg TCDD given 4 times at weekly intervals produced immunotoxicity in the offspring. Immunotoxic endpoints that were unaffected by the highest exposure levels in these studies included B-cell proliferation induced by lipopolysaccharide (LPS) and serum antibody titers to SRBC and bovine gamma globulin (BGG).

Two recent studies have examined immune function in offspring of female mice exposed to TCDD prenatally only[37] or PCB (Kanechlor 500)[38] by cross-fostering the pups to unexposed lactating mice at birth. (It is important to recognize that rodents are born with an immature immune system that matures in the first few weeks following birth. In contrast, the human immune system is considered to be more mature at birth.) C57BL/6 mice exposed to 3.0 μg/kg TCDD on gestational days 6–14 gave birth to offspring (day 21) that had significant thymic atrophy and hypoplasia measured on gestational day 18 or on day 6 postnatally. The thymic effects were no longer apparent by day 14. At 7–8 weeks postnatally, the proliferative responses of T and B cells and the antibody response to SRBC were unaltered while the cytotoxic T lymphocyte response was significantly suppressed relative to controls.[37] These results suggest a selective toxicity of prenatal TCDD exposure on cytotoxic T cells compared with the T helper cells involved in the antibody response to SRBC. In contrast to these results, Takagi *et al.*[38] exposed female C3H mice *per os* to 50 mg/kg Kanechlor 500 (a PCB mixture) twice a week for 4 weeks. The offspring derived from mating to unexposed males had an unaltered antibody response to the T-independent antigen DNP-dextran suggesting no effect on B cell responsiveness. On the other hand, T helper cell activity was significantly suppressed by PCB exposure when assessed in adoptive transfer studies at 4 and 7 weeks after birth, but fully recovered by 11 weeks. Together, these studies confirm prior studies and indicate that T cell function is selectively altered by TCDD or PCB when exposure is prenatal. While both T helper cells and cytotoxic T cells show altered function, T helper cell activity may recover faster than cytotoxic T cell function.

The mechanism for TCDD-induced thymic atrophy has not been fully elucidated. Neither adrenalectomy, hypophysectomy, nor thymosin treatment protected rats from TCDD-induced thymic atrophy, indicating that the effect is not mediated by an effect on the pituitary or adrenal glands, or from reduced production of thymic hormones.[20,39] Although a single report suggested that a high concentration of TCDD *in vitro* caused thymocyte cell death through an apoptotic mechanism,[40] this effect could not be verified in more recent studies.[41] TCDD has been shown to alter thymocyte differentiation *in vitro* in cell cultures[42,43] and organ cultures[44,45] as well as *in vivo* following prenatal exposure to TCDD.[41] This effect appeared to be mediated by alterations in the activity of thymic epithelial cells that make up the thymic stroma rather than by direct effect on the thymocytes. In addition, *in vivo* treatment with TCDD caused a transient suppression of thymocyte proliferation[46] and reduced the migration of lymphocyte stem cells from the fetal liver and bone marrow of mice,[47] suggesting multiple mechanisms for TCDD-induced thymic atrophy.

6. ROLE OF THE *Ah* LOCUS IN HAH IMMUNOTOXICITY

One of the most important advances in the study of HAH toxicity in recent years has been the elucidation of a genetic basis for sensitivity to the toxicity of these chemicals, which may ultimately provide a logical explanation for much of the disparate data in the literature regarding HAH toxicity in different species and in different tissues of the same species. In this regard, many of the biochemical and toxic effects of HAH appear to be mediated via binding to an intracellular protein known as the aromatic hydrocarbon (Ah) or TCDD receptor, in a process similar to steroid hormone receptor-mediated responses.[12,48] Activation of the Ah receptor occurs on binding of the ligand (e.g., TCDD/PCB). The activated receptor–ligand complex is translocated to the nucleus where it binds to DNA at specific sequences called dioxin-response elements (DREs) to activate or repress transcription of the DRE-containing gene. It is hypothesized that the altered transcription leads to the production of or loss of specific protein products that mediate the ultimate biochemical and toxic effects of TCDD. This hypothesis has been proven for the structural genes encoding mRNA for CYP1A1 enzyme activity (i.e., cytochrome P_1450).[49] Differences in toxic potency between various chlorinated cogeners of dioxins and biphenyls generally correlate with differences in Ah receptor binding affinities. The most toxic congeners are approximate stereoisomers of 2,3,7,8-TCDD and are chlorine-substituted in at least three of the four lateral positions in the aromatic ring system.

In mice, allelic variation at the *Ah* locus has been described.[50,51] The different alleles code for Ah receptors that differ in their ability to bind TCDD, and thus help to explain the different sensitivities of various inbred mouse strains to TCDD toxicity. Ah^{bb} C57BL/6 (B6) mice represent the prototypic "responsive" strain and are the most sensitive to TCDD toxicity, while Ah^{dd} DBA/2 (D2) mice represent the prototypic "nonresponsive" strain and require 10- to 20-fold higher doses of TCDD to produce the same toxic effect. Recently, congenic Ah^{dd} mice on a B6 background were derived that differ from conventional B6 mice primarily at the *Ah* locus. The spectrum of biochemical and toxic responses to TCDD exposure was similar in both strains but the doses needed to bring about the responses were significantly higher in congenic mice homozygous for the Ah^{d} allele as compared with mice carrying two Ah^{b} alleles.[52,53]

Two lines of evidence have been used to investigate the Ah receptor dependence of the acute immunotoxicity of TCDD and related HAH: (1) comparative studies using PCDD, PCDF, and PCB congeners that differ in their binding affinity for the Ah receptor and (2) studies using mice of different genetic background known to differ at the *Ah* locus. Vecchi *et al.*[54] were the first to report that the antibody response to SRBC was differentially suppressed by TCDD in B6 mice relative to D2 mice, with D2 mice requiring an approximately 10-fold higher dose to produce the same degree of suppression. Immunosuppression in B6D2F1 and backcross mice supported the role of the *Ah* locus in the expression of TCDD immunotoxicity. 2,3,7,8-TCDF was significantly less potent than TCDD and showed a similar differential immunosuppressive effect in B6 and D2 mice. At the same time, Silkworth and Grabstein[55] reported a B6 versus D2 strain-dependent difference in sensitivity to suppression of the anti-SRBC response by

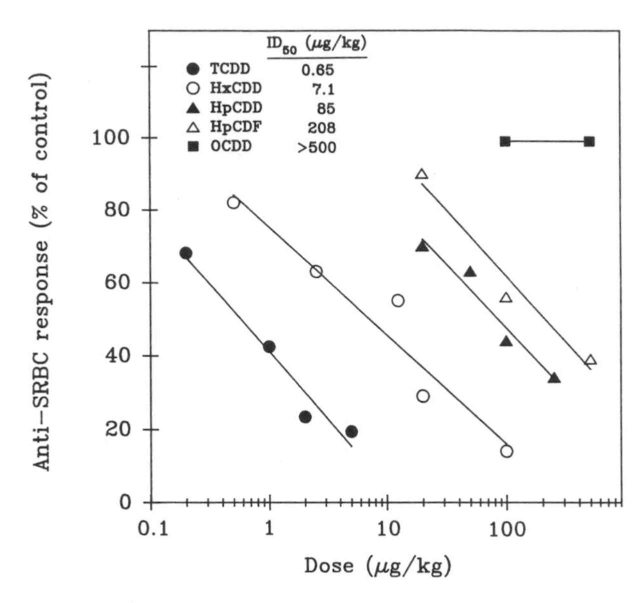

Figure 1. Structure-dependent immunotoxicity of some polychlorinated dioxin and furan isomers. Immunotoxicity was assessed by suppression of the splenic antibody response to SRBC following a single exposure to the purified congener. The ID_{50} represents the dose that resulted in 50% suppression of the antibody response. (Modified from Kerkvliet *et al.*[56])

3,4,3′,4′-tetrachlorobiphenyl, a ligand for the Ah receptor. In comparison, the 2,5,2′,5′-tetrachlorobiphenyl isomer, which lacks affinity for the Ah receptor, was not immunosuppressive in either B6 or D2 mice. Structure–activity relationships were extended by Kerkvliet *et al.*[56] in studies that compared the immunosuppressive potency of the chlorinated dioxin and furan isomers that contaminate technical-grade pentachlorophenol. The 1,2,3,6,7,8-hexachlorodioxin (HxCDD), 1,2,3,4,6,7,8-heptachlorodioxin (HpCDD), and 1,2,3,4,6,7,8-heptachlorofuran (HpCDF) isomers, which bind the receptor, were all significantly immunosuppressive. The dose of each isomer that produced 50% suppression of the antibody response to SRBC (ID_{50}) was 7.1, 85, and 208 μg/kg for HxCDD, HpCDD, and HpCDF, respectively (Fig. 1). The ID_{50} for TCDD was 0.65 μg/kg based on the data of Vecchi *et al.*[57,*] In contrast, octachlorodibenzo-*p*-dioxin (OCDD), which has low affinity for the Ah receptor, was not immunosuppressive at a dose as high as 500 μg/kg.[56] Similar structure–activity relationships have been demonstrated for suppression of the anti-SRBC response by technical-grade PCB mixtures,[58] purified PCB congeners,[59] and PCDF congeners.[60] Likewise, the differential sensitivity of the antibody response to SRBC of congenic Ah^{bb} and Ah^{dd} B6 mice to TCDD-induced suppression supports the concept of an Ah-receptor-dependent mechanism.[53,61]

*The potency of TCDD to suppress the antibody response to SRBC has been reported by several laboratories, with remarkable agreement in the ID_{50} value of 0.7 μg/kg in B6 mice. The ID_{50} in B6C3F1 mice is similar (<1 μg/kg; House *et al.*[24]) or slightly higher (1.2 μg/kg; Holsapple *et al.*[33]) than in B6 mice.

In addition to the antibody response to SRBC, Ah-receptor-dependent immunotoxicity has been demonstrated in mice using other immunologic endpoints. For example, Kerkvliet *et al.*[53] reported that the ID_{50} for suppression of the T-independent antibody response to TNP–LPS in Ah^{bb} B6 mice was 7.0 μg/kg compared with a significantly higher ID_{50} of 30 μg/kg in congenic Ah^{dd} B6 mice. Clark *et al.*[25] were first to report data suggesting that TCDD and PCB isomers suppressed *in vitro* cytotoxic T lymphocyte (CTL) responses of B6 and D2 mice through an Ah-receptor-dependent mechanism. Subsequently, Kerkvliet *et al.*[62] reported that B6 mice congenic at the *Ah* locus showed Ah-dependent sensitivity to suppression of the CTL response following exposure to either TCDD or 3,4,5,3′,4′,5′ hexachlorobiphenyl (HxCB). Furthermore, the potency of TCDD and of three HxCB congeners to suppress the CTL response of Ah^{bb} B6 mice directly correlated with their relative binding affinities for the Ah receptor. The ID_{50} of TCDD for suppression of the CTL response in B6 mice was 7.0 μg/kg.*

In other studies, Luster *et al.*[63] reported that Ah^{bb} B6C3F1 mice were more sensitive than D2 mice to TCDD-induced bone marrow toxicity. In addition, the 2,3,7,8-TCDD isomer was more effective at suppression of CFU-GM colony formation than were compounds that only weakly bind the Ah receptor (2,8-DCDD or OCDD). The ability of TCDD to suppress the cytotoxic responses of activated neutrophils was also shown to segregate with Ah-responsiveness.[64]

In summary, the data relating HAH immunotoxicity to Ah-receptor-dependent events are convincing. However, it should be emphasized that all of the data have been obtained from studies in inbred mice using an acute or subacute exposure regimen. Except for thymic atrophy,[65] structure–immunotoxicity relationships in other species, including rats, have not been established, and inbred strains of other species with defined *Ah* genotype are not currently available. The importance of Ah-receptor-mediated events in chronic, low-level HAH immunotoxicity also remains to be established.

7. SENSITIVE TARGETS FOR HAH IMMUNOTOXICITY

Despite considerable investigation, the cells that are directly altered by HAH exposure leading to suppressed immune function have not been unequivocally identified. The main reason for the lack of definitive progress in this area are the conflicting data reported from different laboratories regarding the ability of TCDD to suppress lymphocyte functions when examined *in vitro*. As discussed in another section of this chapter, the *in vitro* effects of TCDD are greatly influenced by the *in vitro* culture conditions, which may explain the discrepancies in effects observed in different laboratories.

*The dose of TCDD required to suppress the CTL response reported by Kerkvliet *et al.*[62] is significantly greater than that reported by Clark *et al.*[13] who reported CTL suppression following 4 weekly doses of 0.1 μg/kg TCDD. Clark *et al.*[25] also reported that doses of TCDD as low as 4 ng/kg to B6 mice suppressed the *in vitro* generation of CTL, and that the suppression was Ah dependent. The potency of TCDD described in Clark's studies has not been corroborated by other laboratories.

In contrast to *in vitro* studies, the *in vivo* immunotoxicity of TCDD, expressed in terms of suppression of the antibody response to SRBC, is highly reproducible between laboratories. Since the magnitude of the anti-SRBC response depends on the concerted interactions of antigen-presenting cells (e.g., macrophages), regulatory T lymphocytes (e.g., helper and suppressor T cells), and B lymphocytes (e.g., antibody-producing cells), this response has been used most widely to evaluate target cell sensitivity to TCDD and related HAH. In addition, the antibody response to SRBC can be modulated by many nonimmunological factors, including hormonal and nutritional variables, and TCDD is known to affect numerous endocrine and metabolic functions. These latter effects would be manifest only in *in vivo* studies, while only direct effects on macrophage and lymphocyte functions would be evident following *in vitro* exposure. To date, direct *in vitro* effects of TCDD on purified B cell activity have been documented,[66–68] while direct effects on macrophages and T cells have not been described. (The *in vitro* effects of TCDD will be discussed in more detail in Section 8.)

As an *in vivo* approach to evaluate the cellular targets of HpCDD, Kerkvliet and Brauner[31] compared the sensitivity of antibody responses in mice to antigens that differ in their requirements for macrophages and T cells. The T-helper-cell-independent (TI) antigens, DNP–Ficoll and TNP–LPS, were used in these studies. These TI antigens differ from each other in their requirement for macrophages (higher for DNP–Ficoll) and their sensitivity to regulatory (amplifier and suppressor) T cell influence (DNP–Ficoll is sensitive, TNP–LPS is not).[69] Obviously, all antibody responses require B cell differentiation into antibody-secreting plasma cells. Although HpCDD produced dose-dependent suppression of the antibody response to all three antigens, sensitivity to suppression directly correlated with the sensitivity of the response to T cell regulation. The ID_{50}s were 53, 127, and 516 μg/kg for SRBC, DNP–Ficoll, and TNP–LPS, respectively. These results were interpreted as follows: If one assumes that B cell function is targeted in the TNP–LPS response, then regulatory T cells and/or macrophages may represent the more sensitive target in the SRBC and DNP–Ficoll responses. The difference in sensitivity between the SRBC and DNP–Ficoll responses suggests that the T helper cell may be a particularly sensitive target. The differential sensitivity of the antibody responses to TNP–LPS compared with SRBC has been corroborated in TCDD-treated mice.[24,53] Thus, the exquisite *in vivo* sensitivity of the antibody response to SRBC would appear to depend on the T cell and/or macrophage components of the response rather than the B cell, unless the B cells that respond to SRBC are different from the B cells that respond to TNP–LPS. Currently, evidence for such a difference is lacking. The importance of T cells in mediating suppression of the antibody response is supported by studies using congenitally T-cell-deficient "nude" mice, which were significantly less sensitive to HpCDD-induced immune suppression as compared with their T-cell-competent littermates.[31]

On the other hand, the data of Dooley and Holsapple[70] suggest that the B lymphocyte is the cellular target of TCDD. Using separated spleen T cells, B cells, and adherent cells (e.g., macrophages) from vehicle- and TCDD-treated mice, they reported that B cells from TCDD-treated mice were functionally compromised in *in vitro* antibody responses but T cells and macrophages were not. The basis for these different findings *in vitro* has not been established. However, it is possible that T cells are only affected indirectly following antigen exposure (e.g., activated T cells are targeted). Thus, removal of resting T cells from the TCDD-treated animal prior to antigen challenge prevented detection of T cell dysfunction *in vitro*. This interpretation is supported by the findings of Tomar and Kerkvliet[71] that spleen cells taken from TCDD-treated mice were not compromised in their ability to reconstitute the antibody response of lethally irradiated mice. This interpretation is also consistent with the reported lack of direct effects of TCDD and other HAH on T cells *in vitro*.[13,72]

The influence of TCDD exposure on regulatory T cell functions has been addressed in a limited number of studies. Clark *et al.*[13] first proposed that T suppressor cells were induced by TCDD in the thymus that were responsible for the suppressed cytotoxic T cell response. However, increased suppressor cell activity in peripheral lymphoid tissue was not observed in mice exposed to TCDD[73] or HxCB.[74] In terms of T helper cell activity, Tomar and Kerkvliet[71] reported that a dose of 5 μg/kg TCDD suppressed the *in vivo* generation of carrier-specific T helper cells. Lundberg *et al.*[46] reported that thymocytes from B6 mice treated with TCDD (50 μg/kg) were less capable of supporting an *in vitro* anti-SRBC response. These effects on T cells may be related to alterations in the production of the cytokines that are secreted by regulatory T cells. For example, preliminary data from the author's laboratory indicate that the *in vivo* production of several T cell cytokines (IL-2, IL-4, IL-10, γ-interferon) that are produced in control mice in response to allogeneic tumor cells are suppressed in PCB-treated mice.[75,76] The most profound suppression was seen in the production of γ-interferon, a cytokine important for the development of cytotoxic T cells. In contrast, Clark *et al.*[25] reported that T cells from TCDD-treated mice produced normal levels of IL-2 in response to mitogen stimulation *in vitro*.

The direct effect of TCDD exposure on B cell function has been addressed primarily in *in vitro* studies. The issue is difficult to address *in vivo* given that most B cell responses (except perhaps anti-LPS responses) are dependent on interactions with T cells and macrophages. *In vitro* studies have described the direct effects of TCDD on the activation and differentiation of purified B cells.[67,68] These studies suggest that TCDD inhibits the terminal differentiation of B cells via alteration of an early activation event.[68] Increased protein phosphorylation and tyrosine kinase activity in TCDD-treated B cells may underlie this B cell dysfunction.[77,78]

Macrophage functions have also been examined following TCDD exposure and generally found to be resistant to suppression by TCDD when assessed *ex vivo*. Macrophage-mediated phagocytosis, macrophage-mediated tumor cell cytolysis or cytostasis, oxidative reactions of neutrophils and macrophages, and natural killer (NK) cell activity were not suppressed following TCDD exposure, with doses as high as 30 μg/kg failing to suppress NK and macrophage functions.[20,79] A potentially important exception is the reported selective inhibition of phorbol ester-activated antitumor activity of neutrophils by TCDD.[64]

On the other hand, it is interesting to note that the pathology associated with TCDD toxicity often includes neutrophilia and an inflammatory response in liver and skin characterized by activated macrophage and neutrophil accumulation.[10,80–82] While these observations may simply reflect a normal inflammatory response to tissue injury, there is some preliminary experimental evidence that suggests inflammatory cells may be activated by TCDD exposure. For example, Alsharif *et al.*[83] recently reported that TCDD increased superoxide anion production in rat peritoneal macrophages. In addition, it has been shown that TCDD exposure results in an enhanced inflammatory response following SRBC challenge.[84] This effect of TCDD was characterized by a two- to fourfold increase in the number of neutrophils and macrophages locally infiltrating the intraperitoneal site of SRBC injection. However, the time course of the cellular influx was not altered by TCDD. Likewise, the expression of macrophage activation markers (I-A and F4/80) and the antigen-presenting function of the peritoneal exudate cells was unaltered by TCDD. Thus, the effect of TCDD appeared to reflect a quantitative rather than a qualitative change in the inflammatory response. Importantly, TCDD-induced suppression of the anti-SRBC response could not be overcome by increasing the amount of antigen used for sensitization, suggesting that enhanced antigen clearance/degradation by the increased numbers of phagocytic cells (e.g., decreased antigen load) was not responsible for the decreased antibody response in TCDD-treated mice. Thus, the relationship, if any, between the inflammatory and immune effects of TCDD remains to be elucidated.

One mechanism by which TCDD and related HAH may augment inflammatory responses is via enhanced production of inflammatory mediators such as interleukin 1 (IL-1) and tumor necrosis factor (TNF). Recent evidence suggests that the long-recognized hypersusceptibility of TCDD- and PCB-treated animals to endotoxin (LPS)[20,21,85,86] may be related to an increased production of TNF and/or IL-6 in the chemically treated animals.[87–89] The ability of methylprednisolone to reverse the mortality associated with TCDD/endotoxin treatment is also consistent with an inflammatory response.[27] Similarly, increased inflammatory mediator production may underlie the enhanced rat paw edema response to carrageenan and dextran in TCDD-treated rats.[90,91] Limited preliminary data are available to indicate that the production of inflammatory mediators such as TNF[87,88] and IL-6[89] may be increased in HAH-treated animals. Serum complement activity, on the other hand, has been reported to be suppressed in dioxin-treated mice,[22] although enhanced activity was reported at the lowest exposure level when HxCDD was tested. A primary effect of TCDD on IL-1 is supported by the recent findings of Sutter *et al.*[92] that the IL-1β gene may be transcriptionally regulated by TCDD. Likewise, Steppan and Kerkvliet[93] have reported that under some exposure conditions TCDD increased the level of mRNA for IL-1 in TCDD-treated IC21 cells, a macrophage cell line derived from B6 mice. On the other hand, House *et al.*[24] reported that inflammatory macrophages obtained from TCDD-treated mice produced control levels of IL-1 when examined *ex vivo*. Thus, the effect of TCDD on inflammatory mediator production may be a "priming effect" and require coexposure to antigen or LPS. The influence of TCDD on inflammatory mediator production and action is an important area for further study.

Since the rapid influx of phagocytic cells to the site of pathogen invasion is an important factor in host resistance to infection, the ability of TCDD to augment the production of inflammatory chemoattractive mediators would imply that TCDD exposure could result in enhanced host resistance. However, since TCDD exposure is, at the same time, immunosuppressive, resulting in decreased specific immune responses generated by T and B lymphocytes, the overall impact of TCDD exposure on disease susceptibility will likely vary depending on the nature of the pathogen and the major mode of host response to the specific infectious agent. Such effects may in fact help to explain the disparate effects of TCDD in different host resistance models that have been previously reported.

8. *IN VITRO* IMMUNOTOXIC EFFECTS OF HAH

Investigators in the field of TCDD immunotoxicity have long acknowledged the difficulties in consistently demonstrating the immunotoxicity of TCDD when cells from treated animals are tested *ex vivo* or when TCDD is added to culture *in vitro*. While effects following *in vitro* and *ex vivo* exposure to TCDD on lymphocyte functions have been reported,[26,66,68] other laboratories have failed to observe suppression with *in vitro* or *ex vivo* exposure to TCDD or related HAH.[13,46,72] In addition, the effects of TCDD seen *in vitro* are sometimes inconsistent with those observed after *in vivo* assessment of immunotoxicity. For example, the rank order of sensitivity to suppression of T helper cell-dependent and T helper cell-independent antibody responses seen *in vivo*[24,31,53] is not seen *in vitro*[26,66] suggesting that different cellular targets may be affected following *in vitro* exposure to TCDD. More importantly, some data suggest that suppression of the *in vitro* antibody response may occur independent of the Ah receptor. Tucker *et al.*[26] and Holsapple *et al.*[66] reported that direct addition of TCDD *in vitro* suppressed the antibody response to SRBC. However, based on the response of cells from congenic mice as well as a limited structure–activity study, the data of Tucker *et al.*[26] supported an Ah-receptor-dependent suppression while the data of Holsapple *et al.*[66] did not. In the latter study, the magnitude of suppression was comparable using cells from responsive B6C3F1 or congenic heterozygous (B6 Ah^{bd}) mice relative to nonresponsive D2 or homozygous B6 Ah^{dd} mice. In addition, they reported that the 2,7-dichlorodibenzo-*p*-dioxin congener which lacks affinity for the Ah receptor, was equipotent with TCDD in suppressing the *in vitro* response.

In other studies, Davis and Safe[94] compared the *in vitro* structure–immunotoxicity relationships for a series of HAH congeners which show > 14,900-fold difference in *in vivo* immunotoxic potency. Results of these studies indicated that all of the congeners were equipotent *in vitro* and produced a similar concentration-dependent suppression of the *in vitro* anti-SRBC response using cells from either B6 or D2 mice. Coexposure to the Ah-receptor antagonist α-naphthoflavone antagonized the immunosuppression induced by either TCDD or 1,3,6,8-TCDF (a weak Ah-receptor agonist). Collectively, the results supported a mechanism of suppression *in vitro* that was independent of the Ah receptor.

The basis for these variable effects of TCDD *in vitro* is currently unknown. However, recent studies by Morris *et al.*[67] demonstrated that the *in vitro* effects of TCDD on the anti-SRBC response were critically dependent on the type and concentration of the serum used in the *in vitro* culture. Only 3 of 23 lots of serum were able to support a full dose-responsive suppression, and, in serum-free cultures, TCDD caused a 15-fold enhancement of the anti-SRBC response. Thus, differences in media components used in *in vitro* cultures may account for the different effects seen *in vitro* in different laboratories. Other factors such as the TCDD carrier/solvent used, the calcium content of the media, or procedures used for preparation of spleen cell suspensions may all contribute to variable effects of TCDD *in vitro*.

The obvious question relates to the relevance of the *in vitro* findings to the *in vivo* immunotoxicity. In this respect, it is important to note that the concentration of TCDD required for *in vitro* suppression of immune function ($1–30 \times 10^{-9}$ M) of murine lymphocytes is several orders of magnitude higher than the concentration found in lymphoid tissues following exposure *in vivo* to an immunotoxic dose of TCDD.[95] The amount of TCDD associated with isolated spleen cells obtained from mice 2 days following treatment with 5 μg/kg [^{3}H]-TCDD was 2×10^{-15} M per 10^7 spleen cells. Importantly, as much as 50% of the radioactivity associated with whole spleen tissue was recovered in the stromal and/or capsular material (i.e., splenic tissue that resisted passage through the mesh screens used for preparation of spleen cell suspensions). These findings suggest that (1) the most potent effects of TCDD on immune function *in vivo* may be induced indirectly by effects on nonlymphoid cells, or (2) based on the delivered dose of TCDD, this molecule is more toxic than previously thought. Alternatively, TCDD effects *in vivo* on nonlymphoid cells may amplify the direct effects of TCDD on lymphoid tissue. Certainly, additional studies are needed to elucidate the serum components that are permissive for suppression or enhancement of immune responses *in vitro* and to determine their relevance to *in vivo* conditions. Such studies are also likely to provide insight into the mechanisms of TCDD interaction with lymphoid cells.

9. INDIRECT MECHANISMS OF HAH IMMUNOTOXICITY

The difficulty in demonstrating consistent, direct effects of TCDD *in vitro* on lymphocytes, the dependence of those effects on serum components, and the requirement for high concentrations of TCDD are all consistent with an indirect mechanism of TCDD on the immune system. One potentially important indirect mechanism is via effects on the endocrine system. Several endocrine hormones have been shown to regulate immune responses, including glucocorticoids, sex steroids, thyroxine, growth hormone, and prolactin. Importantly, TCDD and other HAH have been shown to alter the activity of all of these hormones.

Kerkvliet *et al.*[62] reported that exposure of mice to HxCB followed by injection of P815 allogeneic tumor cells induced a dose-dependent elevation of serum corticosterone concentrations which correlated with the dose-dependent suppression of the antitumor cytotoxic T cell response. However, since adrenalectomy or treatment with the glucocorticoid receptor antagonist RU38486 failed to protect mice from the immunosuppressive effect of HxCB,[96] a role for the elevated corticosterone in the suppression of the cytotoxic T cell response seems unlikely. Adrenalectomy and hypophysectomy also failed to prevent TCDD-induced thymic atrophy in rats.[39]

Using the P815 allogeneic tumor model, Kerkvliet and Baecher-Steppan[74] reported that male mice were more sensitive than female mice to suppression of the cytotoxic T cell response by HxCB. Castration of male rats partially ameliorated the immunosuppressive effects of HxCB,[96] suggesting a role for testosterone in suppression of the cytotoxic T cell response.

Pazdernik and Rozman[97] suggested that thyroid hormones may play a role in TCDD immunotoxicity based on the finding that radiothyroidectomy prevented the suppression of the anti-SRBC response in rats treated with TCDD. However, since thyroidectomy alone suppressed immune function, the significance of the findings requires further study.

10. IMMUNOTOXICITY OF HAH IN NONHUMAN PRIMATES

A limited number of studies using nonhuman primates as surrogate models for humans have been conducted to assess HAH immunotoxicity. Immunological effects were described in rhesus monkeys and their offspring chronically exposed to TCDD in food at levels of 5 or 25 ppt for 4 years.[98] In the mothers, the total number of T cells increased in monkeys fed 25 ppt TCDD, with a selective increase in $CD8^+$ cells and a decrease in $CD4^+$ cells. However, no significant effect on T cell function was established when assessed as proliferation response to mitogens, alloantigens, or xenoantigens. NK cell activity and production of antibodies to tetanus immunization were normal. In the offspring of TCDD-exposed dams examined 4 years after exposure, a significantly increased antibody response to tetanus toxoid immunization was observed which correlated with TCDD tissue levels. The body burden of TCDD in the offspring ranged from a low of 290 ppt to a high of 1400 ppt.

In other studies, a single injection of TCDD in marmosets (*Callithrix jacchus*) resulted in a delayed decrease in the percentage of $CD4^+$ T cells and $CD20^+$ B cells in the blood and an increase in the percentage of $CD8^+$ cells.[99] The total number of T cells was not significantly altered by TCDD exposure. The $CD4^+$ subset most affected was the $CDw29^+$ "helper-inducer" or "memory" subset, with significant effects observed after a TCDD dose of 10 ng/kg. The no observable effect level (NOEL) for this effect was 3 ng/kg TCDD. Concomitant with suppression of the CDw29 subset in TCDD-treated animals, the percentage of $CD4^+CD45RA^+$ cells increased. This subset has been classified as "suppressor-inducer" or "naive" cells. The changes in the T cell subsets were intensified following *in vitro* culture of the cells with mitogen.[100] Paradoxically, however, a recent study from the same laboratory reported that chronic exposure of young marmosets to very low levels of TCDD (0.3 ng/kg per week for 24 weeks) produced the opposite effect of acute exposure on the $CD4^+CDw29^+$ subset, with TCDD treatment resulting in a significant increase in this population.[101] Concomitantly, the $CD45RA^+$ subset of $CD4^+$ cells decreased. On transfer of the animals to a higher dose of TCDD (1.5 ng/kg per week) for 3 weeks, the enhancing effect was reversed, and suppression of the $CD4^+CDw29^+$ subset was observed. Maximum suppression was seen after 6 weeks of exposure to the higher dose. After discontinuation of dosing, the reduction in the percentage and absolute number of $CD4^+CDw29^+$

cells persisted for 5 weeks, reaching normal range 7 weeks later. These results led the authors to conclude that "extrapolations of the results obtained at higher doses to very low exposures is [*sic*] not justified with respect to the effects induced by TCDD on the immune system of marmosets."

The immunomodulatory effects of chronic low-level PCB exposure in monkeys have also been investigated. In early studies, Thomas and Hinsdill[85] reported that rhesus monkeys fed diets containing 2.5 or 5 ppm of Aroclor 1248 had significantly suppressed antibody response to SRBC but not to tetanus toxoid. These monkeys also had chloracne, alopecia, and facial edema. Similarly, exposure of cynomolgous monkeys to Aroclor 1254 (100 or 400 μg/kg per day) for 3 months suppressed antibody responses to SRBC but not to tetanus toxoid.[102] Suppressive effects on anti-SRBC responses were more severe in cynomolgous monkeys when the PCB mixture contained PCDFs.[103]

Tryphonas *et al.*[104–106] have recently reported results of studies in rhesus monkeys exposed chronically to Aroclor 1254 (5–80 μg/kg per day) for 23 or 55 months. These exposures resulted in steady-state blood PCB levels that ranged from a mean low of 0.01 ± 0.001 ppm in the 5 μg/kg group to a mean high of 0.11 ± 0.01 ppm in the 80 μg/kg group. The only immune parameters that were consistently affected were the primary and anamnestic antibody responses to SRBC which were suppressed in a dose-dependent manner. In contrast, the antibody response to pneumococcus vaccine antigen measured at 55 months of exposure was not significantly altered. At 23 months, the percentage of T helper cells in the blood was significantly decreased in the 80 μg/kg group, and the percentage and absolute number of T suppressor cells was increased; however, these effects were not apparent at 55 months of exposure.[106] Lymphoproliferative responses to PHA and Con A were not significantly altered at 23 months but were dose-dependently suppressed at 55 months. Proliferation to alloantigens was not significantly altered. Likewise, serum immunoglobulin and hydrocortisone levels did not differ between treatment groups. After 55 months, the chemiluminescent response (time to peak) of monocytes was slower in PCB-exposed cells. Also noted at 55 months was a significant elevation in serum hemolytic complement levels, a dose-related increase in NK cell activity, and a dose-related increase in thymosin alpha-1 levels but not thymosin beta-4 levels.[105] Effects on interferon levels were inconsistent, and TNF production was not altered.

The studies in nonhuman primates are important from the standpoint that the antibody response to SRBC emerges as the only immunological parameter consistently suppressed by HAH in several animal species. (A notable exception is the recent report that TCDD does not suppress the antibody response to SRBC in rats.[107]) Other immunological endpoints such as total T cell numbers, percentages of T cell subsets, lymphoproliferative responses, and delayed hypersensitivity responses are inconsistently increased or decreased in various studies. At the present time, it is not clear why the antibody response to SRBC is most consistently altered by HAH exposure in different species. The sensitivity of the anti-SRBC response does not appear to be related solely to the T-cell dependency of the response since antibody responses to other T-dependent antigens (e.g., tetanus toxoid, bovine gamma globulin) are not suppressed and may even be enhanced following HAH exposure. It is possible that the particulate nature of the SRBC antigens is an important factor even though a mechanistic basis for this is not readily apparent. The sensitivity of the technique used to quantify the antibody response may also contribute to apparent increased sensitivity of the SRBC model, which is most often measured as the PFC response rather than serum antibody titers which are usually more variable.

11. IMMUNOTOXICITY OF HAH IN HUMANS

The immunotoxicity of TCDD and related HAH in humans has been the subject of several studies derived from accidental and/or occupational exposures to PCBs, PBBs, and TCDD. Immunological assessment was carried out on patients who consumed acnegenic and hepatotoxic doses of PCDF–PCB-contaminated rice oil in Taiwan in 1979. Clinical symptoms were primarily related to increased frequency of various kinds of infection, especially of the respiratory tract and skin.[108] Immunological effects included decreased serum IgA and IgM but not IgG, decreased percentage of T cells in blood related to decreased $CD4^+$ T helper cells and increased $CD8^+$ T suppressor cells, and suppressed dermal delayed-type hypersensitivity responses to streptokinase/streptodornase and tuberculin antigens.[108] The percentage of anergic patients increased and the degree of induration decreased with increased PCB concentration in the blood. In contrast, lymphoproliferative responses of peripheral blood lymphocytes to T cell mitogens (PHA, PWM, and tuberculin) were significantly augmented in PCB-exposed patients. PCB concentrations in the blood ranged from 3 to 1156 ppb with a mean of 89 $\pm$ 6.9 ppb. The oil was contaminated at PCB concentrations of 4.8 to 204.9 ppm with a mean of 52 $\pm$ 39 ppm. In addition, the oil contained elevated levels of polychlorinated dibenzofurans which likely played a significant role in the toxicity.[109]

Immunotoxic effects were also described in Michigan dairy farmers exposed to polybrominated biphenyls via contaminated dairy products and meat in 1973.[110] Like PCB-exposed patients, the percentage and absolute numbers of T cells in peripheral blood of PBB-exposed farmers were significantly reduced compared with a control group. However, in contrast to PCB, lymphoproliferative responses of T cells were significantly decreased in PBB-exposed persons. Also in contrast to PCB, skin testing using standard recall antigens indicated that PBB-exposed Michigan dairy farmers had significantly increased responses, particularly to candida and varidase. Tissue levels of PBB in the subjects were not determined in these studies.

Webb *et al.*[111] reported the findings from immunological assessment of 41 persons from Missouri with documented adipose tissue levels of TCDD resulting from occupational, recreational, or residential exposure. Of the participants, 16 had tissue TCDD levels less than 20 ppt, 13 had levels between 20 and 60 ppt, and 12 had levels greater than 60 ppt. The highest level was 750 ppt. Data were analyzed by multiple regression based on adipose tissue level and the clinical dependent variable. Increased TCDD levels were correlated with an increased percentage and total number of T lymphocytes. $CD8^+$ and $T11^+$ T cells accounted for the increase, while $CD4^+$ T cells were not altered in percentage or number. Lymphoproliferative responses to T cell mitogens or tetanus toxoid were unaltered as was the cytotoxic T cell response. Serum IgA but not IgG was increased. No adverse clinical disease was associated with TCDD levels in these subjects. Only 2 of the 41 subjects reported a history of chloracne. These findings differ from those reported for the Quail Run Mobile Home Park residents (tissue levels unknown) in which decreased T cell numbers (T3, CD4, and T11) and suppressed cell-mediated immunity were reported.[112] However, subsequent retesting of these anergic subjects failed to confirm the anergy.[113] On the other hand, when serum from some of these individuals was tested for levels of the thymic peptide, thymosin alpha-1, the entire frequency distribution for the TCDD-exposed group was shifted toward lower thymosin alpha-1 levels.[114] The statistically significant difference between the TCDD-exposed persons and controls remained after controlling for age, sex, and socioeconomic status, with a trend of decreasing thymosin alpha-1 levels with increasing number of years of residence in the TCDD-contaminated residential area. The thymosin alpha-1 levels were not correlated with changes in other immune system parameters nor with any increased incidence of clinically diagnosed immune suppression. The decrease in thymosin alpha-1 levels in humans contrasts with the increase in thymosin alpha-1 seen in PCB-treated monkeys.[106]

Finally, Mocarelli *et al.*[115] reported studies on the immune status of 44 children, 20 of whom had chloracne, who were exposed to TCDD following an explosion at a herbicide factory in Seveso, Italy. No abnormalities were found in the following parameters: serum immunoglobulin concentrations, levels of circulating complement, or lymphoproliferative responses to T and B cell mitogens. Interestingly, in a study conducted 6 years after the explosion, a different cohort of TCDD-exposed children exhibited a significant increase in complement protein levels, which correlated with the incidence of chloracne, as well as increased numbers of peripheral blood lymphocytes, and increased lymphoproliferative responses.[116] However, no specific health problems were correlated with dioxin exposure in these children.

It is readily apparent that no clear pattern of immunotoxicity to TCDD, PCB, or PBB emerges from these studies in humans. In some cases T cell numbers or functions increase; in others, they decrease. The findings are not unlike the varied and often conflicting reports found in the literature regarding animal studies of HAH immunotoxicity. The basis for the lack of consistent and/or significant exposure-related effects is unknown and may be dependent on several factors. Most notable in this regard is the generic difficulties in assessing subclinical immunomodulation, particularly in outbred human populations. Most immunological assays have a very broad range of normal responses reducing the sensitivity to detect small changes. Similarly, the assays used to examine immune function in humans exposed to TCDD and related HAH have unfortunately been based to a greater extent on what was clinically "doable" (e.g., lymphocyte phenotype, mitogen responsiveness) rather than on assays that have been shown to be sensitive to TCDD in animal studies. Thus, the lack of consistent and/or significant immunotoxic effects in humans resulting from TCDD exposure may be as much a function of the assays used as the immune status of the cohort. In addition, few studies have examined the immune status of individuals with known, documented exposure to HAH. Rather, cohorts based on presumption of exposure have been studied. There is some evidence to suggest that the lack of consistent, significant effects may sometimes be related to the inclusion of subjects who had little or no actual exposure to TCDD.[111] Likewise, the important role that Ah phenotype plays in TCDD immunotoxicity has not been considered when addressing human sensitivity. Whether there are human equivalents of murine Ah^{bb} and Ah^{dd} types is not known. Finally, in most studies, the assessment of immune function in exposed populations was carried out long after exposure to TCDD ceased. Thus, recovery from the immunotoxic effects of TCDD may have occurred.

Based on the available evidence, it is difficult to draw any significant conclusions regarding the sensitivity of the human immune system to TCDD/PCB toxicity. However, given the current lack of a definitive data base that correlates clinical immunological endpoints with immune status in humans (except in cases of overt immune deficiencies), massive retrospective studies of poorly defined exposure groups cannot be justified to try to "prove" that immune modulation has occurred. Rather, resources would be better directed toward the establishment of a broad data base of normal values for the clinical immunological endpoints that may be of use in immunotoxicity assessments. In conjunction with this effort, research must focus on the definition of sensitive endpoints (i.e., biomarkers) of immune dysfunction in humans so that, in the future, emergency response teams could respond rapidly to accidental exposures to assess the immunological status of the exposed persons. To validate these biomarkers, there is a parallel need for animal research to identify TCDD-sensitive immune endpoints in animal models that can also be measured in humans in order to establish correlative changes in the biomarker and immune function. In particular, it will be important to determine in animal models how well changes in immune function in the lymphoid organs (e.g., spleen, lymph nodes) correlate with changes in the expression of lymphocyte subset/activation markers in peripheral blood since the former are the tissues of choice for laboratory animal studies while the latter is used in human studies. Until such correlations are established, the interpretation of changes observed in subsets/activation markers in human peripheral blood lymphocytes in terms of health risk will be limited to speculation. For example, recent studies in mice have shown that a dose of TCDD that is highly suppressive for the antibody response to SRBC does not alter the frequency of major lymphocyte subsets (i.e., B versus T cells, $CD4^+$ versus $CD8^+$ T cells) in the spleen,[34] demonstrating that functional immunosuppression can occur in the absence of phenotypic alterations in subsets. Research must also continue to develop well-defined animal models using multiple animal species that will lead to an understanding of the underlying mechanisms of HAH immunotoxicity. For example, there is a clear need to document Ah receptor involvement in the immunotoxicity of TCDD and related HAH in species other than mice. These studies need to go beyond descriptive immunotoxicity assessment to determine the mechanistic basis for differences in species sensitivity to TCDD immunotoxicity following both acute and chronic exposure. In the interim, the available data base derived from well-controlled animal studies on HAH immunotoxicity can be used for establishment of no-effect levels and acceptable exposure levels for human risk assessment of TCDD using the same procedures that are used for other noncarcinogenic toxic endpoints. Because the antibody response to SRBC has been shown to be dose-dependently suppressed by TCDD and related HAH in several animal species, this data base would appear to be best suited for current application to risk assessment.

REFERENCES

1. B. Magnusson and A. M. Kligman, The identification of contact allergens by animal assay. The guinea pig maximization test, *J. Invest. Dermatol.* **52,** 268–276 (1985).
2. K. E. Anderson and H. I. Maibach, Guinea pig sensitization assays, an overview, in: *Contact Allergy Predictive Tests in Guinea Pigs* (K. E. Andersen and H. I. Maibach, eds.), pp. 263–290, Karger, Basel (1985).
3. E. V. Buehler, A rationale for the selection of occlusion to induce and elicit delayed contact hypersensitivity in the guinea pig; a prospective test, in: *Contact Allergy Predictive Tests in Guinea Pigs* (K. E. Andersen and H. I. Maibach, eds.), pp. 39–58, Karger, Basel (1985).
4. J. H. Dean, M. I. Luster, G. A. Boorman, and L. D. Lauer, Procedures available to examine the immunotoxicity of chemicals and drugs, *Pharmacol. Rev.* **34,** 137–148 (1982).
5. K. C. Norbury, Immunotoxicology in the pharmaceutical industry, *Environ. Health Perspect.* **43,** 53–59 (1982).
6. P. H. Bick, M. P. Holsapple, and K. L. White, Assessment of the effects of chemicals on the immune system, in: *New Approaches in Toxicity Testing and Their Application in Human Risk Assessment* (A. P. Li, ed.), pp. 165–178, Raven Press, New York (1985).
7. N. I. Kerkvliet, Measurements of immunity and modifications by toxicants, in: *Safety Evaluation of Drugs and Chemicals* (W. E. Lloyd, ed.), pp. 235–256, Hemisphere, Washington, DC (1986).
8. J. H. Exon, L. D. Koller, P. A. Talcott, C. A. O'Reilly, and G. M. Henningsen, Immunotoxicity testing: An economical multiple-assay approach, *Fundam. Appl. Toxicol.* **7,** 387–397 (1986).
9. M. I. Luster, A. E. Munson, P. T. Thomas, M. P. Holsapple, J. D. Fenters, K. L. White, L. D. Lauer, D. R. Germolec, G. J. Rosenthal, and J. H. Dean, Development of a testing battery to assess chemical-induced immunotoxicity: National Toxicology Program's guidelines for immunotoxicity evaluation in mice, *Fundam. Appl. Toxicol.* **10,** 2–19 (1988).
10. J. G. Vos, J. A. Moore, and J. G. Zinkl, Effect of 2,3,7,8-tetrachlorodibenzo-p-dioxin on the immune system of laboratory animals, *Environ. Health Perspect.* **5,** 149–162 (1973).
11. E. E. McConnell, J. A. Moore, J. K. Haseman, and M. W. Harris, The comparative toxicity of chlorinated dibenzo-p-dioxins in mice and guinea pigs, *Toxicol. Appl. Pharmacol.* **44,** 335–356 (1978).
12. A. Poland and J. C. Knutson, 2,3,7,8-Tetrachlorodibenzo-p-dioxin and related halogenated aromatic hydrocarbons: Examination of the mechanism of toxicity, *Annu. Rev. Pharmacol. Toxicol.* **22,** 517–554 (1982).
13. D. A. Clark, J. Gauldie, M. R. Szewczuk, and G. Sweeney, Enhanced suppressor cell activity as a mechanism of immunosuppression by 2,3,7,8-tetrachlorodibenzo-p-dioxin, *Proc. Exp. Biol. Med.* **168,** 290–299 (1981).
14. N. I. Kerkvliet, Halogenated aromatic hydrocarbons (HAH) as immunotoxicants, in: *Chemical Regulation of Immunity in Veterinary Medicine* (M. Kende, J. Gainer, and M. Chirigos, eds.), pp. 369–387, Liss, New York (1984).

15. J. G. Vos and M. I. Luster, Immune alterations, in: *Halogenated Biphenyls, Terphenyls, Naphthalenes, Dibenzodioxins and Related Products* (R. D. Kimbrough and A. A. Jensen, eds.), pp. 295–322, Elsevier, Amsterdam (1989).
16. M. P. Holsapple, D. L. Morris, S. C. Wood, and N. K. Snyder, 2,3,7,8-Tetrachlorodibenzo-p-dioxin-induced changes in immunocompetence: Possible mechanisms, *Annu. Rev. Pharmacol. Toxicol.* **31,** 73–100 (1991).
17. M. P. Holsapple, N. K. Snyder, S. C. Wood, and D. L. Morris, A review of 2,3,7,8-tetrachlorodibenzo-p-dioxin-induced changes in immunocompetence: 1991 update, *Toxicology* **69,** 219–255 (1991).
18. J. E. Thigpen, R. E. Faith, E. E. McConnell, and J. A. Moore, Increased susceptibility to bacterial infection as a sequela of exposure to 2,3,7,8-tetrachlorodibenzo-p-dioxin, *Infect. Immun.* **12,** 1319–1324 (1975).
19. R. D. Hinsdill, D. L. Couch, and R. S. Speirs, Immunosuppression in mice induced by dioxin (TCDD) in feed, *J. Environ. Pathol. Toxicol.* **4,** 401–425 (1980).
20. J. G. Vos, J. G. Kreeftenberg, H. W. B. Engel, A. Minderhoud, and L. M. Van Noorle Jansen, Studies on 2,3,7,8-tetrachlorodibenzo-p-dioxin-induced immune suppression and decreased resistance to infection: Endotoxin hypersensitivity, serum zinc concentrations and effect of thymosin treatment, *Toxicology* **9,** 75–86 (1978).
21. P. T. Thomas and R. D. Hinsdill, The effect of perinatal exposure to tetrachlorodibenzo-p-dioxin on the immune response of young mice, *Drug Chem. Toxicol.* **2,** 77–98 (1979).
22. K. L. White, H. H. Lysy, J. A. McCay, and A. C. Anderson, Modulation of serum complement levels following exposure to polychlorinated dibenzo-p-dioxins, *Toxicol. Appl. Pharmacol.* **84,** 209–219 (1986).
23. M. I. Luster, G. A. Boorman, J. H. Dean, M. W. Harris, R. W. Luebke, M. L. Padarathsingh, and J. A. Moore, Examination of bone marrow, immunologic parameters and host susceptibility following pre- and postnatal exposure to 2,3,7,8-tetrachlorodibenzo-p-dioxin (TCDD), *Int. J. Immunopharmacol.* **2,** 301–310 (1980).
24. R. V. House, L. D. Lauer, M. J. Murray, P. T. Thomas, J. P. Ehrlich, G. R. Burleson, and J. H. Dean, Examination of immune parameters and host resistance mechanisms in B6C3F1 mice following adult exposure to 2,3,7,8-tetrachlorodibenzo-p-dioxin, *J. Toxicol. Environ. Health* **31,** 203–215 (1990).
25. D. A. Clark, G. Sweeney, S. Safe, E. Hancock, D. G. Kilburn, and J. Gauldie, Cellular and genetic basis for suppression of cytotoxic T cell generation by haloaromatic hydrocarbons, *Immunopharmacology* **6,** 143–153 (1983).
26. A. N. Tucker, S. J. Vore, and M. I. Luster, Suppression of B cell differentiation by 2,3,7,8-tetrachlorodibenzo-p-dioxin, *Mol. Pharmacol.* **29,** 372–377 (1986).
27. G. J. Rosenthal, E. Lebetkin, J. E. Thigpen, R. Wilson, A. N. Tucker, and M. I. Luster, Characteristics of 2,3,7,8-tetrachlorodibenzo-p-dioxin induced endotoxin hypersensitivity: Association with hepatotoxicity, *Toxicology* **56,** 239–251 (1989).
28. M. A. Ritter, J. Rozing, and H.- J. Schuurman, The true function of the thymus? *Immunol. Today* **9,** 189–193 (1988).
29. A. G. Clarke and K. A. MacLennan, The many facets of thymic involution, *Immunol. Today* **7,** 204–205 (1986).
30. E. Benjamini and S. Leskowitz, *Immunology: A Short Course,* 2nd ed., Wiley–Liss, New York (1991).
31. N. I. Kerkvliet and J. A. Brauner, Mechanisms of 1,2,3,4,6,7,8-heptachlorodibenzo-p-dioxin (HpCDD)-induced humoral immune suppression: Evidence of primary defect in T cell regulation, *Toxicol. Appl. Pharmacol.* **87,** 18–31 (1987).
32. J. B. Silkworth and L. Antrim, Relationship between Ah receptor-mediated polychlori-

nated biphenyl (PCB)-induced humoral immunosuppression and thymic atrophy, *J. Pharmacol. Exp. Ther.* **235,** 606–611 (1985).
33. M. P. Holsapple, J. A. McCay, and D. W. Barnes, Immunosuppression without liver induction by subchronic exposure to 2,7-dichlorodibenzo-p-dioxin in adult female B6C3F1 mice, *Toxicol. Appl. Pharmacol.* **83,** 445–455 (1986).
34. N. I. Kerkvliet and J. A. Brauner, Flow cytometric analysis of lymphocyte subpopulations in the spleen and thymus of mice exposed to an acute immunosuppressive dose of 2,3,7,8-tetrachlorodibenzo-p-dioxin, *Environ. Res.* **52,** 146–164 (1990).
35. J. G. Vos and J. A. Moore, Suppression of cellular immunity in rats and mice by maternal treatment with 2,3,7,8-tetrachlorodibenzo-p-dioxin, *Int. Arch. Allergy* **47,** 777–794 (1974).
36. R. E. Faith and J. A. Moore, Impairment of thymus-dependent immune functions by exposure of the developing immune system to 2,3,7,8-tetrachlorodibenzo-p-dioxin (TCDD), *J. Toxicol. Environ. Health* **3,** 451–464 (1977).
37. S. D. Holladay, P. Lindstrom, B. L. Blaylock, C. E. Comment, D. R. Germolec, J. J. Heindell, and M. I. Luster, Parinatal thymocyte antigen expression and postnatal immune development altered by gestational exposure to tetrachlorodibenzo-p-dioxin (TCDD), *Teratology* **44,** 385–393 (1991).
38. Y. Takagi, S. Aburada, T. Otake, and N. Ikegami, Effect of polychlorinated biphenyls (PCBs) accumulated in the dam's body on mouse filial immunocompetence, *Arch. Environ. Contam. Toxicol.* **16,** 375–381 (1987).
39. M. J. van Logten, B. N. Gupta, E. E. McConnell, and J. A. Moore, Role of the endocrine system in the action of 2,3,7,8-tetrachlorodibenzo-p-dioxin (TCDD) on the thymus, *Toxicology* **15,** 135–144 (1980).
40. D. J. McConkey, P. I. Hartzell, S. K. Duddy, H. Hakansson, and S. Orrenius, 2,3,7,8-Tetrachlorodibenzo-p-dioxin kills immature thymocytes by Ca^{+2} mediated endonuclease activation, *Science* **242,** 256–259 (1988).
41. B. L. Blaylock, S. D. Holladay, C. E. Comment, J. J. Heindel, and M. I. Luster, Exposure to tetrachlorodibenzo-p-dioxin (TCDD) alters fetal thymocyte maturation, *Toxicol. Appl. Pharmacol.* **112,** 207–213 (1992).
42. W. F. Greenlee, K. M. Dold, R. D. Irons, and R. Osborne, Evidence for direct action of 2,3,7,8-tetrachlorodibenzo-p-dioxin (TCDD) on thymic epithelium, *Toxicol. Appl. Pharmacol.* **79,** 112–120 (1985).
43. J. C. Cook, K. M. Dold, and W. F. Greenlee, An *in vitro* model for studying the toxicity of 2,3,7,8-tetrachlorodibenzo-p-dioxin to human thymus, *Toxicol. Appl. Pharmacol.* **89,** 256–268 (1987).
44. L. Dencker, E. Hassoun, R. d'Argy, and G. Alm, Fetal thymus organ culture as an *in vitro* model for the toxicity of 2,3,7,8-tetrachlorodibenzo-p-dioxin and its congeners, *Mol. Pharmacol.* **28,** 357–363 (1985).
45. R. d'Argy, J. Bergman, and L. Dencker, Effects of immunosuppressive chemicals on lymphoid development in foetal thymus organ cultures, *Pharmacol. Toxicol.* **64,** 33–38 (1989).
46. K. Lundberg, L. Dencker, and K.-O. Gronvik, Effects of 2,3,7,8-tetrachlorodibenzo-p-dioxin (TCDD) treatment *in vivo* on thymocyte functions in mice after activation *in vitro*, *Int. J. Immunopharmacol.* **12,** 459–466 (1990).
47. J. S. Fine, T. A. Gasiewicz, N. C. Fiore, and A. E. Silverstone, Prothymocyte activity is reduced by perinatal 2,3,7,8-tetrachlorodibenzo-p-dioxin exposure, *J. Exp. Pharmacol. Ther.* **255,** 1–5 (1990).
48. S. Cuthill, A. Wilhelmsson, G. G. F. Mason, M. Gillner, L. Poellinger, and J. A.

Gustafsson, The dioxin receptor: A comparison with the glucocorticoid receptor, *J. Steroid Biochem.* **30,** 277–280 (1988).
49. J. P. Whitlock, Genetic and molecular aspects of 2,3,7,8-tetrachlorodibenzo-p-dioxin action, *Annu. Rev. Pharmacol. Toxicol.* **30,** 251–277 (1990).
50. A. Poland, E. Glover, and B. A. Taylor, The murine Ah locus: A new allele and mapping to chromosome 12, *Mol. Pharmacol.* **32,** 471–478 (1987).
51. A. Poland and E. Glover, Characterization and strain distribution pattern of the murine Ah receptor specified by the Ah^{d} and Ah^{b-3} alleles, *Mol. Pharmacol.* **38,** 306–312 (1990).
52. L. S. Birnbaum, M. M. McDonald, P. C. Blair, A. M. Clark, and M. W. Harris, Differential toxicity of 2,3,7,8-tetrachlorodibenzo-p-dioxin (TCDD) in C57B1/6 mice congenic at the Ah locus, *Fundam. Appl. Toxicol.* **15,** 186–200 (1990).
53. N. I. Kerkvliet, L. B. Steppan, J. A. Brauner, J. A. Deyo, M. C. Henderson, R. S. Tomar, and D. R. Buhler, Influence of the Ah locus on the humoral immunotoxicity of 2,3,7,8-tetrachlorodibenzo-p-dioxin (TCDD) immunotoxicity: Evidence for Ah receptor-dependent and Ah receptor-independent mechanisms of immunosuppression, *Toxicol. Appl. Pharmacol.* **105,** 26–36 (1990).
54. A. Vecchi, M. Sironi, M. A. Canegrati, M. Recchis, and S. Garattini, Immunosuppressive effects of 2,3,7,8-tetrachlorodibenzo-p-dioxin in strains of mice with different susceptibility to induction of aryl hydrocarbon hydroxylase, *Toxicol. Appl. Pharmacol.* **68,** 434–441 (1983).
55. J. B. Silkworth and E. M. Grabstein, Polychlorinated biphenyl immunotoxicity: Dependence on isomer planarity and the Ah gene complex, *Toxicol. Appl. Pharmacol.* **65,** 109–115 (1982).
56. N. I. Kerkvliet, J. A. Brauner, and J. P. Matlock, Humoral immunotoxicity of polychlorinated diphenyl ethers, phenoxyphenols, dioxins and furans present as contaminants of technical grade pentachlorophenol, *Toxicology* **36,** 307–324 (1985).
57. A. Vecchi, A. Mantovani, M. Sironi, M. Luini, M. Cairo, and S. Garattini, Effect of acute exposure to 2,3,7,8-tetrachlorodibenzo-p-dioxin on humoral antibody production in mice, *Chem. Biol. Interact.* **30,** 337–341 (1980).
58. D. Davis and S. Safe, Immunosuppressive activities of polychlorinated biphenyls in C57B1/6N mice: Structure–activity relationships as Ah receptor agonists and partial antagonists, *Toxicology* **63,** 97–111 (1990).
59. D. Davis and S. Safe, Dose–response immunotoxicities of commercial polychlorinated biphenyls (PCBs) and their interactions with 2,3,7,8-tetrachlorodibenzo-p-dioxin, *Toxicol. Lett.* **48,** 35–43 (1989).
60. D. Davis and S. Safe, Immunosuppressive activities of polychlorinated dibenzofuran congeners: Quantitative structure–activity relationships and interactive effects, *Toxicol. Appl. Pharmacol.* **94,** 141–149 (1988).
61. J. B. Silkworth, D. S. Cutler, P. W. O'Keefe, and T. Lipinskas, Potentiation and antagonism of 2,3,7,8-tetrachlorodibenzo-p-dioxin effects in a complex environmental mixture, *Toxicol. Appl. Pharmacol.* **119,** 236–247 (1993).
62. N. I. Kerkvliet, L. B. Steppan, B. B. Smith, J. A. Youngberg, M. C. Henderson, and D. R. Buhler, Role of the Ah locus in suppression of cytotoxic T lymphocyte (CTL) activity by halogenated aromatic hydrocarbons (PCBs and TCDD): Structure–activity relationships and effects in C57B1/6 mice, *Fundam. Appl. Toxicol.* **14,** 532–541 (1990).
63. M. I. Luster, L. H. Hong, G. A. Boorman, G. Clark, H. T. Hayes, W. F. Greenlee, K. Dold, and A. N. Tucker, Acute myelotoxic responses in mice exposed to 2,3,7,8-tetrachlorodibenzo-p-dioxin (TCDD), *Toxicol. Appl. Pharmacol.* **81,** 156–165 (1985).
64. M. F. Ackermann, T. A. Gasiewicz, K. R. Lamm, D. R. Germolec, and M. I. Luster,

Selective inhibition of polymorphonuclear neutrophil activity by 2,3,7,8-tetrachlorodibenzo-p-dioxin, *Toxicol. Appl. Pharmacol.* **101,** 470–480 (1989).

65. S. Safe, Polychlorinated biphenyls (PCBs), dibenzo-p-dioxins (PCDDs), dibenzofurans (PCDFs) and related compounds: Environmental and mechanistic considerations which support the development of toxic equivalency factors (TEFs), *Crit. Rev. Toxicol.* **21,** 51–75 (1990).
66. M. P. Holsapple, R. K. Dooley, P. J. McNerney, and J. A. McCay, Direct suppression of antibody responses by chlorinated dibenzodioxins in cultured spleen cells from (C57B1/6 × C3H)F1 and DBA/2 mice, *Immunopharmacology* **12,** 175–186 (1986).
67. D. L. Morris, S. D. Jordan, and M. P. Holsapple, Effects of 2,3,7,8-tetrachlorodibenzo-p-dioxin (TCDD) on humoral immunity: I. Similarities to *Staphylococcus aureus* Cowan Strain I (SAC) in the in vitro T-dependent antibody response, *Immunopharmacology* **21,** 159–170 (1991).
68. M. I. Luster, D. R. Germolec, G. Clark, G. Wiegand, and G. J. Rosenthal, Selective effects of 2,3,7,8-tetrachlorodibenzo-p-dioxin and corticosteroid on in vitro lymphocyte maturation, *J. Immunol.* **140,** 928–935 (1988).
69. H. Braley-Mullen, Differential effect of activated T amplifier cells on B cells responding to thymus-independent type-1 and type-2 antigens, *J. Immunol.* **129,** 484–489 (1982).
70. R. K. Dooley and M. P. Holsapple, Elucidation of cellular targets responsible for tetrachlorodibenzo-p-dioxin (TCDD)-induced suppression of antibody responses: The role of the B lymphocyte, *Immunopharmacology* **16,** 167–180 (1988).
71. R. S. Tomar and N. I. Kerkvliet, Reduced T helper cell function in mice exposed to 2,3,7,8-tetrachlorodibenzo-p-dioxin (TCDD), *Toxicol. Lett.* **57,** 55–64 (1991).
72. N. I. Kerkvliet and L. Baecher-Steppan, Suppression of allograft immunity by 3,4,5,3′,4′,5′-hexachlorobiphenyl. II. Effects of exposure on mixed lymphocyte reactivity in vitro and induction of suppressor cells, *Immunopharmacology* **16,** 13–23 (1988).
73. R. K. Dooley, D. L. Morris, and M. P. Holsapple, Elucidation of cellular targets responsible for tetrachlorodibenzo-p-dioxin (TCDD)-induced suppression of antibody response. 2. Role of the T lymphocyte, *Immunopharmacology* **19,** 47–58 (1990).
74. N. I. Kerkvliet and L. Baecher-Steppan, Suppression of allograft immunity by 3,4,5,3′,4′,5′-hexachlorobiphenyl. I. Effects of exposure on tumor rejection and cytotoxic T cell activity, *Immunopharmacology* **16,** 1–12 (1988).
75. G. K. DeKrey, L. B. Steppan, J. R. Fowles, and N. I. Kerkvliet, 3,4,5,3′,4′,5′-Hexachlorobiphenyl-induced immune suppression: Altered cytokine production by spleen cells during the course of allograft rejection, *J. Immunol.* **150,** 22A (1993).
76. L. B. Steppan, G. K. DeKrey, J. R. Fowles, and N. I. Kerkvliet, Polychlorinated biphenyl induced alterations in the cytokine profile in the peritoneal cavity of mice during the course of P815 tumor rejection, *J. Immunol.* **150,** 134A (1993).
77. C. M. Kramer, K. W. Johnson, R. K. Dooley, and M. P. Holsapple, 2,3,7,8-Tetrachlorodibenzo-p-dioxin (TCDD) enhances antibody production and protein kinase activity in murine B cells, *Biochem. Biophys. Res. Commun.* **145,** 25–32 (1987).
78. G. C. Clark, J. A. Blank, D. R. Germolec, and M. I. Luster, 2,3,7,8-Tetrachlorodibenzo-p-dioxin stimulation of tyrosine phosphorylation in B lymphocytes: Potential role in immunosuppression, *Mol. Pharmacol.* **39,** 495–501 (1991).
79. A. Mantovani, A. Vecchi, W. Luini, M. Sironi, G. P. Candiani, F. Spreafico, and S. Garattini, Effect of 2,3,7,8-tetrachlorodibenzo-p-dioxin on macrophage and natural killer cell mediated cytotoxicity in mice, *Biomedicine* **32,** 200–204 (1980).
80. J. B. Weissberg and J. G. Zinkl, Effects of 2,3,7,8-tetrachlorodibenzo-p-dioxin upon

hemostasis and hematologic function in the rat, *Environ. Health Perspect.* **5,** 119–123 (1973).

81. J. G. Vos, J. A. Moore, and J. G. Zinkl, Toxicity of 2,3,7,8-tetrachlorodibenzo-p-dioxin (TCDD) in C57B1/6 mice, *Toxicol. Appl. Pharmacol.* **29,** 229–241 (1974).
82. S. M. Puhvel and M. Sakamoto, Effect of 2,3,7,8-tetrachlorodibenzo-p-dioxin on murine skin, *J. Invest. Dermatol.* **90,** 354–358 (1988).
83. N. Z. Alsharif, T. Lawson, and S. J. Stohs, TCDD-induced production of superoxide anion and DNA single strand breaks in peritoneal macrophages of rats, *Toxicologist* **10,** 276 (1990).
84. N. I. Kerkvliet and J. A. Oughton, Acute inflammatory response to sheep red blood cell challenge in mice treated with 2,3,7,8-tetrachlorodibenzo-p-dioxin (TCDD): Phenotypic and functional analysis of peritoneal exudate cells, *Toxicol. Appl. Pharmacol.* **119,** 248–257 (1993).
85. P. T. Thomas and R. D. Hinsdill, Effect of polychlorinated biphenyls on the immune responses of rhesus monkeys and mice, *Toxicol. Appl. Pharmacol.* **44,** 41–45 (1978).
86. L. D. Loose, J. B. Silkworth, S. P. Mudzinski, K. A. Pittman, K. F. Benitz, and W. Mueller, Environmental chemical-induced immune dysfunction, *Ecotoxicol. Environ. Safety* **2,** 173–198 (1979).
87. G. C. Clark, M. J. Taylor, A. M. Tritscher, and G. W. Lucier, Tumor necrosis factor involvement in 2,3,7,8-tetrachlorodibenzo-p-dioxin-mediated endotoxin hypersensitivity in C57B1/6 mice congenic at the Ah locus, *Toxicol. Appl. Pharmacol.* **111,** 422–431 (1991).
88. M. J. Taylor, G. C. Clark, Z. Z. Atkins, G. Lucier, and M. I. Luster, 2,3,7,8-Tetrachlorodibenzo-p-dioxin increases the release of tumor necrosis factor-alpha (TNF-α) and induces ethoxyresorufin-o-deethylase (EROD) activity in rat Kupffer's cells (KCs), *Toxicologist* **10,** 276 (1990).
89. N. Hoglen, A. Swim, L. Robertson, and S. Shedlofsky, Effects of xenobiotics on serum tumor necrosis factor (TNF) and interleukin-6 (IL-6) release after LPS in rats, *Toxicologist* **12,** 290 (1992).
90. H. M. Theobald, R. W. Moore, L. B. Katz, R. O. Peiper, and R. E. Peterson, Enhancement of carrageenan and dextran-induced edemas by 2,3,7,8-tetrachlorodibenzo-p-dioxin and related compounds, *J. Pharmacol. Exp. Ther.* **225,** 576–583 (1983).
91. L. B. Katz, H. M. Theobald, R. C. Bookstaff, and R. E. Peterson, Characterization of the enhanced paw edema response to carrageenan and dextran in 2,3,7,8-tetrachlorodibenzo-p-dioxin-treated rats, *J. Pharmacol. Exp. Ther.* **230,** 670–677 (1984).
92. T. R. Sutter, K. Guzman, K. M. Dold, and W. F. Greenlee, Targets for dioxin: Genes for plasminogen activator inhibitor-2 and interleukin-1β, *Science* **254,** 415–418 (1991).
93. L. B. Steppan and N. I. Kerkvliet, Influence of 2,3,7,8-tetrachlorodibenzo-p-dioxin (TCDD) on the production of inflammatory cytokine mRNA by C57B1/6 macrophages, *Toxicologist* **11,** 35 (1991).
94. D. Davis and S. Safe, Halogenated aryl hydrocarbon-induced suppression of the in vitro plaque-forming cell response to sheep red blood cells is not dependent on the Ah receptor, *Immunopharmacology* **21,** 183–190 (1991).
95. C. M. Neumann, L. B. Steppan, and N. I. Kerkvliet, Distribution of 2,3,7,8-tetrachlorodibenzo-p-dioxin (TCDD) in splenic tissue of C57B1/6J mice, *Drug Metab. Dispos.* **20,** 467–469 (1992).
96. G. K. DeKrey, L. Baecher-Steppan, J. A. Deyo, B. B. Smith, and N. I. Kerkvliet, PCB-induced immune suppression: Castration but not adrenalectomy or RU 38486 treatment

partially restores the suppressed cytotoxic T lymphocyte response to alloantigen, *J. Pharmacol. Exp. Ther.* **267,** 308–315 (1993).

97. T. L. Pazdernik and K. K. Rozman, Effect of thyroidectomy and thyroxine on 2,3,7,8-tetrachlorodibenzo-p-dioxin induced immunotoxicity, *Life Sci.* **36,** 695–703 (1985).
98. R. Hong, K. Taylor, and R. Abonour, Immune abnormalities associated with chronic TCDD exposure in rhesus, *Chemosphere* **18,** 313–320 (1989).
99. R. Neubert, U. Jacob-Muller, R. Stahlmann, H. Helge, and D. Neubert, Polyhalogenated dibenzo-p-dioxins and dibenzofurans and the immune system. 1. Effects on peripheral lymphocyte subpopulations of a non-human primate (*Callithrix jacchus*) after treatment with 2,3,7,8-tetrachlorodibenzo-p-dioxin (TCDD), *Arch. Toxicol.* **64,** 345–359 (1990).
100. R. Neubert, U. Jacob-Muller, H. Helge, R. Stahlmann, and D. Neubert, Polyhalogenated dibenzo-p-dioxins and dibenzofurans and the immune system. 2. In vitro effects of 2,3,7,8-tetrachlorodibenzo-p-dioxin (TCDD) on lymphocytes of venous blood from man and a non-human primate (*Callithrix jacchus*), *Arch. Toxicol.* **65,** 213–219 (1991).
101. R. Neubert, G. Golor, R. Stahlmann, H. Helge, and D. Neubert, Polyhalogenated dibenzo-p-dioxins and dibenzofurans and the immune system. 4. Effects of multiple-dose treatment with 2,3,7,8-tetrachlorodibenzo-p-dioxin (TCDD) on peripheral lymphocyte subpopulations of a non-human primate (*Callithrix jacchus*), *Arch. Toxicol.* **66,** 250–259 (1992).
102. J. Truelove, D. Grant, J. Mes, H. Tryphonas, L. Tryphonas, and Z. Zawidzka, Polychlorinated biphenyl toxicity in the pregnant cynomolgus monkey: A pilot study, *Arch. Environ. Contam. Toxicol.* **11,** 583–588 (1982).
103. S. Hori, H. Obana, T. Kashimoto, T. Otake, H. Mishimura, N. Ikegami, N. Kunita, and H. Uda, Effect of polychlorinated biphenyls and polychlorinated quaterphenyls in cynomolgus monkey (*Macaca fascicularis*), *Toxicology* **24,** 123–139 (1982).
104. H. Tryphonas, S. Hayward, L. O'Grady, J. C. K. Loo, D. L. Arnold, F. Bryce, and Z. Z. Zawidzka, Immunotoxicity studies of PCB (Aroclor 1254) in the adult rhesus (*Macaca mulatta*) monkey—preliminary report, *Int. J. Immunopharmacol.* **11,** 199–206 (1989).
105. H. Tryphonas, M. I. Luster, G. Schiffman, L. L. Dawson, M. Hodgen, D. Germolec, S. Hayward, F. Bryce, J. C. K. Loo, F. Mandy, and D. L. Arnold, Effect of chronic exposure of PCB (Aroclor 1254) on specific and nonspecific immune parameters in the rhesus (*Macaca mulatta*) monkey, *Fundam. Appl. Toxicol.* **16,** 773–786 (1991).
106. H. Tryphonas, M. I. Luster, K. L. White, Jr., P. H. Naylor, M. R. Erdos, G. R. Burleson, D. Germolec, M. Hodgen, S. Hayward, and D. L. Arnold, Effects of PCB (Aroclor 1254) on non-specific immune parameters in rhesus (*Macaca mulatta*) monkeys, *Int. J. Immunopharmacol.* **13,** 639–648 (1991).
107. R. Smialowicz, M. Riddle, W. Williams, J. Diliberto, D. Andrews, C. Copeland, and D. Ross, Comparison of the immunotoxicity of 2,3,7,8-tetrachlorodibenzo-p-dioxin (TCDD) in mice and rats, *Toxicologist* **13,** 346 (1993).
108. Y.-C. Lu and Y.-C. Wu, Clinical findings and immunological abnormalities in Yu-Cheng patients, *Environ. Health Perspect.* **59,** 17–29 (1985).
109. T. Kashimoto, H. Miyata, S. Kunita, T. Tung, S. Hsu, K. Chang, S. Tang, G. Ohi, J. Nakagawa, and S. Yamamoto, Role of polychlorinated dibenzofuran in Yusho (PCB poisoning), *Arch. Environ. Health* **36,** 321–326 (1981).
110. J. G. Bekesi, H. A. Anderson, J. P. Roboz, J. Roboz, A. Fischbein, I. J. Selikoff, and J. F. Holland, Immunologic dysfunction among PBB-exposed Michigan dairy farmers, *Ann. N.Y. Acad. Sci.* **320,** 717–728 (1979).
111. K. B. Webb, R. G. Evans, A. P. Knutsen, S. T. Roodman, D. W. Roberts, W. F. Schramm, B. B. Gibson, J. S. Andrews, L. L. Needham, and D. G. Patterson, Medical

evaluation of subjects with known body levels of 2,3,7,8-tetrachlorodibenzo-p-dioxin, *J. Toxicol. Environ. Health* **28,** 183–193 (1989).

112. R. E. Hoffman, P. A. Stehr-Green, K. B. Webb, R. G. Evans, A. P. Knutsen, W. R. F. Schramm, J. L. Staake, B. B. Gibson, and K. K. Steinberg, Health effects of long-term exposure to 2,3,7,8-tetrachlorodibenzo-p-dioxin, *J. Am. Med. Assoc.* **255,** 2031–2038 (1986).
113. R. G. Evans, K. B. Webb, A. P. Knutsen, S. T. Roodman, D. W. Roberts, J. R. Bagby, W. A. Garrett, and J. S. Andrews, A medical follow-up of the health effects of long-term exposure to 2,3,7,8-tetrachlorodibenzo-p-dioxin, *Arch. Environ. Health* **43,** 273–278 (1988).
114. P. A. Stehr-Green, P. H. Naylor, and R. E. Hoffman, Diminished thymosin alpha-1 levels in persons exposed to 2,3,7,8-tetrachlorodibenzo-p-dioxin, *J. Toxicol. Environ. Health* **28,** 285–295 (1989).
115. P. Mocarelli, A. Marocchi, P. Brambilla, P. Gerthoux, D. S. Young, and N. Mantel, Clinical laboratory manifestations of exposure to dioxin in children, a six-year study of the effects of an environmental disaster near Seveso, Italy, *J. Am. Med. Assoc.* **256,** 2687–2695 (1986).
116. G. Tognoni and A. Bonaccorsi, Epidemiological problems with TCDD (A critical view), *Drug Metab. Rev.* **13,** 447–469 (1982).

Chapter 8

Dose–Response Effects of Dioxins

Species Comparison and Implication for Risk Assessment

Angelika M. Tritscher, George C. Clark, and George W. Lucier

1. INTRODUCTION

2,3,7,8-Tetrachlorodibenzo-*p*-dioxin (TCDD) and its structural analogues, polychlorinated aromatics, e.g., chlorinated dibenzodioxins, dibenzofurans, and polychlorinated biphenyls (Fig. 1), are ubiquitous environmental pollutants and produce a broad spectrum of biochemical and toxic effects in animals and humans.[1] Dioxins are produced inadvertently during manufacture of chlorinated phenols and phenoxyherbicides, chlorine bleaching of paper pulp, and combustion of chlorine-containing waste.[2] TCDD, the "prototype" compound of these polychlorinated hydrocarbons, is the most potent and widely studied substance of this class. It is one of the most toxic chemicals tested in laboratory animals and is a potent multisite carcinogen.[3–5] Because of the high acute toxicity, the high carcinogenic potential, and the ubiquitous presence as a trace contaminant in food, water, and soil and therefore continuous human exposure,[6] TCDD and related compounds are of great public health concern. However, there is considerable controversy on how to regulate TCDD and its congeners.[7,8] One of the central issues in the controversy about risk assessment is whether humans are a resistant or a sensitive species to the effects of dioxins. There are wide species differences in the acute toxicity

---

Angelika M. Tritscher, George C. Clark, and George W. Lucier • National Institute of Environmental Health Sciences, Laboratory of Biochemical Risk Analysis, Research Triangle Park, North Carolina 27709.

Dioxins and Health, edited by Arnold Schecter. Plenum Press, New York, 1994.

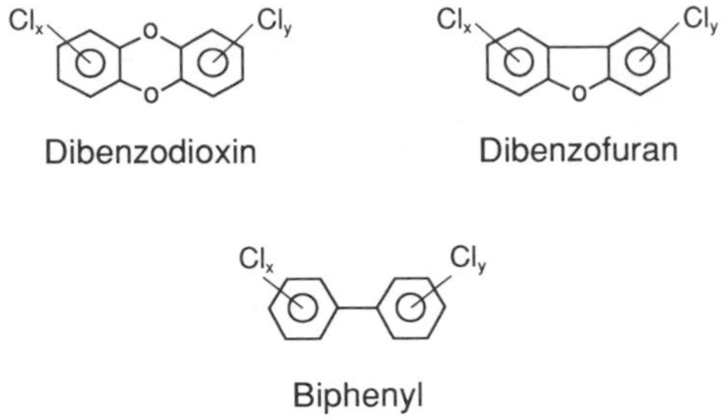

Cl_x Cl_y
Dibenzodioxin

Cl_x Cl_y
Dibenzofuran

Cl_x Cl_y
Biphenyl

Figure 1. Structures of polycyclic halogenated aromatics.

of TCDD,[9] but it is unclear whether carcinogenic responses vary as much as the acute toxicity. Another uncertainty in risk assessment is the lack of knowledge about dose–response relationships especially following chronic low-dose exposure to dioxins.

The "hallmark" of dioxin toxicity is a significant tissue and species specificity. The mechanism of action is not completely understood but most if not all known effects of TCDD seem to involve an initial interaction with an intracellular binding protein, termed the Ah receptor (or dioxin receptor).[10,11] This receptor system is very specific for dioxinlike compounds, it exists in low copy numbers in the cell, but it binds dioxins with a very high affinity (see DeVito and Birnbaum, this volume). Binding of ligand (the dioxinlike compounds) activates the Ah receptor, the receptor/ligand complex then binds to acceptor sites in the nucleus (at DNA) and is thought to alter expression of a subset of genes. Assuming that binding to the receptor results in a biological response, the extent of response should relate to some function of receptor occupancy. The magnitude of response and the shape of the dose–response curve is "dictated" by amount and activity of receptor status in the particular tissue (as well as status of other proteins necessary for the response). Various target genes and responses may have different dose–response curves even within the same tissue, meaning they vary in their sensitivity to dioxin exposure. Also, if a certain number of receptor molecules need to be occupied to result in a measurable biological response, it might imply a practical "threshold" for dioxin exposure. This means there might not be a measurable effect under a certain exposure and thus tissue level. In order to do more accurate human risk assessment, information is needed on background exposure and resulting receptor occupancy.

Because of the great variability in the toxic response in tissues and species, more information on the shape of dose–response curves of dioxin-mediated effects after chronic low-dose exposure as well as their relevance to toxicity/carcinogenicity is needed. Dose–response relationships of biological markers need to be characterized in animal models and compared, when possible, with dose–response relationships in humans. Our research focused on defining the shape of dose–response curves for multiple Ah-receptor-mediated events in an animal model. Together with information on animal-to-human comparison, individual variation, human background exposure, and receptor occupancy, these results will be integrated in the approach of a more biologically (or mechanistically) based risk assessment for humans.

2. DOSE–RESPONSE RELATIONSHIPS

2.1. Biochemical Responses

The most investigated and one of the most sensitive responses of exposure to TCDD and its structural analogues is the induction of enzymes of the mixed-function monooxygenase system. This system contains many different enzymes with the major components being cytochromes P450 which are important for metabolizing endogenous as well as foreign compounds by hydroxylation reactions. The cytochromes are also important for detoxification of foreign substances by hydroxylating them and so increasing their solubility and facilitating their excretion. Some members of the cytochrome P450 family are inducible by foreign compounds. Chemicals that bind to the Ah receptor are known to induce two specific enzymes: cytochrome P450 1A1 (CYP1A1) and cytochrome P450 1A2 (CYP1A2). Studies of the mechanism of induction of CYP1A1 in *in vivo* and *in vitro* systems suggest that TCDD binds to the Ah receptor and following a number of events, dissociation of heat shock protein 90 (HSP 90),[12] association with the ARNT protein, and translocation into the nucleus,[13] this complex binds to regulatory elements on the DNA (DREs, upstream from the start site of the structural gene) resulting in enhanced transcription of the CYP1A1 gene.[14] Other genes are believed to be transcriptionally regulated by the Ah receptor/ligand complex in a way similar to the response in the CYP1A1 gene. These dioxin-responsive genes are reviewed by Sutter and Greenlee.[15]

Other biochemical responses to TCDD exposure are changes in receptor systems which are involved in cell growth and differentiation. TCDD alters binding capacities of epidermal growth factor receptor (EGFR), estrogen receptor (ER), and glucocorticoid receptor (GCR) as well as the Ah receptor itself in animals and humans.

2.1.1. Animals

2.1.1a. Enzyme Induction. The induction of cytochrome P450 enzymes following TCDD exposure occurs *in vivo* and *in vitro* in animals and humans.[1,10,16] The mechanism of induction is widely studied for CYP1A1 and understood so far as described above. Induction of CYP1A2 seems to occur also by transcriptional activation mediated through the Ah receptor. CYP1A2 is induced only in liver whereas CYP1A1 is also induced in extrahepatic tissues, e.g., lymphocytes, lung, placenta, and kidney.

The liver of female rats is a target organ for the carcinogenicity of TCDD and liver tumor incidences are used as the basis for most risk assessment throughout the world.[17] In our studies, dose–response relationships of receptor-mediated events, e.g., the induction of CYP1A1 and CYP1A2, were quantified after chronic exposure to various doses of TCDD in livers of female Sprague–Dawley rats within the framework of a two-stage model for hepatocarcinogenesis.[18] After initiation the animals received bi-

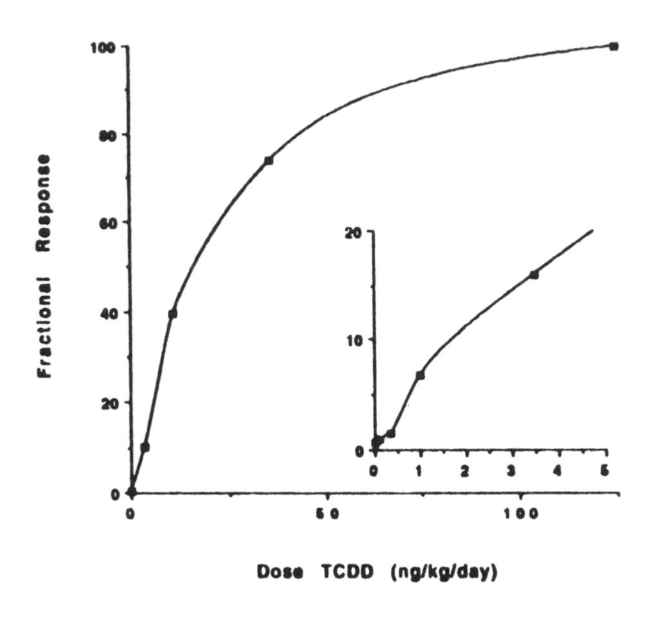

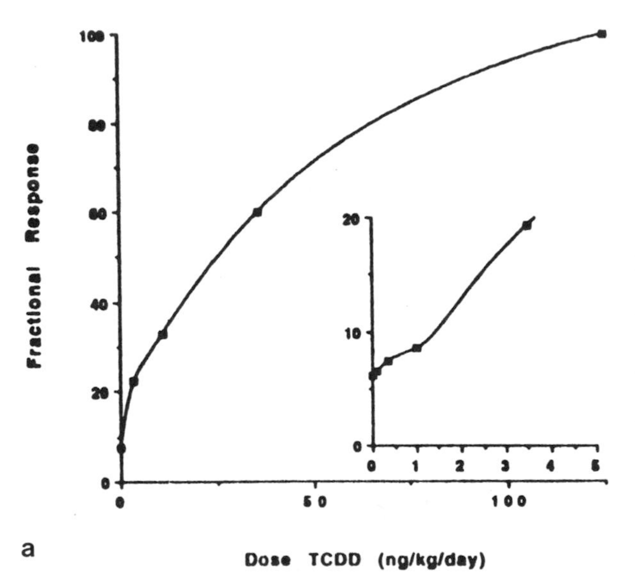

Figure 2. Dose-dependent increase in CYP1A1 and CYP1A2 protein after chronic TCDD treatment. Insets show magnification of the low ends of the curves. (A) Fractional increase in CYP1A1 (top) and CYP1A2 (bottom) as a function of the administered TCDD dose. (B) Increase in CYP1A1 (top) and CYP1A2 (bottom) as a function of the liver TCDD concentration modeled using steady-state Hill kinetics.

weekly oral gavages of TCDD at doses equivalent to 0.1 to 125 ng/kg per day for 30 weeks. CYP1A1 and CYP1A2 were quantified by radioimmunoassay in hepatic microsomes. CYP1A1 was also quantified through its enzyme activity by measuring ethoxyresorufin-*O*-deethylase (EROD) activity (ethoxyresorufin is a substrate for CYP1A1).

Figure 2A shows the fractional increase in both cytochromes. The dose–response

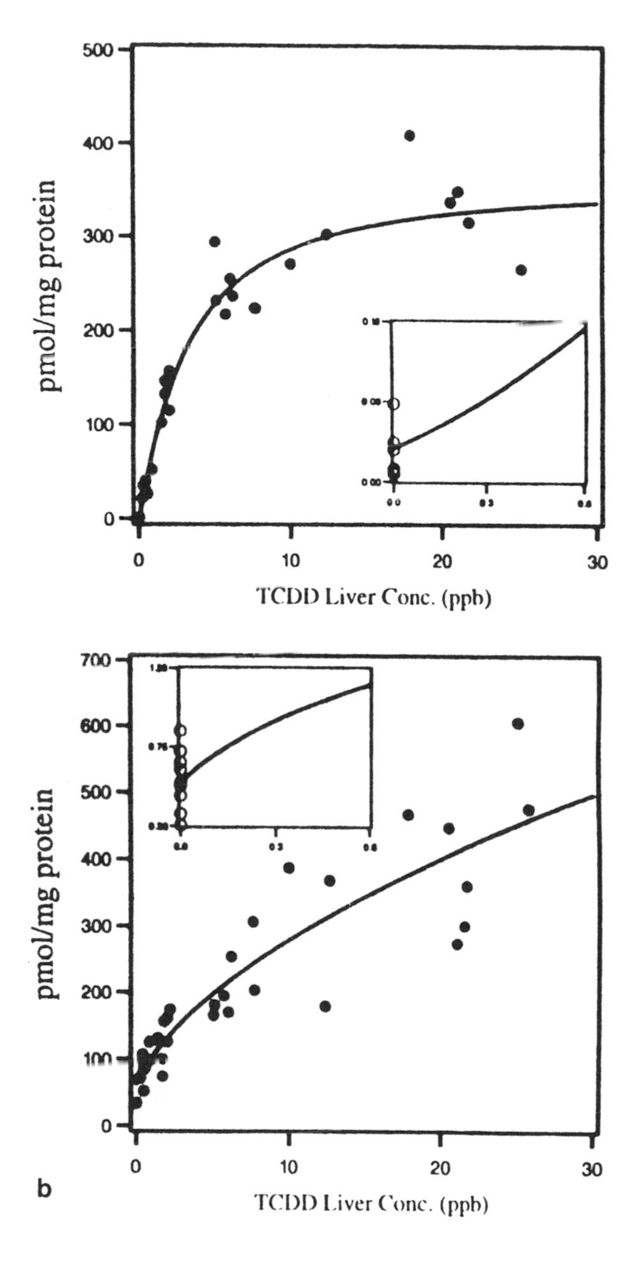

Figure 2. (*Continued*)

trend for both enzymes is highly significant ($p \leq 0.01$) The dose-related increases in CYP1A1 and CYP1A2 resemble a typical receptor-mediated response where there is a proportional relationship between receptor occupancy and response. Maximum induction at the highest dose is approximately 200-fold for CYP1A1 and 10-fold for CYP1A2. CYP1A1-associated enzyme activity (EROD activity) was also measured and showed a strong correlation ($r = 0.85$) with the amount of CYP1A1 protein. This verifies that measuring EROD activity is a good marker for the quantity of CYP1A1 protein. The approximate ED_{50} values for induction of both enzymes and for EROD activity are between 10 and 25 ng/kg per day or, expressed as tissue concentrations, 1.5–4 ppb TCDD. This indicates similar responses to TCDD although CYP1A1 induction (and EROD activity) seems to be the more sensitive marker in the low-dose region. This might be related to the lower constitutive expression of CYP1A1.

The dose–response curves for both enzymes seem to have immediate positive slopes (insets in Fig. 2A) within the dose range of our study. Pharmacodynamic modeling of the data was done using Hill plots assuming that the responses are receptor mediated and that receptor/ligand binding is in equilibrium. This analysis helps to define the shape of dose/response curves and allows extrapolations below the dose range of our study. Hill plots are used to analyze receptor binding data and allow for positive and negative cooperativity among binding sites and the presence of more than one class of binding sites. A Hill coefficient of 1 defines a linear relationship between receptor occupancy and response, a coefficient < 1 supralinearity, and a coefficient $>$ 1 sublinearity. Under the assumption that we have reached steady-state conditions for TCDD binding to the Ah receptor and assuming that TCDD-mediated induction of P450 enzymes occurs by the same mechanism that is responsible for constitutive ("natural") expression of the cytochromes, our models give no indication for a threshold for these TCDD effects. These Hill plots are shown in Fig. 2B and the insets in Fig. 2A and B show a magnification of the low end of the curve to demonstrate the linearity in the low-dose range.

In the same livers the mRNA for CYP1A1 was measured using a highly sensitive quantitative reverse transcriptase polymerase chain reaction (RT-PCR) assay.[19] Hepatic mRNA for CYP1A1 was quantified and the results are shown in Fig. 3. Increase in CYP1A1 mRNA could be detected at lower TCDD concentration than CYP1A1 protein or EROD activity although all three parameters were strongly correlated. The findings are consistent with the idea that increased enzyme induction is related to enhanced transcription of the CYP1A1 gene and that RT-PCR is a sensitive assay for detecting TCDD-mediated changes in CYP1A1 expression.[20]

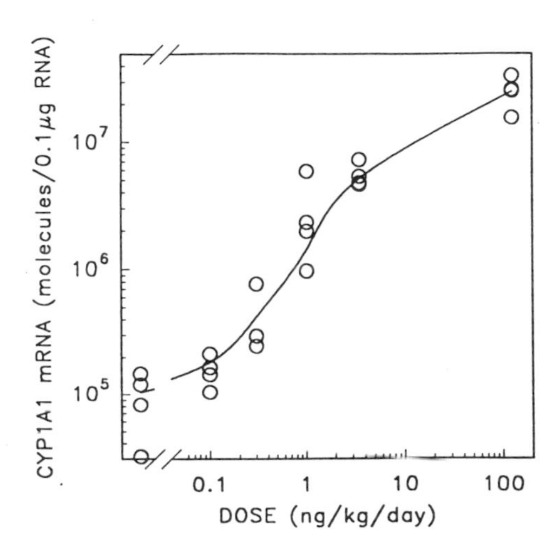

Figure 3. Dose-dependent increase in CYP1A1 mRNA after chronic TCDD treatment as a function of the administered dose.

In another approach, CYP1A1 and CYP1A2 were localized by immunohistochemical techniques in hepatic tissue slices in the same livers analyzed by RIA methods.[18] These studies showed the same localization and distribution pattern for both enzymes. However, it is important to note that the distribution of the enzymes was not uniform throughout the liver lobe. The highest enzyme concentration was observed around central veins at the lower doses. CYP1A1 and CYP1A2 were also detected in the periportal regions in animals receiving higher TCDD doses. This indicates that hepatocytes within the liver lobe vary in their sensitivity to respond to TCDD which is an important factor in evaluation of dose–response relationships since some hepatocytes respond at much lower doses than others.

2.1.1b. Effects on Cellular Receptor Systems. TCDD has been shown to affect a number of receptor systems in the liver which are involved in growth regulation and differentiation. For example, it decreases maximum binding capacity of ER,[21] EGFR,[22,23] and GCR.[24,25] TCDD treatment also increases binding activity of the Ah receptor.[26]

In the same rat liver tumor promotion model used to evaluate dose–response relationships for cytochrome P450 enzymes, we also quantified changes to the EGFR system. EGFR is a transmembrane receptor which is activated through ligand binding, autophosphorylation, and subsequent internalization. These steps are thought to be early events in the generation of a signal for the cell to divide (mitogenic signal). Substances binding and so activating the EGFR (ligands) are EGF, transforming growth factor-α (TGFα), and other EGF-like peptides.

Changes in the EGFR regulatory system may play a role in carcinogenesis,[27] and decrease in EGFR binding has been characterized in our rat liver tumor promotion models by a number of methods.

TCDD treatment decreased EGFR maximum binding capacity in a dose-dependent manner in hepatic membrane preparations in our model (Fig. 4). At the highest dose (125 ng/kg per day) the maximum decrease in EGFR B_{max} was approx-

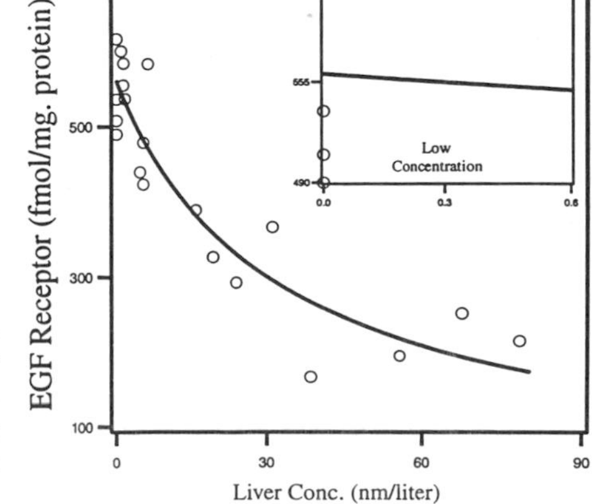

Figure 4. Dose-dependent decrease of membrane EGFR maximum binding capacity as a function of liver TCDD concentration. Points represent individual measured values. The solid line is the best fitting curve modeled using steady-state Hill kinetics. Inset shows magnification of the low end of the curve.

imately 3-fold. The estimated ED_{50} is 10 ng/kg per day or, expressed as tissue levels, 1.5 ppb TCDD. Mathematical modeling similar to that described for the enzyme induction showed no indication of a threshold for this TCDD-mediated biochemical effect[28] (Fig. 4).

Lin *et al.*[23] showed that in mice containing a malfunctional Ah receptor, the effect of TCDD on EGFR ligand binding depends on a functional Ah receptor. This study also indicates that this effect, unlike changes in CYP1A1, does not involve changes in the levels of the mRNA, suggesting that the TCDD-mediated alterations in EGFR binding capacity are on the protein level (posttranscriptional) but they do require the Ah receptor.

The decrease in plasma membrane EGFR was also quantified through EGF-stimulated autophosphorylation of the receptor. Autophosphorylation of the EGFR is an early signal transduction event that follows binding of EGF-like peptides to EGFR. Therefore, it is an indirect measurement of the quantity of the receptor as well as its functionality. The dose–response curve for autophosphorylation is very similar to that of the maximum binding capacity and both measurements show a strong positive correlation ($r = 0.83$).

A mechanistic model was mathematically constructed to predict dose–response curves for TCDD-mediated effects on expression of certain genes in the rat liver.[29] The model assumes that TCDD increases TGFα or other EGF-like peptides which then bind to the EGFR leading to the generation of enhanced signal for the cell to divide. Based on the finding that TCDD increased cell proliferation and decreased plasma membrane EGFR in livers of intact female but not ovariectomized rats,[30] it is also assumed that TCDD's effects on EGF-like peptides are estrogen dependent. This model was used to predict dose–response relationships of TCDD-mediated changes in EGFR, CYP1A1, and CYP1A2 as well as TCDD tissue distribution. The model was able to accurately predict experimental data achieved in our tumor promotion model in female rat liver. Mechanistic models like the one described (developed by Kohn *et al.*[29]) should help to permit more reliable extrapolations of TCDD's effects outside the range of experimental data and may help strengthen the scientific foundation for risk estimates of TCDD's adverse health effects.

Other receptor systems are also affected by TCDD treatment in the liver. Sunahara *et al.*[24] showed that GCR ligand binding capacity is decreased in a dose-dependent manner in the livers of female Sprague-Dawley rats after treatment with single doses of TCDD. GCR is an intracellular binding protein that specifically binds glucocorticoid-like steroids. The ligand-activated receptor undergoes transformation, subsequently translocates into the nucleus and alters transcription of specific genes.

Another steroid hormone receptor affected by TCDD is the ER. Romkes *et al.*[21] reported decreased ER binding capacity after TCDD treatment in the liver and in the uterus of female Long–Evans rats. Single doses of 20 or 80 mg/kg resulted in a downregulation of ER binding capacity up to 50% after 7 and 10 days in the liver. Chronic TCDD treatment (100 ng/kg per day for 30 weeks) of female Sprague–Dawley rats also significantly decreased ER in the liver.[31] The decrease was approximately 60% relative to controls in intact and ovariectomized rats.

In summary, TCDD has significant effects on hormone receptor and growth factor receptor pathways in the livers of female rats, a target organ for the carcinogenic activity of TCDD. These studies suggest that a possible mechanism of TCDD's carcinogenic action is alterations in growth control pathways.

2.1.2. Humans

2.1.2a. Enzyme Induction. (1) *In vitro* exposure. In order to make more valid animal-to-human comparisons of responses to dioxin, we investigated dose–response relationships of enzyme induction to murine and human lymphocytes after *in vitro* exposure to TCDD. Lymphocytes are an easily available tissue and they contain a functional Ah receptor.

The lymphocytes were mitogen stimulated in culture and exposed to various concentrations of TCDD. Exposure to TCDD resulted in a dose-dependent increase in EROD activity. The EC_{50} for murine splenic lymphocytes was 1.3 ± 0.3 nM TCDD and for human peripheral lymphocytes 1.8 ± 0.8 nM TCDD. Although maximum inducibility and baseline values are higher in human lymphocytes as compared with murine lymphocytes, the EC_{50} values are similar. This indicates that human and murine lymphocytes are comparable in their sensitivity to *in vitro* TCDD exposure.[32]

CYP1A1 mRNA was also quantified using the sensitive RT-PCR.[33] The number of mRNA molecules in resting human lymphocytes is very low. After mitogen stimulation and treatment with various amounts of TCDD, the copy number of CYP1A1 mRNA molecules increases in a dose-dependent manner. The magnitude of induction correlates well with induction of EROD activity. This method is a very sensitive approach for detecting dioxin-induced changes in the CYP1A1 gene in accessible human cells and is potentially useful in evaluating effects on dioxin-responsive genes in individuals exposed occupationally or environmentally to dioxin and its structural analogues.

One well-known response to dioxin exposure in humans is the development of chloracne, a persistent form of acne characterized by hyperkeratinization. Dose-dependent increase of 7-ethoxycoumarin-*O*-deethylase activity (ECOD; another substrate for CYP1A1-dependent enzyme activity) was shown in human squamous cell carcinoma (SCC) lines as well as in normal human epidermal cells.[34] The dose–response curves are similar to those for CYP1A1 induction in human lymphocytes.

Since the skin is a sensitive target for dioxin toxicity, human keratinocytes may be a useful model for studying the toxicity of dioxin *in vitro*. Yang *et al.*[35] recently reported that immortalized nontumorigenic human keratinocytes underwent neoplastic transformation after treatment with TCDD. Transformation as well as induction of aryl hydrocarbon hydroxylase activity (AHH; another CYP1A1-associated enzyme activity) were dose dependent. However, it is important to note that neoplastic transformation occurred at lower TCDD doses than enzyme induction. This report is the first evidence for neoplastic transformation in human cells by TCDD *in vitro* and thus might be a useful model for studying TCDD's carcinogenic potential in human cells derived from a sensitive target cell population.

Human thymic epithelial (HuTE) cells were also treated in culture with TCDD and showed dose-dependent increase in ECOD activity.[36] The maximum response varied in cells from individual donors and the EC_{50} values ranged from 0.8 to 2.0 nM TCDD. Thymic involution as well as immunosuppressive response are characteristic responses to TCDD in animals.[37] HuTE cells appear to be a good model for studying mechanistic aspects of TCDD treatment and may also help in making animal-to-human comparison.

(2) *In vivo* exposure. In 1979 a widespread poisoning incident occurred in Taiwan by the accidental ingestion of contaminated rice oil. Most exposed subjects showed characteristic symptoms of polychlorinated biphenyl (PCB) poisoning and the disease was termed *Yu-cheng* (oil disease).[38] Major components of the contamination were polychlorinated dibenzofurans (PCDFs) and dioxinlike PCBs, compounds that are thought to act also through the Ah receptor similar to TCDD. Placental samples were analyzed by GC-MS and Yu-cheng subjects showed elevated levels of specific PCB and PCDF congeners. The amount of exposure for adults was estimated at 1–3 g PCBs and 5 mg PCDFs. Placentas were obtained from nonsmoking women 4–5 years after exposure as well as from unexposed controls. Induction of EROD activity was detectable only in placentas from exposed women, although there was considerable interindividual variation. The values ranged from 1 to 63 pmol/mg protein per min and EROD activity did not correlate with PCDF or PCB concentrations in the tissue of the exposed group.[32]

2.1.2b. Effects on Cellular Receptor Systems. (1) *In vitro* exposure. The effects of TCDD on the EGFR were studied in a human keratinocyte cell line (SCC-12F).[39] This cell line is nontumorigenic and retains many characteristics of normal human epidermal cells, a major target for TCDD toxicity. Pretreatment of SCC-12F cells with TCDD in culture resulted in a concentration-dependent decrease of EGF binding sites. Maximum decrease was 60% as compared with controls and the EC_{50} for this effect was 1.8 nM. The decreased binding was still evident 10 days posttreatment.

(2) *In vivo* exposure. In the 1979 rice oil poisoning incident in Taiwan mentioned above (Yu-cheng), EGFR was characterized in placentas of exposed women and compared with unexposed controls.[40] EGF-stimulated autophosphorylation in exposed subjects was decreased by more than 60% relative to control levels even 4–5 years after exposure. In contrast, the EGFR binding characteristics were not affected, and maximum binding capacity and K_d of the receptor were similar in exposed and control women. The decreased autophosphorylation, however, was significantly correlated with decrease in birth weights. Decreased birth weights are a well-known characteristic of Yu-cheng subjects. There was a significant dose–response relationship between placental EGFR autophosphorylation levels and the PCB concentrations but not the PCDF concentrations.

2.2. Complex Biological Responses

Biochemical responses such as induction of different enzymes are sensitive markers of exposure and reflect interactions of the Ah receptor/TCDD complex with dioxin-responsive genes. However, their relevance to toxic endpoints is often unclear. Complex biological responses involve interactions of multiple events. Measuring a more complex effect might reflect more persistent changes and might correlate better with adverse health effects or toxic endpoints.

TCDD-mediated increase in cell proliferation is an example of a complex biological response and may play a critical role in the development of cancer especially for those chemicals that appear to act through a nongenotoxic mechanism like the dioxins.[41] Although the mechanism of chemically induced mitogenesis and especially the clonal expansion of initiated cells (tumor promotion) is not completely understood, it seems that receptor-mediated events are involved. These concepts have generated the term *receptor-mediated carcinogenesis* and dioxin/Ah receptor interactions are of major interest in this area.[42]

Another effect used to identify tumor-promoting activity in the liver is the development of foci of cellular alteration which are identified through specific marker enzymes like γ-glutamyl transpeptidase (GGT) and placental glutathione-*S*-transferase (PGST).[43]

2.2.1. Animals

2.2.1a. Cell Proliferation. In our rat liver initiation–promotion model, we measured increased cell proliferation through the immunodetection of incorporated bromodeoxyuridine (BrdU). The animals received subcutaneous implants of miniosmotic pumps filled with BrdU 7 days prior to sacrifice. Replicating cells incorporate BrdU in their DNA which can then be detected with specific anti-BrdU antibodies in fixed tissue sections. The results are quantified by counting BrdU-positive nuclei versus total nuclei and expressed as percent cells labeled (labeling index). At least 1000 normal hepatocytes were counted; obvious foci are not included in the determination of this labeling index. The results are presented in Table 1. The dose–response curve shows a decrease at the lower doses (which is significant at 3.5 ng/kg per day) and an increase at higher doses. At the highest dose there is a significant increase (approximately

Table 1
Hepatocellular Changes in Female Sprague–Dawley Rats after Chronic Treatment with TCDD

Dose group	LI (%)	$PGST^{+}$ foci (vol. fract. %)
DEN/corn oil	5.28 ± 2.22	0.57 ± 0.44
DEN/TCDD (3.5 ng/kg)	3.28 ± 1.55*	0.85 ± 0.40
DEN/TCDD (10.7 ng/kg)	3.25 ± 2.91	1.00 ± 1.16
DEN/TCDD (35.7 ng/kg)	6.39 ± 3.62	0.93 ± 0.56
DEN/TCDD (125 ng/kg)	14.35 ± 8.26*	2.23 ± 1.47*

*Statistically different from DEN/corn oil; $p > 0.05$.

3-fold) over controls, although there is a high interindividual variation between animals of the same treatment group. It is important to note that in hepatic focal lesions the labeling index (determined separately) is significantly higher (30–40%) than in surrounding tissue.[44]

2.2.1b. Hepatic Focal Lesions. PGST-positive foci of cellular alterations were quantified in the livers of female rats in our rat liver tumor promotion study (Table 1). DEN-initiated rats showed a dose-related increase in detectable volume fraction of foci but it was statistically significant only at the highest dose. As for cell proliferation there is also a high interindiviual variation (within the same dose group) in this parameter. In general, there is a significant correlation between increases in cell proliferation (replicative DNA synthesis) and enzyme-altered foci data.[30]

3. CANCER DOSE RESPONSE

3.1. Animals

TCDD, the prototypical dioxin, has been tested in numerous long-term carcinogenicity bioassays. All studies yielded positive results and it has been clearly demonstrated that TCDD is a multisite carcinogen in both sexes of mice and rats.[3,4] It is also a carcinogen in the hamster, which is considered the most resistant species to the acute toxicity of TCDD.[45]

Table 2 summarizes tumor sites in various rodents; for a more complete overview see Huff *et al.*[5]

The most cited and currently most used study for estimating the cancer risk of dioxins was published in 1978 by Kociba *et al.*[3] In this lifetime feeding study, female and male Sprague-Dawley rats received TCDD in the diet at doses equivalent to 1, 10, and 100 ng/kg per day. The most significant finding was an increase in hepatocellular hyperplastic nodules and hepatocellular carcinomas in female rats. An increased liver tumor incidence in female rats was detected at 10 ng/kg per day.

Another cancer dose–response study was conducted by NTP[4] in Osborne–Mendel rats and B6C3F1 mice of both sexes. Animals were treated by gavage twice weekly for 2 years at doses equivalent to 1.4, 7.1, and 71 ng/kg per day for female and male rats and male mice. Female mice received 5.7, 28.6, and 286 ng/kg per day. A significant increase in malignant liver tumors was observed at the highest dose in female rats and in both sexes of mice. In male rats the most sensitive response was tumor development in the thyroid gland with an increased incidence at the lowest dose (1.4 ng/kg). Increase in thyroid gland tumors was also detected in high-dose female rats. Several other tumors at various sites were observed at the highest dose in female mice and rats including a significant increase in fibrosarcomas of the subcutaneous tissue in female rats.

The Kociba and NTP studies showed that TCDD is a potent multisite carcinogen that increases neoplasms in rats and mice of both sexes. A greater frequency of liver

Table 2
Sites of Increased Cancer in Animal Bioassays[a]

Rats	
Sprague–Dawley	
Male	Tongue
	Nasal turbinates/hard palate
	Lung
Female	Nasal turbinates/hard palate
	Liver
Osborne–Mendel	
Male	Thyroid
	Adrenal cortex
Female	Liver
	Adrenal cortex
	Subcutaneous fibrosarcoma
Mice	
B6C3F1	
Male	Liver
Female	Subcutaneous fibrosarcoma
	Liver
	Thyroid
B6C3	
Male/female	Thymic lymphomas
	Liver
Hamster	
Syrian golden	
Male	Facial skin carcinoma

[a]Data summarized from Ref. 3, 4, 45.

tumors was observed in female than in male rats, but the male thyroid gland seems to be the most sensitive target tissue.

TCDD-Related Compounds

Few studies have examined the carcinogenicity of other dioxins. NTP[46] tested a mixture of two isomers of hexachlorodibenzo-*p*-dioxin (1,2,3,6,7,8- and 1,2,3,7,8,9-HCDD) in Osborne–Mendel rats and B6C3F1 mice. Animals were treated twice weekly by gavage for 2 years at doses equivalent to 1.25, 2.5, and 5 mg/kg per week for rats and male mice. The doses for female mice were 2.5, 5, and 10 mg/kg per week. HCDD increased liver tumors in both sexes of rats and mice in the higher dose groups. Female rats appear to be more sensitive with a significant increase in tumor incidence at the lowest dose. From this study it can be estimated that HCDD is approximately 1/20th as potent a liver carcinogen as TCDD.

Limited information is available on the carcinogenic potential of other congeners. Flodstrom and Ahlborg[47] tested the relative liver tumor-producing activity of 1,2,3,7,8-pentachlorodibenzo-*p*-dioxin (PeCDD), 2,3,4,7,8-pentachlorodibenzofuran (PeCDF),

3,3′,4,4′,5-pentachlorobiphenyl (PCB126), and 2,3,3′,4,4′-pentachlorobiphenyl (PCB105) after medium-term exposure (20 weeks) in nitrosamine-initiated female Sprague-Dawley rats. The potency of tumor promotion was evaluated by enzyme-altered focus development relative to TCDD. All tested substances acted as promoters of hepatocarcinogenesis with PeCDD being virtually equipotent to TCDD and the other congeners showing about 10-fold less activity in inducing enzyme-altered foci.

In summary, TCDD and HCDD are potent multisite carcinogens in several species, specifically in mice and rats of both sexes. The sensitivity of the carcinogenic response varies between species, target organ, and sex. Carcinogenic effects have been observed at doses that are over two orders of magnitude lower than the maximum tolerated dose with the lowest observable effect at 1 ng/kg per day.[5] Limited information is available on other congeners; however, higher chlorinated compounds (PCDDs and PCDFs) and dioxinlike PCBs bioaccumulate and exhibit toxic effects similar to those of TCDD, and are therefore considered to be carcinogens.[48]

3.2. Humans

Several epidemiologic studies have attempted to determine health effects arising from exposure to phenoxyherbicides and chlorophenols contaminated with dioxins. In the occupational studies, exposure occurred either through the manufacturing process or through application. Based on animal carcinogenicity data, most epidemiologic studies have focused on TCDD exposure. Since measurements of TCDD levels in the serum (rather than milk and adipose tissue) became possible in the late 1980s,[49] more accurate body burden and exposure assessments on a broader basis have enhanced the validity of epidemiologic studies. Observed health effects could now be related to actual tissue levels (rather than estimated levels) and the question of dose–response relationships can at least be partially addressed.

Three recently published studies include various exposure levels and provide evidence for increased cancer in chemical workers with high occupational TCDD exposure. The NIOSH study[50] is a cohort study which includes over 5000 male U.S. chemical workers. This cancer mortality study showed a slight but significant increase in overall cancer, expressed in standard mortality rates (SMR: observed deaths/expected deaths × 100). Because of the typically long latency period (time since first exposure) for the development of chemically induced cancers, a subcohort was identified with a 20-year latency and at least 1 year of exposure. In this subcohort the cancer mortality rate showed an excess of 46% (SMR 146, higher than in the overall cohort SMR 115). In addition to overall cancer, increased cancer at specific sites (soft-tissue sarcoma and respiratory tract tumors) was detected in the long-latency and high-exposure subcohort.

The two other studies are of chemical workers in Germany and they involve smaller cohorts. Manz *et al.*[51] investigated the cancer mortality among 1184 men and 399 women employed between 1952 and 1984 in a chemical plant (Boehringer, Hamburg) which produced herbicides contaminated with TCDD. After an outbreak of chloracne in 1954, the company changed the production process so that TCDD contamination was reduced. Cancer mortality was increased in the overall cohort but was more significant among men employed 20 or more years and among men who began employment before 1955 (high TCDD exposure and long-latency subcohort). As in the NIOSH study, the SMR data indicated an increase in malignant neoplasms in overall sites (SMR 142; see Table 3).

Only 7% of the women worked in locations with expected high exposure and no increased risk of cancer death was observed with the exception of a significant increase in breast cancer mortality (SMR 215, based on nine deaths).

Zober *et al.*[52] reported a slight increase in overall cancer mortality (SMR 117) in a cohort of 247 workers who were exposed to TCDD in a 1953 accident in a chemical plant (BASF). A subcohort was selected based on the development of chloracne and therefore suspected high exposure. The authors reported a twofold excess in cancer mortality in this subcohort.

All three cited studies show an increase in overall cancer mortality after exposure to TCDD, especially after high exposure and long latency which might indicate a dose–response relationship for the development of cancer. Table 3 summarizes the SMRs of the studies overall and for the high-exposure subcohorts. There are several difficulties involved with dose–response effects in epidemiologic studies. The major problems are that exposure to TCDD usually occurs as a part of complex mixtures and that it is difficult to determine control cancer rates because of background exposure to dioxins. And although increasing information is available on human tissue levels,[6] little is known about the impact of background exposure on the occupancy of the Ah receptor. Few epidemiologic studies have actually quantitated dioxin exposure.

There is still controversy whether or not dioxins are a human carcinogen and especially about their carcinogenic potency. In particular, there is a possibility that soft-tissue sarcoma and malignant lymphoma reported in earlier epidemiologic studies are caused by chemicals other than dioxin.[53]

Table 3
Standard Mortality Ratios
in TCDD-Exposed Cohorts of Recently
Published Epidemiologic Studies

	Cancer mortality (SMR)[a]		
	NIOSH[b]	Manz[c]	Zober[d]
Total cohort	115	124	117
Subcohort[e]	146	142	201

[a]Standard mortality ratio: observed deaths/expected deaths × 100.
[b]See Ref. 50.
[c]See Ref. 51.
[d]See Ref. 52.
[e]High exposure and long latency.

Another problem is the lack of site specificity for the carcinogenic action of TCDD in the epidemiologic studies. However, dioxin is a multisite carcinogen in rodents so the lack of site specificity is consistent with animal data. Other important epidemiologic studies published and/or under investigation should help to clarify some of the uncertainties and identify specific risks resulting from dioxin exposure in terms of cancer as well as other adverse health effects such as reproductive toxicity and neurotoxicity. Some of these studies are introduced in other chapters of this book and so are not part of this chapter which deals with dose–response relationships.

4. BIOLOGICALLY BASED RISK ASSESSMENT MODELS

There is considerable controversy on how to regulate TCDD and its structural analogues. Current risk assessment for dioxins in the United States and other countries is based on the carcinogenic activity in experimental animals (rodents). There are currently two models used to estimate safe exposure levels for dioxins.

The first model, used by the U.S. EPA and some other U.S. regulatory agencies, is called the linearized multistage model (LMS). In a multistage model for carcinogenesis (initiation, promotion, progression), the LMS assumes that the mutation rates are a linear function of the dose and that they are constant over time. This means that with a certain confidence level (95% for EPA) the risk of cancer is proportionally related to the dose or exposure levels in the low-dose region of dose–response curves. This model is widely used by regulatory agencies and was developed for genotoxic substances with the assumption that a single mutation may result in carcinogenesis.

The other model is the safety factor approach which assumes the existence of a "safe" exposure level below which no adverse health effect should occur. From a dose with no observable effect in animal experiments it is calculated with the inclusion of safety factors which cover interindividual variation and animal-to-human differences. The combined safety factors set an acceptable daily intake 100–1000 times lower than the no-observable-effect level (NOEL).

In general, countries that use the LMS model have higher risk estimates (lower acceptable exposure limits) than countries that use the safety factor approach. There is still much uncertainty involved in risk estimates for dioxins. This results from a pronounced species and tissue specificity in combination with a lack of information on effects in humans, interindividual variation, background exposure as well as knowledge gaps about the mechanism of action. Both risk assessment models are essentially default positions not necessarily based on mechanism.

A new approach which addresses the need for more scientifically based dioxin risk estimates is the development of a biologically (or mechanistically) based model. This model attempts to utilize mechanistic information, especially the involvement of the Ah receptor.[28,28A] There is controversy about dose–response relationships of receptor-mediated events especially in the low-dose range and it is critical to estimate low-dose exposure risk for humans from high-dose animal data. The assumption that all effects are receptor mediated and that a certain number of receptor molecules need to be occupied for a biological response can imply a practical "threshold" for dioxins below which no toxic effect can be detected in humans. Even if so, considering the overall exposure to dioxinlike compounds, it is of paramount importance to get more information on the background exposure of the general population and the resulting biological consequences. It is important to note, however, that research from our laboratory and presented earlier in this chapter indicates that for some TCDD-mediated effects in a rodent tumor promotion model, no threshold behavior is found.

A biologically based model could help remove some of the uncertainty that is present in LMS or safety factor approaches for estimating dioxin risk. These models need to recognize the mechanisms responsible for the diversity of biological effects that can be elicited by a single ligand interacting with the Ah receptor. They should also attempt to model noncancer endpoints in the risk assessment process since these endpoints are of increasing concern to public health officials.

Dioxin can be used as an example for developing mechanistic models for receptor-mediated carcinogenesis because of the extensive research and the broad data base. Information is available on low-dose behavior and the human data are in reasonable accordance with experimental data in animals.[54] The development of such a model is a dynamic process and with more information the model should provide an increasingly better risk estimate.[8]

5. SUMMARY

In this chapter dose–response relationships of dioxin-induced effects are described in various *in vivo* and *in vitro* systems. Even under the assumption that all effects are mediated through the Ah receptor, it is apparent that the shape of dose–response curves vary for different endpoints or in different target tissues. The possibility of linearity in the low-dose range (nonthreshold behavior) cannot be rejected based on the assumption that dioxin's effect is receptor mediated.

Another conclusion is that humans are very similar in their sensitivity of response to rodents (in terms of biochemical effects). There are several systems available to make animal-to-human comparisons and to identify specific biomarkers for dioxin-induced responses. Characterization of the low-dose behavior of such biomarkers is important for the development of a biologically based dose–response model. A biologically based model should help to clarify some of the uncertainties involved with estimating human risk from dioxin exposure. However, more information is needed on interindividual variation and the possible health consequences of background exposure to dioxin.

ACKNOWLEDGMENTS. The authors wish to thank Dr. J. VandenHeuvel and Mr. C. Sewall for their valuable contribution to this manuscript.

REFERENCES

1. E. K. Silbergeld and T. A. Gasiewicz, Dioxins and the Ah receptor, *Am. J. Ind. Med.* **16,** 455–474 (1989).
2. C. Rappe, in: *Banbury Report 35: Biological Basis for Risk Assessment of Dioxins and Related Compounds* (M. A. Gallo, R. J. Scheuplein, and K. A. van der Heijden, eds.), pp. 121–131, Cold Spring Harbor Laboratory Press, Cold Spring Harbor, NY (1991).
3. R. J. Kociba, D. G. Keyes, J. E. Beyer, R. M. Carreon, C. E. Wade, D. A. Dittenber, R. P. Kalnins, L. E. Frauson, C. Parks, S. D. Barnard, R. A. Hummel, and C. G. Humiston, Results of a two-year chronic toxicity and oncogenicity study of 2,3,7,8-tetrachlorodibenzo-p-dioxin in rats. *Toxicol. Appl. Pharmacol.* **46,** 279–303 (1978).
4. National Toxicology Program, *Bioassay of 2,3,7,8-tetrachlorodibenzo-p-dioxin for possible carcinogenicity (gavage study),* Technical Report Series No. 201, National Toxicology Program, Research Triangle Park, NC (1982).
5. J. Huff, A. Salmon, N. Hooper, and L. Zeise, Long-term carcinogenesis studies on 2,3,7,8-tetrachlorodibenzo-p-dioxin and hexachlorodibenzo-p-dioxins, *Cell Biol. Toxicol.* **7,** 67–94 (1991).
6. A. Schechter, Dioxins and related compounds in humans and in the environment, in: *Banbury Report 35: Biological Basis for Risk Assessment of Dioxins and Related Compounds* (M. A. Gallo, R. J. Scheuplein, and K. A. van der Heijden, eds.), pp. 169–214, Cold Spring Harbor Laboratory Press, Cold Spring Harbor, NY (1991).
7. L. Roberts, Dioxin risks revisited, *Science* **251,** 624–626 (1991).
8. G. W. Lucier, C. J. Portier, and M. A. Gallo, Receptor mechanisms and dose–response models for the effects of dioxins, *Environ. Health Perspect.* **101,** 36–44 (1993).
9. B. Schwetz, J. Norris, G. Sparschu, V. Rowe, P. Gehring, J. Emerson, and C. Gerbig, Toxicology of chlorinated dibenzo-p-dioxins, *Environ. Health Perspect.* **5,** 87–99 (1973).
10. A. Poland and J. C. Knutson, 2,3,7,8-Tetrachlorodibenzo-p-dioxin and related halogenated aromatic hydrocarbons: Examination of the mechanism of toxicity, *Annu. Rev. Pharmacol. Toxicol.* **22,** 517–554 (1982).
11. J. P. Whitlock, Genetic and molecular aspects of 2,3,7,8-tetrachlorodibenzo-p-dioxin action, *Annu. Rev. Pharmacol. Toxicol.* **30,** 251–277 (1990).
12. G. Perdew, Association of the Ah receptor with the 90-kDa heat shock protein, *J. Biol. Chem.* **263,** 13802–13805 (1988).
13. H. Reyes, S. Reisz-Porszasz, an O. Hankinson, Identification of the Ah receptor nuclear translocator protein (Arnt) as a component of the DNA binding form of the Ah receptor, *Science* **256,** 1193–1195 (1992).
14. M. S. Denison, C. L. Phelps, J. Dehoog, H. J. Kim, P. A. Bank, E. F. Yao, and P. A. Harper, in: *Banbury Report 35: Biological Basis for Risk Assessment of Dioxins and Related Compounds* (M. A. Gallo, R. J. Scheuplein, and K. A. van der Heijden, eds.), pp. 337–350, Cold Spring Harbor Laboratory Press, Cold Spring Harbor, NY (1991).
15. T. Sutter and W. Greenlee, Classification of members of the Ah gene battery, *Chemosphere* **25,** 223–226 (1992).
16. J. A. Goldstein and S. Safe, in: *Halogenated Biphenyls, Terphenyls, Naphthalenes, Dibenzodioxins and Related Products* (R. D. Kimbrough and A. A. Jensen, eds.), pp. 239–293, Elsevier, Amsterdam (1989).
17. L. Zeise, J. E. Huff, A. G. Salmon, and N. K. Hooper, Human risks from 2,3,7,8-tetrachlorodibenzo-p-dioxin and hexachlorodibenzo-p-dioxins, *Adv. Mod. Environ. Toxicol.* **17,** 293–342 (1990).

18. A. M. Tritscher, J. A. Goldstein, C. J. Portier, Z. McCoy, G. C. Clark, and G. W. Lucier, Dose–response relationships for chronic exposure to 2,3,7,8-tetrachlorodibenzo-p-dioxin in a rat tumor promotion model: Quantification and immunolocalization of CYP1A1 and CYP1A2 in the liver, *Cancer Res.* **52,** 3436–3442 (1992).
19. J. P. Vanden Heuvel, F. L. Tyson, and D. A. Bell, Construction of recombinant RNA templates for use as internal standard in quantitative RT-PCR, *BioTechniques* **14,** 395–398 (1993).
20. J. P. VandenHeuvel, G. C. Clak, A. M. Tritscher, W. F. Greenlee, G. W. Lucier, and D. A. Bell, Dioxin-responsive genes: Examination of dose–response relationships using quantitative reverse transcriptase-polymerase chain reaction, *Cancer Res.* **54,** 62–68 (1994).
21. M. Romkes, J. Piskorska-Pliezczyneka, and S. Safe, Effects of 2,3,7,8-tetrachlorodibenzo-p-dioxin on hepatic and uterine estrogen receptor levels in rats, *Toxicol. Appl. Pharmacol.* **87,** 306–314 (1987).
22. B. V. Madhukar, D. W. Brewster, and F. Matsumura, Effects of *in vivo*-administered 2,3,7,8-tetrachlorodibenzo-p-dioxin on receptor binding of epidermal growth factor in the hepatic plasma membrane of rat, guinea pig, mouse, and hamster, *Proc. Natl. Acad. Sci USA* **81,** 7407–7411 (1984).
23. C. H. Sewall, G. W. Lucier, A. M. Tritscher, and G. C. Clark, TCDD-mediated changes in hepatic epidermal growth factor receptor may be a critical event in the hepatocarcinogenic action of TCDD, *Carcinogenesis* **14**(9), 1885–1893 (1993).
24. G. I. Sunahara, G.W. Lucier, Z. McCoy, E. H. Bresnick, E. R. Sanchez, and K. G. Nelson, Characterization of 2,3,7,8-tetrachlorodibenzo-p-dioxin-mediated decreases in dexamethasone binding to rat hepatic glucocorticoid receptor, *Mol. Pharmacol.* **36,** 239–247 (1989).
25. F. H. Lin, S. J. Stohs, L. S. Birnbaum, G. Clark, G. W. Lucier, and J. A. Goldstein, The effects of 2,3,7,8-tetrachlorodibenzo-p-dioxin (TCDD) on the hepatic estrogen and glucocorticoid receptors in congenic strains of Ah responsive and Ah non-responsive C57BL/6J mice, *Toxicol. Appl. Pharmacol.* **108,** 129–139 (1991).
26. T. C. Sloop and G. W. Lucier, Dose-dependent elevation of Ah receptor binding by TCDD in rat liver, *Toxicol. Appl. Pharmacol.* **88,** 329–337 (1987).
27. T. Velu, Structure, function and transforming potential of the epidermal growth factor receptor, *Mol. Cell. Endocrinol.* **70,** 205–216 (1990).
28. C. Portier, A. Tritscher, M. Kohn, C. Sewall, G. Clark, L. Edler, D. Hoel, and G. Lucier, Ligand/receptor binding for 2,3,7,8-TCDD: Implications for risk assessment, *Fundam. Appl. Toxicol.* **20,** 48–56 (1993).
28A. M. E. Andersen, J. J. Mills, M. L. Gargas, L. Kedderis, L. S. Birnbaum, D. Neubert, and W. F. Greenlee, Modeling receptor-mediated processes with dioxin: implications for pharmacokinetics and risk assessment, *Risk Analysis* **13**(1), 25–36 (1993).
29. M. C. Kohn, G. W. Lucier, G. C. Clark, C. Sewall, A. M. Tritscher, and C. J. Portier, A mechanistic model of effects of dioxin on gene expression in the rat liver, *Toxicol. Appl. Pharmacol.* **120,** 138–154 (1993).
30. G. W. Lucier, A. Tritscher, T.Goldsworthy, J. Foley, G. Clark, J. Goldstein, and R. Maronpot, Ovarian hormones enhance TCDD-mediated increases in cell proliferation and preneoplastic foci in a two stage model for rat hepatocarcinogenesis, Cancer Res. **51,** 1391–1397 (1991).
31. G. Clark, A. Tritscher, R. Maronpot, J. Foley, and G. Lucier, Tumor promotion by TCDD in female rats, in: *Banbury Report 35: Biological Basis for Risk Assessment of Dioxins and Related Compounds* (M. A. Gallo, R. J. Scheuplein, and K. A. van der Heijden, eds.), pp. 389–404, Cold Spring Harbor Laboratory Press, Cold Spring Harbor, NY (1991).
32. G. C. Clark, A. M. Tritscher, D. A. Bell, and G. W. Lucier, Integrated approach for

evaluating species and interindividual differences in responsiveness to dioxins and structural analogs, *Environ. Health Perspect.* **98,** 125–132 (1992).

33. J. P. VandenHeuvel, G. C. Clark, C. L. Thompson, Z. McCoy, C. R. Miller, G. W. Lucier, and D. A. Bell, CYP1A1 mRNA levels as a human exposure biomarker. The use of quantitative polymerase chain reaction to measure CYP1A1 expression in human blood lymphocytes, *Carcinogenesis* **14**(10), 2003–2006 (1994).
34. W. F. Greenlee, R. Osborne, K. M. Dold, L. G. Hudson, M. J. Young, and J. W. A. Toscano, Altered regulation of epidermal cell proliferation and differentiation by 2,3,7,8-tetrachlorodibenzo-p-dioxin (TCDD), *Rev. Biochem. Toxicol.* **8,** 1–36 (1987).
35. J. Yang, P. Thraves, A. Dritschilo, and J. S. Rhim, Neoplastic transformation of immortalized human keratinocytes by 2,3,7,8-tetrachlorodibenzo-p-dioxin, *Cancer Res.* **52,** 3478–3482 (1992).
36. J. C. Cook, K. M. Dold, and W. F. Greenlee, An *in vitro* model for studying the toxicity of 2,3,7,8-tetrachlorodibenzo-p-dioxin to human thymus, *Toxicol. Appl. Pharmacol.* **89,** 256 (1987).
37. J. Vos and J. Moore, Suppression of cellular immunity in rats and mice by maternal treatment with 2,3,7,8-tetrachlorodibenzo-p-dioxin, *Int. Arch. Allergy Appl. Immunol.* **47,** 777–794 (1974).
38. P. H. Chen, C. Wong, C. Rappe, and M. Nygren, Polychlorinated biphenyls, dibenzofurans and quaterphenyls in toxic rice-bean oil and in the blood and tissues of patients with PCB poisoning (Yu-cheng) in Taiwan, *Environ. Health Perspect.* **59,** 59–65 (1985).
39. L. G. Hudson, W. A. Toscano, and W. F. Greenlee, Regulation of epidermal growth factor binding in human keratinocyte cell line by 2,3,7,8-tetrachlorodibenzo-p-dioxin, *Toxicol. Appl. Pharmacol.* **77,** 251–259 (1985).
40. G. W. Lucier, K. G. Nelson, R. B. Everson, T. K. Wong, R. M. Philpot, T. Tieran, M. Taylor, and G. I. Sunahara, Placental markers of human exposure to polychlorinated biphenyls and polychlorinated dibenzofurans, *Environ. Health Perspect.* **76,** 79–87 (1987).
41. J. Swenberg, F. Richardson, J. Boucheron, F. Deal, S. Belinsky, M. Charbonneau, and B. Short, High- to low-dose extrapolation: Critical determinants involved in the dose response of carcinogenic substances, *Environ. Health Perspect.* **76,** 57–63 (1987).
42. G. W. Lucier, in: *Mechanisms of Carcinogenesis in Risk Identification* (E. H. Vainio, P. N. Magee, D. B. McGregor, and A. J. McMichael, eds.), pp. 87–112, International Agency for Research on Cancer, Lyon, France (1991).
43. H. Pitot, T. Goldsworthy, S. Moran, W. Kennan, H. Glauert, R. Maronpot, and H. Campbell, A method to quantitate the relative initiating and promoting potencies of hepatocarcinogenic agents in their dose–response relationships to altered hepatic foci, *Carcinogenesis* **8,** 1491–1499 (1987).
44. R. R. Maronpot, J. F. Foley, K. Takahashi, T. Goldsworthy, G. Clark, A. M. Tritscher, C. J. Portier, and G. W. Lucier, Dose response for TCDD promotion of hepatocarcinogenesis in rats initiated with DEN: Histologic, biochemical, and cell proliferation endpoints, *Environ. Health Perspect.* **100**(7), 634–642 (1993).
45. M. S. Rao, V. Subbarao, J. D. Prasad, and D. G. Scarpelli, Carcinogenicity of 2,3,7,8-tetrachlorodibenzo-p-dioxin in the Syrian golden hamster, *Carcinogenesis* **9,** 1677–1679 (1988).
46. National Toxicology Program, *Bioassay of a mixture of 1,2,3,6,7,8-hexachlorodibenzo-p-dioxin and 1,2,3,7,8,9-hexachlorodibenzo-p-dioxin for possible carcinogenicity (gavage study),* Technical Report Series No. 198, National Toxicology Program, Research Triangle Park, NC (1980).
47. S. Flodstrom and U. Ahlborg, Relative tumor promoting activity of some polychlorinated

dibenzo-p-dioxin-, dibenzofuran- and biphenyl-congeners in female rats, *Chemosphere* **25,** 169–172 (1992).
48. Environmental Protection Agency Science Advisory Board (EPA SAB), *Review of draft documents "A cancer risk-specific dose estimate for 2,3,7,8-TCDD" and "Estimating risk exposure to 2,3,7,8-TCDD,"* EPA SAB Ad Hoc Dioxin Panel, Washington, DC (1989).
49. D. G. Patterson, Jr., L. Hampton, C. R. Lapeza, Jr., W. T. Belser, V. Green, L. R. Alexander, and L. L. Needham, High-resolution gas chromatographic/high-resolution mass spectrometric analysis of human serum on a whole-weight and lipid base for 2,3,7,8-tetrachlorodibenzo-p-dioxin, *Anal. Chem.* **59,** 2000–2005 (1987).
50. M. A. Fingerhut, W. E. Halperin, D. A. Marlow, L. A. Piacitelli, P. A. Honchar, M. H. Sweeney, A. L. Greife, P. A. Dill, K. Steenland, and A. H. Suruda, Cancer mortality in workers exposed to 2,3,7,8-tetrachlorodibenzo-p-dioxin, *N. Engl. J. Med.* **324,** 212–218 (1991).
51. A. Manz, J. Berger, J. H. Dwyer, D. Flesch-Janys, S. Nagel, and H. Waltsgott, Cancer mortality among workers in chemical plant contaminated with dioxin, *Lancet* **338,** 959–964 (1991).
52. A. Zober, P. Messerer, and P. Huber, Thirty-four-year mortality follow-up of BASF employees exposed to 2,3,7,8-TCDD after an 1953 accident, *Int. Arch. Occup. Environ. Health* **62,** 139–157 (1990).
53. E. S. Johnson, Important aspects of the evidence for 2,3,7,8-tetrachlorodibenzo-p-dioxin (TCDD) carcinogenicity in man, *Environ. Health Perspect.* **99,** 383–390 (1993).
54. G. W. Lucier, Humans are a sensitive species to some of the biochemical effects of structural analogs of dioxin, *Environ. Toxicol. Chem.* **10,** 727–735 (1991).

Chapter 9

Dioxins, Dibenzofurans, PCBs and Colonial, Fish-Eating Water Birds

John P. Giesy, James P. Ludwig, and Donald E. Tillitt

1. INTRODUCTION

Historically, the colonial, fish-eating water birds of the Great Lakes have been exposed to a number of toxic, synthetic, halogenated compounds.[1-10] These exposures have resulted in a number of adverse effects on their reproductive potential,[9-11] such as deformities and lethality of embryos.[9-13] These effects have, in turn, caused declines in populations.[14-16] The most dramatic effect on reproductive performance was the result of eggshell thinning, caused primarily by DDE, the degradation product of DDT.[3,16-21] Since the cessation of the manufacture and use of the most persistent and widespread contaminants, the concentrations of these compounds in fish and birds have decreased.[5,11] Specifically, concentrations of DDE in bird eggs of the North American Great Lakes region have decreased to less than the critical concentration for eggshell thinning. Subsequently, populations of some of the fish-eating water birds have increased rapidly.[22] However, other adverse effects such as localized impairment of reproductive performance[23] and anatomical defects[24-26] have persisted. When we began our studies, it was not known if the observed effects were related to existing contaminants or if these effects would abate as the concentrations decreased further. We speculated that some of the observed effects such as the deformities would

John P. Giesy • Department of Fisheries and Wildlife, Pesticide Research Center and Institute of Environmental Toxicology, Michigan State University, East Lansing, Michigan 48824. James P. Ludwig • The SERE Group, Ltd., Victoria, British Columbia, Canada V8P 3C8. Donald E. Tillitt • National Fisheries Contaminant Research Center, U.S. Fish and Wildlife Service, Columbia, Missouri 65201.

Dioxins and Health, edited by Arnold Schecter. Plenum Press, New York, 1994.

Table 1
Potential Causes of Deformities Observed in Embryos and Chicks of the Great Lakes Region

Nutritional deficiencies related to changes in prey base
New generation pesticides
Continued effects of traditional contaminants
Disease, such as viral infections
Genetic inbreeding
Old, persistent pesticides
Unidentified chemicals, pulp and paper industry

have been observed during the 1950s through the 1970s at equal or greater rates of occurrence, but they were masked by the effects of eggshell thinning. Now that eggs were not being crushed because of thin shells, the more subtle effects could be expressed. Also, since there were only poor correlations between concentrations of the routinely measured contaminants and the observed effects, there was some question as to whether the observed effects were related to these historically significant contaminants or whether they were caused in any way by the presence of contaminants. It has been suggested that since concentrations of all of the known toxic chemicals in the food, tissues, and eggs of birds of the Great Lakes region were declining, the observed effects might not be related to synthetic, organic compounds. There were a number of possible alternative causes to investigate (Table 1). The suite of effects that were observed did not seem to be the types that could all be explained by disease, or nutrition. It had been suggested that the decreased viability of eggs and abnormally great number of birth defects could have been caused by genetic deficiencies caused by a "founder effect." Since the populations of birds had been greatly reduced and for some species almost completely extirpated from the Great Lakes region because of the effects of DDE, the populations were increasing from a few individuals. Thus, it was suggested that there might be an unhealthy low genetic diversity in some of the populations. While this might explain some of the effects like birth deffects, it was not likely that all of the effects could be explained by genetic deficiencies. It was also suggested that the effects might be related to new, nonpersistent pesticides that could have effects at very small concentrations and leave little or no trace in the eggs of afflicted birds. Such effects have not been observed in laboratory or field trials with these types of insecticides so this also seemed unlikely. As we began our investigations in 1987, a prime possibility seemed to be some unidentified compound or compounds, perhaps from the pulp and paper industry, which is known to release a large number of poorly characterized and seldom identified compounds.[27]

Initially we compared the suite of effects that had been observed in wild populations of birds (Table 2) with those that were observed during laboratory exposures to various halogenated hydrocarbons and concluded that this suite of effects as most likely caused by chlorinated dioxins.[7] Therefore, we compared the current, average concentrations of 2,3,7,8-tetrachlorodibenzo-*p*-dioxin (TCDD) or TCDD equivalents (TCDD-EQ) measured in eggs of birds from the region with quantities of TCDD and

Table 2
Effects Observed in Birds of the Great Lakes Region

Eggshell thinning
Deformities
Tumors
Behavioral changes
Immune suppression
Edema
Cardiovascular hemorrhage
Hormonal changes
Enzyme induction P4501A1 and P4502B1
Metabolic changes, wasting syndrome
Depletion of vitamin A
Porphyria

TCDD-EQ known to elicit effects in birds (Fig. 1). Based on the concentrations observed in water birds and their eggs from the Great Lakes region, it seemed unlikely that polychlorinated dibenzodioxins (PCDD) and polychlorinated dibenzofurans (PCDF) were responsible for the observed effects. These same effects have been observed in birds exposed to polychlorinated biphenyls (PCBs) at concentrations similar to those observed.[6,12] However, we observed a fairly poor correlation between the total concentrations of PCBs and egg lethality or birth defects.[23–26] Because the types of effects ob-

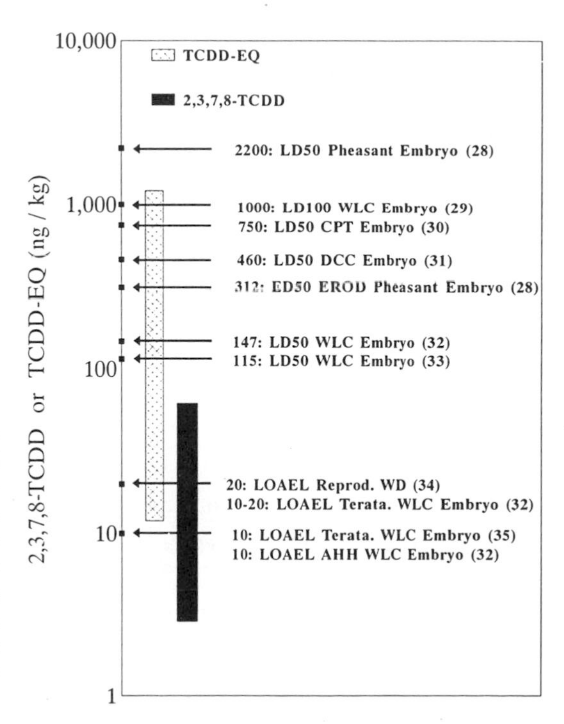

Figure 1. 2,3,7,8-TCDD avian toxicological effects thresholds. The hatched and solid bars represent the relative range of concentrations (pg/g or ng/kg) of TCDD-EQ and 2,3,7,8-TCDD, respectively, that are currently found in the eggs of colonial fish-eating water birds of the Great Lakes. The arrows indicate various toxicity thresholds for avian species which have been tested in laboratory or field studies. The references are given by number in parentheses.

served could be caused by PCBs and the latter chemicals were occurring at the greatest concentrations in the tissues and eggs of birds in the Great Lakes region, before beginning a search for the hypothetical compound(s), we felt that it would be worthwhile to test the hypothesis that the effects were related to the TCDD-like, planar PCB congeners. These compounds are structurally similar and known to cause effects similar to those caused by TCDD. It has been proposed that all of these structurally similar compounds cause their toxicity through the aromatic hydrocarbon receptor (Ah-r) (see Section 5). Thus, we felt that a measure that reflects or incorporates this mechanism of action would be a better technique to test the strength of association between the concentrations of polychlorinated hydrocarbons (PCH), and observed adverse effects.

Here, we describe the current concentrations of TCDD-EQ in tissues of birds of the Great Lakes region and discuss their relationship to egg lethality and birth defects. We also discuss the interactions among some of these Ah-r-active compounds, their relative toxic potencies, the relative importance of the various congeners, and other possible contributors to the total toxicity of complex mixtures. We also describe and compare several methods of assessing the toxic potency of complex mixtures to birds and compare the results of hazard assessments for wildlife with those to protect human health.

2. CHLORINATED HYDROCARBONS IN THE ENVIRONMENT

Many persistent, synthetic PCH have been released to and are widely distributed in the environment.[37–42] Some of the PCH are very persistent,[37,43] and are bioconcentrated and biomagnified.[37,41,43–47] Of these, some classes cause adverse effects at minute concentrations in biota.[46–50] Some of the major groups that are of environmental significance are the PCDDs, PCDFs, and PCBs.[49–51] Together the PCBs, PCDDs, and PCDFs represent 419 individual congeners. Within this overall class are several subclasses defined by the pattern of substitution of the chlorine atoms.[41,45,50,51] One subset of these groups of compounds, those that are both laterally substituted and either non- or mono-*ortho*-substituted, can attain a planar configuration and are referred to as *planar or coplanar congeners* (p-PCH).[50] Our discussion will consider only these p-PCH.

3. COMPLEX MIXTURES

It is difficult to understand or predict potential effects of complex environmental mixtures of PCH on biota not only because there are a great number of compounds, but also because the concentrations of individual components change as a function of space and time. Thus, the mixture to which organisms are exposed at one time or at one location may be very different from that to which they are exposed at other times or locations.[24,52,53] Furthermore, the relative concentrations of the various p-PCH congeners differ from trophic level to trophic level.[54] These differences are caused by environmental "weathering" and the sorting of compounds, based on their solubilities, volatilities, and rates of degradation. The result is mixtures in the environment which not only change spatially and temporally, but which are different from the technical mixtures which were released into the environment. Thus, at this time, it is impossible to use the results of studies that have determined the dose–response relationships of technical mixtures under laboratory conditions to predict effects in real-world wildlife. The study of effects of p-PCH on fish and wildlife was limited for two decades by the fact that it was impossible to assess the toxicological implications of constantly changing mixtures and the need to monitor for total concentrations of p-PCH. Recently, greater understanding of the mechanisms of toxic action of the p-PCH has made it possible to express the potency of mixtures of p-PCH to elicit adverse effects relative to one prototype p-PCH. When this approach has been used, it has been possible to obtain better correlations between observed effects and TCDD-EQ (sometimes referred to as TEQ) than could be obtained for single p-PCH congeners of classes of p-PCH.[50]

4. TOXICITY OF p-PCH

While the suite of effects caused by TCDD may seem diverse and unrelated, all of them are thought to be caused through a common mode of action.[55–70] The mechanism is receptor mediated and involves the binding of particular intracellular receptors which lead to a host of common cellular and subcellular responses and subsequently effects on the whole animal. The similarities of molecular structure and conformation of the p-PCH compounds result in similar toxic effects.[41,47–50] Their primary toxic effects are thought to be exerted through a common mode of action.[56–58] The most widely accepted proposed mechanism of action for p-PCH involves the expression of their biological potency through a specific cytosolic receptor, the Ah-r.[58–60] This receptor binds p-PCH, which are approximately 3×10 Å in size, such as planar PCBs (p-PCBs), PCDDs, and PCDFs, with differing affinities.[50] The resulting receptor–ligand complex is then translocated to the cell nucleus where it elicits specific changes in gene expression.[30,59,61,62] The transformed receptor binds to sequences of DNA called the dioxin-responsive enhancers (DRE). There are a number of DRE in the genome of most animals, thus a number of genes can be affected.[63] Many of the observed toxic effects of the p-PCH are attributable to specific alterations in gene expression.[50,56–58,64–67] The relative toxicity of individual p-PCH is directly proportional to the strength of binding to the Ah-r and the potential to induce cytochrome P4501A isozyme activity.[47–49,60,68,69]

The p-PCH are generally not acutely toxic, but cause chronic toxic responses,[50,62,70,71] including impaired reproductive potential of fish-eating water birds. Because the p-PCH are not generally acutely toxic and accumulate into top predators, such as birds, colonial water birds have been suggested as useful biological monitors for the accumulation and effects of p-PCH in the Great Lakes ecosystem.[23,24,26,72–75]

The most characteristic of these responses include measurement of mixed-function oxidases. These enzymes are part of the general detoxification response of animals to toxic substances. These enzymatic responses are generalized and lead to or are associated with side effects on many critical substrates used in routine metabolism. Levels of hormones, vitamins, and by-products of normal cellular activities are often altered enough to produce a characteristic set of responses now recognized as symptoms of subchronic or chronic exposures to p-PCH (Table 2).[19]

There are a number of pleiotrophic effects of TCDD and p-PCBs on organisms. These effects can be direct or secondary responses to gene regulation. Some of these responses can be directly related to the adverse effects observed, while some are useful as biomarkers, but their relationships to observed adverse effects are less well characterized. The most subtle and important biological effects of TCDD and the dioxinlike p-PCB congeners on wildlife are their effects on endocrine hormones and vitamin homeostasis.[76] TCDD also mimics the effects of thyroxine as a key metamorphosis signal during maturation.[66] TCDD has also been shown to downregulate the epidermal growth factor receptor,[77] which may result in disruption of the patterns of embryonic development at critical stages.

Altered concentrations of thyroid and steroid hormones and vitamin A are frequently reported to co-occur with embryonic abnormalities in wildlife populations exposed to p-PCH.[9] Individuals from these exposed populations have been observed to have altered sexual development,[76,78] sexual dysfunction as adults, and immune system suppression.[59,79] The observations on adult sexual dysfunction are especially significant since young that appear to be normal while raised by intoxicated parents may become reproductively dysfunctional when they mature.[80–84] Poor reproductive efficiencies and adventive, opportunistic diseases are characteristic of the wild animals in these exposed populations of the Great Lakes region.[85] Because of these conserved biochemical mechanisms, concentrations of TCDD-EQ correlated with egg lethality or birth defects in populations of colonial fish-eating water birds while total concentrations of PCB, PCDF, and PCDD did not.[23,86,87]

Vitamin A (retinol) is important in many functions in animals, such as embryonic development, vision, maintenance of the dermally derived tissues, immune competence, hemopoiesis, and reproductive functions.[88–90] Vitamin A is necessary for normal embryonic development[91] and, thus, changes in the status of vitamin A in the plasma or liver may be responsible for the birth defects observed in birds that have been exposed to p-PCH. Laboratory studies have determined that both vitamin A and its storage form in the liver (retinyl palmitate) were depleted in birds exposed to sublethal doses of the dioxinlike, p-PCB congener 77.[92] We have observed an inverse correlation between the concentration of vitamin A in serum and concentrations of p-PCH in tissues of birds from the upper Great Lakes.[73]

p-PCH affect concentrations of vitamin A in both the blood and liver of exposed organisms. These effects are thought to be related at to least two processes. In blood, some of the hydroxylated metabolites of PCBs bind to the carrier protein transthyretin.[89] In the liver, induction of hepatic enzymes such as acyl-CoA:retinol acyltransferase and uridine diphosphate glucuronyl transferase (UDPGT) are thought to alter the metabolic pathways involved in the storage and mobilization of vitamin A, and result in the observed depletion of retinols in the liver.[88,90]

TCDD is known to have effects on both male and female steroid hormones.[92–94]

For instance, TCDD has estrogenic and antiestrogenic effects in different tissues, depending on timing of exposures during development.[46] Furthermore, TCDD is a potent thyroxine (T_4) agonist which may account for its capability to cause wasting syndrome in homeotherms.[50] The induction of the mixed-function monooxygenase system can also reduce the concentrations of circulating steroid hormones, which can have adverse effects on the reproduction of wildlife.[27]

Thyroid hormone, which is an important regulator of development and metabolism, can be influenced by exposure to p-PCH.[66,91–94] There are several possible mechanisms for the observed effects on circulating T_3 and T_4. First, hydroxy-substituted PCB congeners have been observed to displace T_4 from its carrier protein, transthyretin (TTR; prealbumin), which results in effects similar to T_4 deficiency.[93,94] p-PCH can induce UDPGT activity in the liver, which then decreases the concentration of TTR in the blood. Concentrations of TTR are not determined directly in the plasma, but rather T_4 binding capacity is measured. Therefore, it is not possible to distinguish which of the two mechanisms may be causing the observed effects.

There is also evidence that exposure to TCDD can shift normal carbohydrate-dominated metabolism to a fat metabolism, causing afflicted individuals to be unable to utilize a major source of energy. This effect is thought to be related to inhibition of the synthesis of the glucose carrier protein by p-PCH.[26,95]

Methyl sulfone metabolites of p-PCH have also been observed to cause adverse effects. Methyl sulfones accumulate in lung tissue of some marine mammals and are thought to be responsible for some toxic effects.[96]

In addition to the effects of the p-PCBs, it is known that the di-*ortho*-substituted PCBs, which are not very toxic, as a result of effects that are mediated by the Ah-r can cause adverse effects in wildlife. Specifically, these congeners can inhibit dopamine synthesis in the brain, which results in behavioral differences in mammals.[97–104] We have observed behavioral effects in colonial water birds that may be caused by these types of effects,[105] but to date little information is available on this phenomenon. We feel that it is unlikely that the effects of the di-*ortho*-substituted PCBs are responsible for the observed birth defects, but may be important in subtle behavioral shifts.

5. QSAR

The most potent p-PCH identified thus far is TCDD. The relative potency of other p-PCH congeners to cause biochemical or toxicological effects mediated through the Ah-r mechanism is determined by the pattern of substitution of the chlorine atoms on p-PCH.[50] The most potent congeners are those that have at least four chlorine atoms in at least two of the lateral (*meta* and *para*) positions of both of the phenyl rings.[50] The relative potency can be expressed as proportions relative to TCDD and are referred to as toxic equivalency factors (TEF).[49,50,99] The concentrations of individual p-PCH congeners can therefore be measured for Ah-r-mediated potency and reported as TCDD-EQ.[50,51] The potency of complex mixtures, which can include several of the more Ah-active congeners in addition to many less active congeners, can also be expressed relative to TCDD as TCDD-EQ.[50,51] One method involves the quantification of each p-PCH congener in a complex mixture from a biological sample. The potency of the mixture is then calculated by multiplying the concentration of each congener by its TEF value and summing the products.[11,44–51]

6. BIOCHEMICAL RESPONSE ASSAYS

There are a number of limitations associated with the determination of TCDD-EQ in complex mixtures from instrumental analysis and application of TEFs. The large number of p-PCH in environmental samples makes the chemical analysis time-consuming and expensive. In addition, the most important congeners occur at small, but toxicologically relevant concentrations that are sometimes difficult to quantify by routine procedures. The wide range of biological potencies for p-PCH[50] and potential synergistic or antagonistic interactions among p-PCH congeners and other chemicals[48,50,107–110] suggest that an additive model of toxicity is generally adequate but may not always be appropriate.[111] Interactions between or among p-PCH congeners and other compounds in the mixture may further complicate interpretation of the toxicological significance of these mixtures to wildlife. In addition, assessment of possible toxic effects of mixtures on wildlife is complicated by the wide range of reported TEF values for different species and various toxic endpoints used to set TEF values.[50] Depending on the test species and chosen endpoints, the use of different TEF values will result in different calculated TCDD-EQ concentrations.[112,113] Furthermore, TEFs are not available for all biological endpoints and most have been derived from those species amenable to laboratory studies.

The concentrations of TCDD-EQ in complex mixtures of p-PCH may also be determined by use of *in vitro* cell systems in a manner analogous to a chemical detector.[87] The method that uses H4IIE rat hepatoma cells relies on the fact that p-PCH induce specific cytochrome P450-mediated monooxygenase (MO) enzymes through the Ah-r-mediated mechanism.[87,100,101,115,116] Furthermore, relative induction of MO activity by individual p-PCH, as well as by mixtures of these compounds, are correlated with their toxicity to certain species, particularly birds.[48–50,80–84] Therefore, induction of cytochrome P450-mediated MO enzymes *integrates* the concentration and potency of all of the p-PCH congeners present in complex, environmental mixtures. This induction of enzymes has been well characterized *in vitro* with rat hepatoma cells (H4IIE cells) on their exposure to p-PCH-containing extracts of environmental samples.[115–117] This induction measures the potency of an extract[23,87,101,102,113,118–121] which can be expressed as TCDD-EQ by comparing the dose-response of an unknown extract to a standard curve generated with TCDD.[87]

The H4IIE bioassay method of TCDD-EQ determination has both advantages and disadvantages relative to the use of chemical analysis and TEF values. The bioassay is more rapid and less costly than congener-specific chemical analysis. Since the bioassay is a mechanistically based determination of an integrated biochemical response, the results can be expected to be more biologically relevant. The bioassay integrates possible interactions between p-PCH congeners and compounds of other chemical classes thereby providing an integrated measure of potency measured at a cellular site proximate to the site of action. One limitation to the use of bioassay systems is that culture conditions and the use of various carrier solvents or endpoints may make comparison of results among different research groups difficult.[50,87] Also, results obtained using the H4IIE bioassay have only been calibrated against controlled laboratory studies with rodents. This makes it difficult to interpret the toxicological implications of concentrations of TCDD-EQ observed in wildlife species. Only limited correlations are demonstrated in field studies of wildlife species[23] and only a few controlled laboratory studies with individual p-PCH congeners have been conducted.

Until recently only limited comparative data between the instrumental and bioassay approaches have been available for complex mixtures of p-PCHs in samples collected from the environment.[102] More recently we have determined the TCDD-EQ by both instrumental and H4IIE bioassay analysis of a set of p-PCH-containing extracts from birds and their eggs collected at Green Bay, Wisconsin.[112,121] It was found that the TCDD-EQ determined by the instrumental and bioassay techniques were positively correlated, but that the use of the results of the instrumental analyses and application of TEFs in an additive model underestimated the TCDD-EQ measured in the H4IIE bioassay (Fig. 2). The possible reasons for the observed differences were that (1) the mixtures were interacting synergistically or (2) there were compounds present that were not quantified but contributed to the response of the H4IIE cells. It is likely that compounds other than the PCBs, PCDDs, and PCDFs were responsible for a greater concentration of TCDD-EQ in the H4IIE assay than could be accounted for by an additive model which considered the compounds measured. This discrepancy was not caused by the use of inappropriate TEFs or exposure systems, since the TEFs used were derived by using the same techniques in the H4IIE system. This difference is thought to be related to the fact that they assay responds to all of the p-PCH compounds, while we only quantified the PCDD, PCDF, and planar PCB congeners.

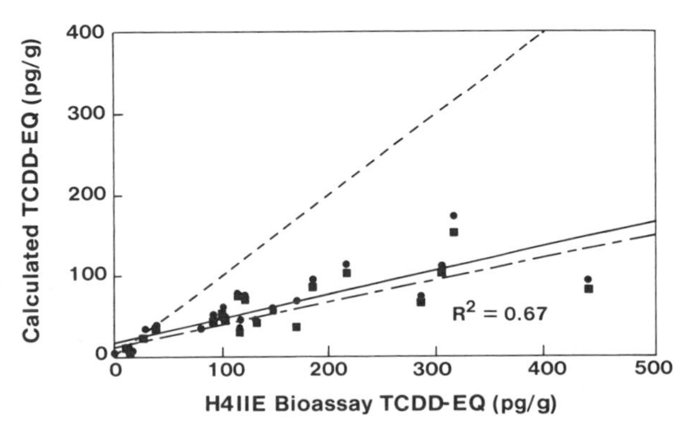

Figure 2. Concentrations of calculated TCDD-EQ as a function of H4IIE bioassay-derived TCDD-EQ. TCDD-EQ were calculated from TEF values using either Tillitt[120] (■) or Tillitt[120] and Safe[50] TEF values (●). Bioassay TCDD-EQ values were determined using the H4IIE bioassay as described in the text. The upper dashed line indicates equality between calculated and bioassay equivalents. The solid line is the linear regression for TCDD-EQ derived from Tillitt–Safe TEFs while the lower dashed line represents the linear regression for the TCDD-EQ derived from Tillitt TEFs. (Reprinted from Jones *et al.*[112] with permission.)

Concentrations of TCDD-EQ determined by calculation using TEFs or H4IIE bioassay have been correlated with adverse effects in birds.[11,23,70,119]

7. RELATIVE CONTRIBUTIONS OF INDIVIDUAL p-PCH

The proportion of TCDD-EQ contributed by PCDD and PCDF congeners in environmental samples from the Great Lakes region is generally small, frequently less than 5% in the fish-eating bird species (Figs. 3 and 4). A similarly great relative importance of p-PCBs has been reported in certain marine mammals.[43] This indicates that marine ecosystems are also strongly influenced by the planar PCBs with toxic responses mediated through the Ah-r mechanism. In contrast to the fish-eating birds in the Great Lakes, the "terrestrial" avian species examined in a Green Bay study contained lesser concentrations of TCDD-EQ[112] and the relative proportion of TCDD-EQ that was contributed by PCDDs and PCDFs in these species ranged from 3.2 to 29% and was greater than for the fish-eating species (Fig. 4). However, the *absolute* concentrations of TCDD-EQ contributed by PCDDs and PCDFs were similar to the piscivorous species.[112,121] This suggests that contamination with PCDD and PCDF can be widespread in all species. In the fish-eating water birds the accumulation of TCDD-EQ from PCB congeners derived from water and food by forage fish is an additional trophic level transfer and thus provides greater bioaccumulation potential than the terrestrial species. Also, there are known local sources of PCBs, whereas the sources of PCDD and PCDF may be more distant and related to atmospheric deposition.

The majority (> 90%) of the TCDD-EQ in the eggs of cormorants and terns in the Great Lakes was related to p-PCBs,[112] rather than PCDD or PCDF, which accounted for between 2 and 9% (12–22 ppt) of TCDD-EQ measured in water bird eggs in Lakes Superior, Huron, and Michigan. The primary contributions to the TCDD-EQ were those of the *dioxinlike* p-PCBs, especially non-*ortho*-chlorinated PCB congeners 126 (3,4,5,3′,4′-PeCB), 77 (3,4,3′,4′-TCB), and 169 (3,4,5,3′,4′,5′-HCB) and

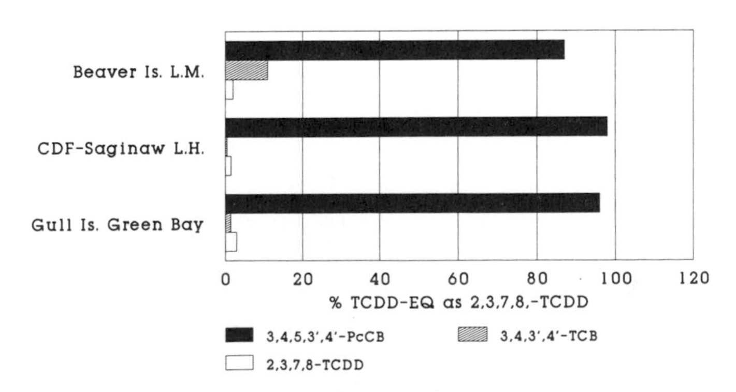

Figure 3. Relative proportions of total TCDD-EQ predicted from instrumental analyses and TEF (EROD induction) values, which are contributed by the p-PCB and 2,3,7,8-TCDD. Samples are of 23-day-old Caspian tern eggs. TEFs used were those of Safe.[50]

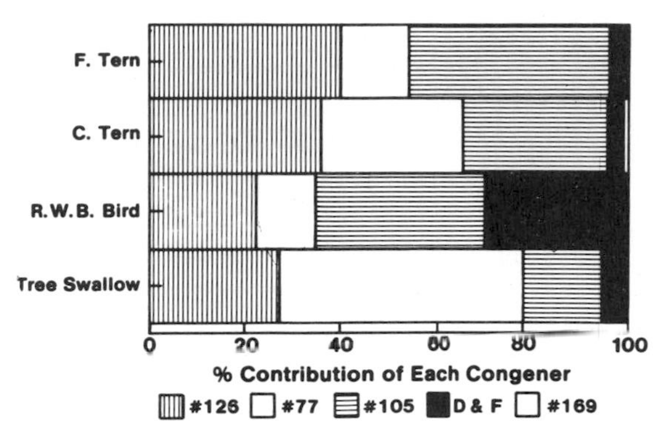

Figure 4. Relative contribution of specific p-PCH congeners to total concentrations of TCDD-EQ. TCDD-EQs were calculated, using Safe[48] TEFs, for individual p-PCH congeners. The contribution of each congener to the total concentration of TCDD-EQ is expressed as a percentage of total calculated TCDD-EQ concentration. Because of the small contributions of equivalents by PCDD and PCDF congeners, the TCDD-EQ contributions of all dioxin and furan congeners were combined (D&F). Values represent species means. F. Tern; Forster's tern; C. Tern, common tern; R.W.B. Bird, red-winged black bird. The segments for each bar graph are in the same order from left to right as shown in the legend. (Reprinted from Jones *et al.*[112] with permission.)

mono-*ortho*-chlorinated congeners 105 (2,3,4,3′,4′-PeCB) and 118 (2,3′,4,4′,5-PeCB). The understanding that dioxinlike bioeffects in fish-eating colonial water birds are related largely to p-PCBs is an emerging consensus worldwide, except near TCDD point sources.[43,52]

Currently, much of the discussion of the safety of consuming fish flesh from the Great Lakes is centered on the concentration of TCDD-EQ contributed by the PCDD and PCDF.[102] However, the concentrations of TCDD-EQ from both PCDD and PCDF are generally in the range of 5 to 10 pptr, wet weight, while concentrations of TCDD-EQ contributed by the PCB congeners can be as great as 250 pptr.[122]

8. INTERACTIONS AMONG p-PCH AND BETWEEN p-PCH AND OTHER COMPOUNDS

There has been much discussion about the possible interaction among individual congeners of p-PCH and between p-PCH classes and other synthetic halogenated compounds in extracts of environmental matrices that contain complex mixtures of p-PCH. An additive model for the prediction of TCDD-EQ is plausible and it is unlikely that the use of this model will result in a great deal of error in predicting the concentrations of TCDD-EQ because of synergisms. While such nonadditive responses could be either greater or less than additive, it seems that the biochemical effects of p-PCH congeners are simply additive.[50] However, there are reports of both infra- and supra-additivity between and among individual p-PCH congeners, complex mixtures of p-PCH and other halogenated hydrocarbons.[50,60,65,67,123]

When nonadditive responses between individual p-PCH congeners and complex mixtures have been observed, they have generally been antagonistic.[80,110,124–126] This interaction is probably related to the fact that the less Ah-r-active congeners still have receptor binding affinities, which when combined with the relatively great concentrations of these less toxic congeners, make them effective competitors for binding sites.[125,126] This reduces the probability of the more toxic p-PCH binding to the Ah-r. However, these less active congeners do not seem to bind with great enough affinity to be effective inducers of EROD activity or cause any of the other adverse effects that are caused by the Ah-r-active congeners.[50] The di-*ortho*-substituted PCB congener 2,2′,4,4′,5,5′ has, under some conditions, been found to be an effective antagonist for 2,3,7,8-TCDD, but some of the results of the studies of interactions among congeners are equivocal. For instance, PCB congener 77 and TCDD caused greater than additive induction of AHH activity in the liver of rainbow trout at doses that corresponded to 0.13 and 0.5 toxic unit (proportion of dose required to elicit a given endpoint). However, at greater doses the mixture was less than additive.[127] Similarly, there was no effect of PCB congener 153 (2,2′,4,4′,5,5′) on the induction of EROD activity by TCDD.[128] PCB congener 153 was found by the same researchers to have an antagonistic effect on the potential of TCDD to induce EROD activity.[128] Pretreatment of mammalian cells results in an increase in the concentration of Ah-r, which could cause a greater response to TCDD, but this is only observed at submaximal exposures to TCDD.

The evidence seems to support the conclusion that complex mixtures of halogenated p-PCH congeners in the extracts from wildlife would be less than additive (antagonistic) rather than more than additive (synergistic). A number of compounds that can bind to the Ah-r, but do not effectively induce the same monooxygenase (P450IA1), enzyme activity as 2,3,7,8-TCDD, are partial antagonists.[107–110] The potency of simple combinations and complex mixtures of PCDD and PCDF to induce EROD activity in hepatocytes or H4IIE cells indicated that the effects were related to the simple additive responses to the 2,3,7,8-substituted congeners.[129] Therefore, it is not likely that nonadditive interactions between or among Ah-r-active or inactive congeners would result in differences between the concentrations of TCDD-EQ derived by instrumental or bioassay methods. In fact, if there were strong antagonisms between the p-PCH and other components of these complex mixtures, one would expect the TCDD-EQ determined in the bioassay to be less, not more than those calculated from the additive model, which was the case when the results of the two methods were compared.[112]

9. SELECTIVE ENRICHMENT OF p-PCH

When studying a complex mixture, such as p-PCH, if the relative proportions of the different congeners change during the bioaccumulation process this change must be accounted for in the hazard assessment. It is difficult to determine safe exposures if concentrations observed in the field cannot be compared with the results of controlled laboratory studies, which have been conducted with the original technical mixtures of the compounds. Chemical weathering resulting from differential solubilities, volatilities, adsorption constants, and degradation rates can result in patterns or relative concentrations of p-PCH congeners which are different from the technical mixtures and also different from one location to another.[130] Furthermore, these patterns can change over time[131] such that the pattern of congeners observed is significantly different from that in the original technical mixtures, which were released to the environment.[130] Also, there are changes in the relative pattern of accumulation in the ecosystem as trophic biomagnification occurs.[112,132,133]

In the Great Lakes system, one way to estimate the relative toxic potency of mixtures is to calculate TCDD-EQ, which measures the total concentration of congeners from PCDD, PCDF, and PCBs and divide this by the total concentration of PCBs.[25,54,112] This relative potency ratio will account for different contributions from the three major classes of congeners and relate it to that fraction that is thought to account for most of the toxic potency, the PCBs.

Selective accumulation of the more toxic PCB congeners can result in a mixture in the tissues of target animals that is more toxic than would be predicted from an estimate of the original Aroclor mixture.[23] This enrichment of the more toxic, non-*ortho*-substituted PCB congeners results in a relative toxic potency of the mixture that is from four to six times greater than the original technical mixture.[23] The toxic potencies determined as the ratio of TCDD-EQ calculated by the H4IIE assay to total concentration of PCB of extracts of double-crested cormorant eggs were 2.5 to 5.24 (mean 3.77) times greater than technical mixtures of Aroclors, which were also measured in the eggs (Fig. 5).[31,86,87]

When a potency ratio was calculated for several of the locations in the Great Lakes, the ratio of toxic potency was indeed found to vary among locations (Fig. 6). Furthermore, the greatest ratio was observed to occur with the least total concentration of PCBs. This relationship is most likely for the greater correlation between adverse

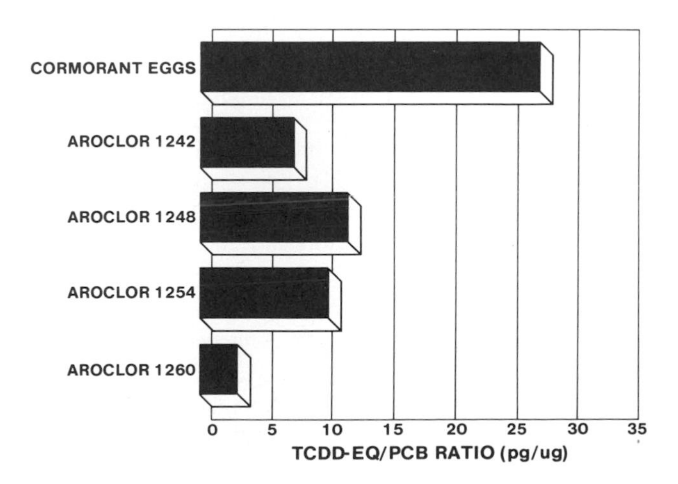

Figure 5. Relative potencies of technical Aroclor mixtures and extracts of double-crested cormorant eggs. The relative potency is the ratio of the concentration of bioassay-derived TCDD-EQ (ng/kg) to total concentration of PCBs (mg/kg). The resulting ratio has units of mg/kg (ppm). (Reprinted from Tillitt *et al.*[23].)

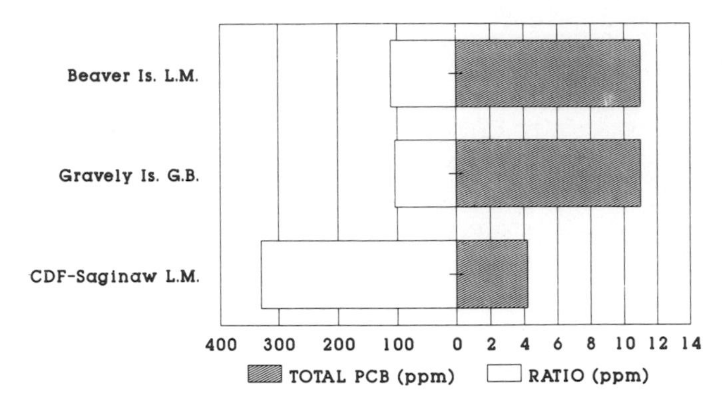

Figure 6. Relative potencies and total concentrations of PCBs in Caspian tern eggs from three locations on the Great Lakes, 1988.

effects and TCDD-EQ than with total concentrations of PCBs (see below). Since the greatest proportion of the TCDD-EQ in birds of the Great Lakes is contributed by the PCBs, there is a general correlation between the concentrations, but there is sufficient variation in the relative potency that a measure of TCDD-EQ gives better prediction of the effects of complex mixtures. The relative potencies (EC_{50} for EROD induction) of extracts from Green Bay ranged from 6 to 56 pg TCDD-EQ/μg PCB, which indicates that total PCB content of a sample is a poor indicator of the biological potency of the toxicity mediated through the Ah-r, even though the measured concentrations of TCDD-EQ were correlated with the total concentration of PCBs.[112] If all of the TCDD-EQ could be attributed to PCB congeners and these congeners were sorted equally in the environment and assimilated and metabolized equally, there would be no significant difference in relative potency among samples, except for the contributions of other compounds.

The smallest PCB-normalized potencies observed in these samples were approximately 10 μg TCDD-EQ/g PCB.[55] This value is similar to the potencies observed for technical-grade PCB preparations in the H4IIE bioassay system.[23,116] However, most of the samples in this study had PCB-normalized potencies that were considerably greater than those of technical-grade PCB preparations.[23]

The greatest enrichment of TCDD-EQ, relative to technical mixtures of Aroclors, is the result of trophic transfers.[54] This is related not only to the presence of non-PCB congeners, but to the enrichment of Ah-r-active congeners, relative to the total concentrations of PCBs. The enrichment of specific p-PCH congeners has previously been demonstrated in water birds.[134–136] This enrichment was greatest for the laterally-substituted congeners which have a relatively great biomagnification potential and are poorly metabolized by most species.[137]

To assess the enrichment of individual PCB congeners within the Great Lakes ecosystem, the relative proportion of the congeners to total concentration of PCBs in bird tissues was compared with that of technical Aroclor mixtures and with samples of fish tissue from Lake Michigan (Table 3).[103–105,122,138] The relative concentrations of

Table 3
Mean, Relative Proportions of Four Planar PCB Congeners in Technical Aroclor Mixtures and Bird and Fish Tissues[a]

Sample	Total PCB (mg/kg)	Relative proportion of congeners (%)			
		77	105	126	169
Bird tissues[112]	4.58	2.28	3.62	0.03	0.00141
Chinook salmon Lake Michigan[122]	1.0	0.22	2.42	0.08	—
Aroclor 1242[103]	—	0.52	—	0.0020	—
Aroclor 1248[103]	—	0.61	—	0.0062	—
Aroclor 1254[103]	—	0.06	—	0.0046	0.08–0.00005
Aroclor 1260[103]	—	0.03	—	0.00080	0.05
Aroclor 1242[138]	—	0.30	0.42	0.003	—
Aroclor 1254[138]	—	0.02	5.49	0.003	—
Aroclor 1260[138]	—	0.001	0.03	0.000	—

[a]Modified from Jones *et al.*[112]

PCB congener 77 (IUPAC) in extracts of bird tissues were the same or greater than those in the technical mixtures of Aroclors 1242, 1248, and 1260. Thus, depending on the relative proportions of these Aroclors in the original mixture released to the environment, this congener may have been enriched, diminished, or stayed the same. Similarly, PCB congener 105 could have been enriched or diminished in the samples, relative to the original technical mixtures depending on the relative proportions of different Aroclors making up the complex mixture. PCB congener 126 was enriched in the extracts of bird tissue regardless of which of the Aroclor mixtures to which they were exposed. The relative contributions of these three congeners to the total mass of PCBs were also greater than those in chinook salmon from Lake Michigan. It is uncertain whether PCB congener 169 was enriched. Values for the relative contribution by weight in Aroclor 1254 ranged from 0.00005 to 0.08. Thus, it is difficult to determine if the value of 0.00141, observed for bird tissues in our study, is an enrichment or diminution (Table 3). The enrichment of these PCB congeners, however, does not explain the discrepancy between the measured and predicted TCDD-EQ, based on the H4IIE TEF values, in the tissue samples. The combined mass contribution of the PCDD and PCDF congeners to the total concentrations of TCDD-EQ was less than 0.5% of the mass of total PCBs present in all samples (Fig. 4).

The relative potencies are different among the original technical Aroclor mixtures (Table 3, Fig. 5). Aroclor 1016, which is Aroclor 1242 with the PCDD, PCDF, and p-PCBs removed, contains essentially no detectable TCDD-EQ. The greatest relative potency is observed in Aroclor 1248 and the least potency is observed in Aroclor 1260 (Fig. 5, Table 3). The relative potency in cormorant eggs from the Great Lakes was more than twice as great as any of the Aroclor mixtures. Thus, the observed enrichment could not be caused by any combination of the original technical Aroclor mixtures. Furthermore, contributions of PCDD and PCDF were not included in these extracts of eggs of fish-eating colonial water birds. Thus, we feel that the greater than predicted potency is primarily related to selective enrichment of the Ah-r-active PCB congeners, but some contribution may be related to the presence of unidentified Ah-r-active compounds as well as PCDD and PCDF congeners.[112]

Biotransformation is the key first step in the elimination of p-PCH congeners and differential biotransformation of PCDD and PCDF congeners by cytochrome P450-mediated pathways has been demonstrated in some species.[139] Both fish and birds have enzymes that are capable of metabolizing PCB congeners.[134,140] However, the activities of cytochrome P450-requiring oxygenase enzymes in fish-eating birds are greater than those in most fishes, but less than those in most mammals.[135,136] Therefore, birds can be expected to eliminate p-PCH, but possibly more slowly than mammals. In general the more substituted congeners tend to be less rapidly metabolized.[134] Also, those congeners that are laterally substituted, such that they do not have two adjacent, unsubstituted carbon atoms are more slowly metabolized[134] and tend to accumulate in animals.[43] PCB congeners with vicinal hydrogen atoms in the *ortho* and *meta* positions with more than one *ortho*-chlorine atom are also resistant to metabolism.[134] The hexachloro biphenyl (2,2′,4,4′,5,5′) was not metabolized by pigeons, rats, or brook trout.[141] More chlorinated congeners can be metabolized if they have adjacent, unsubstituted carbon atoms, whereas those with no vicinal unchlorinated carbons were not metabolized and only slowly excreted.[142–144] This indicates that congeners that were poorly metabolized would not be as readily excreted and would tend to be accumulated selectively in tissues, relative to other congeners, such as the 4,4′-dichloro and 2,2′,5,5′-tetrachloro congeners, which were metabolized to more polar hydroxy metabolites, which could be excreted.

In addition to the effects of metabolism, position in the food chain can affect the relative concentrations of PCB congeners in the tissues and eggs of birds.[145] This is related to many factors, but is primarily related to the relative quantities of different prey items taken by different species or the same species in different locations.

The greater PCB-normalized potency of the p-PCH extracts of avian species from Green Bay, relative to that of Aroclor, indicates that mechanisms of trophic selection or the presence of unidentified Ah-r-active compounds in the extracts could be the cause of the elevated relative potency of the p-PCH mixture. There are two principal ways to correct for changes in potency during a risk assessment: First, each of the active congeners could be quantified and their individual concentrations corrected for relative potencies. Alternatively, an application factor or enrichment factor could be applied to the total concentrations. Since the individual PCB congeners which express the greatest toxic potency have not been measured traditionally and are still seldom considered in regulations, if a constant correction factor can be justified, it could be applied to total concentrations of PCB and correct for the effects of weathering and enrichment through biomagnification. To investigate the assumption that an enrichment factor could be applied to total concentrations of PCBs to correct for different potencies, we calculated water quality criteria (see section below) and then calculated exceedance values for current conditions and compared the exceedances based on total concentrations of PCBs and those based on TCDD-EQ the ratio of the water quality criterion to

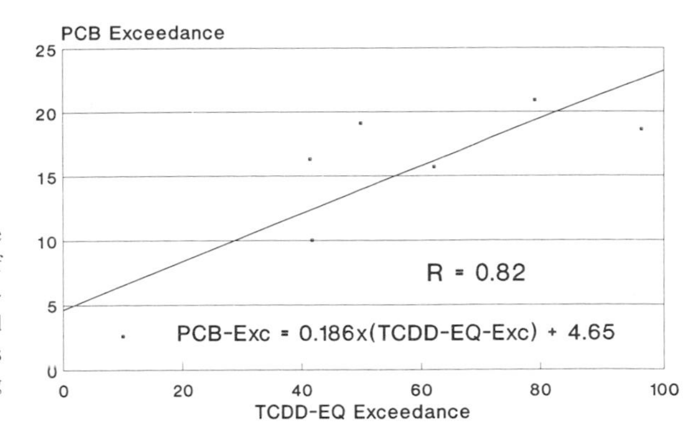

Figure 7. Exceedance values for the effects of total PCBs as a function of the exceedance values in bioassay-derived TCDD-EQ for double-crested cormorant eggs from seven locations in 1988. (Reprinted from Ludwig *et al.*[72]).

protect double-crested cormorants at seven locations, based on total concentrations of PCB and TCDD-EQ (Table 3, Fig. 7).

10. p-PCH OTHER THAN TCDD, TCDF, AND p-PCBs

A number of compounds that are structurally similar to p-PCBs, PCDDs, and PCDFs, and thus should act similarly to these p-PCH, have been observed at concentrations in the environment that, if they have similar toxicological properties, could be of toxicological significance (Table 4). Unfortunately, for many of these compounds, there is currently little known about their toxicological or environmental fate properties. Thus, even if their concentrations in environmental samples were determined instrumentally, it would still be difficult to assess their potential to cause significant adverse effects. It is likely that these compounds do contribute a significant quantity of TCDD-EQ in environmental samples. When TCDD-EQ are measured in the H4IIE bioassay and compared with the TCDD-EQ calculated from instrumental analyses for PCDDs, PCDFs, and PCBs and their TEF values measured in the H4IIE bioassay, we are unable to account for 30 to 50% of the TCDD-EQ measured in the bioassay, especially in samples taken from urbanized and industrialized areas (Fig. 4).

The potency of extracts of peat were found to cause greater induction than could be accounted for by the concentrations of PCDD and PCDF in the extracts.[146] This suggests that there are compounds other than the traditionally measured planar molecules which can be Ah-r-active and thus measured in bioassays of P4501A activity.

Compounds that could contribute to the total concentrations of TCDD-EQ measured in the bioassay also include any or all of the following classes of polychlorinated compounds (Table 4): naphthalenes (PCN), diphenyl ethers (PCDE), diphenyl toluenes (PCDPT), phenoxy anisoles (PCPA), biphenyl anisoles (PCBA), xanthenes (PCXE), xanthones (PCXO), anthracenes (PCAn), fluorenes (PCFl), dihydroanthracenes (PCDHA), diphenylmethanes (PCBM), phenylxylylethanes (PCPXE), dibenzothiophenes (PCDT), quaterphenyls (PCQ), quaterphenyl ethers (PCQE), and biphenylenes

Table 4
Compounds That May, Based on Experimental Evidence or Structure, Be Expected to Have the Potential to Cause Adverse Effects through the Ah-r-Mediated Mechanism of Action

Polycyclic aromatic hydrocarbons
Polychlorinated biphenyls
Polychlorinated dibenzo-p-dioxins
Polychlorinated dibenzo-furans
Polychlorinated naphthalenes
Polychlorinated diphenyltoluenes
Polychlorinated diphenyl ethers
Polychlorinated anisoles
Polychlorinated xanthenes
Polychlorinated xanthones
Polychlorinated anthracenes
Polychlorinated fluorenes
Polychlorinated dihydroanthracenes
Polychlorinated diphenylmethanes
Polychlorinated phenylxylylethanes
Polychlorinated dibenzothiophenes
Polychlorinated quaterphenyls
Polychlorinated quaterphenyl ethers
Polychlorinated biphenylenes
Polybrominated diphenyl ethers
Polychlorinated azoanthracenes

(PCBE). In addition to the chlorinated compounds, brominated and chloro/bromo-substituted analogues of PCDD and PCDF have been found in the environment[147] and are known to induce ethoxyresorufin-*O*-deethylase (EROD) activity *in vivo* and *in vitro*.[50] In addition to the above-mentioned compounds, there are a number of polychlorinated compounds, which are the alkylated forms of these same classes. These include polychlorinated alkylbiphenyls (PCAB), alkylnaphthalenes (PCAN), alkylphenanthrenes (PCAP), and alkyldibenzothiophenes (PCADTh).[148] Alkylated analogues of all of these compounds, including PCDDs and PCDFs, are especially prevalent in sludges and sediments near paper mills that use chlorine in the bleaching process.[149] Polychlorobibenzyls (PCBB) have also been observed in the vicinity of pulp mills.[150]

The PCNs are known to occur in the environment at concentrations that are sufficiently great in some locations that, coupled with their surprisingly high TEF values, could be of toxicological significance similar to that of the PCDD, PCDF, and p-PCBs.[111,151] The TEF values reported by Hanberg[111] for PCN ranged from 0.000007 to 0.002. Total concentrations of PCN in fish from the Great Lakes have been reported to be as great as 5 mg/kg, wet weight. The TEF for the most prevalent PCNs were 0.002. Thus, PCNs could contribute significantly to the total TCDD-EQ. Chlorinated PCNs have been found in great concentrations in the effluents and sludge of pulp and paper processes.[148]

The PCDEs are by-products in the manufacture of chlorinated phenols.[140] Significant concentrations of these compounds have been found in the tissues of humans, fish, and wildlife.[152] PCDEs can be accumulated by fish[153] and induce MFO activity.[154] Furthermore, an unidentified tetrachlorinated PCDE isomer was found at a concentration of approximately 0.9 mg/kg wet wt in the eggs of common terns in Michigan.[155] Concentrations of tri- and tetrachloro-PCDE as great as 3 mg/kg have been found in the tissues of common tern chicks.[156] The PCDEs are similar to PCBs in their dynamics and persistence in animal tissues and induce P-450-type monooxygenase activities. The ED_{50} values for several PCDEs are in the range of 15 to 110 μmol/kg.[152] Unlike PCBs the mono- and dichloro-substituted PCDEs are also potent inducers of EROD activity.[157] It has been postulated that the additional distance between the phenyl moieties reduces the strong effect of *ortho* substitution, which is observed in the PCBs. Thus, PCDEs in the environment may exert a greater effect on the EROD activity than predicted from QSAR relationships developed for PCBs. The PCDEs are rarely monitored in environmental samples because of the paucity of authentic standards and because they occur in the same fraction as the PCDFs and several of the congeners result in the same mass fragments as some of the PCDFs.[152] Thus, since their potencies to induce MFO activities are similar to those for PCBs, PCDEs could account for a significant amount of the induction measured in the H4IIE assay.

The PCDPT are not used in North America, but are manufactured and widely used in Europe as substitutes for PCBs, especially as hydraulic fluids in mining, and have been found in the tissues of aquatic organisms.[158,159] The PCDPT, which are also called diphenylmethanes, are structurally similar to PCBs and cause similar effects.[160] PCDPT are potent inducers of EROD activity in mammals and fish.[159] It is not likely that there are sufficient concentrations of PCDPT in the tissues of North American wildlife to contribute significantly to the TCDD-EQ measured in samples. However, the PCDPT could currently make contributions to the total TCDD-EQ in samples from some locations in Europe[161] and may contribute to the TCDD-EQ measured in North America at some time in the future.

The PCPXEs are known to be potent inducers of EROD activity in the livers of fish,[162] but we are aware of no reports of these compounds occurring in the tissues of fish or birds taken from the environment. Thus, neither the hazard nor the risk presented by these compounds to wildlife can be assessed at this time.

A complete assessment of the possible environmental hazard posed by PCDT is impossible because little is known about their persistence, bioconcentration, or toxicological properties. The PCDT are formed during incineration of organic compounds which contain sulfur and chloro compounds, such as tires.[163–165] The PCDT have a relative potency, as measured in the H4IIE assay, similar to that of the PCDD.[165] There are few reports of the occurrence of PCDT in environmental samples; however, concentrations of the 2,4,6,8-TCDT as great as 8300 and 1000 pg/g wet wt were observed in the tissues of crab and lobster, respectively, from the Elizabeth River, New Jersey. Concentrations of other tetrachlorinated congeners were as great as 500 pg/g in the crabs from the same location. Therefore, it is possible that PCDT could be present locally in sufficiently great concentrations to contribute to TCDD-EQ in environmental samples, including the tissues and eggs of birds from the Great Lakes region of North

America. The TEF for a synthetic mixture of PCDTs (2.4% Cl_2, 74.6% Cl_3, 22.4% Cl_4 and 0.6% Cl_5) was found to be 0.000425 in the H4IIE assay.[120]

PCBE have been little studied, but are structurally similar to the PCDD.[166] PCBE have been found to occur in the environment, generally because of pyrolysis of PCBs.[53] Few authentic standards are available for the PCBE and the concentrations in the environment are generally quite small, compared with those of PCBs. In addition to the chlorinated diphenyl ethers, brominated analogues of these compounds have also been found to occur in the aquatic environment at concentrations that could to toxicologically relevant.[25]

Both PCQs and PCQEs have been found in environmental samples and have been demonstrated to be toxic to animals.[167] However, they are reported to be much less toxic than PCDD or PCDF. Both of these classes of compounds were found in patients who consumed Yusho oil,[167] but it is unlikely that either of these classes of compounds occur in wildlife of the Great Lakes region at concentrations great enough to be of toxicological significance.

In addition to the diaromatic-type compounds, single-ring compounds are known to induce P450-type monooxygenase activity. For instance, hexachlorobenzene (HCB) induces several monooxygenase activities,[56,168] but it is unlikely that it contributes significantly to the concentrations of TCDD-EQ in wildlife since it does not induce the P450IA1-type isozyme which is measured as EROD activity. HCB has been detected in the eggs and tissues of birds in the Great Lakes,[169] but the significance of the concentrations is unknown at this time.

Polybrominated compounds such as PBBs and PBDEs are used in flame retardants such as Bromkal, which contains polybrominated diphenyl ethers (PBDEs).[50] PBBs are known to occur in the environment and can be accumulated into biota.[166] The Bromkal mixture is known to induce EROD activity *in vitro,* but the potency for induction is much less than that of the PCDD, PCDF, or p-PCBs.[111] Since concentrations of these brominated compounds are not generally measured, it is impossible to assess the contribution that these compounds may contribute to the concentrations of TCDD-EQ measured in extracts from biota.

There are a number of polycyclic aromatic hydrocarbons (PAH), compounds that are known to bind to the Ah-r. These compounds may contribute to the toxicity observed in some species. However, it is unlikely that these compounds would be responsible for the effects observed in eggs, since they are rapidly metabolized by vertebrates and thus would not be biomagnified into eggs. Also, since the polar metabolites of these compounds would not be expected to occur in the extracts that are used in the H4IIE assay, it is unlikely that they are responsible for the EROD induction which cannot be accounted for in comparisons of bioassay and instrumental analyses. Because a sufficiently great number of persistent compounds can interact with the Ah-r, it would be useful to use techniques such as the H4IIE bioassay to determine if all of the potential hazardous compounds in complex mixtures have been accounted for.

When the environmental contamination with Ah-r-active compounds is discussed, their impacts are often dismissed because the manufacture of PCBs, which contribute the greatest proportion of the known TCDD-EQ in biological samples, has been discontinued. It is often concluded that nothing can be done about these compounds that have already been released to the environment and that the concentrations of these compounds will eventually decrease to insignificance. While this is partially true, we believe that it is important to remember that a number of other compounds with similar ecotoxicological properties are still manufactured, used, and released into the environment at concentrations that could be of ecological or toxicological significance. Therefore, the use of these compounds should be as tightly regulated as the other p-PCH and their continuing uses reevaluated as more information on their distribution is gained through monitoring. This is especially important if these other classes of compounds are introduced to commerce as substitutes for other members of the more pernicious p-PCH classes.

An implication of these observations for regulators is that discharge limits for a single p-PCH projecting its toxic effects at a given level of discharge must account for the additivity with other p-PCH in the environment to ensure protection of wildlife as well as humans. We know of no case where this policy has been used in the regulatory community which has avoided the issues raised by mixtures and interactions. However, the Great Lakes Initiative developed by the U.S. EPA will take these factors into account.

11. p-PCH EFFECTS ON WILDLIFE IN THE GREAT LAKES REGION

The greatest weakness of all ecoepidemiological wildlife studies when considering cause-and-effect relationships is that all such work is necessarily correlational and nonexperimental. No single chemical cause can be isolated in wildlife and tested for effects, as is done in laboratory situations that can control variables or test each variable singly or in combination. Ideally, one would apply Koch's postulates to wildlife contamination problems in the search for cause-effect relationships (Table 5). However, for studies of wildlife under field conditions, this would be difficult. This is true because it is difficult to conduct controlled laboratory studies with wildlife species and it is difficult to conduct studies with sample sizes that are large enough to allow sufficient statistical power to test hypotheses about effects, such as deformities. Also, it is difficult to conduct laboratory studies of known exposures to the same complex mixtures to which wildlife is exposed under field conditions. We have conducted several studies with animal models, which have been fed fish from the Great Lakes, in

Table 5
Koch's Postulates

Koch's Postulates
1. Observe effect(s)*
2. Correlate to cofactor*
3. Isolate suspected causative agent*
4. Identify suspected causative agent*
5. Introduce suspected causative agent and elicit effect

*Indicates that the postuate has been completed.

an attempt to simulate more natural exposures. However, the logistics for such studies are difficult. Thus, the task is to correlate and then compare with other similarly and differently contaminated populations and species. The best that can be achieved is to consider the weight of evidence and reconcile effects observed in wild populations with controlled laboratory studies done with animal models. These studies can be made more powerful by using both *in vivo* and *in vitro* studies of complex mixtures or fractions into which certain p-PCH have been isolated or enriched. Fox[170] has formalized the six criteria to be used in determination of the validity of ascribing chemical causes to effects in epidemiological studies of wildlife populations. These include consistency of observations, strength of the association, specificity of the association, time sequence, coherence, and the predictive power of the relationship. On these bases, an informed judgment can be made that p-PCH have an influence on populations of wildlife species with a great degree of certainty.[9,25,73] Here we present what we feel is the strongest information supporting the hypothesis that the p-PCBs are the most likely cause of the observed effects on colonial water birds in the Great Lakes region of North America.

All of the symptoms observed in colonial birds of the Great Lakes region are known to be caused by the p-PCH. Thus, even though some of the details of the mechanisms of action remain undescribed, the widespread common effects of the p-PCH on wildlife worldwide, and especially in areas of greatest exposure to p-PCH, suggest largely additive effects of these substances on exposed wild populations. While these phenomena are not yet studied at the ecosystem or community level, it seems likely that at least some of the recent decreases in sizes of populations, extinctions, or shifts in species dominance are at least influenced, if not caused in their entirety, by p-PCH. For example, in the Great Lakes region, the smaller tern species, such as the common (*Sterna hirundo*), Forster's (*S. forsterii*), and black (*Chilodonias niger*), were once more abundant in the region. However, in the last 30 years populations of all three species have decreased and now each is listed as a threatened or endangered species in one or more states. Gulls and cormorants, which have larger body sizes and thus lesser energy requirements per unit body weight, are more tolerant of dioxinlike planar contaminants and their populations have continued to increase since the concentrations of DDE have decreased to below critical levels.[70,119] It is possible that the interspecies dynamics and wildlife populations balance on the Great Lakes are affected as much by contaminants as by other traditional habitat parameters. The p-PCH are prime suspects in these phenomena. These selective effects are subtle and may be difficult to separate from other dynamics of the ecosystem.

12. GLEMEDs

Of all of the adverse effects observed in colonial water birds of the North American Great Lakes region, the most obvious and that which can be most directly related to survival of individuals and population-level effects are embryo lethality and developmental deformities. Most of the embryos or chicks that die during early development

Figure 8. Deformed bill of double-crested cormorant from Green Bay, Wisconsin. (Photo by J. P. Giesy, 1991.)

have been observed to also have developmental deformities,[9,24,26,118] particularly abnormalities that are of ectodermal origin.[171] One of the best documented abnormalities that has been correlated with concentrations of p-PCH in bird eggs is the crossed-bill syndrome (Fig. 8) in North American cormorants.[25] This suite of conditions found in Great Lakes wildlife has been named the GLEMEDs syndrome (Great Lakes embryo mortality, edema, and deformity syndrome)[9]; it mimics chick edema disease caused in offspring of hens exposed to PCDD and PCDF in their feed.[29] The effects observed in birds are similar to those observed in mammals exposed to these chemicals.

The few available case studies illustrating the effects of these chemicals are the reports of the Canadian Wildlife Service on studies performed on herring gulls living on Lake Ontario during the 1970s, particularly between 1974 and 1977. In the late 1960s, anecdotal evidence circulated among field biologists of poor egg hatchability of Lake Ontario herring gull eggs.[13] Official Canadian Wildlife Service surveys began in 1971. Hatchabilities of less than 20% were found at some colonies in Lake Ontario. Productivity was reduced to less than one fledged young per ten nests. Maintenance of a stable population requires fledging rates in the range of five or six chicks per ten nests per year for this species. Initial examination of herring gull eggs and eggs of other species in Lake Ontario documented the presence of DDT and PCBs. However, analytical techniques at the time were insufficient to discriminate among dioxin, furan, and p-PCB congeners or to quantify some of the more toxic p-PCB congeners. Reliable congener-specific chemistry was a decade away. The characteristic symptoms in surviving chicks were similar to those of chick edema disease, which is caused by PCDDs in poultry and had been previously described.[29] Subsequent reanalysis in the late 1980s of a variety of eggs that had been collected from herring gull colonies in Lake Ontario in the 1971–1976 period and archived contained 1–3 ppb of actual TCDD. This TCDD is thought to have been discharged into the Niagara River as a result of herbicide manufacture.[172] This chemically caused epizootic in Lake Ontario is probably the best documented example of *dioxin-caused* effects on wildlife. During the last decade, the symptoms of chick edema disease and GLEMEDS have decreased significantly in the herring gull population of Lake Ontario,[9] but more subtle biochemical effects persist in all species of fish-eating colonial water birds of Lake Ontario and the other Great Lakes.[26,73]

Similar studies of nesting Forster's terns, conducted in Green Bay from 1983 to 1988,[173] and double-crested cormorants and Caspian terns[118] have revealed a similar suite of biological effects, but implicate different p-PCH than TCDD as probable causes. In the Green Bay experience, Kubiak *et al.*[11] found a variety of developmental deformities in the embryos and chicks of Forster's terns including growth deficiencies, deformities, and behavioral differences in parental care of eggs compared with an inland control colony where exposures were significantly less than on Green Bay. Extrinsic adult behavioral abnormalities of inconsistent incubation led to a 4-day-longer incubation period than the reference colony. Reciprocal transplant studies of eggs detected a similar time delay ascribed to toxic substances in the eggs. That study suggested widespread, complex contaminant effects on the reproductive cycle, such as longer incubation times, smaller individuals, and wasting syndrome in those that did hatch. Similar, but less acute problems have been observed in double-crested cormorants and Caspian terns in the upper Green Bay area, where TCDD-EQ calculated from concentrations of p-PCH and TEFs, from 175 up to over 440 ppt have been measured.[24,118]

A study of Caspian terns in Saginaw Bay following the disturbance of sediments in a flood incident has documented concentrations of p-PCB in eggs which was similar to those observed in Green Bay. In this case, between 96 and 98% of the TCDD-EQ was related to the presence of only four p-PCB congeners. These were, in order of relative contribution to the total TCDD-EQ, congeners 77, 126, 169, and 105.[86,118] The developmental effects observed in Caspian terns at Saginaw Bay were severe. Caspian tern eggs contained doses equal to the lethal concentration for 95% of white leghorn chicken eggs for p-PCB congeners 77 and 126, plus 10% of the LD_{50} dose of TCDD. The concentrations of five individual p-PCB congeners reported as TCDD-EQ were more than an order of magnitude greater than that which has been found to cause heart defects in developing chickens.[174] Thus, it is likely that the observed concentrations of TCDD-EQ were sufficient to cause the adverse effects observed. Even though the dioxinlike p-PCB congeners have relative potencies which are less than that of TCDD, they are hundreds to as much as a thousandfold more abundant in the environment and thus can have toxic effects on the wildlife species.

Egg mortality from locations in five geographic regions on and one off the Great Lakes was found to be directly proportional to the concentration of TCDD-EQ (Fig. 9).[23] The current concentrations of both PCBs and TCDD-EQ in cormorant eggs are greater than the estimated no-effect concentrations for this species. The concentrations

Table 6
Concentrations of PCBs (Total), TCDD-EQ, Relative Toxic Potencies, and Egg Lethality for Eggs of Double-Crested Cormorants in the Great Lakes Region

Location	Egg		Relative potency [ng TCDD-EQ/kg / mg PCB/kg]	Potency ratio to Aroclor 1242	Egg lethality % at 23 days
	Total PCB (mg/kg)	TCDD-EQ (ng/kg)	(ng/mg; ppm)		
Pigeon Is. Lk. Ontario	5.5[175]	217[177]	39.5	3.99 : 1	28[170]
Spider Is. Green Bay	6.5[175]	337[177]	51.9	5.34 : 1	37[179]
Little Gull Is. Green Bay	7.3[176]	277[177]	37.9	3.83 : 1	34[179]
Gull Is. N. Lk. Michigan	6.7[176]	175[177]	26.1	2.64 : 1	27[179]
St. Martin's Lk. Huron	5.7[176]	145[177]	25.4	2.57 : 1	20[179]
Tahquamenon Is. Lk. Superior	3.5[176]	146[177]	41.7	4.21 : 1	19[179]
Lk. Winnepegosis Manitoba	0.9[175]	35[177]	38.5	3.89 : 1	8[179]
Mean potency	—	—	37.7 : 1	3.77 : 1	—

of TCDD-EQ were determined by the H4IIE assay in double-crested cormorant eggs from several locations in 1986–1988.[23] The total concentrations of TCDD-EQ, as well as the relative potencies, varied among locations (Table 6, Fig. 6), but were fairly consistant across all of these regions. The death of eggs was found to be directly proportional to the concentration of TCDD-EQ in the eggs (Fig. 9)[23] with a significant positive correlation ($R^2 = 0.703$; $p < 0.0003$). These estimates are based on actual

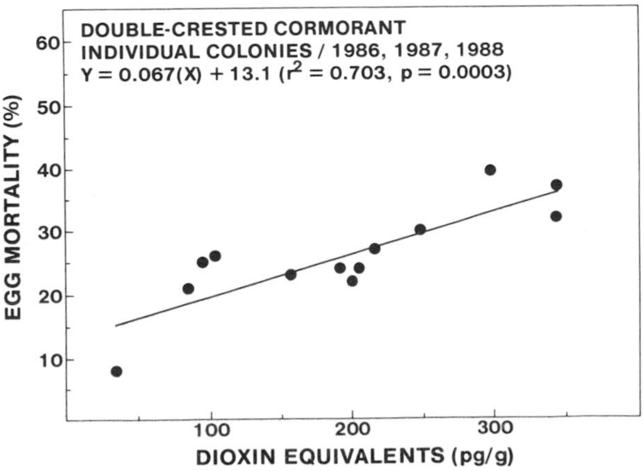

Figure 9. Egg mortality of double-crested cormorants from 12 locations on and one location off of the Great Lakes as a function of TCDD-EQ. (Reprinted with permission from Tillitt *et al.*[23])

Table 7
Common Congenital Deformities Observed in Colonial Fish-Eating Water Birds of the Great Lakes

Crossed bill
Clubfoot
Hip dysplasia
Dwarf appendages
Ascites/edema
Eye deformities
Brain deformities
Skull bones
Gastroschisis

doses in the eggs (Table 6). Thus, there would be less error in predicting these effects than would be expected if they were calculated from the WQC, which include more assumptions.

We investigated the relationships between birth defects observed in both Caspian terns and double-crested cormorants at a number of locations in the Great Lakes. For this study we classified and enumerated the number of various types of deformities in both double-crested cormorants and Caspian terns (Table 7, Fig. 10). We found the rates of abnormalities to range from 2 to 12 per thousand chick embryos and chicks examined (Fig. 11). We observed the greatest rate of deformities in double-crested cormorants to occur in Green Bay (Fig. 11). The greatest percentage of deformities of Caspian terns occurred in the colony on the contained disposal facility (CDF) in Saginaw Bay. When the rates of deformities were correlated with the concentrations of TCDD-EQ, we found very strong, statistically significant correlations (Fig. 12). While this observation alone does not make a cause-and-effect relationship, this correlation is better than that with any of the other contaminants. Concentrations of TCDD-EQ and the relative potencies of the p-PCH mixtures in cormorant eggs varied among locations in the Great Lakes (Figs. 6 and 13), but the greatest concentrations were measured in

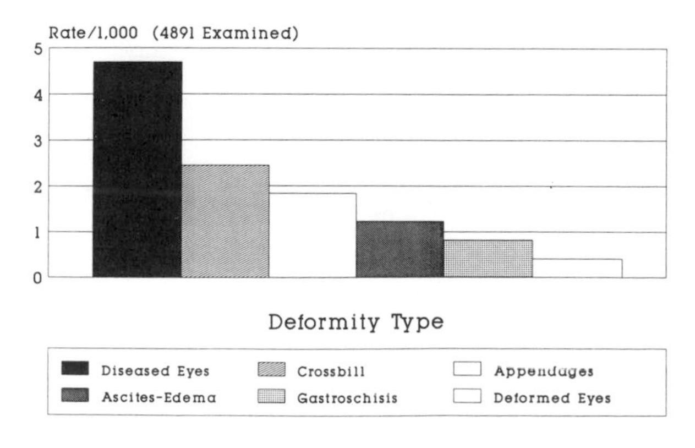

Figure 10. Occurrences (per 1000 chicks examined) of different types of deformities in the embryos and chicks of double-crested cormorants in Green Bay, Lake Michigan, 1986–1989.

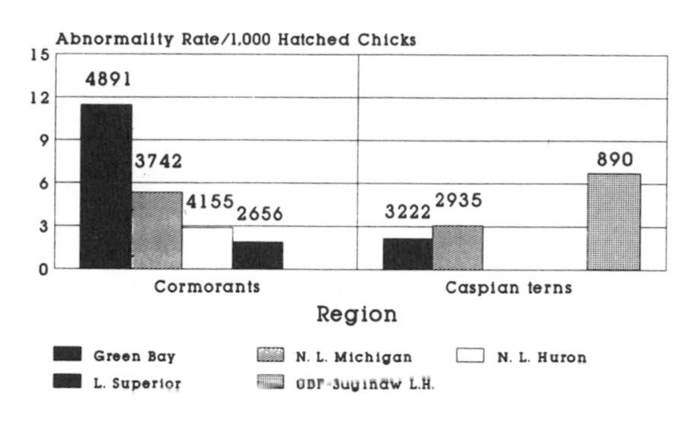

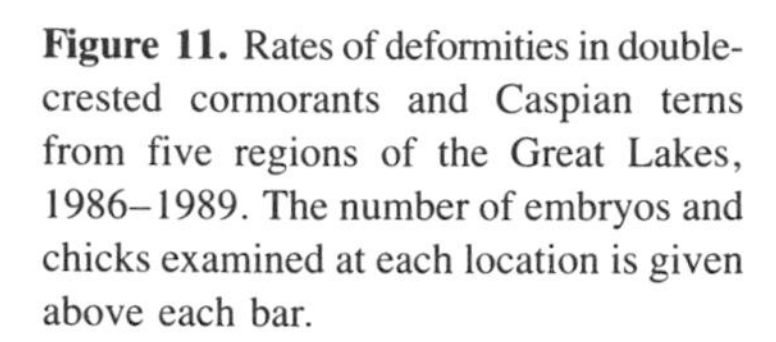

Figure 11. Rates of deformities in double-crested cormorants and Caspian terns from five regions of the Great Lakes, 1986–1989. The number of embryos and chicks examined at each location is given above each bar.

Figure 12. Rates of deformities in double-crested cormorants and Caspian terns in the Great Lakes, 1988, as a function of TCDD-EQ.

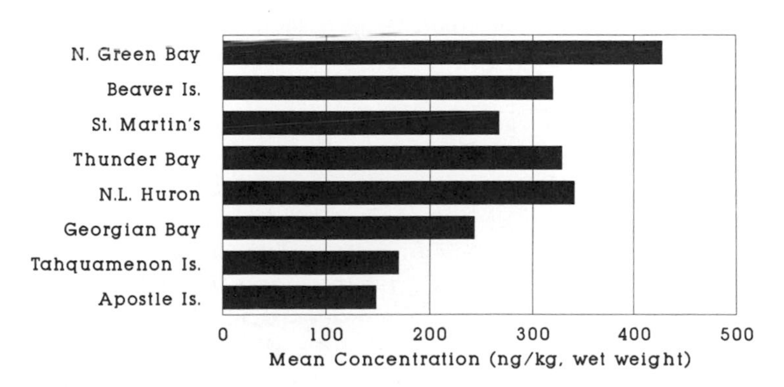

Figure 13. Concentrations of TCDD-EQ in the eggs of double-crested cormorants from eight locations in the Great Lakes, means of 1986, 1988, 1989.

eggs collected from Green Bay,[86] where the greatest rates of deformities and poorest survival of eggs were observed.[25]

13. WILDLIFE HAZARD ASSESSMENTS FOR PCBs

Here we provide a proposed method for conducting wildlife hazard assessments, by using results of field and laboratory studies on target and surrogate domestic species, along with environmental monitoring and chemical analyses, to determine the appropriate water quality criteria (WQC) to protect sensitive wildlife species from the adverse effects of PCBs. We then compare the results of this assessment to an assessment that used human cancer and reproductive effects as endpoints. We based our hazard assessment on PCBs instead of 2,3,7,8-TCDD, because the PCBs have been found to be the primary source of most of the TCDD-EQ for which concentrations in biota have been determined. We conducted wildlife hazard assessments by comparing the threshold for effect with the concentrations of key contaminants that are currently observed in tissues of fish that are eaten by domestic species and eggs of wild birds in the Great Lakes basin. The thresholds for effects, termed the lowest observable adverse effects level (LOAEL), were derived from field observations and correlations with concentrations of key toxicants or from controlled laboratory studies where domestic species of interest were fed known quantities of contaminants in Great Lakes fish. We studied four species of wild birds: herring gulls (*Larus argentatus*), bald eagles (*Haliaeetus leucocephalus*), Caspian terns (*Sterna caspia*), double-crested cormorants (*Phalacrocorax auritus*), and a domesticated surrogate species, white leghorn chickens (*Gallus gallus*)[181] (Tables 8 and 9). The no observable adverse effects level (NOAEL) values were defined to be 10% of the LOAEL [safety factor of 10; Eq. (1)] to correct for the uncertainty in determining the NOAEL from the LOAEL. The LOAEL can be reported

$$\text{NOAEL} = \text{LOAEL} \times 10^{-1} \tag{1}$$

as the reference dose (Ref_t = dose in the target tissue), which is defined as a threshold concentration (mg/kg, wet weight or lipid weight) in a tissue such as that of egg, which causes a defined adverse effect, such as egg lethality or birth defects (Table 10). Alternatively, the dose can be given as the daily intake or by making the appropriate assumptions about food consumption, as the concentration in food (Ref_f = dose in food). Here, LOAEL values are reported as the concentration either in bird eggs or in the food of chickens (Table 10).

The bioconcentration factor (BCF) was used to predict the concentration of p-PCH in water that would result in the reference dose in whole fish tissues, which these animals eat (Table 10). The biomagnification factor (BMF) was used to predict the concentration in water that would result in the reference dose in eggs (Table 10).

Water quality criteria to protect wildlife were derived by dividing the NOAEL by the bioconcentration or biomagnification factor (BCF or BMF) for the species of

Table 8
Lowest Observable Adverse Effect Levels (Concentrations) of PCBs in Bird Eggs or Fish Tissue for Wildlife Species Inhabiting the Great Lakes Region

Species/location	Adverse effect	PCB LOAEL (mg/kg) (wet wt)	Ref.
Bald eagles, Great Lakes Territories	Egg lethality	4.0	180
Herring gulls, Great Lakes	Embryonic deformities egg lethality	5.0	178
Caspian terns, Saginaw Bay, 1988	Egg lethality 21% embryonic deformities	4.2	24
Double-crested cormorants, Lake Superior, 1986–1991	Egg lethality twice as great as control	3.5	23,24
White leghorn chickens, MSU Lab feeding study	Embryonic deformities 42% greater than control	0.294	181

interest (Table 10).[70] For piscivorous fishes we did not attempt to separate the proportion of the contaminants observed in the tissues which was derived directly from the water and that which was obtained from the food. We reasoned that in a food chain exposed to chemicals in water, a ratio that related the concentration of the chemical in the fish to that in the water could be derived. This included both vectors of accumulation, because the prey consumed by the predatory fish would also be in steady state conditions with the concentrations in the waters. We note that, in addition to fish, the

Table 9
Concentrations of 2,3,7,8-TCDD-EQ to Cause Effects in Wildlife Species

Species	Endpoint	TCDD-EQ concentration (pg/g, pptr)
Herring gull—field[182]	LD_{50} egg	2000<>1000
Double-crested cormorant—field[179]	LD_{100} egg	1029
White leghorn chicken—lab[32,81]	LD_{100} adult	1000
Caspian tern—field[179]	LD_{50} egg	750
Herring gull—field[177]	LD_{19} egg	557
Double-crested cormorant—field[23]	LD_{50} egg	460
Caspian tern—field[179]	LD_{35} egg	416
Double-crested cormorant—field[175]	LD_{37} egg	344
Double-crested cormorant—field[179]	LD_{27} egg	217
White leghorn chicken—lab[174]	LD_{50} adult	140
White leghorn chicken—lab[178]	200% increase in heart defects	65
Double-crested cormorant—field[174]	LD_{8} egg	35
White leghorn chicken—lab[178]	LOAEL for heart defects	6.4

Table 10
Reference Doses and Bioconcentration Factors Used to Calculate WQC for PCBs to Protect Great Lakes Wildlife Species

Species (tissue or diet)	Reference dose (NOAEL or LOAEL $\times 10^{-1}$) (mg PCB/kg wet wt)	BCF_c or BMF_c[a]	Water quality criterion (pg/liter, ppq)
Bald eagle (egg)	4.0×10^{-1}	4.0×10^{8} (BMF_c)	1.0
Herring gull (egg)	5.0×10^{-1}	1.6×10^{7} (BMF_c)	31.0
Caspian tern (egg)	4.2×10^{-1}	2.4×10^{7} (BMF_c)	17.0
Double-crested cormorant (egg)	3.5×10^{-1}	1.5×10^{7} (BMF_c)	23.0
White leghorn chicken (diet)	2.9×10^{-2}	1.6×10^{7} (BCF_c)	2.0

[a]BCF_c = BCF from water to fish corrected for selective enrichment of non-*ortho*-substituted PCB congeners. BMF_c = biomagnification factor from fish to bird egg corrected for selective enrichment.

food web includes other fish-eating wildlife. This fact is usually ignored in setting WQC and thus the BMF of toxic chemicals can often be underestimated. We have calculated the reference dose for water (Table 10) by

$$R_w = R_f/BAF_{w\text{-}t} \tag{2}$$

where R_w is the reference dose in water and $BAF_{w\text{-}t}$ is the bioconcentration factor from water to fish tissue corrected for relative potency and biomagnification from one fish to another. WQC were then developed by applying the appropriate application factors. One application factor used in these studies was a 10× factor to estimate the NOAEL from the LOAEL. This factor was applied for the WQC for each species. The final WQC for PCBs was corrected for uncertainty in among-species sensitivities by the application of an additional 10× uncertainty factor. The final WQC for TCDD-EQ was estimated, based on accumulation by eagles, but not corrected for any uncertainty factors.

It is difficult to determine WQC because of the uncertainty associated with estimates of both exposure and the dose–response relationships. Estimates of the factors required to predict the probable accumulation of PCBs or TCDD-EQ from water are particularly uncertain. First of all, the relative proportion of these compounds that are freely dissolved in the water or bioavailable, is impossible to know. Furthermore, the forms of these compounds are continuously changing and never at steady state, which are the conditions under which most predictions are made. We did not use a correction factor for the bioavailable fraction of PCB or TCDD-EQ. We made the conservative assumption that all of the PCB or TCDD-EQ were bioavailable. In reality, the bioavailable fraction is probably from 1 to 10% of the total. For this reason, there is probably at

Table 11
Concentrations of PCBs in Waters of Great Lakes in 1986

Location	Dissolved (ng/liter, pptr)	Particulate (ng/liter, pptr)	Unfiltered water (ng/liter, pptr)
Lk. Superior	0.5[183]	—	0.337[184]
Lk. Michigan	1.4[185]	0.6[185]	1.8[186]
Lk. Huron	0.7[185]	0.3[185]	0.631[183]
Lk. Erie	0.7[185]	0.3[185]	1.378[183]
Lk. Ontario	0.6[185]	0.3[185]	1.41[183]

least a 10- to 100-fold safety factor in our estimates of NOAEL. Since it is difficult to know what portion of the PCB or TCDD is available at any given time. Since there may be an equilibrium between readily available and bound PCB or TCDD-EQ such that there is a continuous source of these compounds in the bioavailable fraction, we chose to not correct for this factor. The concentrations of PCBs measured in the dissolved and particulate fractions of the Great Lakes (Table 11) were positively correlated with the total concentrations of PCBs in the eggs of double-crested cormorants. This indicates that regardless of the relative available fraction the concentrations in the eggs of fish-eating birds are proportional to the concentrations of PCBs in the water. Furthermore, since the BCF and BMF values used in our assessment were derived from field observations, they are more likely to be apparent BCF and BMF values, which take into account the bioavailable fraction.

The values reported for bioaccumulation factors are influenced by bioavailable fraction as well as the physiology and food habits of the species for which BCF or BMF are to be predicted. Laboratory studies with fish generally do not include the accumulation of p-PCH from food and thus must be corrected for what would be expected to be accumulated from food, if water exposures are used to predict the BCF.[179] Better estimates of BMF are available, based on field observations. For an accurate estimate of the exposure from consumption of contaminated food, in the absence of observations of the actual concentrations in food, site-specific estimates of accumulation potentials are needed. Ideally, the exposure dose, expressed as mass of toxicant, per unit mass of organisms per unit time should be known, e.g., pg TCDD-EQ/g body wt per day. For these estimates, in addition to knowledge of the concentrations of the contaminant of interest in each of the dietary items, one would need to have an estimate of the proportion of each item taken in the diet and a conversion value for accumulation efficiency for each compound from each food item. At present, this level of resolution cannot be attained. Therefore, we used average values for accumulation and assumed that the efficiency was the same from all types of food and that the concentrations in food were uniform. Furthermore, we have assumed that the diet consists solely of food items with the specified concentration of toxicant of interest. These assumptions can clearly add uncertainty to the estimates. When possible, it is more accurate to estimate hazard potential from assessments based on concentrations of contaminants in the tissues of prey. We have taken this approach in our hazard assessment, where possible.

The reference doses of complex mixtures in wildlife are difficult to relate to exposures in food or water. The information on toxicity, reported in the literature, is often based on single exposures and may be via injection. Thus, it is difficult to determine the effects of longer-term, continuous exposures in food. For this reason, we have, where possible, used the results of feeding studies with Great Lakes fish. These exposures are confounded by the fact that there are complex mixtures of multiple toxicants in the fish which could influence the response to PCBs or TCDD-EQ. We have chosen to base the hazard assessments on reproductive endpoints, since it is thought that these should be as or more sensitive than effects on adults. For birds we have, thus, based the hazard assessments on the concentrations of PCBs or TCDD-EQ in eggs, which have been associated with observable adverse effects. Alternatively, we have related the effects levels to doses in food, which was fed during longer-term exposures. The hazard assessment for current conditions is reported as exceedance values, which are the ratio of current water concentrations to the WQC:

$$\text{Exceedance} = \text{Concentration in water/WQC} \quad (3)$$

Subsequent to completing the hazard assessment for TCDD-EQ, the concentrations of PCBs in fish tissues and bird eggs were corrected for the change in relative potency, because of a greater concentration of TCDD-EQ per unit PCB in the tissues of fish and bird eggs than technical PCB mixtures (see Section 9). In addition to the total concentration of PCBs, we conducted a hazard assessment with the concentration of TCDD-EQ.

14. WQC FOR PCBs

We calculated the WQC to protect sensitive species of birds from the adverse effects of PCBs to be approximately 1.0 pg PCB/liter (part per quadrillion) (Table 10). In 1986, the year for which the most recent data are available, the total concentrations of PCBs in all of the Great Lakes ranged from 1 to 10 ng/liter (pptr) with the greatest concentrations occurring in Lakes Michigan, Ontario, and Erie (Table 11). The concentrations of PCBs in the water of all five of the Great Lakes in 1986 exceeded our proposed WQC for all four species studied (Table 12). The degree to which the WQC were exceeded varied from a minimum of 11 for herring gulls living on Lake Superior to a maximum of 1800 for bald eagles living on Lake Michigan (Table 12). The exceedances were the least for all species for Lake Superior and tended to be the greatest for Lake Michigan, followed by Lakes Erie and Ontario (Table 12). The exceedence values were calculated for both total PCBs and TCDD-EQ in eggs of double-crested cormorants from seven locations in the Great Lakes (Table 13). In general, the available concentrations of PCB corrected for enrichment by use of the H4IIE enzyme induction bioassay, predicted to be in water were approximately three orders of magnitude greater than the NOAEL.

The proposed WQC to protect wildlife from the adverse effects of PCBs range over five orders of magnitude (Table 14). The value of 17 pg/liter proposed by the Great

Table 12
Exceedance[a] Values for Total Concentrations of PCB for Four Species in the Great Lakes Region Based on Consumption of Fish of Average PCB Content from Each of the Five Great Lakes

Lake	Eagle	Herring gull	Caspian tern	Double-crested cormorant
Superior	340	11	20	15
Michigan	1800	58	110	78
Huron	630	20	37	27
Erie	1400	43	81	60
Ontario	1400	45	83	61

[a]Exceedance = (PCB in water)/WQC: Exceedance values are based on total concentrations of PCBs for each lake (Table 11) and average concentrations in fish resulting from exposure to this concentration.

Lakes Initiative[188] is approximately a factor of 10 greater than what we have proposed as a protective value.

15. WQC FOR TCDD

There is a range of sensitivity to TCDD and TCDD-like compounds among different species of birds.[82] Effect concentrations of TCDD-EQ in birds range from

Table 13
Exceedance[a] Values for Total PCBs and TCDD-EQ in Double-Crested Cormorant Eggs

Location	PCB exceedance	TCDD-EQ exceedance
Pigeon Is. Lk. Ontario	15.7	62.0
Spider Is. Green Bay	18.6	96.3
Little Gull Is. Green Bay	20.9	79.1
Gull Is. N. Lk. Michigan	19.1	50.0
St. Martin's Lk. Michigan	16.3	41.4
Tahquamenon Is. Lk. Superior	10.0	41.7
Lk. Winnepegosis Manitoba	2.6	10.0

[a]Exceedance = concentration of PCB or TCDD-EQ in egg tissue/(Ref_t).

Table 14
Water Quality Criteria for PCBs Recommended by the Great Lakes Initiative[188] Based on Three Different Endpoints

Endpoint	Water quality criterion (pg/liter, ppq)
Human cancer—drinking water	3.0
Human cancer—nondrinking water	3.0
Human health—fish consumption	79
Human health	20
Human health—drinking water	5×10^5
Wildlife	1.7×10^6
Wildlife	1.0×10^3
Wildlife	17

approximately 10 to 500 pg/g body wt.[80,81,187] Clearly, the chicken is a very sensitive species. In fact, when concentrations of TCDD-EQ in the eggs of wild fish-eating birds are compared with the effect concentrations for birds it would be predicted that none of the wildlife species would be able to successfully reproduce in the Great Lakes and that for some species the LC_{99} would be exceeded. At the other end of the spectrum is the pheasant.[28] Since no direct studies of the toxicity of TCDD on wild fish-eating birds were available, we used the information on the toxicity of TCDD to chickens to derive a WQC, which should be sufficient to protect all wildlife species (Table 15). We also used estimates of the no-effect concentrations in birds from our field monitoring studies (Tables 13 and 15).

The WQC was calculated from the bioaccumulation factor and the NOAEL for several species, including the chicken. The product of the BAF used in our hazard assessment was 4.15×10^4. There is a great degree of uncertainty in estimating the

Table 15
Water Quality Criteria for TCDD-EQ Recommended for the Great Lakes, and Concentrations of TCDD-EQ in Fish to Protect Birds and Wildlife

Species (tissue or diet)	Reference dose (NOAEL) (ng/kg, wet wt)	BAF[a]	Water quality criterion (pg/liter, ppq)
Herring gull (egg)	10[b,177]	4.2×10^4	0.3
Caspian tern (egg)	7.5[b,177]	"	0.2
Double-crested cormorant (egg)	4.6[b,177]	"	0.1
Bald eagle (egg)	1.5[173]	8.4×10^6	0.0002
White leghorn chicken (egg)	1.5[c,181]	4.2×10^4	0.04

[a]BAF, bioaccumulation factor (BCF × BMF), where BCF_c = BCF from water to fish for TCDD-EQ and BMF_c = biomagnification factor from fish to bird egg for TCDD-EQ.
[b]$LC_{50}/100$.
[c]LOAEL/10.

accumulation of TCDD from water. In their comprehensive assment of TCDD, Cook *et al.*[187] were able to remove some of the uncertainty in estimates of accumulation of TCDD from water to fish by correcting for organic carbon content of the water and lipid content of the animal to which the TCDD was to be accumulated. Unfortunately, there are few estimates of TCDD concentrations in water or in fish tissue, which provide the necessary information to make these corrections. The concentration of TCDD-EQ in forage fish from Lake Huron is approximately 10 pg TCDD-EQ/g, ww, while that in large lake trout is approximately 350 pg TCDD-EQ/g, ww.[55] Thus, the BAF from forage fish to predators is approximately 10×. This is also approximately the BMF (within a factor of 2) for persistent, neutral organic compounds with a molecular weight similar to TCDD. Since fish-eating birds would be more likely to eat forage fish than the large lake trout and the BAF for lake trout was used in the calculations, no correction was made for biomagnification from fish to bird.

The WQC for TCDD-EQ, based on the white leghorn chicken, would be 0.04 pg TCDD/liter (Table 15). That for the herring gull, which is known to be more tolerant of the effects of TCDD, was estimated to be 0.3 pg TCDD/liter. This level of protection takes into account the potential accumulation of TCDD-EQ into higher trophic levels. For instance, birds that eat larger fish, which have greater concentrations of TCDD-EQ, or other birds would not be adequately protected. Specifically, concentrations of TCDD-EQ in eagle eggs from Thunder Bay, Lake Huron, have been found to be as great as 2000 pg TCDD-EQ/g. Since the concentrations of TCDD-EQ in forage fish from this location contained an average of 10 pg TCDD-EQ/g, ww, the BMF from fish to eagle egg would be approximately 200. Thus, the WQC would need to be approximately 200 times less to protect eagles and colonial fish-eating water birds, if the eagles are approximately as sensitive to the effects of dioxins as are the colonial birds. To be sure of the protection of eagles we have selected the chicken as a surrogate. If the WQC to protect chickens is used and a 200-fold BMF is applied, the WQC to protect eagles would be approximately 0.0002 pg TCDD-EQ/liter.

The values determined to represent a "low" risk to avian species by Cook *et al.*[187] were based on the effects on pheasants and made different assumptions about the degree of availability of the TCDD in water. Even though our assessment used different assumptions, the results of our analysis of colonial fish-eating water birds were similar to those predicted by the U.S. EPA (Table 16).

Table 16
Water Quality Criteria and Fish Quality Proposed by Cook *et al.*[179] to Protect Humans and Wildlife from Adverse Effects of TCDD-EQ in the Great Lakes

Risk	Fish (pg TCDD/g, ww)	Water (pg TCDD/liter)	
		POC[a] = 0.2 mg/liter	POC = 1.0 mg/liter
Low	6	0.07	0.35
High	60	0.7	3.5

[a]POC = organic carbon content of water.

Table 17
Dietary NOAEL, Based on TCDD-EQ in Fish Fed to White Leghorn Chickens[181]

Dietary NOAEL (pg/g, ww)	Concentration of TCDD-EQ in forage fish	Exceedance
0.6	10	16.7

Because of the potential uncertainties associated with determining WQC and because birds accumulate essentially all of their exposure to TCDD-EQ from their food,[195] to assess current conditions we compared the current concentrations of TCDD-EQ in forage fish from Lake Huron[55] to the dietary NOAEL for white leghorn chickens.[181] When this was done, it was found that the current exceedance value is approximately 17 (Table 17). This is a conservative estimate of the current situation, since larger fish, which could also be taken in the diet, contain greater concentrations of TCDD-EQ.

This exercise seems to support the use of the NOAEL calculated from the chicken since the exceedance is in the range that one would expect to see subtle effects on the colonial water birds and more severe effects on birds, such as eagles, which are of a higher trophic leavel. This is, in fact, what is observed: There are subtle effects on survival of embryos and deformities observed in the colonial birds, while eagles, which feed on fish from this location, fail to reproduce.

Based on our field observations, the value of 6 pg TCDD-EQ/g ww of fish, which is given as "low risk" by the U.S. EPA hazard assessment,[187] is probably appropriate for the protection of double-crested cormorants, but would not be sufficient to protect higher trophic levels such as eagles and may not be protective of some of the smaller, more sensitive species of colonial fish-eating birds. We have observed a concentration of approximately 10 pg TCDD-EQ/g ww of forage fish which is similar to the concentration indicated by the EPA[187] to be a small risk to birds. In these areas we have observed adverse effects, such as deformities and embryolethality, but these effects do not seem to be limiting populations of double-crested cormorants. In these areas, wasting syndrome has been observed in Caspian terns and the Caspian terns cannot reproduce normally in some of these areas.[55] Thus, based on our field observations and the hazard assessment, we do not feel that this value would be sufficiently protective of some of the more sensitive species or those, such as eagles, which are a higher level of the food web.

16. WATER QUALITY CRITERIA: HUMAN VERSUS WILDLIFE HEALTH

It is difficult to compare WQC developed for humans and wildlife because humans can restrict consumption of contaminated food items while wildlife cannot. For this reason fish consumption advisories that set an allowable quantity of fish that may be eaten from a particular water body are appropriate for protection of human health but not for the protection of wildlife.[196–198] Similarly, because individuals consume different quantities of fish, establishment of a WQC to protect all humans' health is not appropriate for everyone. The WQC to protect fish and wildlife from the effects of TCDD-EQ, which was proposed by the U.S. FWS, is 5 pg/liter.[6]

The WQC to protect humans from the adverse effects of PCBs, proposed by the Great Lakes Initiative, is 17- and 3-fold greater (Tables 10 and 14) than that proposed by Swain[199,200] to protect humans from the noncancerous developmental, behavioral, and cognitive effects of PCBs. The WQC calculated by Swain[191,192] range from 0.6 to 6.0, with a median of 1.0 pg PCB/liter (based on the McCarthy visual cognition scale) depending on assumptions of exposure.[171,200–207] However, the WQC proposed to protect human health[188] assumed that the hazard of cancer was less than that of other noncancer endpoints, which can occur in a dose-dependent fashion at very small intrauterine or intraegg exposures. For instance, concentrations of PCBs between 1.5 and 2.5 μg PCB/kg ww in blood of human umbilical cords have been correlated with subsequent adverse cognitive effects in human infants.[204] Greater concentrations of PCBs (7.9–12.9 μg PCB/kg) have been observed in the blood of women who ate an average of 23.5 lb (10.6 kg) of Lake Michigan fish per year and gave birth. The WQC proposed to protect humans[188] are similar to the values estimated for the protection of wildlife by Ludwig *et al.*[72] These results indicate that if the most sensitive wildlife species are protected, then humans will also be protected from the most subtle effects. Thus, the adverse effects of p-PCH-type compounds on wild species can serve as an early warning system for potential effects in the human population, but only if WQC for wildlife based on real effects in wild populations are included in the regulatory process.

The WQC that we propose to protect wildlife is approximately 40-fold less than that proposed by the Great Lakes Initiative.[188] Interestingly, the WQC based on both cancer and noncancer endpoints for the protection of human health from the effects of TCDD are greater than those proposed to protect wildlife (Table 18). This is probably justified, since humans eat less fish in their diet. We feel that the lesser value should be

Table 18
Water Quality Criteria Proposed by Several Governmental Agencies for the Protection of Wildlife and Humans from TCDD

Endpoint	Water quality criterion (pg TCDD/liter, ppq)	Ref.
Human health—cancer: drinking water	1.0×10^{-2}	180
Human health—noncancer: drinking water	1.0×10^{-1}	180
Human health—cancer: nondrinking water	1.0×10^{-2}	180
Human health—noncancer: nondrinking water	1.0×10^{-1}	180
Human health	1.3×10^{-2}	186
Human health	1.4×10^{-2}	186
Wildlife	8.5×10^{-3}	180

adopted since this would protect both wildlife and humans. Thus, in the case of TCDD, it appears that if wildlife are protected, humans will also be. Thus, wildlife would be a good sentinel species.

There is ongoing controversy over the relative sensitivity of humans to the effects of p-PCH[208] including both PCBs[209] and PCDDs.[210–213] There is even some controversy over the potency of TCDD as a mammalian carcinogen.[214,215] This uncertainty is leading to a reassessment of the reference doses and uncertainty factors used in human health hazard and risk assessments *based on cancer* for these types of compounds. It is difficult to conduct controlled studies on the effects of chemicals on humans. For that reason, the chronic effects of chemicals on humans are estimated from short-term, high-level exposures of animal models, which may be more or less sensitive to the carcinogenicity of p-PCH than humans. A safety factor of 10-fold is generally added to the assessment process to correct for among-species differences in sensitivity. This assumes that humans are more sensitive to chronic effects than are shorter-lived animals.[196,197,216,217]

Recent epidemiological evidence and the results of studies of the mechanism and modes of action of p-PCH indicate that humans may be much less sensitive to the Ah-r-mediated toxic effects of p-PCH than other species.[208–211,213] This is particularly true for carcinogenesis.[211–213] The U.S. EPA has stated that there is inadequate evidence of carcinogenicity of PCBs in humans.[218] We agree. The reason for this greater resistance or tolerance is unknown, but may be related to the interaction of the Ah-r with the p-PCH or with the DNA.[208] Long-term studies on the effects on humans exposed to p-PCH from accidental or industrial exposure for over 30 years generally have failed to exhibit rates of cancer greater than expected for the population.[208] Exposure to p-PCH, such as PCDD, at Seveso, Italy, or to PCB in electrical workers have not resulted in measurable increases in the rates of cancers in these exposed groups. These observations have stimulated a reassessment of the risk of cancer posed by these compounds to humans[208] and a possible change in the proposed reference doses for cancer that could lead to a relaxation of proposed WQC.[211] Currently, the standards for environmental concentrations of PCDD in Europe and Japan are 170 and 1700 times less stringent than those in the United States.[213]

If the environmental standards for p-PCH, based on human exposure, and the cancer endpoint are relaxed, then the WQC will not be sufficient to protect either wildlife or humans from the noncancer adverse effects of these compounds. Many subtle effects in humans will be ignored. Currently, the Science Advisory Board of the U.S. EPA is reexamining the model that will be used to set environmental standards for p-PCH.[213] The first phase of the reassessment will be to develop a better model of human health effects, which is based on the current state of knowledge about the Ah-r-mediated mechanism of action of the p-PCH.[208] Subsequently, an assessment of aquatic ecological risk of p-PCH will be conducted. The results of our analysis indicate that the protection of wildlife species will require more stringent regulations than for the protection of humans from cancer. We suggest that the protection of wildlife populations and humans from subtle noncancer effects should be given equal priority to that of protecting human populations from cancer. We have dcrived WQC based on the responses of wildlife rather than cancer in humans.[70] Our analysis indicates that WQC based on human cancer effects are not adequate to protect sensitive, wildlife species. Additionally, the method for deriving WQC has profound implications for remediation, litigation, and damage assessments. If wildlife are protected for the most sensitive endpoints, the most relevant of which seems to be reproduction, all components of the ecosystem including human health should be protected adequately.

The proposed WQC are very protective and if these criteria are met we would not expect to see any adverse effects related to these compounds in wildlife. This does not mean that adverse effects would be expected if the proposed criteria are exceeded. Also, since the proposed criteria do not consider bioavailability, it would be expected that the actual safe concentration in water could be as much as 10 to 100 times as great without causing adverse effects. Also, the proposed criteria assume that the species of interest consume only the identified prey with a specified BCF or BMF. This too provides some degree of safety. The 14 ng PCB/liter of total PCBs, which was suggested in the 1980 WQC document,[219] might not protect some species of wildlife since we have documented effects at a concentration of less than 2 ng PCB/liter in areas of Lake Michigan.

There is a great deal of uncertainty in the risk assessment process (Table 19). WQC developed as we did here are probably no better than ± 10 to 100×. For this reason it does not seem very worthwhile to argue about the exact WQC, especially when it will be impossible to directly validate models of concentrations in the water. For compounds that are already in the environment such as PCBs and TCDD-EQ, we advocate monitoring of wildlife species or their diet. For new compounds not already released into the environment, we advocate a conservative approach to allowed releases of bioaccumulable compounds.

WQC that are used to establish water pollution standards and permissible loadings of substances to public waters can be derived by several methods. These techniques generally involve hazard and risk assessment procedures. For nonpersistent, non-biomagnified compounds and elements, WQC are derived from the acute and chronic toxicity to aquatic organisms. For persistent organic compounds which are bioaccumulated and biomagnified, the effects on organisms higher in the food web must be considered. In the Great Lakes region of North America, the primary emphasis has

Table 19
Uncertainties Associated with Parameters Used in Hazard Assessment

Factor	Range of uncertainty
NOAEL for a species	10×
Species sensitivity	10×
Lipid content	5×
Trophic level	2×
Available fraction	10–100×
BMF	3×
BCF	10–100×

been on the potential for adverse effects to humans who eat fish. The primary endpoint considered in hazard and risk assessments has been cancer. Other endpoints such as teratogenicity, intellectual performance, and immune suppression and reproductive impairment are seldom considered. By hazard and risk assessments, regulators have endeavored to predict safe concentrations of potentially toxic chemicals which could be allowed to enter the environment. Once hazardous materials have entered the environment, these same criteria can be used as target values to determine if environmental damage has occurred, and the degree of remediation required to restore an ecosystem. For aquatic environments, these take the form of WQC. These WQC consider many environmental processes, such as dissipation, bioaccumulation, biomagnification, degradation, and dilution. For the most widespread and hazardous, synthetic, organic chemicals, these regulatory decisions are preoccupied by potential effects on the health of humans, particularly the risk for additional cancers in the population. Regulatory actions assume that ecosystems have assimilative capacities for persistent chemicals[9] and that risk of human cancer is the most relevant and sensitive endpoint. For that reason, long-term effects of persistent, organic chemicals that have the potential to biomagnify and cause *chronic, population-level* effects on wildlife have received scant attention. Humans have the option of restricting consumption of contaminated food. In fact, most fish consumption advisories are established to recommend a safe quantity of fish to be consumed in some specified time period by a person of average size, age, and sensitivity. On the other hand, wildlife cannot avoid contaminated food supplies and thus are at greater risk.

Cancer in humans may not be the most sensitive endpoint to measure.[208,220,221] There are physiological and developmental effects in a number of species, including humans, that are more sensitive, immediate, and demonstrative endpoints than cancer.[9,221] Subtle effects such as growth retardation[171] or altered development,[201–203] immune system suppression and elevated rates of disease,[171] wasting syndromes,[222,223] and behavioral changes in both adults and juveniles[11] have been observed in human and wildlife populations exposed to synthetic p-PCHs such as PCBs. Furthermore, the epidemiological evidence supports the contention that humans in the Great Lakes basin are expressing subtle, chronic effects related to the exposure to PCBs and similar compounds.[199,200]

Not only are there endpoints in humans that are more sensitive than cancer, but humans may not be the most sensitive species to the effects of p-PCH such as PCBs,[221,224] PCDDs, and PCDFs.[50,215] There are many wild species that may be inherently more sensitive to some classes of contaminants or receive greater exposures because they do not have the varied diets of humans. When wildlife species have been considered in the derivation of WQC, generally only direct, acute effects on aquatic organisms such as death have been considered. Some states in the Great Lakes region have even based their criteria for wildlife only on the water they consume, ignoring the much more important route of exposure for p-PCH represented by consumption of fish. Even though some of the methods for deriving WQC are very complex and include mechanisms to predict chronic effects and differences in sensitivities among species, they generally do not include the wildlife species, such as mink, eagles, and other fish-eating water birds, that may eat contaminated aquatic organisms exclusively. The currently used risk assessment procedures do not account for exposures to complex mixtures of potentially hazardous compounds that can cause subtle, long-term, chronic effects on the dynamics of wildlife populations. Typically, these agents cause effects on the offspring of contaminated adults through decreased fecundity or even reproductive failure. These phenomena can affect wildlife populations directly over longer periods of time.[9] Documented adverse effects in wildlife populations are consistent with the types and frequencies of effects observed in laboratory studies of the effects of p-PCH. These effects have been observed in fish,[225] birds,[9,23,220,226] mink,[227] turtles,[228] and humans.[199,200]

The use of nonconventional *real world* bioeffects and reproductive endpoints to establish LOAEL values and NOAEL for persistent, lipophilic, toxic contaminants will require new testing protocols and data on the reproductive outcomes of serial exposed generations of human and wildlife populations. Although recommended by several authors,[68,69] the effectiveness of regulations to protect wildlife, such as birds, has rarely been assessed by conducting field studies. Regulations have generally been set based on modeled expected results. The longitudinal studies of behavioral effects of contaminants on children of parents who have been exposed to PCBs through eating contaminated Great Lakes fish are providing some of these data for humans.[201–203] Long-term studies of herring gulls in the Great Lakes region have produced a 20-year record of bioeffects which have been linked to chemicals.[178] Similar studies conducted for shorter periods of time have demonstrated the same types of effects in other species of colonial fish-eating water birds.[23,72,170]

17. WILDLIFE AS ENVIRONMENTAL SENTINELS

Ecotoxicology and especially *wildlife toxicology* are relatively new fields of endeavor.[229] However, even in ancient times humans exhibited an awareness of the condition of birds, and hence the Greek maxim *a bad crow lays a bad egg*.[229] Even though there are records of human-caused episodes of toxicological effects in populations of wildlife species from ancient times, only with the widespread use of synthetic organic chemicals since World War II have large-scale chemically-induced wildlife epizootics occurred. The ability to describe these effects and to understand the role of toxic chemicals has required developments in a number of scientific fields, including environmental, analytical chemistry and wildlife biochemical toxicology. Limited knowledge of the basic biochemistry, physiology, and natural histories of wildlife species has also limited the ability to document and understand the effects of contaminants on wildlife populations. The analytical tools of these multidisciplinary fields are providing a comprehensive picture of the effects of trace concentrations of toxic synthetic hydrocarbons in wildlife species.[9,73,170] The study of effects of toxic chemicals on wildlife populations is limited by the complexity of a large number of species interacting with each other as well as their natural habitat and human-caused physical changes to their environment along with the effects of synthetic organic chemicals. However, through the efforts of multidisciplinary research teams of experts in environmental chemistry, chemodynamics, toxicology, biochemistry, pathology, and ecology, rapid progress is being made. The ability to establish the cause–effect relationships between concentrations of chemicals in complex mixtures with adverse effects and to be able to predict the fates and effects of these chemicals in the ecosystem is developing rapidly.[170]

In our efforts to investigate the linkages between certain synthetic halogenated chemicals in the Great Lakes and effects on populations of wild birds, we have followed Koch's postulates (Table 5). To date, we have observed adverse effects, including egg mortality and deformities in chicks. These effects have been correlated with the concentrations of several compounds, but the strongest correlations are with the concentrations of TCDD-EQ. The types of effects observed are the same as those that can be caused by these compounds in laboratory studies of animal models. Furthermore, the concentrations required to elicit the responses are in the same range that would be expected from laboratory studies. Thus, we feel as if we have completed the first four of the postulates. It is likely that the effects observed are caused by the TCDD-EQ contributed by PCDD, PCDF, and p-PCBs. This conclusion is further supported by the fact that when the p-PCH fraction was removed by selective carbon column chromatography, we were able to remove the fraction that caused the induction in the H4IIE bioassay.

We have not completed the fifth postulate *in vivo* because it is difficult to conduct studies of this type in the laboratory with the same species that occur in the wild. This is because of many logistical problems, such as the fact that no uncontaminated source of organisms is available. Even if uncontaminated wildlife species were available, our techniques to rear them in the laboratory are not sufficiently well developed to allow valid laboratory comparisons. Instead, we have fed fish from the Great Lakes to chickens.[173] The results of these studies indicated that the same types of effects observed under field conditions could be induced in these species under laboratory conditions. Finally, relative to studies of the potential for these compounds to cause deformities, it is difficult to demonstrate statistically significant effects because of the small rates of deformities observed and the small sample sizes that are possible in laboratory studies.

A method for conducting wildlife hazard assessments that uses results of field and laboratory studies on target and surrogate domestic species, along with environmental monitoring and chemical analyses, to determine the appropriate WQC to protect sensitive wildlife species from the adverse effects of p-PCH has been developed.[70] When compared with an assessment that used human cancer risk assessments and visual cognition and memory effects as endpoints, the WQC to protect humans and wildlife were found to be similar in magnitude.[200]

When possible it is best to use actual concentrations of PCBs or TCDD-EQ in tissues of fish or birds or their eggs, as the most proximate measure of exposure. If it is not possible to make measurements directly on the target organisms, the next best estimate of exposure is to measure the concentrations of compounds in their food. Use of concentrations in water is the least accurate estimate of exposure and thus includes more safety factors to assure protection of all species from any adverse effects.

Here, we have established that the types of effects observed in wildlife populations are similar to those observed in mammals, which are used as models of the potential effects on humans, and that environmental criteria to protect wildlife are similar to those predicted to protect human health. The allowable effects in wildlife are greater than those in humans, but their exposure is greater, since they cannot restrict their intake. The result is that if wildlife species, such as birds, which are at the top of the food chain are protected, it is likely that human health in the same ecosystem will also be protected. We do not imply that if effects are observed in wildlife, there will necessarily be similar effects in humans. Conversely, if wildlife are protected from adverse effects resulting from environmental exposures to contaminants, it is likely that humans will be protected from the same sources of exposure. However, it must be remembered that there may be other significant exposures to toxicants, such as household and occupational exposures. Thus, while wildlife biomonitoring cannot protect humans from all exposures, monitoring their responses can be a useful procedure. For this reason, wildlife, especially colonial water birds of the Great Lakes region, have been advocated as sentinels of environmental exposures.[2,16,73,224,226]

18. REMEDIATION AND ENVIRONMENTAL MANAGEMENT

The greatest contribution to the TCDD-EQ in biological samples from the Great Lakes is from the p-PCB congeners. Thus, it has been concluded that the toxicity of these chemicals is not a relevant issue because all that can be done from a regulatory standpoint has been accomplished. This situation could lead to the conclusion that all that can be done is to wait for the concentrations of these compounds to decrease to below the threshold for adverse effect. This is partly true: However, because the rate of loss of these compounds through burial and degradation is slow[132] and environmental sorting mechanisms[230] tend to keep the more toxic Ah-active p-PCH in circulation, these toxicants will be active in the ecosystem for many years. For this reason, and because, once released, they cannot be effectively removed from the environment, every effort must be made to minimize any further release of these compounds. This means immediate cleanup of "hot spots," in rivers, bays, and along the shorelines, landfills and spill areas which continue to be a source of these chemicals and adoption of sufficiently protective WQC to minimize any further releases. Only in this way will the exposure to wildlife and humans be reduced effectively. Simply because the manufacture of PCBs has been stopped does not mean that nothing more can be done to protect wildlife from the effects of p-PCH. Currently, PCDD and PCDF contribute between 2 and 30% of the TCDD-EQ measured in the tissues of Great Lakes fish and wildlife, according to proposed TEF values.[24,112,122,177] However, these compounds come from diverse point and nonpoint sources and occur as by-products of industrial processes and are contaminants in other products. Therefore, because these compounds are very persistent in the environment and have the greatest toxic potency of all of the p-PCH, every effort should be made to minimize their formation and entry into the ecosystem. Even if all possible steps are taken to minimize releases of p-PCH into the environment, those that are already in the environment will continue to be important in management decisions. Some such decisions include: (1) the setting of fish consumption advisories sufficient to protect humans from chronic subtle effects of p-PCH. (2) Opening of certain bodies of water so that exposure of wildlife or humans to p-PCH is increased. For example, the removal of dams and weirs on some rivers may result in greater exposure of upstream wildlife, which cannot selectively avoid or limit consumption of contaminated fish, and of humans who will have greater access to the contaminated fish. (3) Planting fish, which may become attractive nuisances such as polyploid salmon in the Great Lakes. These genetically altered fish will grow larger, be exposed longer and result in greater concentrations of p-PCH in their tissues, thus causing greater exposure to both humans and wildlife who may consume them. (4) If sensitive species are to be reintroduced, then one must be certain that the appropriate habitat include sufficiently small concentrations of toxic chemicals to allow the reintroduced species to reproduce. For instance, it would not seem appropriate to spend vast resources to reintroduce peregrine falcons, bald eagles, mink, otters, lake trout, or any other sensitive species into a contaminated environment where they cannot sustain viable populations until concentrations were reduced. Clean food supplies in the habitat are essential for reintroduction programs of threatened or endangered species if self-sustaining populations are the goal.

Currently, the concentrations of PCBs and other p-PCH in all of the Great Lakes exceed the WQC to protect even the most tolerant species, such as herring gulls and double-crested cormorants, but the concentrations are approaching the threshold for subtle effects in the more tolerant species. This explains why the populations of herring gulls and double-crested cormorants are increasing in most areas of the Great Lakes after being almost extirpated, as a result of the toxic effects of other synthetic, halogenated hydrocarbons such as the DDT group.[72,74] The concentrations of PCBs currently greatly exceed the WQC to protect bald eagles living on all five of the Great Lakes, which partly explains why the productivities of individuals that eat Great Lakes fish and other fish-eating water birds are less than optimal in all locations and less than replacement in some locations on the Great Lakes.[180] The WQC to protect mink from the effects of PCBs is currently exceeded in all of the Great Lakes except Superior. We have determined that the concentration of total PCBs in Saginaw River water is currently 300 times greater than the concentration that would result in the dietary no-effect concentration for mink.[177] Wildlife are as sensitive or more so to the effects of persistent, synthetic organic toxicants, such as PCBs, as are humans. Because they do not have the option to restrict consumption of contaminated food items, wild animals are often at similar or greater risk than humans. Therefore, wildlife hazard assessments should be included in the process to devise WQC. We recommend a WQC for total concentrations of PCBs of 0.1 pg PCB/liter (100 fg PCB/liter; 100 parts per quintillion, ppq) to fully protect the most sensitive wildlife species. This WQC would be protective of bald eagles (WQC = 1.0 ppq total PCBs) and the white leghorn chicken (WQC = 2 ppq), which are species known to be sensitive to the effects of PCB-type compounds and provide a safety factor of approximately 10, which would allow for the effects of all those p-PCH that act through the same mechanism of action. For TCDD-EQ, we suggest a WQC of 0.2 fg TCDD-EQ/liter (ppq).

The WQC suggested here can be useful target values to supply in modeling scenarios to determine necessary reductions in loadings and to predict the impact of various remedial actions. However, since the proposed values are both well below the currently attainable method detection limits for total PCBs and TCDD, it will be impossible to verify the models or the WQC. Furthermore, it will not be possible to use monitoring of water to assess changes in concentrations with time. Thus, we suggest that monitoring programs that focus on colonial fish-eating water birds would be an appropriate alternative.

Many synthetic p-PCH from both point and nonpoint sources, which can cause effects similar to those of PCBs, are neither considered in hazard assessments nor regulated. Based on our knowledge of what has occurred with PCBs, every effort should be made to keep these compounds out of the ecosystem. Even if this value is adopted, it is possible that subtle, adverse effects may be observed in some sensitive species or individuals, especially tertiary predators. If this WQC is established and remedial actions taken to expedite attainment of this level, humans who eat Great Lakes fish may still be at risk and fish consumption advisories will still be necessary for some time.

Monitoring programs generally focus on a few contaminants, while many potentially important p-PCH are not quantified, even if they are observed in the sample extract. We further suggest that all of the p-PCH with are manufactured should be monitored, including the chlorinated compounds formed in the paper manufacturing process. Monitoring should focus on wildlife as the most sensitive indicators.[69]

Because there are a large number of known and potential p-PCH, which can express toxicity through the Ah-r-mediated mechanism of action and the chemical quantification of these compounds is difficult, time-consuming, and expensive, we recommend the use of biochemical detector systems to measure the functional response of biota to these complex mixtures. Such bioassays have the potential to be automated and to provide rapid, sensitive, inexpensive measures of the toxic potency of complex mixtures of p-PCH. Also, these functional bioassays can account for the interactions between and among p-PCH congeners as well as other synthetic hydrocarbons, which may act synergistically or antagonistically. Because these assays measure the interaction of the p-PCH with its biochemical receptor, assays of this type give better predictability of the toxic effects of p-PCH.

The environmental concentrations of p-PCH are not decreasing very rapidly because of continued, constant releases from land stocks, contaminated soils and aquatic sediments, and a large mass of these compounds in the atmosphere. PCBs are the toxicant class of primary toxicological significance in the Great Lakes ecosystem. Even though PCBs are no longer manufactured, they are still in use in some applications and enter the aquatic ecosystem from both point and diffuse sources. Most PCBs manufactured are still in use or held in situations such as poorly built landfills that make them likely to enter the environment in the future. For this reason, decisions pertaining to human health and wildlife management will need to be made against this continuing level of background contamination for the foreseeable future. For the human health community, this means learning to recognize those effects that could be caused by p-PCH in sensitive populations or those that are at greater risk because of occupational or lifestyle exposure.

REFERENCES

1. A. P. Gillman, G. A. Fox, D. B. Peakall, S. M. Teeple, T. R. Carroll, and G. T. Haymes, Reproductive parameters and egg contaminant levels of Great Lakes herring gulls, *J. Wildl. Manage.* **41,** 458–468 (1977).
2. D. V. Weseloh, P. Mineau, and D. J. Hallett, Organochlorine contaminants and trends in reproduction in Great Lakes herring gulls, 1974–1978, *Proc. 44th North Am. Wildl. Nat. Resour. Conf.*, pp. 543–557, Wildlife Management Inst., Washington, DC (1979).
3. D. V. Weseloh, S. M. Teeple, and M. Gilbertson, Double-crested cormorants of the Great Lakes: Egg laying parameters, reproductive failure and contaminant residues in eggs, Lake Huron 1972–1973, *Can. J. Zool.* **61,** 427–436 (1983).
4. D. V. Weseloh, T. W. Custer, and B. M. Braune, Organochlorine contaminants in eggs of common terns from the Canadian Great Lakes, 1981, *Environ. Pollut.* **59,** 141–160 (1989).
5. R. J. Allan, A. J. Ball, V. W. Cairns, G. A. Fox, A. P. Gilman, A. P. Peakall, D. A. Piekarz, J. C. Van Oosdam, D. C. Villeneuve, and D. T. Williams, *Toxic Chemicals in the Great Lakes and Associated Effects,* Environment Canada, Dept. Fisheries & Oceans, Health & Welfare Canada, Vols. I and II (1991).
6. R. Eisler, *Polychlorinated biphenyl hazards to fish, wildlife and invertebrates: A synoptic review,* USDI, USFWS Contaminant Hazard Review No. 7 (1986).
7. J. P. Ludwig, J. P. Giesy, C. L. W. Bowerman, S. Heaton, R. Aulerich, S. Bursian, H. J. Auman, P. D. Jones, L. L. Williams, D. E. Tillitt, and M. Gilbertson, A comparison of water quality criteria in the Great Lakes Basin based on human or wildlife health, *J. Great Lakes Res.* **19,** 789–807 (1994).
8. R. Eisler, *Dioxin hazards to fish, wildlife and invertebrates: A synoptic review,* USDI, USFWS Contaminant Hazard Review No. 8 (1986).
9. M. Gilbertson, T. J. Kubiak, J. P. Ludwig, and G. Fox, Great Lakes embryo mortality, edema, and deformities syndrome (GLEMEDS) in colonial fish-eating birds: Similarity to chick edema disease, *J. Toxicol. Environ. Health* **33,** 455–520 (1991).
10. M. Gilbertson, Etiology of chick edema disease in herring gulls in the lower Great Lakes, *Chemosphere* **12,** 357–370 (1983).
11. T. J. Kubiak, H. J. Harris, L. M. Smith, T. R. Schwartz, D. L. Stalling, J. A. Trick, L. Sileo, D. E. Docherty, and T. C. Erdman, Microcontaminants and reproductive impairment of the Forster's tern on Green Bay, Lake Michigan—1983, *Arch. Environ. Contam. Toxicol.* **18,** 706–727 (1989).
12. C. F. Tumasonis, B. Bush, and F. D. Baker, PCB levels in egg yolks associated with embryonic mortality and deformity of hatched chicks, *Arch. Environ. Contam. Toxciol.* **1,** 312–324 (1973).
13. M. Gilbertson, R. D. Morriss, and R. A. Hunter, Abnormal chicks and PCB residue levels in eggs of colonial birds on the lower Great Lakes (1971–1973), *Auk* **93,** 435–442 (1976).
14. D. B. Peakall, Accumulation and effects on birds, in: *PCBs in the Environment* (J. S. Aaid, ed.), Vol. II, pp. 31–47, CRC Press, Boca Raton, FL (1986).
15. D. B. Peakall, Known effects of pollutants on fish-eating birds in the Great Lakes of North America, in: *Toxic Contamination in Large Lakes* (N. W. Schmidtke, ed.), Vol. II, pp. 39–54, Lewis Publishers, Chelsea, MI (1988).
16. D. B. Peakall and G. A. Fox, Toxicological investigations of pollutant-related effects in Great Lakes gulls, *Environ. Health. Perspect.* **71,** 187–193 (1987).

17. J. Struger and D. V. Weseloh, Great Lakes Caspian terns: Egg contaminants and biological implications, *Colon. Water Birds* **8,** 142–149 (1985).
18. J. Struger, D. V. Weseloh, D. J. Hallett, and P. Mineau, Organochlorine contaminants in herring gull eggs from the Detroit and Niagara Rivers and Saginaw Bay (1978–1982): Contaminant discriminants, *J. Great Lakes Res.* **11,** 223–230 (1985).
19. D. W. Anderson and J. J. Hickey, Eggshell changes in certain North American birds, *Proc. XVth Int. Ornithol. Congr.* pp. 514–540 (1969).
20. J. E. Elliott, R. J. Norstrom, and J. A. Keith, Organochlorines and eggshell thinning in northern gannets (*Sula bassanus*) from eastern Canada, *Environ. Pollut.* **52,** 81–102 (1988).
21. P. C. Baumann and D. M. Whittle, The status of selected organics in the Laurentian Great Lakes: An overview of DDT, PCBs, dioxins, furans and aromatic hydrocarbons, *Aquat. Toxicol.* **11,** 241–257 (1988).
22. I. M. Price and D. V. Weseloh, Increased numbers and productivity of double-crested cormorants *Phalacrocorax auritus*) on Lake Ontario, *Can. Field Nat.* **100,** 474–482 (1986).
23. D. E. Tillitt, G. T. Ankley, J. P. Giesy, J. P. Ludwig, H. Kurita-Matsuba, D. V. Weseloh, P. S. Ross, C. Bishop, L. Sileo, K. L. Stromberg, J. Larson, and T. J. Kubiak, Polychlorinated biphenyls residues and egg mortality in double-crested cormorants from the Great Lakes, *Environ. Toxicol. Chem.* **11,** 1281–1288 (1992).
24. N. Yamashita, S. Tanabe, J. P. Ludwig, H. Kurita, M. E. Ludwig, and R. Tatsukawa, Embryonic abnormalities and organochlorine contamination in double-crested cormorants (*Phalacrocorax auritus*) and Caspian terns (*Hydroprogne caspia*) from the Upper Great Lakes, collected in 1988, *Environ. Pollut.* **79,** 163–173 (1992).
25. U. Seleström, B. Jansson, A. Kierkegaard, C. de Wit, T. Odsjö, and M. Olson, Polybrominated diphenyl ethers (PBDE) in biological samples from the Swedish environment, *Chemosphere* **26,** 1703–1718 (1993).
26. G. A. Fox, D. V. Weseloh, T. J. Kubiak, and T. C. Erdman, Reproductive outcomes in colonial fish-eating birds: A biomarker for developmental toxicants in Great Lakes food chains, *J. Great Lakes Res.* **17,** 153–157 (1991).
27. P. V. Hodson, M. McWhirther, K. Ralph, B. Gray, D. Thivierge, J. H. Carey, and M. C. Levesque, Effects of bleached Kraft mill effluent on fish in the St. Maurice river, Quebec, *Environ. Toxicol. Chem.* **11,** 1635–1651 (1992).
28. J. A. Nosek, J. R. Sullivan, T. E. Amundson, S. R. Craven, L. M. Miller, A. G. Fitspatrick, M. E. Cook, and R. E. Peterson, Embryotoxicity of 2,3,7,8-tetrachlorodibenzo-p-dioxin in ring-necked pheasants, *Environ. Toxicol. Chem.* **12,** 1215–1222 (1992).
29. G. R. Higgenbotham, A. Huang, D. Firestone, J. Verrett, J. Reese, and A. D. Campbell, Chemical and toxicological evaluations of isolated and synthetic chloro-derivatives of dibenzo-*p*-dioxin, *Nature* **220,** 702–703 (1986).
30. J. P. Ludwig, SERE Group, unpublished data.
31. D. E. Tillitt, *Characterization studies of the H4IIE bioassay for Assessment of Planar Halogenated Hydrocarbons in Fish and Wildlife,* Ph.D. dissertation, p. 152, Michigan State University, East Lansing (1989).
32. M. J. Verrett, US Food & Drug Administration Memorandum to D. Firestone, dated 8 June (1976).
33. D. S. Henschel, LD50 and teratogenicity studies of the effects of TCDD on chicken embryos, Society of Environmental Toxicology and Chemistry, Annual Meeting, Houston, Texas, 1993, p. 280.

34. D. H. White, J. T. Seginak, and D. J. Hoffman, Dioxins and furans linked to reproductive impairment in Arkansas wood ducks, *J. Wildl. Manage.* **58,** 100–106 (1994).
35. D. S. Henshel, B. M. Hehn, M. T. Vo, and J. D. Steeves, A short-term test for dioxin teratogenicity using chicken embryos, in: *Environmental Toxicology and Risk Assessment,* Vol. 2, ASTM STP 1216 (J. W. Gorsuch, F. J. Dwyer, C. G. Ingesoll, and T. W. LaPoint, eds.), ASTM, Philadelphia, (in press).
36. A. Poland, and E. Glover, Comparison of 2,3,7,8-tetrachlorodibenzo-p-dioxin, a potent inducer of aryl hydrocarbon hydroxylase, with 3-methylcholanthrene, *Mol. Pharmacol.* **10,** 349–359 (1974).
37. S. Tanabe, N. Kannan, M. Fukushima, T. Okamoto, T. Wakimoto, and R. Tatsukawa, Persistent organochlorines in Japanese coastal waters: An introspective summary from a Far East developed nation, *Mar. Pollut. Bull.* **20,** 344–352 (1989).
38. R. J. Norstrom, Bioaccumulation of polychlorinated biphenyls in Canadian wildlife, in: *Hazards, Decontamination and Replacement of PCBs,* (J.-P. Crine, ed.), pp. 1–16, Plenum Press, New York (1987).
39. R. J. Norstrom, M. Simon, D. C. G. Muir, and R. E. Schweinsburg, Organochlorine contaminants in arctic marine food chains: Identification, geographical distribution and temporal trends in polar bears, *Environ. Sci. Technol.* **22,** 1063–1071 (1988).
40. P. De Voogt and U. A. T. Brinkman, Production, properties and usage of polychlorinated biphenyls, in: *Halogenated Biphenyls, Terphenyls, Naphthalenes, Dibenzodioxins and Related Products* (R. D. Kimbrough and A. A. Jensen, eds.), pp. 3–45, Elsevier, Amsterdam (1989).
41. P. De Voogt, J. W. M. Wegener, J. C. Klamer, G. A. Van Zijl, and H. Govers, Prediction of environmental fate and effects of heteroatomic polycyclic aromatics by QSARs: The position of n-octanol/water partition coefficients, *Biomed. Environ. Sci.* **1,** 194–209 (1988).
42. P. De Voogt, D. E. Wells, L. Reutergardh, and U. A. T. Brinkman, Biological activity, determination and occurrence of planar, mono- and di-ortho PCBs, *Int. J. Environ. Chem.* **40,** 1–46 (1990).
43. S. Tanabe, N. Kannan, A. Subramanian, S. Watanabe, and R. Tatsukawa, Highly toxic coplanar PCBs: Occurrence, source, persistency and toxic implications to wildlife and humans, *Environ. Pollut.* **47,** 147–163 (1987).
44. L. S. Birnbaum, The role of structure in the disposition of halogenated aromatic xenobiotics, *Environ. Health Perspect.* **61,** 11–20 (1985).
45. D. C. G. Muir, R. J. Norstrom, and M. Simon, Organochlorine contaminants in arctic marine food chains: Accumulation of specific polychlorinated biphenyls and chlordane-related compounds, *Environ. Sci. Technol.* **22,** 1071–1079 (1988).
46. A. Parkinson and S. Safe, Mammalian biologic and toxic effects of PCBs, in: *Polychlorinated Biphenyls (PCBs): Mammalian and Environmental Toxicology* (S. Safe and O. Hutzinger, eds.), pp. 49–75, Springer-Verlag, Berlin (1989).
47. S. Tanabe, N. Kannan, T. Wakimoto, R. Tatsukawa, T. Okamoto, and Y. Masuda, Isomer-specific determination and toxic evaluation of potentially hazardous coplanar PCBs, dibenzofurans and dioxins in the tissues of "Yusho" PCB poisoning victim and in the causal oil, *Toxicol. Environ. Chem.* **24,** 215–231 (1989).
48. S. Safe, S. Bandiera, T. Sawyer, B. Zmudzka, G. Mason, M. Romkes, M. A. Denomme, J. Sparling, A. B. Okey, and T. Fujita, Effects of structure on binding to the 2,3,7,8-TCDD receptor protein and AHH induction-halogenated biphenyls, *Environ. Health Perspect.* **61,** 21–33 (1985).
49. S. Safe, Determination of 2,3,7,8-TCDD toxic equivalent factors (TEFs): Support for the use of the *in vitro* AHH induction assay, *Chemosphere* **16,** 791–802 (1987).

50. S. Safe, Polychlorinated biphenyls (PCBs), dibenzo-p-dioxins (PCDDs), dibenzofurans (PCDFs), and related compounds: Environmental and mechanistic considerations which support the development of toxic equivalency factors (TEFs), *Crit. Rev. Toxicol.* **21,** 51–88 (1990).
51. V. A. McFarland and J. U. Clarke, Environmental occurrence, abundance, and potential toxicity of polychlorinated biphenyl congeners: Considerations for a congener-specific analysis, *Environ. Health Perspect.* **81,** 225–239 (1989).
52. L. M. Smith, T. R. Schwartz, K. Feltz, and T. J. Kubiak, Determination and occurrence of AHH-active polychlorinated biphenyls, 2,3,7,8 TCDD and 2,3,7,8 TCDF in Lake Michigan sediment and biota: The question of their relative toxicological significance, *Chemosphere* **21,** 1063–1085 (1990).
53. L. M. Smith, D. L. Stalling, and L. Johnson, Determination of part-per-trillion levels of polychlorinated dibenzofurans and dioxins in environmental samples, *Anal. Chem.* **56,** 1830–1842 (1984).
54. P. D. Jones, G. T. Ankley, D. A. Best, R. Crawford, N. Krishnan, J. P. Giesy, T. J. Kubiak, J. P. Ludwig, J. L. Newsted, D. E. Tillitt, and D. A. Verbrugge, Biomagnification of bioassay-derived 2,3,7,8-tetrachlor-dibenzo-p-dioxin equivalents, *Chemosphere* **26,** 1203–1212 (1993).
55. J. P. Whitlock, The regulation of gene expression by 2,3,7,8-tetrachlordibenzo-p-dioxin, *Pharmacol. Rev.* **39,** 147–161 (1987).
56. J. A. Goldstein, P. Linko, J. N. Huckins, and D. L. Stalling, Structure–activity relationships of chlorinated benzenes as indicators of different forms of cytochrome P_{450} in rat liver, *Chem. Biol. Interact.* **41,** 131–139 (1982).
57. J. A. Goldstein, Structure–activity relationships for the biochemical effects and the relationship to toxicity, in: *Halogenated Biphenyls, Terphenyls, Naphthalenes, Dibenzodioxins and Related Products,* Vol. 4 (R. D. Kimbrough and A. A. Jensen, eds.), pp. 151–190, Elsevier, Amsterdam (1980).
58. A. Poland and C. Knutson, 2,3,7,8-Tetrachlorodibenzo-p-dioxin and related halogenated aromatic hydrocarbons: Examination of the mechanism of toxicity, *Annu. Rev. Pharmacol. Toxicol.* **22,** 517–554 (1982).
59. D. W. Nebert, The Ah locus: Genetic differences in toxicity, cancer, mutation, and birth defects, *Crit. Rev. Toxicol.* **20,** 153–174 (1990).
60. J. P. Landers and N. J. Bunce, The Ah-receptor and the mechanism of dioxin toxicity, *Biochem J.* **276,** 273–287 (1991).
61. M. S. Denison, P. A. Bank, and E. F. Yao, DNA sequence-specific binding of transformed Ah receptor to a dioxin responsive enhancer: Looks aren't everything, *Chemosphere* **25,** 33–36 (1992).
62. F. J. Gonzalez, R. H. Tukey, and D. W. Nebert, Structural gene products of the Ah locus. Transcriptional regulation of cytochrome P1-450 and P3-450 mRNA levels by 3-methylcholanthrene, *Mol. Pharmacol.* **26,** 117–121 (1984).
63. M. S. Denison, J. M. Fisher, and J. P. Whitlock, Protein–DNA interactions at recognition sites for the dioxin–Ah receptor complex, *J. Biol. Chem.* **264,** 16478–16482 (1989).
64. R. M. Pratt, Receptor-dependent mechanisms of glucocorticoid and dioxin-induced cleft palate, *Environ. Health Perspect.* **61,** 35–40 (1985).
65. J. B. Silkworth, D. S. Cutler, and G. Sack, Immuno-toxicity of 2,3,7,8-tetrachlorodibenzo-p-dioxin in a complex environmental mixture from the Love Canal, *Fundam. Appl. Toxicol.* **12,** 303–312 (1989).
66. J. D. McKinney, J. Fawkes, S. Jordan, K. Chae, S. Oatley, R. E. Coleman, and W. Briner, 2,3,7,8-Tetrachlorodibenzo-p-dioxin (TCDD) as a potent and persistent thyroxine agonist:

A mechanistic model for toxicity based on molecular reactivity, *Environ. Health Perspect.* **61,** 41–53 (1985).
67. J. B. Silkworth, D. S. Cutler, P. W. O'Keefe, and T. Lipniskas, Potentiation and antagonism of 2,3,7,8-tetrachlor-p-dibenzo dioxin effects in a complex environmental mixture, *Toxicol. Appl. Pharmacol.* **119,** 236–247 (1993).
68. C. F. Mason, *Biology of Freshwater Pollution,* 2nd ed., Wiley, New York (1991).
69. C. F. Mason, G. T. Sawyer, B. Keys, S. Bandiera, M. Romkes, J. Piskorska-Pliszcynska, B. Zmudzka, and S. Safe, Polychlorinated dibenzofurans (PCDFs): Correlation between in vivo and in vitro structure–activity relationships, *Toxicology* **37,** 1–12 (1985).
70. E. E. McConnell, Acute and chronic toxicity and carcinogenesis in animals, in: *Halogenated Biphenyls, Terphenyls, Naphthalenes, Dibenzodioxins and Related Products* (R. D. Kimbrough and A. A. Jensen, eds.), pp. 161–193, Elsevier, Amsterdam (1989).
71. R. E. Morrissey and B. A. Schwetz, Reproductive and developmental toxicity in animals, in: *Halogenated Biphenyls, Terphenyls, Naphthalenes, Dibenzodioxins and Related Products* (R. D. Kimbrough and A. Jensen, eds.), pp. 195–225, Elsevier, Amsterdam (1989).
72. J. P. Ludwig, M. E. Ludwig, and H. J. Auman, Uptake of chemicals from Great Lakes fish by double-crested cormorants and herring gull chicks, *Abstracts of the Cause–Effects Linkages II Symposium, September 27–28, 1991, Traverse City, Michigan,* pp. 23–34, (1991).
73. G. A. Fox, M. Gilbertson, A. P. Gillman, and T. J. Kubiak, A rationale for the use of colonial, fish-eating birds to monitor the presence of developmental toxicants in Great Lakes fish, *J. Great Lakes Res.* **17,** 151–152 (1991).
74. G. A. Fox, Eggshell quality: Its ecological and physiological significance in a DDT-contaminated common tern population, *Wilson Bull.* **88,** 459–477 (1976).
75. K. Stromberg, U.S. Fish and Wildlife Service, Green Bay, WI, unpublished data.
76. T. Colborn and C. Clement, *Chemically-Induced Alterations in Sexual Development: The Wildlife/Human Connection,* Princeton Scientific Publishing Co., Princeton, NJ (1992).
77. J. L. Newsted and J. P. Giesy, Effects of 2,3,7,8-tetrachlordibenzo-p-dioxin (TCDD) on epidermal growth factor binding and protein kinase activity in the RTH149 rainbow trout hepatoma cell line, *Aquat. Toxicol.* **23,** 119–135 (1992).
78. D. M. Fry and T. K. Toon, DDT-induced feminization of gull embryos, *Science* **213,** 922–924 (1981).
79. A. P. Brouwer, *Interference of 2,3,7,8-tetrachlorbiphenyl in vitamin A (retinols) metabolism: Possible implications for toxicity and carcinogenicity of polyhalogenated aromatic hydrocarbons,* Radiobiological Inst., Health Res. TNO, Rijswijk, The Netherlands (1987).
80. B. Brunström and J. Lund, Differences between chick and turkey embryos in sensitivity to 3,3′,4,4′-tetrachlor-biphenyl and in concentration/affinity of the hepatic receptor for 2,3,7,8-tetrachloro-p-dioxin, *Comp. Biochem.* **91C,** 507–512 (1988).
81. B. Brunström and L. Andersson, Toxicity and 7-ethoxyresorufin O-deethylase-inducing potency of coplanar polychlorinated biphenyls (PCBs) in chick embryos, *Arch. Toxicol.* **62,** 263–266 (1988).
82. B. Brunström, Sensitivity of embryos from duck, goose, herring gull and various chicken breeds to 3,3′,4,4′-tetrachlorobiphenyl, *Poult. Sci.* **67,** 52–57 (1987).
83. B. Brunström, Mono-ortho-chlorinated chlorobiphenyls: Toxicity and induction of 7-ethoxyresorufin O-deethylase (EROD) activity in chick embryos, *Arch. Toxicol.* **64,** 188–192 (1990).
84. B. Brunström, Toxicity and EROD-inducing potency of polychlorinated biphenyls (PCBs) and polycyclic aromatic hydrocarbons (PAHs) in avian embryos, *Comp. Biochem. Toxicol.* **100C,** 241–243 (1991).

85. P. Beland, S. Deguise, C. Girard, A. Lagase, D. Martineau, R. Michaud, D. Muir, R. Norstrom, E. Pelletier, and L. Shugart, Toxic compounds, health and reproductive effects in St. Lawrence beluga whales, *J. Great Lakes Res.* **19,** 766–777 (1993).
86. D. E. Tillitt, J. P. Giesy, and G. T. Ankley, Characterization of the H4IIE rat hepatoma cell bioassay as a tool for assessing toxic potency of planar halogenated hydrocarbons in environmental samples, *Environ. Sci. Technol.* **25,** 87–92 (1989).
87. D. E. Tillitt, G. T. Ankley, D. A. Verbrugge, J. P. Giesy, J. P. Ludwig, and T. J. Kubiak, H4IIE rat hepatoma cell bioassay-derived 2,3,7,8-tetrachloro-*p*-dioxin equivalents in colonial, fish-eating water birds from the Great Lakes, *Arch. Environ. Toxicol.* **21,** 91–101 (1991).
88. A. Brouwer, Inhibition of thyroid hormone transport in plasma of rats by polychlorinated biphenyls, *Arch. Toxicol. Suppl.* **13,** 440–445 (1989).
89. A. P. Brouwer, Binding of a metabolite of 3,4,3′,4′-tetrachlorobiphenyl to transthyretin reduces serum vitamin A transport by inhibiting the formation of the protein complex carrying both retinol and thyroxine, *Toxicol. Appl. Pharmacol.* **85,** 301–312 (1989).
90. A. P. Brouwer, J. H. Reijnders, and J. H. Koeman, Polychlorinated biphenyl (PCB) contaminated fish induces vitamin A and thyroid hormone deficiency in the common seal (*Phoca vitulina*), *Aquat. Toxicol.* **15,** 99–106 (1989).
91. M. H. Zile, Vitamin A homeostasis endangered by environmental pollutants, *Proc. Soc. Exp. Biol. Med.* **201,** 141–153 (1992).
92. T. A. Mably, R. W. Moore, and R. E. Peterson, *In utero* and lactational exposure of male rats to 2,3,7,8-tetrachlorodibenzo-p-dioxin: 1. Effects on androgenic status, *Toxicol. Appl. Pharmacol.* **114,** 97–107 (1992).
93. P. A. Spear and T. W. Moon, Low dietary iodine and thyroid anomalies in ring doves, *Streptopelia risoria,* exposed to 3,3′,4,4′ tetrachlorobiphenyl, *Arch. Environ. Contam. Toxicol.* **14,** 547–553 (1985).
94. T. A. Mably, R. W. Moore, R. W. Goy, and R. E. Peterson, *In utero* and lactational exposure of male rats to 2,3,7,8-tetrachlorodibenzo-p-dioxin: 2. Effects on sexual behavior and the regulation of luteinizing hormone secretion in adulthood, *Toxicol. Appl. Pharmacol.* **114,** 108–117 (1992).
95. E. Enan, P. C. C. Liu, and F. Matsumura, 2,3,7,8-Tetrachlordibenzo-p-dioxin causes reduction of glucose transporting activities in the plasma membranes of adipose tissue and pancreas from the guinea pig, *J. Biol. Chem.* **267,** 19785–19791 (1992).
96. S. Jensen and B. Jansson, Anthropogenic substances in seal from the Baltic: Methyl sulphone metabolites of PCb and DDT, *Ambio* **5,** 257–260 (1976).
97. R. F. Seegal, B. Bush, and K. O. Brosch, Sub-chronic exposure of the adult rat to Aroclor® 1254 yields regionally-specific changes in central dopaminergic function, *Neurotoxicology* **12,** 55–66 (1991).
98. R. F. Seegal, B. Bush, and K. O. Brosch, Comparison of effects of Aroclors® 1016 and 1260 on non-human primate catecholamine function, *Toxicology* **66,** 145–163 (1991).
99. J. A. van Zorge, J. H. van Wijnen, R. M. C. Theelen, K. Olie, and M. van den Berg, Assessment of the toxicity of mixtures of halogenated dibenzo-p-dioxins and dibenzofurans by use of toxicity equivalency factors (TEF), *Chemosphere* **19,** 1881–1895 (1989).
100. R. F. Seegal, B. Bush, and W. Shain, Lightly chlorinated ortho-substituted PCB congeners decrease dopamine in nonhuman primate brain and in tissue culture, *Toxicol. Appl. Pharmacol.* **106,** 136–144 (1990).
101. T. Zacharewski, T. Harris, S. Safe, H. Thoma, and O. Hutzinger, Applications of the *in vitro* aryl hydrocarbon hydroxylase induction assay for determining "2,3,7,8-

tetrachlorodibenzo-*p*-dioxin equivalents". Pyrolyzed brominated flame retardants, *Toxicology* **51,** 177–189 (1988).
102. T. Zacharewski, L. Safe, S. Safe, B. Chittim, D. DeVault, K. Wiberg, P. A. Berquist, and C. Rappe, Comparative analysis of polychlorinated dibenzo-*p*-dioxin and dibenzofuran congeners in Great Lakes fish extracts by gas chromatography–mass spectrometry and *in vitro* enzyme induction activities, *Environ. Sci. Technol.* **23,** 730–735 (1989).
103. N. Kannan, S. Tanabe, T. Wakimoto, and R. Tatsukawa, Coplanar polychlorinated biphenyls in Aroclor and Kanechlor mixtures, *J. Off. Assoc. Anal. Chem.* **70,** 451–454 (1987).
104. W. Shain, B. Bush, and R. Seegal, Neurotoxicity of polychlorinated biphenyls: Structure–activity relationship of individual congeners, *Toxicol. Appl. Pharmacol.* **111,** 33–42 (1991).
105. M. Mora, H. J. Auman, J. P. Ludwig, J. P. Giesy, D. A. Verbrugge, and M. E. Ludwig, PCBs and chlorinated insecticides in plasma of Caspian terns: Relationships with age, productivity and colony site tenacity, *Arch. Environ. Toxicol. Contam.* **24,** 320–331 (1992).
106. N. Kannan, S. Tanabe, and R. Tatsukawa, Toxic potential of non-ortho and mono-ortho coplanar PCBs in commercial PCB preparations: 2,3,7,8-T4 CDD toxicity equivalence factors approach, *Bull. Environ. Contam. Toxicol.* **41,** 267–276 (1988).
107. R. Bannister, D. Davis, T. Zacharewski, I. Tizard, and S. Safe, Aroclor 1254 as a 2,3,7,8-tetrachlorodibenzo-p-dioxin antagonist: Effects on enzyme induction and immunology, *Toxicology* **46,** 29–42 (1987).
108. R. Stahlmann, T. Schultz-Schalge, M. Korte, C. Renschler, I. Baumann-Wilschke, and D. Neubert, Clinical chemistry and haematology in rats after short-term and after long-term exposure to 2,3,7,8-tetrachlorodibenzo-p-dioxin (TCDD), *Chemosphere* **25,** 1207–1214 (1992).
109. R. Bannister, M. Kelly, and S. Safe, Synergistic interactions of 2,3,7,8-TCDD and 2,2′,4,4′,5,5′-hexachlorobiphenyl in C57BL/Gj and DBA/2J mice: Role of the Ah receptor, *Toxicology* **44,** 159–169 (1987).
110. R. Bannister, L. Biegal, D. Davis, B. Astroff, and S. Safe, 6-Methyl-1.3.8-trichlorodibenzofuran (MCDF) as a 2,3,7,8-tetrachlorodibenzo-p-dioxin antagonist in C57BL/6 mice, *Toxicology* **54,** 139–154 (1989).
111. A. Hanberg, M. Stahlberg, A. Georgellis, C. deWit, and V. G. Ahlborg, Swedish dioxin survey: Evaluation of the H4IIE bioassay for screening for dioxin-like enzyme induction, *Pharmacol. Toxicol.* **69,** 442–449 (1991).
112. P. D. Jones, J. P. Giesy, J. L. Newsted, A. A. Verbrugge, D., L. Beaver, G. T. Ankley, D. E. Tillitt, and K. B. Lodge, 2,3,7,8-Tetrachlordibenzo-p-dioxin equivalents in tissues of birds at Green Bay, Wisconsin, USA, *Arch. Environ. Toxicol. Safety* **24,** 345–354 (1993).
113. D. E. Tillitt, T. J. Kubiak, G. T. Ankley, and J. P. Giesy, Dioxin-like toxic potency in Forster's tern eggs from Green Bay, Lake Michigan, North America, *Chemosphere* **26,** 2079–2084 (1993).
114. D. E. Tillitt and J. P. Giesy, Michigan State University, East Lansing, unpublished data.
115. J. A. Bradlaw and J. L. Casterline, Induction of enzymes in cell cultures: A rapid screen for the detection of planar chlorinated organic compounds. *J. Assoc. Off. Anal. Chem.* **62,** 904–916 (1979).
116. T. W. Sawyer, A. D. Vatcher, and S. Safe, Comparative aryl hydrocarbon hydroxylase induction activities of commercial PCBs in Wistar rats and rat hepatoma H-4-II-E cells in culture, *Chemosphere* **13,** 695–701 (1984).
117. J. L. Casterline, J. A. Bradlaw, B. J. Puma, and Y. Ku, Screening of fresh water fish

extracts for enzyme-inducing substances by an aryl hydrocarbon hydroxylase induction bioassay technique, *J. Assoc. Off. Anal. Chem.* **66,** 1136–1139 (1983).
118. G. T. Ankley, D. E. Tillitt, and J. P. Giesy, Maternal transfer of bio-active polychlorinated aromatic hydrocarbons in spawning chinook salmon (*Oncorhynchus tschawytscha*), *Mar. Environ. Res.* **28,** 231–234 (1989).
119. J. P. Ludwig, H. J. Auman, H. Kurita-Matsuba, M. E. Ludwig, L. M. Campbell, J. P. Giesy, D. E. Tillitt, P. Jones, N. Yamashita, S. Tanabe, and R. Tatsukawa, Caspian tern reproduction in the Saginaw Bay ecosystem following a 100-year flood event, *J. Great Lakes Res.* **19,** 96–108 (1993).
120. G. T. Ankley, D. E. Tillitt, J. P. Giesy, P. D. Jones, and D. A. Verbrugge, Bioassay-derived 2,3,7,8-tetrachlorodibenzo-p-dioxin equivalents (TCDD-EQ) in the flesh and eggs of Lake Michigan chinook salmon and possible implications for reproduction, *Can. J. Fish. Aquat. Sci.* **48,** 1685–1690 (1991).
121. G. T. Ankley, G. T. Niemi, K. B. Lodge, H. J. Harris, D. L. Beaver, D. E. Tillitt, T. R. Schwartz, J. P. Giesy, P. D. Jones, and C. Hagley, Uptake of planar polychlorinated biphenyls and 2,3,7,8-substituted polychlorinated dibenzofurans and dibenzo-p-dioxins by birds nesting in the lower Fox River/Green Bay, Wisconsin, *Arch. Environ. Contam. Toxicol.* **24,** 332–334 (1993).
122. L. L. Williams, J. P. Giesy, N. DeGalan, D. A. Verbrugge, D. E. Tillitt, and G. T. Ankley, Prediction of concentrations of 2,3,7,8-TCDD equivalents (TCDD-EQ) in trimmed, chinook salmon filets from Lake Michigan from total concentrations of PCBs and fish size, *Environ. Sci. Technol.* **26,** 1151–1159 (1992).
123. L. S. Birnbaum, Distribution and excretion of 2,3,7,8-tetrachlordibenzo-p-dioxin in congenic strains of mice which differ at the Ah locus, *Drug. Metab. Dispos.* **14,** 34–40 (1986).
124. J. Haake, S. Safe, K. Mayura, and T. D. Phillips, Aroclor 1254 as an antagonist of the teratogenicity of 2,3,7,8-tetrachlorodibenzo-p-dioxin, *Toxicol. Lett.* **38,** 299–306 (1987).
125. B. T. Astroff, T. Zacharewski, S. Safe, M. P. Arlatto, and A. Parkinson, 6-Methyl-1,3,8-trichlorodibenzofuran as a 2,3,7,8-tetrachlorodibenzo-p-dioxin antagonist: Inhibition of the induction of rat cytochrome P-450 isozymes and related monooxygenase activities, *Mol. Pharmacol.* **33,** 231–236 (1988).
126. L. M. Biegel, M. Harris, D. Davis, R. Rosengren, L. Safe, and S. Safe, 2,2′,3,3′,5,5′-Hexachlorobiphenyl as a 2,3,7,8-tetrachlorodibenzo-p-dioxin antagonist in C57BL/6j mice, *Toxicol. Appl. Pharmacol.* **97,** 561–571 (1989).
127. D. M. Janz and C. D. Metcalf, Nonadditive interactions of mixtures of 2,3,7,8-TCDD and 3,3′,4,4′-tetrachlorobiphenyl on aryl hydrocarbon hydroxylase induction in rainbow trout (*Oncorhynchus mykiss*), *Chemosphere* **23,** 467–472 (1991).
128. J. van der Kolk, A. P. J. M. van Birgelen, H. Poiger, and C. Schlatter, Interactions of 2,2′,4,4′,5,5′-hexachlorobiphenyl and 2,3,7,8-tetrachlorodibenzo-p-dioxin in a subchronic feeding study in the rat, *Chemosphere* **25,** 2023–2027 (1992).
129. D. Schrenk, H.-P. Lipp, T. Wiesmüller, H. Hagenmaier, and K. W. Bock, Assessment of biological activities of mixtures of polychlorinated dibenzo-p-dioxins: Comparison between defined mixtures and their constituents, *Arch. Toxicol.* **65,** 114–118 (1990).
130. T. R. Schwartz and D. L. Stalling, Chemometric comparison of polychlorinated biphenyl residues and toxicologically active polychlorinated biphenyl congeners in the eggs of Forster's terns (*Sterna fosteri*), *Arch. Environ. Contam. Toxicol.* **20,** 183–199 (1991).
131. B. G. Oliver and A. Niimi, Trophodynamic analysis of polychlorinated biphenyl congeners and other chlorinated hydrocarbons in the Lake Ontario ecosystem, *Environ. Sci. Technol.* **22,** 388–397 (1988).
132. B. G. Oliver, M. N. Charlton, and R. W. Durham, Distribution, redistribution, and

geochemistry of polychlorinated biphenyl congeners and other chlorinated hydrocarbons in Lake Ontario sediments, *Environ. Sci. Technol.* **23,** 200–208 (1989).
133. J. F. Brown, Metabolic alterations of PCB residues in aquatic fauna: Distributions of cytochrome P4501A- P4502B-like activities, *Mar. Environ. Res.* **34,** 261–266 (1992).
134. J. P. Boon, F. Eijgenraam, J. M. Everaats, and J. C. Duinker, A structure–activity relationship (SAR) approach towards metabolism of PCBs in marine animals from different trophic levels, *Mar. Environ. Res.* **27,** 159–176 (1989).
135. J. T. Borlakoglu, J. P. G. Wilkins, and C. H. Walker, Polychlorinated biphenyls in fish-eating sea birds—Molecular features and metabolic interpretations, *Mar. Environ. Res.* **24,** 15–19 (1988).
136. C. H. Walker, Persistent pollutants in fish-eating sea birds—Bioaccumulation, metabolism and effects, *Aquat. Toxicol.* **17,** 293–324 (1990).
137. D. Broman, C. Näf, C. Rolff, Y. Zebühr, B. Fry, and J. Hobbie, Using ratios of stable nitrogen isotopes to estimate bioaccumulation and flux of polychlorinated dibenzo-p-dioxins (PCDDs) in two food chains from the northern Baltic, *Environ. Toxicol. Chem.* **11,** 331–345 (1992).
138. R. J. Pruell, R. D. Bowen, S. J. Fluck, J. A. LiVosi, D. J. Cobb, and J. L. Lake, *PCB congeners in American lobster, Homarus americanus and winter flounder, Pseudopleuronectes americanus, from New Bedford Harbor, Massachusetts,* USEPA Environmental Assessment Group Report, Washington, DC (1988).
139. D. T. H. M. Sijm, A. L. Yarechewski, D. C. G. Muir, G. R. B. Webster, W. Seinen, and A. Opperhuizen, Biotransformation and tissue distribution of 1,2,3,7-tetrachlorodibenzo-p-dioxin, 1,2,3,4,7-pentachlorodibenzo-p-dioxin and 2,3,4,7,8-pentachlorodibenzofuran in rainbow trout, *Chemosphere* **21,** 845–866 (1990).
140. G. Sundström and O. Hutzinger, The synthesis of chlorinated diphenyl ethers, *Chemosphere* **5,** 305–312 (1976).
141. O. Hutzinger, D. M. Nash, S. Safe, A. S. W. DeFreitas, R. J. Norstrom, D. J. Wildish, and V. Zitko, Polychlorinated biphenyls: Metabolic behavior of pure isomers in pigeons, rats and brook trout, *Science* **178,** 312–314 (1972).
142. J. C. Gage and S. Holm, The influence of molecular structure on the retention and excretion of polychlorinated biphenyls by the mouse, *Toxicol. Appl. Pharmacol.* **36,** 555–560 (1976).
143. H. B. Matthews and M. W. Anderson, Effect of chlorination on the distribution and excretion of polychlorinated biphenyls, *Drug. Metab. Dispos.* **3,** 371–380 (1975).
144. H. B. Matthews and D. B. Tuey, The effect of chlorine position on the distribution and excretion of four hexachlorobiphenyl isomers, *Toxicol. Appl. Pharmacol.* **53,** 377–388 (1980).
145. S. Focardi, C. Leonzio, and C. Fossi, Variations in polychlorinated biphenyl congener composition in eggs of Mediterranean water birds in relation to their position in the food chain, *Environ. Pollut.* **52,** 243–255 (1988).
146. P. Kopponen, J. Tarhanen, J. Ruuskanen, R. Törrönen, and S. Kärenlampi, Peat induces cytochrome P4501A1 in Hepa-1 cell line, comparison with fly ashes from combustion of peat coal, heavy fuel oil and hazardous waste, *Chemosphere* **26,** 1499–1506 (1993).
147. P. Haglund, K. E. Egebäck, and B. Jansson, Analysis of polybrominated dioxins and furans in vehicle exhaust, *Chemosphere* **17,** 2129–2140 (1988).
148. J. Koistinen and T. Nevalainen, Identification and level estimation of aromatic coelutes of polychlorinated dibenzo-p-dioxins and dibenzofurans in pulp mill products and wastes, *Environ. Sci. Technol.* **26,** 2499–2507 (1920).
149. H. R. Buser, L. O. Kjeller, S. E. Swanson, and C. Rappe, Methyl-, polymethyl and

alkylpolychlorodibenzofurans identified in pulp mill sludge and sediments, *Environ. Sci. Technol.* **20,** 404–408 (1989).
150. T. Nevalainen and J. Koistinen, Model compound synthesis for the structure determination of new unknown planar aromatic compounds originating from pulp mill, *Chemosphere* **23,** 1581–1589 (1991).
151. U. Järnberg, L. Asplund, C. de Wit, A.-K. Grafström, P. Haglund, B. Jansson, K. Lexén, M. Strandell, M. Olson, and B. Johnson, Polychlorinated biphenyls and polychlorinated naphthalenes in Swedish sediment and biota: Levels, patterns and time trend, *Environ. Sci. Technol.* **27,** 1364–1374 (1993).
152. M. Becker, T. Phillips, and S. Safe, Polychlorinated diphenyl ethers: A review, *Toxicol Environ. Chem.* **33,** 189–200 (1991).
153. Y. C. Chui, R. F. Addison, and F. C. P. Law, Acute toxicity and toxicokinetics of chlorinated diphenyl ethers in trout, *Xenobiotica* **20,** 489–499 (1990).
154. Y. C. Chui, M. M. Hansell, R. F. Addison, and F. C. P. Law, Effects of chlorinated diphenyl ethers on the mixed-function oxidases and ultrastructure of rat and trout liver, *Xenobiotica* **20,** 489–494 (1985).
155. C. J. Stafford, Halogenated diphenyl ethers identified in avian tissues and eggs by GC/MS, *Chemosphere* **12,** 1487–1495 (1983).
156. T. W. Custer, C. M. Bunck, and C. J. Stafford, Organochlorine concentrations in prefledging common terns at three Rhode Island colonies, *Colon. Water Birds* **8,** 150–153 (1985).
157. L. Howie, R. Dickerson, D. Davis, and S. Safe, Immunosuppressive and monooxygenase induction activities of polychlorinated diphenyl ether congeners in C57BL/6N mice: Quantitative structure–activity relationships, *Toxicol. Appl. Pharmacol.* **105,** 254–263 (1990).
158. A. J. Murk, J. H. J. van den Berg, J. H. Koeman, and A. Brouwer, The toxicity of tetrachlorobenzyltoluenes (Ugilec® 141) and polychlorinated biphenyls (Aroclor® 1254 and PCB-77) compared to A*h*-responsive and Ah-non-responsive mice, *Environ. Pollut.* **72,** 46–67 (1991).
159. R. F. Addison, M. E. Zink, D. E. Willis, and J. J. Wrench, Induction of hepatic mixed function oxidase activity in trout (*Salvelinus fontinalis*) by Aroclor 1254 and some aromatic hydrocarbon replacements, *Toxicol. Appl. Pharmacol.* **63,** 166–172 (1982).
160. H. Friege, W. Stock, J. Alberti, A. Poppe, I. Juhnke, J. Knie, and W. Schiller, Environmental behavior of polychlorinated mono-methyl-substituted diphenyl-methanes (Me-PCDMs) in comparison with polychlorinated biphenyls (PCBs) II: Environmental residues and aquatic toxicity, *Chemosphere* **18,** 1367–1378 (1989).
161. P. G. Wester and F. van der Valk, Tetrachlorobenzyltoluenes in eel from the Netherlands, *Bull. Environ. Contam. Toxicol.* **45,** 69–73 (1990).
162. R. F. Addison, P. D. Hansen, H.-J. Pluta, and D. E. Willis, Effects of Ugilec-141, PCB substitute based on tetrachlorobenzyltoluenes, on hepatic mono-oxygenase induction in estuarine fish, *Mar. Environ. Res.* **31,** 137–144 (1991).
163. H. R. Buser, and C. Rappe, Determination of polychlorodibenzothiophenes, the sulfur analogues of polychlorodibenzofurans, using various gas chromatographic/mass spectrometric techniques, *Anal. Chem.* **63,** 1210–1217 (1991).
164. H. R. Buser, I. S. Dolezal, M. Wolfensberger, and C. Rappe, Polychlorodibenzothiophenes, the sulfur analogues of the polychlorodibenzofurans identified in incineration samples, *Environ. Sci. Technol.* **25,** 1637–1643 (1991).
165. D. R. Hilker, K. M. Aldous, R. M. Smith, P. W. O'Keefe, J. F. Gierthy, J. Jurusik, S. W. Hibbons, D. Spink, and R. J. Parillo, Detection of sulfur analog of 2,3,7,8-TCDD in the environment, *Chemosphere* **14,** 1275–1284 (1985).

166. E. R. Barnhart, D. G. Patterson, J. A. H. MacBride, L. R. Alexander, C. A. Alley, and W. E. Turner, Polychlorinated biphenylene production for quantitative reference material, in: *Chlorinated Dioxins and Dibenzofurans in Perspective* (C. Rappe, G. Choudhary, and L. H. Keith, eds.), pp. 501–510, Lewis Publishers, Chelsea, MI (1986).
167. T. Kashimoto and H. Miyata, Differences between Yusho and other kinds of poisonings involving only PCBs, in: *PCBs in the Environment* (J. W. Waid, ed.), Vol. 3, pp. 1–26, CRC Press, Boca Raton, FL (1987).
168. K. Lundgren, and C. Rappe, Detection of alkylated polychlorodibenzo furans and alkylated polychlorodibenzo-p-dioxins by tandem mass spectrometry for the analysis of crustacean samples, *Chemosphere* **23,** 1591–1604 (1991).
169. G. J. Niemi, T. E. Davis, G. D. Veith, and B. Vieux, Organochlorine chemical residues in herring gulls, ring-billed gulls and common terns of western Lake Superior, *Arch. Environ. Contam. Toxicol.* **15,** 313–320 (1986).
170. G. A. Fox, Practical causal inference for ecoepidemiologists, *J. Toxicol. Environ. Health* **33,** 359–373 (1991).
171. W. J. Rogan, B. C. Gladen, K. L. Hung, S. L. Koong, L. Y. Shih, J. S. Taylor, Y. C. Wu, D. Yang, U. D. Yang, N. B. Ragan, and C. C. Hsu, Congenital poisoning by polychlorinated biphenyls and their contaminants in Taiwan. *Science* **241,** 334–336 (1988).
172. M. Gilbertson, The Niagara labyrinth—The human ecology of producing organochlorine chemicals, *Can. J. Fish. Aquat. Sci.* **42,** 1681–1692 (1985).
173. D. J. Hoffman, B. A. Rattner, L. Sileo, D. Docherty, and T. J. Kubiak, Embryo-toxicity, teratogenicity, and aryl hydrocarbon hydroxylase activity in Forster's terns on Green Bay, Lake Michigan, *Environ. Res.* **42,** 176–184 (1987).
174. M. O. Cheung, E. F. Gilbert, and R. E. Peterson, Cardiovascular teratogenicity of 2,3,7,8-tetrachlorodibenzo-p-dioxin in the chick embryo, *Toxicol. Appl. Pharmacol.* **61,** 197–204 (1981).
175. T. R. Schwartz and T. J. Kubiak, U.S. Fish and Wildlife Service, unpublished data.
176. S. Tanabe, Ehime University, Matsuga, Japan, unpublished data.
177. J. P. Giesy, Michigan State University, unpublished data.
178. D. V. Weseloh, C. A. Bishop, R. J. Norstrom, and G. A. Fox, Monitoring levels and effects of contaminants in herring gull eggs on the Great Lakes 1974–1990, *Abstracts of the Cause–Effects Linkages II Symposium,* Traverse City, MI, September 27–28, 1991, Michigan Audubon Society, pp. 29–31, (1991).
179. J. P. Ludwig and H. Kurita, EPA, Ann Arbor, unpublished data.
180. W. W. Bowerman, D. A. Best, E. D. Evans, S. Postupalsky, M. S. Martel, K. Kozie, R. L. Welch, R. H. Schell, K. F. Darling, J. C. Rogers, T. J. Kubiak, D. E. Tillitt, T. R. Schwartz, P. D. Jones, and J. P. Giesy, PCB concentration in plasma of nesting bald eagles from the Great Lakes Basin, North America, in: *Organohalogen Compounds* (O. Hutzinger and H. Fiedler, eds.), Vol. 1, pp. 203–206, Ecoinforma Press, Bayreuth, Germany (1991).
181. C. L. Summer, *An Avian Ecosystem Health Indicator: The Reproductive Effects Induced by Feeding Great Lakes Fish to White Leghorn Laying Hens,* M.S. thesis, Michigan State University (1992).
182. M. Gilbertson, International Joint Commission, Windsor, Ontario (1989).
183. R. J. Stevens and M. A. Nielson, Inter- and intra-lake distributions of trace organic contaminants in surface waters of the Great Lakes, *J. Great Lakes Res.* **15,** 377–393 (1989).
184. J. E. Baker and S. J. Eisenreich, PCBs and PAHs as tracers of particulate dynamics in large lakes, *J. Great Lakes Res.* **15,** 84–103 (1989).

185. W. M. Strachan and S. J. Eisenrich, Mass balance of toxic chemicals in the Great Lakes: The role of atmospheric deposition, in: Appendix I from the Workshop on the Estimation of Atmospheric Loadings of Toxic Chemicals to the Great Lakes Basin, October 29–31, 1986, Scarborough Ontario, International Joint Commission, Windsor, Ontario (1988).
186. D. M. Swakhammer and D. E. Armstrong, Estimation of the atmospheric and non-atmospheric contributions and losses of PCBs for Lake Michigan on the basis of sediment records of remote lakes, *Environ. Sci. Technol.* **20,** 879–883 (1986).
187. P. M. Cook, R. J. Erickson, R. L. Spehar, S. P. Bradbury, and G. T. Ankley, *Interim report on the assessment of 2,3,7,8-tetrachlordibenzo-p-dioxin risk to aquatic life and associated wildlife,* Environmental Protection Agency, Washington, DC (1993).
188. U.S. EPA, *Great Lakes Water Quality Initiative: Procedure for Deriving Criteria for the Protection of Wildlife,* U.S. EPA Office of Water Regulations and Standards, *Federal Register,* March, Washington, DC (1993).
189. Anon, *Proposed Water Quality Guidance for the Great Lakes,* EPA, *Federal Register* March, **45,** 79339, Washington, DC (1985), (1993).
190. Michigan Department of Natural Resources, Water Resources Commission, *Guidelines for Part 4, Water Quality Standards (Rule 323.1057)* (1985).
191. Michigan Department of Natural Resources, Water Resources Commission.
192. U.S. EPA, *Maximum Contaminant Level, Safe Drinking Water Act* (1976).
193. International Joint Commission, *Great Lakes Water Quality Agreement of 1978 Revised. An IJC Agreement with Annexes and Terms of Reference, between the U.S. and Canada, signed at Ottawa, November 22 and Phosphorus Load Reduction Supplement Signed October 16, 1993 as Amended by Protocol Signed November 18, 1987* (1989).
194. U.S. EPA *Federal Register* **45,** 5831 (1989).
195. C. A. F. M. Romijn, R. Luttik, D. van de Meent, W. Slooff, and J. H. Canton, *Ecotoxicol. Environ. Safety* **26,** 61–85 (1993).
196. C. Cox, A. Vaillancourt, and A. F. Johnson, *A Method for Determining the Intake of Various Contaminants Through the Consumption of Ontario Sport Fish,* Environment Ontario (1989).
197. P. A. Cunningham, J. M. McCarthy, and D. Zeitlin, *Results of the 1989 census of the state fish/shellfish consumption advisory programs,* Report to U.S. EPA Office of Water Regulations and Standards, Washington, DC (1989).
198. M. L. Dourson and J. M. Clark, Fish consumption advisories: Toward a unified, scientifically credible approach, *Regul. Toxicol. Pharmacol.* **12,** 161–178 (1990).
199. W. R. Swain, Effects of organochlorine chemicals on the reproductive outcome of humans who consumed contaminated Great Lakes fish: An epidemiologic consideration, *J. Toxicol. Environ. Health* **33,** 587–639 (1991).
200. W. R. Swain, A review of research on the effects of human exposure to orally ingested polychlorinated biphenyls: Reference dose and exposure assessment, *Abstracts of the papers given at the Cause–Effects Linkages II Symposium, September 27–28, 1991, Traverse City, Michigan,* pp. 32–33 (1991).
201. J. L. Jacobson, S. W. Jacobson, and H. E. B. Humphrey, Effects of exposure to PCBs and related compounds on growth and activity in children, *Neurotoxicol. Teratol.* **12,** 319–326 (1990).
202. J. L. Jacobson, S. W. Jacobson, and H. E. B. Humphrey, Effects of *in utero* exposure to polychlorinated biphenyls and related contaminants on cognitive functioning in young children, *J. Pediatr.* **113,** 38–45 (1990).
203. J. L. Jacobson, S. W. Jacobson, and H. E. B. Humphrey, Follow-up on children from the

Michigan fish-eaters cohort study: Performance at age four, *J. Great Lakes Res.* **19,** 776–778 (1993).
204. T. Colborn, Epidemiology of Great Lakes bald eagles, *J. Environ. Toxicol. Health* **33,** 395–453 (1991).
205. T. Colborn, *Nontraditional evaluation of risk from fish contaminants,* Report for the committee on Evaluation of the Safety of Fishery Products, Food and Nutrition Board, Institute of Medicine, National Academy of Science, Washington, DC (1991).
206. T. Colborn, Background paper–An overview of the toxic substances and their effects in the Great Lakes basin ecosystem, International Joint Commission, pp. 1–24 (1989).
207. H. A. Tilson, J. L. Jacobson, and W. J. Rogan, Polychlorinated biphenyls and the developing nervous system: Cross-species comparisons, *Neurotoxicol. Teratol.* **12,** 239–248 (1990).
208. D. J. Hanson, Dioxin toxicity: New studies prompt debate, regulatory action, *Chem. Eng. News* August **12,** 7–14 (1991).
209. P. H. Abelson, Excessive fear of PCBs, *Science* 261 (1991).
210. C. Gorman, The double take on dioxin, *Time* August **26,** 42 (1991).
211. Anon, Reassessment of dioxin based on new science, *Chemecology* November, 1–2 (1991).
212. Anon, Dioxin re-examined, *Economist* March (1991).
213. Anon, Science Scope, 10-25-1991, p. 507 (1991).
214. R. E. Keenan, R. J. Wenning, A. H. Parsons, and D. J. Paustenbach, A re-evaluation of the tumor histopathology of Kociba et al. (1978) using 1990 criteria: Implications for the risk assessment of 2,3,7,8-TCDD using the linearized computer multistage model, in: *Organohalogen Compounds* (O. Hutzinger and H. Fiedler, eds.), Vol. 1, pp. 549–554, Ecoinforma Press, Bayreuth, Germany (1991).
215. Anon, *Tier I Wildlife Criteria for the Great Lakes Water Quality Initiative,* Working Group for Wildlife, Wisconsin Department of Natural Resources, Madison, October 18 (1991).
216. M. W. Layard, D. J. Paustenbach, R. J. Wenning and R. E. Keenan, Risk assessment of 2,3,7,8-TCDD using a biologically-based cancer model: A re-evaluation of the Kociba et al. (1978) bioassay, in: *Organohalogen Compounds* (O. Hutzinger and H. Fiedler, eds), Vol. 1, pp. 549–554, Ecoinforma Press, Bayreuth, Germany (1991).
217. U.S. EPA, Assessing human health risks from chemically contaminated fish and shellfish, Report No. EPA 503/8-89-002, Office of Water Regulations and Standards (WH-552), Washington, DC (1989).
218. U.S. EPA, *Federal Register* January 30 (1991).
219. U.S. EPA, *Ambient water quality criteria for Polychlorinated Biphenyls,* Report No. EPA 440/5-5-80-068, Environmental Protection Agency, Office of Water Regulations and Standards, Washington, DC (1980).
220. T. Colborn, Epidemiology of Great Lakes eagles, *J. Toxicol. Environ. Health* **33,** 395–453 (1991).
221. T. Colborn, H. A. Bern, P. Blair, S. Brasseur, G. R. Cunha, W. Davis, K. D. Dohler G. Fox, M. Fry, E. Gray, R. Green, M. Hines, T. J. Kubiak, J. McLochlan, J. P. Meyers, R. E. Peterson, P. J. H. Reyners, A. Sota, G. van der Kraak, F. Vom Saal, and P. Whitten, *Chemically Induced Alterations in Sexual Development: The Wildlife/Human Connection,* Elsevier, Amsterdam (1992).
222. B. Leece, M. A. Dinomme, R. Yowner, A. Li, and J. Landers, Nonadditive interactive effects of polychlorinated biphenyl congeners in rats: Role of the 2.3.7.8-tetrachlorodibenzo-p-dioxin receptor, *Can. J. Physiol. Pharmacol.* **65,** 1908–1912 (1987).
223. H. J. Harris, Persistent toxic substances and birds and mammals in the Great Lakes, in:

Toxic Contaminants and Ecosystem Health: A Great Lakes Focus (M. S. Evans, eds.), pp. 557–559, Wiley, New York (1988).

224. T. Clark, K. Clark, S. Patterson, D. Mackay, and R. Norstrom, Wildlife monitoring, modeling and fugacity, *Environ. Sci. Technol.* **22,** 120–127 (1988).
225. J. F. Brown, R. W. Lawton, M. R. Ross, and J. Feingold, in: *Organohalogen Compounds* (O. Hutzinger and H. Fiedler, eds.), pp. 283–286, Ecoinforma Press, Bayreuth, Germany (1991).
226. P. Mineau, G. A. Fox, R. J. Norstrom, D. V. Weseloh, D. J. Hallet, and J. A. Ellenton, in: *Toxic Contaminants in the Great Lakes* (J. O. Nriagu and M. S. Simons, eds.), pp. 426–452, Wiley, New York (1989).
227. C. D. Wren, Cause–effect linkages between chemicals and populations of mink (*Mustella vison*) and otter (*Lutra canadensis*) in the Great Lakes Basin, *J. Toxicol. Environ. Health* **33,** 549–585 (1991).
228. C. A. Bishop, R. J. Brooks, J. H. Carey, P. Ng, R. J. Norstrom, and D. R. S. Lean, The case for a cause–effect linkage between environmental contamination and development in eggs of the common snapping turtle (*Chelydra serpentina*) from Ontario, Canada, *J. Toxicol. Environ. Health* **33,** 521–547 (1991).
229. D. J. Hoffman, B. A. Rattner, and R. J. Hall, Wildlife toxicology, *Environ. Sci. Technol.* **24,** 276–283 (1991).
230. L. P. Burkhard, D. E. Armstrong, and A. W. Andren, Partitioning behavior of polychlorinated biphenyls, *Chemosphere* **14,** 1703–1716 (1985).

Chapter 10

Developmental and Reproductive Toxicity of Dioxins and Other Ah Receptor Agonists

H. Michael Theobald and Richard E. Peterson

1. INTRODUCTION

A well-established principle of developmental biology is that ontogeny recapitulates phylogeny. The embryo/fetus of all vertebrate species undergoes a morphologic transition from a single cell embryo to a neonate.[1] During this transition there are stages in which a casual observer would find it difficult to distinguish between the embryos of different vertebrate species, even if they were embryos of species in different vertebrate classes—fish, amphibians, reptiles, birds, and mammals. It is interesting to note that the form of toxicity in the neonate of one species can resemble that which occurs in adults of another species that is considered to be of a lower phylogenetic order. Human infants exposed perinatally to complex mixtures of halogenated aromatic hydrocarbons exhibit signs of toxicity similar to those of adult monkeys exposed only to 2,3,7,8-tetrachlorodibenzo-*p*-dioxin (TCDD).[2–12] Thus, the signs of toxicity in human infants appear to be more similar to those that occur in adult monkeys than they are to those that have been most commonly reported in adult humans.

There are four main sections in this chapter. They are (1) a section on the general principles of developmental and reproductive toxicity related specifically to the TCDD-like halogenated aromatic hydrocarbons, (2) a description of the developmental toxicity exerted by these chemicals in mammals, (3) a section on the reproductive toxicity of these chemicals in mammals, and (4) an overall summary. Relevant infor-

H. Michael Theobald • School of Pharmacy, University of Wisconsin, Madison, Wisconsin 53706. Richard E. Peterson • School of Pharmacy and Environmental Toxicology Center, University of Wisconsin, Madison, Wisconsin 53706.

Dioxins and Health, edited by Arnold Schecter. Plenum Press, New York, 1994.

mation on the developmental toxicity of TCDD-like halogenated aromatic hydrocarbons in humans will be described and emphasized. However, the bulk of information will be derived from the results of experimental studies using laboratory animals. Developmental toxicity to fish and bird species will be mentioned, although the focus will be on mammals.

Four major recurrent themes will be highlighted. Two of these are fairly well-established principles of developmental toxicology, and the remainder are hypothetical themes that relate specifically to exposure of the mammalian embryo–fetus to TCDD-like halogenated aromatic hydrocarbons. The recurrent themes are (1) that the embryo or fetus of any species is generally more susceptible to adverse effects following toxic exposure than are adult animals of the same species, (2) that chemically induced developmental toxicity is produced only when the embryo or fetus is exposed during critical periods of development, (3) the hypothesis that all adverse developmental effects of TCDD are mediated by the aryl hydrocarbon (Ah) receptor, and (4) the hypothesis that a clustering of developmental effects related to TCDD toxicity in organs that are derived from embryonic ectoderm includes toxicity to the developing nervous system. The omission of reproductive toxicity in the third theme on Ah receptor involvement is the result of the paucity of information that is available to evaluate this hypothesis with respect to reproductive toxicity in mammals. Information on biologic mechanisms will be presented where appropriate as an aid in developing the two hypothetical themes. A more in-depth discussion of potential mechanisms can be found in Peterson *et al.*[13]

2. GENERAL PRINCIPLES

2.1. The Class of Chemicals

TCDD is one of 75 chlorinated dibenzo-*p*-dioxin (PCDD) congeners. Other members of this class of halogenated aromatic hydrocarbons include the polybrominated dibenzo-*p*-dioxins (PBDDs), polychlorinated and polybrominated dibenzofurans (PCDFs and PBDFs), as well as the polychlorinated biphenyls (PCBs) and polybrominated biphenyls (PBBs). A feature common to all members of this class of chemicals is binding to the Ah receptor which is responsible for induction of aryl hydrocarbon hydroxylase (AHH), a cytochrome P-450 enzyme.[14–16] Among all members of this class of chemicals, TCDD has one of the greatest binding affinities for the Ah receptor, and is one of the most potent members at producing toxic responses.[16,17] Thus, TCDD is the prototype congener for studying developmental and reproductive toxicity induced by exposure to this class of chemicals. Developmental and reproductive toxicity following exposure to TCDD is related to the parent compound. There is no evidence that metabolites of TCDD are involved in producing developmental and reproductive toxicity.

PCDD and PBDD congeners with decreased lateral (numbered positions 2, 3, 7, and 8 on the dibenzo-*p*-dioxin tricyclic system), or increased nonlateral chlorine or bromine substitutions, bind to the Ah receptor with less affinity than that of TCDD.[16] Such congeners are also less potent than TCDD at producing developmental toxicity.[18–21] PCB congeners with zero or one chlorine atom *ortho* to the bond that links the two phenyl rings, and at least one chlorine atom in a *meta* position, can assume a coplanar conformation sterically similar to that of TCDD. In contrast, steric hindrance prevents PCB congeners with two or more *ortho* chlorines from assuming a coplanar conformation. These di-, tri-, and tetra-*ortho* congeners do not bind with high affinity to the Ah receptor and they generally do not produce the same toxicity as TCDD.[22]

The existence of two PCB classes, those that are Ah receptor agonists and those that are not, complicates the interpretation of data related to developmental and reproductive toxicity following human exposure. This is because the existing instances of large-scale human exposure have most commonly involved complex chemical mixtures rather than single agents. Experimental studies using laboratory animals, on the other hand, have been designed to study the developmental and reproductive toxicity of individual congeners. In general, the rank order potencies of the various congeners for producing developmental toxicity in laboratory animals follow those that would be predicted by their rank order affinity for binding to the Ah receptor.[13,16,18–20]

2.2. Developmental Toxicity Defined

Adverse effects of chemical exposure to the embryo or fetus can have four possible outcomes.[23]: death, retardation of growth, structural malformations, and organ system dysfunction. Exposure of the human embryo/fetus to complex mixtures of halogenated aromatic hydrocarbons during critical developmental periods can produce each of these four outcomes. In this chapter, the four possible outcomes of developmental toxicity will be divided into the following three categories: (1) death/growth/clinical signs, (2) structural malformations, and (3) functional alterations.

In nonmammalian vertebrate species, the term *embryotoxic* denotes all transient or permanent adverse effects induced by exposure of the embryo to chemicals regardless of mechanism. Similarly, the term *embryomortality* refers to death of the nonmammalian embryo at any stage of development.

For the mammalian embryo/fetus the terms *prenatal toxicity* and *prenatal mortality* will be used to denote the corresponding concepts of toxicity—short of lethality and lethality, respectively. Prenatal mortality is quantified as the percentage of nonviable embryos or fetuses at any given maternal dose or level of exposure.

Structural malformations and teratogenicity refer to abnormalities that originate from the impairment of an event that is typical of fetal development, but that is largely irreversible. Like the event itself, the impairment must be irreversible if it is to be classified as a structural malformation. Structural malformations are macroscopic in nature and they are generally apparent soon after birth. Structural malformations have been observed following exposure to TCDD-like congeners in only a few mammalian species. In most mammalian species, structural malformations following TCDD exposure during gestation do not occur at all, or if they do, relatively high levels of exposure to TCDD-like congeners are required.

Functional alterations include any organ system dysfunction that can be related to toxic exposure of the embryo/fetus. Unlike structural malformations, functional alterations are caused by developmental irregularities that are often microscopic in nature. An additional contrast is that functional alterations, unlike structural malformations, more frequently do not become apparent until well after birth. Functional alterations in mammalian species following exposure to TCDD-like Ah receptor agonists include, but are not limited to, transient or permanent dysfunctions of the nervous system.

2.3. Reproductive Toxicity Defined

Toxic effects on reproduction and reproductive behavior following exposure of sexually mature mammals to TCDD-like congeners will be considered separately for male and females. In mammals, the clinical or experimental evaluation of male reproductive toxicity would include a description of effects on the following parameters: spermatogenesis, sperm motility, sperm morphology, serum or plasma levels of important male reproductive hormones such as the gonadotropins, luteinizing hormone (LH), and follicle-stimulating hormone (FSH), as well as testosterone and other androgens, target organ responsiveness to androgens, the birth weight of live offspring, and sexual behavior and reproductive success in adulthood. In addition, when evaluating reproductive toxicity in males, it is important to quantify the prenatal death rate and occurrence of developmental toxicity in offspring sired by the affected males.[24] In females, the clinical or experimental evaluation of reproductive toxicity would include effects on the duration of the menstrual/estrous cycle, the length and intensity of menstruation, the frequency of anovulatory menstrual cycles, serum or plasma levels of important hormones such as FSH, LH, estrogen, and progesterone, sexual behavior, ability to conceive, and ability to carry pregnancy to term. Effects of TCDD-like congeners on reproduction in the female have not been as well characterized as have been the effects of these compounds on reproduction in the male. Three-generation studies can be used to assess mammalian reproduction in both males and females.[23]

2.4. Phylogenetic Aspects of Developmental Toxicity

Developmental toxicity caused by exposure to TCDD-like congeners has been extensively studied in fish, birds, and mammals. When different species, even within the same vertebrate class, are compared, there are differences with respect to the developmental effects observed. For example, liver lesions, edema, and thymic hypoplasia are hallmark signs of toxicity resulting from injection of TCDD-like congeners into chicken eggs, whereas these effects are not observed following similar injections into turkey eggs.[25] In addition, the incidence of cardiac malformations is increased in chicken embryos exposed to TCDD,[26,27] but there is no TCDD-induced increase in the incidence of these heart malformations in embryos of the ring-necked pheasant or eastern bluebird.[28,29] In mammals, cleft palate formation can result from prenatal exposure to TCDD in mice.[30] However, the mouse (relative to other mammals), like the chicken (relative to other birds), is uniquely sensitive to structural malformations caused by TCDD-like congeners.[31] In most animal species exposed to TCDD, structural malformations are not observed. The only effects of early life stage TCDD exposure common to all fish, bird, and mammalian species are decreased growth and embryomortality or prenatal mortality.

While different adverse developmental outcomes can result from exposure to TCDD-like congeners in different species, within a given species Ah receptor agonists generally produce the same pattern of developmental toxicity. Furthermore within a given species, TCDD-like congeners vary widely in the dosage or level of exposure required to produce a given developmental effect.[31] This difference in potency of TCDD-like congeners (within a given species) can be the result of pharmacokinetic dissimilarities between congeners and/or differences in their Ah receptor binding affinity and intrinsic activity.[31] Between species, however, pharmacokinetic differences do not seem sufficient to explain the species differences in susceptibility to individual TCDD-like congeners.[32]

For a given congener, factors intrinsic to the affected organs or tissues themselves, within different species, may play a role in determining their different sensitivities to toxicity. For example, isolated palatal shelves from the mouse, rat, and human respond similarly to TCDD in organ culture resulting in an organ culture equivalent of a cleft palate.[33–35] The key difference is that in isolated palatal shelves from species other than the mouse, much higher concentrations of TCDD are required to elicit the same response, and these concentrations cannot be achieved *in vivo* without causing prenatal mortality. This suggests that species differences in the susceptibility to cleft palate formation are the result of differences in the interaction between TCDD and the developing palatal shelves themselves. The net effect is that other developmental effects such as prenatal mortality predominate over cleft palate formation in mammalian species other than the mouse.

One aspect of developmental toxicity that is common to many species, even those in different vertebrate classes, is that the embryo or fetus is more susceptible to mortality induced by TCDD-like congeners than are adult animals of the same species. The LD_{50} of TCDD in rainbow trout sac fry is 25 times less than that in juvenile rainbow trout.[36,37] The LD_{50} of TCDD in the chicken embryo is 100–200 times less than the TCDD dose that causes mortality in adult chickens,[38,39] whereas fertilized ring-necked pheasant eggs are 14–23 times more sensitive to TCDD-induced lethality than are ring-necked hen pheasants.[29,40] The LD_{50} in the hamster embryo/fetus is 64–280 times less than that in adult hamsters.[41] However, there is at least one exception to the generalization of relatively greater TCDD sensitivity of the embryo/fetus compared with the adult. Bullfrog tadpoles (*Rana catesbeiana*) and adult bullfrogs are both relatively insensitive to TCDD-induced toxicity.[42]

When exposed to TCDD during adulthood, mammalian species display wide differences in the LD_{50} for TCDD. On the other hand, when exposure occurs during the developmental period, the lethal potency of TCDD tends to be more similar across species. Adult hamsters (LD_{50} 1157–5051 μg/kg) are three orders of magnitude more resistant to TCDD-induced lethality than are adult guinea pigs (LD_{50} 1 μg/kg). Yet, a maternal dose of 18 μg TCDD/kg can increase the incidence of prenatal mortality in the hamster. Since this dose is only 12-fold higher than the maternally toxic dose (1.5 μg/kg) that killed one of five guinea pig dams and increased the incidence of prenatal mortality in the guinea pig, the hamster embryo/fetus approaches that of other mammalian species in its sensitivity to TCDD-induced lethality.[43,44] Levels of maternal TCDD exposure as low as 1.5 and 1 μg/kg can produce prenatal mortality in the guinea pig and monkey, respectively. This indicates that monkeys are at least as sensitive to TCDD-induced prenatal mortality as is the most sensitive rodent species. While these levels of exposure in guinea pigs and monkeys can produce maternal lethality, the range of doses that cause prenatal mortality across mammalian species is narrower than the range of doses that cause lethality to adult animals. This indicates that the magnitude of the species differences in lethal potency of TCDD is affected by whether exposure occurs during early development or in adulthood.

PCDD and PCDF congeners that are laterally substituted, and are therefore some of the most potent Ah receptor agonists, bioaccumulate preferentially in fish, reptiles, birds, and mammals compared with congeners that lack chlorine substitution in the 2, 3, 7, and 8 positions.[45–47] Coplanar PCBs and/or mono-*ortho* chlorine-substituted analogues of the coplanar PCBs are also known to bioaccumulate in fish and wildlife as well as in humans.[48–52] The combined effects of the laterally substituted PCDD, PBDD, PCDF, PBDF, PCB, and PBB congeners acting through the Ah receptor-mediated mechanism have the potential to decrease feral fish and wildlife populations secondary to developmental toxicity.

Humans do not appear to be exempt from the occurrence of developmental toxicity induced by complex halogenated aromatic hydrocarbon mixtures. Such mixtures which contain both TCDD-like and non-TCDD-like congeners have been implicated in causing developmental toxicity in the Yusho and Yu-cheng poisoning episodes in Japan and Taiwan.[7,10,11] In these two episodes of human poisoning the mothers of the affected infants consumed rice oil that was accidentally contaminated with PCBS, PCDFs, and polychlorinated quaterphenyls (PCQs). The potency of TCDD in producing mortality is not known in either the adult human or the human embryo/fetus. However, the phylogenetic aspects of developmental toxicity suggest that the human embryo/fetus is likely to be more susceptible to such lethal effects than are human adults. Furthermore, it is likely that susceptibility of the human embryo/fetus to prenatal mortality induced by TCDD-like congeners is not substantially different from that of other mammalian species.

2.5. Critical Periods of Exposure

As already mentioned, the embryo or fetus in most vertebrate species is more sensitive to the lethal effects of TCDD-like congeners than are adult animals of the same species. In mammals the embryo/fetus, itself, is likely to be exposed to only a small fraction of the total maternal dose. The potency of TCDD for producing developmental toxicity, in the mouse for example, is clearly evident when one considers that after dosing the pregnant dam with TCDD on gestational day 11 only about 0.0003% of a maternally administered TCDD dose can be isolated from the palatal shelves or urinary tract of the embryo/fetus on gestational day 14.[33] In the mouse embryo/fetus this level of TCDD exposure in the palatal shelves and kidney results in the structural malformations cleft palate and hydronephrosis, respectively. The potency of TCDD for producing developmental toxicity in the mouse is further evidenced by the fact that cleft palate and hydronephrosis can be induced by levels of maternal TCDD exposure that produce no overtly toxic effects on the pregnant dam.[31,53] Another important feature of developmental toxicity in the mouse is that the potency of TCDD depends on the developmental stage at which exposure occurs. A difference of one gestational day can be critically important. This critical window concept will be illustrated by using the mouse as an example.

Prenatal mortality is increased in NMRI mice when a single dose of 45 μg TCDD/kg is administered to pregnant dams on gestational day 6.[54] Similarly, a single dose of 24 μg TCDD/kg administered to pregnant C57BL/6 mice on gestational day 6 increases prenatal mortality in the C57BL/6 mouse strain.[55] The incidence of prenatal mortality in these two mouse strains is less when similar doses of TCDD are given to pregnant dams between gestational day 7 and 15.[54,55] On the other hand, when daily cumulative doses of TCDD are administered to pregnant CD-1 mice on gestational days 7–16, not including day 6, it requires a daily dose of approximately 200 μg/kg to affect a significant increase in prenatal mortality.[56] It is important to note that CD-1, C57BL/6, and NMRI mice appear to be similarly sensitive to cleft palate and hydronephrosis induced by TCDD.[18,30,54,57] However, the results in CD-1 mice might suggest that this strain is less susceptible than the others with respect to TCDD-induced prenatal mortality.

The concept of a critical window for TCDD-induced prenatal mortality provides a potential explanation for the apparent insensitivity of the CD-1 embryo/fetus exposed to daily doses of TCDD on gestational days 7–16. It could very well be that the critical window for prenatal mortality in the mouse occurs on or before gestational day 6. If the embryo/fetus is not exposed to TCDD by gestational day 6, larger doses of TCDD are required to produce prenatal mortality. Given that TCDD exposure of the pregnant CD-1 dams did not begin until gestational day 7, this interpretation is consistent with the ability of a single 24 μg TCDD/kg dose to increase the incidence of prenatal mortality when administered to pregnant C57BL/6 mice on gestational day 6, but not when administered on gestational day 8, 10, 12, or 14.[55] Similar to the results with C57BL/6 mice, the largest increase in prenatal mortality in NMRI mice occurs when the mice are exposed to a single dose of TCDD on gestational day 6.[54]

Developmental effects in the mouse other than prenatal mortality are also affected by a critical window. Cleft palate can be induced by TCDD exposure when the TCDD is administered to pregnant dams on gestational days 6–12. After day 12 the ability of TCDD to cause cleft palate diminishes with time such that there is no palatal clefting when TCDD is administered after day 13 even though birth normally occurs between days 19 and 21 in the mouse.[55] Hydronephrosis can be induced when TCDD is administered at any time during the interval between gestational days 6 and 14. Unlike cleft palate, however, a peak period for the development of hydronephrosis occurs after birth. Postnatal day 1 was identified as the peak postnatal period of susceptibility to TCDD-induced hydronephrosis in neonatal mice exposed to TCDD only by suckling.[58]

There are a few general comments that can be made regarding prenatal mortality during development of the human embryo/fetus. These are that (1) death of the human embryo/fetus during the early stages of gestation, weeks 1–4, results in its resorption by the maternal system, (2) death of the human embryo/fetus during weeks 4–8 results in heavy maternal bleeding which is frequently undetected as a fetal death by the mother, and (3) death of the human embryo/fetus after 8 weeks results in expulsion of the uterine contents and spontaneous abortion.[23] The indicated time periods suggest that chemically induced prenatal mortality of the human embryo/fetus prior to the eighth week of pregnancy might not be recognized and attributed as such. Day 6, an apparent critical period for prenatal mortality in the mouse, roughly marks the onset of organogenesis in that species. The corresponding time in human pregnancy might be approximately day 13 after fertilization.[59] Given that both dose and time of exposure are important determinants of developmental toxicity, it is important to add that the minimal levels of TCDD exposure required to produce developmental toxicity in the human embryo/fetus have not been determined. However, for any given level of exposure it is likely that the time of exposure would determine what developmental toxicities, if any, could potentially be observed. Adverse effects of chemical exposure during weeks 1 and 2 of a human pregnancy result predominantly in prenatal death although other manifestations of developmental toxicity are possible. Exposure during weeks 3 through 7 results predominantly in major morphologic abnormalities and prenatal death, and exposure during weeks 8 through 36 results predominantly in functional alterations and minor structural abnormalities.[23]

2.6. Ectoderm as a Target Tissue for TCDD-Induced Developmental Toxicity

Organs that originate from embryonic ectoderm are well-known targets for toxicity following exposure of adult animals to TCDD-like congeners. For example, treatment of adult monkeys with TCDD results in effects involving the skin, meibomian glands, and nails.[2] In addition, a response similar to the chloracne that develops in adult humans[60–62] has been observed following exposure of adult monkeys to TCDD.[63] Similarly, a hallmark sign of fetal/neonatal toxicity in the human Yusho and Yu-cheng poisoning episodes is a clustering of effects that can be referred to as an ectodermal dysplasia syndrome. The effects that characterize this syndrome consist of hyperpigmentation of the skin and mucous membranes, hyperpigmentation and deformation of finger- and toenails, hypersecretion of the meibomian gland, conjunctivitis, gingival hyperplasia, presence of erupted teeth in newborn infants and altered eruption of permanent teeth, and abnormally shaped tooth roots.[3–12] Additional effects on human infants that are not related to ectoderm, but resemble effects that have been observed following exposure of adult monkeys to TCDD, include subcutaneous edema of the face and eyelids.[2,8,9,64,65] In addition, accelerated tooth eruption has been observed in newborn mice exposed to TCDD by lactation,[66] as well as in the exposed human infants mentioned above. The similarities between some of the effects observed in human infants exposed to complex mixtures of halogenated aromatic hydrocarbons with those in adult monkeys and newborn mice exposed only TCDD suggest that at least some of the effects in human infants exposed during the Yusho and Yu-cheng incidents may have been caused by exposure to TCDD-like congeners. This possibility is important given the fact that the affected human infants were perinatally exposed to contaminated rice oil that contained a complex mixture of PCBs, PCDFs, and PCQs that was ingested by the mothers of these infants. Some but not all of these congeners have TCDD-like Ah receptor binding affinity.

Chloracne, which is the most often cited effect of TCDD exposure, involving the skin in adult humans, has an animal correlate in the hairless mouse, and can be studied by using a mouse teratoma cell line in tissue culture.[17] However, it has rarely been recognized in the dioxin literature that the nervous system, like the skin, is derived from embryonic ectoderm.[1] As will be described in the section on developmental toxicity in mammals, neurobehavioral effects occur following transplacental and lactational exposure to TCDD-like congeners in mice,[67–71] as well as perinatal exposure to TCDD itself in monkeys.[72,73] In some of the Yu-cheng children who were exposed transplacentally to PCBs, PCDFs, and PCQs there was a clinical impression of a developmental delay and psychomotor delay including impairment of intellectual development.[9,74] As there is a clustering of effects of TCDD in ectoderm-derived organs, it is possible to speculate that effects of TCDD-like congeners on the nervous system are responsible for some of the neurobehavioral effects observed in these children. Further research is required, however, to characterize and elucidate the mechanisms by which TCDD affects the nervous system. In addition, it is important to maintain the perspective that some but not all toxic effects resulting from exposure to TCDD-like congeners occur in organs that are derived from ectoderm.

2.7. Antiestrogenic Actions

In mammals, estrogens are necessary for normal uterine development and for maintenance of the adult uterus. The cyclic production of estrogens partially regulates the cyclic production of FSH and LH that results in estrous cycling. In addition, estrogens are necessary for the maintenance of pregnancy. Any insult that causes a decrease in circulating levels of estrogen or target cell estrogen responsiveness in adult females can alter normal hormonal balance and action, and potentially result in reproductive toxicity.

During development in mammals, androgens secreted by the testes are important for imprinting male sexual characteristics and sexual behavior, since in the absence of androgens the genetic development of genetic males is feminized.[75–78] However, for some developmental effects in the central nervous system, androgens must first be aromatized to form estrogen prior to exerting an effect.[76,79] Numerous studies have shown that the administration of an antiestrogen during the critical period in which sexual dimorphism of the CNS develops can cause profound effects in the brain of the male rat. For example, the antiestrogen tamoxifen demasculinizes the sexually dimorphic nucleus located in the preoptic area of the hypothalamus so that it is smaller in size, and resembles that of the female rat.[80] In addition, perinatal exposure to this compound can inhibit subsequent testicular development so that the testes are aspermatozoic.[81] In contrast, the administration of estrogen to neonatal female rats causes masculinization and defeminization of the CNS similar to the effects of aromatizable androgens.[82]

Gonadectomy in bird embryos results in the retention of masculine characteristics. Thus, in bird embryos, unlike mammals, the default sex is male and gonadal steroids are required to promote the development of feminized characteristics.[75,83] For this reason antiestrogenic compounds might result in the retention of masculine characteristics in bird embryos, whereas they cause feminization and demasculinization of the developing nervous system in mammals. A number of man-made environmental contaminants have estrogenic activity and have caused developmental irregularities in the reproductive tracts of various bird species.[84] However, the effects of TCDD on the reproductive tracts of male and female bird embryos have not been established. The results in mammals suggest that TCDD is an antiestrogen rather than an estrogen. Therefore, it would not be expected to produce the same effects on sexual development and function in birds as those that are produced by estrogenic substances such as DDT, Methoxychlor, mirex, and their metabolites.

In female mice and rats, and in cultured cell lines, exposure to TCDD results in antiestrogenic actions. TCDD treatment of female mice decreases uterine weight and antagonizes the ability of exogenous 17β-estradiol to increase uterine weight.[85] Similarly in rats, TCDD treatment results in decreased uterine weight, decreased uterine peroxidase activity, and decreased tissue concentration of estrogen receptors within the uterus and liver.[86] When TCDD and 17β-estradiol are co-administered to the same female rat, TCDD diminishes or prevents the 17β-estradiol-induced increases in uterine weight, peroxidase activity, progesterone receptor concentration, and expression of EGF receptor mRNA.[86–88] In MCF-7 human breast adenoma cells treated with TCDD *in vitro,* the antiestrogenic effects of TCDD include a reduction of the 17β-estradiol-induced secretion of certain estrogen-dependent proteins, an inhibition of the estrogen-induced postconfluent cell proliferation, and a reduction in the number of occupied nuclear estrogen receptors.[89–92] In addition, TCDD treatment of MCF-7 cells causes induction of cytochrome P-450 drug metabolizing enzymes and results in increased metabolism of estrogens.[93,94]

It is important to note that the relative potencies of TCDD-like congeners in producing antiestrogenic actions are consistent with their relative binding affinities for the Ah receptor both *in vivo*[87] and *in vitro.*[89,95] Thus, the mechanism by which TCDD-like congeners produce antiestrogenic effects probably involves an interaction with the Ah receptor. While the potential developmental toxicities that could be caused by the antiestrogenic effects of TCDD-like congeners in birds and mammals have not yet been fully evaluated, there are functional reproductive system alterations in adult male rats that were exposed to TCDD perinatally.[96–99] One potential cause of these functional alterations is that sexually dimorphic characteristics within the brain fail to be imprinted because of the TCDD-induced antiestrogenic effects during development.

3. DEVELOPMENTAL TOXICITY IN MAMMALS

3.1. Death/Growth/Clinical Signs

Gestational exposure to TCDD produces a characteristic pattern of toxicity in most laboratory mammals that consists of thymic hypoplasia, subcutaneous edema, decreased prenatal growth, and prenatal mortality.[53] In addition to these common prenatal toxicities resulting from TCDD exposure are other effects that are highly species specific. Examples of the latter are cleft palate formation in the mouse,[30,31,57] an increased incidence of extra ribs in the rabbit,[100,101] and intestinal hemorrhage in the rat.[102,103] At large enough doses, TCDD exposure can result in prenatal mortality which may be detectable only by staining for resorption sites on dissection of the uterus. Prenatal toxicity and prenatal mortality can, in some cases, be accompanied by maternal toxicity. The most sensitive indicators of overt maternal toxicity might include decreased body weight gain or marked edema compared with vehicle-dosed control animals.

The degree to which prenatal mortality is associated with maternal toxicity can depend on the species. In the guinea pig, rabbit, rat, and mouse, the dose–response relationship for maternal toxicity is essentially the same as that for increased prenatal mortality.[44,56,100,103] Even in the hamster where overt maternal toxicity is less severe at doses that cause prenatal mortality than it is in other species, fetuses exhibit increases in neutrophilic metamyelocytes and bands, while increases in leukocyte number and bands are also found in maternal blood.[44] More recently, it has been shown that TCDD exposure in the mouse causes rupture of the embryo–maternal vascular barrier which results in hemorrhage of fetal blood into the maternal circulation.[104] It is not known whether these extraembryonic hematologic changes are contributory or coincidental with developmental toxicity in these species. However, their occurrence reinforces the concept that prenatal mortality can be associated with maternal toxicity.

In rhesus monkeys, prenatal mortality in a particular pregnancy may or may not be associated with maternal toxicity. Some rhesus monkeys fed diets that contained 25 or 50 ppt TCDD before and during pregnancy were unable to carry their pregnancies to term even though overt toxicity was not apparent in the mothers.[2,105–107] In rhesus monkeys exposed to a total cumulative dose of 1 μg TCDD/kg during the first trimester of pregnancy, 13 of 16 pregnancies resulted in prenatal mortality. Within 20–147 days after aborting, 8 of the 13 females that had aborted showed signs of maternal toxicity and 3 of these 8 monkeys died.[108,109] Thus, 5 of 13 instances of prenatal mortality occurred in the absence of overt maternal toxicity.

In humans, decreased autophosphorylation of the EGF receptor in the placenta is associated with decreased birth weight in infants born to exposed mothers 4 years after the initial Yu-cheng exposure incident.[110] While a cause-and-effect relationship was not established, this result supports the conclusion that careful study is needed to define the relationship between maternal toxicity, placental toxicity, and developmental toxicity in humans. In addition, this relationship needs to be better described in rodents and monkeys as well.

It has been suggested that the embryotoxic effects in rats are caused by an indirect action, presumably maternal toxicity.[111] Yet, structural malformations produced in mice after TCDD exposure are apparently caused by direct effects on the palate[33–35] and ureter.[112] Thus, while some developmental effects may eventually be found to be indirectly produced through maternal toxicity and/or placental toxicity, other such effects may be caused by direct actions of TCDD exerted within the embryo/fetus itself.

3.2. Structural Malformations

In inbred strains of mice the teratogenic response is characterized by (1) altered cellular proliferation, (2) metaplasia, and (3) modified terminal differentiation.[17] This response in the mouse is extremely organ specific such that it occurs only in the palate, kidney, and thymus.[31] In addition, the degree to which each of the three characteristics affects the developmental process depends on the organ affected.

The medial edge epithelium in the developing palatal shelves is an ectoderm that retains the ability to transform into mesenchyme. On completion of the epithelial-to-mesenchyme transformation process, the once separate and apposing palatal shelves are fused so that a single continuous tissue is formed.[113,114] When isolated palatal shelves are exposed to TCDD as explants *in vitro,* the epithelial-to-mesenchyme transformation of the basal epithelial cells is prevented by TCDD exposure. Instead, there is a differentiation into a stratified squamous epithelium such that the cells originating from the basal epithelium eventually resemble the squamous keratinizing cells that normally occur within the oral epithelium.[33–35] These results indicate that characteristics 2 and 3 (metaplasia and modified terminal differentiation) play predominant roles in producing a palatal cleft. On the other hand, hydronephrosis is caused by hyperplasia of the luminal epithelium within the developing ureter which results in obstruction and leads to damage of the renal papilla resulting from back pressure.[115] Therefore, characteristic 1 (altered cellular proliferation) may be more predominant in producing hydronephrosis than it is in producing cleft palate formation.

3.2.1. Cleft Palate

It is interesting that the medial edge epithelium in the developing palate originates from ectoderm,[113] because cleft palate formation in TCDD-exposed mice then brings to mind the ectodermal dysplasia syndrome that was previously described for adult monkeys and human infants. At large enough concentrations of TCDD, isolated rat and human palatal shelves respond to TCDD in explant culture the same way as do isolated palatal shelves from mice.[33–35] However, in the rat embryo/fetus, very large maternally toxic doses that cause a substantial incidence of prenatal mortality are required to produce cleft palate formation.[116] In addition, no defects or clefts of the secondary palate have been reported in monkey infants[109,117] or human infants.[31,118] Therefore, it would seem that the palatal shelves in mice are uniquely sensitive to TCDD exposure, whereas other aspects of the ectodermal dysplasia syndrome or prenatal mortality are more predominant in other species.

The incidence of cleft palate formation was examined in ten inbred strains of mice, five strains with Ah receptors that have a relatively high affinity for TCDD, and five strains with lower affinity Ah receptors.[119] In the five latter strains, there was only a 0–3% incidence of cleft palate formation, whereas four of the five strains with high-affinity Ah receptors developed an incidence of at least 50%. The one strain with high-affinity Ah receptors that is resistant to TCDD-induced cleft palate formation is also resistant to cleft palate formation induced by glucocorticoids, and therefore may be a strain in which cleft palates are simply difficult to obtain.

Subsequently, Hassoun *et al.*[120] evaluated co-segregation of the *Ah* locus and 2,3,7,8-tetrachlorodibenzofuran (TCDF)-induced cleft palate formation in a series of recombinant mouse strains called BXD mice. After matings with eight different BXD strains with high-affinity Ah receptors, the incidence of TCDF-induced cleft palate formation was greater than 85%. After similar matings within eight different BXD strains with lower-affinity Ah receptors, the incidence of TCDF-induced cleft palate formation was less than 2%. These results corroborate those of Poland and Glover,[119] and they indicate that susceptibility to cleft palate formation induced by TCDD-like congeners is a trait that segregates with the *Ah* locus. Therefore, the *Ah* locus and the Ah receptor are involved in the formation of palatal clefts that are induced by TCDD-like congeners. In addition, the results suggest that the *Ah* locus may be involved in producing other effects that are part of the ectodermal dysplasia syndrome in other species.

As TCDD-induced cleft palate formation in mice is mediated by the Ah receptor, structure–activity relationships based on Ah receptor binding should predict the relative potencies of different TCDD-like congeners for producing cleft palate formation. It turns out that the rank order potency of various congeners for producing this effect is generally similar to their affinity for binding to the Ah receptor.[16,18–21]

It is consistent with the structure–activity relationships for Ah receptor binding that TCDD-like PCDD, PBDD, PCDF, PBDF, and PCB congeners are effective agonists for producing cleft palate formation.[18–21] It is also consistent with the structure–activity relationships for binding to the Ah receptor that a number of hexachlorobiphenyls do not induce cleft palate formation. These molecules either lack sufficient lateral halogen substitution, or are substituted in such a manner that they cannot achieve a coplanar conformation. Included in this category are the di- and tetra-*ortho*-chlorine-substituted 2,2′,3,3′,5,5′-, 2,2′,3,3′,6,6′- 2,2′,4,4′,5,5′-, and 2,2′,4,4′,6,6′-hexachlorobiphenyls (HCBs).[121] The structure–activity relationships for Ah receptor binding predict that the mono-*ortho*-chlorine-substituted 2,3,3′,4,4′,5-HCB is a weak teratogen. It is 3×10^5 times less potent than TCDD in cleft palate formation and 5×10^6 times less potent than TCDD in Ah receptor binding affinity.[49]

A result that would not be expected according to the structure–activity relationships for binding to the Ah receptor is that the di-*ortho*-chlorine-substituted 2,2′,3,3′,4,4′-HCB causes cleft palate formation and hydronephrosis in mice.[121] However, another di-*ortho*-chlorine-substituted PCB, 2,2′,4,4′,5,5′-HCB, can also cause hydronephrosis and is a very weak inducer of ethoxyresorufin-*o*-deethylase (EROD) activity.[122,123]

2,2′,4,4′,5,5′-HCB appears to be a partial Ah receptor agonist, because at large enough doses it can competitively displace TCDD from the murine hepatic cytosolic Ah receptor and inhibit TCDD-induced cleft palate formation and immunotoxicity in C57BL/6 mice.[122,123] These latter results suggest that PCB congeners do not have to be in a strictly coplanar configuration to bind to the Ah receptor.

3.2.2. Hydronephrosis

Hydronephrosis is the most sensitive developmental response elicited by TCDD in mice, as it is produced by maternal doses of TCDD that do not cause palatal clefting or overt maternal toxicity.[18,30,57,115,124,125] Isolated ureters from day 12 embryos exposed to 1×10^{-10} M TCDD *in vitro* display evidence of epithelial cell hyperplasia.[112] This demonstrates that epithelial cell hyperplasia, the hallmark sign of TCDD-induced hydronephrosis *in vivo,* is caused by a direct action of TCDD on the ureteric epithelium itself. As was the case with cleft palate formation, the results of genetic crosses with different strains of BXD mice show that susceptibility to hydronephrosis segregates with the *Ah* locus.[120] In addition, the structure–activity relationships for TCDD-like congeners in producing hydronephrosis are consistent with those for Ah receptor binding.[16,18–21] Therefore, the *Ah* locus and the Ah receptor are involved in producing hydronephrosis in mice. However, the urinary tract is derived from embryonic mesoderm rather than ectoderm. Therefore, effects involving the Ah receptor can be produced in mice even though these effects are not part of the ectodermal dysplasia syndrome.

Hydronephrosis has been reported after administration of low maternal doses of TCDD to rats and hamsters. Possibly because of the small number of fetuses examined, the observed incidences of hydronephrosis in rats after exposure to cumulative maternal doses of less than 2 μg TCDD/kg have not been statistically significant.[30,126] On the other hand, following a single dose of 1.5 μg TCDD/kg, administered on gestational day 7 or 9, the incidence of hydronephrosis in hamster fetuses was 11 and 4.2%, respectively. This is in contrast to an incidence of less than 1% in control hamster fetuses. Accordingly, hydronephrosis is one of the most sensitive indicators of prenatal dioxin toxicity in hamsters.[44]

3.2.3. Other Structural Malformations

While the mouse seems uniquely sensitive to cleft palate formation, and hydronephrosis has been described in the mouse, rat, and hamster, most other mammalian species seem to be unaffected by structural malformations following prenatal exposure to TCDD-like congeners. However, the few other forms of structural malformation that have been described also seem to be highly species specific. These include extra ribs produced after TCDD exposure in the rabbit,[100,101] and rocker bottom heel produced after exposure to a complex mixture of halogenated aromatic hydrocarbons in human infants affected by the Yusho poisoning incident.[8] As previously noted, however, the contaminated rice oil contained a mixture of PCBs and PCDFs some of which have

TCDD-like Ah receptor binding activity. Therefore, the rocker bottom heel effect observed in human infants cannot be linked to TCDD-like congeners with certainty. Considering all mammalian species examined, structural malformations related to TCDD exposure are not common.

3.3. Functional Alterations

3.3.1. Male Reproductive System

Testosterone and/or its metabolite dihydrotestosterone (DHT) are essential prenatally and/or early postnatally for imprinting and development of accessory sex organs,[127–129] and for initiation of spermatogenesis.[130] For the accessory sex organs the concept of imprinting essentially means that there is a critical period, which starts before birth and in some cases lasts through puberty, during which adequate concentrations of androgens are required for normal growth, function, and hormone responsiveness to occur in these organs when the animal becomes sexually mature.[127–129,131,132] If perinatal imprinting has failed to occur in the accessory sex organs of a neonatal male rat, the result could be diminished reproductive capacity when the animal becomes sexually mature. This is suggested by the results of surgical ablation experiments which indicate that the seminal vesicles and dorsal lateral lobes of the prostate are essential for fertility in the rat.[133] On the other hand, the degree to which a perinatal androgen deficiency can mimic the effects of surgical ablation of these organs is not known.

Sexual differentiation of the CNS in mammalian males is dependent on the presence of androgens during early development. In rats the critical period of sexual differentiation within the CNS extends from late fetal life to the first week of postnatal life.[134] When adequate circulating levels of testicular androgens are absent during this time, adult male rats display high levels of feminine sexual behavior (e.g., lordosis), low levels of masculine sexual behavior, and a feminized pattern of LH secretion in adulthood.[135,136] In contrast, perinatal exposure of rats to androgens will result in the masculinization of sexually dimorphic neural parameters including reproductive behaviors, regulation of LH secretion, and several morphologic indices such as the size of certain brain nuclei.[137,138]

Because of the metabolism of androgens within the CNS, the mechanisms by which these hormones cause sexual differentiation of the CNS is not completely understood. 17β-Estradiol can be formed within the CNS by the local aromatization of testosterone. In rats, mice, and hamsters, 17β-estradiol biosynthesized within the brain from testosterone may be the major steroid responsible for mediating sexual differentiation of the CNS.[76,139] In rats, 5α-DHT, unlike testosterone or 17β-estradiol, is not able to promote either masculinization of male sexual behavior or suppression of female sexual behavior. On the other hand, prenatal administration of DHT to guinea pigs and monkeys can facilitate some effects on masculine sexual differentiation.[139] This last result would indicate that androgens *per se* are important mediators of sexual differentiation in guinea pigs and monkeys because DHT cannot be aromatized to estrogens. Even in rats, androgens themselves appear to be required for the normal differentiation of certain nonreproductive male behaviors, as well as spinal reflexes involved in the copulatory response in males.[134,140] It is notable that such spinal reflexes can be activated in the rat by DHT, whereas 17β-estradiol does not have any appreciable effect.[140] Thus, the degree to which estrogens, rather than androgens, play the predominant role in mediating sexual differentiation of the CNS is species dependent, but in many species both are involved.

Since TCDD can be transferred from mother to young during lactation,[141,142] and it can cause decreased plasma concentrations of testosterone and DHT in adult male rats,[143] it was postulated that TCDD exposure during development might impact on the male reproductive system. Effects on perinatal TCDD exposure on the male reproductive system have been described, but the extent to which changes in plasma testosterone concentrations in fetal rats can explain the observed effects is not known.[96–99] The following section describes the effects of low-level, perinatal TCDD exposure on prenatal and postnatal hormone levels, secondary sex organs, spermatogenesis, and sexual behavior in the rat. It turns out that the maternal dosage levels required to produce some of these effects in rats are so low that the effects themselves represent some of the most sensitive indicators of TCDD toxicity in any mammalian species.

3.3.1a. Prenatal Hormone Levels. Prenatal exposure to TCDD results in reduced plasma testosterone concentrations in fetal rats surgically removed from their mother and evaluated prior to birth. In unexposed male fetuses, mean plasma testosterone concentrations are greater than those in unexposed female fetuses on days 17–21 of gestation (birth normally occurs between days 20 and 22 in the rat). This sex-based difference is greatest during the time of the prenatal testosterone surge which occurs in male fetuses on days 17–19.[97] Following exposure to a maternal dose of 1 μg TCDD/kg administered on day 15 of gestation, plasma testosterone concentrations are reduced in male fetuses such that the magnitude of the sex-based difference is diminished on gestational days 18–21.[96,97]

3.3.1b. Postnatal Hormone Levels. In unexposed male rats, there is a postnatal peak in plasma testosterone concentrations that occurs approximately 2 hr after birth.[97,144] In male rats prenatally exposed to a maternal dose of 1 μg TCDD/kg on day 15 of gestation, the postnatal surge in plasma testosterone concentration is only one-half as large as that which occurs in unexposed male rats. In addition, the peak concentration is delayed in TCDD-exposed male rats such that it does not reach its zenith until 4 hr after birth.[96,97]

Dams exposed to a single dose of TCDD (1 μg/kg) while pregnant were allowed to nurse their pups, and after weaning, plasma testosterone concentrations were measured in male pups that had been exposed to TCDD *in utero* and via lactation. On days 32, 49, and 63 after birth, mean plasma testosterone concentrations were reduced by up to 69% after perinatal TCDD exposure. However, these differences were not statistically significant. At 120 days after birth, there was essentially no difference between TCDD-exposed and unexposed male rats with respect to their plasma testosterone concentrations.[97] When TCDD-exposed male rats were compared with unexposed animals, results similar to those described for testosterone were obtained for plasma DHT concentrations on postnatal days 49, 63, and 120. The mean plasma concentration of DHT in TCDD-exposed male rats was decreased by up to 59% relative to that in the unexposed rats, although this difference was not statistically significant. At 120 days the mean plasma DHT concentration in the male rats exposed perinatally to TCDD was not different from that of the control rats.

When measured 32 days after birth there are modest depressant effects of perinatal TCDD exposure on the plasma concentrations of LH[97] and follicle-stimulating hormone (FSH).[99] On postnatal day 32 these affects appear to be dose related with 1 μg TCDD/kg being the lowest maternal dose that results in a significant depression of the plasma concentrations of either hormone. However, the effect of perinatal TCDD exposure on these gonadotropin concentrations is transitory because no effect on either hormone concentration is observed at 49, 63, or 120 days after birth.[99]

3.3.1c. Nonhormonal Indices of Androgenic Status. While plasma androgen concentrations can give one-time, snapshot indications of androgen status, certain measurements of androgen target organs can be used to produce an index of androgenic status that is integrated over time. These measurements, which are dependent on circulating androgen concentrations and androgenic responsiveness, include anogenital distance,[145] time of testis descent,[146] and changes in the growth and wet weight of accessory sex organs. A single maternal TCDD dose as low as 0.16 μg/kg administered on day 15 of gestation reduces anogenital distance in 1- and 4-day-old male pups, even when slight decreases in body length are considered.[97] In addition, testis descent is an androgen-mediated developmental event that normally occurs in rats between 20 and 25 days of age. In male rats exposed perinatally to a single maternal dose of up to 1.0 μg TCDD/kg, testis descent is delayed in a dose-dependent manner. The maximal extent of this delay is 1.7 days.[97]

When changes in feed consumption and body weight are accounted for, the lowest maternal dose of TCDD that affected an index of androgenic status in the rat was 0.16 μg/kg. This dose results in a significantly depressed ventral prostate weight expressed relative to body weight in 32-day-old pups. Maternal doses of TCDD up to 1 μg/kg result in dose-dependent, though modest, reductions in seminal vesicle, ventral prostate, and cauda epididymis weights in perinatally exposed rats that are 32, 49, 63, and 120 days old.[97,99] In this experiment, the two lowest maternal doses of TCDD used, 0.064 and 0.16 μg/kg, did not affect feed consumption or body weight in either the pregnant dams or the male pups. However, at a maternal dose of 1 μg TCDD/kg there was a statistically significant body weight decrease of no more than 22% in the male offspring. Nevertheless, the spectrum of effects on indices of androgenic status, which is caused by perinatal TCDD exposure to doses as low as 0.16 μg TCDD/kg given on gestational day 15, demonstrates that perinatal TCDD exposure can decrease androgenic status in male rats from the fetal stage into adulthood.[96] Certain of these effects tend to diminish with time, though not completely as the rat matures.[96,97] Even so, statistically significant decreases in ventral prostate weight were obtained in 120-day-old rats that had been exposed perinatally to TCDD.[97]

3.3.1d. Spermatogenesis. Decreased spermatogenesis is among the most sensitive responses of the male rat reproductive system to perinatal TCDD exposure.[96,99] Testis and epididymis weights and indices of spermatogenesis were determined on postnatal days 32, 49, 63, and 120. Perinatal TCDD exposure caused dose-related decreases in testis and epididymis weights. Weights of the caudal portion of the epididymis where mature sperm are stored prior to ejaculation were decreased the most, by approximately 45%. The number of sperm per cauda epididymis was decreased by 75 and 65% on days 63 and 120, respectively, and appeared to be the most sensitive effect of perinatal TCDD exposure on the male reproductive system. Daily sperm production was decreased by up to 43% at puberty (day 49), but the decrease was less at sexual maturity (day 120). Seminiferous tubule diameter was decreased at all four developmental stages: juvenile, pubertal, postpubertal, and mature. Each effect of TCDD was dose-related and in all cases except for testis weight, a significant decrease was seen in response to the lowest maternal TCDD dose tested (0.064 μg/kg) during at least one stage of sexual development. In general, the magnitude of the decreases recovered with time, though not completely, suggesting that perinatal TCDD exposure delays sexual maturation. The lowest dose of TCDD that caused a significant decrease in testis weight was 0.40 μg/kg when the effect was measured in 32-day-old rats. Thus, effects of *in utero* TCDD exposure on spermatogenesis in 63- and 120-day-old rats were caused by maternal doses of TCDD that were too low to affect testis weight.

In normal rats, daily sperm production does not reach a maximum until 100–125 days of age,[147] but in rats perinatally exposed to less than 1 μg TCDD/kg it takes longer for sperm production to reach the adult level. Furthermore, the length of the delay is directly related to maternal TCDD dose.[99] If the dose is 1 μg TCDD/kg or greater, the reduction in spermatogenesis may be permanent as daily sperm production is reduced in male rat offspring at 300 days of age.[148] The mechanism by which perinatal TCDD exposure decreases spermatogenesis in adulthood is unknown. Therefore, it is unclear whether the irreversible effect at the largest maternal dose (1 μg TCDD/kg) is caused by the same mechanism as that at smaller maternal doses. The largest dose results in a transient decrease in feed consumption and body weight gain, whereas the smallest doses do not have this effect. At these smaller doses it is not yet known if the male offspring eventually recover their full spermatogenic capacity.

A key observation for postulating mechanisms by which perinatal TCDD exposure reduces spermatogenesis in adulthood is the finding that the ratio of leptotene spermatocytes per Sertoli cell in the testes of 49-, 63-, and 120-day-old rats is not affected by *in utero* and lactational TCDD exposure even though daily sperm production is reduced.[99] Since Sertoli cells provide spermatogenic cells with functional and structural support[149] and the upper limit of daily sperm production in adult rats is directly dependent on the number of Sertoli cells per testis,[150] at least three possible mechanisms for the decrease in daily sperm production can be postulated. TCDD could increase the degeneration of cells intermediate in development between leptotene spermatocytes and terminal-stage spermatids (the cell type used to calculate daily sperm production); decrease postleptotene spermatocyte cell division (meiosis); and/or decrease the number of Sertoli cells per testis.[151] Elucidating the mechanism by which perinatal TCDD exposure decreases spermatogenesis is important because it is one of the most sensitive responses of the male reproductive system to TCDD.

3.3.1e. Reproductive Capability. To assess reproductive capability, male rats born to dams given TCDD (0.064, 0.16, 0.40, or 1.0 μg/kg) or vehicle on day 15 of gestation were mated with control virgin females when the males were about 70 and 120 days of age.[96,99] Fertility index of the males is defined as numbers of males impregnating females divided by number of males mated. The two highest maternal TCDD doses decreased the fertility index of the male offspring by 11 and 22%, respectively. However, these decreases were not statistically significant, and at lower doses the fertility index was not reduced. Gestation index, defined as the percentage of control dams mated with TCDD-exposed males that delivered at least one live offspring, was also not affected by perinatal TCDD exposure. With respect to progeny of these matings, there was no effect on litter size, live birth index, or 21-day survival index. When perinatal TCDD-exposed males were mated again at 120 days of age, there was no effect on any of these same parameters. Thus, despite pronounced reductions in cauda epididymal sperm reserves, when the TCDD-treated males were mated, perinatal TCDD exposure had little or no effect on fertility of male rats or on survival and growth of their offspring.[96,99]

Since rats produce and ejaculate ten times more sperm than are necessary for normal fertility and litter size,[152,153] the absence of a reduction in fertility of male rats exposed perinatally to TCDD is not inconsistent with the substantial reductions in testicular spermatogenesis and epididymal sperm reserves. In contrast, reproductive efficiency in human males is lower than that in the rat, the number of sperm per ejaculation being close to that required for fertility.[154] Thus, measures of fertility using rats do not appear to be appropriate for low-dose extrapolation in humans.[155] A percent reduction in daily sperm production in humans, similar in magnitude to that observed in rats,[96,99] could reduce fertility in men.

3.3.1f. Demasculinization of Sexual Behavior. Mably *et al.*[96,98] assessed sexually dimorphic functions in male rats born to dams given graded doses of TCDD or vehicle on day 15 of gestation. Masculine sexual behavior was assessed in male offspring at 60, 75, and 115 days of age by placing a male rat in a cage with a receptive control female and observing the first ejaculatory series and subsequent postejaculatory interval. The number of mounts and intromissions (mounts with vaginal penetration) before ejaculation were increased by a maternal TCDD dose of 1.0 μg/kg. The same males exhibited 12- and 11-fold increases in mount and intromission latencies, respectively, and a 2-fold increase in ejaculation latency. All latency effects were dose-related and significant at a maternal TCDD dose as low as 0.064 μg/kg (intromission latency) and 0.16 μg/kg (mount and ejaculation latencies). Copulatory rates (number of mounts + intromissions/time from first mount to ejaculation) were decreased to less than 43% of the control rate. This effect on copulatory rates was dose-related, and a statistically significant effect was observed at maternal TCDD doses as low as 0.16 μg/kg. Post-ejaculatory intervals were increased 35% above the control interval and a statistically significant effect was observed at maternal doses of TCDD as low as 0.40 μg/kg. Collectively, these results demonstrate that perinatal TCDD exposure demasculinizes sexual behavior.

Since perinatal exposure to a maternal TCDD dose of 1.0 μg/kg has no effect on the open field locomotor activity of adult male rats,[156] the increased mount, intromission, and ejaculation latencies appear to be specific for these masculine sexual behaviors, not secondary to a depressant effect of TCDD on motor activity. Postpubertal plasma testosterone and DHT concentrations in littermates of the rats evaluated for masculine sexual behavior were as low as 56 and 62%, respectively, of control.[96,97] However, plasma testosterone concentrations that are only 33% of control are still sufficient to masculinize sexual behavior of adult male rats.[157] Therefore, the modest reductions in adult plasma androgen concentrations following perinatal TCDD exposure were not of sufficient magnitude to demasculinize sexual behavior.

3.3.1g. Feminization of Sexual Behavior. Mably *et al.*[96,98] studied whether the potential of adult male rats to display feminine sexual behavior was altered by perinatal TCDD exposure. Male offspring of dams treated on day 15 of gestation with various doses of TCDD up to 1 μg/kg or vehicle were castrated at about 120 days of age and beginning at approximately 160 days of age were injected weekly for 3 weeks with 17β-estradiol benzoate, followed 42 hr later by progesterone. Four to six hours after the progesterone injection on weeks 2 and 3, the male was placed in a cage with a sexually excited control stud male. The frequency of lordosis in response to being mounted by the stud male was increased from 18% (control males) to 54% by the highest maternal TCDD dose, 1.0 μg/kg. Lordosis intensity scored after Hardy and DeBold[158] as (1) for light lordosis, (2) for moderate lordosis, and (3) full spinal dorsoflexion in response to a mount was increased in male rats by perinatal TCDD exposure. Both effects on lordosis behavior in males were dose-related and significant at maternal TCDD doses as low as 0.16 μg/kg (increased lordotic frequency) and 0.40 μg/kg (increased lordotic intensity). Together they indicate a feminization of sexual behavior in these animals. Although severe undernutrition from 5 to 45 days after birth potentiates the display of lordosis behavior in adult male rats,[159] the increased frequency of lordotic behavior was seen at a maternal TCDD dose, 0.16 μg/kg, which had no effect on feed intake or body weight. It was concluded that perinatal TCDD exposure feminizes sexual behavior in adult male rats independent of undernutrition.

3.3.1h. Feminization of the Regulation of LH Secretion. The effect of perinatal TCDD exposure on regulation of LH secretion by ovarian steroids was determined in male offspring at about 270 days of age.[98] There is normally a distinct sexual dimorphism in the regulation of LH secretion. In adult rats gonadectomized and primed with estrogen, progesterone injection in females causes a marked increase in plasma LH concentration, whereas no increase is seen in males similarly primed with estrogen and injected with progesterone.[160] LH secretion is not increased by progesterone in estrogen-primed control males. However, LH secretion is feminized (increased) in male rats that had been perinatally exposed to TCDD. Single maternal doses of TCDD, as low as 0.4 μg/kg, resulted in significant increases in LH secretion in male rats evaluated when they were sexually mature.[96,98] Thus, perinatal TCDD exposure increases pituitary and/or hypothalamic responsiveness of male rats to ovarian steroids in adulthood, suggesting that feminization of the sexually dimorphic pattern of LH secretion may be permanent.

3.3.1i. Comparison to Other Ah-Receptor-Mediated Responses. The induction of hepatic cytochrome P-4501A1 and its associated EROD activity are extremely sensitive Ah-receptor-mediated responses to TCDD exposure. Yet in 120-day-old male rats that had been exposed to TCDD perinatally, alterations in sexual behavior, regulation of LH secretion, and spermatogenesis were observed when induction of hepatic EROD activity could no longer be detected.[96–99] These results suggest that TCDD affects sexual behavior, gonadotropin regulation, and spermatogenesis when virtually no TCDD remains in the body. Furthermore, these effects may be irreversible.[98,99]

3.3.1j. Possible Mechanisms and Significance. The most plausible explanation for the demasculinization of sexual behavior and feminization of sexual behavior and LH secretion is that perinatal exposure to TCDD impairs sexual differentiation of the CNS. Neither undernutrition, altered locomotor activity, reduced sensitivity of the penis to sexual stimulation, nor modest reductions in adult plasma androgen concentrations of the male offspring can account for these effects.[98] On the other hand, exposure of the developing brain to testosterone, conversion of testosterone to 17β-estradiol within the brain, and events initiated by the binding of 17β-estradiol to its receptor are all critical for sexual differentiation of the CNS and have the potential to be modulated by TCDD. If TCDD interferes with any of these processes during late gestation and/or early neonatal life, it could irreversibly demasculinize and feminize sexual behavior[79,161,162] and feminize the regulation of LH secretion[163,164] in male rats in adulthood.

Treatment of dams on day 15 of gestation with 1.0 μg TCDD/kg significantly decreases plasma testosterone concentrations in male rat fetuses on days 18 and 20 of gestation and in male rat pups 2 hr postpartum.[97] Thus, the ability of maternal TCDD exposure to reduce prenatal and early postnatal plasma testosterone concentrations may account, in part, for the impaired sexual differentiation of male rats exposed perinatally to TCDD. Other mechanisms that may cause the TCDD-induced impairment in CNS sexual differentiation are: a decrease in the formation of 17β-estradiol from testosterone within the CNS (that is independent of the decrease in plasma testosterone concentrations) and/or a reduction in responsiveness of the CNS to estrogen during the critical period of sexual differentiation. The latter mechanism is consistent with the Ah-receptor-mediated antiestrogenic action of TCDD.

In utero and/or lactational exposure to TCDD may cause similar effects in other animal species, including nonhuman primates,[165–167] in which sexual differentiation is under androgenic control. In humans there is evidence that social factors account for much of the variation in sexually dimorphic behavior; however, there is also evidence that prenatal androgenization influences both the sexual differentiation of such behavior and brain hypothalamic structure.[78,168,169]

3.3.2. Neurobehavioral Effects

Following administration of [^{14}C]-TCDD in the rat the highest concentrations of TCDD-derived radioactivity in the CNS are found in the hypothalamus. Much lower concentrations are found in the cerebral cortex and cerebellum.[170] Only limited information with respect to the presence of Ah receptors at these sites is available. It has been reported that Ah receptors found in brain tissue may be associated with glial cells rather than neurons.[171,172] While Ah receptors could not be detected in human frontal cortex,[172] whole rat or mouse brain, they were detected in hamster cerebrum and guinea pig cerebellum.[173] Thus, TCDD can enter the brain and Ah receptors may be present in brain tissue. These results support the hypothesis that the nervous system is a site of action of TCDD and TCDD-like congeners, and that functional effects on the nervous system are part of the TCDD-induced ectodermal dysplasia syndrome. Included are the effects of perinatal TCDD exposure on sexual differentiation of the CNS in male rats, as well as the neurobehavioral effects of perinatal TCDD exposure in monkeys and transplacental exposure to a mixture of PCBs, PCDFs, and PCQs in Yucheng children.

3.3.2a. Mice. CD-1 mice transplacentally exposed and NMRI mice postnatally exposed to the Ah receptor agonist, 3,3′,4,4′-TCB, exhibit neurobehavioral and neurochemical alterations in adulthood.[67–71] The neurobehavioral effects in transplacentally exposed CD-1 mice consist of circling, head bobbing, hyperactivity, impaired forelimb grip strength, impaired ability to traverse a wire rod, impaired visual placement responding, and impaired learning of a one-way avoidance task.[67] The brain pathology in adult mice exhibiting this syndrome includes decreases in dopamine levels and dopamine receptor binding in the corpus striatum, both of which are associated with elevated levels of motor activity.[68,69] In another set of tests, NMRI mice dosed postnatally with 3,3′,4,4′-TCB were significantly less active than controls at the onset of testing, but more active than controls at the end of the test period. This pattern of effects can be interpreted as a failure of the treated mice to habituate to the test apparatus. In addition, it is associated with significantly altered muscarinic receptor, rather than dopamine receptor levels in the hippocampus. No effects on muscarinic receptor levels were found in the cortex.[70,71] The so-called "spinning syndrome" described for transplacentally exposed CD-1 mice was not observed in the postnatally exposed NMRI mice. Taken together these results using CD-1 and NMRI mice show that developmental neurobehavioral effects are observed following exposure to 3,3′,4,4′-TCB, although strain differences and/or the stage of development of the mouse at the time of exposure may contribute to differences in the specific effects observed. It must be noted, however, that while 3,3′,4,4′-TCB itself is an Ah receptor agonist, it can be metabolized (far more rapidly than TCDD) to metabolites that do not have Ah receptor binding activity. Therefore, these results with 3,3′,4,4′-TCB in mice cannot be linked with certainty to the Ah receptor until they are demonstrated following perinatal exposure to TCDD and other TCDD-like congeners.

3.3.2b. Monkeys. Schantz and Bowman[72] and Bowman *et al.*[73] have conducted a series of studies on the long-term neurobehavioral effects of perinatal TCDD exposure in monkeys. Because these were the first studies to evaluate the behavioral teratology of TCDD, monkeys exposed to TCDD via the mother during gestation and lactation were screened on a large number of behavioral tests at various stages of development.[73] At the dose studied (5 or 25 ppt in the maternal diet), TCDD did not affect reflex development, visual exploration, locomotor activity, or fine motor controls in any consistent manner.[174] However, perinatal TCDD exposure did produce a specific, replicable deficit in cognitive function.[72] TCDD-exposed offspring were impaired on object learning, but were unimpaired on spatial learning. TCDD exposure also produced changes in the social interactions of mother–infant dyads.[175] TCDD-exposed infants spent more time in close physical contact with their mothers. This pattern of effects was similar to that seen in lead-exposed infants and suggested that mothers were providing increased care to the TCDD-exposed infants.[175]

3.3.2c. Humans. The intellectual and behavioral development of Yu-cheng children transplacentally exposed to PCBs, PCDFs, and PCQs was studied through 1985 by Rogan *et al.*[9] In Yu-cheng children, matched to unexposed children of similar age, area of residence, and socioeconomic status, there was a clinical impression of developmental or psychomotor delay in 12 (10%) Yu-cheng children compared with 3 (3%) control children, and of a speech problem in 8 (7%) Yu-cheng children versus 3 (3%) control children. Also, except for verbal IQ on the Wechsler Intelligence Scale for Children, Yu-cheng children scored lower than control children on three developmental and cognitive tests.[9] Neurobehavioral data on Yu-cheng children obtained after 1985 shows that the intellectual development of these children continues to lag somewhat behind that of matched control children.[34] Also, Yu-cheng children are rated by their parents and teachers to have a higher activity level, more health, habit, and behavioral problems, and to have a temperamental clustering closer to that of a "difficult child."[74] It is concluded that in humans transplacental exposure to halogenated aromatic hydrocarbons can affect CNS function postnatally. However, which congeners, TCDD-like versus non-TCDD-like, are responsible for the neurotoxicity is unknown.

4. REPRODUCTIVE TOXICITY IN MAMMALS

4.1. Male Reproductive System

When given to adult animals in sufficient dosage to cause overt toxicity, TCDD-like congeners can decrease plasma androgen concentrations, cause abnormal testicular morphology, decrease spermatogenesis, and reduce fertility. Certain of these effects have been reported in rhesus monkeys, rats, guinea pigs, and mice treated with either TCDD, TCDD-like congeners, or toxic fat that was later discovered to contain TCDD.[103,176–181] Effects of TCDD on spermatogenesis are characterized by degenerating spermatocytes and mature spermatozoa within the seminiferous tubules, and a reduction in the number of tubules that contain mature spermatozoa.[176,179,180]

In contrast to the highly sensitive effects of perinatal TCDD exposure on the reproductive system of adult male rats, exposure to TCDD during adulthood is not nearly as sensitive in causing male reproductive toxicity. On the other hand, plasma testosterone and DHT concentrations are decreased to a greater extent following adult TCDD exposure, compared with the relatively small decrease in plasma androgen concentrations that occurs following perinatal TCDD exposure.[97,143] This last point emphasizes the vital importance of adequate serum androgen concentrations during male reproductive system development. The mechanism by which TCDD decreases serum androgen concentrations in adult rats is thought to involve effects on steroid synthesis in the testis,[143,182,183] and effects on the regulation of LH secretion by the hypothalamus/pituitary axis. In the testis of rats exposed in adulthood, TCDD inhibits testosterone production by decreasing the rate-limiting step (cholesterol mobilization) in the testosterone biosynthetic pathway.[184] This contributes to the fall in plasma testosterone concentrations. However, even in the face of decreased serum androgen concentrations, there is no compensatory increase in serum LH concentrations.[183–187] This occurs because TCDD increases the potency of androgens and estrogens as feedback inhibitors of pituitary LH secretions. Thus, the hypothalamic/pituitary axis is a target for TCDD toxicity in adulthood contributing to the androgenic deficiency by failing to increase LH secretion when plasma androgen concentrations are low.[186,187]

4.2. Female Reproductive System

The most significant effects of TCDD on the female reproductive system in rats and monkeys appear to be decreased fertility, inability to maintain pregnancy for the full gestational period in monkeys, and decreased litter size in rats. In some studies, reduced plasma concentrations of estrogen and progesterone, and signs of ovarian dysfunction such as anovulation and suppression of the estrous cycle have been reported.[105,106,177] Unfortunately, only limited attention has been given to the effects of TCDD on the female reproductive system, especially in the nonpregnant state. The following sections review effects of TCDD on female reproduction.

4.2.1. Reproductive Function/Fertility

The three-generation study of Murray *et al.*[188] examined the reproductive effects of exposure of male and female rats to relatively low levels of TCDD (0, 0.001, 0.01, and 0.1 μg/kg per day). The results showed exposure-related effects on fertility, an increased time between first cohabitation and delivery, and a decrease in litter size. The effects on fertility and litter size were observed at 0.1 μg/kg per day in the F_0 generation and at 0.01 μg/kg per day in the F_1 and F_2 generations. Additionally, in a 13-week exposure to 1–2 μg/kg per day of TCDD in nonpregnant female rats, Kociba *et al.*[177] reported anovulation and signs of ovarian dysfunction, as well as suppression of the estrous cycle. However, at daily exposures of 0.001–0.01 μg/kg in a 2-year study, Kociba *et al.*[189] reported no effects on the female reproductive system. The NOAEL and LOAEL for TCDD toxicity were respectively 0.001 and 0.01 μg/kg per day in both the Kociba *et al.*[189] and Murray *et al.*[188] studies. However, Murray *et al.*[188] emphasized reproductive endpoints, whereas Kociba *et al.*[189] emphasized oncogenicity.

Allen and his colleagues investigated the effects of TCDD on reproduction in the monkey.[2,72,105–107,174] In a series of studies, female rhesus monkeys were fed 25, 50, or 500 ppt TCDD for up to 9 months. Females exposed to 500 ppt TCDD showed obvious clinical signs of toxicity and lost weight throughout the study. Five of the eight monkeys died within 1 year after exposure was initiated. Following 7 months of exposure to 500 ppt TCDD, seven of the eight females were bred to unexposed males. Only three of these females were able to conceive and of these, only one was able to carry her infant to term.[106] When females exposed to 50 ppt TCDD in the diet were bred at 7 months, two of eight females did not conceive and four of the six that did conceive could not carry their pregnancies to term. As one monkey delivered a stillborn infant, only one conception resulted in a live birth.[107] At the 25 ppt dietary exposure level, only one of the eight mated females gave birth to a viable infant. As described in an abstracted summary the results at 50 and 500 ppt TCDD are compared with a group of monkeys given a dietary exposure to PBB (0.3 ppm, Firemaster FF-1) in which seven of seven exposed females were able to conceive, five gave birth to live infants, and one gave birth to a stillborn infant.[105] Similar results were obtained in a later study in a group of monkeys fed diets that contained no TCDD.[72,174] While the effects at 500 ppt TCDD may be associated with significant maternal toxicity, this would not appear to be the case at the lower doses. After 25 and 50 ppt TCDD, no overt effects on maternal health were described, but the ability to conceive and maintain pregnancy was reduced.[72,105,174]

McNulty[108] examined the effect of TCDD exposure during the first trimester of pregnancy (gestational age 25–40 days) in the rhesus monkey. At a total dose of 1 μg/kg given in nine divided doses, three of four pregnancies ended in abortion, two of these in animals that demonstrated no maternal toxicity. At a total dose of 0.2 μg/kg, one of four pregnancies ended in abortion. This did not appear different from the control population, but the low number of animals per group did not permit statistical analysis. McNulty[108] also administered single 1 μg/kg doses of TCDD on gestational day 25, 30, 35, or 40. The number of animals per group was limited to three, but it appeared that the most sensitive periods were the earlier days of pregnancy, days 25 and 30, and that both maternal toxicity and fetotoxicity were reduced when TCDD was given on later gestational days. For all days at which a single 1 μg TCDD/kg dose was given (gestational day 25, 30, 35, or 40), 10 of 12 pregnancies terminated in spontaneous abortion. Thus, of 16 monkeys given 1 μg TCDD/kg in single or divided doses between days 25 and 40 of pregnancy, there were only three normal births.[108,109]

4.2.2. Alterations in Hormone Levels

The potential for TCDD to alter circulating female hormone levels has been examined, but only to a very limited extent. In monkeys fed a diet that contained 500 ppt TCDD for up to 9 months, the length of the menstrual cycle, as well as the intensity and duration of menstruation were not appreciably affected by TCDD exposure.[106] However, there was a decrease in serum estradiol and progesterone in five of the eight exposed monkeys, and in two of these animals the reduced steroid concentrations were consistent with anovulatory menstrual cycles. In summary form, Allen *et al.*[105] described the effects of dietary exposure of female monkeys to 50 ppt TCDD. After 6 months of exposure to this lower dietary level of TCDD, there was no effect on the serum estradiol and progesterone concentrations of these monkeys. Thus, the presence of these hormonal alterations is dependent on the level of dietary TCDD exposure. Shiverick and Muther[190] reported that there was no change in circulating levels of estradiol in the rat after exposure to 1 μg/kg per day on gestational days 4–15. Thus, the effect of TCDD exposure on circulating female hormone levels may depend both on species and on level of exposure. It appears that any significant effect is only seen at relatively high exposure levels, but more research needs to be done.

5. SUMMARY

1. TCDD-like halogenated aromatic hydrocarbons cause developmental toxicity in various species of fish, birds, and mammals. In addition, they cause reproductive toxicity in mammals.

2. In fish, birds, and mammals, the embryo/fetus is generally more susceptible to TCDD-induced lethality than is the adult animal. The range of lethal potencies for TCDD across species is narrower during early development than is the range of LD_{50}s for adult exposures across the same species.

3. For certain endpoints of developmental toxicity (e.g., cleft palate, hydronephrosis), genetic studies and/or structure–activity relationships indicate that the Ah receptor plays a role in mediating the toxicity. However, it has not been shown that the Ah receptor is involved in all endpoints of developmental toxicity.

4. The particular toxic effects observed vary between different species of fish, birds, and mammals. However, within a given species, different Ah receptor agonists generally produce the same spectrum of toxic effects.

5. In birds and mammals, structural malformations occur in relatively few species. Functional alterations, on the other hand, may be the most sensitive signs of TCDD-induced toxicity in mammals.

6. In mammals, the role that maternal toxicity plays in producing adverse developmental effects is not completely understood. In some cases, however, developmental toxicity results from TCDD-induced metaplasia, modified terminal differentiation, and/or altered proliferation within the embryo/fetus itself.

7. Functional alterations following TCDD exposure in mammalian species can

lead to deficits in cognitive function in monkeys, and toxicity to the male reproductive system in rats.

8. The most sensitive toxic endpoints following perinatal TCDD exposure in male rats include decreased spermatogenesis, and demasculinization and feminization of sexual behavior after the animal becomes mature.

9. Certain effects in human infants exposed perinatally to complex mixtures resemble effects that can be produced in ectoderm-derived organs of neonatal mice and adult monkeys exposed to TCDD. This clustering of effects has been called an ectoderm dysplasia syndrome and it may include effects of perinatal exposure to TCDD-like congeners on the CNS in monkeys and human infants.

10. Developmental toxicity in human infants has been produced following accidental exposure to complex mixtures that contain Ah receptor agonists as well as other substances. Exposed human infants have been affected by developmental and/or psychomotor delays. Even though it is not known which components in the mixtures are responsible for these effects, the occurrence of neurobehavioral effects in other mammals exposed to TCDD increases the likelihood that the TCDD-like congeners present are at least partly responsible for the observed effects in human infants.

ACKNOWLEDGMENTS. The authors acknowledge Donald J. Bjerke for his constructive comments and criticisms of the manuscript. This work was supported in part by NIH grant ES01332.

REFERENCES

1. B. I. Balinski, *An Introduction to Embryology,* pp. 367–423, Saunders, Philadelphia (1970).
2. J. R. Allen, D. A. Barsotti, J. P. Van Miller, L. J. Abrahamson, and J. J. Lalich, Morphological changes in monkeys consuming a diet containing low levels of 2,3,7,8-tetrachlorodibenzo-*p*-dioxin, *Food Cosmet. Toxicol.* **15,** 401–410 (1977).
3. I. Taki, S. Hisanaga, and Y. Amagase, Report on Yusho (chlorobiphenyl poisoning) pregnant women and their fetuses, *Fukuoka Acta Med.* **60,** 471–474 (1969). [in Japanese]
4. A. Yamaguchi, T. Yoshimura, and M. Kuratsune, A survey on pregnant women having consumed rice oil contaminated with chlorobiphenyls and their babies, *Fukuoka Acta Med.* **62,** 117–121 (1971). [in Japanese]
5. I. Funatsu, F. Yamashita, T. Yosikane, T. Funatsu, Y. Ito, and S. Tsugawa, A chlorobiphenyl induced fetopathy, *Fukuoka Acta Med.* **62,** 139–149 (1971).
6. K. C. Wong and M. Y. Hwang, Children born to BCP poisoning mothers, *Clin. Med. (Taipai)* **7,** 83–87 (1981). [in Chinese]
7. S. T. Hsu, C. I. Ma, S. K. H. Hsu, S. S. Wu, N. H. M. Hsu, C. C. Yeh, and S. B. Wu, Discovery and epidemiology of PCB poisoning in Taiwan: A four-year followup, *Environ. Health Perspect.* **59,** 5–10 (1985).
8. F. Yamashita and M. Hayashi, Fetal BCP syndrome: Clinical features, intrauterine growth retardation and possible alteration in calcium metabolism, *Environ. Health Perspect.* **59,** 41–45 (1985).
9. W. J. Rogan, B. C. Gladen, K.-L. Hung, S.-L. Koong, L.-Y. Shih, J. S. Taylor, Y.-C.

Wu, D. Yang, N. B. Ragan, and C.-C. Hsu, Congenital poisoning by polychlorinated biphenyls and their contaminants in Taiwan, *Science* **241,** 334–338 (1988).
10. M. Kuratsune, Yusho, with reference to Yu-cheng, in: *Halogenated Biphenyls, Terphenyls, Naphthalenes, Dibenzodioxins and Related Products,* 2nd ed. (R. D. Kimbrough and A. A. Jensen, eds.), pp. 381–400, Elsevier, Amsterdam (1989).
11. W. J. Rogan, Yu-cheng, in: *Halogenated Biphenyls, Terphenyls, Naphthalenes, Dibenzodioxins and Related Products,* 2nd ed. (R. D. Kimbrough and A. A. Jensen, eds.), pp. 401–415, Elsevier, Amsterdam (1989).
12. S.-J. Lan, Y.-Y. Yen, Y.-C. Ko, and E.-R. Chin, Growth and development of permanent teeth germ of transplacental Yu-cheng babies in Taiwan, *Bull. Environ. Contam. Toxicol.* **42,** 931–934 (1989).
13. R. E. Peterson, H. M. Theobald, and G. L. Kimmel, Developmental and reproductive toxicity of dioxins and related compounds, *CRC Crit. Rev. Toxicol.* **23,** 283–335. (1993).
14. D. W. Nebert and J. E. Gielen, Genetic regulation of aryl hydrocarbon hydroxylase induction in the mouse, *Fed. Proc.* **31,** 1315–1325 (1972).
15. P. E. Thomas, R. E. Kouri, and J. J. Hutton, The genetics of aryl hydrocarbon hydroxylase induction in mice: A single gene difference between C57BL/6J and DBA/2J, *Biochem. Genet.* **6,** 157–168 (1972).
16. S. Safe, Polychlorinated biphenyls (PCBs), dibenzo-*p*-dioxins (PCDDs), dibenzofurans (PCDFs), and related compounds: Environmental and mechanistic considerations which support the development of toxic equivalency factors (TEFs), *Crit. Rev. Toxicol.* **21,** 51–88 (1990).
17. A. Poland and J. C. Knutson, 2,3,7,8-Tetrachlorodibenzo-*p*-dioxin and related halogenated aromatic hydrocarbons: Examination of the mechanism of toxicity, *Annu. Rev. Pharmacol. Toxicol.* **22,** 517–554 (1982).
18. H. Weber, M. W. Harris, J. K. Haseman, and L. S. Birnbaum, Teratogenic potency of TCDD, TCDF and TCDD–TCDF combinations in C57BL/6N mice, *Toxicol. Lett.* **26,** 159–167 (1985).
19. L. S. Birnbaum, M. W. Harris, E. R. Barnhart, and R. E. Morrissey, Teratogenicity of three polychlorinated dibenzofurans in C57BL/6N mice, *Toxicol. Appl. Pharmacol.* **90,** 206–216 (1987).
20. L. S. Birnbaum, M. W. Harris, D. D. Crawford, and R. E. Morrissey, Teratogenic effects of polychlorinated dibenzofurans in combination in C57BL/6N mice, *Toxicol. Appl. Pharmacol.* **91,** 246–255 (1987).
21. L. S. Birnbaum, R. E. Morrissey, and M. W. Harris, Teratogenic effects of 2,3,7,8-tetrabromodibenzo-*p*-dioxin and three polybrominated dibenzofurans in C57BL/6N mice, *Toxicol. Appl. Pharmacol.* **107,** 141–152 (1991).
22. J. A. Goldstein, The structure–activity relationsihps of halogenated biphenyls as enzyme inducers, *Ann. N.Y. Acad. Sci.* **320,** 164–178 (1979).
23. B. S. Shane, Human reproductive hazards. Evaluation and chemical etiology, *Environ. Sci. Technol.* **23,** 1187–1195 (1989).
24. J. H. Pearn, Teratogenesis and the male. An analysis with special reference to herbicide exposure, *Med. J. Aust.* **2,** 16–20 (1983).
25. B. Brunstrom and J. Lund, Differences between chick and turkey embryos in sensitivity to 3,3′,4,4′-tetrachlorobiphenyl and in concentration affinity of the hepatic receptor of 2,3,7,8-tetrachlorodibenzo-*p*-dioxin, *Comp. Biochem. Physiol.* **91C,** 507–512 (1988).
26. M. O. Cheung, E. F. Gilbert, and R. E. Peterson, Cardiovascular teratogenicity of 2,3,7,8-tetrachlorodibenzo-*p*-dioxin in the chick embryo, *Toxicol. Appl. Pharmacol.* **61,** 197–204 (1981).

27. M. O. Cheung, E. F. Gilbert, and R. E. Peterson, Cardiovascular teratogenesis in chick embryos treated with 2,3,7,8-tetrachlorodibenzo-*p*-dioxin, in: *Toxicology of Halogenated Hydrocarbons, Health and Ecological Effects* (M. A. Q. Khan and R. H. Stanton, eds.), pp. 202–208, Pergamon Press, Elmsford, NY (1981).
28. D. A. Thiel, S. G. Martin, J. W. Duncan, M. J. Lemke, W. R. Lance, and R. E. Peterson, Evaluation of the effects of dioxin-contaminated sludge on wild birds, TAPPI Proceedings, 1988 Environmental Conference, pp. 487–506 (1988).
29. J. A. Nosek, J. R. Sullinvan, S. R. Craven, A. Gendon-Fitzpatrick, and R. E. Peterson, Embryotoxicity of 2,3,7,8-tetrachlorodibenzo-*p*-dioxin in ring-necked pheasants, *Environ. Toxicol. Chem.* **17,** 1215–1222 (1993).
30. K. D. Courtney and J. A. Moore, Teratology studies with 2,4,5-trichlorophenoxyacetic acid and 2,3,7,8-tetrachlorodibenzo-*p*-dioxin, *Toxicol. Appl. Pharmacol.* **20,** 396–403 (1971).
31. L. S. Birnbaum, Developmental toxicity of TCDD and related compounds: Species sensitivities and differences, in: *Biological Basis for Risk Assessment of Dioxins and Related Compounds* (M. A. Gallo, R. J. Scheuplein and K. A. van der Heijden, eds.), pp. 51–68, Banbury Report 35, Cold Spring Harbor Laboratory Press, Cold Spring Harbor, NY (1991).
32. T. A. Gasiewicz, L. E. Gieger, G. Rucci, and R. A. Neal, Distribution, excretion, and metabolism of 2,3,7,8-tetrachlorodibenzo-*p*-dioxin in C57BL/6J, DBA/2J and B6D2F1/J mice, *Drug Metab. Dispos.* **11,** 497–503 (1983).
33. B. D. Abbott, J. J. Diliberto, and L. S. Birnbaum, 2,3,7,8-Tetrachlorodibenzo-*p*-dioxin alters embryonic palatial medial epithelial cell differentition *in vitro*. *Toxicol. Appl. Pharmacol.* **100,** 119–131 (1989).
34. B. D. Abbott and L. S. Birnbaum, Rat embryonic palatial shelves respond to TCDD in organ culture, *Toxicol. Appl. Pharmacol.* **103,** 441–451 (1990).
35. B. D. Abbott and L. S. Birnbaum, TCDD exposure of human embryonic palatial shelves in organ culture alters the differentiation of medial epithelial cells, *Teratology* **43,** 119–132 (1991).
36. M. K. Walker and R. E. Peterson, Potencies of polychlorinated dibenzo-*p*-dioxins, dibenzofurans, and biphenyl congeners for producing early life stage mortality in rainbow trout, (Oncorhynchus mykiss), *Aquat. Toxicol.* **21,** 219–238 (1991).
37. J. M. Kleeman, J. R. Olson, and R. E. Peterson, Species differences in 2,3,7,8-tetrachlorodibenzo-*p*-dioxin toxicity and biotransformation in fish, *Fundam. Appl. Toxicol.* **10,** 206–213 (1988).
38. J. B. Greig, G. Jones, W. H. Butler, and J. M. Barnes, Toxic effects of 2,3,7,8-tetrachlorodibenzo-*p*-dioxin, *Food Cosmet. Toxicol.* **11,** 585–595 (1973).
39. P. M. Allred and J. R. Strange, The effects of 2,4,5-trichlorophenoxyacetic acid and 2,3,7,8-tetrachlorodibenzo-*p*-dioxin on developing chicken embryos, *Arch. Environ. Contam. Toxicol.* **5,** 483–489 (1977).
40. J. A. Nosek, J. R. Sullivan, S. S. Hurley, S. R. Craven, and R. E. Peterson, Toxicity and reproductive effects of 2,3,7,8-tetrachlorodibenzo-*p*-dioxin in ring-necked pheasant hens, *J. Toxicol. Environ. Health* **35,** 187–198 (1991).
41. J. R. Olson, M. A. Holscher, and R. A. Neal, Toxicity of 2,3,7,8-tetrachlorodibenzo-*p*-dioxin in the golden Syrian hamster, *Toxicol. Appl. Pharmacol.* **55,** 67–78 (1980).
42. P. W. Beatty, M. A. Holscher, and R. A. Neal, Toxicity of 2,3,7,8-tetrachlorodibenzo-*p*-dioxin in larval and adult forms of Rana catesbeiana, *Bull. Environ. Contam. Toxicol.* **16,** 578–581 (1976).
43. J. R. Olson and B. P. McGarrigle, Characterization of the developmental toxicity of 2,3,7,8-TCDD in the golden Syrian hamster, *Toxicologist* **10,** 313 (abstr.) (1990).

44. J. R. Olson and B. P. McGarrigle, Comparative developmental toxicity of 2,3,7,8-tetrachlorodibenzo-*p*-dioxin (TCDD), *Chemosphere* **25,** 71–74 (1991).
45. D. L. Stalling, L. M. Smith, J. D. Petty, J. W. Hogan, J. L. Johnson, C. Rappe, and H. R. Busser, Residues of polychlorinated dibenzo-*p*-dioxins and dibenzofurans in Laurentian Great Lakes fish, in: *Human and Environmental Risks of Chlorinated Dioxins and Related Compounds* (R. E. Tucker, A. L. Young, and A. P. Gray, eds.), pp. 221–240, Plenum Press, New York (1983).
46. P. M. Cook, M. K. Walker, D. W. Kuehl, and R. E. Peterson, Bioaccumulation and toxicity of TCDD and related compounds in aquatic ecosystems, in: *Biological Basis for Risk Assessment of Dioxins and Related Compounds* (M. A. Gallo, R. J. Scheuplein, and K. A. van der Heijden, eds.), pp. 143–168, Banbury Report 35, Cold Spring Harbor Laboratory Press, Cold Spring Harbor, NY (1991).
47. U.S. EPA, Bioaccumulation of Selected Pollutants in Fish, Vol. 1, A National Study, Office of Water Regulations and Standards, Washington, DC, EPA 506/6-90/001a (1991).
48. S. Tanabe, PCB problems in the future: Foresight from current knowledge, *Environ. Pollut.* **50,** 5–28 (1988).
49. N. Kannan, S. Tanabe, and R. Tatsukawa, Potentially hazardous residues of non-*ortho* chlorine substituted coplanar PCBs in human adipose tissue, *Arch. Environ. Health* **43,** 11–14 (1988).
50. M. J. Mac, T. R. Schwartz, and C. C. Edsall, Correlating PCB effects on fish reproduction using dioxin equivalents, *Soc. Environ. Toxicol. Chem., 9th Annu. Meet. Abstr.*, p. 116 (1988).
51. T. J. Kubiak, H. J. Harris, L. M. Smith, T. R. Schwartz, D. L. Stalling, J. A. Trick, L. Sileo, D. E. Docherty, and T. C. Erdman, Microcontaminants and reproductive impairment of the Forster's tern on Green Bay, Lake Michigan—1983, *Arch. Environ. Contam. Toxicol.* **18,** 706–727 (1989).
52. L. M. Smith, T. R. Schwartz, K. Feltz, and T. J. Kubiak, Determination and occurrences of AHH-active polychlorinated biphenyls, 2,3,7,8-tetrachlorodibenzo-*p*-dioxin and 2,3,7,8-tetrachlorodibenzofuran in Lake Michigan sediment and biota. The question of their relative toxicological significance, *Chemosphere* **21,** 1063–1085 (1990).
53. L. A. Couture, B. D. Abbott, and L. S. Birnbaum, A critical review of the developmental toxicity and teratogenicity of 2,3,7,8-tetrachlorodibenzo-*p*-dioxin: Recent advances toward understanding the mechanism, *Teratology* **42,** 619–627 (1990).
54. D. Neubert and I. Dillman, Embryotoxic effects in mice treated with 2,4,5-trichlorophenoxy acetic acid and 2,3,7,8-tetrachlorodibenzo-*p*-dioxin, *N.S. Arch. Pharmacol.* **272,** 243–264 (1972).
55. L. A. Couture, M. W. Harris, and L. S. Birnbaum, Characteristics of the peak period of sensitivity for the induction of hydronephrosis in C57BL/6N mice following exposure to 2,3,7,8-tetrachlorodibenzo-*p*-dioxin, *Fundam. Appl. Toxicol.* **15,** 142–150 (1990).
56. K. D. Courtney, Mouse teratology studies with chlorodibenzo-*p*-dioxins, *Bull. Environ. Contam. Toxicol.* **16,** 674–681 (1976).
57. J. A. Moore, B. N. Gupta, J. G. Zinkl, and J. G. Voss, Postnatal effects of maternal exposure to 2,3,7,8-tetrachlorodibenzo-*p*-dioxin (TCDD), *Environ. Health Perspect.* **5,** 81–85 (1973).
58. L. Couture-Haws, M. W. Harris, M. M. McDonald, A. C. Lockhart, and L. S. Birnbaum, Hydronephrosis in mice exposed to TCDD-contaminated breast milk, identification of the peak period of sensitivity and assessment of potential recovery, *Toxicol. Appl. Pharmacol.* **107,** 413–428 (1991).
59. D. R. Dunnihoo, *Fundamentals of Gynecology and Obstetrics*, p. 283, Lippincott, Philadelphia (1990).

60. K. D. Crow, Chloracne, *Trans. St. Johns Hosp. Dermatol. Soc.* **56,** 77–99 (1970).
61. R. D. Kimbrough, The toxicity of polychlorinated polycyclic compounds and related chemicals, *CRC Crit. Rev. Toxicol.* **2,** 445–489 (1974).
62. J. S. Taylor, Chloracne—a continuing problem, *Cutis* **13,** 585–591 (1974).
63. E. E. McConnell, J. A. Moore, and D. W. Dalgard, Toxicity of 2,3,7,8-tetrchlorodibenzo-*p*-dioxin in rhesus monkeys (Macaca mulatta) following a single oral dose, *Toxicol. Appl. Pharmacol.* **43,** 175–187 (1978).
64. J. A. Moore, E. E. McConnell, D. W. Dalgard, and M. W. Harris, Comparative toxicity of three halogenated dibenzofurans in guinea pigs, mice, and rhesus monkeys, *Ann. N.Y. Acad. Sci.* **320,** 151–163 (1979).
65. K. L. Law, B. T. Hwang, and I. S. Shaio, PCB poisoning in newborn twins, *Clin. Med. (Taipei)* **7,** 88–91 (1981). [in Chinese]
66. B. V. Madhukar, D. W. Brewster, and F. Matsumura, Effects of *in vivo*-administered 2,3,7,8-tetrachlorodibenzo-*p*-dioxin on receptor binding of epidermal growth factor in the hepatic plasma membrane of rat, guinea pig, mouse, and hamster, *Proc. Natl. Acad. Sci. USA* **81,** 7407–7411 (1984).
67. H. A. Tilson, G. J. Davis, J. A. McLachlan, and G. W. Lucier, The effects of polychlorinated biphenyls given prenatally on the neurobehavioral development of mice, *Environ. Res.* **18,** 466–474 (1979).
68. S. M. Chou, T. Miike, W. M. Payne, and G. L. Davis, Neuropathology of "spinning syndrome" induced by prenatal intoxication with a PCB in mice, *Ann. N.Y. Acad. Sci.* **320,** 373–395 (1979).
69. A. K. Agrawal, H. A. Tilson, and S. C. Bondy, 3,4,3′,4′-Tetrachlorobiphenyl given to mice prenatally produces long-term decreases in striatal dopamine and receptor binding sites in the caudate nucleus, *Toxicol. Lett.* **7,** 417–424 (1981).
70. P. Eriksson, Effects of 3,3′,4,4′-tetrachlorbiphenyl in the brain of the neonatal mouse, *Toxicology* **49,** 43–48 (1988).
71. P. Eriksson, U. Lundkvist, and A. Fredriksson, Neonatal exposure to 3,3′,4,4′-tetrachlorbiphenyl: Changes in spontaneous behavior and cholinergic muscarinic receptors in the adult mouse, *Toxicology* **69,** 27–34 (1991).
72. S. L. Schantz and R. E. Bowman, Learning in monkeys exposed perinatally to 2,3,7,8-tetrachlorodibenzo-*p*-dioxin (TCDD), *Neurotox. Teratol.* **11,** 13–19 (1989).
73. R. E. Bowman, S. L. Schantz, M. L. Gross, and S. A. Ferguson, Behavioral effects in monkeys exposed to 2,3,7,8-TCDD transmitted maternally during gestation and for four months of nursing, *Chemosphere* **18,** 235–242 (1989).
74. Y. C. Chen, Y. L. Guo, C. C. Hsu, and W. J. Rogan, Cognitive development of Yu-Cheng ("oil" disease) in children prenatally exposed to heat-degraded PCBs, *J. Am. Med. Assoc.* **268,** 3213–3218 (1992).
75. A. Jost, Basic sexual trends in the development of vertebrates, *CIBA Found. Symp.* **62,** 3–18 (1978).
76. B. S. McEwen, Sexual maturation and differentiation: The role of the gonadal steroids, *Prog. Brain Res.* **48,** 281–307 (1978).
77. J. D. Wilson, F. W. George, and J. F. Griffin, The hormonal control of sexual development, *Science* **211,** 1278–1284 (1981).
78. A. A. Ehrhardt, and F. L. Meyer-Bahlburg, Effects of prenatal sex hormones on gender-related behavior, *Science* **211,** 1312–1317 (1981).
79. B. S. McEwen, I. Lieberburg, C. Chaptal, and L. C. Krey, Aromatization: Important for sexual differentiation of the neonatal rat brain, *Horm. Behav.* **9,** 249–263 (1977).
80. K. D. Dohler, S. S. Srivastava, J. E. Shryne, B. Jarzab, A. Sipos, and R. A. Gorski, Differentiation of the sexually dimorphic nucleus in the preoptic area of the rat brain is

inhibited by postnatal treatment with an estrogen antagonist, *Neuroendocrinology* **38,** 297–301 (1984).
81. K. D. Dohler, A. Coquelin, F. Davies, M. Hines, J. E. Shryne, P. N. Sickmoller, B. Jarzab, and R. A. Gorski, Pre- and postnatal influence of an estrogen antagonist on differentiation of the sexually dimorphic nucleus of the preoptic area of male and female rats, *Neuroendocrinology* **42,** 443–448 (1986).
82. L. E. Gray, Jr., Chemical-induced alterations of sexual differentiation: A review of effects in humans and rodents, in: *Advances in Modern Environmental Toxicology* (M. A. Mehlman, ed.), pp. 203–230, Princeton Scientific, Princeton, NJ (1992).
83. E. Wolff and E. Wolff, The effects of castration on bird embryos, *J. Exp. Zool.* **116,** 59–97 (1951).
84. G. A. Fox, Epidemiological and pathobiological evidence of contaminant-induced alterations in sexual development in free-living wildlife, in: *Advances in Modern Environmental Toxicology* (M. A. Mehlman, ed.), pp. 203–230, Princeton Scientific, Princeton, NJ (1992).
85. M. A. Gallo, E. J. Hesse, G. J. McDonald, and T. H. Umbreit, Interactive effects of estradiol and 2,3,7,8-tetrachlorodibenzo-*p*-dioxin on hepatic cytochrome P-450 and mouse uterus, *Toxicol. Lett.* **32,** 123–132 (1986).
86. S. Safe, B. Astroff, M. Harris, T. Zacharewski, R. Dickerson, M. Romkes, and L. Biegel, 2,3,7,8-Tetrachlorodibenzo-*p*-dioxin (TCDD) and related compounds as antiestrogens: Characterization and mechanism of action, *Pharmacol. Toxicol.* **69,** 400–409 (1991).
87. B. Astroff and S. Safe, 2,3,7,8-Tetarchlorodibenzo-*p*-dioxin as an antiestrogen: Effect on rat uterine peroxidase activity, *Biochem. Pharmacol.* **39,** 485–488 (1990).
88. B. Astroff, C. Rowlands, R. Dickerson, and S. Safe, 2,3,7,8-Tetrachlorodibenzo-*p*-dioxin inhibition of 17β-estradiol-induced increases in rat uterine epidermal growth factor receptor binding activity and gene expression, *Mol. Cell. Endocrinol.* **72,** 247–252 (1990).
89. J. F. Gierthy, D. W. Lincoln II, M. B. Gillespie, J. I. Seeger, H. L. Martinez, H. W. Dickerman, and S. A. Kumar, Suppression of estrogen-regulatred extracellular tissue plasminogen activator activity of MCF-7 cells by 2,3,7,8-tetrachlorodibenzo-*p*-dioxin, *Cancer Res.* **47,** 6198–6203 (1987).
90. J. F. Gierthy and D. W. Lincoln II, Inhibition of postconfluent focus production in cultures of MCF-7 human breast cancer cells by 2,3,7,8-tetrachlorodibenzo-*p*-dioxin, *Breast Cancer Res. Treat.* **12,** 227–233 (1988).
91. M. Harris, J. Piskorska-Pliszczynska, M. Romkes, and S. Safe, Structure-dependent induction of aryl hydrocarbon hydroxylase in human breast cancer cell lines and characterization of the Ah receptor, *Cancer Res.* **49,** 4531–4535 (1989).
92. L. Biegel and S. Safe, Effects of 2,3,7,8-tetrachlorodibenzo-*p*-dioxin (TCDD) on cell growth and the secretion of the estrogen-induced 34-, 52-, and 160-kDa proteins in human breast cancer cells, *J. Steroid Biochem. Mol. Biol.* **37,** 725–732 (1990).
93. D. C. Spink, D. W. Lincoln II, H. W. Dickerman, and J. F. Gierthy, 2,3,7,8-Tetrachlorodibenzo-*p*-dioxin causes an extensive alteration of 17β-Estradiol metabolism in MCF-7 breast tumor cells, *Proc. Natl. Acad. Sci. USA* **87,** 6917–6921 (1990).
94. D. C. Spink, H. P. Eugster, D. W. Lincoln II, J. D. Schuetz, E. G. Schuetz, J. A. Johnson, L. A. Kaminsky, and J. F. Gierthy, 17β-estradiol hydroxylation catalyzed by human cytochrome P450 1A1: A comparison of the activities induced by 2,3,7,8-tetrachlorodibenzo-*p*-dioxin in MCF-7 cells with those from heterologous expression of the cDNA, *Arch. Biochem. Biophys.* **293,** 342–348 (1992).
95. M. Harris, T. Zacharewski, and S. Safe, Effects of 2,3,7,8-tetrachlorodibenzo-*p*-dioxin and related compounds on the occupied nuclear estrogen receptor in MCF-7 human breast cancer cells, *Cancer Res.* **50,** 3579–3584 (1990).

96. T. A. Mably, R. W. Moore, D. L. Bjerke, and R. E. Peterson, The male reproductive system is highly sensitive to *in utero* and lactational 2,3,7,8-tetrachlorodibenzo-*p*-dioxin exposure, in: *Biological Basis for Risk Assessment of Dioxins and Related Compounds* (M. A. Gallo, R. J. Scheuplein and K. A. van der Heijden, eds.), pp. 69–78, Banbury Report 35, Cold Spring Harbor Laboratory Press, Cold Spring Harbor, NY (1991).
97. T. A. Mably, R. W. Moore, and R. E. Peterson, *In utero* and lactational exposure of male rats to 2,3,7,8-tetrachlorodibenzo-*p*-dioxin: 1. Effects on androgenic status, *Toxicol. Appl. Pharmacol.* **114,** 97–107 (1992).
98. T. A. Mably, R. W. Moore, R. W. Goy, and R. E. Peterson, *In utero* and lactational exposure of male rats to 2,3,7,8-tetrachlorodibenzo-*p*-dioxin. 2. Effects on sexual behavior and the regulation of luteinizing hormone secretion in adulthood, *Toxicol. Appl. Pharmacol.* **114,** 108–117 (1992).
99. T. A. Mably, D. L. Bjerke, R. W. Moore, A. Gendron-Fitzpatrick, and R. E. Peterson, *In utero* and lactational exposure of male rats to 2,3,7,8-tetrachlorodibenzo-*p*-dioxin. 3. Effects on spermatogenesis and reproductive capability, *Toxicol. Appl. Pharmacol.* **114,** 118–126 (1992).
100. E. M. Giavini, M. Prati, and C. Vismara, Effects of 2,3,7,8-tetrachlorodibenzo-*p*-dioxin administered to pregnant rats during the preimplantation period, *Environ. Res.* **29,** 185–189 (1982).
101. E. M. Giavini, M. Prati, and C. Vismara, Rabbit teratology studies with 2,3,7,8-tetrachlorodibenzo-*p*-dioxin, *Environ. Res.* **27,** 74–78 (1982).
102. G. L. Sparschu, F. L. Dunn, and V. K. Rowe, Study of the teratogenicity of 2,3,7,8-tetrachlorodibenzo-*p*-dioxin in the rat, *Food Cosmet. Toxicol.* **9,** 405–412 (1971).
103. K. S. Khera and J. A. Ruddick, Polychlorodibenzo-*p*-dioxins: Perinatal effects and the dominant lethal test in Wistar rats, in: *Chlorodioxins—Origin and Fate* (E. H. Blair, ed.), pp. 70–84, American Chemical Society, Washington, DC (1973).
104. K. S. Khera, Extraembryonic tissue changes induced by 2,3,7,8-tetrachloro-*p*-dioxin and 2,3,7,8-pentachlorodibenzofuran with a note on direction of maternal blood flow in the labyrinth of C57BL/6N mice, *Teratology* **45,** 611–627 (1992).
105. J. R. Allen, D. A. Barsotti, L. K. Lambrecht, and J. P. Van Miller, Reproductive effects of halogenated aromatic hydrocarbons on nonhuman primates, *Ann. N.Y. Acad. Sci.* **320,** 419–425 (1979).
106. D. A. Barsotti, L. J. Abrahamson, and J. R. Allen, Hormonal alterations in female rhesus monkeys fed a diet containing 2,3,7,8-tetrachlorodibenzo-*p*-dioxin, *Bull. Environ. Contam. Toxicol.* **21,** 463–469 (1979).
107. S. L. Schantz, D. A. Barsotti, and J. R. Allen, Toxicological effects produced in nonhuman primates chronically exposed to fifty parts per trillion 2,3,7,8-tetrachlorodibenzo-*p*-dioxin (TCDD), *Toxicol. Appl. Pharmacol.* **48**(2), A180 (abstr.) (1979).
108. W. P. McNulty, Fetotoxicity of 2,3,7,8-tetrachlorodibenzo-*p*-dioxin (TCDD) for rhesus macaques (Macaca mulatta), *Am. J. Primatol.* **6,** 41–47 (1984).
109. W. P. McNulty, Toxicity and fetotoxicity of TCDD, TCDF and PCB isomers in rhesus macaques (Macaca mulatta), *Environ. Health Perspect.* **60,** 77–88 (1985).
110. G. I. Sunahara, K. G. Nelson, T. K. Wong, and G. W. Lucier, Decreased human birth weights after *in utero* exposure to PCBs and PCDFs are associated with decreased placental EGF-stimulated receptor autophosphorylation capacity, *Mol. Pharmacol.* **32,** 572–578 (1987).
111. R. Krowke, G. Franz, and D. Neubert, Embryotoxicity. Is the TCDD-induced embryotoxicity in rats due to maternal toxicity? *Chemosphere* **18,** 291–298 (1989).
112. B. D. Abbott and L. S. Birnbaum, Effects of TCDD on embryonic ureteric epithelial EGF receptor expression and cell proliferation, *Teratology* **41,** 71–84 (1990).

113. J. E. Fitchett and E. D. Hay, Medial edge epithelium transforms to mesenchyme after embryonic palatal shelves fuse, *Dev. Biol.* **131,** 455–474 (1989).
114. C. F. Shuler, D. E. Halpern, Y. Guo, and A. C. Sank, Medial edge epithelium (MEE) fate traced by cell linkage analysis during epithelial–mesenchymal transformation *in vivo. J. Cell Biol.* **115,** 147a (abstr.) (1991).
115. B. D. Abbott, L. S. Birnbaum, and R. M. Pratt, TCDD-induced hyperplasia of the ureteral epithelium produces hydronephrosis in murine fetuses, *Teratology* **35,** 329–334 (1987).
116. B. A. Schwetz, J. M. Norris, G. L. Sparschu, V. K. Rowe, P. J. Gehring, J. L. Emerson, and C. G. Gerbig, Toxicology of chlorinated dibenzo-*p*-dioxins, *Environ. Health Perspect.* **5,** 87–99 (1973).
117. M. R. Zingeser, Anomalous development of the soft palate in rhesus macaques (Macaca mulatta) prenatally exposed to 3,4,7,8-tetrachlorodibenzo-*p*-dioxin, *Teratology* **19,** 54A (abstr.) (1979).
118. G. M. Fara and G. Del Corno, Pregnancy outcome in the Seveso area after TCDD contamination, in: *Prevention of Physical and Mental Congenital Defects, Part B, Epidemiology, Early Detection and Therapy, and Environmental Factors,* pp. 279–285, Liss, New York (1985).
119. A. Poland and E. Glover, 2,3,7,8-Tetrachlorodibenzo-*p*-dioxin: Segregation of toxicity with the Ah locus, *Mol. Pharmacol.* **17,** 86–94 (1980).
120. E. Hassoun, R. d'Argy, L. Dencker, L.-G. Lundin, and P. Borwell, Teratogenicity of 2,3,7,8-tetrachlorodibenzofuran in BXD recombinant inbred strains, *Toxicol. Lett.* **23,** 37–42 (1984).
121. T. A. Marks and R. E. Staples, Teratogenic evaluation of the symmetrical isomers of hexachlorbiphenyl (HCB) in the mouse, in: *Proc. 20th Ann. Meet. Teratol. Soc., Portsmouth, NH,* June, p. 54A (abstr.) (1980).
122. L. Biegel, M. Harris, D. Davis, R. Rosengren, L. Safe, and S. Safe, 2,2′,4,4′,5,5′-Hexachlorobiphenyl as a 2,3,7,8-tetrachlorodibenzo-*p*-dioxin antagonist in C57BL/6 mice, *Toxicol. Appl. Pharmacol.* **97,** 561–571 (1989).
123. R. E. Morrissey, H. W. Harris, J. J. Diliberto, and L. S. Birnbaum, Limited PCB antagonism of TCDD-induced malformations in mice, *Toxicol. Lett.* **60,** 19–25 (1992).
124. L. S. Birnbaum, H. Weber, M. W. Harris, J. C. Lamb IV, and J. D. McKinney, Toxic interaction of specific polychlorinated biphenyls and 2,3,7,8-tetrachlorodibenzo-*p*-dioxin: Increased incidence of cleft palate in mice, *Toxicol. Appl. Pharmacol.* **77,** 292–302 (1985).
125. B. D. Abbott, K. S. Morgan, L. S. Birnbaum, and R. M. Pratt, TCDD alters the extracellular matrix and basal lamina of the fetal mouse kidney, *Teratology* **35,** 335–344 (1987).
126. E. M. Giavini, M. Prati, and C. Vismara, Embryotoxic effects of 2,3,7,8-tetrachlorodibenzo-*p*-dioxin administered to female rats before mating, *Environ. Res.* **31,** 105–110 (1983).
127. L. W. K. Chung and G. Raymond, Neonatal imprinting of the accessory glands and hepatic monooxygenases in adulthood, *Fed. Proc.* **35,** 686 (1976).
128. J. Rajfer and D. S. Coffey, Effects of neonatal steroids on male sex tissues, *Invest. Urol.* **17,** 3–8 (1979).
129. D. S. Coffey, Androgen action and the sex accessory tissues, in: *The Physiology of Reproduction* (E. Knobil and J. Neill, eds.), pp. 1081–1119, Raven Press, New York (1988).
130. E. Steinberger and A. Steinberger, Hormonal control of spermatogenesis, in: *Endocrinology,* 2nd ed. (L. J. DeGroot, ed.), pp. 2132–2136, Saunders, Philadelphia (1989).
131. C. Desjardins and R. A. Jones, Differential sensitivity of rat accessory-sex-tissues to androgen following neonatal castration or androgen treatment, *Anat. Rec.* **166,** 299 (1970).

132. L. W. K. Chung and G. Ferland-Raymond, Differences among rat sex accessory glands in their neonatal androgen dependency, *Endocrinology* **97,** 145–153 (1975).
133. K. Queen, C. B. Dhabuwala, and C. G. Pierrpoint, The effect of the removal of the various accessory sex glands on the fertility of male rats, *J. Reprod. Fertil.* **62,** 423–426 (1981).
134. N. J. MacLusky and F. Naftolin, Sexual differentiation of the central nervous system, *Science* **211,** 1294–1303 (1981).
135. R. A. Gorski, The neuroendocrine regulation of sexual behavior, in: *Advances in Psychobiology* (G. Newton and A. H. Riessen, eds.), Vol. 2, pp. 1–58, Wiley, New York (1974).
136. C. A. Barraclough, Sex differentiation of cyclic gonadotropin secretion, in: *Advances in the Biosciences* (A. M. Kaye and M. Kaye, eds.), Vol. 25, pp. 433–450, Pergamon Press, Elmsford, NY (1980).
137. G. Raisman and P. M. Field, Sexual dimorphism in the neuropil of the preoptic area of the rat and its dependence on neonatal androgen, *Brain Res.* **54,** 1–29 (1973).
138. R. A. Gorski, J. H. Gordon, J. E. Shryne, and A. M. Southam, Evidence for a morphological sex difference within the medial preoptic area of the rat brain, *Brain Res.* **148,** 333–346 (1978).
139. R. W. Goy and B. S. McEwen, *Sexual Differentiation of the Brain,* pp. 137–139, MIT Press, Cambridge, MA (1980).
140. B. L. Hart, Activation of sexual reflexes of male rats by dihydrotestosterone but not estrogen, *Physiol. Behav.* **23,** 107–110 (1979).
141. M. Van den Berg, C. Heeremans, E. Veenhoven, and K. Olie, Transfer of polychlorinated dibenzo-*p*-dioxins and dibenzofurans to fetal and neonatal rats, *Fundam. Appl. Toxicol.* **9,** 635–644 (1987).
142. J. A. Moore, M. W. Harris, and P. W. Albro, Tissue distribution of [^{14}C] tetrachlorodibenzo-*p*-dioxin in pregnant and neonatal rats, *Toxicol. Appl. Pharmacol.* **37,** 146–147 (1976).
143. R. W. Moore, C. L. Potter, H. M. Theobald, J. A. Robinson, and R. E. Peterson, Androgenic deficiency in male rats treated with 2,3,7,8-tetrachlorodibenzo-*p*-dioxin, *Toxicol. Appl. Pharmacol.* **79,** 99–111 (1985).
144. J. Weisz and I. L. Ward, Plasma testosterone and progesterone titers of pregnant rats, their male and female fetuses, and neonatal offspring, *Endocrinology* **106,** 306–316 (1980).
145. F. Neumann, R. von Berswordt-Wallrabe, W. Elger, H. Steinbeck, J. D. Hahn, and M. Kramer, Aspects of androgen-dependent events as studied by antiandrogens, *Recent Prog. Horm. Res.* **26,** 337–410 (1970).
146. J. Rajfer and P. C. Walsh, Hormonal regulation of testicular descent: Experimental and clinical observations, *Urology* **118,** 985–990 (1977).
147. G. W. Robb, R. P. Amann, and G. J. Killian, Daily sperm production and epididymal sperm reserves of pubertal and adults rats, *J. Reprod. Fertil.* **54,** 103–107 (1978).
148. R. W. Moore, T. A. Mably, D. L. Bjerke, and R. E. Peterson, In utero and lactational 2,3,7,8-tetrachlorodibenzo-*p*-dioxin (TCDD) exposure decreases androgenic responsiveness of male sex organs and permanently inhibits spermatogenesis and demasculinizes sexual behavior in rats, *Toxicologist* **12,** 81 (abstr.) (1992).
149. C. W. Bardin, C. Y. Cheng, N. A. Mustow, and G. L. Gunsalus, The Sertoli cell, in: *The Physiology of Reproduction* (E. Knobil and J. D. Neill, eds.), pp. 933–974, Raven Press, New York (1988).
150. L. D. Russell and R. N. Peterson, Determination of the elongate spermatid–Sertoli cell ratio in various mammals, *J. Reprod. Fertil.* **70,** 635–641 (1984).
151. J. M. Orth, G. L. Gunsalus, and A. A. Lamperti, Evidence from Sertoli cell-depleted rats

indicates that spermatid number in adults depends on numbers of Sertoli cells produced during perinatal development, *Endocrinology* **122,** 787–794 (1988).
152. J. H. Aafjes, J. M. Vels, and E. Schenck, Fertility of rats with artificial oligozoospermia, *J. Reprod. Fertil.* **58,** 345–351 (1980).
153. R. P. Amann, Use of animal models for detecting specific alterations in reproduction, *Fundam. Appl. Toxicol.* **2,** 13–26 (1982).
154. P. K. Working, Male reproductive toxicology: Comparison of the human to animal models, *Environ. Health Perspect.* **77,** 37–44 (1988).
155. M. L. Meistrich, A method of quantitative assessment of reproductive risks to the human male, *Fundam. Appl. Toxicol.* **18,** 479–490 (1992).
156. S. L. Schantz, T. A. Mably, and R. E. Peterson, Effects of perinatal exposure to 2,3,7,8-tetrachlorodibenzo-*p*-dioxin (TCDD) on spatial learning and memory and locomotor activity in rats, *Teratology* **43,** 497 (1991).
157. D. A. Demassa, E. R. Smith, B. Tennent, and J. M. Davidson, The relationship between circulating testosterone levels and male sexual behavior in rats, *Horm. Behav.* **8,** 275–286 (1977).
158. D. F. Hardy and J. F. DeBold, Effects of coital stimulation upon behavior of the female rat, *J. Comp. Physiol. Psychol.* **78,** 400–408 (1972).
159. G. Forsberg, K. Abrahamsson, P. Södersten, and P. Eneroth, Effects of restricted maternal contact in neonatal rats on sexual behavior in the adult, *J. Endocrinol.* **104,** 427–431 (1985).
160. S. Taleisnik, L. Caligaris, and J. J. Astrada, Sex difference in the release of luteinizing hormone evoked by progesterone, *J. Endocrinol.* **44,** 313–321 (1969).
161. B. L. Hart, Manipulation of neonatal androgen: Effects on sexual responses and penile development in male rats, *Physiol. Behav.* **8,** 841–845 (1972).
162. R. E. Whalen and K. L. Olsen, Role of aromatization in sexual differentiation: Effects of prenatal ATD treatment and neonatal castration, *Horm. Behav.* **15,** 107–122 (1981).
163. F. Gogan, I. A. Beattie, M. Hery, E. Laplante, and C. Kordon, Effect of neonatal administration of steroids or gonadectomy upon oestradiol-induced luteinizing hormone release in rats of both sexes, *J. Endocrinol.* **85,** 69–74 (1980).
164. F. Gogan, A. Slama, B. Bizzini-Koutznetzova, F. Dray, and C. Kordon, Importance of perinatal testosterone in sexual differentiation in the male rat, *J. Endocrinol.* **91,** 75–79 (1981).
165. S. M. Pomerantz, R. W. Goy, and M. M. Roy, Expression of male-typical behavior in adult female psuedohermaphroditic rhesus: Comparisons with normal males and neonatally gonadectomized males and females, *Horm. Behav.* **20,** 483–500 (1986).
166. J. Thornton and R. W. Goy, Female-typical sexual behavior of rhesus and defeminization by androgens given prenatally, *Horm. Behav.* **20,** 129–147 (1986).
167. R. W. Goy, F. B. Bercovitch, and M. C. McBrair, Behavior masculinization is independent of genital masculinization in prenatally androgenized female rhesus macaques, *Horm. Behav.* **22,** 552–571 (1988).
168. M. Hines, Prenatal gonadal hormones and sex differences in human behavior, *Psychol. Bull.* **92,** 56–80 (1982).
169. S. LeVay, A difference in hypothalamic structure between heterosexual and homosexual men, *Science* **253,** 1034–1037 (1991).
170. R. Pohjanvirta, T. Vartiainen, A. Uusi-Rauva, J. Monkkonen, and J. Tuomisto, Tissue distribution, metabolism and excretion of ^{14}C-TCDD in a TCDD-susceptible and a TCDD-resistant rat strain, *Pharmacol. Toxicol.* **66,** 93–100 (1990).

171. J. B. Carlstedt-Duke, Tissue distribution of the receptor for 2,3,7,8-tetrachlorodibenzo-*p*-dioxin in the rat, *Cancer Res.* **39,** 3172–3176 (1979).
172. E. K. Silbergeld, Dioxin: Distribution of Ah receptor binding in neurons and glia from rat and human brain, *Toxicologist* **12,** 196 (abstr.) (1992).
173. T. A. Gasiewicz, Receptors for 2,3,7,8-tetrachlorodibenzo-*p*-dioxin: Their inter- and intraspecies distribution and relationship to the toxicity of this compound, in: *Proceedings of the Thirteenth Annual Conference on Environmental Toxicology,* AFAMRL-TR-82-101, Air Force Aerospace Medical Laboratory, Wright-Patterson AFB, Ohio, pp. 250–269 (1983).
174. R. E. Bowman, S. L. Schantz, L., N. C. A. Weerasinghe, M. Gross, and D. Barsotti, Chronic dietary intake of 2,3,7,8-tetrachlorodibenzo-*p*-dioxin (TCDD) at 5 or 25 parts per trillion in the monkey: TCDD kinetics and dose–effect estimate of reproductive toxicity, *Chemosphere* **18,** 243–252 (1989).
175. S. L. Schantz, M. K. Laughlin, H. C. Van Valkenberg, and R. E. Bowman, Maternal care by rhesus monkeys of infant monkeys exposed to either lead or 2,3,7,8-tetrachlorodibenzo-*p*-dioxin (TCDD), *Neurotoxicology* **7,** 641–654 (1986).
176. J. R. Allen and L. A. Carstens, Light and electron microscopic observations in Macaca mulatta monkeys fed toxic fat, *Am. J. Vet. Res.* **28,** 1513–1526 (1967).
177. R. J. Kociba, P. A. Keeler, G. N. Park, and P. J. Gehring, 2,3,7,8-Tetrachlorodibenzo-*p*-dioxin (TCDD): Results of a 13 week oral toxicity study in rats, *Toxicol. Appl. Pharmacol.* **35,** 533–574 (1976).
178. J. P. Van Miller, J. J. Lalich, and J. R. Allen, Increased incidence of neoplasms in rats exposed to low levels of 2,3,7,8-tetrachlorodibenzo-*p*-dioxin, *Chemosphere* **6,** 537–544 (1977).
179. E. E. McConnell, J. A. Moore, J. K. Haseman, and M. W. Harris, The comparative toxicity of chlorinated dibenzo-*p*-dioxins in mice and guinea pigs, *Toxicol. Appl. Pharmacol.* **44,** 335–356 (1978).
180. I. Chahoud, R. Krowke, A. Schimmel, H. Merker, and D. Neubert, Reprodutive toxicity and pharmacokinetics of 2,3,7,8-tetrachlorodibenzo-*p*-dioxin. I. Effects of high doses on the fertility of male rats, *Arch. Toxicol.* **63,** 432–439 (1989).
181. R. E. Morrissey and B. A. Schwetz, Reproductive and developmental toxicity in animals, in: *Halogenated Biphenyls, Terphenyls, Naphthalenes, Dibenzodioxins and Related Products,* 2nd ed. (R. D. Kimbrough and A. A. Jensen, eds.), pp. 195–225, Elsevier, Amsterdam (1989).
182. R. W. Moore, C. R. Jefcoate, and R. E. Peterson, 2,3,7,8-Tetrachorodibenzo-*p*-dioxin inhibits steroidogenesis in the rat testis by inhibiting the mobilization of cholesterol to cytochome $P450_{scc}$, *Toxicol. Appl. Pharmacol.* **109,** 85–97 (1991).
183. S. Ruangwises, L. L. Bestervelt, D. W. Piper, C. J., Nolan, W. W. Piper, Human chorionic gonadotropin treatment prevents depressed 17α-hydrolase/C_{17-20} lyase activities and serum testosterone concentrations in 2,3,7,8-tetrachlorodibenzo-*p*-dioxin treated rats, *Biol. Reprod.* **45,** 143–150 (1991).
184. J. M. Kleeman, R. W. Moore, and R. E. Peteson, Inhibition of testicular steroidogenesis in 2,3,7,8-tetrachlorodibenzo-*p*-dioxin-treated rats: Evidence that the key lesion occurs prior to or during pregnenolone formation, *Toxicol. Appl. Pharmacol.* **106,** 112–125 (1990).
185. R. W. Moore, J. A. Parsons, R. C. Bookstaff, and R. E. Peterson, Plasma concentrations of pituitary hormones in 2,3,7,8-tetrachlorodibenzo-*p*-dioxin-treated male rats, *J. Biochem. Toxicol.* **4,** 165–172 (1989).
186. H. C. Bookstaff, R. W. Moore, and R. E. Peterson, 2,3,7,8-Tetrachlorodibenzo-*p*-dioxin

increases the potency of androgens and estrogens as feedback inhibitors of luteinizing hormone secretion in male rats, *Toxicol. Appl. Pharmacol.* **104,** 212–224 (1990).

187. R. C. Bookstaff, F. Kamel, R. W. Moore, D. L. Bjerke, and R. E. Peterson, Altered regulation of pituitary gonadotropin-releasing hormone (GnRH) receptor number and pituitary responsiveness to GnRH in 2,3,7,8-tetrachlorodibenzo-*p*-dioxin-treated male rats, *Toxicol. Appl. Pharmacol.* **105,** 78–92 (1990).
188. F. J. Murray, F. A. Smith, K. D. Nitschke, C. G. Humiston, R. J. Kociba, and B. A. Schwetz, Three-generation reproduction study of rats given 2,3,7,8-tetrachlorodibenzo-*p*-dioxin (TCDD) in the diet, *Toxicol. Appl. Pharmacol.* **50,** 241–252 (1979).
189. R. J. Kociba, D. G. Keyes, J. E. Beyer, R. M. Carreon, C. E. Wade, D. A. Dittenber, R. P. Kalnins, L. E. Frauson, D. N. Park, S. D. Barnard, R. A. Hummel, and C. G. Humiston,
Results of a two-year chronic toxicity and oncogenicity study of 2,3,7,8-tetrachlorodibenzo-*p*-dioxin in rats, *Toxicol. Appl. Pharmacol.* **46,** 279–303 (1978).
190. K. T. Shiverick and T. F. Muther, 2,3,7,8-Tetrachlorodibenzo-*p*-dioxin (TCDD) effects on hepatic microsomal steroid metabolism and serum estradiol of pregnant rats, *Biochem. Pharmacol.* **32,** 991–995 (1983).

Chapter 11

Aquatic Toxicity of Dioxins and Related Chemicals

Mary K. Walker and Richard E. Peterson

1. GENERAL TOXICOLOGICAL CHARACTERISTICS

Polychlorinated dibenzo-*p*-dioxins (PCDDs), dibenzofurans (PCDFs), and biphenyls (PCBs) belong to a family of lipophilic halogenated aromatic hydrocarbons that have similar structures, resist chemical and biological degradation, and persist in the environment posing a potential risk to fish, wildlife, and human health. There are more than 400 possible polychlorinated dioxins, dibenzofuran, and biphenyl congeners; however, only 21 are considered highly toxic.[1,2] The more potent congeners are planar or coplanar molecules with lateral chlorine substitutions, approximate isostereomers of 2,3,7,8-tetrachlorodibenzo-*p*-dioxin (TCDD), the most potent PCDD, PCDF, or PCB congener.[3] In mammals, TCDD and TCDD-like dioxin, dibenzofuran, and biphenyl congeners evoke similar patterns of toxic responses within a given species, and share a common mechanism of action mediated by binding the cellular aryl hydrocarbon (Ah) receptor and altering gene transcription.[3,4] The ability of TCDD and TCDD-like congeners to produce toxicity in mammalian species correlates with their Ah receptor binding affinity and their ability to induce cytochrome P450IA enzyme activity, a gene subfamily that catalyzes monooxygenase reactions and whose expression is regulated by the Ah receptor.[4–6]

Mary K. Walker • School of Pharmacy, University of Wisconsin, Madison, Wisconsin 53706. Richard E. Peterson • School of Pharmacy and Environmental Toxicology Center, University of Wisconsin, Madison, Wisconsin 53706.

Dioxins and Health, edited by Arnold Schecter. Plenum Press, New York, 1994.

1.1. Toxicological Characteristics in Fish

Certain lines of evidence suggest that TCDD and TCDD-like PCDD, PCDF, and PCB congeners may also manifest toxicity in fish via an Ah receptor-mediated mechanism.

1.1.1. Presence of the Ah Receptor and P450IA1 Inducibility

The ability of mammalian Ah receptor agonists to induce hepatic cytochrome P450IA1 activity in a variety of fish species correlates with the presence of the Ah receptor in those species. The Ah receptor has been detected in advanced classes of fish, including two species of elasmobranchs, smooth dogfish (*Mustelus canis*)[7] and spiny dogfish (*Squalus acanthias*)[7]; and five species of teleosts, scup (*Stenotomus chrysops*),[7] winter flounder (*Pseudopleuronectes americanus*),[7] killifish (*Fundulus heteroclitus*),[7] rainbow trout (*Oncorhynchus mykiss*),[7,8] and brown trout (*Salmo trutta*),[7] but has not been detected in Atlantic hagfish (*Myxine glutinosa*)[7] or sea lamprey (*Petromyzon marinus*),[7] species belonging to a primitive class of jawless aquatic vertebrates, the agnathans. TCDD or other mammalian Ah receptor agonists induced hepatic cytochrome P450IA1 activity in the seven fish species in which the Ah receptor was detected,[7–13] but these same receptor agonists did not induce P450IA1 activity in hagfish or sea lamprey, the two species in which the Ah receptor was not detected.[7,14] Although the assay methods used may not be suitable for detecting the Ah receptor in hagfish and sea lamprey, the inability of Ah receptor agonists to induce P450IA1 enzyme activity in these species strongly suggests that the Ah receptor is not present in hagfish or sea lamprey, and that in fish, as in birds and mammals, Ah receptor binding is required for P450IA1 induction.

The Ah receptor has also been detected in fish cells in culture including RTH-149 rainbow trout hepatoma cells,[15] PLHC-1 desert topminnow (*Poeciliopsis lucida*) hepatoma cells,[7] and RTG-2 rainbow trout embryonic gonad cells.[16] Mammalian Ah receptor agonists induced cytochrome P450IA1 activity in both hepatoma cell lines,[7,15] and competitively bound to the putative Ah receptor in RTG-2 embryonic gonad cells with a rank order of agonist affinity identical to that for the murine Ah receptor.[16]

1.1.2. Characterization of the Fish Ah Receptor

In mammals the Ah receptor is a ligand-dependent, DNA-binding protein. Following specific binding of the ligand to the Ah receptor, the ligand–receptor complex binds to specific DNA sequences termed dioxin-responsive enhancers (DREs).[17] DREs are adjacent to genes, such as cytochrome P450IA1, whose transcription is regulated by Ah receptor ligands, such as TCDD. Recent evidence has shown that following specific binding of TCDD to the putative hepatic Ah receptor in rainbow trout and killifish, the ligand–receptor complex can bind an oligonucleotide sequence containing a DRE upstream of the mouse cytochrome P450IA1 gene.[18] This finding suggests that the recognition sequence for Ah receptor responsive genes is conserved among fish and mammals.

1.1.3. Structure–Activity Relationships

TCDD is the most potent PCDD, PCDF, or PCB congener in producing toxicity in fish,[19,20] and is one of the most potent PCDD, PCDF, or PCB congeners for inducing cytochrome P450IA1 enzyme activity in fish *in vivo* and *in vitro*.[21–25] Although the absolute potencies of TCDD-like PCDD, PCDF, and PCB congeners for producing toxicity in fish and mammals differ,[20,26] the rank order potency of these same congeners is similar, suggesting that toxicity is mediated by a common mechanism.

1.2. Summary

The strongest evidence to date that TCDD and TCDD-like PCDD, PCDF, and PCB congeners manifest toxicity in fish via an Ah receptor-mediated mechanism includes: (1) presence of the Ah receptor in fish, (2) induction of cytochrome P450IA1 enzymes by mammalian Ah receptor agonists in fish species possessing the Ah receptor, (3) evolutionary conservation of Ah receptor recognition sequences upstream of genes regulated by Ah receptor agonists, and (4) similar structure–activity relationships for toxicity in fish and mammals. Although these data provide evidence that the Ah receptor may mediate toxicity for PCDDs, PCDFs, and PCBs in fish, the role of the Ah receptor in producing all signs of toxicity is not firmly established. Additional research that would provide even stronger evidence that the Ah receptor mediates toxicity of TCDD and TCDD-like congeners in fish includes (1) structure–activity evidence showing that the rank order binding affinities of TCDD and TCDD-like congeners for the Ah receptor in fish correlate with their rank order potencies for a toxic endpoint and (2) genetic evidence showing that toxicity of TCDD segregates with the Ah receptor locus.

2. ENVIRONMENTAL EXPOSURE OF FISH

2.1. Historical Contamination

2.1.1. PCDDs and PCDFs

TCDD was first reported in 1970 in fish collected near areas that had been heavily exposed to the defoliant 2,4,5-trichlorophenoxyacetic acid (2,4,5-T) during the Vietnam war.[27] Historical data of PCDDs and PCDFs in fish prior to 1970 are limited; however, historical data on PCDD and PCDF contamination can be found in lake sediments.[28] Dated sediment cores from Lake Huron showed that PCDD and PCDF contamination abruptly increased between 1930 and 1940, then steadily rose to current levels, paralleling the production of chlorinated aromatic hydrocarbons.[28] Recent determination of congener-specific concentrations of PCDDs and PCDFs in fish in the United States has established valuable baseline information for future comparisons.[29–31]

2.1.2. PCBs

PCBs were first analytically detected in fish in 1966 in Sweden.[32] However, retrospective analysis of fish museum specimens collected from the Great Lakes shows that PCB contamination first appeared in fish in 1949 and increased steadily through the 1960s.[33] PCB concentrations in fish declined between 1970 and 1980, reflecting the stringent controls on point sources of PCBs in 1972 and the final ban of commercial production in the United States in 1977.[34] Currently, PCB concentrations in fish appear to be fluctuating around a new lower level which may result from continued input of PCBs from nonpoint sources and from the compartmental cycling of PCBs already present in the environment.[34–38]

2.2. Exposure and Bioaccumulation of PCDDs, PCDFs, and PCBs

Adult fish are exposed to PCDDs, PCDFs, and PCBs, in the water, sediment, and food, and bioaccumulation is dependent on the physical and chemical characteristics of individual PCDD, PCDF, and PCB congeners and on the biotransformation and elimination rates of individual congeners in fish.[39–42] As a result of these factors, fish preferentially bioaccumulate TCDD and TCDD-like PCDD and PCDF congeners,[29–31,43] and preferentially bioaccumulate PCB congeners with higher chlorine content (i.e., penta-, hexa-, and heptachlorinated biphenyls).[44,45]

During fish early development, the major route of exposure to TCDD and TCDD-like congeners is not from contaminated food and water; rather, early life stages of fish (eggs and sac fry) are exposed to PCDDs, PCDFs, and PCBs by the deposition of these lipophilic chemicals from maternal tissues to the oocytes during vitellogenesis.[42,46–51] PCDDs, PCDFs, and PCBs have been detected in the eggs of a variety of fish species from the Great Lakes to the Baltic Sea.[52–62]

3. TOXICITY OF PCDDS, PCDFS, AND PCBS TO FISH

The adverse effects of PCDDs, PCDFs, and PCBs on fish have been studied in two primary ways: (1) laboratory exposures of fish (i.e., waterborne, dietary, intraperitoneal injection) to single congeners or commercial mixtures and (2) the correlation of PCDD, PCDF, and/or PCB concentrations in the environment with abnormalities in fish populations, such as mortality during early development,[61,63,64] or thyroid hyperplasia in adult fish.[65] While field research can integrate the impact of multiple environmental contamination on fish, laboratory exposures can identify the specific responses associated with exposure to a single toxicant, and determine the dose–response relationships for those responses. Thus, laboratory research can elucidate whether the body burden of a particular contaminant in fish in the environment is likely to produce the abnormalities observed in feral fish populations. Both field and laboratory research are vital to understanding the toxicity of PCDDs, PCDFs, and PCBs to fish and in predicting the risk that these compounds pose to fish in the environment. This chapter will focus primarily on the toxic responses characterized in laboratory studies.

3.1. General Characteristics of Toxicity

Given the variety of exposure routes, exposure durations, and experimental designs used to study the toxicity of TCDD and TCDD-related congeners to fish, it is difficult to compare the sensitivity of different fish species and the sensitivity of different toxic endpoints following exposure. In general, the exposure concentration, exposure duration, and exposure route are not as important as the actual body burden accumulated during exposure. In addition, TCDD and TCDD-like PCDD, PCDF, and PCB congeners produce adverse effects which are delayed in onset occurring days, weeks, or even months after exposure. Therefore, acute toxicity tests with a short observation period (i.e., 96 hr LC_{50}) will not accurately reflect the sensitivity of a given fish species or the toxicity of a particular congener. The most meaningful comparisons can be made when toxic responses are based on the body burden of TCDD and TCDD-like congeners and when the fish are observed for many weeks or months after exposure.

3.2. Exposure to Individual Congeners

3.2.1. TCDD and TCDD-Like Congeners

TCDD is the most extensively studied congener in avian, mammalian, and piscine species and serves as a prototype for studying the toxicity of other TCDD-like PCDD, PCDF, and PCB congeners in fish.

3.2.1a. Lethality. Both mammalian and piscine species vary in their sensitivity to the lethal potency of TCDD, and fish are among the most sensitive vertebrates (Table 1).[3,66,67] The guinea pig[66] and the hamster[68] are the most and least sensitive mammals, respectively, with LD_{50}s of 1.0 μg TCDD/kg body weight and 5000 μg/kg. Juvenile fish approach the sensitivity of guinea pigs with TCDD LD_{50}s from 3 μg/kg in yellow perch (*Perca flavescens*)[67] and common carp (*Cyprinus carpio*)[67] to 16 μg/kg in bluegill (*Lepomis macrochirus*).[67]

Table 1
Lethal Potency of TCDD in Mammals and Juvenile Fish

	TCDD LD_{50} (μg/kg body wt)	Ref.
Mammals		
Guinea pig	0.6–2.1	66
Rat	22–45	66
Hamster	5000	68
Juvenile fish		
Yellow perch	3 (2–4)[a]	67
Common carp	3 (2–4)	67
Black bullhead	5 (4–8)	67
Rainbow trout	10 (7–15)	67
Largemouth bass	11 (9–14)	67
Bluegill	16 (12–23)	67

[a]Numbers in parentheses represent 95% confidence interval.

Fish are more sensitive to the lethal potency of TCDD during early development than as juveniles or adults (Fig. 1).[20,67,69–72] Rainbow trout swim-up fry[69] (21 days posthatch, 0.38 g) are approximately 10 times more sensitive to the lethal potency of TCDD than juvenile rainbow trout[67] (1 year old, 35 g). A body burden of 0.99 μg TCDD/kg in rainbow trout swim-up fry produced 45% mortality (LD_{45}),[69] while 10 μg/kg in juveniles produced 50% mortality (LD_{50}).[67] Rainbow trout are even more sensitive when exposed as newly fertilized eggs.[72] Following exposure of rainbow trout eggs to waterborne or injected TCDD, LD_{50}s ranged from 0.230 to 0.488 μg TCDD/kg egg wet weight,[72] 20 times more sensitive than juveniles.[67]

The most sensitive fish species and most sensitive developmental stage to TCDD-induced lethality are lake trout (*Salvelinus namaycush*) during early development.[70,71] When exposed as newly fertilized eggs, lake trout egg burdens as low as 0.040 μg

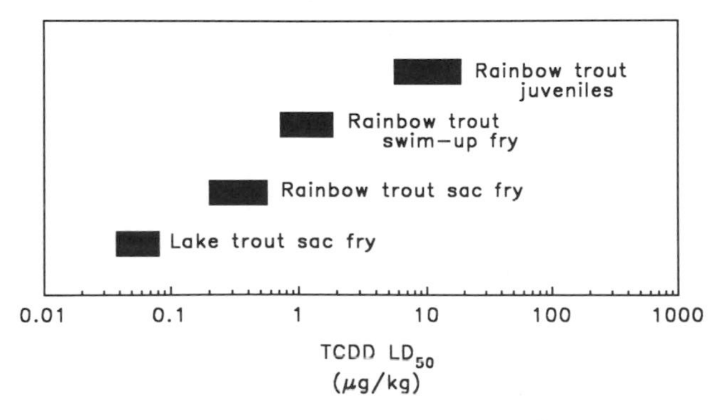

Figure 1. TCDD LD_{50}s showing the effect of developmental stages of fish to the lethal potency of TCDD.

TCDD/kg egg significantly increased sac fry mortality[70] and TCDD LD_{50}s ranged from 0.047 to 0.065 μg/kg egg,[71,72] three to eight times more sensitive than the same life stages of rainbow trout following the same routes of exposure.[72]

3.2.1b. Acute Toxicity of TCDD to Juvenile Fish. Acute toxicity of TCDD to juvenile fish has been shown to be dose-dependent and occur as a function of the TCDD body burden. Following exposure of juvenile fish to TCDD via water, sediment, diet, or intraperitoneal (i.p.) injection, overt toxicity of TCDD is characteristically manifested by decreased feed consumption, decreased body weight gain (wasting), and delayed mortality (Table 2). Species that exhibit these toxic responses to TCDD exposure include coho salmon (*Oncorhynchus kisutch*),[73,74] rainbow trout,[67,69,75–79] fathead minnow (*Pimephales promelas*),[81] common carp,[42,67] channel catfish (*Ictalurus punctatus*),[81] black bullhead (*Ictalurus melas*),[67] guppy (*Poecilia reticulata*),[73–75] zebrafish (*Danio rerio*),[51] largemouth bass (*Micropterus salmoides*),[67] bluegill,[67] and yellow perch.[67,82]

Many studies have observed decreases in feed consumption following exposure of juvenile fish to TCDD,[67,69,73–76,78,78,82] but few have quantitated this response.[73,74] Miller and co-workers[73,74] measured the percentage of feed consumed in 20 min by coho salmon after exposure to waterborne TCDD for 24, 48, or 96 hr. The percent feed consumed decreased with increasing dose of TCDD, increasing duration of exposure, and increasing time after exposure. By 15 days after a 48-hr exposure to 5.6 and 56.0 ng/liter, coho salmon consumed 89 and 55% of available feed, respectively, while fish exposed to these same doses for 96 hr consumed 78 and 21%, respectively, compared with 96% for controls. By 30 days after exposure, all fish exposed to 56.0 ng/liter had stopped eating (0% consumed).

Decreased feed consumption following exposure to TCDD subsequently results in decreased body weight gain or wasting. Kleeman *et al.*[67] observed decreases in body weight gain of yellow perch, common carp, black bullhead, rainbow trout, largemouth bass, and bluegill after treatment with TCDD; however, a significant decrease, relative to controls, was only observed in rainbow trout (5–125 μg/kg), bluegill (25 and 125 μg/kg), and yellow perch (5 μg/kg). Decreased feed intake and wasting are not likely to be the ultimate cause of TCDD-induced lethality in fish, however. At a sufficiently high dose of TCDD (LD_{90}), fish die before reducing their feed intake and before losing weight.[67] Rainbow trout showed wasting prior to death following a single i.p. dose of 5, 25, or 125 μg TCDD/kg. Yellow perch, however, exhibited wasting only after treatment with 5 μg/kg, while fish dosed with 25 or 125 μg/kg did not lose weight prior to death.[67]

Following acute exposure of juvenile fish to TCDD, mortality is delayed in onset, and the length of the delay is dose-dependent with lower lethal doses of TCDD producing a longer delay. Following treatment of common carp with 5 and 125 μg TCDD/kg, the time required to produce 50% mortality was 10 and 6 weeks, respectively, while at these same doses in yellow perch, 50% mortality occurred at 10 and 1 week, respectively.[67] Time to onset of mortality is not dependent on species sensitivity (i.e., mortality does not occur faster in more sensitive species). At 125 μg/kg, yellow

Table 2
Acute Toxicity of TCDD in Juvenile Fish

Species	Exposure			Organism		Toxicity		Ref.
	Route	Duration	Nominal dose	Body burden	Observation[a]	LOAEL[b]	Signs	
Coho salmon *Oncorhynchus kisutch*								
3.5 g	Water static	24, 48, 96 hr	0, 0.056, 0.56, 5.6, 56 ng/L	N.D.[c]	56 days	5.6 ng/L 24–96 hr All effects	Decreased feed intake, wasting, delayed mortality, fin necrosis	73
3.5 g	Water static	24, 48, 96 hr	0, 0.056, 0.56, 5.6, 56 ng/L	0, 0, 0.125, 2.17, N.D. μg/kg[d]	114 days	5.6 ng/L 24–96 hr All effects	Decreased feed intake, wasting, delayed mortality, fin necrosis	74
7.25 g	Water static	24, 48, 96 hr	0, 56, 100, 560, 1000 ng/L	N.D.	80 days	5.6 ng/L 24–96 hr All effects	Decreased feed intake, wasting, delayed mortality, skin discoloration, fin necrosis	73
Rainbow trout *Oncorhynchus mykiss*								
0.38 g	Water flow-thru	28 days	0, 0.038, 0.079, 0.176, 0.382, 0.789 ng/L	0, 0.99, N.D., 4.52, 10.95, 15.41 μg/kg[e]	28 days	0.99 μg/kg All effects	≥ 0.99 μg/kg: Delayed mortality, wasting, decreased feed intake, lethargy, fin necrosis ≥ 10.95 μg/kg: 100% mortality	69
0.85 g	Water static	64 hr 16 hr renewal	0, 10, 100 ng/L	N.D.	70 days	10 ng/L Wasting	≥ 10 ng/L: Wasting, ascites, exophthalmia 100 ng/L: Delayed mortality, 100% mortality by day 27	85
2–2.5 g[f]	Diet	15 weeks 6 days/week	0, 0.0023, 2.3, 2300 μg/kg feed	0.48, 0.06, 1.6, 1380 μg/kg[g]	None	1380 μg/kg All effects	1380 μg/kg: Decreased feed intake, wasting, delayed mortality, fin necrosis, 100% mortality by day 71	76

6–10 g	i.p. injection	Single 2.6 ml/kg		0, 0.1, 1, 10 µg/kg	35 days	10 µg/kg All effects	Fin hyperpigmentation and necrosis	78
35 g	Water static	6 hr	0, 285 ng/L	0, 2.6 µg/kg[h]	139 days	2.6 µg/kg All effects	Wasting, delayed mortality, fin necrosis	77
25–45 g	i.p. injection	Single 2.6 ml/kg		0, 1, 5, 25, 125 µg/kg	80 days	5 µg/kg Wasting	≥ 5 µg/kg: Wasting, delayed mortality	67
25–55 g	i.p. injection	Single 20 µl	N.R.[i]	0, 0.01, 0.05, 0.1, 0.5, 1, 5 µg/kg	12 weeks	5 µg/kg All effects	Decreased feed intake, wasting, fin necrosis, abnormal swimming, delayed mortality	79
Fathead minnows *Pimephales promelas*								
0.5–1.0 g	Water static	28 days 4 day renewal	0, 1.0, 10, 100 ng/L	0, 70, 320, 2000 µg/kg[j]	44 days	70 µg/kg Mortality	Delayed mortality, 100% mortality in all groups by day 72	80
1.0–2.0 g	Water static	24, 48, 72, & 96 hr	0, 0.1, 1.0, 10, 100 ng/L	N.D.	60 days	10.0 ng/L 24–96 hr Mortality	Delayed mortality observed as a function of dose and exposure duration	80
Common carp *Cyprinus carpio*								
13–27 g	i.p. injection	Single 2.6 ml/kg		0, 1, 5, 25, 125 µg/kg	80 days	5 µg/kg Mortality	≥ [illegible] µg/kg: Hyperpigmentation ≥ 5 µg/kg: Delayed mortality ≥ 25 µg/kg: Fin necrosis, cutaneous hemorrhage	67
15 g	Water flow-thru	71 days	0, 0.060 ng/L	0, 2.2 µg/kg	61 days	2.2 µg/kg All effects	Fin hyperpigmentation and necrosis, hemorrhages, edema, abnormal swimming, exophthalmia, delayed mortality	42
Channel catfish *Ictalurus punctatus*								
Fingerling	Sediment	15 days	100 µg/kg sediment	5.9 µg/kg[k]	5 days	5.9 µg/kg Mortality	Fin necrosis, hemorrhages, erratic swimming, delayed mortality; 100% mortality by day 20	81

(continued)

Table 2 (*Continued*)

Species	Exposure			Organism		Toxicity		Ref.
	Route	Duration	Nominal dose	Body burden	Observation[a]	LOAEL[b]	Signs	
Black bullhead *Ictalurus melas*								
4–8 g	i.p. injection	Single 2.6 ml/kg		0, 1, 5, 25, 125 μg/kg	80 days	5 μg/kg Mortality	≥ 5 μg/kg: Delayed mortality ≥ 25 μg/kg: Fin necrosis	67
Guppy *Poecilia reticulata*								
8–12 mm	Water static	24 hr	0, 0.01, 0.1, 1, 10 ng/L	N.D.	69 days	1 ng/L Fin necrosis	≥ 1 ng/L: Fin necrosis at 42 days 10 ng/L: Delayed mortality	74
9–40 mm	Water static	5 days	0, 0.1, 1, 10 ng/L	N.D.	37 days	0.1 ng/L Mortality	Decreased feed intake, fin necrosis, decreased swimming activity, delayed mortality; 100% mortality in all groups by day 37	75
10–40 mm	Water static	5 days	0, 100, 1000, 10,000 ng/L	N.D.	37 days	N.R.	Decreased feed intake, skin discoloration, fin necrosis, erosion of upper jaw, delayed mortality	73
Zebrafish *Danio rerio*								
Adult[m], 0.5–0.7 g	Diet	1 day 20 mg/fish	0, 50, 250, 500, 1000 μg/kg feed	N.D.	ca. 35 days		≥ 250 μg/kg feed: Decreased body weight, lethargy 1000 μg/kg feed: Skin lesions and discoloration	51
Mosquito fish *Gambusia affinis*								

Adult[m]	Sediment	15–17 days	100 μg/kg sediment	11.7 μg/kg[n]	None	11.7 μg/kg Mortality	Hemorrhaging, lethargic swimming, delayed mortality; 100% mortality by day 15	81
Largemouth bass *Micropterus salmoides*								
5–9 g	i.p. injection	Single 2.6 ml/kg		0, 1, 5, 25, 125 μg/kg	80 days	25 μg/kg Mortality	≥ 5 μg/kg: Hyperpigmentation ≥ 25 μg/kg: Delayed mortality, fin necrosis	67
Bluegill *Lepomis macrochirus*								
26–34 g	i.p. injection	Single 2.6 ml/kg		0, 1, 5, 25, 125 μg/kg	80 days	25 μg/kg All effects	Wasting, delayed mortality, fin necrosis, cutaneous hemorrhage	67
Yellow perch *Perca flavescens*								
34–46 g	i.p. injection	Single 2.6 ml/kg		0, 1, 5, 25, 125 μg/kg	80 days	5 μg/kg Wasting	≥ 5 μg/kg: Wasting, delayed mortality ≥ 25 μg/kg: Fin necrosis, cutaneous hemorrhage	67

[a]Duration of observation postexposure.
[b]LOAEL, lowest observable adverse effect level of TCDD.
[c]N.D., not determined.
[d]Body burden of TCDD 114 days postexposure.
[e]Body burdens of TCDD were determined on days 0, 7, 14, 21, and 28 of exposure. Nominal water concentrations of 0, 0.038, 0.176, 0.382, and 0.789 ng TCDD/liter produced peak body burdens of 0.027, 0.99, 4.52, 10.95, and 15.41 μg/kg on days 28, 21, 28, 28, and 21 of exposure, respectively.[69]
[f]Freeze-dried weight.
[g]Body burden of TCDD of one fish randomly selected from each dietary exposure concentration on day 65 of exposure.
[h]Body burden of TCDD determined at end of 6-hr exposure.
[i]N.R., not reported.
[j]Mean body burden of TCDD for fathead minnows that died during 28-day exposure.
[k]Mean body burden of TCDD determined on day 15 of exposure.
[m]Sexually mature.
[n]Peak body burden of TCDD determined on day 7 of 15-day exposure.

perch reached 50% mortality faster than common carp (1 versus 6 weeks); however, the two species have equivalent LD_{50}s (3 μg/kg).[67]

Chronic exposure of fish to TCDD also produces delayed mortality,[42,69,75,76,80] and mortality is not necessarily averted if the exposure ends prior to the onset of lethality.[42,69,75,80] Rainbow trout fed 2300 μg TCDD/kg feed for 105 days reached 50 and 88% mortality on days 61 and 71, respectively.[76] In another study, rainbow trout, exposed to waterborne TCDD for 28 days, did not exhibit mortality at exposure concentrations of 0.038 and 0.079 ng TCDD/liter; however, mortality reached 45 and 83%, respectively, 28 days after the exposure ended.[69]

Another common feature of acute TCDD toxicity in juvenile fish is fin necrosis. Fin margins become necrotic and fin rays fragment[67,78,82] often leading to increased susceptibility to fungal infection.[74,76] Fin necrosis was the most sensitive acute effect of TCDD exposure in guppies exposed to 1 ng/liter for 5 days,[74] while most fish species exhibited fin necrosis at doses of TCDD that also produced significant lethality and wasting.[42,51,67,69,73,75–79,81,82] In addition, fin necrosis in guppies has been shown to be reversible.[74] Following exposure to 10 ng TCDD/liter for 24 hr, 87% of the guppies exhibited fin necrosis 42 days postexposure, but by 69 days postexposure, only 21% exhibited fin necrosis.[74] The decreased incidence of fin necrosis could not completely be explained by differential mortality of only those guppies with fin necrosis.

Other responses of juvenile fish to acute TCDD exposure are more species-specific, such as cutaneous hyperpigmentation, cutaneous hemorrhage, and ascites. Cutaneous hyperpigmentation was observed in TCDD-exposed rainbow trout, common carp, and largemouth bass.[42,67,78] Common carp exhibited increased pigmentation at doses as low as 1 μg TCDD/kg and increasing doses produced graded increases in pigmentation.[67] Cutaneous hemorrhage have been described in TCDD-exposed yellow perch, bluegill, common carp, mosquito fish, and channel catfish.[42,67,78,81,82] Hemorrhages in yellow perch,[67,82] bluegill,[67] and common carp[67] were primarily associated with necrotic fin margins; however, yellow perch[82] also exhibited hemorrhages around the electroreceptor sensory pits of the head, and common carp exhibited hemorrhages along the lateral body wall.[42] Nasal hemorrhages were observed in mosquito fish[81] prior to death, and anal and lower jaw hemorrhages accompanied mortality of channel catfish.[81] Ascites have been described in both rainbow trout[78] and yellow perch[82] at doses as low as 10 and 25 μg/kg, respectively, while edema has been described in common carp at a body burden of 2.2 μg/kg.[42]

Lastly, two studies have investigated the acute toxicity of TCDD-like congeners to juvenile fish. Following exposure of juvenile rainbow trout to 3,3′,4,4′-tetrachlorobiphenyl[84] (3,3′,4,4′-TCB, IUPAC #77[83]) or 2,3,7,8-tetrachlorodibenzofuran (2,3,7,8-TCDF)[69], toxicity of both congeners was manifested by the same pattern of toxic responses as TCDD (Table 3). Rainbow trout, exposed to 100–1400 ng 3,3′,4,4′-TCB/liter for 50 days, exhibited decreased feed consumption, decreased body weight gain, and were disoriented.[84] In addition, delayed mortality was observed 28 days after exposure in the highest treatment group.[84] In a separate study, juvenile rainbow trout, exposed to 0.41–8.78 ng 2,3,7,8-TCDF/liter for 28 days, exhibited decreased feed consumption, wasting, delayed mortality, and lethargy.[69] Fin necrosis was not noted in either study.

Table 3
Acute Toxicity of TCDD-Like Dibenzofuran and Biphenyl Congeners in Juvenile Rainbow Trout

	Exposure			Organism		Toxicity		
Congener	Route	Duration	Nominal concentration	Body burden	Observation[a]	LOAEL[b]	Signs	Ref.
3,3′,4,4′-TCB[c]	Water flow-thru	50 days	0, 100, 400, 700, 1400 ng/L	0, 800, 1900, 5100, 12,000 μg/kg[d]	28 days	800 μg/kg Wasting	≥ 800 μg/kg: Decreased feed intake, wasting, disoriented 12,000 μg/kg: Delayed mortality	84
2,3,7,8-TCDF[e]	Water flow-thru	28 days	0, 0.41, 0.90, 1.79, 3.93, 8.78 ng/L	0, 2.48, N.D.,[f] N.D., 11.9, N.D. μg/kg[g]	28 days	0.90 ng/L Wasting	≥ 0.90 ng/L: Wasting ≥ 1 79 ng/L: Decreased feed intake, lethargy ≥ 3 93 ng/L (11.9 μg/kg): Delayed mortality	69

[a]Duration of observation postexposure.
[b]LOAEL, lowest observable adverse effect level for 3,3′,4,4′-TCB or 2,3,7,8-TCDF.
[c]3,3′,4,4′-Tetrachlorobiphenyl (IUPAC #77[83]).
[d]Mean body burdens of 3,3′,4,4′-TCB measured on day 50 exposure.
[e]2,3,7,8-Tetrachlorodibenzofuran.
[f]N.D., not determined.
[g]Peak body burdens of 2,3,7,8-TCDF during 28-day exposure.

Table 4
Histopathologic Lesions Induced by TCDD in Juvenile Fish

Tissue	Lesion	Species	Dose, route, duration	Ref.
Epithelial				
Liver	Glycogen depletion and cytoplasmic vacuolation	Rainbow trout	2300 μg/kg feed, 105 days	76
			100 ng/L, water, 96 hr	85
			≥ 10 μg/kg, i.p., single	78
		Yellow perch	≥ 25 μg/kg, i.p., single	82
	Hypertrophy and lipidosis	Yellow perch	≥ 1 μg/kg, i.p., single	82
Gills	Epithelial hyperplasia	Yellow perch	≥ 25 μg/kg, i.p., single	82
	Fusion of secondary lamellae	Rainbow trout	≥ 10 μg/kg, i.p., single	78
Stomach	Submucosal edema	Yellow perch	≥ 25 μg/kg, i.p., single	82
	Necrosis and hyperplasia of serous mucosal glands	Rainbow trout	≥ 10 μg/kg, i.p., single	78
			100 ng/L, water, 96 hr	85
Pancreas	Acinar cell cytoplasmic vacuoles containing eosinophilic material	Rainbow trout	≥ 10 μg/kg, i.p., single	78
			100 ng/L, water, 96 hr	85
Lymphomyeloid				
Thymus	Lymphoid depletion, thymic atrophy	Rainbow trout	≥ 10 μg/kg, i.p., single	78
		Yellow perch	≥ 25 μg/kg, i.p., single	82
Spleen	Lymphoid depletion	Rainbow trout	≥ 10 μg/kg, i.p., single	78
			5 μg/kg, i.p., single	79
		Yellow perch	≥ 25 μg/kg, i.p., single	82
Head kidney	Lymphoid and hematopoietic depletion, decreased percentage of blast cells	Rainbow trout	≥ 10 μg/kg, i.p., single	78
		Yellow perch	≥ 25 μg/kg, i.p., single	82
Hematology	Leukocytopenia and thrombocytopenia	Rainbow trout	≥ 1 μg/kg, i.p., single	78
Cardiovascular				
Heart	Myocyte necrosis; pericardial mesothelial hypertrophy and hyperplasia; fibrinous pericarditis	Yellow perch	≥ 25 μg/kg, i.p., single	82

3.2.1c. Histopathologic Effects. Histologically, TCDD-exposed juvenile fish typically exhibit lesions in epithelial and lymphomyeloid tissues (Table 4). The liver has been the epithelial tissue most frequently examined for TCDD-induced lesions.[76,78,79,82,85] Juvenile rainbow trout, exposed to dietary (2300 μg/kg feed),[76] waterborne (100 ng/liter),[85] or i.p. injected TCDD (10–125 μg/kg body weight),[78] exhibited hepatocellular glycogen depletion and cytoplasmic vacuolation. Yellow perch[82] exhibited similar hepatic changes following 25 and 125 μg/kg, but also showed hepatocellular hypertrophy and lipidosis at doses as low as 1 μg/kg.

Other TCDD-induced epithelial lesions tend to be species-specific and include effects on gills, gastric mucosa, and exocrine pancreas. Yellow perch[82] exhibited epithelial hyperplasia of gill filaments and secondary lamellae at 25 and 125 μg TCDD/kg, while rainbow trout[78] exhibited fusion of secondary gill lamellae at doses as low as 10 μg/kg. Gastric lesions included significant submucosal edema of the stomach in yellow perch at 25 and 125 μg/kg,[82] and hyperplasia and necrosis of the serous gastric mucosal glands in rainbow trout following treatment with ≥ 10 μg/kg[78] and following waterborne exposure to 100 ng/liter.[85] TCDD-induced pancreatic lesions have only been described in rainbow trout.[78,85] Following i.p. treatment[78] or waterborne exposure,[85] rainbow trout exhibited cytoplasmic vacuoles containing eosinophilic material in the acinar cells of the pancreas.

TCDD induced lymphomyeloid lesions in juvenile fish in the thymus, spleen, and head kidney. Both yellow perch[82] and rainbow trout[78,79] exhibited thymic and splenic lymphoid depletion, and hypocellularity of hematopoietic tissues in the head kidney. In addition, the head kidney in both species showed a decreased percentage of blast cells and an increased percentage of differentiated lymphomyeloid cells.[78,82] Rainbow trout[78] were more sensitive than yellow perch[82] to TCDD-induced lymphomyeloid lesions with doses of 10 and 25 μg/kg, respectively, producing histopathologic effects. In rainbow trout,[78] decreased hematopoiesis in the head kidney resulted in a decrease in circulating leukocytes and thrombocytes at doses as low as 1 μg TCDD/kg, representing the lowest observable adverse effect level (LOAEL) for TCDD-induced lesions in juvenile fish.

Cardiotoxicity is observed in chicken embryos following TCDD exposure,[86-88] but has only been described in one juvenile fish species. Juvenile yellow perch,[82] exposed to 25 or 125 μg TCDD/kg, exhibited myocyte necrosis; hypertrophy and hyperplasia of pericardial mesothelium; and fibrinous pericarditis. Furthermore, TCDD-induced cardiac lesions were more severe in smaller perch (20 g) than in larger perch (40 g).[82]

3.2.1d. Biochemical Effects. TCDD and TCDD-like congeners have been shown to produce a variety of biochemical effects in mammals including enzyme induction and effects on various membrane and cytosolic receptors.[89] In fish, the induction of cytochrome P450IA1 enzyme activity by TCDD and related compounds has been extensively studied and reviews on this subject can be found elsewhere.[90-93] In general, the induction of P450IA1 enzyme activity by TCDD and TCDD-like congeners in fish has been used as an index of contaminant exposure, rather than an index of toxicity, and the role of P450IA1 enzyme activity in TCDD-induced toxicity in fish is unclear.

Research on the biochemical effects of TCDD in fish, other than P450IA1 enzyme induction, is limited. Newsted and Giesy[94,95] investigated the effects of TCDD on the epidermal growth factor receptor (EGF-R) in rainbow trout. In mammals, TCDD decreases the number of hepatic EGF-R *in vivo*[96] and *in vitro*.[97-99] In rainbow trout, TCDD produced a dose- and time-dependent decrease in the number of hepatic EGF-R *in vivo*,[94] and produced a similar pattern of responses *in vitro*[94,95] in the RTH-149 rainbow trout hepatoma cell line. EGF and its embryonic counterpart, transforming growth factor-α (TGF-α), both EGF-R agonists, play a role in differentiation and proliferation of epithelial tissue.[100,101] While changes in the number of EGF-R by TCDD may contribute to epithelial lesions, such as fin necrosis, thymic and splenic involution, and gastric hyperplasia, there is no evidence as yet for or against this possibility in fish.

3.2.1e. Behavioral Effects. The effects of TCDD on fish behavior, such as predator–prey interactions, reproductive habits, or feeding, have not been studied. However, anecdotal observations of lethargy or a generalized decrease in swimming activity following exposure of fish to TCDD have been described by many investigators.[42,51,67,69,75,81] Mehrle and co-workers[69] observed the behavioral responses of rainbow trout swim-up fry during and after a 28-day exposure to waterborne TCDD. Rainbow trout exhibited abnormal head-up swimming, lethargic swimming, and lack of response to external stimuli in all TCDD treatment groups.[69] Furthermore, the abnormal behavior persisted until the end of the study 28 days postexposure.[69] In a parallel study, Mehrle and co-workers[69] observed similar behavioral changes in rainbow trout swim-up fry exposed to waterborne 2,3,7,8-TCDF.

Recent studies in mammals have shown that *in utero* exposure to TCDD caused sexual behavior changes in adulthood that may have resulted from altered sexual differentiation of the central nervous system perinatally.[102] The effect of TCDD on the central nervous system during fish early development, on sexual behavior, and behavior in general in fish is an area open for future research.

3.2.1f. Immunotoxicity. TCDD alters immune function in mammalian species at doses well below those that produce overt toxicity, and can increase the susceptibility of mammals to bacterial pathogens.[103] Although TCDD causes severe lymphoid depletion in the thymus, spleen, and hematopoietic tissue in fish at lethal doses, adverse effects of TCDD on immune responses in fish at sublethal doses have not been documented.[104,105] Following treatment of juvenile rainbow trout with 0, 0.1, or 1 μg TCDD/kg, cellular and humoral immune responses were unaltered as measured by response to sheep red blood cells (SRBC) in the Jerne plaque assay, complement-mediated lysis of SRBC in a chromium release assay, and mitogenic stimulation by concanavalin A and pokeweed mitogen.[104] Resistance to disease was also unaffected in fish by sublethal doses of TCDD. Rainbow trout, exposed to infectious hematopoietic necrosis virus (IHNV) 2 weeks after treatment with 0.01, 0.1, or 1.0 μg TCDD/kg, showed an increase in the severity of IHNV histopathologic lesions, but did not exhibit increased mortality compared with fish exposed to IHNV without TCDD treatment.[105] This evidence suggests that rainbow trout are less sensitive than mammals to the immunotoxic action of TCDD; however, since TCDD severely depletes circulating thrombocytes and leukocytes at doses as low as 1 μg/kg in rainbow trout,[82] further research in this area is warranted.

3.2.1g. Reproductive Toxicity. Research on the effects of TCDD on fish reproduction is limited to one study.[51] Individual breeding female zebrafish were fed 20 mg of feed on one day containing 0, 50, 250, 500, and 1000 μg TCDD/kg and were then observed during five subsequent spawning cycles (ca. 35 days).[51] At 50 μg TCDD/kg feed, there were no signs of overt toxicity in the adults, nor were there effects on spawning, embryo viability, or ovarian histology. At 250–1000 μg TCDD/kg feed, the number of eggs produced per female per spawning decreased to ca. 100, compared with 200–400 in control fish, and within one or two spawning cycles (7–14 days) egg production ceased. Histologic examination of the ovaries 35 days after exposure revealed a decrease in the number of mature, vitellogenic follicles, an increase in the number of immature, previtellogenic oocytes, and an increase in the number of atretic follicles. Larvae that hatched from eggs spawned from TCDD-treated females (250–1000 μg/kg feed) developed cranial and peritoneal edema after hatching and died within 2–3 days. The effects of TCDD on reproduction observed in this study occurred at doses that also produced overt toxicity in the adult; adult zebrafish exhibited weight loss, lethargy, and skin lesions at doses of 250–1000 μg/kg feed.[51]

The reproductive toxicity of two TCDD-like congeners, 2,3,4,7,8-pentachlorodibenzofuran (2,3,4,7,8-PeCDF)[106] and 3,3′,4,4′-TCB,[107] have recently been studied in fish and preliminary results indicate that the most sensitive effects occur during early life stage development, and reproductive toxicity in the adults occurs only at concentrations at or above those that produce toxicity during fish early development. Lake trout from the Experimental Lakes Area in Ontario, Canada, were injected i.p. with 1 μg/kg 2,3,4,7,8-PeCDF in 1988, and recaptured and spawned in 1990 and 1991.[106] Adults successfully spawned; fertilization rates were unaffected; however, there was an increased incidence of lethal deformities in lake trout embryos.[106] In a separate study, male and female white perch were given i.p. injections of 0, 0.2, 1, and 5 mg 3,3′,4,4′-TCB/kg every 3 weeks for a total of 9 weeks beginning in December and were spawned the following March.[107] While the highest dose of 5 mg/kg decreased the number of females reaching gonadal maturity, a dose as low as 1 mg/kg significantly increased larval mortality between hatching and 7 days posthatch without apparent effects on adult reproduction.[107]

Studies in mammals have shown that TCDD can significantly affect reproduction at doses that do not produce overt toxicity in the adults.[108–111] Whether TCDD-induced reproductive toxicity in fish is more sensitive than acute effects on the adult fish remains to be determined. Based on current information, however, mortality during fish early development is a more sensitive endpoint of TCDD toxicity than effects on reproduction. The lack of research on the effects of TCDD on fish reproduction represents a large information gap. Definitive studies that chronically expose fish throughout the entire reproductive cycle are needed, as well as research on the long-term reproductive success of offspring exposed to TCDD via maternal transfer. This basic lack of knowledge significantly increases the uncertainty in predicting the risk that TCDD and related compounds pose to fish populations in the environment.

3.2.1h. Developmental Toxicity. Although juvenile fish are among the most sensitive vertebrates to TCDD-induced lethality, the most sensitive adverse effects of TCDD in fish occur during early development. Fish embryos exhibit toxicity at egg burdens less than 0.300 μg TCDD/kg egg[113] and in some species as low as 0.040 μg/kg egg,[70] 25 times more sensitive than the LOAEL in juvenile fish (throm-

Table 5
Effects of TCDD on Fish Early Development following Oocyte or Fertilized Egg Exposure

	Exposure			Organism		Toxicity		
Species	Route	Duration	Nominal dose	Egg burden	Observation[a]	LOAEL[b]	Signs	Ref.
Rainbow trout *Oncorhynchus mykiss*	Water static	96 hr 24 hr renewal	0, 0.1, 1, 10 ng/L	N.D.[c]	Post-swim-up	1 ng/L Growth	Decreased growth rate, sac fry mortality, yolk sac and pericardial edema, exophthalmia, shortened maxillas, opercular defects	85
	Water static	48 hr 12 hr renewal	0, 25, 50, 75, 100, 150 ng/L	0, 0.279, 0.428, 0.466, 0.720, 1.62 μg/kg	7 days post-swim-up	0.279 μg/kg egg Sac fry mortality	Hatching and sac fry mortality, yolk sac and pericardial edema, exophthalmia, and subcutaneous hemorrhages	72
	Injection	Single 0.2 μl/egg	N.R.[d]	0, 0.194, 0.291, 0.437, 0.656, 0.983 μg/kg	7 days post-swim-up	0.291 μg/kg egg Sac fry mortality	Hatching and sac fry mortality, yolk sac and pericardial edema, exophthalmia, and subcutaneous hemorrhages	72
Lake trout *Salvelinus namaycush*	Water static	48 hr 12 hr renewal	0, 0.1, 1, 10, 100 ng/L	0, <0.015, <0.015, 0.040, 0.400 μg/kg	60 days post-swim-up	0.040 μg/kg egg Sac fry mortality	Hatching and sac fry mortality, yolk sac edema, circulatory shutdown, necrosis of retina, brain, and spinal cord	70
	Water static	48 hr 12 hr renewal	0, 10, 20, 40, 62, 100 ng/L	0, 0.034, 0.055, 0.121, 0.226, 0.302 μg/kg	60 days post-swim-up	0.055 μg/kg egg Sac fry mortality and edema	Hatching and sac fry mortality, yolk sac edema, exophthalmia, and subcutaneous hemorrhages	71
	Injection	Single 0.2 μl/egg	N.R.	0, 0.024, 0.033, 0.044, 0.058,	7 days post-swim-up	0.044 μg/kg egg	Hatching and sac fry mortality, yolk sac edema,	72

				0.077, 0.103 µg/kg		Sac fry mortality	exophthalmia, and subcutaneous hemorrhages	
	Maternal transfer	2 months prespawning	Dietary exposure of adults	0, 0.001–0.387 µg/kg	Swim-up	0.050 µg/kg egg Sac fry mortality	Hatching and sac fry mortality, yolk sac edema, exophthalmia, and subcutaneous hemorrhages	117
Northern pike *Esox lucius*	Water static	96 hr 48 hr renewal	0, 0.1, 1, 10 ng/L	N.D.	Post-swim-up	0.1 ng/L Growth	Decreased growth rate, hatching and sac fry mortality, generalized edema and hemorrhages	112
Japanese medaka *Oryzias latipes*	Water static	Fertilization to hatch	0, 0.5, 2.4, 7, 12, 33.5, 57.9 ng/L	0, <0.1, 0.3, 0.8, 1.2, 3.3, 4.8 µg/kg	3 days posthatch	0.3 µg/kg egg Edema and hemorrhages	Embryos showed a sequence of effects: Vascular hemorrhage, pericardial edema, collapse of yolk sac, circulatory shutdown, brain necrosis, and death by 3 days posthatch	113
	Maternal transfer	1 hr[e]	0, 12, 52 ng/L[e]	0, 0.2, 0.64–5[f] µg/kg	3 days posthatch	0.2 µg/kg egg All effects	Pericardial edema, hemorrhages, skeletal deformities, severe anemia, mortality 3 days posthatch	50
Mummichog *Fundulus heteroclitus*	Water static	Fertilization to hatch	0, 0.25–200 ng/L	N.D.	Posthatch	200 ng/L Mortality	Mortality, tube heart, collapse of yolk sac	114
Zebrafish *Danio rerio*	Maternal transfer	1 day 20 mg/fish[g]	0, 50, 250, 500, 1000 µg/kg feed[g]	N.D.	5–6 days posthatch	250 µg/kg feed All effects	No significant egg mortality, 100% mortality after hatch associated with cranial and peritoneal edema	51

[a]End of observation during early life stage development.
[b]LOAEL, lowest observable adverse effect level of TCDD.
[c]N.D., not determined.
[d]N.R., not reported.
[e]Waterbone exposure of adults to TCDD for 1 hr.
[f]Egg burden of TCDD from adults exposed to 52 ng TCDD/liter decline over a 28-day period.
[g]Dietary exposure of adults for 1 day.

bocytopenia and leukocytopenia, 1 μg/kg, juvenile rainbow trout).[78] Following exposure of newly fertilized fish eggs to TCDD in water or by direct injection, or exposure of developing oocytes by maternal deposition, toxicity is characteristically manifested by subcutaneous edema, hemorrhages, and mortality (Table 5). These responses have been observed in lake trout,[70–72] rainbow trout,[20,72,85] northern pike (*Esox lucius*),[112] Japanese medaka (*Oryzias latipes*),[50,113] mummichog (*Fundulus heteroclitus*),[114] and zebrafish.[51]

Rainbow trout, lake trout, northern pike, Japanese medaka, mummichog, and zebrafish are called sac fry or yolk-sac larvae at hatching because they continue to absorb nutrients and energy from their yolk. At swim-up, when exogenous feeding begins, the yolk sac has been absorbed and fish are termed fry. The length of time between hatching and exogenous feeding is temperature- and species-dependent; Japanese medaka begin exogenously feeding 1 day posthatch at 23°C,[115] while northern pike begin feeding 16–18 days posthatch at 10°C.[116]

The time course of TCDD toxicity during early development of lake trout, rainbow trout, northern pike, and zebrafish is similar; toxicity is primarily manifested after hatching. At elevated egg burdens of TCDD (3–4 times LD_{99}), embryos die prior to or during hatching; however, at lower egg burdens of TCDD (< 3 times LD_{99}), toxicity is manifested after hatching during the sac fry stage prior to swim-up. Lake trout eggs, containing 0.400 μg TCDD/kg egg (3.5 times the LD_{100} of 0.121 μg/kg), developed normally until 1 week prior to hatch when embryos exhibited subcutaneous hemorrhages; 54% of these embryos died prior to or during hatching.[70] Lake trout with egg burdens of 0.040–0.302 μg/kg (< 2.5 times LD_{100} of 0.121 μg/kg) did not exhibit increased mortality prior to hatch; however, beginning at hatch toxicity was manifested by fluid accumulation beneath the yolk sac epithelial membrane, subcutaneous hemorrhages, and mortality.[70,71] Lake trout sac fry typically died with severe yolk sac edema, exophthalmia, petechial hemorrhages, disruption of the vitelline vasculature, cessation of blood circulation to the tail, head, and gills, and arrested development of skeletal and soft tissues (Fig. 2).[70,71] Histologically, moribund lake trout sac fry also exhibited intraocular hemorrhage, and severe congestion and hemorrhages in the capillary bed behind the eye.[70] In addition, when the toxicity of TCDD is compared among different routes of egg exposure, waterborne exposure, injection, or maternal transfer, the TCDD egg burdens that produced 50% mortality were not significantly different, ranging from 0.047 to 0.065 μg TCDD/kg egg.[71,72,117]

In lake trout, TCDD appears to impact the cardiovascular system first causing congestion of vascular beds in a variety of sites leading to severe fluid accumulation and cessation of blood circulation in the yolk sac and body.[70] Further study has shown that increasing egg concentrations of TCDD produced dose-dependent increases in the volume, protein content, and eosinophil content of the yolk sac edema fluid.[118] In addition, polyacrylamide gel electrophoresis (PAGE) of edema fluid proteins revealed a pattern of protein banding similar to that produced by PAGE of adult lake trout serum, suggesting that the yolk sac edema is an ultrafiltrate of blood.[118] Based on immunohistochemistry, Guiney and co-workers[119] have recently shown that exposure of fertilized eggs to TCDD induced cytochrome P450IA1 enzymes in vascular endothelial cells of developing lake trout. While it is possible that TCDD may cause

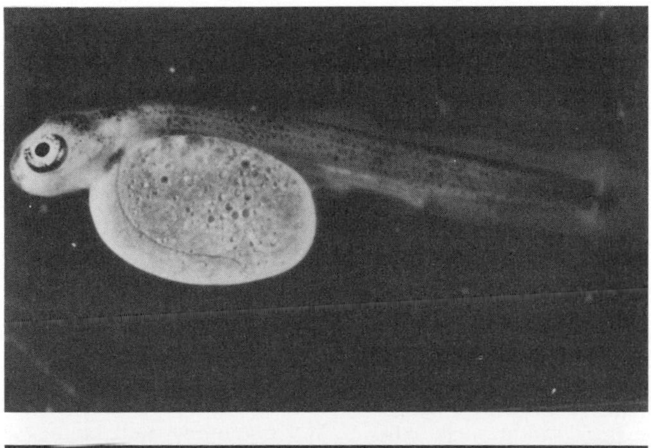

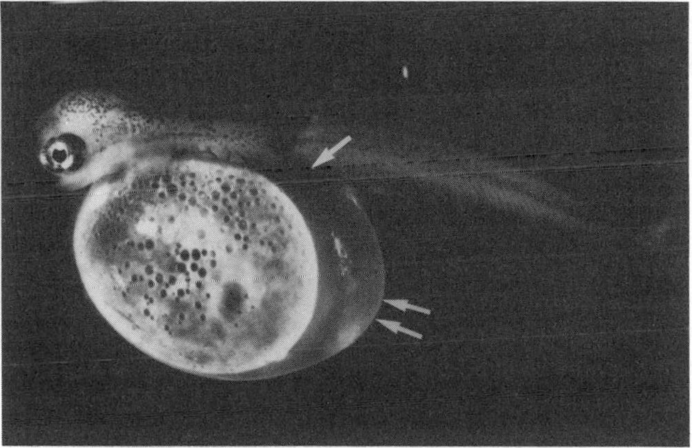

Figure 2. Representative lake trout sac fry (4 weeks posthatch) exposed as a fertilized egg to either vehicle (top) or TCDD (bottom panel, egg TCDD concentration of 0.400 μg/kg). TCDD manifested toxicity after hatching by fluid accumulation beneath the yolk sac epithelial membrane (double arrow) and subcutaneous hemorrhages (single arrow) prior to death. Reprinted with permission.[70]

hemodynamic and/or vascular permeability changes during lake trout early development secondary to cytochrome P450IA1 enzyme induction in the vascular endothelium, ultimately leading to edema and mortality,[119] there is no evidence to support or refute this hypothesis. A final point is that Walker and co-workers[71] have shown that the edema in lake trout sac fry becomes so severe that it increases sac fry wet weight. One explanation for this increase in wet weight may be that loss of fluid from the bloodstream in producing yolk sac edema causes the affected sac fry to take up excessive water from the external environment in an attempt to restore blood volume to normal.

Rainbow trout[72,85] and northern pike,[112] exposed to TCDD as eggs, exhibit the same pattern of toxic responses as lake trout,[71] sac fry mortality associated with edema. Helder[85,112] exposed rainbow trout and northern pike eggs to waterborne

TCDD (0, 0.1, 1.0, and 10.0 ng/liter) for 96 hr. Rainbow trout manifested toxicity after hatching by reduced growth rate, yolk sac and pericardial edema, exophthalmia, and sac fry mortality.[85] Northern pike exhibited some hatching mortality, but primarily sac fry mortality associated with generalized edema and hemorrhages.[112] Histologically, rainbow trout and northern pike exposed to TCDD showed disruption of blood capillaries,[85,112] and as with lake trout,[71] TCDD-induced mortality in rainbow trout and northern pike occurred prior to swim-up.

Walker and co-workers[72] found that regardless of the route of egg exposure, waterborne or injection, TCDD manifested toxicity during rainbow trout early development by sac fry edema and mortality. Furthermore, the TCDD LD_{50}s following waterborne exposure or direct injection of rainbow trout eggs were nearly identical, 0.421 (0.331–0.489) and 0.439 (0.346–0.516) μg TCDD/kg egg (95% fiducial limits), respectively.[72] In addition, rainbow trout showed significant strain differences in sensitivity to TCDD-induced lethality.[20] McConaughy strain rainbow trout were the most sensitive strain tested with an LD_{50} of 0.230 μg/kg egg, while Eagle Lake strain rainbow trout were the most resistant with an LD_{50} of 0.488 μg/kg egg.[20]

Morphologically, TCDD-induced lesions during rainbow trout and lake trout early development resemble blue-sac disease,[70–72,85] an edematous syndrome observed in some hatchery-raised salmonids.[120,121] The etiology of blue-sac disease is poorly understood, but many exogenous factors also induce the edematous syndrome including reduced water flow, increased water ammonia concentrations, and low dissolved oxygen levels.[120,122]

Zebrafish[51] eggs and yolk-sac larvae, exposed as oocytes to maternally derived TCDD, exhibited a similar course of toxicity and pattern of responses as lake trout, rainbow trout, and northern pike. Egg mortality was not affected by TCDD exposure, and toxicity was manifested after hatching by cranial and peritoneal edema, and subsequent mortality of yolk-sac larvae 2–3 days posthatch.[51]

In the Japanese medaka, the time course of TCDD-induced toxicity during early development is different than that for rainbow trout, lake trout, northern pike, or zebrafish. TCDD-induced lesions in Japanese medaka develop after organogenesis but prior to hatching.[113] Following exposure of eggs to TCDD on the day of fertilization, embryos developed normally until day 4 or 5 of development when blood flow in the caudal vein was decreased. Subcutaneous hemorrhages and pericardial edema subsequently developed, followed by collapse of the yolk sac, failure of heart chamber formation, circulatory shutdown, and mortality.[113]

The sensitivity of medaka to TCDD is reduced when embryos are exposed after the formation of the liver rudiment (days 4–5 postfertilization).[113] When Japanese medaka were exposed to TCDD between fertilization and day 5 of development, egg burdens of 9–17 μg TCDD/kg egg produced severe edema, hemorrhages, and mortality. When exposed on day 6 of development, percent mortality and the percentage of embryos with severe lesions (edema and hemorrhage) were reduced at egg burdens of 10 μg/kg egg. When exposed on day 7 or 8, percent mortality was reduced further and severe lesions were absent at egg burdens of 9–10 μg/kg egg, and if exposure began on day 9, toxicity was not observed at egg burdens of 7 μg/kg egg. This evidence suggests that the sensitivity of medaka to TCDD toxicity depends on a specific event during embryonic development occurring between days 5 and 9 postfertilization.[113]

The pattern of lesions induced by TCDD during fish early development, characterized by cardiovascular and circulatory changes, edema, and hemorrhages, resembles TCDD-induced lesions in other vertebrate species including chickens, mice, and monkeys.[66,123–125] Chickens[66,123] exposed to TCDD exhibit "chick edema disease" which is characterized by ascites and by pericardial, subcutaneous, and pulmonary edema. Chicken embryos,[88] injected with TCDD as newly fertilized eggs, develop abnormalities in the vitelline vasculature including a decrease in the area of yolk vascularized and short, abnormally bent vitelline blood vessels. Mice[124] treated with acute doses of TCDD exhibited subcutaneous edema, ascites, fluid accumulation in the thoracic cavity, and submucosal edema in the stomach and small and large intestine. Mice also showed intraorbital hemorrhage and subcutaneous hemorrhages in the eyelids following TCDD exposure.[124] Monkeys[125] typically developed periorbital and facial edema, ascites, subcutaneous edema in the lower abdomen, as well as pericardial hemorrhages, focal hemorrhages in the lungs, and petechial hemorrhages over the entire body surface following TCDD exposure. Although rats do not exhibit edema following TCDD treatment, TCDD treatment enhances the response of rats to edemagenic agents,[126,127] suggesting that TCDD may affect the vascular endothelium and/or mediators of vascular permeability.

Early life stages of fish may provide a valuable model for delineating the mechanism of action of TCDD as a result of their extreme sensitivity, the specific time course of toxicity, and the similarities of TCDD-induced toxicity with higher vertebrates. Early life stages of fish have already provided a useful model for studying the quantitative structure–*toxicity* relationships of TCDD-like PCDD, PCDF, and PCB congeners.[20,128]

3.2.1i. Structure–Toxicity Relationships during Early Development. In fish, as in birds and mammals, (1) the toxicity of TCDD-like PCDD, PCDF, and PCB congeners is manifested in a manner characteristic of the most potent congener, TCDD, and (2) the potencies of PCDD, PCDF, and PCB congeners vary in a predictable manner characteristic of the class of chemical, dioxin, furan, or biphenyl, and the chlorine substitution pattern in that class. Quantitative structure–toxicity relationships of TCDD and TCDD-like congeners in fish have only been studied during fish early development.[19,20,128] Wisk and Cooper[19] exposed Japanese medaka eggs to graded waterborne concentrations of TCDD, 1,2,3,7,8-PeCDD, 1,2,3,4,7,8-HxCDD, or 2,3,7,8-TCDF. Toxicity of all TCDD-like congeners was manifested in a manner identical to TCDD, consisting of subcutaneous hemorrhages, pericardial edema, circulatory shutdown, and mortality.[19,113] In a study by Walker and Peterson,[20] newly fertilized rainbow trout eggs were injected with graded doses of TCDD, or the following TCDD-like congeners: 1,2,3,7,8-PeCDD, 1,2,3,4,7,8-HxCDD, 2,3,7,8-TCDF, 1,2,3,7,8-PeCDF, 2,3,4,7,8-PeCDF, 1,2,3,4,7,8-HxCDF, 3,3′,4,4′-TCB (#77), or 3,3′,4,4′,5-PeCB (#126). Toxicity of all eight TCDD-like congeners was manifested in a manner characteristic of TCDD, dose-related increases in sac fry mortality associated with subcutaneous hemorrhages and yolk sac edema.[20,72]

In fish, as in mammals, the toxic potency of TCDD-like PCDD and PCDF congeners is dependent on the chemical structure of the congener. In Japanese medaka[19] and rainbow trout,[20,128] potency of TCDD-like PCDD congeners decreased with increasing nonlateral chlorine substitution, and the rank order toxic potency was TCDD > 1,2,3,7,8-PeCDD > 1,2,3,4,7,8-HxCDD. For TCDD-like PCDF congeners, potency decreased with loss of chlorine at the C(4) position and did not substantially change with the addition of chlorine at C(1), and the rank order toxic potency was 2,3,4,7,8-PeCDF = 1,2,4,7,8-HxCDF > 1,2,3,7,8-PeCDF = 2,3,7,8-TCDF.[20,128] The quantitative structure–activity relationship for early life stage mortality caused by TCDD and TCDD-like PCDD and PCDF congeners in Japanese medaka[19] and rainbow trout[20,128] is essentially the same as that for both Ah receptor binding affinity and acute toxicity in rats.[129,130]

The toxic potency of TCDD-like coplanar PCB congeners, those lacking *ortho*-chlorine substitution, is also structure-dependent, and, based on a limited amount of information on the toxicity of these congeners to mammals, the quantitative structure–toxicity relationships for these congeners appear to differ between fish and mammals. In both fish[20,128] and mammals,[131,132] the toxic potency of 3,3′,4,4′,5-PeCB (#126) is greater than 3,3′,4,4′-TCB (#77). However, the potencies of both of these PCB congeners for producing rainbow trout early life stage mortality[20,128] were significantly lower than their potencies proposed for human health risk assessment.[26] The potencies proposed for human health risk assessment are based on acute toxicity data and P450IA1 enzyme induction in mammals.[26] In addition, the induction potency of both 3,3′,4,4′,5-PeCB (#126) and 3,3′,4,4′-TCB (#77) for cytochrome P450IA1 activity is also lower in fish[21] compared with their induction potency in mammals.[131]

The toxic potency of TCDD-like, mono-*ortho*-chlorinated PCB congeners also appears to differ between fish and mammals. 2,3,3′,4,4′-PeCB (#105) and 2,3′,4,4′,5-PeCB (#118) bind the Ah receptor, induce cytochrome P450IA1 enzyme activity, and produce toxicity in mammals.[131–133] However, neither of these two congeners produced toxicity during rainbow trout early development when injected up to egg doses of 7000 μg/kg egg.[20] Further studies have indicated that neither 2,3,3′,4,4′-PeCB (#105) nor 2,3,4,4′,5-PeCB (#118) produce toxicity when injected into rainbow trout eggs up to 140,000 μg/kg egg.[134] In addition, neither 2,3′,4,4′,5-PeCB (#118) nor 2,3,3′,4,4′-PeCB (#105) induced hepatic cytochrome P450IA1 enzyme activity in scup[135]; however, 2,3′,4,4′,5-PeCB (#118) was shown to induce P450IA1 activity in rainbow trout at a dose six times higher than that administered to scup.[136]

The reason for the potential differences in the toxic potency of TCDD-like PCB congeners between fish and mammals is unknown. If these TCDD-like PCB congeners were more rapidly eliminated than TCDD during fish early development, then their toxic potency would be greatly reduced compared with TCDD. However, Zabel and co-workers[137] have found that elimination of 3,3′,4,4′,5-PeCB (#126), 3,3′,4,4′-TCB (#77), 2,3,3′,4,4′-PeCB (#105), and 2,3′,4,4′,5-PeCB (#118) during the egg and sac fry stages of rainbow trout was only 13–25%[137] compared with 0% for TCDD.[72,137] Nevertheless, the 13–25% elimination of these four PCB congeners cannot explain the 10 to 100-fold lower potency of these coplanar and mono-*ortho*-chlorine-substituted PCB congeners in fish[20] compared with mammals.[26] Alternative explanations for the lower potency of TCDD-like PCB congeners in fish are that (1) the ligand binding site on the Ah receptor has lower binding affinity for PCB congeners in fish than in mammals and (2) the mono-*ortho*-chlorinated analogues of the coplanar PCBs are unable to attain the coplanar structure required to bind the Ah receptor because of the lower temperatures at which fish studies are conducted. It must be noted, however, that the high toxic potencies of TCDD-like PCBs in mammals have been recently questioned.[138] Evidence suggests that the relative potencies of the PCB congeners estimated for mammals are too high. If this is supported in other studies, it would mean the toxic potencies in fish and mammals are not as different as is currently thought.

3.2.2. Summary

Fish are among the most sensitive vertebrates to the lethal potency of TCDD, and early life stages of fish represent the most sensitive developmental stage, one to two orders of magnitude more sensitive than juveniles, with egg burdens of TCDD in the low parts per trillion range producing mortality. Signs of toxicity and histopathologic lesions produced by TCDD in juvenile fish are similar to those seen in higher vertebrates and include decreased food intake, wasting, delayed mortality, and lesions in epithelial and lymphomyeloid tissues. Signs of toxicity and histopathologic lesions produced by TCDD during fish early development, characterized primarily by cardiovascular and circulatory changes, edema, hemorrhages, and mortality, also resemble TCDD-induced toxicity in higher vertebrates such as chickens, mice, and monkeys. Quantitative structure–toxicity relationships for TCDD-like PCDD and PCDF congeners are similar between fish and mammals, while toxic potencies of coplanar and mono-*ortho*-chlorine-substituted PCB congeners appear lower in fish than in mammals, and the reasons for this difference remain to be explained. Finally, the biochemical, behavioral, and reproductive toxicity of TCDD in fish remain to be adequately studied.

3.3. Exposure to Commercial PCB Mixtures

Commercial PCB mixtures were manufactured by batch chlorination of the biphenyl ring structure, and the mixture of congeners produced in a given formulation included congeners *with* and *without* a chlorine substitution pattern resembling that of TCDD.[139–141] In addition, PCDFs were found as trace contaminants in some commercial PCB products.[142] Therefore, results of toxicity studies using commercial PCB formulations represent the combined toxicity of TCDD-like PCBs and PCDFs (Ah receptor agonists) and non-TCDD-like PCBs (non-Ah receptor agonists), which include a class of lightly chlorinated noncoplanar PCB congeners that are neurotoxic in mammals.[143] Although research on the toxicity of commercial PCB mixtures to fish will not be comprehensively reviewed in this chapter, the primary similarities and key differences between TCDD-induced toxicity and toxicity of commercial PCB mixtures to fish will be highlighted.

3.3.1. Aroclor and Clophen

The majority of toxicity studies of PCB commercial mixtures have used Aroclor formulations. Most of the commercial PCBs manufactured in the United States were sold under the trade name Aroclor followed by a four-digit number, such as Aroclor 1254, where the "12" represents the biphenyl molecule and "54" the weight percentage of chlorine in the formulation. One exception was Aroclor 1016, a distillate of Aroclor 1242, which contained ca. 40–42% chlorine, but had a higher percentage of di- and trichlorobiphenyls and lower percentage of tetra- and pentachlorobiphenyls than Aroclor 1242.[140] The PCB congener composition of Clophen A50, manufactured in the Federal Republic of Germany, resembles that of Aroclor 1254[140] and has been used in a few pertinent toxicity studies in fish.

3.3.1a. Acute Toxicity. Many of the acute effects produced by Aroclor and Clophen PCB formulations in fish resemble acute effects produced by TCDD. Following acute or chronic exposure of juvenile fish to dietary Clophen A50, or waterborne or dietary Aroclor 1016, 1221, 1232, 1242, 1248, 1254, and 1260, toxicity is typically manifested by decreased feed intake, wasting, delayed mortality, and fin necrosis.[144–153] Following exposure of juvenile pinfish[144] (*Lagodon rhomboides*) and spot[144] (*Leiostomus xanthurus*) to 5 μg Aroclor 1254/liter for up to 45 days, both species exhibited delayed mortality, and fish continued to die even after being placed in PCB-free water. Prior to death, pinfish developed fungal infections and hemorrhages, while spot typically ceased feeding, exhibited wasting, and developed fin and epithelial lesions.[144] In a different study, adult minnows (*Phoxinus phoxinus*),[153] exposed to dietary Clophen A50 (0, 20, 200, or 2000 mg PCB/kg feed) for 40 days, exhibited decreased feed intake, wasting, lethargy, fin necrosis, and delayed mortality.

When the acute toxicity of different Aroclor formulations are compared, generally, the more highly chlorinated PCB mixtures are less potent than the lower chlorinated ones.[148,151] Following exposure of juvenile cutthroat trout (*Salmo clarki*) to graded water concentrations of Aroclor 1221, 1232, 1242, 1248, 1254, 1260, 1262, and 1268, the 96-hr LD_{50}s were 1.2, 2.5, 5.4, 5.7, 42, 61, 50, and 50 mg/liter, respectively.[148,151] Thirty-day chronic exposure of rainbow trout, bluegill, and channel catfish showed a similar relationship for Aroclor 1242, 1248, 1254, and 1260, although Aroclor 1248 was slightly more toxic than 1242 to all species tested.[151]

Following acute and chronic exposure of juvenile fish to Aroclor formulations, other signs of toxicity included lethargy, cutaneous hyperpigmentation, and uncoordinated swimming.[145–147]

3.3.1b. Reproductive Toxicity. The reproductive toxicity of commercial PCB formulations in fish has been more thoroughly studied than TCDD-induced reproductive toxicity. In general, following exposure of adult fish to Aroclor formulations, the most sensitive adverse effects occur during early life stage development, and reproductive toxicity in the adults occurs only at PCB concentrations at or above those that produce early life stage mortality.[145,150,152] Adult minnows[153] (*Phoxinus phoxinus*), exposed to dietary Clophen A50 (0, 20, 200, or 2000 mg PCB/kg feed) for 40 days, were observed at spawning, 195 days postexposure. Spawning of fish in the 200 and 2000 mg/kg treatment groups was delayed by 1 and 3 weeks, respectively, compared with controls. Egg production and fertilization were unaffected by Clophen exposure, while egg and hatching mortality were significantly increased in the highest exposure group. Adult sheepshead minnows (*Cyprinodon variegatus*),[145] exposed to waterborne Aroclor 1254 (0.1, 0.32, 1.0, 3.2, or 10 μg/liter) for 4 weeks, exhibited overt toxicity including decreased feed intake and fin necrosis at 10 μg/liter. Egg production and percent fertilization were unaffected at 10 μg/liter, while egg and hatching mortality were significantly increased.

The effects of Aroclor formulations on sex hormone levels in fish have been studied in a variety of species. In general, the levels of circulating steroid hormones are reduced by exposure to commercial PCB mixtures; however, studies have not linked decreased plasma sex hormone concentrations with functional reproductive deficits in fish. Circulating progesterone and 17β-estradiol were decreased in female common carp following exposure to Aroclor 1248, and circulating testosterone was decreased in males.[154] Similarly, plasma estrogen and androgen concentrations were decreased in female and male rainbow trout, respectively, following exposure to Aroclor 1254.[155] Aroclor 1254 decreased gonadosomatic index, plasma 17β-estradiol, hepatic estrogen receptor concentration, and pituitary secretion of gonadotropin in female Atlantic croaker (*Micropogonias undulatus*).[156]

3.3.1c. Toxicity during Early Development. As seen with TCDD, the early developmental stages of fish are more sensitive than juveniles or adults to the lethal potency of commercial PCB mixtures. This has been observed for fathead minnows,[150,152] sheepshead minnows,[145,149] and flagfish (*Jordanella floridae*).[150] Newly hatched fathead minnow yolk-sac larvae were 20 times more sensitive to the lethal potency of waterborne Aroclor 1242 than 3-month-old juvenile fathead minnows with 96 hr LC_{50}s of 15 and 300 μg/liter, respectively.[150]

Furthermore, during fish early development Aroclor toxicity is manifested in a manner similar to TCDD; egg mortality occurs at extremely high egg doses, but at lower egg doses toxicity is primarily manifested by mortality after hatching and prior to exogenous feeding. Eggs, obtained from adult sheepshead minnows[145] following exposure to waterborne Aroclor 1254 (0.1, 0.32, 1.0, 3.2, and 10 μg/liter), showed egg and hatching mortality only in the highest exposure group (10 μg/liter maternal exposure), while toxicity in lower exposure groups (0.32–3.2 μg/liter) was primarily manifested by dose-related increases in mortality of yolk-sac larvae prior to the onset of exogenous feeding. Similarly, eggs, obtained from adult minnows (*Phoxinus phoxinus*)[150] exposed to Clophen A50, exhibited increased egg and hatching mortality in the highest treatment group (2000 mg/kg feed, parental exposure), while toxicity of Clophen A50 in a lower treatment group (200 mg/kg feed) was manifested after hatching by yolk-sac larval mortality associated with pericardial edema.

3.3.1d. Differences in Toxicity of TCDD and Commercial PCB Mixtures. Differences in the manner in which toxicity of commercial PCB mixtures are manifested in fish compared with TCDD have been repeatedly observed. Two such differences are growth stimulation and decreased time-to-hatch. Dietary exposure of juvenile brown trout[157] and minnows (*Phoxinus phoxinus*)[158] to Clophen A50 has been shown to stimulate growth (i.e., increased weight gain), rather than produce a wasting syndrome (i.e., decreased weight gain). Furthermore, the increases in growth could not be explained by differences in gonadal maturation, levels of extractable fat, or edema.[157,158] Interestingly, juvenile coho salmon,[159] exposed to a mixture of three non-TCDD-like PCB congeners—2,2′,5,5′-TCB (#52), 2,2′,4,5,5′-PeCB (#101), and 2,2′,4,4′,5,5′-HxCB (#153)—for 72 days, exhibited a time-related increase in body weight gain, relative to controls.

A second difference between TCDD-induced toxicity and toxicity of PCB mixtures is the effects of length of time between fertilization and hatching. Following exposure of fish eggs to TCDD, time-to-hatch is unaffected.[70,71,85,112,113] However, coho salmon eggs[160] and minnow (*Phoxinus phoxinus*)[158] eggs exposed to Aroclor 1254 and Clophen A50, respectively, hatched significantly earlier than controls. Embryos that hatched early were not advanced in development; rather, embryos hatched prematurely.[158,160] Decreased time-to-hatch, however, has not been consistently observed in developmental toxicity studies of commercial PCB mixtures. Time-to-hatch was unaffected following exposure of sheepshead minnow eggs to Aroclor 1016[147] and 1254,[149] and following exposure of brook trout (*Salvelinus fontinalis*) eggs to Aroclor 1254.[161]

3.3.2 Summary

The similarities between the acute, reproductive, and developmental toxicity of TCDD and commercial PCB formulations to fish are noteworthy, suggesting that the toxicity of commercial PCB formulations to fish is primarily manifested by Ah receptor agonists (i.e., TCDD-like PCB and PCDF congeners). The reason for differences in acute and developmental toxicity of PCB mixtures compared with TCDD may be related to (1) the presence of PCB congeners with chlorination patterns that do not resemble TCDD, (2) interactions between mixture components, and/or (3) the presence of other unidentified contaminants.

4. RISK POSED BY TCDD AND TCDD-LIKE PCDD, PCDF, AND PCB CONGENERS TO FISH IN THE ENVIRONMENT

4.1. TCDD

The concentration of TCDD in fish in the environment is too low to produce overt toxicity in juvenile or adult fish.[29–31,42,67,162] The highest concentration of TCDD in fish sampled from 388 sites across the United States was 0.203 μg/kg in smallmouth buffalo (*Ictiobus bubalus*) from a site in Arkansas,[31] while concentrations in lake trout collected from the Great Lakes in 1984 ranged from 0.001 μg/kg in Lake Superior to 0.049 μg/kg in Lake Ontario.[29] While species sensitivity to TCDD toxicity varies, the LOAEL in juvenile fish is 1.0 μg TCDD/kg,[78] 5- to 1000-fold higher than concentrations of TCDD detected in fish in the environment.

Early life stages of fish, however, are at higher risk because they are up to 50 times more sensitive than adults to TCDD toxicity.[20,70–72,85,112,113,128] The concentration of TCDD in fish eggs in the environment is only beginning to be examined. Striped bass eggs from the Hudson River contained 0.029 μg TCDD/kg egg,[60] approaching the LOAEL during fish early development of 0.040 μg TCDD/kg egg.[70]

4.2. Environmental Mixtures of TCDD and TCDD-Like Congeners

Although TCDD is the most potent PCDD, PCDF, or PCB congener, it is not the only one that has bioaccumulated in fish in the environment. To assess the risk that complex mixtures of TCDD and TCDD-like PCDD, PCDF, and PCB congeners pose to feral fish, the toxic potency of the mixture must be determined. The toxic potency of a mixture of PCDDs, PCDFs, and PCBs can be expressed as an equivalent concentration of TCDD by taking into account the potencies of the individual TCDD-like congeners, relative to TCDD. This TCDD equivalent concentration (TEC) can be determined using chemical data or biological data.[164] Calculation of TEC from chemical data involves determining the concentration of each TCDD and TCDD-like PCDD, PCDF, and PCB congener in an environmental sample, and applying congener-specific toxic equivalency factors (TEFs). TEF, or toxicity potency of a PCDD, PCDF, and PCB congener, relative to TCDD, is defined as the ratio TCDD ED_{50}/congener ED_{50}. Multiplication of the concentration of a congener in a mixture by its respective TEF yields a TEC for that congener. The toxic potency of the entire mixture can then be expressed as an equivalent concentration of TCDD by addition of the TECs for all of the individual congeners in the mixture.[2,164] Alternatively, calculation of TEC from biological data involves extracting the PCDDs, PCDFs, and PCBs from the environmental sample, and measuring the ability of the extract, relative to TCDD, to produce a TCDD-like response.[163] P450IA1 enzyme induction in the H4IIE rat hepatoma cell line is the most commonly used TCDD response and *in vitro* system, respectively, for comparing the potency of an environmental extract with the potency of TCDD.[163]

Both methods have their limitations. Using chemical data and congener-specific TEFs assumes that TCDD and TCDD-like congeners act additively to produce toxicity, and this assumption has not been unequivocally determined. Alternatively, H4IIE cells have been shown to overestimate the toxic potency of coplanar PCBs to fish by 8 to 100 fold.[20,133,165] Since PCBs occur in fish in the environment at concentrations often 1000 times greater than the combined concentrations of PCDDs and PCDFs,[60,62] H4IIE cells will overestimate TCDD equivalents concentration of an environment mixture of PCDDs, PCDFs, and PCBs with respect to fish early life stage mortality.[20]

Because early life stages of fish are more sensitive than adults to TCDD toxicity, they are also at higher risk of toxicity from complex mixtures of TCDD-like congeners. Using fish-specific TEFs and determining the concentration of PCDD, PCDF, and

PCB congeners in feral lake trout eggs, Walker and co-workers[62] calculated TECs for lake trout eggs in Lakes Ontario and Michigan. Fish-specific TEFs were based on the ability of TCDD and TCDD-like PCDD, PCDF, and PCB congeners, injected into rainbow trout eggs, to produce early life stage mortality and were defined as TCDD LD_{50}/congener LD_{50}.[20] If a TEC calculated for lake trout eggs exceeded the LOAEL for TCDD-induced mortality during lake trout early development of 0.040 μg TCDD/kg egg,[70] then the combined presence of TCDD-like PCDDs, PCDFs, and PCBs in feral lake trout eggs would increase the risk of early life stage mortality. TECs for lake trout eggs collected from Lake Michigan and Lake Ontario in 1988 were calculated to be 0.008 and 0.029 μg/kg egg, respectively.[62] These values approach but do not exceed the TCDD LOAEL for lake trout early life stage mortality of 0.040 μg/kg egg, suggesting that in 1988 the combined presence of PCDDs, PCDFs, and PCBs in Lake Michigan and Lake Ontario lake trout eggs did not pose a significant risk to lake trout early life stage survival.[62]

Field studies of feral fish populations provide evidence that TCDD and TCDD-like PCDDs, PCDFs, and PCBs may be impacting fish early life stage survival in the environment. Swedish scientists have described an edematous syndrome, M74-syndrome, during early development of Atlantic salmon (*Salmo salar*).[166] Sac fry typically exhibited edema and vascular necrosis prior to death.[166,167] PCDDs and PCDFs transferred from the maternal tissues to the eggs are suspected to be one of the causative agents. In the United States, fisheries biologists have described an increased incidence of blue-sac disease in lake trout sac fry from Lake Ontario.[168] Lake trout eggs collected from Lake Ontario adults exhibited increased mortality during the egg and sac fry stages of development from 1977 to 1983, and sac fry mortality was associated with yolk sac edema and hemorrhages, identical to blue-sac disease.[168] Although PCDDs and PCDFs have not been unequivocally identified as the causative agents, a retrospective risk assessment currently being undertaken indicates that the concentrations of PCDDs and PCDFs were likely high enough in the late 1970s to impact lake trout early life stage survival.[169]

Future biomonitoring efforts need to determine congener-specific PCDD, PCDF, and PCB concentrations in feral fish eggs; however, these analyses are very costly and labor intensive. Future efforts to predict the risk that TCDD and TCDD-like PCDD, PCDF, and PCB congeners pose to fish early life stage development should focus chemical analyses on those PCDD, PCDF, and PCB congeners in feral fish eggs that contribute the largest percentage to a TEC and pose the greatest risk to fish early life stage mortality. Based on current data, those congeners include TCDD, 1,2,3,7,8-PeCDD, 2,3,7,8-TCDF, 2,3,4,7,8-PeCDF, and 3,3′,4,4′,5-PeCB (#126).[62,169]

4.3. Conclusions

To reliably predict the risk that TCDD and TCDD-like PCDD, PCDF, and PCB congeners pose to fish in the environment, additional information is needed such as (1) the LOAEL for TCDD-induced sublethal effects during fish early development,

(2) validation that TCDD-like congeners interact additively to produce toxicity, and (3) an understanding of the role that other environmental contaminants play in modulating TCDD toxicity. The LOAEL for TCDD during fish early development is based on mortality, and TCDD may produce adverse effects during fish early development at concentrations lower than those that produce mortality. In addition, the assumption that TCDD and TCDD-like PCDD, PCDF, and PCB congeners act additively to produce toxicity has not been unequivocally determined; nor has the manner in which TCDD-like congeners and non-TCDD-like congeners interact to produce toxicity been studied. Given the fact that early life stages of fish in the environment are exposed to maternally derived TCDD-like PCDD, PCDF, and PCB congeners, and that early life stages of fish are extremely sensitive to these chemicals, the combined presence of PCDDs, PCDFs, and PCBs in feral fish eggs likely increase the risk of adverse effects during early development in many fish species which may ultimately impact recruitment.

To predict the risk that TCDD and TCDD-like congeners pose to fish in the environment, future research should focus on (1) measuring the concentration of these contaminants in fish and fish eggs in the environment, (2) studying the interactive effects of PCDD, PCDF, and PCB congeners present in complex environmental mixtures, and (3) studying the sublethal effects of TCDD in fish, such as adverse effects on early development, behavior, reproduction, and biochemical and cellular mechanisms.

ACKNOWLEDGMENTS. We thank Mike Hornung, Erik Zabel, Linda Damos, and Dorothy Nesbit for their expert assistance. This work was supported in part by the University of Wisconsin Sea Grant Institute under a grant R/MW-52 from the National Sea Grant College Program, National Oceanic and Atmospheric Administration, U.S. Department of Commerce and from the State of Wisconsin; and Great Lakes Protection Fund grant FG6901038, and EPA Cooperative Agreement CR819065-01-0, Mod. 1.

REFERENCES

1. S. Safe, S. Bandiera, T. Sawyer, L. Robertson, L. Safe, A. Parkinson, P. E. Thomas, D. E. Ryan, L. M. Reik, W. Levin, M. A. Denomme, and T. Fujita, PCBs: Structure–function relationships and mechanisms of action, *Environ. Health Perspect.* **60,** 47–66 (1985).
2. J. S. Bellin and D. G. Barnes, Interim procedures for estimating risks associated with exposure to mixtures of chlorinated dibenzo-*p*-dioxins and -dibenzofurans (CDDs and CDFs), EPA/625/3-89/016, Risk Assessment Forum, Washington, DC (1989).
3. A. Poland and J. C. Knutson, 2,3,7,8-Tetrachlorodibenzo-*p*-dioxin and related halogenated aromatic hydrocarbons: Examination of the mechanism of toxicity, *Annu. Rev. Pharmacol. Toxicol.* **22,** 517–554 (1982).
4. J. P. Whitlock, Jr., Genetic and molecular aspects of 2,3,7,8-tetrachlorodibenzo-*p*-dioxin action, *Annu. Rev. Pharmacol. Toxicol.* **30,** 251–277 (1990).
5. A. Poland, E. Glover, and A. S. Kende, Stereospecific, high affinity binding of 2,3,7,8-tetrachlorodibenzo-*p*-dioxin by hepatic cytosol, *J. Biol. Chem.* **251,** 4936–4946 (1976).

6. A. Poland and E. Glover, Chlorinated biphenyl induction of aryl hydrocarbon hydroxylase activity: A study of the structure–activity relationship, *Mol. Pharmacol.* **13,** 924–938 (1977).
7. M. E. Hahn, A. Poland, E. Glover, and J. J. Stegeman, The Ah receptor in marine animals: Phylogenetic distribution and relationship to cytochrome P450IA inducibility, *Mar. Environ. Res.* **34,** 87–92 (1992).
8. L. J. Heilmann, Y.-Y. Sheen, S. W. Bigelow, and D. W. Nebert, Trout P450IA1: cDNA and deduced protein sequence, expression in liver, and evolutionary significance, *DNA* **7,** 379–387 (1988).
9. R. L. Binder and J. J. Stegeman, Induction of aryl hydrocarbon hydroxylase activity in embryos of an estuarine fish, *Biochem. Pharmacol.* **29,** 949–951 (1980).
10. M. J. Vodicnik, C. R. Elcombe, and J. J. Lech, The effect of various types of inducing agents on hepatic microsomal monooxygenase activity in rainbow trout, *Toxicol. Appl. Pharmacol.* **59,** 364–374 (1981).
11. R. J. Schwen and G. J. Mannering, Hepatic cytochrome P-450-dependent monooxygenase systems of the trout, frog and snake—III. Induction, *Comp. Biochem. Physiol.* **71B,** 445–453 (1982).
12. E. Monosson and J. J. Stegeman, Cytochrome P450E (P450IA) induction and inhibition in winter flounder by 3,3′,4,4′-tetrachlorobiphenyl: Comparison of response in fish from Georges Bank and Narragansett Bay, *Environ. Toxicol. Chem.* **10,** 765–774 (1991).
13. R. M. Smolowitz, M. E. Hahn, and J. J. Stegeman, Immunohistochemical localization of cytochrome P-450IA1 induced by 3,3′,4,4′-tetrachlorobiphenyl and by 2,3,7,8-tetrachlorodibenzofuran in liver and extrahepatic tissues of the teleost *Stenotomus chrysops* (Scup), *Drug Metab. Dispo.* **19,** 113–123 (1991).
14. A. Goksoyr, T. Andersson, D. R. Buhler, J. J. Stegeman, D. E. Williams, and L. Forlin, Immunochemical cross-reactivity of β-naphthoflavone-inducible cytochrome P450 (P450IA) in liver microsomes from different fish species and rat, *Fish Physiol. Biochem.* **9,** 1–13 (1991).
15. A. Lorenzen and A. B. Okey, Detection and characterization of [^{3}H]2,3,7,8-tetrachlorodibenzo-*p*-dioxin binding to Ah receptor in a rainbow trout hepatoma cell line, *Toxicol. Appl. Pharmacol.* **106,** 53–62 (1990).
16. H. I. Swanson and G. H. Perdew, Detection of the Ah receptor in rainbow trout: Use of 2-azido-3-[^{125}I]-7,8-dibromodibenzo-*p*-dioxin in cell culture, *Toxicol. Lett.* **58,** 85–95 (1991).
17. M. S. Denison, The molecular mechanism of action of 2,3,7,8-tetrachlorodibenzo-*p*-dioxin and related halogenated aromatic hydrocarbons, *Chemosphere* **23,** 1825–1830 (1991).
18. M. E. Hahn, J. J. Stegeman, A. Poland, and M. S. Denison, The Ah receptor in aquatic species: Presence, DNA binding, and implications for toxicity of dioxins, and related compounds, *Soc. Environ. Toxicol. Chem. Abstr.* **13,** 44 (1992).
19. J. D. Wisk and K. R. Cooper, Comparison of the toxicity of several polychlorinated dibenzo-*p*-dioxins and 2,3,7,8-tetrachlorodibenzofuran in embryos of the Japanese medaka (*Oryzias latipes*), *Chemosphere* **20,** 361–377 (1990).
20. M. K. Walker and R. E. Peterson, Potencies of polychlorinated dibenzo-*p*-dioxins, dibenzofurans, and biphenyls, relative to 2,3,7,8-tetrachlorodibenzo-*p*-dioxin, for producing early life stage mortality in rainbow trout (*Oncorhynchus mykiss*), *Aquat. Toxicol.* **21,** 219–238 (1991).
21. D. M. Janz and C. D. Metcalfe, Relative induction of aryl hydrocarbon hydroxylase by 2,3,7,8-TCDD and two coplanar PCBs in rainbow trout (*Oncorhynchus mykiss*), *Environ. Toxicol. Chem.* **10,** 917–923 (1991).

22. J. L. Parrott, P. V. Hodson, D. G. Dixon, and M. R. Servos, Toxic equivalent factors (TEFs) for dioxins in fish, *Soc. Environ. Toxicol. Abstr.* **12,** 30 (1991).
23. J. L. Parrott, P. V. Hodson, D. G. Dixon, and M. R. Servos, Liver EROD induction by several 2,3,7,8-chlorinated dibenzofurans in rainbow trout, *Soc. Environ. Toxicol. Chem. Abstr.* **13,** 113 (1992).
24. J. H. Clemons, M. R. vanden Heuvel, N. C. Bols, and D. G. Dixon, Toxic equivalent factors for selected congeners of PCDDs, PCDFs, and PCBs using a rainbow tout liver cell line (RTL-W1), *Soc. Environ. Toxicol. Chem. Abstr.* **13,** 201 (1992).
25. D. E. Tillitt and S. M. Cantrell, Planar halogenated hydrocarbon (PHH) structure activity relationship in a teleost (PLHC) cell line, *Soc. Environ. Toxicol. Chem. Abstr.* **13,** 45 (1992).
26. S. Safe, Polychlorinated biphenyls (PCBs), dibenzo-*p*-dioxins (PCDDs), dibenzofurans (PCDFs), and related compounds: Environmental and mechanistic considerations which support the development of toxic equivalency factors (TEFs), *Crit. Rev. Toxicol.* **21,** 51–88 (1990).
27. R. Baughman and M. Meselson, An analytical method for detecting TCDD (dioxin): Levels of TCDD in samples from Vietnam, *Environ. Health Perspect.* **5,** 27–35 (1973).
28. J. M. Czuczwa and R. A. Hites, Environmental fate of combustion-generated polychlorinated dioxins and furans, *Environ. Sci. Technol.* **18,** 444–450 (1984).
29. D. DeVault, W. Dunn, P. Bergqvist, K. Wiberg, and C. Rappe, Polychlorinated dibenzofurans and polychlorinated dibenzo-*p*-dioxins in Great Lakes fish: A baseline and interlake comparison, *Environ. Toxicol. Chem.* **8,** 1013–1022 (1989).
30. T. Zacharewski, L. Safe, S. Safe, B. Chittim, and D. DeVault, Comparative analysis of polychlorinated dibenzo-*p*-dioxin and dibenzofuran congeners in Great Lakes fish extracts by gas chromatography-mass spectrometry and *in vitro* enzyme induction activities, *Environ. Sci. Technol.* **23,** 730–735 (1989).
31. Environmental Protection Agency (EPA), Bioaccumulation of Selected Pollutants in Fish, U.S. EPA, Office of Water Regulations and Standards, Washington, DC 506/6-90/001b (April, 1991).
32. S. Jensen, Report of a new chemical hazard, *New Sci.* **32,** 612 (1966).
33. W. J. Neidermyer and J. J. Hickey, Chronology of organochlorine compounds in Lake Michigan fish, 1929–1966, *Pestic. Monit. J.* **10,** 92–95 (1976).
34. F. M. D'Itri, in: *Toxic Contamination in Large Lakes* (N. W. Schmidtke, ed.), Vol. 2, pp. 51–84, Lewis Publishers, Chelsea, MI (1988).
35. R. V. Thomann, J. P. Connolly, and N. A. Thomas, in: *PCBs and the Environment* (J. S. Waid, ed.), Vol. 3, pp. 153–180, CRC Press, Boca Raton, FL (1986).
36. D. J. Hallett, in: *Solving Hazardous Waste Problems, Learning from Dioxins* (J. H. Exner, ed.), pp. 94–104, American Chemical Society, Washington, DC (1987).
37. P. C. Baumann and D. M. Whittle, The status of selected organics in the Laurentian Great Lakes: An overview of DDT, PCBs, dioxins, furans, and aromatic hydrocarbons, *Aquat. Toxicol.* **11,** 241–257 (1988).
38. U. Borgman and D. M. Whittle, DDE, PCB, and mercury concentration trends in Lake Ontario rainbow smelt (*Osmerus mordax*) and slimy sculpin (*Cottus cognatus*): 1977 to 1988, *J. Great Lakes Res.* **18,** 298–308 (1992).
39. G. R. Shaw and D. W. Connell, in: *PCBs and the Environment* (J. S. Waid, ed.), Vol. I, pp. 121–133, CRC Press, Boca Raton, FL (1986).
40. M. Craig Barber, L. A. Suarez, and R. R. Lassiter, Modelling bioaccumulation of organic pollutants in fish with an application to PCBs in Lake Ontario salmonids, *Can. J. Fish. Aquat. Sci.* **48,** 318–337 (1991).
41. A. Opperhuizen and D. T. H. M. Sijm, Bioaccumulation and biotransformation of poly-

chlorinated dibenzo-*p*-dioxins and dibenzofurans in fish, *Environ. Toxicol. Chem.* **9,** 175–186 (1990).

42. P. M. Cook, M. K. Walker, D. W. Kuehl, and R. E. Peterson, in: *Biological Basis for Risk Assessment of Dioxins and Related Compounds* (M. A. Gallo, R. J. Scheuplein, and C. A. van der Heijden, eds.), Banbury Report 35, pp. 143–167, Cold Spring Harbor Laboratory Press, Cold Spring Harbor, NY (1991).
43. D. W. Kuehl, P. M. Cook, and A. R. Batterman, Uptake and depuration studies of PCDDs and PCDFs in freshwater fish, *Chemosphere* **15,** 2023–2026 (1986).
44. B. G. Oliver and A. J. Niimi, Trophodynamic analysis of polychlorinated biphenyl congeners and other chlorinated hydrocarbons in the Lake Ontario ecosystem, *Environ. Sci. Technol.* **22,** 388–397 (1988).
45. M. S. Evans, G. E. Noguchi, and C. P. Rice, The biomagnification of polychlorinated biphenyls, toxaphene, and DDT compounds in a Lake Michigan offshore food web, *Arch. Environ. Contam. Toxicol.* **20,** 87–93 (1991).
46. P. D. Guiney, M. J. Melancon, Jr., J. J. Lech, and R. E. Peterson, Effects of egg and sperm maturation and spawning on the distribution and elimination of a polychlorinated biphenyl in rainbow trout (*Salmo gairdneri*), *Toxicol. Appl. Pharmacol.* **47,** 261–272 (1979).
47. A. J. Niimi, Biological and toxicological effects of environmental contaminants in fish and their eggs, *Can. J. Fish. Aquat. Sci.* **40,** 306–312 (1983).
48. M. J. Vodicnik and R. E. Peterson, The enhancing effect of spawning on elimination of a persistent polychlorinated biphenyl from female yellow perch, *Fundam. Appl. Toxicol.* **5,** 770–776 (1985).
49. G. T. Ankley, D. E. Tillett, and J. P. Giesy, Maternal transfer of bioactive polychlorinated aromatic hydrocarbons in spawning chinook salmon (*Oncorhynchus tshawytscha*), *Mar. Environ. Res.* **28,** 231–234 (1989).
50. R. Prince and K. R. Cooper, Biological effects in and deposition to eggs produced by female Japanese medaka (*Oryzias latipes*) exposed to 2,3,7,8-tetrachlorodibenzo-*p*-dioxin (TCDD), *Toxicologist* **10,** 314 (1990).
51. R. Wannemacher, A. Rebstock, E. Kulzer, D. Schrenk, and K. W. Bock, Effects of 2,3,7,8-tetrachlorodibenzo-*p*-dioxin on reproduction and oogenesis in zebrafish (*Brachydanio rerio*), *Chemosphere* **24,** 1361–1368 (1992).
52. N. Johansson, PCB—Indications of effects on fish, PCB Conference, Swedish Salmon Research Institute, National Swedish Environment Protection Board, Stockholm, pp. 58–67 (December, 1970).
53. H. von Westernhagen, H. Rosenthal, V. Dethlefsen, W. Ernst, U. Harms, and P.-D. Hansen, Bioaccumulating substances and reproductive success in Baltic flounder (*Platichthys flesus*), *Aquat. Toxicol.* **1,** 85–99 (1981).
54. P.-D. Hansen, H. von Westernhagen, and H. Rosenthal, Chlorinated hydrocarbons and hatching success in Baltic herring spring spawners, *Mar. Environ. Res.* **15,** 59–76 (1985).
55. G. Monod, Egg mortality of Lake Geneva char (*Salvelinus alpinus* L.) contaminated by PCB and DDT derivatives, *Bull. Environ. Contam. Toxicol.* **35,** 531–536 (1985).
56. P. F. Morrison, J. F. Leatherland, and R. A. Sonstegard, Proximate composition and organochlorine and heavy metal contamination of eggs from Lake Ontario, Lake Erie and Lake Michigan coho salmon (*Oncorhynchus kisutch* Walbaum) in relation to egg survival, *Aquat. Toxicol.* **6,** 73–86 (1985).
57. P. Cameron, H. von Westernhagen, V. Dethlefsen, and D. Janssen, Chlorinated hydrocarbons in North Sea whiting (*Merlangius merlangus*) and effects on reproduction, International Council for the Exploration of the Sea, Marine Environmental Quality Committee, E:25, pp. 1–10 (1986).

58. J. P. Giesey, J. L. Newsted, and D. L. Garling, Relationships between chlorinated hydrocarbon concentrations and rearing mortality of chinook salmon (*Oncorhynchus tshawytscha*) eggs from Lake Michigan, *J. Great Lakes Res.* **12,** 82–98 (1986).
59. R. B. Spies and D. W. Rice, Jr., Effects of organic contaminants on reproduction of the starry flounder (*Platichthys stellatus*) in San Francisco Bay, *Mar. Biol.* **98,** 191–200 (1988).
60. L. M. Smith, T. R. Schwartz, K. Feltz, and T. J. Kubiak, Determination and occurrence of AHH-active polychlorinated biphenyls, 2,3,7,8-tetrachlorodibenzo-*p*-dioxin and 2,3,7,8-tetrachlorodibenzofuran in Lake Michigan sediment and biota. The question of their relative toxicological significance, *Chemosphere* **21,** 1063–1085 (1990).
61. M. J. Mac and C. C. Edsall, Environmental contaminants and the reproductive success of lake trout in the Great Lakes: An epidemiological approach, *J. Toxicol. Environ. Health* **33,** 375–394 (1991).
62. M. K. Walker, E. Zabel, R. E. Peterson, P. M. Cook, and P. Marquis, Risk posed by polychlorinated dibenzo-*p*-dioxins (PCDDs), dibenzofurans (PCDFs), and biphenyls (PCBs) to lake trout early life stage survival in the Great Lakes, *Soc. Environ. Toxicol. Chem. Abstr.* **13,** 45 (1992).
63. M. J. Mac, in: *Toxic Contaminants and Ecosystem Health: A Great Lakes Focus* (M. S. Evans and J. E. Gannon, eds.), pp. 389–401, Wiley, New York (1988).
64. L. L. Williams and J. P. Giesy, Relationships among concentrations of individual polychlorinated biphenyl (PCB) congeners, 2,3,7,8-tetrachlorodibenzo-*p*-dioxin equivalents (TCDD-EQ), and rearing mortality of chinook salmon (*Oncorhynchus tshawytscha*) eggs from Lake Michigan, *J. Great Lakes Res.* **18,** 108–124 (1992).
65. J. F. Leatherland and R. A. Sonstegard, in: *Contaminant Effects on Fisheries* (V. W. Cairns, P. V. Hodson, and J. O. Nriagu, eds.), Vol. 16, pp. 115–149, Wiley, New York (1984).
66. B. A. Schwetz, J. M. Norris, G. L. Sparschu, V. K. Rowe, P. J. Gehring, J. L. Emerson, and C. G. Gerbig, Toxicology of chlorinated dibenzo-*p*-dioxins, *Environ. Health Perspect.* **5,** 87–99 (1973).
67. J. M. Kleeman, J. R. Olson, and R. E. Peterson, Species differences in 2,3,7,8-tetrachlorodibenzo-*p*-dioxin toxicity and biotransformation in fish, *Fundam. Appl. Toxicol.* **10,** 206–213 (1988).
68. J. R. Olson, M. A. Holscher, and R. A. Neal, Toxicity of 2,3,7,8-tetrachlorodibenzo-*p*-dioxin in the golden Syrian hamster, *Toxicol. Appl. Pharmacol.* **55,** 67–78 (1980).
69. P. M. Mehrle, D. R. Buckler, E. E. Little, L. M. Smith, J. D. Petty, P. H. Peterman, D. L. Stalling, G. M. DeGraeve, J. J. Coyle, and W. J. Adams, Toxicity and bioconcentration of 2,3,7,8-tetrachlorodibenzo-*p*-dioxin and 2,3,7,8-tetrachlorodibenzofuran in rainbow trout, *Environ. Toxicol. Chem.* **7,** 47–62 (1988).
70. J. M. Spitsbergen, M. K. Walker, J. R. Olson, and R. E. Peterson, Pathologic alterations in early life stages of lake trout (*Salvelinus namaycush*), exposed to 2,3,7,8-tetrachlorodibenzo-*p*-dioxin as fertilized eggs, *Aquat. Toxicol.* **19,** 41–72 (1991).
71. M. K. Walker, J. M. Spitsbergen, J. R. Olson, and R. E. Peterson, 2,3,7,8-Tetrachlorodibenzo-*p*-dioxin (TCDD) toxicity during early life stage development of lake trout (*Salvelinus namaycush*), *Can. J. Fish. Aquat. Sci.* **48,** 875–883 (1991).
72. M. K. Walker, L. C. Hufnagle, Jr., M. K. Clayton, and R. E. Peterson, Development of an egg injection method for assessing the early life stage toxicity of halogenated aromatic hydrocarbons in rainbow trout (*Oncorhynchus mykiss*), *Aquat. Toxicol.* **22,** 15–38 (1992).
73. R. A. Miller, L. A. Norris, and C. L. Hawkes, Toxicity of 2,3,7,8-tetrachlorodibenzo-*p*-dioxin (TCDD) in aquatic organisms, *Environ. Health Perspect.* **5,** 177–186 (1973).
74. R. A. Miller, L. A. Norris, and B. R. Loper, The response of coho salmon and guppies to

2,3,7,8-tetrachlorodibenzo-*p*-dioxin (TCDD) in water, *Trans. Am. Fish. Soc.* **108,** 401–407 (1979).

75. L. A. Norris and R. A. Miller, The toxicity of 2,3,7,8-tetrachlorodibenzo-*p*-dioxin (TCDD) in guppies (*Poecilia reticulatus* Peters), *Bull. Environ. Contam. Toxicol.* **12,** 76–80 (1974).
76. C. L. Hawkes and L. A. Norris, Chronic oral toxicity of 2,3,7,8-tetrachlorodibenzo-*p*-dioxin (TCDD) to rainbow trout, *Trans. Am. Fish. Soc.* **106,** 641–645 (1977).
77. D. R. Branson, I. T. Takahashi, W. M. Parker, and G. E. Blau, Bioconcentration kinetics of 2,3,7,8-tetrachlorodibenzo-*p*-dioxin in rainbow trout, *Environ. Toxicol. Chem.* **4,** 779–788 (1985).
78. J. M. Spitsbergen, J. M. Kleeman, and R. E. Peterson, Morphologic lesions and acute toxicity in rainbow trout (*Salmo gairdneri*) treated with 2,3,7,8-tetrachlorodibenzo-*p*-dioxin, *J. Toxicol. Environ. Health* **23,** 333–358 (1988).
79. M. E. J. van der Weiden, J. van der Kolk, A. H. Penninks, W. Seinen, and M. van der Berg, A dose/response study with 2,3,7,8-TCDD in the rainbow trout, *Chemosphere* **20,** 1053–1058 (1990).
80. W. J. Adams, G. M. DeGraeve, T. D. Sabourin, J. D. Cooney, and G. M. Mosher, Toxicity and bioconcentration of 2,3,7,8-TCDD to fathead minnows (*Pimephales promelas*), *Chemosphere* **15,** 1503–1511 (1986).
81. R. S. Yockim, A. R. Isensee, and G. E. Jones, Distribution and toxicity of TCDD and 2,4,5-T in an aquatic model ecosystem, *Chemosphere* **7,** 215–220 (1978).
82. J. M. Spitsbergen, J. M. Kleeman, and R. E. Peterson, 2,3,7,8-Tetrachlorodibenzo-*p*-dioxin toxicity in yellow perch (*Perca flavescens*), *J. Toxicol. Environ. Health* **23,** 359–383 (1988).
83. K. Ballschmiter and M. Zell, Analysis of polychlorinated biphenyls by capillary gas chromatography, *Fresenius J. Anal. Chem.* **302,** 20–31 (1980).
84. D. L. Stalling, N. J. Huckins, J. D. Petty, J. L. Johnson, and H. O. Sanders, An expanded approach to the study and measurement of PCBs and selected planar halogenated aromatic environmental pollutants, *Ann. N.Y. Acad. Sci.* **320,** 48–59 (1979).
85. T. Helder, Effects of 2,3,7,8-tetrachlorodibenzo-*p*-dioxin (TCDD) on early life stages of rainbow trout (*Salmo gairdneri,* Richardson), *Toxicology* **19,** 101–112 (1981).
86. M. O. Cheung, E. F. Gilbert, and R. E. Peterson, Cardiovascular teratogenicity of 2,3,7,8-tetrachlorodibenzo-*p*-dioxin in the chicken embryo, *Toxicol. Appl. Pharmacol.* **61,** 197–204 (1981).
87. A. B. Rifkind, Y. Hattori, R. Levi, M. J. Hughes, C. Quilley, and D. R. Alonso, in: *Biological Mechanisms of Dioxin Action* (A. Poland and R. D. Kimbrough, eds.), pp. 255–266, Cold Spring Harbor Laboratory Press, Cold Spring Harbor, NY (1984).
88. D. S. Henschel, B. M. Hehn, M. T. Vo, and J. D. Steeves, in: *Environmental Toxicology and Risk Assessment* (J. W. Gorsuch, F. J. Dwyer, C. G. Ingersoll, and T. W. LaPoint, eds.), pp. 159–174, Vol. 2, American Society for Testing and Materials, Philadelphia (1993).
89. J. A. Goldstein and S. Safe, in: *Halogenated Biphenyls, Terphenyls, Naphthalenes, Dibenzodioxins and Related Products* (R. D. Kimbrough and A. A. Jensen, eds.), pp. 239–293, Elsevier, Amsterdam (1989).
90. J. J. Lech, M. J. Vodicnik, and C. R. Elcombe, in: *Aquatic Toxicology* (L. J. Weber, ed.), pp. 107–148, Raven Press, New York (1982).
91. M. J. Melancon, R. L. Binder, and J. J. Lech, in: *Toxic Contaminants and Ecosystem Health: A Great Lakes Focus* (M. S. Evans and J. E. Gannon, eds.), pp. 215–236, Wiley, New York (1988).

92. J. J. Stegeman, Cytochrome P450 forms in fish: Catalytic, immunological, and sequence similarities, *Xenobiotica* **19,** 1093–1110 (1989).
93. A. Goksoyr and L. Forlin, The cytochrome P-450 system in fish, aquatic toxicology and environmental monitoring, *Aquat. Toxicol.* **22,** 287–312 (1992).
94. J. L. Newsted and J. P. Giesy, Effect of 2,3,7,8-tetrachlorodibenzo-*p*-dioxin (TCDD) on the epidermal growth factor receptor in hepatic plasma membranes of rainbow trout, *Toxicol. Appl. Pharmacol.* **118,** 119–130 (1993).
95. J. L. Newsted and J. P. Giesy, The effects of 2,3,7,8-tetrachlorodibenzo-*p*-dioxin on epidermal growth factor binding and protein kinase activity in the RTH-149 rainbow trout hepatoma cell line, *Aquat. Toxicol.* **23,** 119–135 (1992).
96. B. V. Madhukar, D. W. Brewster, and F. Matsumura, Effects of *in vivo*-administered 2,3,7,8-tetrachlorodibenzo-*p*-dioxin on receptor binding of epidermal growth factor in the hepatic plasma membrane of rat, guinea pig, mouse, and hamster, *Proc. Natl. Acad. Sci. USA* **81,** 7407–7411 (1984).
97. R. Osborne and W. F. Greenlee, 2,3,7,8-Tetrachlorodibenzo-*p*-dioxin (TCDD) enhances terminal differentiation of cultured human epidermal cells, *Toxicol. Appl. Pharmacol.* **77,** 434–440 (1985).
98. L. G. Hudson, W. A. Toscano, Jr., and W. F. Greenlee, Regulation of epidermal growth factor binding in a human keratinocyte cell line by 2,3,7,8-tetrachlorodibenzo-*p*-dioxin (TCDD), *Toxicol. Appl. Pharmacol.* **77,** 251–259 (1985).
99. L. G. Hudson, W. A. Toscano, Jr., and W. F. Greenlee, 2,3,7,8-Tetrachlorodibenzo-*p*-dioxin (TCDD) modulate epidermal growth factor (EGF) binding to basal cells from a human keratinocyte cell line, *Toxicol. Appl. Pharmacol.* **82,** 481–492 (1986).
100. E. D. Adamson, in: *Growth Factors and Development* (M. Nilsen-Hamilton, ed.), Vol. 24, pp. 1–29, Academic Press, New York (1990).
101. A. Partanen, in: *Growth Factors and Development* (M. Nilsen-Hamilton, ed.), Vol. 24, pp. 31–55, Academic Press, New York (1990).
102. T. A. Mably, R. W. Moore, R. W. Goy, and R. E. Peterson, *In utero* and lactational exposure of male rats to 2,3,7,8-tetrachlorodibenzo-*p*-dioxin. 2. Effects on sexual behavior and the regulation of luteinizing hormone secretion in adulthood, *Toxicol. Appl. Pharmacol.* **114,** 118–126 (1992).
103. J. G. Vos and M. I. Luster, in: *Halogenated Biphenyls, Terphenyls, Naphthalenes, Dibenzodioxins and Related Products* (R. D. Kimbrough and A. A. Jensen, eds.), pp. 295–322, Elsevier, Amsterdam (1989).
104. J. M. Spitsbergen, K. A. Schat, J. M. Kleeman, and R. E. Peterson, Interactions of 2,3,7,8-tetrachlorodibenzo-*p*-dioxin (TCDD) with immune responses of rainbow trout, *Vet. Immunol. Immunopathol.* **12,** 263–280 (1986).
105. J. M. Spitsbergen, K. A. Schat, J. M. Kleeman, and R. E. Peterson, Effects of 2,3,7,8-tetrachlorodibenzo-*p*-dioxin (TCDD) or Aroclor 1254 on the resistance of rainbow trout (*Salmo gairdneri* Richardson), to infectious hematopoietic necrosis virus, *J. Fish. Dis.* **11,** 73–83 (1988).
106. P. D. Delorme, F. J. Ward, W. L. Lockhart, and D. C. G. Muir, Reproductive success of feral lake trout and white suckers treated with toxaphene, chlordane or 2,3,4,7,8-pentachlorodibenzofuran, *Soc. Environ. Toxicol. Chem. Abstr.* **13,** 138 (1992).
107. E. Monosson, Effects of 3,3′,4,4′-tetrachlorobiphenyl on reproductive processes in white perch, *Soc. Environ. Toxicol. Chem. Abstr.* **13,** 259 (1992).
108. R. J. Kociba, P. A. Keeler, G. N. Park, and P. J. Gehring, 2,3,7,8-Tetrachlorodibenzo-*p*-dioxin (TCDD): Results of a 13 week oral toxicity study in rats, *Toxicol. Appl. Pharmacol.* **35,** 553–574 (1976).

109. F. J. Murray, F. A. Smith, K. D. Nitschke, C. G. Huniston, R. J. Kociba, and B. A. Schwetz, Three-generation reproduction study of rats given 2,3,7,8-tetrachlodordibenzo-*p*-dioxin (TCDD) in the diet, *Toxicol. Appl. Pharmacol.* **50,** 241–252 (1979).
110. W. P. McNulty, Fetotoxicity of 2,3,7,8-tetrchlorodibenzo-*p*-dioxin (TCDD) for rhesus macaques (*Macaca mulatta*), *Am. J. Primatol.* **6,** 41–47 (1984).
111. W. P. McNulty, Toxicity and fetotoxicity of TCDD, TCDF, and PCB isomers in rhesus macaques (*Macada mulatta*), *Environ. Health Perspect.* **60,** 77–88 (1985).
112. T. Helder, Effects of 2,3,7,8-tetrachlorodibenzo-*p*-dioxin (TCDD) on early life stages of the pike (*Esox lucius* L.), *Sci. Total Environ.* **14,** 255–264 (1980).
113. J. D. Wisk and K. R. Cooper, The stage specific toxicity of 2,3,7,8-tetrachlorodibenzo-*p*-dioxin in embryos of the Japanese medaka (*Oryzias latipes*), *Environ. Toxicol. Chem.* **9,** 1159–1169 (1990).
114. R. Prince and K. R. Cooper, Differential embryo sensitivity to 2,3,7,8-tetrachlorodibenzo-*p*-dioxin (TCDD) in *Fundulus heteroclitus, Toxicologist* **9,** 43 (1989).
115. R. V. Kirchen and W. R. West, *The Japanese Medaka, Its Care and Development,* pp. 1–36, Carolina Biological Co., Burlington, NC (1976).
116. H. Westers, in: *Culture of Nonsalmonid Freshwater Fishes* (R. R. Stickney, ed.), pp. 91–101, CRC Press, Boca Raton, FL (1990).
117. M. K. Walker, P. M. Cook, A. Batterman, B. Butterworth, C. Bernini, J. J. Libal, L. C. Hufnagle, and R. E. Peterson, Translocation of 2,3,7,8-tetrachlorodibenzo-*p*-dioxin from adult female lake trout (*Salvelinus namaycush*) to oocytes: Effects on early life stage development and sac fry survival, *Can. J. Fish. Aquat. Sci.,* **51,** (in press).
118. P. D. Guiney, M. K. Walker, and R. E. Peterson, Edema in TCDD exposed lake trout sac fry is an ultrafiltrate of blood, *Soc. Environ. Toxicol. Chem. Abstr.* **11,** 96 (1990).
119. P. D. Guiney, J. J. Stegeman, R. M. Smolowitz, M. K. Walker, and R. E. Peterson, Localization of 2,3,7,8-tetrachlorodibenzo-*p*-dioxin (TCDD)-induced cytochrome P450IA1 in vascular endothelium of early life stages (ELS) of lake trout, *Soc. Environ. Toxicol. Chem. Abstr.* **13,** 45 (1992).
120. K. Wolf, Blue-sac disease of fish, U.S. Fish Wildlife Service Fish Disease Leaflet #15, pp. 1–4 (1969).
121. R. J. Roberts and C. J. Shepherd, in: *Handbook of Trout and Salmon Diseases* (R. J. Roberts and C. J. Shepherd, eds.), 2nd ed., pp. 94–101, Fishing News Books Ltd., Farnham, Surrey, England (1986).
122. D. E. Burkhalter and C. M. Kaya, Effects of prolonged exposure to ammonia on fertilized eggs and sac fry of rainbow trout (*Salmo gairdneri*), *Trans. Am. Fish. Soc.* **106,** 470–475 (1977).
123. D. Firestone, Etiology of chick edema disease, *Environ. Health Perspect.* **5,** 59–66 (1973).
124. J. G. Vos, J. A. Moore, and J. G. Zinkl, Toxicity of 2,3,7,8-tetrachlorodibenzo-*p*-dioxin (TCDD) in C57B1/6 mice, *Toxicol. Appl. Pharmacol.* **29,** 229–241 (1974).
125. J. R. Allen, D. A. Barsotti, J. P. Van Miller, L. J. Abrahamson, and J. J. Lalich, Morphological changes in monkeys consuming a diet containing low levels of 2,3,7,8-tetrachlorodibenzo-*p*-dioxin, *Food Cosmet. Toxicol.* **15,** 401–410 (1977).
126. H. M. Theobald, R. W. Moore, L. B. Katz, R. O. Pieper, and R. E. Peterson, Enhancement of carrageenan and dextran-induced edema by 2,3,7,8-tetrachlorodibenzo-*p*-dioxin and related compounds, *J. Pharmacol. Exp. Ther.* **225,** 576–583 (1983).
127. L. B. Katz, H. M. Theobald, R. C. Bookstaff, and R. E. Peterson, Characterization of the

enhanced paw edema response to carrageenan and dextran in 2,3,7,8-tetrachlorodibenzo-*p*-dioxin-treated rats, *J. Pharmacol. Exp. Ther.* **230,** 670–677 (1984).
128. M. K. Walker and R. E. Peterson, in: *Chemically Induced Alterations in Sexual and Functional Development: The Wildlife/Human Connection* (T. Colborn and C. Clement, eds.), pp. 195–202, Princeton Scientific Publishing Co., Princeton, NJ (1992).
129. G. Mason, T. Sawyer, B. Keys, S. Bandiera, M. Romkes, J. Piskorska-Pliszczynska, B. Zmudzka, and S. Safe, Polychlorinated dibenzofurans (PCDFs): Correlation between *in vivo* and *in vitro* structure–activity relationships, *Toxicology* **37,** 1–12 (1985).
130. G. Mason, K. Farrell, B. Keys, J. Piskorska-Pliszczynska, L. Safe, and S. Safe, Polychlorinated dibenzo-*p*-dioxins: Quantitative *in vitro* and *in vivo* structure–activity relationships, *Toxicology* **41,** 21–31 (1986).
131. S. Bandiera, S. Safe, and A. B. Okey, Binding of polychlorinated biphenyls classified as either phenobarbitone-, 3-methylcholanthrene- or mixed-type inducers to cytosolic Ah receptor, *Chem. Biol. Interact.* **39,** 259–277 (1982).
132. S. Safe, Determination of 2,3,7,8-TCDD toxic equivalent factors (TEFs): Support for the use of the *in vitro* AHH induction assay, *Chemosphere* **16,** 791–802 (1987).
133. B. Leece, M. A. Denomme, R. Towner, S. M. A. Li, and S. Safe, Polychlorinated biphenyls: Correlation between *in vivo* and *in vitro* quantitative structure–activity relationships (QSARs), *J. Toxicol. Environ. Health* **16,** 379–388 (1985).
134. E. W. Zabel, P. M. Cook, and R. E. Peterson, Toxic equivalency factors of polychlorinated dibenzo-*p*-dioxin, dibenzofuran, and biphenyl congeners based on early life stage mortality in rainbow trout (*Oncorhynchus mykiss*), *Aquat. Toxicol.* (in press).
135. J. W. Gooch, A. A. Elskus, P. J. Kloepper-Sams, M. E. Hahn, and J. J. Stegeman, Effects of *ortho* and non-*ortho*-substituted polychlorinated biphenyl congeners on the hepatic monooxygenase system in scup (*Stenotomus chrysops*), *Toxicol. Appl. Pharmacol.* **98,** 422–433 (1989).
136. J. U. Skaare, E. G. Jensen, A. Goksoyr, and E. Egaas, Response of xenobiotic metabolizing enzymes of rainbow trout (*Oncorhynchus mykiss*) to the mono-*ortho* substituted polychlorinated biphenyl congener 2,3′,4,4′,5-pentachlorobiphenyl, PCB-118, detected by enzyme activities and immunochemical methods, *Arch. Environ. Contam. Toxicol.* **20,** 349–352 (1991).
137. E. W. Zabel, R. E. Peterson, P. M. Cook, and P. Marquis, Elimination of polychlorinated biphenyl (PCB), dibenzo-*p*-dioxin (PCDD), and dibenzofuran (PCDF) congeners during early life stage (ELS) development of rainbow trout, *Soc. Environ. Toxicol. Chem. Abstr.* **13,** 229 (1992).
138. M. J. DeVito, W. E. Maier, J. J. Diliberto, and L. S. Birnbaum, Comparative ability of various PCBs, PCDFs, and TCDD to induce cytochrome P-450 1A1 and 1A2 activity following 4 weeks of treatment, *Fundam. Appl. Toxicol.* **20,** 125–130 (1993).
139. J. N. Huckins, D. L. Stalling, and J. D. Petty, Carbon-foam chromatographic separation of non-*o,o′*-chlorine substituted PCBs from Aroclor mixtures, *J. Assoc. Off. Anal. Chem.* **63,** 750–755 (1980).
140. P. de Voogt and U. A. T. Brinkman, in: *Halogenated Biphenyls, Terphenyls, Naphthalenes, Dibenzodioxins and Related Products* (R. D. Kimbrough and A. A. Jensen, eds.), pp. 3–45, Elsevier, Amsterdam (1989).
141. D. E. Schulz, G. Petrick, and J. C. Duinker, Complete characterization of polychlorinated biphenyl congeners in commercial Aroclor and Clophen mixtures by multidimensional gas chromatography–electron capture detection, *Environ. Sci. Technol.* **23,** 852–859 (1989).
142. G. W. Bowes, M. J. Mulvihill, B. R. T. Simoneit, A. L. Burlingame, and R. W. Ris-

ebrough, Identification of chlorinated dibenzofurans in American polychlorinated biphenyls, *Nature* **256,** 305–307 (1975).
143. W. J. Rogan and B. C. Gladen, Neurotoxicology of PCBs and related compounds, *Neurotoxicology* **13,** 27–35 (1992).
144. D. J. Hansen, P. R. Parrish, J. I. Lowe, A. J. Wilson, Jr., and P. D. Wilson, Chronic toxicity, uptake, and retention of Aroclor® 1254 in two estuarine fishes, *Bull. Environ. Contam. Toxicol.* **6,** 113–119 (1971).
145. D. J. Hansen, S. C. Schimmel, and J. Forester, Aroclor® 1254 in eggs of sheepshead minnows: Effects on fertilization success and survival of embryos and fry, *Proc. 27th Annu. Conf. Southeastern Assoc. Game Fish. Comm.*, pp. 420–423 (1973).
146. D. J. Hansen, P. R. Parrish, and J. Forester, Aroclor 1016: Toxicity to and uptake by estuarine animals, *Environ. Res.* **7,** 363–373 (1974).
147. D. J. Hansen, S. C. Schimmel, and J. Forester, Effects of Aroclor® 1016 on embryos, fry, juveniles, and adults of sheepshead minnows (*Cyprinodon variegatus*), *Trans. Am. Fish. Soc.* **104,** 584–588 (1975).
148. D. L. Stalling and F. L. Mayer, Jr., Toxicities of PCBs to fish and environmental residues, *Environ. Health Perspect.* **1,** 159–164 (1972).
149. S. C. Schimmel, D. J. Hansen, and J. Forester, Effects of Aroclor® 1254 on laboratory-reared embryos and fry of sheepshead minnows (*Cyprinodon variegatus*), *Trans. Am. Fish. Soc.* **103,** 582–586 (1974).
150. A. V. Nebeker, F. A. Puglisi, and D. L. DeFoe, Effect of polychlorinated biphenyl compounds on survival and reproduction of the fathead minnow and flagfish, *Trans. Am. Fish. Soc.* **103,** 562–568 (1974).
151. F. L. Mayer, P. M. Mehrle, and H. O. Sanders, Residue dynamics and biological effects of polychlorinated biphenyls in aquatic organisms, *Arch. Environ. Contam. Toxicol.* **5,** 501–511 (1977).
152. D. L. DeFoe, G. D. Veith, and R. W. Carlson, Effects of Aroclor® 1248 and 1260 on the fathead minnow (*Pimephales promelas*), *J. Fish. Res. Board Can.* **35,** 997–1002 (1978).
153. B.-E. Bengtsson, Long-term effects of PCB (Clophen A50) on growth, reproduction and swimming performance in the minnow (*Phoxinus phoxinus*), *Water Res.* **14,** 681–687 (1980).
154. T. Yano and H. Matsuyama, Stimulatory effect of PCB on the metabolism of sex hormones in carp hepatopancreas, *Bull. Jpn. Soc. Sci. Fish.* **52,** 1847–1852 (1986).
155. K. Sivarajah, C. S. Franklin, and W. P. Williams, The effects of polychlorinated biphenyls on plasma steroid levels and hepatic microsomal enzymes in fish, *J. Fish Biol.* **13,** 401–409 (1978).
156. P. Thomas, Effects of Aroclor 1254 and cadmium on reproductive endocrine function and ovarian growth in Atlantic croaker, *Mar. Environ. Res.* **28,** 499–503 (1989).
157. N. Johanson, A. Larsson, and K. Lewander, Metabolic effects of PCB (polychlorinated biphenyls) on the brown trout (*Salmo trutta*), *Comp. Gen. Pharmacol.* **3,** 310–314 (1972).
158. B.-E. Bengtsson, Increased growth in minnows exposed to PCBs, *Ambio* **8,** 169–170 (1979).
159. E. H. Gruger, Jr., T. Hruby, and N. L. Karrick, Sublethal effects of structurally related tetrachloro-, pentachloro-, and hexachlorobiphenyl on juvenile coho salmon, *Environ. Sci. Technol.* **10,** 1033–1037 (1976).
160. M. T. Halter and H. E. Johnson, Acute toxicities of a polychlorinated biphenyl (PCB) and DDT alone and in combination to early life stages of coho salmon (*Oncorhynchus kisutch*), *J. Fish Res. Board Can.* **31,** 1543–1547 (1974).
161. W. L. Mauck, P. M. Mehrle, and F. L. Mayer, Effects of the polychlorinated biphenyl

Aroclor® 1254 on growth, survival, and bone development in brook trout (*Salvelinus fontinalis*), *J. Fish. Res. Board Can.* **35,** 1084–1088 (1978).

162. C. J. Schmitt, J. L. Zajicek, and P. H. Peterman, National contaminant biomonitoring program: Residues of organochlorine chemicals in U.S. freshwater fish, 1976–1984, *Arch. Environ. Contam. Toxicol.* **19,** 748–781 (1990).
163. G. Eadon, L. Kaminsky, J. Silkworth, K. Aldous, D. Hilker, P. O'Keefe, R. Smith, J. Gierthy, J. Hawley, N. Kim, and A. DeCaprio, Calculation of 2,3,7,8-TCDD equivalent concentrations of complex environmental contaminant mixtures, *Environ. Health Perspect.* **70,** 221–227 (1986).
164. J. L. Casterline, Jr., J. A. Bradlaw, B. J. Puma, and Y. Ku, Screening of freshwater fish extracts for enzyme-inducing substances by an aryl hydrocarbon hydroxylase induction bioassay technique, *J. Assoc. Off. Anal. Chem.* **66,** 1136–1139 (1983).
165. D. E. Tillitt, J. P. Giesy, and G. T. Ankley, Characterization of the H4IIE rat hepatoma cell bioassay as a tool for assessing toxic potency of planar halogenated hydrocarbons in environmental samples, *Environ. Sci. Technol.* **25,** 87–92 (1991).
166. L. Norrgren, T. Andersson, P.-A. Bergqvist, and I. Bjorklund, Chemical, physiological and morphological studies of feral Baltic salmon (*Salmo salar*) suffering from abnormal fry mortality, *Environ. Toxicol. Chem.* **12,** 2065–2075 (1993).
167. K. R. Cooper, H. Liu, P.-A. Bergqvist, and C. Rappe, Evaluation of Baltic herring and Icelandic cod liver oil for embryo toxicity, using the Japanese medaka (*Oryzias latipes*) embryo larval assay, *Environ. Toxicol. Chem.* **10,** 707–714 (1991).
168. J. C. Skea, J. Symula, J. Miccoli, and G. Aylesworth, Summary of salmonid reproductive studies in several New York State lakes and hatcheries, New York State Department of Environmental Conservation (1986).
169. P. M. Cook, personal communication.

Chapter 12

Dioxins and Mammalian Carcinogenesis

James Huff

1. INTRODUCTION

First notice of an adverse human health effect (occupational chloracne) of (TCDD) became known in 1957,[1] evidence of teratogenicity in animals was reported in 1971,[2] and the first international symposium was held in 1973.[3] World public concern was raised in July of 1976 when a chemical company producing trichlorophenol in Seveso, Italy, exploded and contaminated the countryside with TCDD and other reaction products.[4–8] Since then, other "accidents" or large-scale occupational and environmental exposures have been divulged or have occurred.[7,9]

The first evidence implicating TCDD as being carcinogenic in experimental animals appeared in 1977.[10] Since then, several studies on TCDD and related chemicals confirmed and extended these early observations,[11–13] and these form the prime basis of this chapter. Currently only one albeit limited experimental carcinogenesis study appears to be ongoing[14]; this likely reflects a lack of investigator reporting, since considerable research effort continues regarding TCDD carcinogenesis.

Moreover, much research has been done using TCDD as a model compound to explore mechanisms of carcinogenesis.[15–20] TCDD has been detected in human breast milk and fat[21,22]; anecdotal remarks are common that "no human being on earth escapes having detectable levels of dioxins."

Epidemiological data from occupationally exposed workers now show accumulating and convincing evidence that exposures to TCDD are associated with several cancers in humans: respiratory, lung, thyroid gland, connective and soft tissue sarcoma, hematopoietic system, liver, and all cancers.[20,23–30,101]

James Huff • Environmental Carcinogenesis Program, National Institute of Environmental Health Sciences, Research Triangle Park, North Carolina 27709.

Dioxins and Health, edited by Arnold Schecter. Plenum Press, New York, 1994.

TCDD is but one of an expanding number of chemicals first discovered as being carcinogenic in experimental animals, and some years or decades later being confirmed as carcinogenic to humans.[31–34] Notably, all chemicals or exposure circumstances known to cause cancer in humans have likewise been shown to induce cancer in laboratory animals; that is, for those that can and have been studied adequately. Thus, experimental carcinogenesis results must continue to be taken seriously.

2. PURPOSE OF THIS CHAPTER

In this overview paper the available data from carcinogenesis experiments of TCDD are summarized, select findings are detailed (see composite and individual tables), evidence for mechanism of complete carcinogenicity is presented, environmental contamination is mentioned, documentation of human carcinogenicity is given, supplemental experiments are posed, and further implications for potential cancer hazards to human health are offered.

For more details on any of these issues, readers are urged to select and obtain appropriate sources listed in the References. Any of mine can be obtained by request from me.

3. ENVIRONMENTAL CONTAMINATION

Ubiquitous contamination of the air, land, and waters by dioxins—and similar bioaccumulated and virtually nondegradable chemicals like DDT/metabolites and PCBs/PBBs—appears on the increase despite the fact that (except for DDT prior to the early 1970s) none of these typically organochlorine chemicals were purposefully used or "allowed" in the environment. Yet, globally the evidence of universal contamination undergirds the view that "dioxins are everywhere." In water environments for instance, TCDD associates with sediments, biota, and the organic fraction of ambient waters.[35,36] Because of profound lipophilicity, and low chemical and biological degradation in aquatic environs, TCDD accumulates in biota; then characteristically it is transported up the aquatic food chain and to the local avian and mammalian genera, and eventually to humans.

Gardner *et al.*[37] demonstrated that sediment-bound carcinogens are associated with induced cancers in winter flounder, and that they are trophically transferred up the food chain from sediment-exposed and contaminated blue mussels to the flounder. These experiments support the view that sediment-bound carcinogens and other toxic chemicals are biologically available via the food chain.[38,39] Perhaps extended experiments could be undertaken whereby laboratory rodents are given contaminated sediments taken from areas inhabited by fish or other aquatic life having tumors, like those experiments of McConnell *et al.*,[40] Lucier *et al.*,[41] and Umbreit *et al.*,[42] who exposed rodents to dioxin-containing soils. These collective results showed clearly that "bound dioxin" was readily available biologically, like sediment and mussel-containing carcinogens. For liver tumors in particular there seems to be a strong association between those fish species affected and habitat existence with contaminated sediments, and "in fact, no studies of liver neoplasia in non-bottom-dwelling fish species have been reported."[43]

The U.S. EPA for instance issued in November of 1992 a national survey of chemical residues in fish: 99% of captured fish were found to contain DDE (a metabolite of DDT, not permitted for use since the early 1970s), mercury was found in 92% of fish, PCBs in 91%, and so on. Both whole-body and fish fillet samples were analyzed for 60 selected chemicals: 15 "dioxins" and "furans," 10 polychlorinated biphenyls, 21 pesticides and herbicides, mercury, biphenyl, and 12 other organic compounds; found in greater than 50% of fish were seven "dioxins," six pesticides, PCBs, and other industrial organic chemicals, and mercury; and all chemicals were detected in at least one of the 388 locations.[35,36] PCBs, essentially banned since the late 1970s and early 1980s, were detected at the highest concentrations: a maximum of 124,000 ppb, and an average of 1900 ppb; for a somewhat hypothetical and perhaps perspective translation of these concentrations, Chriswell[44] calculated various products one might obtain at a concentration of 1 ppb for several substances from a large reservoir (see Table 1). In this context one begins to better appreciate and understand that relatively miniscule concentrations can and do take on considerable significance, especially since these exposures are in reality forever.

Recognized sources of dioxin exposures include fossil fuel burning, waste (municipal, hazardous, biomedical) combustion, phenoxy acid herbicide spraying, disposal of herbicide production wastes, and diesel truck emissions. For the latter, Jones[45] opines that heavy-duty diesel trucks are a significant source of human exposures to dioxins, now and in the future. Moreover, mobile-source dioxin emissions are responsible for more releases than the combined discharges from waste incineration sources.

Table 1
Perspective on Parts per Billion Using Several Selected Contaminants[a]

Substance	One ppb in 3,000,000 acre-feet[b]/year would amount to enough of "substance" to:
Chromium	Plate 50,000 car bumpers
Gold	Run the Federal government for nearly 20 minutes or support 50 average families for eternity
Herbicides	Kill all of the dandelions in 100,000 lawns
Insecticides	Fill 5,000,000 aerosol cans of bug killer
Lead	Cast 1,000,000 bullets
Mercury	Fill 4,000,000 rectal thermometers
Phenols	Produce 250,000 bottles of Lysol

[a]From Chriswell (1977).[44]
[b]Amount of water used in the Los Angeles area per year.

Table 2
TCDD Carcinogenicity Experiments in Animals

1977	Van Miller, Lalich, and Allen[10]
1978	Kociba, Keyes, Beyer, Carreon, *et al.*[72]
1979	Toth, Somfai-Relle, Sugar, and Bence[98]
1982	National Cancer Institute and National Toxicology Program[65,66]
1987	Della Porta, Dragani, and Sozzi[99]
1988	Rao, Subbarao, Prasad, and Scarpelli[100]
1990	Zeise, Huff, Salmon, and Hooper[11]

4. EXPERIMENTAL CARCINOGENESIS DATA

Available studies on TCDD carcinogenesis (Table 2) have been reviewed,[11–13,17,19,20] and these with abbreviated results are summarized chronologically in Tables 3–8 and are presented collectively as well in Table 9. Within the seven cited papers in Table 2 there are 18 separate sex–species experiments, five routes of exposure, TCDD concentrations ranging from 0.001 to 100 μg/kg body weight, exposure durations of 5 weeks to 2 years, and experiments lasting from 12 months to the life span. Under these varied experimental conditions, TCDD in all cases uniformly caused tumors. Both non-neoplastic (toxic) and neoplastic (benign and malignant tumors) effects were observed in various and multiple tissue/organ sites.

In every species so far exposed to TCDD, either long term or short term followed by an observation period, and by every route of exposure, clear carcinogenic responses

Table 3
Van Miller, Lalich, and Allen (1977)[10]

Species/strain	Sprague–Dawley male rats 10/group
Exposures and duration	Feed: 0.001–1000 ppb, 78 weeks; 17 weeks with no exposure
Cancer results	Total tumors increased 30–70% in 0.005–5 ppb groups

Table 4
Kociba, Keyes, Beyer, Carreon, *et al.* (1978)[72]

Species/strain	Sprague–Dawley male and female rats, 86 controls, 50/exposed group
Exposures and duration	Feed: 21–2200 ppt (= 0.02–2 ppb) (0.001–0.1 μg/kg/day); 2 years
Cancer results	Male: tongue, nose, palate Female: lung, liver, nose, palate

Table 5
Toth, Somfai-Relle, Sugar, and Bence (1979)[98]

Species/strain	Swiss male mice, 100 controls 45/exposed group
Exposures and duration	Gavage: 0.007–7.0 μg/kg/week, 1 year exposure, life span
Cancer results	Liver tumors: 0.7 μg group only

Table 6
National Cancer Institute-National Toxicology Program (1982)[65,66]

Species/strain	Osborne–Mendel male and female rats, 75 controls, 50/exposed group
Exposures and duration	Gavage: 0.0014–0.071 μg/kg/day, 2 years
Cancer results	Male: thyroid, liver?, adrenal Female: liver, skin, adrenal
Species/strain	B6C3F1 male and female mice, 75 controls 50/exposed group
Exposures and duration	Gavage: male 0.0014–0.071, female 0.0057–0.29 μg/kg/day, 2 years
Cancer results	Male: lung, liver Female: liver, thyroid, skin, lymphoma
Species/strain	Swiss–Webster male and female mice, 45 controls, 30/exposed group
Exposures and duration	Dermal: 0.001–0.005 mg/app., 3×/week, 2 years
Cancer results	Male: skin (fibrosarcoma) (?) Female: skin (fibrosarcoma)

Table 7
Della Porta, Dragani, and Sozzi (1987)[99]

Species/strain	B6C3 male and female mice 42–50/group
Exposures and duration	Gavage: 2.5–5.0 μg/kg/week × 52 weeks, observe until 78 weeks
Cancer results	All: liver tumors
Species/strain	B6C3 and B6C male and female mice, 89–106/group
Exposures and duration	i.p.: 1–30 μg/kg/week, for 5 weeks, observe until 78 weeks
Cancer results	All lymphoma M & F B6C3: liver tumors

Table 8
Rao, Subbarao, Prasad, and Scarpelli (1988)[100]

Species/strain	Syrian golden male hamsters
Exposure and duration	i.p.: 100 μg/kg 2–6 ×, one/4 weeks s.c.: 50–100 μg/kg 6 ×, one/4 weeks 12–13 months
Cancer results	Facial skin: squamous cell carcinoma

have been found. Tumors of various types in nine organ systems (including total tumors) have been induced, and these rarely appeared to occur at the site of application; for example, there were no tumors of the forestomach or stomach after oral (gavage or diet) administration, or of injection site areas. Most tumors in the single dermal study were located at or near the site of application (yet skin tumors were reported in mice after gavage exposure), and one might consider the tumors of the tongue and palate to be in the path of the diet exposure. Thus, the majority of cancer sites are distant from the exposure route.

Table 9
Summary of Carcinogenicity Experiments of TCDD

Authors	Species/strain[a]	Routes and exposures	Cancer results
Van Miller *et al.* (1977)[10]	Male Sprague–Dawley rats, 10/grp	Feed: 0.001–1000 ppb, 78 wk exp. + 17 wk obs	Total tumors >in all grps but 0.001 ppb
Kociba *et al.* (1978)[72]	M & F Sprague–Dawley rats, 86 con, 50 exp	Feed: 21–2200 ppt (0.001–0.1 μg/kg/d), 2 yr	M: tongue, nose, palate; F: lung, liver, nose, palate
Toth *et al.* (1979)[98]	Male Swiss mice, 100 con, 45 exp	Gavage: 0.007–7.0 μg/kg/wk, 1 yr exp, life-span obs	Liver tumors in 0.7 μg grp; none in 0.007; higher doses died
NTP (1982)[65]	M & F Osborne–Mendel rats 75 con, 50 exp	Gavage: 0.0014–0.071 μg/kg/d, 2 yr	M: Thyroid, liver?, adrenal; F: liver, skin, adrenal
	M & F B6C3F1 mice, 75 con, 50 exp, 2 yr	Gavage: M—0.0014–0.071, F—0.0057–0.29 μg/kg/d	M: lung, liver; F: liver, thyroid, skin, lymphoma
NTP (1982)[66]	Swiss–Webster mice, 45 con, 30 exp	Dermal: 0.001–0.005 mg/appl, 3×/wk, 2 yr	F: skin fibrosarcoma; M: same?
Della Porta *et al.* (1987)[99]	M & F B6C3 mice, 42–50/grp	Gavage: 2.5–5.0 μg/kg/wk × 52 wk, obs until 78 wk	M: liver; F: liver
	M & F B6C3 & B6C mice, 89–106/grp	i.p.: 1–30 μg/kg/wk × 5 wk, obs until 78 wk	All: lymphoma; M & F B6C3: liver
Rao *et al.* (1988)[100]	M Syrian golden hamsters, 12–13 months	i.p.: 100 μg/kg, 2–6×, one/4 wk; s.c.: 50–100 μg/kg, 6×, one/4 wk	Both routes: facial skin, squamous cell carcinoma

Abbreviations: con, controls; exp, exposed; F, female; M, male.

In rats, TCDD induced neoplasms in the lung, oral and nasal cavities, thyroid and adrenal glands, and liver (see discussion of gender-specific liver tumor induction in Section 7, point 3). In mice, TCDD induced neoplasms in the liver, subcutaneous tissue, thyroid gland, lung, and lymphopoietic tissue (lymphomas). In hamsters, TCDD produced squamous cell carcinomas of the facial skin.

Tumors of the integumentary system have been reported after oral (rats and mice), i.p. (hamsters), and dermal (mice) exposures. TCDD is a transspecies (rat, hamster, mouse), transstrain, transsex, multisite, complete carcinogen. The most potent of the identified chemical carcinogens, TCDD induced carcinogenic effects in laboratory animals with exposures as low as 0.001 μg/kg body wt per day.[11–13] Thus, even under these relatively crude experimental paradigms, no threshold level has so far been convincingly found for TCDD carcinogenic activity.

5. GENETIC TOXICOLOGY

TCDD is not considered to be genotoxic,[46,47] although it enhanced transformation of C3H 10T1/2 cells induced by MNNG[48] and was mutagenic to mouse lymphoma cells.[49] In four *Salmonella* tester strains exposed to the components of Agent Orange—2,4-D, 2,4,5-T, and esters, TCDD, DCDD—none exhibited any mutagenic activity.[50] In a host-mediated *in vivo*/*in vitro* assay using peritoneal macrophages, TCDD was reported to transform these cells dose-dependently.[51,52] And the TCDD cell transforming potential was seven times more active than TBrDD. As with the reported long-term carcinogenicity experiments, Agent Orange *per se* as a composite mixture has not been studied for genotoxicity.

6. MECHANISM OF ACTION

Mechanisms of carcinogenesis—and in particular chemically associated carcinogenicity—have attracted considerable scientific and public attention in the last decade. Much insight has been gained that can only lead to more reasoned and better prevention, intervention, and treatment for the reduction of environmentally caused cancers.[53–55] However, there seems to be an exaggerated tendency to embrace "mechanisms" not yet fully characterized, completely tested, proven, and accepted.[56]

A major difficulty confronting researchers and regulators centers on what is actually meant by the oft-used term *mechanism:* to many this term connotes simply pharmacokinetics, whereas to others the word means molecular bases of action. Some have other notions with respect to genotoxicity, metabolism, toxicity, and so forth. In chemical carcinogenesis, for example, much discussion and research center on the role exogenously enhanced cell proliferation has on the carcinogenesis process.

Proposed mechanisms of TCDD carcinogenesis do not yet account for the many relatively rare and varied types of tumors found in exposed animals. Receptor-mediated carcinogenesis may prove most relevant.[16,18] Further, differences of responses among sexes may signal hormonal influences regarding site specificity. In two-stage liver and skin models, TCDD exhibits considerable promotional activity.[e.g.,57,58] Yet, both liver and skin as well as other organs respond to the carcinogenic effects of TCDD alone; that is, without exogenous initiation and in organs with no or few naturally occurring tumors. Studies that should allow further clarification as to cellular and molecular mechanisms of TCDD-induced cancers include protooncogene activation and tumor suppressor gene inactivation patterns in control and TCDD-induced tumors. Importantly this should be accomplished not only for liver, but also for tumors in other organs such as lung, thyroid gland, and lymphopoietic system, because there appears to be concordance between tumors at these sites in animals and in humans and "dioxin" exposures.

As another example of species commonality of responsiveness, the toxic mode of action of TCDD is remarkably similar in rainbow trout and in mammals,[59] and aquatic contamination with TCDD is widespread.[36]

As with most chemical carcinogens, attempts to restrictively "classify" TCDD as an initiator, a promoter, or a complete carcinogen[11–13,17,20,60] can only confuse the public health and regulatory efforts toward prevention of "dioxin"-associated diseases. These "operational terms" are useful for scientific communication, but may be misunderstood or misused by some, and could lead to an actual hindrance of scientific progress.[61]

Interestingly enough, as we learn more and more about the intricacies and multiplicities of "mechanisms of carcinogenesis" and carcinogenesis processes, we have not yet been able to conclusively and unequivocally describe—after only four or five decades of intense searching—the complete mechanistic process for a single chemical carcinogen. Undoubtedly significant gains have been and are being made, but we still await a solitary and inarguable mechanism of action of carcinogenicity for even one chemical. For a few we are quite close. Some believe, including myself, that "a mechanism" or some "unified theory" of carcinogenesis may not be congruent with available information; perhaps more logically, we must continue to approach each chemical (or "class of chemicals") as if it exhibits a distinct mechanism or mechanisms of cancer induction (e.g., individual benzene metabolites; dioxin and receptor-mediated effects; DES and hormonal perturbations), and may indeed function differently among different species or sexes. Of course this hypothesis may be disproven as well, as we gather further information and more knowledge.

7. SUPPLEMENTAL EXPERIMENTS

Other practical (and relatively easy) experiments that may help answer some additional questions about TCDD and carcinogenicity could include the following:

1. "Reread" histology tissue sections from the long-term studies conducted by the National Cancer Institute/National Toxicology Program. In the past, several individuals/groups examined these sections and came to different diagnostic and numerical conclusions. Thus, for use in quantitative risk assessment, one must have consensus incidence data.

2. Because more current diagnostic criteria have been developed and used by the NTP in recent years for tumors of the liver in rats,[62] especially in an attempt to end the use of the older (and confusing) nomenclature "neoplastic nodule," liver sections should be reexamined and rediagnosed.

3. Further, some difference of opinion surrounds the liver tumor incidence in rats, and centers on whether TCDD induces liver tumors in only one sex; in the NTP study,[12] results for male Osborne–Mendel rats were 0/74 versus 0/50, 0/50, 3/50 [one had both a neoplastic nodule and a carcinoma; the trend statistic for males was significant ($p = 0.005$), and the controls versus top dose was marginal ($p = 0.06$)]; for female Osborne–Mendel rats, 5/75, 7% versus 1/49, 2%; 3/50, 6%; 14/49, 29% (trend, $p < 0.001$; controls versus top dose, $p = 0.001$).

Control incidence rates for liver tumors in Osborne–Mendel rats are typically low; neoplastic nodules: males = 7/975, 0.7%; females = 16/970, 1.6%; heptocellular carcinoma: males = 2/975, 0.2%; females = 4/970, 0.4%.[63] Given that only 9 liver tumors have been seen in nearly 1000 contemporary male control O–B rats, the 4 liver tumors observed in this experiment are clearly related to TCDD exposure; the observed carcinoma further supports this causality. Focal cellular alterations were increased in both sexes: male, 0/74 versus 3/50, 3/50, 6/50; female, 0/75, 0/49, 8/50, 16/49. Interestingly, hepatic cell proliferation (e.g., DNA synthesis measured after 1 and 2 weeks' TCDD exposure to Sprague–Dawley rats) either was not increased or was significantly reduced, more so in female than in male rats.[64]

To address this possible gender–response issue, since even in female rats only three of the tumors in the top dose group were carcinomas, one needs to (1) rediagnose these tumors using the current nomenclature and (2) take additional sections of the remaining liver to get a closer indication of the "true" incidence of these tumors.

Additional evidence for association with liver tumors comes from TCDD-caused liver tumors in both male and female mice, and hexachlorodibenzo-*p*-dioxin induced liver tumors in both male (0/74 versus 0/49, 1/50, 4/48) and female (5/75 versus 10/50, 12/50, 30/50) O–B rats and in both sexes of B6C3F1 mice.[11–13,65,66] Historically, NCI/NTP has identified only four chemicals causing liver tumors in O–B rats: 1,2-dibromoethane (female), 1,4-dioxane (female), hexachlorodibenzo-*p*-dioxin (male equivocal, female), and 2,3,7,8-TCDD (female). In my view both "dioxins" were actually positive for male rat liver as well. Nonetheless, the male O–B rat does appear to be considerably less responsive to chemicals and liver cancer than does the female O–B rat, and compared with the Fischer rat[67,68] relatively few chemicals induced liver tumors in O–B rats: 55 chemicals were evaluated and only 4 induced liver tumors.

4. Long-term exposure experiments in Fischer rats should be conducted, not only for comparisons to most other chemical studies using Fischer 344 rats in the NCI/NTP data base, but also to evaluate the influence of cell proliferation[56,64,69,70] and oncogene activation on the carcinogenesis process.[54,71] Studies have been done mainly in mice, but only Osborne–Mendel[65] and Sprague–Dawley[72] rats have been used.

5. Additional long-term exposure groups might include:

a. A single TCDD exposure or relatively few exposures group, and observe the animals for life
b. A combination-chemicals exposure group, such as several "dioxins" or "fur-

ans," TCDD and PCBs/PBBs, TCDD and phenoxy acid herbicides, or several other chemicals humans are exposed to routinely

c. "Agent orange"—50:50 mixture of *n*-butyl esters of 2,4,5-trichlorophenoxyacetic acid (contaminated with TCDD) and 2,4-dichlorophenoxyacetic acid—has surprisingly never been studied for carcinogenic activity

d. A multigenerational exposure regimen to include preconception, conception, gestation, lactation, and long-term (lifetime)

8. DIOXINS, HERBICIDES, AND HUMAN CANCERS

Debate continues about whether TCDD, chlorophenoxy herbicides, or a combination of these and other attendant industrial chemicals and solvents are the ultimate cause of associated cancers and other adverse health effects.[7,11,23–30,73–85,101] This tends to centralize once again the important and neglected issue of evaluating mixtures of chemicals for toxic effects, including cancer.[86–88] Obviously humans do not get exposed to single discrete agents, but are bombarded by a menagerie of chemicals, mixtures of exposures, and varied life-style, environmental, and occupational circumstances that impact collectively.[89]

A consistent finding from the latest cohort studies shows a 54% (95% CI = 31–81%) collective SMR increase in cancer mortality 20 or more years after exposure to TCDD (H. Vainio, personal communication, 1993). These "total cancer" increases in TCDD-cohort mortality certainly fit the current mechanistically based carcinogenesis epitome. The debate continues over whether or not these studies report "true" associations between "dioxin" exposure and cancer in humans. Silbergeld[90] believes, as I do, that "at present, the balance of the epidemiological studies suggests that dioxins are carcinogenic to humans." Findings in humans were predated from experiments in laboratory animals,[34,89,91] and support the reasoning that TCDD is a multispecies carcinogen.

According to the International Agency for Research on Cancer,[92] nine epidemiology studies involving TCDD-exposed cohorts are ongoing: two centered in France (paper and pulp workers; registry of exposures to phenoxy acid herbicides/contaminants), one in Germany (chemical industry phenoxy acid herbicide workers), one in Italy (persons exposed in 1976 Seveso accident), one in the Netherlands (phenoxy acid herbicide manufacturers), one in Sweden (fishermen), three in the United States (metropolitan Atlanta; polychlorinated hydrocarbon exposures; rare cancers as sentinel events). Additionally, the NIOSH dioxin registry and cohort mortality study continues together with the IARC worldwide dioxin registry effort.

The accumulating experimental,[13] mechanistic,[20,93,94] and epidemiological evidence (see above) supports the conclusion that TCDD alone, and likely chlorophenoxy herbicides as well, are indeed associated with human cancers.[77,95,96] These instances add to the growing list of human carcinogens that were first shown to induce cancers in experimental animals and subsequently in humans.[31–34,97] Thus, public health ethics dictate that all potential exposures to this and other dioxins and herbicides should be minimized, and where feasible eliminated.

Table 10
Summary Conclusions of Experimental Carcinogenesis Studies in Laboratory Animals

TCDD causes cancer in
Multiple species
Multiple strains
Both sexes
Multiple organs and tissues
TCDD causes cancer by
Multiple routes
Various durations of exposure
Ranges of exposure concentrations
Species and strains
Mice
Swiss
B6C3F1
B6C
Rats
Sprague–Dawley
Osborne–Mendel
Hamsters
Syrian golden
Humans
Workers
Routes and durations of exposure
Feed, oral (gavage), dermal, injection
5, 16, 52, 78, and 104 weeks, life span

9. SUMMARY AND CONCLUSIONS

Long-term carcinogenesis studies using laboratory animals exposed to TCDD have shown conclusively that this chemical is a complete carcinogen. In a series of independent studies from the first reported in 1977 to the latest in 1988, TCDD has consistently induced cancers in a variety of organs and systems. The "receptor-

Table 11
Sites of TCDD Carcinogenicity[a]

Nose, palate, tongue
(Lung), liver, (skin)
(Thyroid) and adrenal glands
Hematopoietic (lymphoma, multiple myeloma, leukemia)
(Connective and sift tissue)
(Liver)
(Respiratory system)
(Total tumors)

[a]Probable sites in humans are shown in parentheses.

Table 12
TCDD (and Other Dioxin and Dioxin-like) Exposures and Human Health

Question:	Is "dioxin" exposure hazardous to humans?
Facts:	TCDD is extremely toxic.
	TCDD is an enzyme (MFO) inducer.
	TCDD is immunotoxic.
	TCDD is fetotoxic and teratogenic.
	TCDD impairs reproductive performance.
	TCDD is not "mutagenic" or "genotoxic."
	TCDD is carcinogenic in animals.
	All humans have dioxin body burdens.
	Biological half-life in humans is ~7 years.
	TCDD is chloracnegenic in humans.
	TCDD is associated with cancer in humans.

mediated" mechanism of carcinogenic activity surely supports the findings of multiple sites of carcinogenesis. TCDD-induced neoplastic lesions occurred in each of 18 individual sex–species experiments.

In rats, TCDD induced neoplasms in lung, oral/nasal cavities, thyroid and adrenal glands, and liver. In mice, TCDD caused neoplasms in the liver, subcutaneous tissue, thyroid gland, lung, and lymphopoietic system (lymphomas). In hamsters, TCDD prouced squamous cell carcinomas of the facial skin. TCDD is a transspecies (rat, mouse, and hamster), trans-strain (Sprague–Dawley and Osborne–Mendel rats; B6C3F1, Swiss–Webster, and B6C mice), transsex, multisite, complete carcinogen. The multiple sites and types of induced cancers in animals and in humans argue in favor of complete carcinogenicity, although many "mechanistic" studies—often concentratedly those of the two-stage liver or skin models—do strongly support a "promoter-type" mode.

The most potent of the identified chemical carcinogens, TCDD induced carcinogenic effects in laboratory animals with exposures as low as 0.001 μg/kg body weight per day. Proposed mechanism(s) of TCDD-induced carcinogenesis have yet to account for the many relatively rare types of tumors observed in exposed animals. In two-stage liver or skin models, TCDD exhibits considerable tumor promotion activity.

Table 13
TCDD Mechanism(s) of Carcinogenesis?

1. Ah receptor–cytochrome P-450 activity enzyme inducer
2. Regulation of cellular differentiation and/or cell division
3. Promoter or enhancer of initiated cells
4. Receptor-mediated tumor promoter
5. Initiation activity
6. Classical "nongenotoxic" complete carcinogen
7. Combination of mechanisms

Table 14
Evidence for Complete Carcinogen Activity of TCDD

1. Results from whole animal studies: species, strains, sexes, potency
2. Induction of uncommonly occurring and typically nonpromotable site-specific cancers
3. Multiple-site carcinogenicity
4. Irreversibility of neoplasia
5. No demonstrated threshold
6. Semantic dogma

With respect to carcinogenic hazards to humans, these experimental data alone indicate that TCDD should be considered for practical and public health purposes as being likely carcinogenic to humans; this species extrapolative conclusion gains growing support from epidemiological studies showing clear associations between exposure to dioxin and increases in cancers of the respiratory system (particularly lung), soft-tissue sarcomas, and total cancers. Other risks appear to be elevated for cancers of the testicle, thyroid gland, and other endocrine glands, largely in workers exposed to chlorophenoxy herbicides.

Thus, using the collective experimental and epidemiological evidence, the prudent course of action would be to minimize to the greatest extent possible all potential exposures to this and other dioxins as well as to phenoxyacid herbicides, and where technologically and agriculturally feasible exposures should be eliminated.

Regarding the proposed concept of thresholds for dioxins, or for that matter the generic issue of threshold carcinogenesis, one need only to view the carcinogenesis results from long-term experiments in laboratory animals to decide that a threshold does not pertain or sustain for dioxins: that is, in every adequate study done so far has induced carcinogenic effects, and at exposure levels as low as 0.001 μg/kg body wt per day.[13] Lucier *et al.*[17,18] and Portier *et al.*[93] confirmed and extended the "lack-of-threshold" truth using molecular studies measuring cytochrome P 450 polymorphisms.

Table 15
Conclusions from Long-Term Carcinogenesis Studies

The experimental carcinogenesis data show that TCDD:

1. Is carcinogenic at extremely low exposures
2. Has no apparent "threshold" exposure level
3. Exhibits "promoter" activity
4. Demonstrates evidence of "initiator" activity
5. Possesses complete carcinogenic activity
6. Should be regarded as likely carcinogenic to humans
7. Associated with lung, (thyroid gland), skin, and total tumor increases in humans

In conclusion, the evidence that TCDD is a complete carcinogen comes from the overwhelming results in whole animal experiments whereby (1) all species, strains, and sexes so far studied exhibit TCDD-induced cancers, (2) uncommonly occurring and typically nonpromotable organ site cancers are induced, (3) dose-related and multiple-site carcinogenesis have been observed, and (4) adequate evidence of thresholds for cancer development has not been demonstrated. As posed by Bailar,[73] "the hypothesis that low exposures are entirely safe for humans is distinctly less tenable now than before."

ACKNOWLEDGMENTS. For reading and offering meaningful comments on this chapter, I thank Dr. Lauren Zeise (California Environmental Protection Agency), Dr. George Lucier (National Institute of Environmental Health Sciences), and Dr. Jerrold Ward (National Cancer Institute). Additionally, I appreciate the invitation from Dr. Arnold Schecter (College of Medicine, Binghamton, NY) to prepare this overview chapter on TCDD and mammalian carcinogenesis. Ms. Donna Hebert Ratcliff was most helpful in coordinating various administrative aspects.

REFERENCES

1. J. Kimmig and K. H. Schulz, Occupational acne (so-called chloracne) due to chlorinated aromatic cyclic ethers [in German], *Dermatologia* **115,** 540–546 (1957).
2. K. D. Courtney and J. A. Moore, Teratology studies with 2,4,5-trichlorophenoxyacetic acid and 2,3,7,8-tetrachlorodibenzo-p-dioxin, *Toxicol. Appl. Pharmacol.* **20,** 396–403 (1971).
3. EHP, Perspective on Chlorinated Dibenzodioxins and Dibenzofurans, *Environ. Health Perspect.* **5,** pp. 1–312 (1973).
4. J. E. Huff and J. S. Wassom, Chlorinated dibenzodioxins and dibenzofurans. An annotated literature collection 1934–1973, *Environ. Health Perspect.* **5,** 283–312 (1973).
5. J. E. Huff and J. S. Wassom, Health hazards from chemical impurities: Chlorinated dibenzodioxins and dibenzofurans, *Int. J. Environ. Stud.* **6,** 13–17 (1974).
6. J. E. Huff and J. S. Wassom, Hazardous contaminants: Chlorinated dibenzodioxins and dibenzofurans, in: *Environmental Chemicals: Human and Animal Health* (E. Savage, ed.), Vol. II, pp. 175–197, Colorado State University, Fort Collins (1974).
7. J. E. Huff, J. Moore, R. Saracci, and L. Tomatis, Long-term hazards of polychlorinated dibenzodioxins and polychlorinated dibenzofurans, *Environ. Health Perspect.* **36,** 221–240 (1980).
8. IPCS, Polychlorinated Dibenzo-para-dioxins and Dibenzofurans, Environmental Health Criteria No. 88, International Programme on Chemical Safety, World Health Organization, Geneva (1989).
9. T. Kauppinen, M. Kogevinas, E. Johnson, H. Becher, P. A. Bertazzi, H. B. Bueno De Mesquita, D. Coggon, L. M. Green, M. Littorin, E. Lynge, J. D. Mathews, M. Neuberger, J. Osman, B. Pannett, N. Pearce, R. Winkleman, and R. Saracci, Chemical exposure in manufacture of phenoxy herbicides and chlorophenols and in spraying of phenoxy herbicides, *Am. J. Ind. Med.* **23,** 903–920 (1993).
10. J. P. Van Miller, J. J. Lalich, and J. R. Allen, Increased incidence of neoplasms in rats exposed to low levels of tetrachlorodibenzo-p-dioxin, *Chemosphere* **6,** 537–544 (1977).

11. L. Zeise, J. Huff, A. Salmon, and K. Hooper, Human risks from 2,3,7,8-tetrachlorodibenzo-p-dioxin and hexachlorodibenzo-p-dioxins, *Adv. Mod. Environ. Toxicol.* **17,** 293–342 (1990).
12. J. E. Huff, A. Salmon, K. Hooper, and L. Zeise, Long-term carcinogenesis studies on 2,3,7,8-tetrachlorodibenzo-p-dioxin and hexachlorodibenzo-p-dioxins, *Cell Biol. Toxicol.* **7,** 67–94 (1991).
13. J. E. Huff, 2,3,7,8-TCDD: A potent and complete carcinogen in experimental animals, *Chemosphere* **25,** 173–176 (1992).
14. IARC, Directory of agents being tested for carcinogenicity, No. 16, International Agency for Research on Cancer, Lyon (1994).
15. A. Poland and J. Knutson, 2,3,7,8-TCDD and related halogenated aromatic hydrocarbons: Examination of the mechanisms of toxicity, *Annu. Rev. Pharmacol.* **22,** 517–554 (1982).
16. G. W. Lucier, Receptor-mediated carcinogenesis, in: *Mechanisms of Carcinogenesis in Risk Evaluation* (H. Vainio, P. Magee, D. McGregor, and A. J. McMichael, eds.), IARC Sci. Publ. 116, International Agency for Research on Cancer, Lyon (1992).
17. G. W. Lucier, G. C. Clark, A. M. Tritscher, J. Foley, and R. R. Maronpot, Mechanisms of dioxin tumor promotion: Implications for risk assessment, *Chemosphere* **25,** 177–180 (1992).
18. G. W. Lucier, C. J. Portier, and M. A. Gallo, Receptor mechanisms and dose–response models for the effects of dioxins, *Environ. Health Perspect.* **101,** 36–44 (1993).
19. G. W. Lucier, G. C. Clark, C. Hiermath, A. M. Tritscher, C. H. Sewall, and J. E. Huff, Carcinogenicity of TCDD in laboratory animals: Implications for risk assessment, *Toxicol. Indust. Health* **9,** 631–668.
20. J. E. Huff, G. W. Lucier, and A. M. Tritscher, Carcinogenicity of TCDD: Experimental, mechanistic, and epidemiologic evidence, *Annu. Rev. Pharmacol. Toxicol.* **34,** 343–372 (1994).
21. F. A. Gunter and J. D. Gunter (eds.), *Residues of Pesticides and Other Contaminants in the Total Environment,* Springer-Verlag, Berlin (1983).
22. H. M. H. Ip and D. J. H. Phillips, Organochlorine chemicals in human breast milk in Hong Kong, *Arch. Environ. Contam. Toxicol.* **18,** 490–494 (1989).
23. M. Eriksson, L. Hardell, and H.-O. Adami, Exposure to dioxins as a risk factor for soft tissue sarcoma: A population-based case–control study, *J. Natl. Cancer Inst.* **82,** 486–490 (1990).
24. A. Zober, P. Messerer, and P. Huber, Thirty-four-year mortality follow-up of BASF employees exposed to 2,3,7,8-TCDD after the 1953 accident, *Int. Arch. Occup. Environ. Health* **62,** 139–157 (1990).
25. M. A. Fingerhut, W. E. Halpern, D. A. Marlow, L. A. Piacitelli, P. A. Honchar, M. H. Sweeney, A. L. Greife, P. A. Dill, K. Steenland and A. J. Suruda, Cancer mortality in workers exposed to 2,3,7,8-tetrachlorodibenzo-p-dioxin, *N. Engl. J. Med.* **324,** 212–218 (1991).
26. A. Manz, J. Berger, J. H. Dwyer, D. Flesch-Janys, S. Nagel, and H. Waltsgott, Cancer mortality among workers in chemical plant contaminated with dioxin, *Lancet* **338,** 959–964 (1991).
27. R. Saracci, M. Kogevinas, P. A. Bertazzi, H. B. Bueno De Mesquita, D. Coggon, L. M. Green, T. Kauppinen, K. A. L'Abbe, M. Littorin, E. Lynge, J. D. Mathews, M. Neuberger, J. Osman, N. Pearce, and R. Winkleman, Cancer mortality in an international cohort of workers exposed to chlorophenoxy herbicides and chlorophenols, *Lancet* **338,** 1027–1032 (1991).

28. A. H. Smith and M. L. Warner, Biologically measured human exposure to 2,3,7,8-tetrachlorodibenzo-p-dioxin and human cancer, *Chemosphere* **25,** 219–222 (1992).
29. L. Hardell, Phenoxy herbicides, chlorophenols, soft-tissue sarcoma (STS) and malignant lymphoma, *Br. J. Cancer* **67,** 1154–1155 (1993).
30. E. Lynge, Cancer in phenoxy herbicide manufacturing workers in Denmark, 1947–87, an update, *Cancer Causes Contr.* **4,** 261–272 (1993).
31. L. Tomatis, The predictive value of rodent carcinogenicity tests in the evaluation of human risks, *Annu. Rev. Pharmacol. Toxicol.* **19,** 511–530 (1979).
32. L. Tomatis, A. Aitio, J. Wilbourn and L. Shuker, Human carcinogens so far identified, *Jpn. J. Cancer Res.* **80,** 795–807 (1989).
33. J. E. Huff and D. P. Rall, Relevance to humans of carcinogenesis results from laboratory animal toxicology studies, in: *Maxcy-Rosenau's Public Health and Preventive Medicine,* 13th ed. (J. M. Last, ed.), pp. 433–440, 453–457, Appleton–Century–Crofts, New York (1992).
34. J. E. Huff, Chemicals and cancer in humans: First evidence in experimental animals, *Environ. Health Perspect.* **100,** 201–210 (1993).
35. EPA, National Study of Chemical Residues in Fish, Vols. I and II, EPA 823-R-92-008a and 008b, Office of Science and Technology, U.S. Environmental Protection Agency, Washington, DC (1992).
36. EPA, Interim report on data and methods for assessment of 2,3,7,8-tetrachlorodibenzo-p-dioxin risks to aquatic life and associated wildlife, EPA/600/R-93/055, Office of Research and Development, U.S. Environmental Protection Agency, Washington, DC (1993).
37. G. R. Gardner, P. P. Yevich, J. C. Harshbarger, and A. R. Malcolm, Carcinogenicity of Black Rock Harbor sediment to the Easter oyster and trophic transfer of Black Rock Harbor carcinogens from the blue mussel to the winter flounder, *Environ. Health Perspect.* **90,** 53–66 (1991).
38. J. C. Harshbarger and J. B. Clark, Epizootiology of neoplasms in bony fish of North America, *Sci. Total Environ.* **94,** 1–32 (1990).
39. E. Casillas, D. Weber, C. Haley, and S. Sol, Comparison of growth mortality in juvenile sand dollars (Dendraster excentricus) as indicators of contaminated marine sediments, *Environ. Toxicol. Chem.* **11,** 559–569 (1992).
40. E. E. McConnell, G. W. Lucier, R. C. Rumbaugh, P. W. Albro, D. J. Harvan, J. R. Hass, and M. W. Harris, Dioxin in soil: Bioavailability after ingestion by rats and guinea pigs, *Science* **223,** 1077–1079 (1984).
41. G. W. Lucier, R. C. Rumbaugh, R. Z. McCoy, R. Hass, D. Harvan, and P. Albro, Ingestion of soil contaminated with 2,3,7,8-tetrachlorodibenzo-p-dioxin (TCDD) alters hepatic enzyme activities in rats, *Fundam. Appl. Toxicol.* **6,** 364–371 (1986).
42. T. H. Umbreit, E. J. Hesse, and M. A. Gallo, Bioavailability of dioxin in soil from a 2,4,5-T manufacturing plant, *Science* **232,** 497–499 (1986).
43. E. B. May, R. Lukacovic, H. King, and M. M. Lipsky, Hyperplastic and neoplastic alterations in the livers of white perch (Morone americana) fron the Chesapeake Bay, *J. Natl. Cancer Inst.* **79,** 137–143 (1987).
44. C. D. Chriswell, How much is a part per trillion, anyway? Chemecology (letter), November (1977).
45. K. H. Jones, Diesel truck emissions, an unrecognized source of PCDD/PCDF exposure in the United States, *Risk Anal.* **13,** 245–252 (1993).
46. J. S. Wassom, J. E. Huff, and N. Loprieno, A review of the genetic toxicology of chlorinated dibenzo-p-dioxins, *Mutat. Res.* **47,** 141–160 (1977).
47. IARC, 2,3,7,8-Tetrachlorodibenzo-para-dioxin (TCDD), in: *Genetic and Related Effects: An Updating of Selected IARC Monographs from Volumes 1 to 42,* Suppl. 6, pp. 508–510,

IARC Monographs on the Evaluation of Carcinogenic Risks to Humans, International Agency for Research on Cancer, Lyon (1987).
48. D. J. Abernethy, W. F. Greenlee, J. C. Huband, and C. J. Boreiko, 2,3,7,8-Tetrachlorodibenzo-p-dioxin [TCDD] promotes the transformation of C3H/10T1/2 cells, *Carcinogenesis* **6,** 651–653 (1985).
49. A. M. Rogers, M. E. Andersen, and K. C. Black, Mutagenicity of 2,3,7,8-tetrachlorodibenzo-p-dioxin and perfluoro-N-decanoic acid in L5178Y mouse-lymphoma cells, *Mutat. Res.* **105,** 445–449 (1982).
50. K. Mortelmans, S. Haworth, Speck, and E. Zeiger, Mutagenicity testing of Agent Orange components and related chemicals, *Toxicol. Appl. Pharmacol.* **75,** 137–146 (1984).
51. T. Massa, Esmaeili, H. Fortmeyer, B. Schlatterer, H. Hagenmaier, and P. Chandra, Cell transforming and oncogenic activity of 2,3,7,8-tetrachlorodibenzo-p-dioxin and 2,3,7,8-tetrabromodibenzo-p-dioxin, *Anticancer Res.* **12,** 2053–2060 (1992).
52. T. Massa, Esmaeili, H. Fortmeyer, B. Schlatterer, H. Hagenmaier and P. Chandra, Carcinogenic and co-carcinogenic potential of 2,3,7,8-tetrachlorodibenzo-p-dioxin in a host-mediated in vivo/in vitro assay, *Chemosphere* **25,** 1085–1090 (1992).
53. J. C. Barrett, Mechanism of action of known human carcinogens, in: *Mechanisms of Carcinogenesis in Risk Identification* (H. Vainio, P. Magee, D. McGregor, and A. McMichael, eds.), IARC Sci. Publ. 116, pp. 115–134. International Agency for Research on Cancer, Lyon (1992).
54. J. C. Barrett and R. W. Wiseman, Molecular carcinogenesis in humans and rodents, in: *Comparative Molecular Carcinogenesis* (A. Klein-Szanto, M. Anderson, J. C. Barrett, and T. J. Slaga, eds.), Vol. 376, pp. 1–30, Wiley, New York (1992).
55. J. C. Barrett, Mechanisms of multistep carcinogenesis and risk assessment, *Environ. Health Perspect.* **100,** 9–20 (1993).
56. J. E. Huff, Mechanisms, chemical carcinogenesis, and risk assessment: Cell proliferation and cancer, *Am. J. Ind. Med.* in press. [Earlier version in *Ramazzini Newsletter* **3,** 47–50 (1992).]
57. Y. P. Dragan, X.-H. Xu, T. L. Goldsworthy, H. A. Campbell, R. R. Maronpot, and H. C. Pitot, Characterization of the promotion of altered hepatic foci by 2,3,7,8-tetrachlorodibenzo-p-dioxin in the female rat, *Carcinogenesis* **13,** 1389–1395 (1992).
58. S. Flodstrom and U. G. Ahlborg, Relative liver tumor promoting activity of some polychlorinated dibenzo-p-dioxin-, dibenzofuran- and biphenyl-congeners in female rats, *Chemosphere* **25,** 169–172 (1992).
59. J. L. Newsted and J. P. Giesy, Effect of 2,3,7,8-tetrachlorodibenzo-p-dioxin (TCDD) on the epidermal growth factor receptor in hepatic plasma membranes of rainbow trout (Oncorhynchus mykiss), *Toxicol. Appl. Pharmacol.* **119,** 41–51 (1993).
60. J. W. Holder and H. M. Menzel, Analysis of 2,3,7,8-TCDD tumor promotion activity and its relationship to cancer, *Chemosphere* **19,** 861–868 (1989).
61. H. Yamasaki, Multistage carcinogenesis: Implication for risk evaluation, *Cancer Metast. Rev.* **7,** 5–18 (1988).
62. R. R. Maronpot, C. A. Montgomery, G. A. Boorman, and E. E. McConnell, National Toxicology Program nomenclature for hepatoproliferative lesions of rats, *J. Toxicol. Pathol.* **14,** 263–273 (1986).
63. D. G. Goodman, J. M. Ward, R. A. Squire, M. B. Paxton, W. D. Reichardt, K. C. Chu, and M. S. Linhart, Neoplastic and nonneoplastic lesions in aging Osborne–Mendel rats, *Toxicol. Appl. Pharmacol.* **55,** 433–447 (1980).
64. T. R. Fox, L. L. Best, S. M. Goldsworthy, J. J. Mills, and T. L. Goldsworthy, Gene expression and cell proliferation in rat liver after 2,3,7,8-tetrachlorodibenzo-p-dioxin exposure, *Cancer Res.* **53,** 2265–2271 (1993).

65. NTP, Carcinogenesis Bioassay of 2,3,7,8-Tetrachlorodibenzo-p-dioxin (Cas No. 1746-01-6) in Osborne–Mendel Rats and B6C3F1 Mice (Gavage Study), Tech. Rep. Ser. No. 209, pp. 1–195, National Toxicology Program, Research Triangle Park, NC (1982).
66. NTP, Carcinogenesis Bioassay of 2,3,7,8-Tetrachlorodibenzo-p-dioxin (Cas No. 1746-01-6) in Swiss–Webster Mice (Dermal Study), Tech. Rep. Ser. No. 201, pp. 1–113, National Toxicology Program, Research Triangle Park, NC (1982).
67. J. E. Huff, J. Cirvello, J. K. Haseman, and J. R. Bucher, Chemicals associated with site-specific neoplasia in 1394 long-term carcinogenesis experiments in laboratory rodents, *Environ. Health Perspect.* **93,** 247–271 (1991).
68. J. E. Huff, and J. K. Haseman, Long-term chemical carcinogenesis experiments for identifying potential human cancer hazards. Collective data base of the National Cancer Institute and National Toxicology Program (1976–1991), *Environ. Health Perspect.* **96,** 23–31 (1991).
69. R. L. Melnick, J. E. Huff, J. C. Barrett, R. R. Maronpot, G. Lucier, and C. J. Portier, Cell proliferation and chemical carcinogenesis: A symposium overview, *Mol. Carcinogen.* **7,** 135–138 (1993). [Reprinted in *Environ. Health Perspect.* **101** (Suppl. 5) 3–8 (1993).]
70. R. L. Melnick and J. E. Huff, Liver carcinogenesis is not a predicted outcome of chemically induced hepatocyte proliferation, *Toxicol. Ind. Health* **9,** 415–438 (1993). Reprinted in: *Identification and Public Health Control of Environmental and Occupational Diseases* (M. A. Mehlman and A. Upton, eds.), Princeton Sci. Publ. Princeton, NJ, in press.
71. J. A. Boyd and J. C. Barrett, Genetic and cellular basis of multistep carcinogenesis, *Pharmacol. Ther.* **46,** 469–486 (1990).
72. R. J. Kociba, D. G. Keyes, J. E. Beyer, R. M. Carreon, C. E. Wade, D. A. Dittenber, R. P. Kalnins, L. E. Frauson, C. N. Park, S. D. Barnard, R. A. Hummel, and C. G. Humiston, Results of a two-year chronic toxicity study and oncogenicity study of 2,3,7,8-tetrachlorodibenzo-p-dioxin in rats, *Toxicol. Appl. Pharmacol.* **46,** 279–303 (1978).
73. J. C. Bailar, Editorial: How dangerous is dioxin? *N. Engl. J. Med.* **423,** 260–262 (1991).
74. A. Blair and S. H. Zahm, Patterns of pesticide use among farmers: Implications for epidemiologic research, *Epidemiology* **4,** 55–62 (1993).
75. A. Blair, A. Linos, P. A. Stewart, L. F. Burnmeister, R. Gibson, G. Everett, L. Schuman, and K. P. Cantor, Evaluation of risks for non-Hodgkin's lymphoma by occupation and industry exposures from a case–control study, *Am. J. Ind. Med.* **23,** 301–312 (1993).
76. S. K. Hoar, A. Blair, F. F. Holmes, C. D. Biysen, R. J. Robel, R. Hoover, and J. J. Fraumeni, Agricultural herbicide use and risk of lymphoma and soft-tissue sarcoma, *J. Am. Med. Assoc.* **256,** 1141–1147 (1986).
77. IARC, 2,3,7,8-Tetrachlorodibenzo-para-dioxin [TCDD], in: *Overall Evaluations of Carcinogenicity: An Updating of IARC Monographs Volumes 1 to 42,* Suppl. 7, pp. 350–354, IARC Monographs on the Evaluation of Carcinogenic Risks to Humans, International Agency for Research on Cancer, Lyon (1987).
78. E. S. Johnson, Human exposure to 2,3,7,8-TCDD and risk of cancer, *Crit. Rev. Toxicol.* **21,** 451–463 (1992).
79. E. S. Johnson, W. Parsons, C. R. Weinberg, D. L. Shore, J. Mathews, D. G. Patterson, and L. L. Needham, Current serum levels of 2,3,7,8-tetrachlorodibenzo-p-dioxin in phenoxy acid herbicide applicators and characterization of historical levels, *J. Natl. Cancer Inst.* **84,** 1648–1653 (1992).
80. E. S. Johnson, Important aspects of the evidence for TCDD carcinogenicity in man, *Environ. Health Perspect.* **99,** 383–390 (1993).
81. S. H. Zahm and A. Blair, Pesticides and non-Hodgkin's lymphoma, *Cancer Res.* (Suppl.) **52,** 5485s–5488s (1992).
82. H. I. Morrison, K. Wilkins, R. Semensiw, Y. Mao, and D. Wigle, Herbicides and cancer, *J. Natl. Cancer Inst.* **84,** 1866–1874 (1992).

83. A. C. Pesatori, D. Consonni, A. Tironi, M. T. Landi, C. Zocchetti, and P. A. Bertazzi, Cancer morbidity in the Seveso area, 1976–1986, *Chemosphere* **25,** 209–212 (1992).
84. T. Sinks, Misclassified sarcomas and confounded dioxin exposure, *Epidemiology* **4,** 3–6 (1993).
85. A. J. Suruda, E. M. Ward, and M. A. Fingerhut, Identification of soft tissue sarcoma deaths in cohorts exposed to dioxin and to chlorinated naphthalenes, *Epidemiology* **4,** 14–19 (1993).
86. R. S. H. Yang and NTP, Toxicity studies of pesticide/fertilizer mixtures administered in drinking water to F344/N rats and B6C3F1 mice, Toxicity Rep. Ser. No. 36, National Toxicology Program, Research Triangle Park, North Carolina (1993).
87. R. S. H. Yang, Strategy for studying health effects of pesticides/fertilizer mixtures in groundwater, *Rev. Environ. Contam. Toxicol.* **127,** 1–22 (1992).
88. A. K. Chaturvedi, Toxicological evaluation of mixtures of ten widely used pesticides, *J. Appl. Toxicol.* **13,** 183–188 (1993).
89. J. E. Huff, J. K. Haseman, and D. P. Rall, Scientific concepts, value, and significance of chemical carcinogenesis studies, *Annu. Rev. Pharmacol. Toxicol.* **31,** 621–652 (1991).
90. E. K. Silbergeld, Carcinogenicity of dioxins, *J. Natl. Cancer Inst.* **83,** 1198–1199 (1991).
91. D. P. Rall, M. D. Hogan, J. E. Huff, B. A. Schwetz, and T. W. Tennant, Alternatives to using human experience in assessing health risks, *Annu. Rev. Public Health* **8,** 355–385 (1987).
92. M. Coleman, J. Wahrendorf, and E. Demaret, Directory of on-going research in cancer epidemiology 1992, IARC Sci. Publ. No. 117, #281, #294, #370, #483, #591, #690, #891, #1032, #1106, International Agency for Research on Cancer, Lyon (1992).
93. C. Portier, A. Tritscher, M. Kohn, C. Sewall, G. Clark, L. Elder, D. Hoel, and L. Lucier, Ligand/receptor binding for 2,3,7,8-TCDD: Implications for risk assessment, *Fundam. Appl. Toxicol.* **20,** 48–56 (1993).
94. M. C. Kohn, G. W. Lucier, and C. J. Portier, A mechanistic model of effects of dioxin on gene expression in the rat liver, *Toxicol. Appl. Pharmacol.* **120,** 138–154 (1993).
95. IARC, Occupational exposures in spraying and application of insecticides, in: *Occupational Exposures in Insecticides Application,* IARC Monographs on the Evaluation of Carcinogenic Risks to Humans, Vol. 53, pp. 45–92, International Agency for Research on Cancer, Lyon (1991).
96. DHHS/NTP, 2,3,7,8-Tetrachlorodibenzo-p-dioxin [TCDD], Seventh Annual Report on Carcinogens, Vol. 7, pp. 794–800, National Toxicology Program, Research Triangle Park, NC (1993).
97. J. E. Huff, Chemicals causally associated with cancers in humans and in laboratory animals—A perfect concordance, in: *Carcinogenesis* (M. P. Waalkes and J. M. Ward, eds.), pp. 25–37, Raven Press, New York (1994).
98. K. Toth, S. Somfai-Relle, and J. Sugar, and J. Bence, Carcinogenicity of herbicide 2,4,5-trichlorophenoxyethanol containing dioxin and of pure dioxin in Swiss mice, *Nature* **278,** 548–549 (1979).
99. G. Della Porta, T. A. Dragani, and G. Sozzi, Carcinogenic effects of infantile and long-term 2,3,7,8-tetrachlorodibenzo-p-dioxin treatment in the mouse, *Tumori* **73,** 99–107 (1987).
100. M. S. Rao, V. Subbarao, J. D. Prasad, and D. G. Scarpelli, Carcinogenicity of 2,3,7,8-tetrachlorodibenzo-p-dioxin in the Syrian golden hamster, *Carcinogenesis* **9,** 1677–1679 (1988).
101. P. A. Bertazzi, A. C. Pesatori, D. Consonni, A. Tironi, M. T. Landi, and C. Zochetti, Cancer incidence in a population accidentally exposed to 2,3,7,8-tetrachloro-para-dioxin, *Epidemiology* **4,** 398–406.

Chapter 13

Neurochemical and Behavioral Sequelae of Exposure to Dioxins and PCBs

Richard F. Seegal and Susan L. Schantz

1. INTRODUCTION

In a previous review of the literature related to the toxicity of polychlorinated biphenyls (PCBs), one of the present authors[1] reviewed the experimental evidence demonstrating that one of the major classes of PCBs—the *ortho*-substituted congeners—were capable of altering neurological function in a wide variety of experimental preparations. In this chapter we broaden the review to include a discussion of the behavioral effects of exposure as well as contributions that coplanar PCBs, dioxins (PCDDs), and dibenzofurans (PCDFs) make to the overall neurotoxic risk to humans from exposure to complex environmental mixtures.

PCBs, PCDDs, and PCDFs are members of a class of widely dispersed, environmentally persistent organic compounds known as halogenated aromatic hydrocarbons (HAHs). PCBs were first manufactured in the 1930s[2] and, because of their physical and chemical properties,[3] gained widespread use in the electrical industry, as heat-exchange fluids in the preparation of edible oils, and as sealants in concrete silos.[4] Based on the number of sites that can be chlorinated on the biphenyl moiety, there are 209 theoretically possible congeners.[5] However, only 135 of these congeners were produced in commercial synthesis and have actually been identified in the environment.[6] The most desirable characteristic of the PCBs (i.e., their thermal and physical

---

Richard F. Seegal • New York State Department of Health, Wadsworth Center for Laboratories and Research, Albany, New York 12201, and School of Public Health, University at Albany, State University of New York, Albany, New York 12203. Susan L. Schantz • Institute for Environmental Studies, University of Illinois at Urbana–Champaign, Urbana, Illinois 61801.

Dioxins and Health, edited by Arnold Schecter. Plenum Press, New York, 1994.

stability) greatly impedes their breakdown when they are released into the environment.[7,8] This situation is complicated by the fact that differences in the physical properties and biodegradation characteristics of individual PCB congeners have led to differential accumulation of certain PCB congeners in the environment. As a result of these and other processes, environmental samples rarely reflect the congener makeup of the original commercial mixtures. For example, recent data have shown that, under optimal conditions of PCB concentration, temperature, and time, certain strains of anaerobic bacteria may break down complex mixtures of highly chlorinated PCBs to more lightly chlorinated congeners.[9,10]

Because PCBs are both stable and highly lipophilic, they bioaccumulate in wildlife and foodstuffs.[11] Thus, one of the major routes of exposure for humans to PCBs is via consumption of contaminated foodstuffs.

PCDDs are highly toxic compounds that have no known industrial use and are produced as by-products during the manufacture of herbicides and defoliants (e.g., Agent Orange), and wood preservatives (e.g., pentachlorophenol); and as unwanted contaminants during the incineration of municipal wastes and during the bleaching of paper products produced from wood pulp.[12] Elevated levels of PCDDs are found near some European municipal waste incinerators; contamination from these incinerators has resulted in milk and meat products being withdrawn from the market.[13] Elevated levels of PCDDs have also been found in fish and other wildlife living downstream from wood pulp processing plants[14] and efforts are now being made to alter the paper bleaching process to minimize the amounts of PCDDs that are generated.[15]

PCDFs are structurally related to PCDDs and are unwanted by-products of the iron-catalyzed synthesis of PCBs.[3] PCDFs are also produced by the combustion of PCBs in oxygen-poor environments such as occurs during fires in PCB-containing transformers and during the heating of PCBs used as heat-exchange fluids.[16,17] PCDFs were a major source of contamination following the well-known transformer fire in 1981 at the Binghamton State Office Building in New York and in other similar fires.[18,19] In spite of massive efforts and the expenditure of more than $45 million, the Binghamton State Office Building has not yet reopened because of residual contamination!

Unfortunately, PCBs, PCDDs, and PCDFS exist in the environment as complex mixtures, making the task very difficult of identifying which of these putative neurotoxicants are responsible for inducing changes in human central nervous system (CNS) function. To further complicate the task, these HAHs are often found in combination with other toxicants such as methyl mercury, lead, and organic pesticides.[20] Furthermore, because of physical and biological processes (e.g., photolysis and bacterial dehalogenation[4,21]) that may occur when HAHs are released into the environment, and because of the metabolic processes that take place when the compounds are bioaccumulated by wildlife and eventually consumed by humans, the HAH congener patterns found in human tissues and fluids bear little resemblance to the patterns seen in the original commercial mixtures that entered the environment.[22] Thus, one of the most daunting tasks facing environmental neurotoxicologists is the need to determine which of the above agents or congeners, or mixtures of them, are responsible for the reported alterations in human CNS function and the possible degree of interaction between these classes of HAH congeners and other toxicants.

Comparison of Toxicological Effects of *Ortho*-Substituted and Coplanar PCB Congeners

Although HAHs consist of three major classes of compounds (PCBs, PCDDs, and PCDFs), each of which contains many individual congeners, they can be divided into two major subgroups based on their chemical structures and toxicological properties.

The first group consists of PCDDs, PCDFs and coplanar PCBs, all of which appear to exert their toxic efforts (as hepatotoxicants, immunotoxicants, and procarcinogens) by interacting with a cytosolic receptor protein known as the Ah receptor. This receptor exhibits saturable and high-affinity binding for the planar HAHs.[23] In turn, the ligand/receptor complexes interact with specific nuclear DNA sequences resulting in the induction of several cytochrome P-450 isozymes[24] including P4501A1 and P4501A2 and associated microsomal oxygenases including aryl hydrocarbon hydroxylase (AHH) and ethoxyresorufin *O*-deethylase (EROD).[25] All coplanar HAHs bind to the Ah receptor and induce AHH and EROD activity in both tissue culture and whole animal preparations but the individual congeners differ in their relative potencies.[26] 2, 3, 7, 8-TCDD (TCDD) is the most active, while coplanar PCB congeners have approximately 1% of the potency of TCDD. The presence of a single *ortho* chlorine substitution (mono-*ortho* coplanars) markedly reduces, but does not eliminate, the ability of the congener to interact with the Ah receptor and induce enzyme activity and the related toxicological effects.

In an attempt to simplify the task of determining the toxicity of complex mixtures of HAHs, Safe[24] has developed a toxic equivalency factor (TEF) based on the differential ability of the coplanar HAHs to induce AHH activity. In this schema the most toxic dioxin congener, 2,3,7,8-TCDD, was assigned a value of 1, with other less potent congeners having values of less than 1; e.g., the TEF for 2,3,7,8-TCDF is 0.1 and the TEF for the most dioxinlike PCB congener, 3,4,5,3′, 4′, is 0.01. TEFs have also been calculated for commercial PCB mixtures which contain coplanar and mono-*ortho*-coplanar congeners. For example, Aroclor 1248 has been assigned an overall TEF of approximately 0.000017. Using this approach it is theoretically possible to compare doses of different HAH mixtures based on their calculated TEFs.

The second major class of HAHs consists of PCB congeners with two or more *ortho*-substituted congeners. These congeners do not interact with the Ah receptor because the *ortho* chlorines prevent the molecule from assuming a planar conformation and they are therefore assigned a TEF value of zero.

Although TEFs may be useful for predicting toxic responses associated with compounds that bind to the Ah receptor, there are significant limitations to this approach because TEFs ignore the potential contribution that *ortho*-substituted congeners make to the overall toxicity of complex mixtures. Seegal and co-workers, and more recently Tilson and co-workers, have demonstrated in *in vitro* preparations[27–30] that *ortho*-substituted PCB congeners possess significant neurochemical activity. Furthermore, Schantz *et al.*[31] have shown that perinatal exposure of nonhuman primates to Aroclor 1016, which is composed almost entirely of *ortho*-substituted congeners and is devoid of planar, dioxinlike congeners, induces changes in neurobehavioral function that persist until adulthood. Finally, nonhuman primates, exposed perinatally to equivalent TEF doses of Aroclor 1248 and TCDD, exhibit qualitatively and quantitatively different behavioral responses.[32,33] These findings emphasize the shortcomings of the TEF concept for estimating the risk of neurological dysfunctions caused by environmental exposure to mixtures of HAHs. Indeed, both the data and current opinion[34,35] suggest that the neurological risk of exposure to HAHs would be severely underestimated if the contribution of the *ortho*-substituted congeners were ignored (i.e., if TCDD TEFs were used to estimate the risk of exposure to complex environmental mixtures of HAHs).

2. EPIDEMIOLOGICAL STUDIES

One of the primary motivations for the laboratory-based study of the possible neurological effects of HAHs stems from a series of epidemiological studies conducted in Japan, Taiwan, and the United States. These studies have been extensively reviewed elsewhere and will only by briefly discussed here.

In 1968 in Japan and in 1978 in Taiwan, approximately 4000 Japanese and Taiwanese were exposed, via consumption of contaminated rice oil, to PCBs, PCDFs, and polychlorinated naphthalenes. In Japan this incident was known as Yusho,[36] while in Taiwan it became known as Yu-cheng.[37] In both countries the exposure resulted in dermatological and neurological abnormalities in adults and infants, including "cola"-colored changes in skin pigmentation, decreases in nerve conduction velocities in adults, and evidence of developmental delays and impaired cognitive function in children exposed during the perinatal period.[38–40]

Studies by Hsu and Rogan in Taiwan support the earlier findings from Japan and suggested long-term decreases in cognitive function in perinatally exposed children.[41,42] These studies suggest that exposure to high concentrations of HAHs during early development may have long-term consequences on the health of children, including changes in CNS function. However, because of the relatively large number of putative neurotoxicants to which the populations were exposed, it has not been possible to unequivocally determine which are responsible for the developmental dysfunctions.

There have also been two epidemiological studies conducted in the United States that have examined the behavioral consequences of chronic, low-level exposure to PCBs. The first study was conducted in Michigan and examined the neurological consequences of perinatal exposure to PCBs (and other contaminants) via maternal consumption of contaminated fish from Lake Michigan.[43] The infants in this study were underresponsive and hyporeflexive at birth.[44] When they were tested at 7 months of age, they showed a dose-dependent deficit in visual recognition memory[45] and in a follow-up study conducted when the children were 4 years of age, deficits in short-term memory processing were still evident.[46,47] Higher PCB exposure was also associated with reduced activity levels at 4 years of age.[48] Schantz *et al.*[49] have also reported reduced activity levels in a cohort of children whose mothers were exposed to PCBs from eating contaminated beef and dairy products.

The second epidemiological study was carried out by Rogan and Gladen, who examined the effects of low-level, perinatal exposure to PCBs in a cohort of children from North Carolina. The North Carolina infants were also underresponsive and hyporeflexive at birth[50] and continued to show delays in psychomotor development at 6 and 12 months of age.[51] However, unlike the children in the Michigan cohort, the North Carolina children showed no evidence of continued behavioral impairments in follow-up testing at 3 to 5 years of age.[52] Disparities between the two U.S. studies are not fully understood but may include differences in exposure (the Michigan cohort was exposed to contaminants present in Great Lakes fish in addition to PCBs) and differences in age of the children at the time of assessment as well as differences in behavioral testing methodologies. Additional epidemiological studies examining the behavioral sequelae of exposure to toxicants in Great Lakes fish are presently being conducted in New York by Helen Daly and in Wisconsin by John Dellinger. Hopefully, results from these two additional studies will help resolve the differences between the Michigan and North Carolina studies.

3. EXPERIMENTAL METHODS TO QUANTIFY CHANGES IN NERVOUS SYSTEM FUNCTION FOLLOWING EXPOSURE TO ENVIRONMENTAL TOXICANTS

Experimental techniques from four major subdisciplines of neuroscience are available for the study of possible alterations in CNS function induced by exposure to HAHs. The first approach, electrophysiology, evolved from clinical neurology and measures the electrical activity of the nervous system. Measures include peripheral nerve conduction velocities, which are decreased in Yusho patients, electroencephalograms (EEGs), which provide an overall estimate of the electrical activity of the brain,[53] and sensory evoked potentials (SEPs), which provide data on the status of both peripheral and central neural pathways involved in the processing of sensory information.[54]

Because EEGs provide a relatively nonspecific measure of neuronal activity,[55] they have been largely supplanted by SEPs, which provide a more specific index of electrical activity. In this procedure, the electrical activity of the brain is recorded following stimulation of specific sensory pathways including the somatosensory (stimulation of peripheral nerves), visual (either flashing lights or checkerboard patterns of light are presented), and auditory (tones of different amplitude and frequency are presented) systems. These procedures have been used to determine the effects of strokes and lesions of the CNS,[56] exposure to electric and magnetic fields,[57] and perinatal exposure to inorganic lead.[58a] Recently, Chen and Hsu[58b] have demonstrated that there are significant increases in the latencies of the auditory-evoked P300 potential and significant decreases in the P300 amplitude in children born to Yu-Cheng mothers. These electrophysiological findings, seen in children exposed perinatally to contaminated rice oil, suggest deficits in cognitive function and the ability to process auditory information. Because these procedures are largely noninvasive, and can be used in both humans and animals, significant effort should be made to expand the use of these techniques to aid in determining the effects of exposure to HAHs on integrative functions of the CNS.

The second major methodology, known as neurobehavioral toxicology, involves the study of behavioral changes induced following exposure to putative neurotoxicants including HAHs. Tests have been developed that measure sensory function, motor and reflex development, and learning or cognitive function.[59,60] Indeed, the National Toxicology Program has developed a standardized series of tests that are capable of quantifying changes in behavior to a wide spectrum of potentially neurotoxic agents.[61] Although considerable data have been generated concerning behavioral changes following surgical ablation or chemical lesioning of specific nuclei in the brain,[62,63] it is generally not possible, using only behavioral procedures, to determine the site(s) of action of a toxicant or the underlying physiological or biochemical mechanisms responsible for the observed behavioral dysfunctions. Nevertheless, neurobehavioral toxicology is an extremely important subdiscipline of neurotoxicology because it allows the experimenter to determine whether underlying biochemical or morphological changes are of functional significance to the organism. Furthermore, advances have been made in developing test procedures that facilitate extrapolation from animals to humans, thus allowing comparisons of results derived from laboratory and epidemiological studies.[59] A major portion of this chapter is devoted to a review of the behavioral effects of PCBs and dioxins.

A third major component of neurosciences that is available to the neurotoxicologist is measurement of neuroanatomical changes following exposure to putative neurotoxicants. Indeed, one of the earliest reports of PCB exposure describes changes in spinal cord morphology.[64] In addition, changes in the status of the blood–brain barrier and evidence of toxicant-induced neuropathology have been noted following exposure to lead and methyl mercury.[65] The recent development of immunocytochemical techniques using antibodies raised against specific neurochemicals [e.g., changes in neuronal concentrations of tyrosine hydroxylase, glial fibrillary acidic protein (GFAP)] coupled with improved light microscopic and morphometric procedures has allowed study of the effects of neurotoxicants on changes in neuronal and glial cell number, morphometry, and reactivity to insult.[66]

The fourth major component of neurotoxicology is neurochemical toxicology. Most often, changes in concentrations of neurotransmitters, their metabolites and precursors, and activity of synthesizing enzymes are measured following exposure to known and suspected neurotoxicants. Because the concentrations of neurotransmitters in the CNS are highly regulated,[67] changes in concentrations provide an objective measure of changes in CNS function.[68] In addition, because the majority of neurotransmitters and their regulatory enzymes are present in lower mammals as well as humans,[69,70] these measures, to a large extent, avoid the potential problems that may arise when comparisons are made between different species, e.g., interpretation of species-typical behaviors. Finally, alterations in central neurochemical function can often be estimated by measuring concentrations of metabolites of the neurotransmitters in biological fluids such as plasma[71] and urine.[72] Thus, similar measures can be collected in epidemiological studies and the results compared with identical measures collected in the laboratory. For example, urinary concentrations of homovanillic acid (HVA), a major metabolite of the biogenic amine neurotransmitter dopamine (DA), are elevated in humans and test animals following exposure to inorganic lead[73] and to commercial mixtures of PCBs.[74]

Although each of these approaches has been considered separately, the soundest approach to understanding the consequences and mechanisms by which putative neurotoxicants alter CNS function is one that correlates electrophysiological, biochemical, and neuroanatomical data with information on observable and quantifiable functional changes (e.g., behavioral dysfunctions). Although this goal may not always be attainable, efforts should be made to combine as many of the above approaches as possible, whenever possible.

3.1. Dopamine Biochemistry

The neurotransmitter that has been most extensively investigated following exposure to PCBs and related HAHs is DA.[1] This neurotransmitter is involved in the initiation and control of motor behavior,[63,75–77] in the regulation of endocrine function,[78,79] and in learning and memory.[80–82] Thus, changes in concentrations of DA and/or its metabolites may mediate many of the behavioral changes that have been shown to be altered following exposure to PCBs and related HAHs.

To better understand the potential mechanisms by which environmental neurotoxicants can affect dopaminergic neurons, Fig. 1 is a schematic diagram of the synthesis and catabolism of DA. DA is normally stored in vesicles, where it is protected against catabolism by the major degradative enzymes catechol-*O*-methyltransferase (COMT) and monoamine oxidase (MAO) until it is released into the synaptic cleft following depolarization of the nerve terminal. Newly synthesized DA is stored in a physiologically releasable pool while the remaining DA is stored in essentially nonreleasable pools.[83] Following release, DA can be metabolized to 3-methoxytyramine (a minor short-lived metabolite) by extraneuronal COMT. However, the majority of the DA is taken up into the neuron by a high-affinity uptake transporter, whereupon it is susceptible to degradation to HVA or 3,4-dihydroxyphenylacetic acid (DOPAC).[84] Thus, neurotransmitter concentrations are closely regulated[67]; under physiological conditions, neuronal concentrations of DA do not vary. Estimates of dopaminergic neuronal activity, in the absence of changes in steady-state concentrations, are normally determined by measuring concentrations of its metabolites (i.e., HVA and DOPAC) and expressing the results as a ratio of HVA or DOPAC/DA. Elevations in this ratio are indicative of increased activity while decreases indicate decreased neuronal activity.[71]

Under certain conditions (e.g., prolonged exposure to environmental neurotoxicants), brain DA concentrations may be altered. These decreases may be related to changes in availability of the precursor, tyrosine[85]; decreases in the activity or concentrations of the synthetic enzymes, tyrosine hydroxylase (EC 1.14.3.a) or dopa decar-

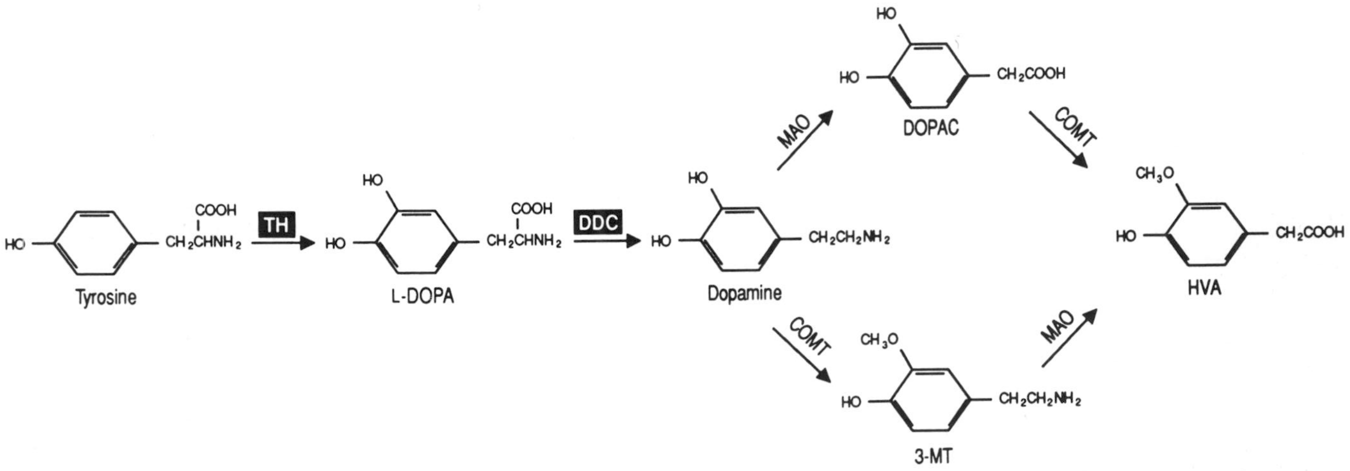

Figure 1. Diagram showing the pathway for the synthesis and metabolism of dopamine highlighting the major intermediate products and enzymes involved. Abbreviations: TH, tyrosine hydroxylase; DDC, dopa decarboxylase; MAO, monoamine oxidase; COMT, catechol-*O*-methyltransferase; L-DOPA, L-3,4-dihydroxyphenylalanine; DOPAC, 3,4-dihydroxyphenylacetic acid; 3-MT, 3-methoxytyramine; HVA, 4-hydroxy-3-methoxyphenylacetic acid.

boxylase (EC 4.1.1.26); or increases in the rates of release of DA from vesicular storage sites which would result in increased concentrations of HVA and/or DOPAC. Discussion of the possible mechanisms by which HAHs alter dopaminergic function will take place in the sections describing the experimental effects of *in vivo* and *in vitro* exposure to these compounds.

3.2. Measurement of Dopamine and Its Metabolites

The most widely used experimental technique for the measurement of concentrations of DA and its metabolites in a wide variety of biological tissues and fluids is high-performance liquid chromatography (HPLC) with electrochemical detection. This procedure combines relative ease of measurement (e.g., minimal sample preparation) with a high degree of sensitivity. Most modern systems are capable of measuring picomolar concentrations of DA and its metabolites.[86,87] Furthermore, in a single analytical run it is possible to measure concentrations of other biogenic amine neurotransmitters (e.g., epinephrine, norepinephrine, and serotonin) as well as concentrations of their metabolites (e.g., 5-hydroxyindoleacetic acid, 3-methoxy-4-hydroxyphenylglycol), thus simplifying the task of estimating neuronal activity by determining the metabolite/neurotransmitter ratios as previously discussed.

Biogenic amine neurotransmitters and their metabolites have been measured in brain tissue,[88,89] in extraneuronal fluid using *in vivo* microdialysis,[90,91] in cerebrospinal fluid,[92] in urine,[74] and in plasma.[93,94] Furthermore, these analytical procedures are ideal for measurement of biogenic amine neurotransmitters and metabolites in *in vitro* preparations such as pheochromocytoma (PC12) cells.[27]

4. NEUROCHEMICAL RESULTS: COMPARISON OF DIOXINLIKE COPLANAR AND *ORTHO*-SUBSTITUTED PCB CONGENERS

4.1. *In Vivo* Results

4.1.1. Adult Exposure

In order to determine the pharmacological or toxicological responses of the CNS to PCBs without confounding these events with additional changes that occur during development (e.g., possible alterations in maternal behavior, nutritional status), Seegal *et al.* have conducted a number of studies examining the neurochemical sequelae of exposure of the adult animal to PCBs.

In the first series of studies, adult rats were orally gavaged with a single dose of Aroclor 1254 (500 or 1000 mg/kg) and sacrificed 1, 3, 7, or 14 days after exposure to determine the relationship between brain PCB concentrations and possible alterations in biogenic amine neurotransmitter concentrations.[95–97] All biogenic amine neurotransmitter concentrations were reduced on postexposure days 1 and 3 and gradually returned to preexposure levels by day 14. Brain PCB concentrations were inversely related to neurotransmitter concentrations (i.e., they were high immediately following exposure and decreased to near preexposure levels by day 14), suggesting that, following acute exposure, relatively high levels of PCBs must be present in brain to reduce biogenic amine concentrations.

In a more recent study, adult rats were orally exposed for 30 days to powdered lab chow adulterated with either 500 or 1000 ppm of Aroclor 1254.[88] Following exposure the animals were sacrificed and regional brain concentrations of biogenic amine neurotransmitters and individual PCB congeners were determined. Brain concentrations of PCBs ranged from 55 to 82 ppm with no significant differences in either PCB concentrations or congener distribution between brain regions. Although Aroclor 1254 contains a number of coplanar and mono-*ortho*-coplanar congeners,[6] all congeners that were found in brain were di-*ortho*-substituted. Brain PCB concentrations were similar to those seen in the acute exposure experiment described above and yet neurochemical changes were limited to decreases in concentrations of DA and its metabolites in the lateral olfactory tract and the caudate nucleus. This study was the first to demonstrate neurotransmitter- and region-specific effects following exposure to PCBs and suggested that central dopaminergic neurons may be differentially sensitive to PCBs.

Similar regional and neurochemical specificities were again found in adult nonhuman primates following 20-week exposure to either Aroclor 1016 or Aroclor 1260 at doses of 0.8, 1.6, or 3.2 mg/kg per day.[89] This exposure paradigm neither affected the body weight of the animals nor induced any signs of chlorobiphenyl poisoning. However, both Aroclor mixtures significantly reduced DA concentrations in the caudate, putamen, substantia nigra, and hypothalamus. No changes were detected in concentrations or activity of other biogenic amine neurotransmitters. As seen in the subchronic rodent study, all PCB congeners in brain were either mono- or di-*ortho*-substituted.

Subsequent to this study we exposed pheochromocytoma (PC12) cells, a continuous cell line derived from a rat adrenal gland tumor that synthesizes, stores, releases, and metabolizes DA in a manner similar to that of the mammalian CNS,[98,99] to the congeners that accumulated in nonhuman primate brain following exposure to Aroclor 1016 (i.e., 2,4,4′,2,4,2′,4′, and 2,5,2′,5′). All congeners, when presented either singly or as a mixture that reflected the congener ratios in nonhuman primate brain, significantly reduced cellular DA concentrations.[28] These results suggest that the PCB congeners found in the brains of the Aroclor 1016-exposed animals are capable of reducing cellular DA content and may be responsible for the observed *in vivo* decreases in brain DA concentrations.

An additional study carried out in nonhuman primates was designed to determine whether the observed decreases in brain DA concentrations persisted following removal of the animals from PCBs. Seegal *et al.*[100] exposed animals, on a daily basis, to 3.2 mg/kg of either Aroclor 1016 or 1260 for 20 weeks. PCB exposure was then terminated and serum PCB concentrations were measured for an additional 24 weeks postexposure to provide an index of changes in PCB body burden. At the time of sacrifice, brain PCB and DA concentrations were determined. Both serum and brain PCB concentrations decreased by 50–75% (Fig. 2) although brain DA concentrations

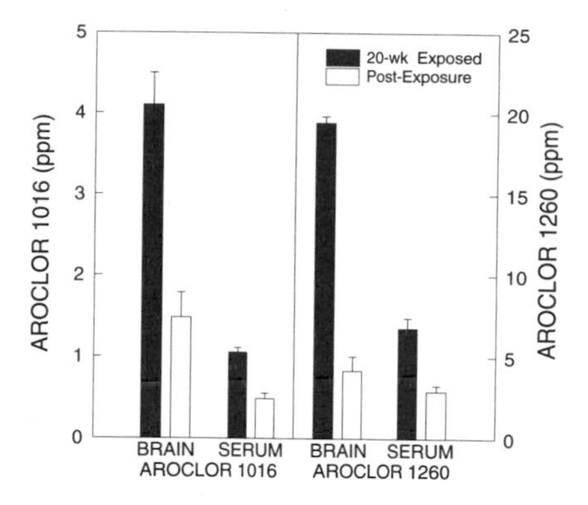

Figure 2. PCB concentrations in brain and serum from nonhuman primates sacrificed either immediately following 20-week exposure to 3.2 mg/kg per day of either Aroclor 1016 or 1260 or 24 weeks following a similar 20-week exposure period. N = 4–5 animals per treatment group.

were indistinguishable from those seen in the animals sacrificed immediately following exposure, providing strong evidence that subchronic exposure of the adult animal to PCBs results in long-term, if not permanent, decreases in DA content.

4.1.2. Perinatal Studies

The earliest study suggesting that PCB exposure alters CNS neurochemical function was that of Chou *et al.*[64] In that study mice were exposed on gestational days 10–16 to 3,4,3′,4′-tetrachlorobiphenyl (TCB), an archetypical coplanar congener, at a dose of 32 mg/kg per day. Several toxic effects, including increased pup mortality and an increase in the number of stillborn pups, were observed. In the surviving offspring, approximately half of the animals exhibited a syndrome of rapid spinning and head bobbing and were designated *spinners*. Of immediate interest to this review, administration of *d*-amphetamine, an indirect DA agonist, to hyperactive TCB-exposed mice decreased their activity level, a paradoxical effect similar to that observed in neonatally 6-OHDA-lesioned rats.[101,102] These early findings were important because they were the first to suggest altered DA function as a possible mechanism for the PCB-induced behavioral effects.

Additional studies of perinatal exposure of mice to TCB have been carried out by Tilson *et al.* and Agrawal *et al.* Tilson *et al.*[103] determined that perinatal exposure to TCB altered the behavior of the offspring while Agrawal *et al.*[104] noted decreases in both striatal DA concentrations and the density of DA receptor binding sites that were detected up to 1 year postexposure. More recently, Eriksson[105] exposed 10-day-old mice to a single intraperitoneal injection of either 0.41 or 41 mg/kg of TCB, which resulted in significant decreases in the density of muscarinic acetylcholine receptors in hippocampus measured using the muscarinic antagonist [^{3}H]-QNB. In a later study, Eriksson *et al.*[106] found altered patterns of motor activity in adult mice that had been exposed to the same doses of TCB on postnatal day 10. However, unlike the earlier findings of hyperactivity reported by Chou *et al.*,[64] Tilson *et al.*,[103] and Agrawal *et al.*,[104] Eriksson and co-workers reported that the mice were initially hypoactive and failed to habituate to the test apparatus over time. Schantz *et al.*[107] have also reported hypoactivity in rats whose dams were exposed to 2 or 8 mg/kg per day of TCB on days 10–16 of gestation.

The above studies have demonstrated that perinatal exposure of mice to TCB induces behavioral and neurochemical changes in the offspring that persist into adulthood. Because both dopaminergic and cholinergic effects were observed, it is likely that perinatal exposure to TCB affects at least two neurochemical sites of action, one affecting dopaminergic function and the other decreasing the density of muscarinic acetylcholine receptors. Additional efforts should be devoted to determining whether other neurotransmitters are altered and whether these changes are secondary to changes in DA neurochemistry.

A note of caution concerning the dose levels of TCB used in these studies must be considered. Both Chou and Tilson reported large numbers of deaths and stillbirths following perinatal exposure to TCB, indicating that this dose of TCB was fetotoxic. Furthermore, in an attempt to replicate the above findings in rats, Seegal *et al.* (unpublished observations) exposed pregnant rat dams to 32 mg/kg of TCB during gestational days 10–16 and also noted very high rates of pup mortality. Because of the fetotoxicity (and perhaps associated maternal toxicity), combined with the bias in the selection of pups for testing as a result of the high rate of TCB-induced pup mortality, it is difficult to conclude that the behavioral and neurochemical sequelae reported in the early Chou, Tilson, and Agrawal studies were direct effects of the TCB. Both Schantz's and Eriksson's behavioral effects were in the opposite direction from those reported in the earlier studies (i.e., hypo- rather than hyperactivity), but they used lower, nonfetotoxic doses of TCB. Interestingly, the later findings are more consistent with the activity changes reported in PCB-exposed children.[48]

In spite of the persistent alterations in nervous system function observed following perinatal exposure to TCB, Tilson *et al.*[103] were unable to detect significant changes in either behavior or neurochemistry in the adult animal, regardless of the dose of TCB they used. These results suggest that TCB, and perhaps other coplanar HAHs, are neuroteratogens, i.e., compounds that alter the development and function of the nervous system when administered during development but are without effect when administered to adult animals.

More recently, Seegal[108] conducted a series of studies examining the effects of perinatal (gestational, lactational, and gestational/lactational) exposure of rats to a commercial mixture of PCBs (Aroclor 1016). Dams were exposed to adulterated powdered chow containing 30, 100, or 300 ppm of Aroclor 1016. No significant effects on either pup or maternal body weight or mortality were noted. Offspring were sacrificed on postnatal days 25, 35, 60, and 90; biogenic amine concentrations were determined in the caudate nucleus, nucleus accumbens, frontal cortex, and hippocampus. Unlike the effects observed following either oral exposure of the adult nonhuman primate to Aroclor 1016 or intrastriatal injection of individual *ortho* PCB congeners into the adult rat, which resulted in decreases in regional DA concentrations (see below), perinatal exposure of the rat to Aroclor 1016 resulted in increases in neurotransmitter content, including elevations in DA, 5-HT, and norepinephrine (Fig. 3). Because Aroclor 1016 is composed primarily of lightly chlorinated *ortho*-substituted congeners,[6] we assumed that the differences in direction of change in neurotransmitter concentrations between

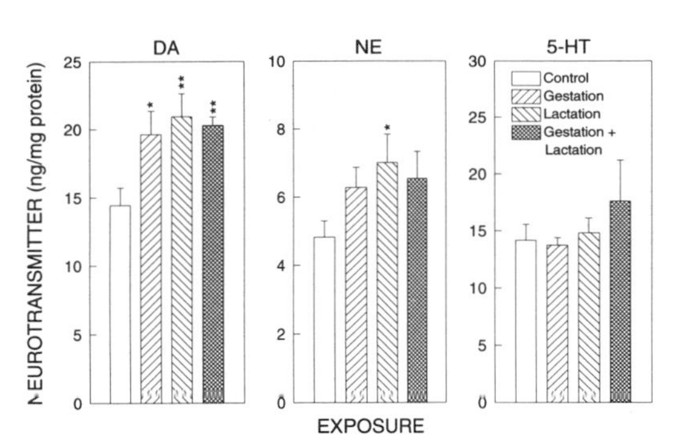

Figure 3. Dopamine, norepinephrine, and serotonin concentrations in the substantia nigra of female offspring perinatally exposed to Aroclor 1016. Dams were exposed to 100 ppm adulterated chow during gestation (GD 8–21), lactation (D 0–21), or gestation + lactation; offspring were sacrificed on postnatal day 60. N = 6–10 animals per treatment group. Significantly different from control: $*p \leq 0.05$, $**p \leq 0.01$.

adult-exposed and perinatally exposed animals were related to the congeners to which the animals were exposed rather than the age of the animal at the time of exposure, species differences, or differences in the route of administration of the PCBs (e.g., *ortho*-substituted versus the coplanar PCB results of Agrawal *et al.*[104] However, recent data from our laboratory, as well as data from Brouwer and colleagues (personal communication), suggest that perinatal exposure to low concentrations of TCB (1–10 mg/kg) also results in elevations in brain DA concentrations. These findings suggest that it may be the timing of exposure that is important—with perinatal exposure resulting in increased neurotransmitter concentrations and adult exposure resulting in decreased concentrations regardless of the class of congeners to which the offspring are exposed. We are continuing to investigate the interactions between the congeners to which the animals are exposed and the age of the animal at the time of exposure.

The above results not only underscore the importance of using several doses of a neurotoxicant (large doses of TCB decrease DA while low doses may increase DA) but also demonstrate the differences that may exist between the mature and developing nervous system in their responses to the same neurotoxicants. Possible mechanisms by which these compounds differentially affect neurotransmitter function in the developing and adult animal will be discussed in a later section.

4.2. *In Vitro* Results

The above series of studies have clearly demonstrated that exposure of either the developing or adult nervous system of rodents and nonhuman primates to PCBs results in long-term changes in CNS concentrations of DA. However, because of the complexity of the mammalian nervous system (e.g., the presence of a blood–brain barrier that may actively inhibit the movement of certain compounds into brain, together with the presence of many different cell types and different neurotransmitter systems) and the fact that orally administered PCBs are subject to hepatic metabolism, it is difficult to determine whether the changes in DA function observed *in vivo* are: (1) related to the action of the parent congener(s), (2) related to the action of metabolites of PCBs, or (3) mediated by changes in other neurotransmitter or endocrine systems that then interact with DA.

One approach to this problem is to use an *in vitro* system composed of a single cell type that expresses a single phenotype—in this instance the synthesis, storage, release, and metabolism of catecholamines, and in particular DA. Tissue-culture procedures provide a means of directly assessing the action of PCBs since there is no blood–brain barrier to exclude access of PCBs to the cells in culture nor evidence of substantial metabolism of PCBs *in vitro*. Finally, assays are performed in a closed environment where only small volumes of test solution are required and where there are no mechanisms for the removal of DA or its metabolites from the culture dish. Thus, it is a fairly simple task to experimentally determine whether the observed decreases in cellular DA concentrations are related to: (1) increased release of DA into the media, (2) increased catabolism of DA to its metabolites HVA and DOPAC, or (3) inhibition of the synthesis of DA.

Pheochromocytoma (PC12) cells, derived from a rat adrenal gland tumor,[99,109] represent such a system. They have been used by Seegal and Shain in a series of studies to investigate the regulation of catecholamine synthesis and metabolism, the structure–activity relationship of PCB congeners and the potential mechanisms by which PCBs alter dopaminergic function.

Seegal *et al.*[27] first exposed PC12 cells to Aroclor 1254, a commercial mixture composed primarily of tetra- to hepta-chlorinated PCB congeners. Aroclor 1254 resulted in dose- and time-dependent decreases in cellular DA content that resembled the decreases we had previously observed *in vivo*. Accompanying these decreases in cellular DA content were parallel decreases in media concentrations of DA and its metabolites (i.e., HVA and DOPAC) (Table 1). These results, in addition to demonstrating a direct effect of PCBs on cellular DA content, strongly suggest that the PCB-induced decreases are not related to increased release of DA from the cells or enhanced metabolism of DA, but instead are related to a decrease in the rate of synthesis of DA.

Although the above results clearly demonstrate that exposure of PC12 cells to a commercial mixture of PCB congeners results in significant decreases in cellular DA content, the experiments provided no information on possible differences in the activity of structurally disparate PCB congeners. This information is vitally important because PCBs of environmental origin exist as complex mixtures that differ from the commercial mixtures in their congener makeup. Thus, information on the differential toxicity of individual PCB congeners will allow a more realistic estimate of the risk to humans following consumption of foodstuffs contaminated with structurally disparate PCB

Table 1
Neurotransmitter and Metabolite Concentrations in PC12 Cells and Growth Media (Expressed as a Percent of DMSO Control)

Aroclor 1254 Exposure	Cellular Dopamine	Media		
		Dopamine	DOPAC	HVA
10 ppm	60	51	63	64
100 ppm	28	50	56	63

Table 2
IC_{50} Values (μM) for *Ortho*-Substituted and Coplanar PCB Congener-Mediated Decrease in Cellular Dopamine Content

Ortho-substituted congener	IC_{50}	Coplanar congener	IC_{50}
2,2′	64	4,4′	>1000
2,4,6,2′	71	3,4,3′,4′	>1000
2,5,2′	83	3,4,5,3′,4′	>1000
2,4,6,3′	90		
2,5,2′,5′	92		

congeners, as well as providing insights to the possible mechanisms by which PCBs alter cellular DA content.

Shain *et al.*[110] have recently carried out experiments to determine the structure–activity relationships of more than 50 PCB congeners. These results demonstrate that *ortho*-substituted congeners (Table 2) are the most potent, while *meta*- and *para*-substituted, dioxinlike congeners are totally inactive and suggest that the mechanisms by which PCB congeners alter DA function *in vitro* differ from those that result in tumor promotion,[111,112] hepatic[113] or immunotoxic effects[114,115] since these effects require the interaction of laterally substituted, dioxinlike congeners with the Ah receptor.[24]

Seegal *et al.*[116] have begun to investigate the locus and mechanisms by which the most potent PCB congener (2, 2′) inhibits the synthesis of DA *in vitro* using neuroblastoma cells (N1E-N115). These cells lack the enzyme, aromatic amino acid decarboxylase, needed for the conversion of *l*-Dopa to DA and hence produce large quantities of *l*-Dopa. Because *l*-Dopa is not stored in vesicles, it therefore "leaks" into the growth medium. If PCBs inhibit DA synthesis by inhibiting the rate-limiting enzyme, tyrosine hydroxylase, medium concentrations of *l*-Dopa should be decreased. When the neuroblastoma cells were exposed to 2, 2′, there were significant decreases in medium concentrations of *l*-Dopa (Fig. 4) suggesting that, at least in tissue culture, PCBs appear to decrease cellular DA concentrations by inhibiting the activity of the rate-limiting enzyme, tyrosine hydroxylase.

One of us (R.F.S.) is also undertaking experiments in whole animals to determine whether the structure–activity relationships and the mechanisms by which PCBs decrease brain DA concentrations resemble those we have observed in tissue culture.

Although the above results, obtained in both *in vivo* and *in vitro* preparations, provide insights into the potential mechanisms by which PCBs alter CNS function, it is also necessary to determine whether these changes are sufficient to induce behavioral changes in assessing the relative risks to humans from exposure to these compounds. The following section provides a selective review of the current literature on the behavioral effects of perinatal exposure of nonhuman primates to Aroclor mixtures and dioxins. In combination with the above biochemical data, the behavioral findings provide an important set of data that support the epidemiological findings that PCBs

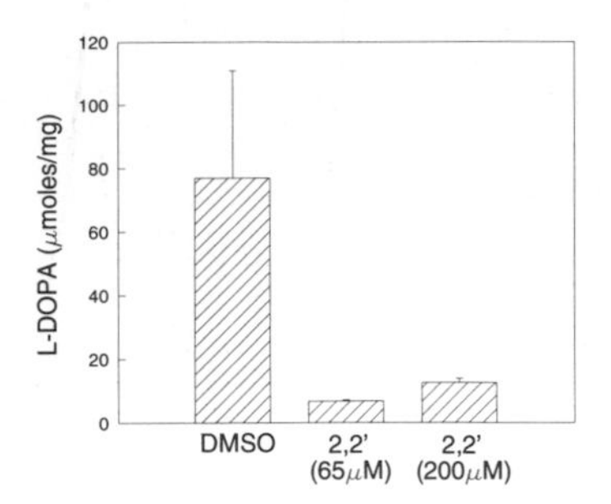

Figure 4. *l*-Dopa production and release into media by N1E-N115 cells in culture following 1-hr exposure to media containing varying concentrations of 2,2′-dichlorobiphenyl. DMSO served as the vehicle at a final concentration of 0.1%. $N = 4$ wells per treatment condition.

and related HAHs may be human neurotoxicants as well as providing a basis for understanding the alterations in human CNS function.

5. BEHAVIORAL RESULTS: COMPARISON OF PCBs AND TCDD

5.1. Groups and Exposure

Schantz and colleagues have conducted a series of laboratory studies assessing the long-term effects of perinatal exposure to HAHs on behavioral function in monkeys.[32,33,117–123] In these experiments, adult female rhesus monkeys were fed experimental diets containing either PCBs (Aroclor 1016 or Aroclor 1248) or TCDD for a period of months and then bred to produce offspring. Control subjects were age-matched monkeys born to unexposed mothers. The PCB and TCDD exposure groups discussed in this chapter are summarized in Table 3. These groups have been described in more detail in other published reports.[121,124]

As shown in Table 3, some experimental groups were born and nursed concurrently with maternal PCB or TCDD exposure, whereas others were born and nursed after the maternal exposure had ended. In all cases, the offspring were raised with their biological mothers until weaning. Thus, exposure to PCBs or TCDD was both transplacental and transmammary. Because an important goal of this review is to compare and contrast the behavioral effects of PCBs and TCDD, we have limited our discussion to those PCB- and TCDD-exposed groups that were tested on the same series of behavioral tests.

Given this goal, it is instructive to estimate how the maternal PCB and TCDD doses used in these studies compare in terms of TCDD toxic equivalency factors (TEFs). TEFs have been estimated for many of the commercial PCB mixtures. The two PCB mixtures used in our monkey studies differ substantially in congener makeup and in TEFs. Aroclor 1016 is made up entirely of lightly chlorinated, *ortho*-substituted congeners and is assumed to have a TEF of zero, with little or no TCDD-like toxicity. In contrast, Aroclor 1248 is a more highly chlorinated mixture with a high percentage of tetra- and penta-chlorinated congeners, including some that have Ah activity. Various authors have cited TEFs of around 0.00002 for Aroclor 1248.[26,125] Using the TEF

Table 3
Summary of PCB and TCDD Exposure Groups

Maternal exposure condition	Duation of exposure (months) (mean ± S.D.)	Total intake (mean ± S.D.)	Maternal exposure status at time of birth	Live births	Subjects tested
PCB		mg/kg			
2.5 ppm Aroclor 1248 (1)	18.2 ± 1.7	52.90 ± 9.30	1.5 years postexposure	7/8	4
2.5 ppm Aroclor 1248 (2)	18.2 ± 1.7	48.55 ± 3.24	3 years postexposure	3/7	3
0.25 ppm Aroclor 1016	21.8 ± 2.2	4.52 ± 0.56	Concurrent with exposure	7/7	7
1.0 ppm Aroclor 1016	21.8 ± 2.2	18.41 ± 3.64	Concurrent with exposure	7/7	6
Control 1	—	—	Unexposed	7/7	5
Control 2	—	—	Unexposed	6/6	6
TCDD		ng/kg			
5.0 ppt TCDD (1)	16.2 ± 0.4	59.6 ± 5.0	Concurrent with exposure	6/7	6
5.0 ppt TCDD (2)	36.3 ± 0.2	141.0 ± 9.0	Concurrent with exposure	5/6	5
5.0 ppt TCDD (3)	42.0 ± 0.0	163.0 ± 8.0	1.5 years postexposure	7/8	7
25.0 ppt TCDD	48.0 ± 0.0	938.0 ± 36.0	1.5 years postexposure	3/4	3
Control 1	—	—	Unexposed	7/7	7
Control 2	—	—	Unexposed	3/4	3
Control 3	—	—	Unexposed	7/8	6

of 0.0000173 cited by Smith *et al.*,[125] we calculate that the 2.5-ppm Aroclor 1248 dose we used is roughly equivalent to a dose of 43 ppt TCDD. Thus, in terms of concentrations of the contaminants in the maternal diets, our 2.5-ppm Aroclor 1248 dose group was quite close to our 25-ppt TCDD dose group. Another way to calculate the equivalent TCDD dose, taking into account the duration of exposure, is based on the total PCBs consumed by the animals rather than the concentration of PCBs in the maternal diet. When TEFs were calculated based on the total amount of PCBs consumed, the TCDD equivalent dose for the Aroclor 1248 mothers is approximately 90 ng/kg, closer to the total amount of TCDD consumed by the 5-ppt TCDD mothers (Table 3). Using either method of calculation, the Aroclor 1248 exposure group was quite similar to the TCDD exposure groups in terms of TEFs, suggesting that comparisons of the behavioral effects seen in the Aroclor 1248-exposed and TCDD-exposed offspring can legitimately be made. If the TEF concept is valid for predicting the degree of nervous system impairment induced by complex PCB mixtures, effects should be qualitatively and quantitatively similar in the Aroclor 1248- and TCDD-exposed offspring, and absent altogether in the Aroclor 1016-exposed groups. However, the TEF calculations ignore the potential contribution the *ortho*-substituted PCBs may make to the overall risk of CNS dysfunction. The present comparison of behavioral changes seen following exposure to TCDD, Aroclor 1248, and Aroclor 1016 (which contains only *ortho*-substituted congeners) should aid in determining the utility of the TEF concept for assessing HAH-induced CNS changes.

5.2. Behavioral Tests

5.2.1. Tests and Apparatus

The learning ability of the PCB- and TCDD-exposed monkey offspring described in Table 3 was evaluated in a commonly used monkey test apparatus known as the Wisconsin General Testing Apparatus (WGTA).[126] As shown in Fig. 5A and B, the WGTA consists of a table with movable screens at each end and a movable tray that can be rolled across the table between the two screens. The monkey is placed in a cage behind an opaque screen at one end of the apparatus and the operator sits behind a screen with a one-way mirror at the other end. Stimulus objects are placed on the tray and a food reward is concealed beneath the correct object. On each test trial, the opaque screen is raised, the tray is moved to within the monkey's reach, and the monkey is allowed to displace one stimulus object and obtain the food reward, if the response is appropriate. The screen in front of the monkey is then lowered, the tray is drawn back, and the correct object is baited for the next trial.

The most common type of problems used with the WGTA are spatial and object discrimination and reversal problems (RL problems). Figure 5A shows a monkey presented with a spatial RL problem. The tray holds two identical square blocks and either the right or the left one is correct: the monkey must determine which stimulus is correct. Figure 5B shows a monkey presented with an object RL problem. Either the

Figure 5. (A) Monkey doing a simple spatial discrimination in a WGTA. (B) Monkey doing an object discrimination problem in a WGTA. (Both reprinted with permission from Levin *et al.*[127])

circle or the plus is correct and the monkey must determine which is correct. Other object quality discriminations use color or pattern as the salient cues. Stimulus dimensions can also be present, but unrelated to the solution of the problem. For example, the monkey could be presented with a problem in which the same spatial position is rewarded on every trial, but the objects shown in Fig. 5B could be used. The position of the objects would be varied from trial to trial so that each is associated with the correct spatial position 50% of the time. The presence of irrelevant stimulus dimensions generally makes the problems more difficult to solve because the monkey must first determine which dimension (i.e., object or spatial position) is rewarded, and then, within the correct dimension, it must determine which choice (i.e., circle or plus/left or right) is correct.

In discrimination reversal learning, the monkey is usually trained on a particular spatial or object discrimination task until it makes 9 or 10 consecutive correct responses. Then a reversal is made. That is, the previously incorrect position or object becomes correct. Typically a series of these reversals are given. Each time the monkey achieves the learning criterion of 9 or 10 correct, the other position or object becomes correct. Typically the first reversal is the most difficult for the monkey to learn. Performance becomes progressively better over subsequent reversals and nearly error-free performance is achieved after 5–10 reversals.

Another type of learning problem that can be presented in the WGTA is alternation training. Conceptually, alternation training is the logical extreme of reversal learning, where the correct position or object is alternated on every trial. In our PCB and TCDD experiments we used a spatial alternation procedure [delayed spatial alternation (DSA)] in which the location of the reward was alternated on every trial. First the left position was rewarded, then the right, then the left again, and so on. Alternation paradigms offer the opportunity to test short-term memory function. This is done by imposing delays of varying lengths between the trials and examining percent correct performance over progressively longer delays. Typically there is a gradual decline in correct performance as the intertrial delays get longer.

5.2.2. Advantages of Using RL and DSA in Neurotoxicology

Although training monkeys on complex learning tasks such as RL and DSA is time-consuming and difficult, there are several advantages to using these tests in neurotoxicology testing. First, both tests have been used extensively in studies designed to assess the role of various discrete brain structures in learning and memory.[127] As a result of these studies, the brain areas and neural circuits critical for accurate performance on RL and DSA are quite well understood. Decrements in RL performance are seen with specific lesions of the frontal and temporal lobes[128,129] and the mammillary bodies.[130] On the other hand, decrements in DSA are related to specific prefrontal cortical lesions[128] and to disruption of dopaminergic innervation of the prefrontal cortex.[81] Decrements in RL and DSA are also seen following lesions to subcortical nuclei such as the medial dorsal nucleus of the thalamus and the caudate nucleus that have direct connections with the prefrontal cortex.[131,132] Because the pattern of deficits seen on these tasks varies depending on the site of the lesion, the pattern seen after toxicant exposure may be useful in differentiating which brain regions or neural circuits have been altered. Alternatively, if neurochemical or neuropathological effects have been identified in discrete brain regions, RL or DSA may be useful in determining whether the damage is severe enough to be of functional significance to the animal.

Second, these tasks assess complex behavioral functions that are not species-specific. For example, both paradigms have been used successfully in a wide range of species including birds, rodents, monkeys, and humans.[133,134] Furthermore, if RL performance is expressed as a ratio score which relates reversal performance to original learning,[135,136] it is possible to avoid the problem of trying to extrapolate absolute performance levels across divergent species. Third, human tests of cognitive function exist that are remarkably similar to RL and DSA. For example, the Wisconsin Card Sorting Test (WCST) is a widely used standard clinical test for humans which has task requirements that are nearly identical to those of RL as it is presented in the WGTA.[134] In the WCST, the subject must match cards from a deck to a set of four stimulus cards based on the color, shape, or number of items printed on the card, with no external cues to guide responding. The examiner simply replies "right" or "wrong" after each card is placed, and the subject must determine which dimension (color, shape, or number) is correct by trial and error. After he or she has learned which is correct and has made 10 correct card placements, the correct dimension changes without notice and the subject must again use trial and error to determine the correct response. As with RL, a series of these "reversals" are given. Just as RL deficits have been related to frontal cortical damage in monkeys, impaired performance on the WCST has been shown to be a sensitive indicator of frontal lobe dysfunction in humans.[134] Currently, the WCST is being used by Jacobson and Jacobson in a long-term follow-up of their PCB-exposed cohort of children in Michigan. The similarity of the WCST to RL should greatly facilitate comparisons of our laboratory results in monkeys to epidemiological findings in humans exposed perinatally to PCBs and other contaminants found in Lake Michigan fish.

5.3. Behavioral Results

5.3.1. Discrimination–Reversal Learning

At approximately 14 months of age (10 months after PCB or TCDD exposure ended), the Aroclor 1016-, Aroclor 1248-, and TCDD-exposed monkeys were trained on a series of four RL problems, including spatial discrimination tasks with and without presentation of irrelevant color and shape cues, a color discrimination task, and a shape discrimination task. For each problem, original learning was followed by a series of reversals.

5.3.1a. PCB-Exposed Groups. As illustrated in Fig. 6A, monkey offspring whose mothers were fed a diet containing 1.0 ppm Aroclor 1016 were impaired in their ability to learn the simple left–right spatial RL task.[121] They required more trials than did their age-matched controls to learn the original discrimination and the first reversal, but were not impaired on later reversals. Offspring whose mothers were fed a lower dose of Aroclor 1016 and offspring whose mothers had a prior history of exposure to 2.5 ppm Aroclor 1248 were not impaired on spatial RL. In an earlier study, offspring that were born concurrent with the maternal exposure to the 2.5-ppm Aroclor 1248 diet

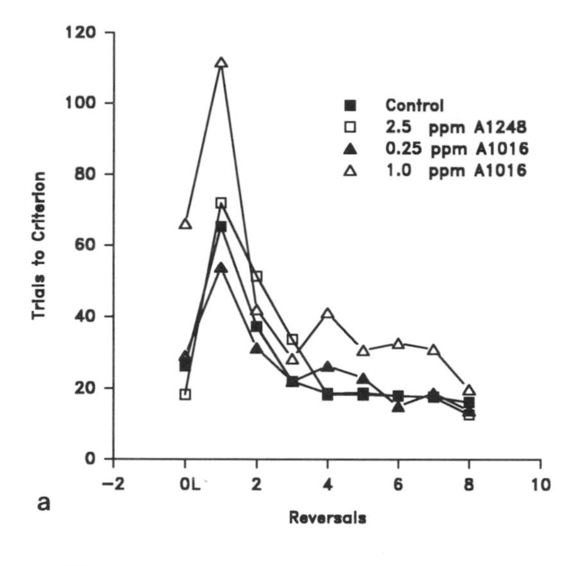

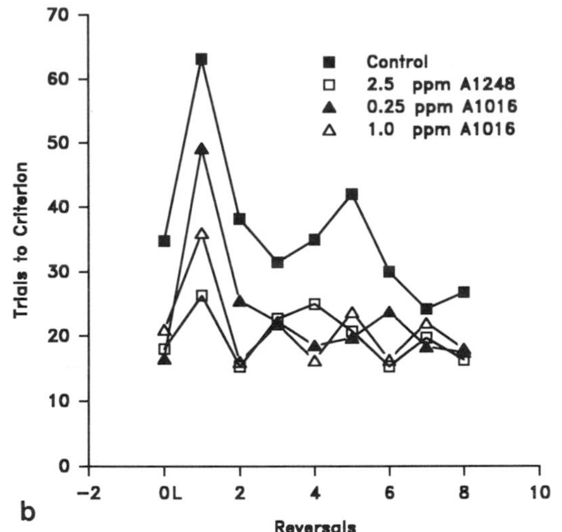

Figure 6. Performance of Aroclor 1016- and Aroclor 1248-exposed monkeys on (A) spatial discrimination–reversal learning and (B) object discrimination–reversal learning. Graphs show trials to criterion on original learning and eight reversals.

showed significant mortality. Infants that survived postweaning were tested on discrimination–reversal learning and were found to be profoundly impaired on spatial RL.[117]

Paradoxically, after training on a second spatial task which included irrelevant color and shape cues, both the Aroclor 1016-exposed offspring and the offspring whose mothers had a prior history of exposure to 2.5 ppm Aroclor 1248 learned object RL problems in fewer trials than age-matched controls.[121] Figure 6B shows trials to criterion for the shape RL. The results were similar for the color RL, but did not quite reach statistical significance.

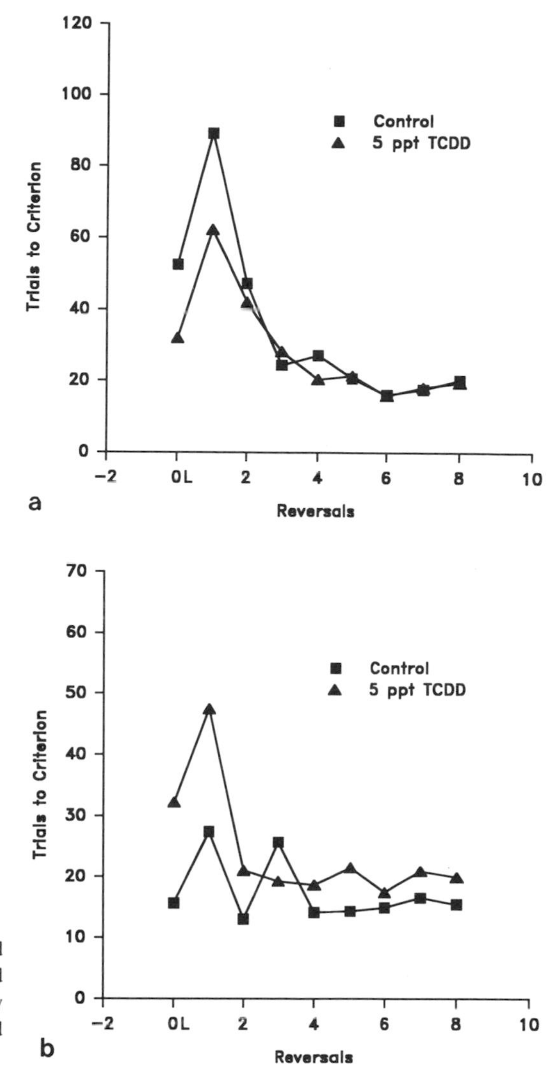

Figure 7. Performance of TCDD-exposed monkeys on (A) spatial reversal learning and (B) object reversal learning. Graphs show trials to criterion on original learning and eight reversals.

5.3.1b. TCDD-Exposed Groups. Figure 7 shows the effects of 5 ppt TCDD on the same spatial and object RL tasks. As Fig. 7 (see also Table 4) illustrates, the effects seen in the TCDD-exposed monkeys were qualitatively different from those seen in the PCB-exposed monkeys. In fact, the pattern was just the reverse of that seen in the PCB-exposed monkeys! TCDD-exposed monkeys were facilitated on simple spatial RL and impaired on object RL. In contrast, the pattern *was* similar to that for PCBs in the sense that these effects were present on original learning and the first reversal, but not on later reversals.

The 5-ppt TCDD monkeys born after maternal exposure were not different from

Table 4
Comparison of Aroclor 1016-, Aroclor 1248-, and TCDD-Exposed Monkeys on RL and DSA

	Spatial RL	Object RL	DSA
Aroclor 1016	↓	↑	↓
Aroclor 1248	—	↑	↓
TCDD	↑	↓	—

controls on either spatial or object RL. However, the 25-ppt monkeys born after maternal exposure had ended did show the expected pattern of effects. They were facilitated on spatial RL and impaired on object RL.

5.3.2. Delayed Spatial Alternation

The PCB- and TCDD-exposed monkeys have also been tested on DSA. The monkeys were presented with two identical objects (e.g., red square blocks) and the spatial position of the reward was alternated back and forth between the two. Delays of 5, 10, 20, or 40 sec were interposed between trials in counterbalanced order. The monkeys were tested on DSA for 80 test sessions when they were 4–6 years of age.

The DSA results for the Aroclor 1016-exposed monkeys were subtle. Neither PCB-exposed group differed significantly from controls. However, the two PCB-exposed groups did differ significantly from each other. The difference between the 0.25- and 1.0-ppm Aroclor 1016 groups on the DSA task was related to a combination of facilitated performance in the 0.25-ppm group and impaired performance in the 1.0-ppm group. Interestingly, this kind of biphasic effect is consistent with what has been seen in lead-exposed monkeys on the DSA task. Monkeys exposed to low levels of lead perinatally exhibit somewhat facilitated DSA performance,[137] whereas animals exposed to higher levels of lead show deficits on DSA.[138]

In contrast to Aroclor 1016, monkey offspring born to mothers with a prior history of exposure to Aroclor 1248 exhibited a pronounced deficit in DSA choice accuracy when they were tested at 4–6 years of age (Fig. 8A).[33] The figure shows percent correct performance averaged across all four intertrial delays. Even after prolonged testing, the PCB-exposed monkeys never reached control levels of performance.

This profound deficit in DSA performance is particularly striking when one considers that the monkeys were 4 to 6 years of age at the time of testing and had not been exposed to PCBs since they were weaned at 4 months of age. Given that the monkeys were young adults at the time of testing, it is quite likely that the deficit represents a permanent consequence of early PCB exposure. It is also striking that these effects were detected in monkey offspring born 1.5 to 3.0 years after maternal PCB exposure ended. These data suggest that, although some recovery of learning function may occur, offspring remain at risk for certain types of neurobehavioral effects long after maternal PCB exposure has ended.

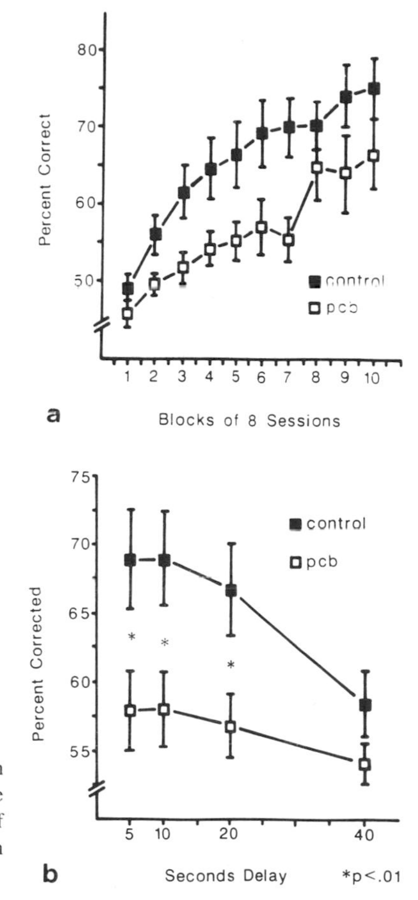

Figure 8. Performance of Aroclor 1248-exposed monkeys on delayed spatial alternation. (A) Percent correct performance over test sessions; (B) percent correct performance at each of the four different intertrial delays. (Reprinted with permission from Levin *et al.*[33])

Figure 8B shows percent correct performance for the Aroclor 1248 monkeys at each of four different intertrial delays. Significant PCB-related deficits were observed when there was a 5-, 10-, or 20-sec delay between trials, but not when there was a 40-sec delay between trials. Interestingly, the deficit was just as pronounced at the 5-sec delay as it was at the 10- and 20-sec delays. The lack of an effect at the 40-sec delay was probably related to the fact that control performance fell to near-chance levels when there was a 40-sec delay between trials.

In contrast to the PCB results, TCDD does not seem to impair DSA performance (Table 4). In general, TCDD-exposed animals performed slightly, though not significantly, better than controls. These results are consistent with the facilitated spatial RL observed in TCDD-exposed animals, and serve to further highlight the difference in behavioral effects in TCDD- and PCB-exposed animals. Given that the Aroclor 1248 dose used was similar to the TCDD doses used on a TEF basis, these findings of severely impaired DSA performance in Aroclor 1248-exposed animals and little or no effect in TCDD-exposed animals suggest that the PCB-related deficit in behavioral function may not be Ah receptor-mediated. The qualitative differences between the PCB mixtures and TCDD on RL further support the thesis that PCBs and TCDD may alter behavioral function through different mechanisms. If so, the concept of TCDD TEFs may not be appropriate for estimating neurobehavioral outcomes from exposure to complex PCB mixtures.

5.4. Discussion of Behavioral Data

Monkeys with lesions to the dorsolateral area of the prefrontal cortex (DLPFC) show a pattern of deficits on spatial RL very similar to that observed in the 1.0-ppm Aroclor 1016-exposed offspring.[128] Lesioned monkeys are impaired on the original discrimination and the early reversals, but are unimpaired on later reversals. In contrast, animals with damage to other brain areas such as the temporal lobe or mammillary bodies are impaired across all of the reversals of spatial RL problems.[129,130] The similarity of effects in PCB-exposed and DLPFC-lesioned animals suggests that PCBs may alter the function of the DLPFC or of one of the input or output pathways that comprise the neural circuitry of the DLPFC. An important brain stem projection to the DLPFC is the mesocortical dopamine pathway. As discussed previously, brain stem dopamine input to the DLPFC has an important influence on spatial learning and memory. Specific chemical lesions that disrupt dopamine input to the DLPFC have been shown to cause a specific pattern of spatial learning deficits nearly identical to those seen following surgical ablation of the DLPFC itself.[81,134,139] Since PCBs alter dopamine levels and dopamine turnover in the brain stem and in other brain areas, it is likely that PCBs may cause deficits in spatial RL by altering the dopamine input to the DLPFC.

The improved object RL performance seen in the PCB-exposed monkeys is also consistent with this interpretation since improved object RL has also been reported in monkeys with lesions of the DLPFC.[128] Because animals with damage to this brain region do not attend well to spatial cues, they are less distracted by the spatial positions of the objects and are thus able to focus more exclusively on other dimensions such as shape or color and can learn to discriminate them more readily.

The pattern of effects on RL was different for TCDD. Spatial RL was not impaired, whereas object RL was. The prefrontal cortex would appear to be a possible site of action for TCDD, because the deficit was present only on original learning and the first several reversals, which resembles the changes seen with damage to that area. However, with TCDD, the orbital prefrontal cortex, which is important for object learning,[140,141] would be the more likely site of action. Our current knowledge of the anatomical connections of the orbital prefrontal area is less detailed and precise than our knowledge of the dorsolateral area, so it is more difficult to speculate how TCDD might differentially affect the functioning of this area. However, it is clear that the orbital and dorsolateral prefrontal areas are anatomically and functionally distinct,[134] while one interesting difference between the two is in their descending projections. The orbital cortex projects to basal forebrain cholinergic areas,[142] whereas the dorsolateral area projects to brain stem catecholaminergic areas.[143] Unfortunately, we also know very little about the neurochemical effects of TCDD in the brain, so it is very difficult to form any hypotheses about potential mechanisms at this point.

The DSA results also suggested the DFPLC as a site of damage. Studies of brain function in both rats and monkeys have shown that the particular pattern of DSA deficit that we observed in PCB-exposed monkeys, in which the effect is as pronounced at very short intertrial delays as it is at longer delays, is a sensitive indicator of dysfunction in the DLPFC.[128,139] Furthermore, specific chemical lesioning of the mesocortical dopaminergic projection to the DLPFC impairs DSA nearly as much as cortical ablation does,[81,139] strongly suggesting a functional role of the mesocortical dopamine system in control of DSA. DSA deficits are also seen in monkeys after dopamine receptor blockade via haloperidol administration.[144] The findings specifically linking DSA deficits to altered dopamine function in the DLPFC strengthen the hypothesis stated above that PCBs may cause deficits in spatial learning and memory by altering dopamine input to the DLPFC. Again, Seegal and colleagues' findings of altered brain dopamine levels in rats and monkeys following PCB exposure also support this interpretation.[1]

5.5. Behavioral Conclusions

In these studies, gravid monkeys were fed PCB and TCDD diets that were similar in dose based on TEFs, and their offspring were later studied for behavioral effects. An important finding was that the behavioral effects of complex PCB mixtures differed qualitatively from those of TCDD. PCB-exposed monkeys were impaired on two spatial learning tasks, but showed facilitated performance on an object learning task. TCDD-exposed monkeys showed just the opposite pattern of effects on learning tasks: they were facilitated on spatial learning tasks and impaired on object learning tasks. These qualitative differences suggest that the behavioral effects observed in the PCB-exposed monkeys may not be mediated through Ah receptor binding, and suggest that TEFs may not be appropriate for estimating neurobehavioral outcomes following perinatal PCB exposure. Our results highlight the need to investigate other potential mechanisms for PCB-related neurobehavioral effects. Seegal and Shain's[1] data suggest one possible mechanism that could be a direct effect of *ortho*-substituted congeners on dopamine synthesis.

The learning effects seen in both PCB- and TCDD-exposed monkeys are suggestive of altered functioning in the prefrontal cortex. However, the pattern of deficits seen in the PCB-exposed monkeys suggests altered function of the dorsolateral region, which is important for spatial learning, whereas the pattern seen in the TCDD-exposed monkeys is suggestive of alterations in the orbital region, which is important for object learning. Although it is unlikely that PCBs or TCDD would target a specific area of the cerebral cortex, leaving the rest of the brain unaffected, it does seem likely that the prefrontal cortex, and/or some part of its neural circuitry, is one area that is affected, particularly with PCB exposure. Given the findings of Seegal and colleagues, and the important role DLPFC dopamine plays in mediating spatial RL and DSA, it seems likely that the PCB-related spatial learning deficits that we observed could be the result of altered dopamine input to the DLPFC. It is more difficult to formulate a hypothesis for the TCDD-related effects based on the limited knowledge of the neurochemical effects of TCDD exposure. Since neuropathological and neurochemical examinations were not carried out on the brains of these monkeys, there are currently no data to confirm or disprove our hypothesis of prefrontal effects. The prefrontal cortex, as well as its input and output pathways, will be an important focus for future neuropathological and neurochemical investigation.

The findings of spatial learning deficits in the Aroclor 1016-exposed monkeys are significant because they suggest that lightly chlorinated, *ortho*-substituted PCB congeners are behaviorally toxic. As discussed earlier in this chapter, Seegal *et al.*[28] have assessed neurochemical function in the brains of monkeys exposed to Aroclor 1016. Long-term reductions in brain DA content were observed. Interestingly, only three of the more than 50 PCB congeners that make up Aroclor 1016 were detected in the brains of the exposed monkeys (i.e., 2,4,4′-trichlorobiphenyl, 2,4,2′,4′-tetrachlorobiphenyl, and 2,5,2′,5′-tetrachlorobiphenyl). The monkey offspring assessed in our study showed a preferential accumulation of the same three *ortho*-substituted PCB congeners in body fat.[31,145] The Seegal *et al.* findings suggest that the cognitive deficits observed in Aroclor 1016-exposed monkey offspring may be related to altered brain DA function and implicate three specific congeners that were shown to preferentially accumulate in body fat as the possible source of the effects. However, until the three congeners are tested for behavioral effects, both alone and in combination, and until behavior and DA function are assessed in the same animals, no firm conclusions can be drawn. Rodent studies of the behavioral and neurochemical effects of these and other individual PCB congeners are currently under way in both laboratories, and should soon shed further light on this important issue.

6. SUMMARY AND CONCLUSIONS

Taken together, the neurochemical and neurobehavioral data reviewed in this chapter strongly suggest that PCBs and related HAHs constitute an important class of neurotoxicants. However, the neuronal mechanisms responsible for the changes in CNS function remain largely unknown. Until the mechanisms by which this diverse class of environmental contaminants alter CNS function are understood, it will be difficult to progress beyond descriptive studies.

A number of factors may be responsible for this lack of information. First is the relatively short period of time that the neural and behavioral effects of PCBs and HAHs have been studied. The first clinical studies describing peripheral neurological effects of exposure were published in the late 1960s,[38] the first papers describing the neural effects of PCBs in experimental animals were published in the early 1970s,[146–148] and the first epidemiological/behavioral study of exposure to environmental levels of these contaminants was not published until 1984.[44] The majority of the earlier studies were, of necessity, descriptive in nature. Additionally, except for the rice-oil poisoning incidents in Japan and Taiwan, the effects of PCBs appear to be more subtle in nature than other neurotoxicants such as methyl mercury or lead; HAHs have thus not yet acquired the stigma (and hence the urgency for research) that has driven lead research.

Second, PCBs and HAHs constitute a large class of compounds that differ in their structure and toxicological activity, thus complicating the task of determining which congeners should be studied and limiting the ability to understand the mechanisms by which these compounds alter CNS function. Indeed, until recently the *ortho*-substituted congeners were thought to be biologically inactive because they did not interact with the Ah receptor. Initial studies either exposed animals to complex commercial mixtures of PCB congeners or used individual congeners that may not have been environmentally relevant. More recently, investigators[107,149] have turned to describing the behavioral, endocrinological, and neurological effects of perinatal exposure to individual PCBs that reflect the congener makeup found in environmental samples. Although these studies will eventually lead to a classification of HAHs based on their behavioral and neurological activity, they are labor-intensive and the list of environmentally relevant congeners is long!

Third, particularly for the developmental studies, there have been few attempts to standardize the protocols for exposure, perhaps because the potential mechanisms of action are poorly understood. Each investigator appears to have a favorite time period for exposing the animals (i.e., gestational days 10–16,[64,103,104] gestational days 6–15,[150] gestational day 6 to postweaning day 21[108]). Thus, in addition to differences in congener exposure, differences in the duration of exposure to the putative neurotoxicants may affect the outcomes, limiting the ability to generalize results from different laboratories.

Finally, the available data, gathered in the adult and developing organism as well as in tissue-culture studies, suggest that the neurological effects and potential mechanisms of action of the various HAH compounds may depend on the developmental status of the organism at the time of exposure. Thus, the task of describing the risk of exposure to HAHs depends not only on which class of compounds the organism is exposed to, but also on the age at exposure. Nevertheless, certain facts exist that may aid in understanding the mechanisms of action and hence the risk to humans following exposure to these compounds.

Although there are 209 theoretically possible PCB congeners, 135 TCDF congeners, and 75 TCDD congeners, it may be possible, as we have previously suggested, to differentiate these 419 compounds into two major classes: coplanar compounds (including the TCDDs, TCDFs, and dioxinlike PCB congeners) that interact with the Ah receptor and differ from each other only in terms of their potency, and di-*ortho*-substituted congeners, which do not interact with the Ah receptor.

On the basis of data that we have gathered in adult rodents and nonhuman primates,[88,89] as well as statements made by Tilson *et al.*[103] that "signs of TCB neurotoxicity have not been observed in adult animals receiving large doses," we suggest that *ortho*-substituted congeners are the major DA neurotoxicant in the adult and may reduce brain DA concentrations by inhibiting the activity of tyrosine hydroxylase, the rate-limiting enzyme in the synthesis of DA.

The problem of determining the mechanisms by which HAHs alter CNS function in the developing organism is much more complex than attempting to understand the mechanisms responsible for altering function in the adult nervous system. On the one hand, coplanar PCBs appear to alter neurological function in the neonate, but not in the adult. Thus, the dioxinlike congeners may be neuroteratogens, altering CNS structure and/or function when administered during development, but are devoid of activity when administered to the mature organism. On the other hand, perinatal exposure to Aroclor 1016 increases brain DA concentrations. Clearly the developing nervous system is quite different from the adult nervous system and major efforts must be made to understand the multiple mechanisms by which the two different classes of HAHs alter CNS function in the neonate.

We suggest that future avenues of research should involve study of the effects of HAHs on endocrine function because both coplanar and *ortho*-substituted PCBs have been shown to affect both estrogen activity and thyroid hormones.[149] In turn, changes in these hormones would have profound and long-term consequences on the developing nervous system[151–153] while having either no effect or only a temporary effect in the adult. Future efforts will aid in understanding whether alterations in these hormones, induced by exposure to PCBs, will induce persistent changes in the developing nervous system.

Regardless of the mechanisms by which PCBs alter nervous system function, one common factor in the majority of studies appears to be alterations in brain DA concentrations. As stated previously, changes in either brain DA concentrations or the number of dopaminergic receptors are correlated with deficits in cognitive behaviors that are reminiscent of the changes observed in nonhuman primates and humans exposed perinatally to PCBs. However, recent work by Eriksson *et al.*[105,106] has demonstrated the importance of examining other neurotransmitter systems since he has shown that early postnatal exposure to PCBs also alters cholinergic and muscarinic neurochemical function as well as the behavior of the exposed animal. However, at this point, cholinergic function has not been evaluated in the adult animal. If neurotoxicology is to provide *relevant* information to public health officials that can be used both to set guidelines for exposure and to suggest pharmacologically and behaviorally based programs of intervention, it is obvious that additional work is needed to determine the biochemical underpinnings of the altered behavioral responses seen in perinatally exposed humans and nonhuman primates.

ACKNOWLEDGMENTS. Supported in part by the National Institute of Environmental Health Sciences and the U.S. Environmental Protection Agency.

REFERENCES

1. R. F. Seegal and W. Shain, Neurotoxicity of polychlorinated biphenyls: The role of ortho-substituted congeners in altering neurochemical function, in: *The Vulnerable Brain and Environmental Risks,* Volume 2, *Toxins in Food* (R. L. Isaacson and K. F. Jensen, eds.), pp. 169–191, Plenum Press, New York (1992).

2. National Research Council, *Polychlorinated Biphenyls,* National Academy of Sciences, Washington, DC (1979).
3. M. D. Erickson, *Analytical Chemistry of PCBs,* pp. 15–23, Butterworths, Boston (1986).
4. O. Hutzinger, S. Safe, and V. Zitko, *The Chemistry of PCBs,* CRC Press, Cleveland (1974).
5. M. D. Mullin, C. M. Pochini, S. McCrindle, M. Romkes, S. H. Safe, and L. M. Safe, High-resolution PCB analysis: Synthesis and chromatographic properties of all 209 PCB congeners, *Environ. Sci. Technol.* **18,** 468–476 (1984).
6. M. D. Erickson, *Analytical Chemistry of PCBs,* pp. 24–34, Butterworths, Boston (1986).
7. S. Safe, L. Safe, and M. Mullin, Polychlorinated biphenyls: Environmental occurrence and analysis, in: *Polychlorinated Biphenyls (PCBs): Mammalian and Environmental Toxicology* (S. Safe and O. Hutzinger, eds.), pp. 1–13, Springer-Verlag, Berlin (1987).
8. L. G. Hansen, Environmental toxicology of polychlorinated biphenyls, in: *Polychlorinated Biphenyls (PCBs): Mammalian and Environmental Toxicology* (S. Safe and O. Hutzinger, eds.), pp. 15–48, Springer-Verlag, New York (1987).
9. D. O. Abramowicz, Aerobic and anaerobic biodegradation of PCBs: A review, *Crit. Rev. Biotechnol.* **10,** 241–251 (1990).
10. M. R. Harkness, J. B. McDermott, D. A. Abramowicz, J. J. Salvo, W. P. Flanagan, M. L. Stephens, F. J. Mondello, R. J. May, J. H. Lobos, K. M. Carroll, M. J. Brennan, A. A. Bracco, K. M. Fish, G. L. Warner, P. R. Wilson, D. K. Dietrich, D. T. Lin, C. B. Morgan, and W. L. Gately, In situ stimulation of aerobic PCB biodegradation in Hudson River sediments, *Science* **259,** 503–507 (1993).
11. M. S. Evans, G. E. Noguchi, and C. P. Rice, The biomagnification of polychlorinated biphenyls, toxaphene and DDT compounds in a Lake Michigan offshore food web, *Arch. Environ. Contam. Toxicol.* **20,** 87–93 (1991).
12. C. Rappe and H. R. Buser, Chemical and physical properties, analytical methods, sources and environmental levels of halogenated dibenzodioxins and dibenzofurans, in: *Halogenated Biphenyls, Terphenyls, Naphthalenes and Related Products* (R. D. Kimbrough and A. A. Jensen, eds.), pp. 71–102, Elsevier, Amsterdam (1989).
13. A. K. D. Liem, R. Hoogerbrugge, P. R. Kootstra, A. P. J. M. de Jong, J. A. Marsman, A. C. den Boer, R. S. den Hartog, G. S. Groenemeijer, and H. A. van deKlooster, Levels and patterns of dioxins in cow's milk in the vicinity of municipal waste incinerators and metal reclamation plants in the Netherlands, in: *Organohalogen Compounds,* Volume 1: *Dioxin '90* (O. Hutzinger and H. Fiedler, eds.), pp. 567–570, Ecoinforma Press, Bayreuth (1990).
14. D. W. Kuehl, B. C. Butterworth, A. McBride, S. Kroner, and D. Bahnick, Contamination of fish by 2,3,7,8-tetrachlorodibenzo-*p*-dioxin: A survey of fish from major watersheds in the United States, *Chemosphere* **18,** 1997–2014 (1989).
15. P. Axegard and L. Renberg, The influence of bleaching chemicals and lignin content on the formation of polychlorinated dioxins and dibenzofurans, *Chemosphere* **19,** 661–668 (1989).
16. H. R. Buser, H. P. Bosshardt, and C. Rappe, Formation of polychlorinated dibenzofurans (PCDFs) from the pyrolysis of PCBs, *Chemosphere* **1,** 109–119 (1978).
17. M. Morita, J. Nakagawa, K. Akiyama, S. Mimura, and N. Isono, Detailed examination of polychlorinated dibenzofurans in PCB preparations and Kanemi Yusho oil, *Bull. Environ. Contam. Toxicol.* **18,** 67–73 (1977).
18. P. W. O'Keefe, J. B. Silkworth, J. F. Gierthy, R. M. Smith, A. P. DeCaprio, J. N. Turner, G. Eadon, D. R. Hilker, K. M. Aldous, L. S. Kaminsky, and D. N. Collins, Chemical and biological investigations of a transformer accident in Binghamton, NY, *Environ. Health Perspect.* **60,** 201–209 (1985).

19. P. W. O'Keefe and R. M. Smith, PCB capacitor/transformer accidents, in: *Halogenated Biphenyls, Terphenyls, Naphthalenes, Dibenzodioxins and Related products* (R. D. Kimbrough and A. A. Jensen, eds.), pp. 417–444, Elsevier, Amsterdam (1989).
20. J. R. Clark, D. DeVault, R. J. Bowden, and J. Weishaar, Contaminant analysis of fillets from Great Lakes coho salmon, 1980, *J. Great Lakes Res.* **10,** 38–48 (1984).
21. D. L. Bedard, R. Unterman, L. H. Bopp, M. J. Brennan, M. L. Haberl, and C. Johnson, Rapid assay for screening and characterizing microorganisms for the ability to degrade polychlorinated biphenyls, *Appl. Environ. Microbiol.* **51,** 761–768 (1986).
22. B. Bush, J. Snow, S. Connor, and R. Koblintz, Polychlorinated biphenyl congeners (PCBs), *p,p'*-DDE and hexachlorobenzene in human milk in three areas of upstate New York, *Arch. Environ. Contam. Toxicol.* **14,** 443–450 (1985).
23. K. Farrell, L. Safe, and S. Safe, Synthesis and aryl hydrocarbon receptor binding properties of radiolabeled polychlorinated dibenzofuran congeners, *Arch. Biochem. Biophys.* **259,** 185–195 (1987).
24. S. Safe, Polychlorinated biphenyls (PCBs), dibenzo-*p*-dioxins (PCDDs), dibenzofurans (PCDFs), and related compounds: Environmental and mechanistic considerations which support the development of toxic equivalency factors (TEFs), *CRC Crit. Rev. Toxicol.* **21,** 51–88 (1990).
25. A. Parkinson and S. Safe, Mammalian biologic and toxic effects of PCBs, in: *Polychlorinated Biphenyls (PCBs): Mammalian and Environmental Toxicology* (S. Safe and O. Hutzinger, eds.), pp. 49–75, Springer-Verlag, New York (1987).
26. T. W. Sawyer, A. D. Vatcher, and S. Safe, Comparative aryl hydrocarbon hydroxylase induction activities of commercial PCBs in Wistar rats and rat hepatoma H-4-II E cells in culture, *Chemosphere* **13,** 695–701 (1984).
27. R. F. Seegal, K. Brosch, B. Bush, M. Ritz, and W. Shain, Effects of Aroclor 1254 on dopamine and norepinephrine concentrations in pheochromocytoma (PC-12) cells, *Neurotoxicology* **10,** 757–764 (1989).
28. R. F. Seegal, B. Bush, and W. Shain, Lightly chlorinated ortho-substituted PCB congeners decrease dopamine in nonhuman primate brain and in tissue culture, *Toxicol. Appl. Pharmacol.* **106,** 136–144 (1990).
29. W. E. Maier, P. R. S. Kodavanti, and H. A. Tilson, In vitro effects of polychlorinated biphenyl congeners in Na^+/K^+-ATPase in selected rat brain regions, *Toxicologist* **13,** 213 (1993).
30. P. R. S. Kodavanti, D. Shin, H. A. Tilson, and G. J. Harry, Changes in cellular calcium homeostasis and toxicity of 2,2'-dichlorobiphenyl in rat cerebellar granule cells, *Soc. Neurosci. Abstr.* **18,** 1606 (1992).
31. S. L. Schantz, E. D. Levin, and R. E. Bowman, Long-term neurobehavioral effects of perinatal polychlorinated biphenyl (PCB) exposure in monkeys, *Environ. Toxicol. Chem.* **10,** 747–756 (1991).
32. S. L. Schantz and R. E. Bowman, Learning in monkeys exposed perinatally to 2,3,7,8-tetrachlorodibenzo-*p*-dioxin (TCDD), *Neurotoxicol. Teratol.* **11,** 13–19 (1989).
33. E. D. Levin, S. L. Schantz, and R. E. Bowman, Delayed spatial alteration deficits resulting from perinatal PCB exposure in monkeys, *Arch. Toxicol.* **62,** 267–273 (1988).
34. U. G. Ahlborg, A. Brouwer, M. A. Fingerhut, J. L. Jacobson, S. W. Jacobson, S. W. Kennedy, A. A. F. Kettrup, J. H. Koeman, H. Poiger, C. Rappe, S. H. Safe, R. F. Seegal, J. Tuomisto, and M. van den Berg, Impact of polychlorinated dibenzo-p-dioxins, dibenzofurans, and biphenyls on human and environmental health, with special emphasis on application of the toxic equivalency factor concept, *Eur. J. Pharmacol.* **228,** 179–199 (1992).
35. Risk Assessment Forum, *Workshop report on toxicity equivalency factors for polychlori-*

nated biphenyl congeners, U.S. Environmental Protection Agency, EPA/625/3-91/020, Washington, DC (1991).
36. M. Kuratsune, T. Youshimara, J. Matsuzaka, and A. Yamaguchi, Epidemiologic study on Yusho: A poisoning caused by ingestion of rice oil contaminated with a commercial brand of polychlorinated biphenyls, *Environ. Health Perspect.* **1,** 119–128 (1972).
37. S.-T. Hsu, C.-I. Ma, S. K.-H. Hsu, S.-S. Wu, N. H.-M. Hsu, C.-C. Yeh, and S.-B. Wu, Discovery and epidemiology of PCB poisoning in Taiwan: A four-year followup, *Environ. Health Perspect.* **59,** 5–10 (1985).
38. Y. Kuroiwa, Y. Murai, and T. Santa, Neurological and nerve conduction velocity studies on 23 patients with chlorobiphenyls poisoning, *Fukuoka Igaku Zasshi* **60,** 446–462 (1969).
39. Y.-C. Lü and P.-N. Wong, Dermatological, medical, and laboratory findings of patients in Taiwan and their treatments, in: *PCB Poisoning in Japan and Taiwan* (M. Kuratsune and R. E. Shapiro, eds.), Alan R. Liss, New York (1984).
40. M. Harada, Intrauterine poisoning: Clinical and epidemiological studies of the problem, *Bull. Inst. Const. Med.* **25,** 1–60 (1976).
41. W. J. Rogan, B. C. Gladen, K. L. Hung, S. L. Koong, L. Y. Shih, J. S. Taylor, Y. C. Wu, D. Yang, N. B. Rogan, and C. C. Hsu, Congenital poisoning by polychlorinated biphenyls and their contaminants in Taiwan, *Science* **241,** 334–336 (1988).
42. Y.-C. J. Chen, Y.-L. Guo, C.-C. Hsu, and W. J. Rogan, Cognitive development of Yu-cheng ('oil disease') children prenatally exposed to heat-degraded PCBs, *J. Am. Med. Assoc.* **268,** 3213–3218 (1992).
43. P. M. Schwartz, S. W. Jacobson, G. G. Fein, and J. L. Jacobson, Lake Michigan fish consumption as a source of polychlorinated biphenyls in human cord serum, maternal serum, and milk, *Am. J. Public Health* **73,** 293–296 (1983).
44. J. L. Jacobson, S. W. Jacobson, P. M. Schwartz, G. G. Fein, and J. K. Dowler, Prenatal exposure to an environmental toxin: A test of the multiple effects model, *Dev. Psychol.* **20**(4), 523–532 (1984).
45. S. W. Jacobson, G. G. Fein, J. L. Jacobson, P. M. Schwartz, and J. K. Dowler, The effect of intrauterine PCB exposure on visual recognition memory, *Child Dev.* **56,** 853–860 (1985).
46. J. L. Jacobson, S. W. Jacobson, and H. E. B. Humphrey, Effects of in utero exposure to polychlorinated biphenyls and related contaminants on cognitive functioning in young children, *J. Pediatr.* **116,** 38–45 (1990).
47. J. L. Jacobson, S. W. Jacobson, R. J. Padgett, G. A. Brumitt, and R. L. Billings, Effects of prenatal PCB exposure on cognitive processing efficiency and sustained attention, *Dev. Psychol.* **28,** 297–306 (1992).
48. J. L. Jacobson, S. W. Jacobson, and H. E. B. Humphrey, Effects of exposure to PCBs and related compounds on growth and activity in children, *Neurotoxicol. Teratol.* **12,** 319–326 (1990).
49. S. L. Schantz, J. L. Jacobson, S. W. Jacobson, and H. E. B. Humphrey, Behavioral correlates of polychlorinated biphenyl (PCB) body burden in school-aged children, *Toxicologist* **10,** 303 (1990).
50. W. J. Rogan, B. C. Gladen, J. D. McKinney, N. Carreras, P. Hardy, J. D. Thullen, J. Tinglestad, and M. Tully, Neonatal effects of transplacental exposure to PCBs and DDE, *J. Pediatr.* **109,** 335–341 (1986).
51. B. Gladen and W. Rogan, Decrements on six-month and one-year Bayley scores and prenatal polychlorinated biphenyls (PCB) exposure, *Am. J. Epidemiol.* **128,** 912 (1988).
52. B. C. Gladen and W. J. Rogan, Effects of perinatal polychlorinated biphenyls and dichlorodiphenyl dichloroethane on later development, *J. Pediatr.* **119,** 58–63 (1991).
53. E.-J. Speckmann and C. E. Elger, Introduction to the neurophysiological basis of the EEG

and DC potentials, in: *Electroencephalography: Basic principles, Clinical Applications and Related Fields* (E. Niedermeyer and F. Lopes da Silva, eds.), pp. 1–13, Urban & Schwarzenberg, Baltimore (1987).
54. C. W. Erwin, M. P. Rozear, R. A. Radtke, and A. C. Erwin, Somatosensory evoked potentials, in: *Electroencephalography: Basic Principles, Clinical Applications and Related Fields* (E. Niedermeyer and F. Lopes da Silva, eds.), pp. 817–833, Urban & Schwarzenberg, Baltimore (1987).
55. F. H. Lopes Da Silva, Dynamics of EEGs as signals of neuronal populations: Models and theoretical considerations, in: *Electroencephalography: Basic Principles, Clinical Applications and Related Fields* (E. Niedermeyer and F. Lopes da Silva, eds.), pp. 15–28, Urban & Schwarzenberg, Baltimore (1987).
56. R. Spehlmann, *Evoked Potential Primer: Visual, Auditory, and Somatosensory Evoked Potentials in Clinical Diagnosis,* Butterworths, Boston (1985).
57. R. Dowman, J. R. Wolpaw, R. F. Seegal, and S. Satya-Murti, Chronic exposure to 60-Hz electric and magnetic fields: II. Neurophysiologic effects, *Bioelectromagnetics* **10,** 302–318 (1989).
58. a. D. Otto, V. Benigmus, K. Muller, C. Barton, K. Seiple, J. Prah, and S. Schroeder, Effects of low to moderate lead exposure on slow cortical potentials in young children: Two year follow-up study, *Neurobehav. Toxicol. Teratol.* **4,** 733–737 (1981); b. Y.-J. Chcn and C.-C. Hsu, Effects of prenatal exposure to PCBs on the neurological function of children: A neuropsychological and neurophysiological study, *Devel. Med. and Child Neurol.* **36,** 312–320 (1994).
59. M. E. Stanton and L. P. Spear, Workshop on the qualitative and quantitative comparability of human and animal developmental neurotoxicity, Work Group I report: Comparability of measures of developmental neurotoxicity in humans and laboratory animals, *Neurotoxicol. Teratol.* **12,** 261–267 (1990).
60. J. Buelke-Sam and C. F. Mactutus, Workshop on the qualitative and quantitative comparability of human and animal developmental neurotoxicity, Work Group II report: Testing methods in developmental neurotoxicity for use in human risk assessment, *Neurotoxicol. Teratol.* **12,** 269–274 (1990).
61. H. A. Tilson, Animal neurobehavioral test battery in NTP assessment, in: *Advances in Neurobehavioral Toxicology: Applications in Environmental and Occupational Health* (B. L. Johnson, ed.), pp. 403–418, Lewis Publishers, Chelsea, MI (1990).
62. J. B. Cavanagh, Lesion localisation: Implications for the study of functional effects and mechanisms of action, *Toxicology* **49,** 131–136 (1988).
63. T. Archer, W. Danysz, A. Fredriksson, G. Jonsson, J. Luthman, L. Sundstrom, and A. Teiling, Neonatal 6-hydroxydopamine-induced dopamine depletions: Motor activity and performance in maze learning, *Pharmacol. Biochem. Behav.* **31,** 357–364 (1988).
64. S. M. Chou, T. Miike, W. M. Payne, and G. J. Davis, Neuropathology of "spinning syndrome" induced by prenatal intoxication with a PCB in mice, *Ann. N.Y. Acad. Sci.* **320,** 373–396 (1979).
65. Z. Annau, Organometals and brain development, *Prog. Brain Res.* **73,** 295–303 (1988).
66. J. P. O'Callaghan, D. B. Miller, and J. F. Reinhard, Jr., Characterization of the origins of astrocyte response to injury using the dopaminergic neurotoxicant, 1-methyl-4-phenyl-1,2,3,6-tetrahydropyridine, *Brain Res.* **521,** 73–80 (1990).
67. M. E. Wolf and R. H. Roth, Autoreceptor regulation of dopamine synthesis, *Ann. N.Y. Acad. Sci.* **604,** 323–343 (1990).
68. I. J. Kopin, J. H. White, and K. Bankiewicz, A new approach to biochemical evaluation of brain dopamine metabolism, *Cell. Mol. Neurobiol.* **8,** 171–179 (1988).

69. C. R. Clark, G. M. Geffen, and L. B. Geffen, Catecholamines and attention. I. Animal and clinical studies, *Neurosci. Biobehav. Rev.* **11,** 341–352 (1987).
70. G. Venturini, F. Stocchi, V. Margotta, S. Ruggieri, D. Bravi, P. Bellantuono, and G. Palladini, A pharmacological study of dopaminergic receptors in planaria, *Neuropharmacology* **28,** 1377–1382 (1989).
71. R. H. Roth, Neuroleptics: Functional neurochemistry, in: *Neuroleptics: Neurochemical, Behavioral and Clinical Perspectives* (J. T. Coyle and S. J. Enna, eds.), pp. 119–156, Raven Press, New York (1983).
72. I. J. Kopin, K. Bankiewicz, and J. Harvey-White, Effect of MPTP-induced parkinsonism in monkeys on the urinary excretion of HVA and MHPG during debrisoquin administration, *Life Sci.* **43,** 133–141 (1988).
73. E. K. Silbergeld and J. J. Chisolm, Lead poisoning: Altered urinary catecholamine metabolites as indicators of intoxication in mice and children, *Science* **192,** 153–155 (1976).
74. R. F. Seegal, K. O. Brosch, and R. Okoniewski, The degree of PCB chlorination determines whether the rise in urinary homovanillic acid production in rats is peripheral or central in origin, *Toxicol. Appl. Pharmacol.* **96,** 560–564 (1988).
75. J. A. Dominic and K. E. Moore, Acute effects of α-methyltyrosine on brain catecholamines and on spontaneous and amphetamine stimulated motor activity in mice, *Arch. Int. Pharmacodyn. Ther.* **178,** 166–176 (1969).
76. P. H. Kelly, Drug induced motor behavior, in: *Handbook of Psychopharmacology* (L. L. Iversen, S. D. Iversen, and S. H. Snyder, eds.), pp. 295–332, Plenum Press, New York (1977).
77. T. Archer, W. Danysz, A. Fredriksson, G. Jonsson, J. Luthman, E. Sundström, and A. Teiling, Neonatal 6-hydroxydopamine-induced dopamine depletions: Motor activity and performance in maze learning, *Pharmacol. Biochem. Behav.* **31,** 357–364 (1988).
78. M. Goiny, S. Cekan, and K. Uvnas-Moberg, Effects of dopaminergic drugs on plasma levels of steroid hormones in conscious dogs, *Life Sci.* **38,** 2293–2300 (1986).
79. T. L. Sourkes, Neural and endocrine functions of dopamine, *Psychoneuroendocrinology* **1,** 69–78 (1975).
80. T. Sawaguchi, M. Matsumura, and K. Kubota, Dopamine enhances the neuronal activity of spatial short-term memory task in the primate prefrontal cortex, *Neurosci. Res.* **5,** 465–473 (1988).
81. T. J. Brozoski, R. M. Brown, H. E. Rosvold, and P. S. Goldman, Cognitive deficit caused by regional depletion of dopamine in prefrontal cortex of rhesus monkey, *Science* **205,** 929–931 (1979).
82. J. S. Schneider and D. F. Roeltgen, Delayed matching-to-sample, object retrieval, and discrimination reversal deficits in chronic low dose MPTP-treated monkeys, *Brain Res.* **615,** 351–354 (1993).
83. H. Nissbrandt, E. Sundström, G. Jonsson, S. Hjorth, and A. Carlsson, Synthesis and release of dopamine in rat brain: Comparison between substantia nigra pars compacta, pars reticulata, and striatum, *J. Neurochem.* **52,** 1170–1182 (1989).
84. P. L. McGeer, J. C. Eccles, and E. G. McGeer, Catecholamine neurons, in: *Molecular Neurobiology of the Mammalian Brain* pp. 233–293, Plenum Press, New York (1978).
85. W. T. Chance, T. Foley-Nelson, J. L. Nelson, and J. E. Fischer, Tyrosine loading increases dopamine metabolite concentrations in the brain, *Pharmacol. Biochem. Behav.* **35,** 195–199 (1990).
86. R. F. Seegal, K. O. Brosch, and B. Bush, High-performance liquid chromatography of biogenic amines and metabolites in brain, cerebrospinal fluid, urine and plasma, *J. Chromatogr.* **377,** 131–144 (1986).

87. P. Herregodts, B. Velkeniers, G. Ebinger, Y. Michotte, L. Vanhaelst, and E. Hooghe-Peters, Development of monoaminergic neurotransmitters in fetal and postnatal rat brain: Analysis by HPLC with electrochemical detection, *J. Neurochem.* **55,** 774–779 (1990).
88. R. F. Seegal, B. Bush, and K. O. Brosch, Sub-chronic exposure of the adult rat to Aroclor 1254 yields regionally-specific changes in central dopaminergic function, *Neurotoxicology* **12,** 55–66 (1991).
89. R. F. Seegal, B. Bush, and K. O. Brosch, Comparison of effects of Aroclors 1016 and 1260 on nonhuman primate catecholamine function, *Toxicology* **66,** 145–163 (1991).
90. E. Castañeda, I. Q. Whishaw, L. Lermer, and T. E. Robinson, Dopamine depletion in neonatal rats: Effects on behavior and striatal dopamine release assessed by intracerebral microdialysis during adulthood, *Brain Res.* **508,** 30–39 (1990).
91. B. H. C. Westerink, J. B. De Vries, and R. Duran, Use of microdialysis for monitoring tyrosine hydroxylase activity in the brain of conscious rats, *J. Neurochem.* **54,** 381–387 (1990).
92. R. F. Seegal, Lumbar cerebrospinal fluid homovanillic acid concentrations are higher in female than male non-human primates, *Brain Res.* **334,** 375–379 (1985).
93. J. D. Elsworth, D. J. Leahy, R. H. Roth, and D. E. Redmond, Jr., Homovanillic acid concentrations in brain, CSF and plasma as indicators of central dopamine function in primates, *J. Neural Transm.* **68,** 51–62 (1987).
94. N. M. Munoz, C. Tutins, and A. R. Leff, Highly sensitive determination of catecholamine and serotonin concentrations in plasma by liquid chromatography-electrochemistry, *J. Chromatogr.* **493,** 157–163 (1989).
95. R. F. Seegal, B. Bush, and K. O. Brosch, Polychlorinated biphenyls induce regional changes in brain norepinephrine concentrations in adult rats, *Neurotoxicology* **6,** 13–24 (1985).
96. R. F. Seegal, K. O. Brosch, and B. Bush, Regional alterations in serotonin metabolism induced by oral exposure of rats to polychlorinated biphenyls, *Neurotoxicology* **7,** 155–166 (1986).
97. R. F. Seegal, K. O. Brosch, and B. Bush, Polychlorinated biphenyls produce regional alterations of dopamine metabolism in rat brain, *Toxicol. Lett.* **30,** 197–202 (1986).
98. L. A. Greene and G. Rein, Release, storage and uptake of catecholamines by a clonal cell line of nerve growth factor (NGF) responsive pheochromocytoma cells, *Brain Res.* **129,** 247–263 (1977).
99. B. Kittner, M. Brautigam, and H. Herken, PC12 cells: A model system for studying drug effects on dopamine synthesis and release, *Arch. Int. Pharmacodyn. Ther.* **286,** 181–194 (1987).
100. R. F. Seegal, B. Bush, and K. O. Brosch, Decreases in dopamine concentrations in adult non-human primate brain persist following removal from polychlorinated biphenyls, *Toxicology* **86,** 71–87 (1994).
101. B. A. Shaywitz, J. H. Klopper, R. D. Yager, and J. W. Gordon, Paradoxical response to amphetamine in developing rats treated with 6-hydroxydopamine, *Nature* **261,** 153–155 (1976).
102. J. Luthman, A. Fredriksson, T. Lewander, G. Jonsson, and T. Archer, Effects of d-amphetmine and methylphenidate on hyperactivity produced by neonatal 6-hydroxydopamine treatment, *Psychopharmacology* **99,** 550–557 (1989).
103. H. A. Tilson, G. J. Davis, J. A. Mclachlan, and G. W. Lucier, The effects of polychlorinated biphenyls given prenatally on the neurobehavioral development of mice, *Environ. Res.* **18,** 466–474 (1979).

104. A. K. Agrawal, H. A. Tilson, and S. C. Bondy, 3,4,3′,4′-Tetrachlorobiphenyl given to mice prenatally produces long term decreases in striatal dopamine and receptor binding sites in the caudate nucleus, *Toxicol. Lett.* **7,** 417–424 (1981).
105. P. Eriksson, Effects of 3,3′,4,4′-tetrachlorobiphenyl in the brain of the neonatal mouse, *Toxicology* **49,** 43–48 (1988).
106. P. Eriksson, U. Lundkvist, and A. Fredriksson, Neonatal exposure to 3,3′,4,4′-tetrachlorobiphenyl: Changes in spontaneous behaviour and cholinergic muscarinic receptors in the adult mouse, *Toxicology* **69,** 27–34 (1991).
107. S. L. Schantz, J. Moshtaghian, and D. K. Ness, Long-term effects of perinatal exposure to PCB congeners and mixtures on locomotor activity of rats, *Teratology* **45,** 524–525 (1992).
108. R. F. Seegal, Perinatal exposure to Aroclor 1016 elevates brain dopamine concentrations in the rat, *Toxicologist* **12,** 320 (1992).
109. L. A. Greene and A. S. Tischler, Establishment of a noradrenergic clonal line of rat adrenal pheochromocytoma cells which respond to nerve growth factor, *Proc. Natl. Acad. Sci. USA* **73,** 2424–2428 (1976).
110. W. Shain, B. Bush, and R. F. Seegal, Neurotoxicity of polychlorinated biphenyls: Structure–activity relationship of individual congeners, *Toxicol. Appl. Pharmacol.* **111,** 33–42 (1991).
111. M. A. Hayes, Carcinogenic and mutagenic effects of PCBs, in: *Polychlorinated Biphenyls (PCBs): Mammalian and Environmental Toxicology* (S. Safe and O. Hutzinger, eds.), pp. 77–95, Springer-Verlag, New York (1987).
112. S. Safe, Polychlorinated biphenyls (PCBs): Mutagenicity and carcinogenicity, *Mutat. Res.* **220,** 31–47 (1989).
113. C. V. Rao and S. A. Banerji, Effect of feeding polychlorinated biphenyl (Aroclor 1260) on hepatic enzymes of rats, *Indian J. Exp. Biol.* **28,** 149–151 (1990).
114. J. B. Silkworth, L. Antrim, and G. Sack, *Ah* receptor mediated suppression of the antibody response in mice is primarily dependent on the *Ah* phenotype of lymphoid tissue, *Toxicol. Appl. Pharmacol.* **86,** 380–390 (1986).
115. D. Davis and S. Safe, Immunosuppressive activities of polychlorinated biphenyls in C57BL/6N mice: Structure–activity relationships as Ah receptor agonists and partial antagonists, *Toxicology* **63,** 97–111 (1990).
116. R. F. Seegal, B. Bush, and W. Shain, Neurotoxicology of ortho-substituted polychlorinated biphenyls, *Chemosphere* **23,** 1941–1949 (1991).
117. R. E. Bowman, M. P. Heironimus, and J. R. Allen, Correlation of PCB body burden with behavioral toxicology in monkeys, *Pharmacol. Biochem. Behav.* **9,** 49–56 (1978).
118. R. E. Bowman and M. P. Heironimus, Hypoactivity in adolescent monkeys perinatally exposed to PCBs and hyperactive as juveniles, *Neurobehav. Toxicol. Teratol.* **3,** 15–18 (1981).
119. R. E. Bowman, M. P. Heironimus, and D. A. Barsotti, Locomotor hyperactivity in PCB-exposed rhesus monkeys, *Neurotoxicology* **2,** 251–268 (1981).
120. S. L. Schantz, N.K. Laughlin, H. C. Van Valkenberg, and R. E. Bowman, Maternal care by rhesus monkeys of infant monkeys exposed to either lead or 2,3,7,8-tetrachlorodibenzo-*p*-dioxin, *Neurotoxicology* **7,** 637–650 (1986).
121. S. L. Schantz, E. D. Levin, R. E. Bowman, M. P. Heironimus, and N. K. Laughlin, Effects of perinatal PCB exposure on discrimination-reversal learning in monkeys, *Neurotoxicol. Teratol.* **11,** 243–250 (1989).
122. R. E. Bowman, S. L. Schantz, S. A. Ferguson, H. Y. Tong, and M. L. Gross, Controlled

exposure of female rhesus monkeys to 2,3,7,8-TCDD: Cognitive behavioral effects in their offspring, *Chemosphere* **20,** 1103–1108 (1990).

123. S. L. Schantz, S. A. Ferguson, and R. E. Bowman, Effects of 2,3,7,8-tetrachlorodibenzo-*p*-dioxin (TCDD) on behavior of monkeys in peer groups, *Neurotoxicol. Teratol.* **14,** 433–446 (1992).
124. R. E. Bowman, S. L. Schantz, M.L. Gross, and S. A. Ferguson, Behavioral effects in monkeys exposed to 2,3,7,8-TCDD transmitted maternally during gestation and for four months of nursing, *Chemosphere* **18,** 235–242 (1989).
125. L. M. Smith, T. R. Schwartz, K. Feitz, and T. J. Kubiak, Determination and occurrence of AHH-active polychlorinated biphenyls, 2,3,7,8-tetrachloro-*p*-dioxin and 2,3,7,8-tetrachlorodibenzofuran in Lake Michigan sediment and biota. The question of their relative toxicological significance, *Chemosphere* **21,** 1063–1085 (1990).
126. H. F. Harlow and J. A. Bromer, A test apparatus for monkeys, *Psychol. Rep.* **2,** 434–436 (1938).
127. E. D. Levin, S. L. Schantz, and R. E. Bowman, Use of the lesion model for examining toxicant effects on cognitive behavior, *Neurotoxicol. Teratol.* **14,** 131–141 (1992).
128. P. S. Goldman, H. E. Rosvold, B. Vest, and T. W. Galkin, Analysis of the delayed-alternation deficit produced by dorsolateral prefrontal lesions in the rhesus monkey, *J. Comp. Physiol. Psychol.* **77,** 212–220 (1971).
129. H. Mahut, Spatial and object reversal learning in monkeys with partial temporal lobe ablations, *Neuropsychologia* **9,** 409–424 (1971).
130. E. J. Holmes, N. Butters, S. Jacobson, and B. M. Stein, An examination of the effects of mammillary-body lesions on reversal learning sets in monkeys, *Physiol. Psychol.* **11,** 159–165 (1983).
131. A. Isseroff, H. E. Rosvold, T. W. Galkin, and P. S. Goldman-Rakic, Spatial memory impairments following damage to the mediodorsal nucleus of the thalamus in rhesus monkeys, *Brain Res.* **232,** 97–113 (1982).
132. K. Battig, H. E. Rosvold, and M. Mishkin, Comparison of the effects of frontal and caudate lesions on discrimination learning in monkeys, *J. Comp. Physiol. Psychol.* **55,** 458–463 (1962).
133. J. M. Warren, Primate learning in comparative perspective, in: *Behavior of Nonhuman Primates: Modern Research Trends* (A. M. Schrier, H. F. Harlow, and F. Stolintz,, eds.), pp. 249–281, Academic Press, New York (1965).
134. P. S. Goldman-Rakic, Circuitry of primate prefrontal cortex and regulation of behavior by representational memory, in: *Handbook of Physiology—The Nervous System* (F. Plum and V. Mountcastle, eds.), pp. 373–417, American Physiological Society, Bethesda (1987).
135. D. M. Rumbaugh and M. A. Jeeves, A comparison of two discrimination-reversal indices intended for use with diverse groups of organisms, *Psychon. Sci.* **6,** 1–2 (1966).
136. J. M. Warren, An assessment of the reversal index, *Anim. Behav.* **15,** 493–498 (1967).
137. E. D. Levin and R. E. Bowman, Long-term effects of chronic postnatal lead exposure on delayed spatial alternation in monkeys, *Neurotoxicol. Teratol.* **10,** 505–510 (1989).
138. E. D. Levin and R. E. Bowman, Long-term effects on the Hamilton search task and delayed alternation in monkeys, *Neurobehav. Toxicol. Teratol.* **8,** 219–224 (1986).
139. M. Bubser and W. J. Schmidt, 6-Hydroxydopamine lesion of the rat prefrontal cortex increases locomotor activity, impairs acquisition of delayed alternation tasks, but does not affect uninterrupted tasks in the radial maze, *Behav. Brain Res.* **37,** 157–168 (1990).
140. P. S. Goldman, H. E. Rosvold, and M. Mishkin, Evidence for behavioral impairment following prefrontal lobectomy in the infant monkey, *J. Comp. Physiol. Psychol.* **70,** 454–462 (1970).

141. B. Jones and M. Mishkin, Limbic lesions and the problem of stimulus–reinforcement associations, *Exp. Neurol.* **36,** 362–377 (1972).
142. M.-M. Mesulam and E. J. Mufson, Neural inputs into the nucleus basalis of the substantia innominata (CH4) in the rhesus monkey, *Brain* **107,** 253–274 (1984).
143. A. F. T. Arnsten and P. S. Goldman-Rakic, Alpha adrenergic mechanisms in prefrontal cortex associated with cognitive decline in aged nonhuman primates, *Science* **230,** 1273–1276 (1985).
144. R. T Bartus, Short-term memory in the rhesus monkey. Effects of dopamine blockade via acute haloperidol administration, *Pharmacol. Biochem. Behav.* **9,** 353–357 (1978).
145. D. A. Barsotti and J. P. Van Miller, Accumulation of a commercial polychlorinated biphenyl mixture (Aroclor 1016) in adult rhesus monkeys and their nursing infants, *Toxicology* **30,** 31–44 (1984).
146. M. Ogawa, Electrophysiological and histochemical studies of experimental chlorobiphenyl poisoning, *Fukuoka Igaku Zasshi* **62,** 74–78 (1971).
147. N. Suenaga, K. Yamada, T. Hidaka, and T. Fukuda, Influences of PCB on the brain catecholamine levels in rats, *Fukuoka Igaku Zasshi* **66,** 589–592 (1975).
148. K. Shiota, Postnatal behavioral effects of prenatal treatment with PCBs (polychlorinated biphenyls) in rats, *Okajimas Folia Anat. Jpn.* **53,** 105–114 (1976).
149. D. K. Ness, S. L. Schantz, J. Mostaghian, and L. G. Hansen, Effects of perinatal exposure to specific PCB congeners on thyroid hormone concentrations and thyroid histology in the rat, *Toxicol. Lett.* **68,** 311–323 (1993).
150. G. Pantaleoni, D. Fanini, A. M. Sponta, G. Palumbo, R. Giorgi, and P. M. Adams, Effects of maternal exposure to polychlorinated biphenyls (PCBs) on F1 generation behavior in the rat, *Fundam. Appl. Toxicol.* **11,** 440–449 (1988).
151. R. A. Gorski, Steroid hormones and brain function: Progress, principles, and problems, in: *Steroid Hormones and Brain Function* (C. H. Sawyer and R. A. Gorski, eds.), pp. 1–26, University of California Press, Los Angeles (1971).
152. T. O. Fox, C. C. Vito, and S. J. Wieland, Estrogen and androgen receptor proteins in embryonic and neonatal brain: Hypothesis for roles in sexual differentiation and behavior, *Am. Zool.* **18,** 525–537 (1978).
153. J. P. O'Callaghan, D. B. Miller, and J. F. Reinhard, Jr., 1-Methyl-4-phenyl-1,2,3,6-tetrahydropyridine (MPTP)-induced damage of striatal dopaminergic fibers attenuates subsequent astrocyte response to MPTP, *Neurosci. Lett.* **117,** 228–233 (1990).

Chapter 14

Exposure Assessment

Measurement of Dioxins and Related Chemicals in Human Tissues

Arnold Schecter

1. INTRODUCTION

Toxic chemical exposure assessment in humans has been approached in a variety of ways. Environmental measurements, such as of air levels in a workplace, have been used to estimate exposure. For example, useful indirect estimates of exposure have been calculated from the amount of time spent in a workplace known to be contaminated with asbestos. Packs of cigarettes smoked per day for a given number of years has been used to estimate dose of the carcinogens from cigarette smoke. Direct measurement of chemical levels in blood or other tissues is a more recent approach. The latter presupposes some knowledge of the metabolism of the compound of concern as well as a good exposure history. In recent years, it has become possible to measure, in a sensitive and specific fashion, the highly toxic, lipid-soluble, synthetic polychlorinated dioxins (PCDDs), dibenzofurans (PCDFs), and biphenyls (PCBs) in human tissue. This chapter will primarily review our own work in developing and using human tissue measurement to document exposure to the halogenated dioxins and structurally related compounds, including the chlorinated dibenzofurans and PCBs.

Chlorinated dioxins, as noted in other chapters in this volume, are ubiquitous and persistent environmental contaminants, particularly in industrialized countries. The primary source of exposure to the general public is from food, especially meat, milk, and fish,[1–5A] which account for well over 90% of the dioxins found in humans. Herbicides, municipal waste incinerators, PCB fires, fungicides, and wood preservatives are some of the sources of these chemicals which are primarily synthetic in origin.[6–13]

Arnold Schecter • Department of Preventive Medicine, College of Medicine, State University of New York Health Science Center–Syracuse, Binghamton, New York 13903.

Dioxins and Health, edited by Arnold Schecter. Plenum Press, New York, 1994.

Analytic techniques developed during the past few decades make it possible to measure polychlorinated dibenzo-*p*-dioxins, PCDFs, and PCBs down to the parts per trillion (ppt) and, on occasion, parts per quadrillion (ppq) levels in human tissues. Worldwide, approximately two to three dozen laboratories are currently capable of performing these measurements in a reliable, valid, and reproducible fashion using known standards and employing improved capillary columns followed by high-resolution gas chromatography combined with high-resolution mass spectrometry. The increased sensitivity and specificity of these newer analytical techniques makes it possible to measure the approximately 16 chlorinated dioxin and dibenzofuran congeners that are found in tissues of most adults from industrialized countries. These congeners are frequently present in levels and patterns that are typical for given broad geographical regions.

Some dioxins have shown remarkable persistence in human tissues. Increased 2,3,7,8,-tetrachlorodibenzodioxin (2,3,7,8-TCDD) body burden from relatively high occupational or environmental exposures has been documented more than 30 years after exposure.[14,15] In cases of lower exposures or those involving some of the less persistent dioxins, dibenzofurans, or PCBs, human tissue levels often return to general population levels in a shorter time, ranging from several months to several years.[16–20] Elevated tissue levels in characteristic patterns have been used to document intake of dioxins or dibenzofurans from exposure to Agent Orange, chlorophenols, and municipal incinerators.[21–26]

Background

The first measurement of dioxins in biological tissue (human and fish) was performed by Robert Baughman in the early 1970s.[23,24] He found what was calculated to be up to 1450 ppt of 2,3,7,8-TCDD in the lipid portion of breast milk from South Vietnamese nursing mothers who had been exposed to Agent Orange, a mixture of phenoxyherbicides contaminated with 2,3,7,8-TCDD. John Constable, a surgeon from Harvard Medical School; Matthew Meselson, a biochemist at Harvard University and Baughman's then Ph.D. advisor; and others collected the milk. Validation of these results by analysis of archived aliquots of the original specimens was later reported by Schecter and Ryan; Ryan's analyses, employing modern techniques, found 2,3,7,8-TCDD levels very similar to those originally reported.[27]

Later in the 1970s Rappe, Masuda, and others documented elevated dioxin and dibenzofuran congeners in tissues from persons occupationally or environmentally exposed.[28–30] In the early 1980s Gross and colleagues were the first to measure elevated 2,3,7,8-TCDD in adipose tissue from U.S. Vietnam veterans who had been classified as "exposed" to Agent Orange by U.S. government investigators.[31]

In a project organized at the Binghamton Clinical Campus of the State University of New York Health Science Center–Syracuse, a collaborating group of scientists first began measuring adipose tissue and blood in U.S. workers exposed to dioxins, dibenzofurans, and PCBs after a PCB transformer fire in Binghamton, New York, in

1981.[17,32–35] This study was the first to report substantial dioxin and dibenzofuran levels in blood and adipose tissue from a general population comparison group. At that time, the general population, having no special dioxin exposure, was expected to have no measurable dioxin body burden. The extent of environmental and food dioxin contamination was yet to be discovered.

A conceptual basis for estimating the combined toxicity of mixtures of PCDDs and PCDFs was developed when New York State Department of Health officials and scientists were required to set guidelines for reentry of the contaminated Binghamton State Office Building. Based on *in vivo* and *in vitro* studies as well as theoretical considerations, toxic PCDD and PCDF congeners were assigned a relative toxicity factor or weighting compared with 2,3,7,8-TCDD. The measured level of each congener was then multiplied by its "dioxin toxic equivalency factor" (TEF) to give the dioxin "toxic equivalents" (TEq). The TEFs, subject to modification as new data become available, vary from 1.0 for 2,3,7,8-TCDD (by definition) to 0.001 in the case of OCDD. The sum of these calculated 2,3,7,8-TCDD toxic equivalents gives an approximation of total dioxinlike toxicity. Non-2,3,7,8-chlorine-substituted dioxins and dibenzofurans, rarely found in human tissues, are given a weighting of zero.[36–39]

2. METHODOLOGY

Tissue, such as fat, milk, or blood, is first collected in chemically cleaned containers (free of dioxins, dibenzofurans, and PCBs), and then frozen. Specimens are kept frozen at between −20 and −70°C until analyzed.

Analytical methodology involving capillary column separation and high-resolution gas chromatography–mass spectrometry, with the use of known dioxin and dibenzofuran standards, has been previously described and is also considered elsewhere in this book.[40,41] The majority of laboratories involved in the studies reported in this chapter have successfully participated in U.S. Environmental Protection Agency–National Institutes of Health (EPA–NIH) and/or World Health Organization (WHO) interlaboratory validation studies.[42–45]

3. SELECTED CASE STUDIES

3.1. Occupational and Other Exposures

3.1.1. The Binghamton State Office Building, Binghamton, New York

Table 1 is adapted from a paper that reported the first PCDD/F tissue level measurements in exposed U.S. workers and the general population on a wet weight basis.[32,46] In 1981, firefighters responding to a PCB transformer fire at the State Office Building in Binghamton, New York, and later cleanup workers were exposed to a

Table 1
PCDD and PCDF Levels and Dioxin Toxic Equivalents (TEq) from Workers Involved in a PCB Transformer Incident (adipose tissue, ppt, lipid)[a]

	Controls (mean of 8)	Exposed workers 1	2	3	4
2378-TCDD	10.2	19.3	43.5	19.6	13.7
23478-PnCDF	20.3	ND[b]	37.4	55.2	88.1
Total HxCDF	44.4	ND	20.8	117.4	308.0
1234678-HpCDF	23.4	12.5	22.3	21.4	46.4
Total PCDDs	1395	464	2611	1793	1175
Total PCDFs	88	13	81	193	441
Total PCDD/Fs	1483	477	2692	1986	1616
Total PCDD TEq	35	32	88	51	34
Total PCDF TEq	15	0.1	21	39	75
Total PCDD/Fs TEq	50	32	109	90	109

[a]Totals are rounded.
[b]ND, not detected.

mixture of PCBs, PCDFs, and PCDDs.[47] The mean value of the control samples, surgically obtained fat tissue specimens from eight adults undergoing surgery at a Binghamton hospital, showed surprisingly high levels of total dioxins and dibenzofurans (~1483 ppt, lipid) in this first congener-specific measurement of tissue from the U.S. general population. Preliminary findings were confirmed by repeating tissue analyses from these workers in later studies by Schecter and Ryan.[35]

Human tissue dioxin levels were first reported on a "wet" or "whole weight" basis, as these were. Later, when it became clear that even fat (adipose) tissue varied in lipid content, these measurements came to be more commonly reported on a lipid basis. The TEqs were calculated using the "International Toxicity Equivalency Factors" that are currently widely accepted and used by the U.S. Environmental Protection Agency and similar agencies in some European countries.[37–39]

In the samples analyzed, chlorinated dibenzofurans were found at higher levels than dioxins, presumably reflecting a higher concentration of PCBs than chlorinated benzenes in the transformer fluid. In the presence of oxygen and heat, PCBs are converted mainly to dibenzofurans, whereas chlorinated benzenes primarily yield dioxins.

Worker number 2 showed an especially elevated 2,3,7,8-TCDD level, and elevated 2,3,4,7,8-pentachlorodibenzofuran (PnCDF) and hexachlorodibenzofuran (HxCDF) levels were particularly noticeable in workers 3 and 4. Buser reported these PCDF congeners as well as 2,3,7,8-TCDD and others among those detected in the soot from the fire.[47] Worker 1 had PCDD/F levels lower than the general population comparison group, including low levels for those congeners that were found in the soot, except for 2,3,7,8-TCDD. It is difficult to characterize this worker's intake without knowing his tissue levels prior to exposure; he may have had a relatively small

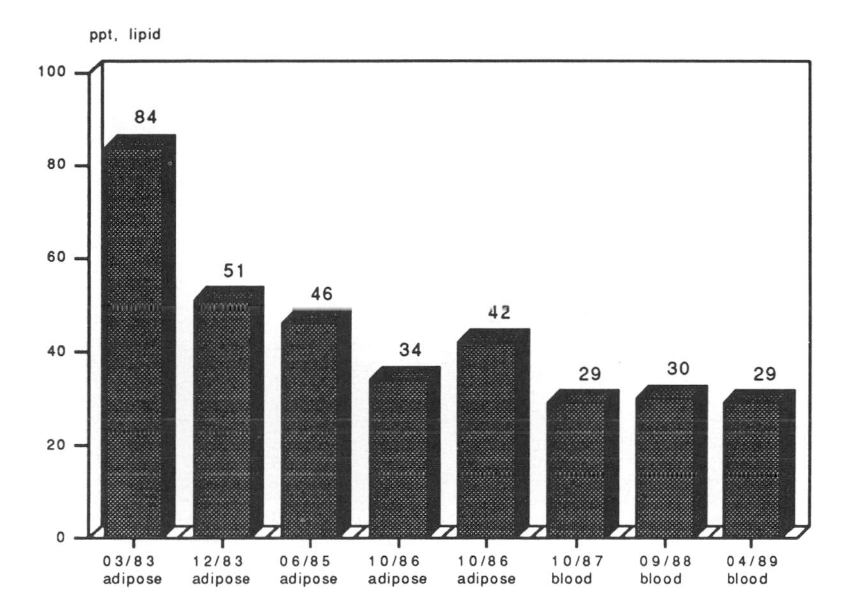

Figure 1. 2,3,4,7,8-PnCDF serial samples from a cleanup worker.

intake, he may have eliminated the PCDD/Fs more rapidly than the other workers, and/or he may have had relatively low levels before exposure. This illustrates the importance of combining tissue measurement with exposure history, since this worker's exposure was similar to workers 3 and 4. The limitations of using tissue measurements alone to determine exposure to dioxins and related chemicals were pointed out in a recent National Academy of Sciences report.[48]

Figures 1 and 2 show serial 2,3,4,7,8-PnCDF and HxCDF levels in one worker, a supervising engineer for the cleanup, from seven samples, fat or whole blood, the first taken 2 years after exposure, and the rest during the following 6 years.[18,40] These congeners were also found in the soot from the fire.[47] The decrease over time in this patient's tissue levels and others' have demonstrated that dibenzofurans have a shorter half-life of elimination than 2,3,7,8-TCDD.[49] Since dibenzofuran tissue levels return to baseline levels sooner than if the exposure were to 2,3,7,8-TCDD, tissue samples should be collected as soon as possible after suspected PCDF exposure in order to document exposure.

For the patients' convenience, the use of fat tissue biopsies to obtain tissue for dioxin analysis has generally been replaced by the use of blood, as was the case in the three most recent serial samples shown here. There is, however, typically a slight difference between blood and adipose tissue dioxin and dibenzofuran levels in the higher chlorinated congeners (some of the hexa-, hepta-, and octa- congeners), where higher levels may be found in whole blood lipid than in fat or adipose lipid.[50–52]

Figure 3 reports wet weight serum PCB levels in parts per billion (ppb) from seven firefighters who were involved in putting out the fire in the Binghamton State Office

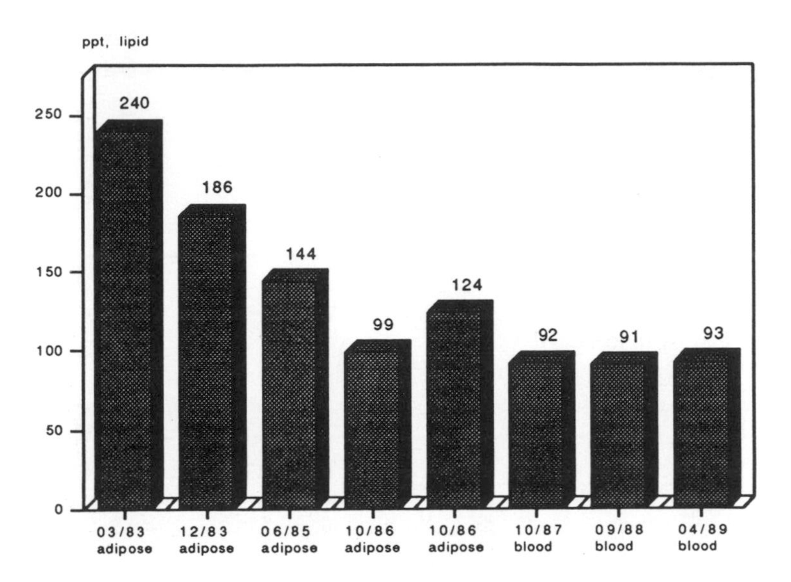

Figure 2. 1,2,3,4,7,8- and 1,2,3,6,7,8-HxCDF serial samples from a cleanup worker.

Building.[20] Blood samples were obtained within days after the fire, and were analyzed using a packed column technique. Serum PCB levels were also obtained 10 months later. The use of serial values documented a decrease in PCB levels in the 10-month period for these seven firefighters, although some of their initial levels were within the usual range found in adults' serum in New York State, 5 to 10 ppb, wet weight. This decrease was very rapid relative to the decrease in dibenzofurans for the worker shown

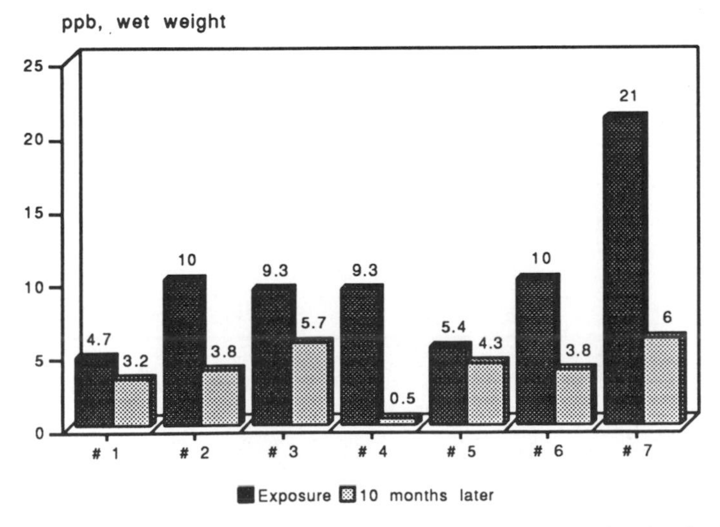

Figure 3. Serial serum PCB levels in exposed firefighters. PCBs reported as Aroclor 1254.

in Figs. 1 and 2. If serial measurements had not been utilized, or if blood had first been collected 10 months after the fire, it would have been difficult to document intake of these chemicals from the incident from blood values alone, although an occupational medical history would have provided convincing medical evidence of PCB exposure and almost certain intake.

3.1.2. BASF Factory, Ludwigshafen, Germany

A second incident illustrates the remarkable persistence that 2,3,7,8-TCDD can show in human tissue and the usefulness of tissue measurement to document high exposure.[14] Exposed workers from the BASF factory in Ludwigshafen, Germany, exhibited elevated adipose tissue levels when tested over three decades after exposure following a dioxin contamination incident. An uncontrolled reaction at the BASF factory exposed these workers to TCDD 32 years before their tissues were obtained for dioxin measurement.[14] Figure 4 shows elevated 2,3,7,8-TCDD ranging from 11 to 141 ppt in these six workers. German general population adult adipose tissue 2,3,7,8-TCDD levels are approximately 3 to 6 ppt, similar to U.S. levels.

Many of these workers had chloracne, an easy-to-observe, although relatively insensitive, nonspecific, and rare biomarker of exposure to chlorinated organic chemicals. Some felt quite ill at the time of exposure, others become ill in a manner consistent with dioxin exposure at a later time.

Approximate fat tissue TCDD levels at the time of exposure, calculated using a 5-year half-life of elimination as our estimate,[53,54] ranged from about 900 to 12,000 ppt. Current estimates of TCDD half life are closer to 11 years, but a biphasic elimination is believed to be more probable, with initial elimination more rapid. These estimated levels at the time of exposure are similar to measured 2,3,7,8-TCDD blood

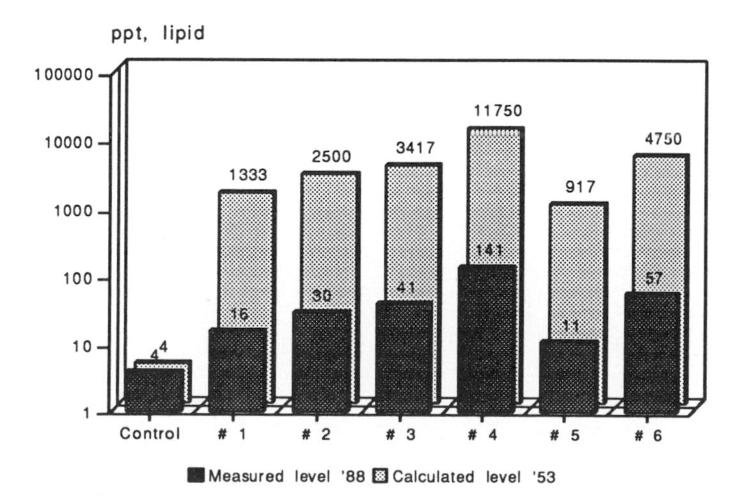

Figure 4. 2,3,7,8-TCDD in six exposed workers and controls from Germany. Measured 32 years after exposure; calculated for time of exposure with 5-year half-life assumed.

levels in exposed persons from samples taken in 1976 at time of exposure to the dioxin incident in Seveso, Italy. Chloracne was usually, but not always, present in exposed persons when blood levels were over 10,000 ppt, and was usually absent below this level.[55] The Seveso blood was sampled very soon after acute exposure and intake, not years later (as in the Ludwigshafen incident), when more of an equilibrium would be expected.

3.1.3. Massachusetts Vietnam Veteran Agent Orange Pilot Study

Table 2 presents selected data from the Massachusetts Vietnam Veteran Agent Orange Pilot Study.[21,51,56,57] The goal of the study was to determine whether U.S. troops, including ground troops, could be found with elevated 2,3,7,8-TCDD from Agent Orange exposure several decades after Vietnam service. Over 12 million gallons of Agent Orange, a half-and-half mixture of the *n*-butyl esters of 2,4-dichlorophenoxy-acetic acid and 2,4,5-trichlorophenoxyacetic acid herbicides, was sprayed in Vietnam, for the most part between 1962 and 1971, in a defoliation project named "Operation Ranch Hand." The name "Agent Orange" came from the orange band on the 55-gallon drums in which the herbicide was stored. The average 2,3,7,8-TCDD content of Agent Orange is believed to have been approximately 2 ppm, although levels as high as 30 ppt were reported.[58]

Twenty years after exposure, levels of 2,3,7,8-TCDD measured in adipose lipid of eight veteran members of Operation Ranch Hand and two other veterans who had

Table 2
2,3,7,8-TCDD in Adipose Tissue of U.S. Vietnam Veterans (ppt, wet weight)

Duty	Concentration of TCDD	1969 TCDD level	
		a	*b*
Flight engineer	55	440	610
Mechanic	41	330	460
Pilot	36	290	400
Navigator	19	150	210
Mechanic	17	140	200
Crew chief	17	140	200
Mechanic	12	100	140
Crew chief	10	80	110
Mechanic	8	70	100
Pilot	7	60	80
Bulldozer operator	17	140	200
Crew of spray helicopter	10	80	110

[a]Three half lives assumed.
[b]This level corrected for fat increase with age, by 1.4. The first eight were members of Operation Ranch Hand. Lipid averaged 97%.

contact with Agent Orange ranged from 7 to 55 ppt, with an average of 21 ppt, as shown in Table 2.[57] Extrapolating back to estimated levels at time of exposure in Vietnam, and correcting for dilution of 2,3,7,8-TCDD from increased body fat with age,[59] we estimate approximate adipose tissue levels at time of exposure to have ranged from 80 to 610 ppt. In comparison, 3 to 6 ppt of 2,3,7,8-TCDD is usually observed in fat, milk, or blood lipid from U.S. general population adults.[25,60–62]

The estimated level of 610 ppt in adipose tissue lipid for the flight engineer is similar to 616 ppt (serum lipid), the highest value actually measured in a U.S. Air Force Vietnam veteran involved in Agent Orange spraying closer to time of exposure in a larger U.S. government study.[63] This estimated level is also similar to the 200 to 1450 ppt of TCDD measured in 1970 samples of human milk lipid from Vietnamese women living in Agent Orange-sprayed areas.[23,27]

3.1.4. Michigan Vietnam Veteran Agent Orange Study

The Massachusetts veteran study confirmed that, decades after service in Vietnam, some U.S. Air Force Agent Orange sprayers and some Army ground troops still had elevated TCDD levels, confirming Agent Orange exposure and intake. In the Michigan Vietnam Veteran Agent Orange Study it was decided to measure dioxins, dibenzofurans, and the dioxinlike PCBs in blood as well as dioxins in semen to better characterize total dioxin and dioxinlike chemical tissue levels.[22,64–66]

Table 3 presents measured TCDD levels in whole blood lipid in 1991 and 1992, 24 to 25 years after exposure to Agent Orange in Vietnam, and estimated levels in the late 1960s, at time of exposure, for six veterans from the Michigan study. Of the 50 veterans selected as having had possible Agent Orange exposure in Vietnam, 6 showed elevated TCDD, ranging from 21 to 131 ppt, while the remaining 44 had an average whole blood lipid level of 4.1 ppt. Assuming a 5-year half-life of elimination, single-

Table 3
TCDD in Vietnam Veterans' Blood and Estimated Level at Time of Exposure

Veteran #	Years since exposure	Measured level 1991–92 (ppt)	TCDD at time of exposure	
			a	*b*
1105	24	31.0	1133	219
1115	24	21.3	631	124
1107	24	131.0	5840	1112
0493	24	22.9	636	124
0500	23	54.5	1596	328
0111	24	20.4	645	127
Mean of 44	23–24	4.1	—	—

[a]Assuming 5-year half-life.
[b]Assuming 10-year half-life.

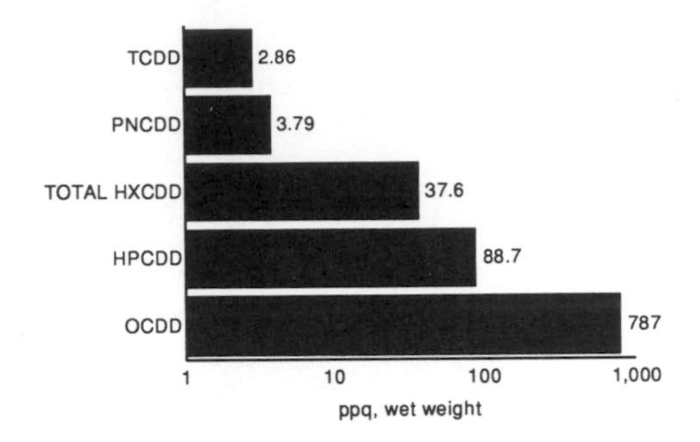

Figure 5. Mean dioxin levels in composite semen samples ($N = 17$).

order kinetics, and a single-compartment model, estimated levels at time of exposure would have varied from 636 to 5840, and assuming a 10-year half-life of elimination, 124 to 1112 ppt.

Since only the elevation of 2,3,7,8-TCDD can be related to Agent Orange exposure during Vietnam service, the other dioxin, dibenzofuran, and PCB levels for these veterans should be similar to adult general population levels. These data therefore documented substantially higher total levels and dioxin toxic equivalents from dioxins, dibenzofurans, and the dioxinlike PCBs than was previously suspected.

For the first time, PCDDs and PCDFs were identified in human semen samples from 17 veterans which were pooled in order to have sufficient sample size.[64,65] Figures 5 and 6 present these semen PCDD and PCDF levels on a wet weight basis in parts per quadrillion. The low lipid content of the semen samples did not allow accurate lipid measurements, so lipid-adjusted data are not presented. The pattern of the dioxins in semen resembles that seen in other tissues, with levels increasing as chlorination increases. However, the pattern of dibenzofurans does not appear to follow the usual pattern in human tissue because of the large amount of OCDF present in this sample. The detection of dioxins and dibenzofurans in semen is consistent with the hypothesis that male-mediated dioxin reproductive toxicity might sometimes involve dioxin transfer from semen to egg or zygote.[67]

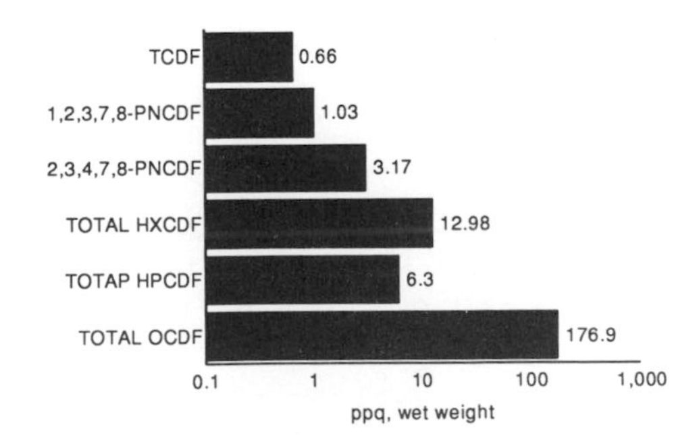

Figure 6. Mean dibenzofuran levels in composite semen samples ($N = 17$).

3.1.5. Vietnamese Exposed to Agent Orange

Although studies of human exposure to Agent Orange contaminated with 2,3,7,8-TCDD usually focus on U.S. or, more rarely, Australian or Korean veterans,[21,56,57,62,64,68–73] those persons with the highest and longest exposure to Agent Orange are the Vietnamese living in the south of Vietnam. It was there that Agent Orange was sprayed over about 10% of the forest and farmland of South Vietnam between 1962 and 1971, with the heaviest spraying occurring between 1967 and 1970.[27,58,74–84]

Current 2,3,7,8-TCDD tissue levels in Vietnamese are shown in Table 4 from samples collected in the 1980s and 1990s.[77] 2,3,7,8-TCDD varies in this series of pooled blood and adipose specimens from 1.2 ppt in whole blood from nonsprayed Hanoi, the capital of Vietnam located in the north, to 28 ppt in a pooled blood sample from Bien Hoa in the south, an Agent Orange-sprayed area. By converting to dioxin toxic equivalents,[37–39] we obtain estimated total dioxin toxicity values of from 10 in the nonsprayed north to 77 ppt in the sprayed southern area of Da Nang, from all chlorinated dioxins and dibenzofurans, not only from the TCDD which presumably is partly from Agent Orange. TCDD contributes 18 ppt to the highest total TEq value of 77 from Da Nang blood while the Bien Hoa sample with a lower total TEq of 47 has a higher contribution from TCDD, 28 ppt. The Ho Chi Minh City (formerly Saigon) blood shown here was obtained from young women at the obstetrics hospital in that city and reflects low TCDD levels that are characteristic of younger persons who were born some time after Agent Orange spraying had ended. The higher levels of dioxins and dibenzofurans other than TCDD in southern Vietnamese reflect industrialization and possibly the use of chlorophenols in various agricultural uses.

It is also interesting to note that 2,3,7,8-TCDD is elevated in a pooled sample of fat tissue in veterans of the North Vietnam Army as compared with the general population in the north, 8.1 ppt versus 1.4 ppt, respectively. This is consistent with uptake of 2,3,7,8-TCDD from Agent Orange from serving for many years in the south where Agent Orange was sprayed, and from living on or near the Ho Chi Minh Trail, which ran from the north to the south of Vietnam and was also heavily sprayed to discourage troop movement. However, not all of the elevated dioxins are from Agent Orange; total PCDD and PCDF levels in these former soldiers are also elevated and contribute to the higher dioxin toxic equivalent level.

It should be noted that the highest Vietnamese adipose tissue level, 103 ppt of TCDD, was found in an individual adult sample collected in the 1980s and presumably indicated previous Agent Orange exposure.[84] Because of the scarcity of qualified dioxin laboratories and the high cost of analyses, pooled or combined blood samples were used to provide average values for a given geographic region in our more recent studies of Vietnam.

The elevated tissue TCDD levels in Table 4 were found over two decades after Agent Orange was initially sprayed, again demonstrating the persistence of TCDD in human tissue. To illustrate the levels of 2,3,7,8-TCDD found at the time of exposure, Table 5 shows some of the higher measured breast milk levels in 1970 specimens

Table 4
Dioxins and Dibenzofurans in Human Blood and Adipose Tissue from Vietnam (ppt, lipid)[a]

	Southern Vietnam					Central Vietnam				North Vietnam		
	Ho Chi Minh Blood $N = 50$	Dong Nai Blood $N = 33$	Tay Bien Blood $N = 50$	Bien Hoa Blood $N = 50$	Ma Da Forest Blood $N = 50$	Quang Tri Blood $N = 50$	Hue Blood $N = 50$	Da Nang Blood $N = 50$	A Luoi Blood $N = 33$	Hanoi Blood $N = 32$	Hanoi Adi.[b] $N = 10$	Soldiers Adi. $N = 10$
TCDD	3.4	12	6.8	28	12	9.5	11	18	15	1.2	1.4	8.1
Total PCDDs	1087	1938	682	855	286	767	995	1529	174	126	274	685
Total PCDFs	82	133	26	75	29	130	265	315	61	41	31	56
Total PCDD/Fs	1169	2071	708	930	315	897	1260	1844	235	167	305	741
PCDD TEq	14.2	31	12.02	38.8	16.2	23.1	29.8	46.2	16.4	6	4	20.9
PCDF TEq	14.2	18	3.4	8.0	2.7	10.9	26.7	30.6	5.3	6	6	9.4
PCDD/F TEq	28	49	15	47	19	34	57	77	22	12	10	30

[a] Totals are rounded.
[b] Adi., adipose.

Table 5

PCDD and PCDF Levels in Human Milk from Vietnam (ppt, lipid basis)

	South											North
	Tan Uyen[a] Village $N = 1$		Tan Uyen						Da Nang pool $N = 11$	HCMC pool $N = 38$	Song Be pool $N = 12$	Dong Nai pool $N = 11$	Hanoi pool $N = 30$
			3 individual analyses, 1973			3 pools of $N = 2$, 1985–88							
	1970	1973							1985–90	1985–90	1985–90	1985–90	1985–90
2,3,7,8-TCDD	1450	400	77	100	230	2.9	5.2	11	5.6	7.1	17	10	2.1
Total PCDDs	1450*	400*	547*	187*	530*	307	167	164	406	302	277	180	104
Total PCDFs	NA[b]	NA	23*	NA	27*	183	14	11	129	33	45	54	23
Total PCDD/Fs	NA	NA	570*	187*	557*	490	181	175	535	335	322	234	127
Total TEq	1450*	400*	89*	100*	244*	26	9	20	35	19	32	26	9

[a]See Ref. 1.

[b]NA, not available, only 2,3,7,8-TCDD was quantified.

*Levels would have been higher if other congeners had been analyzed.

expressed on a lipid basis, which were collected while Agent Orange was still being sprayed. These are compared with 1973 samples, collected 2 to 3 years after spraying ended, and more recent samples.[23,27,77] The 1970 breast milk lipid-adjusted value of 1450 ppt is the highest measured TCDD level in milk following Agent Orange exposure, and also the highest dioxin level reported in human breast milk to date.

These values may also serve as approximations of TCDD levels for others exposed in the past to large amounts of the once commonly used phenoxyherbicide 2,4,5-T, which was contaminated with the most toxic dioxin, 2,3,7,8-TCDD, in the parts per million range. These very high dioxin levels from women nursing in two Vietnamese villages suggest that studying the children as well as the women who were unfortunately exposed to these high dioxin levels should be useful in characterizing the potential health consequences of dioxin exposure. In the rice oil poisonings with chlorinated dibenzofurans and PCBs in Japan and Taiwan, health problems were documented in children exposed from transplacental transfer and presumably also from nursing; this is reported in some detail elsewhere in this volume.[29,85–87] Levels of PCBs in U.S. mothers from the general population, lower than those found in the rice oil poisonings, have also been associated with neurobehavioral and neurodevelopmental pathology in children.[88,89] Serious nervous system and endocrine alterations in male rodents born to dams exposed to even small doses of TCDD during pregnancy have also been documented, and are described in some detail elsewhere in this volume.[90–92]

3.1.6. Municipal Incinerator Worker Dioxin Exposure

Municipal incinerators, commonly used to dispose of household waste in industrial countries, have been found to generate dioxins and dibenzofurans during combustion.[93–95] For that reason, the question of hazards to workers and the general public has become an issue of concern. Because incinerator workers have the closest contact with incinerator ash, it was decided to measure dioxins in the blood of incinerator workers from an older, and presumably less environmentally “safe,” New York City incineration facility, to determine whether bioavailability of dioxins from incinerator ash could be documented in humans.

Table 6 summarizes selected congeners in whole blood dioxin measurements from two municipal incinerator worker cohorts, one American and one German. The first was exposed to boiler ash containing PCDD/Fs at a municipal incinerator located in New York City.[96] Dioxin levels in pooled blood were compared with control pooled blood of New York City residents, matched by sex and age. The elevated total dibenzofuran level is especially striking: 103 ppt in workers versus 47 in controls. Total dioxins also were somewhat higher in workers than controls, 904 versus 700 ppt, respectively. Congeners paralleling those found in the incinerator ash were elevated in the workers’ blood. These findings led to the implementation of more stringent worker protection measures.

The second worker column represents mean dioxin levels from 10 individual blood samples from workers at an old, non-state-of-the-art, German municipal waste incinerator. Their levels are compared with those of 25 matched controls. Statistically

Table 6
Selected Congeners and Total PCDD and PCDF Levels in Pooled Blood of Municipal Incinerator Workers and Matched Controls (ppt, lipid)

	U.S. matched controls $n = 14$	U.S. workers $n = 56$	German matched controls $n = 25$	German workers $n = 10$
1,2,3,7,8-PnCDD	5.2	7.7	14.1	11.0
Total HxCDD	65.1	74.9	93.7	85.9
OCDD	531	695	601	1051
2,3,7,8-TCDF	3.5	8.1	3.3	2.7
Total HxCDF	20.3	33.6	30.2	52.3
Total HpCDF	18	50	22.9	43.9
Total PCDDs	700	904	829	1262
Total PCDFs	47	103	91	133
Total PCDD/Fs	747	1007	920	1395

significant elevations ($p < 0.05$) are seen in the incinerator workers' blood for OCDD, HxCDF, HpCDF, and total PCDD and PCDF. A recent study of workers from a modern German municipal incinerator, where worker protection and incinerator technology are at a higher level, did not find elevated dioxins in workers' blood.

3.1.7. Dioxin Chemists with Potential Exposure

Chemists handling dioxins are another group at risk of occupational dioxin exposure.[15,17,97,98] Table 7 gives the results of dioxin tissue measurements in three dioxin chemists. The U.S. chemist #1 synthesized the brominated 2,3,7,8-tetrabromodibenzo-*p*-dioxin (2,3,7,8-TBrDD) as well as chlorinated 2,3,7,8-TCDD.[98] A 2,3,7,8-TBrDD level of 625 ppt was found 36 years after exposure, whereas no 2,3,7,8-TBrDD could be found in general population blood or in several North American blood specimens (with a detection limit of 3 to 6 ppt). The 2,3,7,8-TCDD was also somewhat elevated, 18 ppt, 36 years after exposure, as compared with the usual 3 to 6 ppt found in blood from the general U.S. population.[25] After the synthesis of 2,3,7,8-TCDD, chemist #1 developed chloracne, headaches, backaches, and severe pain in both legs on exertion and was hospitalized for observation. Six months prior to synthesizing the 2,3,7,8-TCDD, he synthesized the 2,3,7,8-TBDD, which is thought to be about as toxic as 2,3,7,8-TCDD, and experienced minor signs and symptoms of exposure, including very mild chloracne. We estimate that his total dioxin level at time of exposure was between 1928 and 148,011 ppt, lipid, using 10- and 5-year half lives of elimination, respectively. If this was the case, total dioxin toxic equivalents were between 673 and 146,756 ppt, lipid.

The second U.S. chemist, who had potential contact with dioxins from his research on chick edema disease, and the European chemist shown in Table 7 have dioxin blood levels similar to the general population levels except for HpCDD and OCDD, which are elevated in the European chemist. Since we do not have data prior to

Table 7
Chlorinated and Brominated PCDDs and PCDFs in Whole Blood—Two U.S. Chemists, U.S. General Population (G.P.), a European Chemist, and German General Population (ppt, lipid)

	U.S. G.P. $N = 100$		U.S. Chem.[b]		U.S. Chem.		German G.P. $N = 85$		European Chem.	
	PCDD/F	TEq	PCDD/F	TEq	PCDD/F	TEq	PCDD/F	TEq	PCDD/F	TEq
2378-TBDD[c]	NA	NA	625	625	—	—	—	—	ND(3)	ND
2378-TBDD[d]			142628	142628	—	—	—	—	—	—
2378-TCDD[c]	5.2	5.2	18	18	3.4	3.4	3.6	3.6	4.2	4.2
2378-TCDD[d]			4098	4098	—	—	—	—	—	—
12378-PnCDD	21	10.5	18	9	11	5.5	14	7	11	5.5
Total HxCDD	112	11.2	87	8.7	79.3	7.9	81.1	8.11	70	7
1234678-HpCDD	187	1.87	181	1.81	87	0.87	93.8	0.938	181	1.81
OCDD	1174	1174	912	0.912	828	0.828	596	0.596	1215	1.21
2378-TCDF	3.1	0.31	ND (2)	ND(0.1)	2	0.2	2.5	0.25	ND(2)	ND
23478-PnCDF	13.0	6.5	13	6.5	7.4	3.7	36.8	18.4	24	0.24
Total HxCDF	32.6	3.26	23	2.9	14	1.4	31.6	3.16	24	0.21
1234678-HpCDF	36	0.36	43	0.43	14	0.14	21.8	0.218	21	0.021
OCDF	4.2	0.0042	NA	NA	ND(3)	ND(1.5)	5.5	0.0055	NA	NA
Total PCDDs	1499	30	1216	38	1009	19	788	20	1481	20
Total PCDFs	89	10	87	10	37	5	98	22	69	14
Total PCDD/Fs	1588	40	1303	48	1046	24	886	42	1550	34
Total PBCDD/Fs[c]	—	—	1928	673	—	—	—	—	—	—
Total PBCDD/Fs[d]	—	—	148011	146756	—	—	—	—	—	—

[a]Totals are rounded. NA, not available; ND, not detected, detection limit in parentheses. Chem., chemist.
[b]Chemist was exposed to TCDD and TBDD during synthesis, blood was drawn in 1990 (see Ref. 22). Other chemists were potentially exposed to multiple PCDD/Fs.
[c]Calculated level at time of exposure assuming a 10-year half-life.
[d]Calculated level at time of exposure assuming a 5-year half-life.

exposure, nor specific information concerning metabolism-kinetics for these congeners (an area where further research is indicated), we cannot rule out intake of dioxins from these findings. However, these tissue levels do not document intake of dioxins from workplace exposure. The European chemist still works with dioxins, in contrast to the first U.S. chemist, whose last exposure was decades prior to the time his blood was sampled.

3.1.8. Dioxin Levels in AIDS Patients

An interesting finding, reported in a New York State-based pilot study, was that of elevated levels of dioxins and dibenzofurans, known immunosuppressants, in clinically symptomatic AIDS patients as compared with asymptomatic but human immunodeficiency virus (HIV)-infected patients who had below-average dioxin levels.[99] Pooled whole blood from 100 persons in each group was analyzed. The AIDS patients who were sick with opportunistic infections had elevated PCDD/F blood lipid levels and the asymptomatic HIV-positive persons had levels below the general population levels.

Figure 7 presents the PCDD and PCDF dioxin toxic equivalents for these three groups. These findings are consistent with the hypothesis that dioxins and dibenzofurans may sometimes be cofactors in expression of the disease syndrome, AIDS, in HIV-positive (infected) persons. If this hypothesis is correct, populations with higher levels could become ill more rapidly whereas those with lower dioxin body burdens might develop the disease state at a slower rate or to a lesser extent. Although illness or treatment might have caused elevated blood lipid levels of dioxins in the AIDS patient group (there is no published work on this issue at present), this would not account for the below-average level of PCDD/Fs in the asymptomatic HIV^+ persons. Further studies involving Individual analyses of blood and prospective monitoring of

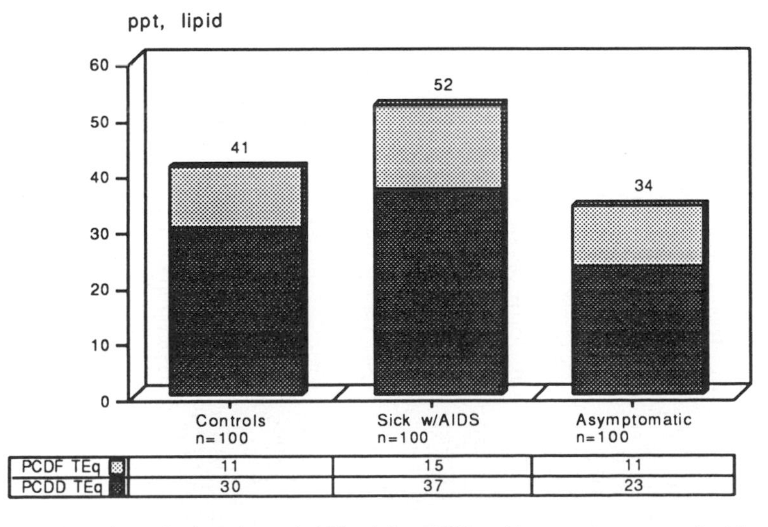

	Controls n=100	Sick w/AIDS n=100	Asymptomatic n=100
PCDF TEq	11	15	11
PCDD TEq	30	37	23

Figure 7. Dioxin toxic equivalents in pooled blood from HIV-positive persons compared with controls.

persons with HIV infection would be a first step in providing data for validation or refutation of the hypothesis that the immunosuppressant dioxins and dibenzofurans may sometimes be cofactors in the expression of clinical disease in HIV-infected persons.

3.1.9. Chemical Workers from Ufa, Russia, and Their Children

The Chimprom Manufacturing Complex in Ufa, Russia, a large chemical company manufacturing many chlorinated compounds, produced 2,4,5-T from 1965 to 1967. As noted previously, 2,4,5-T is characteristically contaminated with 2,3,7,8-TCDD. During that time several hundred factory production workers were believed to have been directly exposed to the 2,4,5-T with its dioxin contamination. Many employees sought medical assistance at the time of exposure for evaluation of skin rashes and other medical problems; 137 employees out of approximately 231 involved in 2,4,5-T production were registered at the Occupational Medicine Institute in Ufa as having chloracne, or acne caused by chlorinated organic chemicals. In 1991, blood from workers who had been employed at the plant during these years was collected and analyzed; their 2,3,7,8-TCDD blood lipid levels 25 years after exposure ranged from 36 to 291 ppt, whereas we had previously found 2 to 5 ppt of TCDD in blood from the Russian general population.[100,101]

In 1992, 25 years after exposure to 2,3,7,8-TCDD from the production of 2,4,5-T, individual blood samples were collected from 65 residents of Ufa, including exposed workers and their children.[101] Table 8 presents mean TCDD levels of four

Table 8
PCDD and PCDF Levels for Ufa Russia Chemical Workers, Their Children, and the General Population (whole blood)

	Control pool $N = 100$	Mothers' mean $N = 4$	Children's mean $N = 6$
2,3,7,8-TCDD	12	168	49
1,2,3,7,8-PnCDD	9.5	28	23
1,2,3,4,7,8/1,2,3,6,7,8-HxCDD	6	11	11
1,2,3,7,8,9-HxCDD	ND(3)	3	4
1,2,3,4,6,7,8-HpCDD	9.7	11	31
OCDD	73	170	279
2,3,7,8-TCDF	ND(2)	1	3
2,3,4,7,8-PnCDF	8	11	10
1,2,3,4,7,8/1,2,3,6,7,8-HxCDF	7	13	11
2,3,4,6,7,8-HxCDF	ND(3)	2	4
1,2,3,4,6,7,8-HpCDF	8.4	16	33
Total PCDDs	112	391	397
Total PCDFs	26	43	61
Total PCDD/Fs	138	434	458

mothers, mean levels of six of their children, and the results of a pooled sample (n = 100) from residents of Ufa who had not worked in the factory. Other than transplacental transfer *in utero,* intake from nursing, or possible exposure to contaminated clothing, these children had no other known special exposure to TCDD, yet they exhibit tissue TCDD levels decades after their mother's exposure.

The most probable route of exposure for these children was through nursing, although this has not yet been confirmed. Little data exist on the transmission of dioxins and dibenzofurans from exposed persons to family members. A previous report of two case studies found that unexposed mates' PCB blood patterns matched their occupationally exposed husbands' blood patterns, the intake occurring presumably from contact with or while washing contaminated clothing.[102] Two children of a male Chimprom factory worker with a TCDD blood level of 108 ppt had TCDD blood levels similar to the Ufa controls, consistent with transplacental and nursing intake for the children of female workers. Further research is needed to determine possible health effects on these workers and their children.

3.1.10. Pentachlorophenol Dioxin Exposure in China

Certain rural areas in China have been and are being sprayed with dioxin- and dibenzofuran-contaminated sodium pentachlorophenol (Na-PCP) to control the spread of snailborne schistosomiasis. In Table 9 a sample of Na-PCP from China is shown to have a total dioxin and dibenzofuran content of just under 1000 ppb and a TEq of 29.9 ppb. The Na-PCP is contaminated with very low levels of 2,3,7,8-TCDD, 1,2,3,7,8-PnCDD, 2,3,7,8-TCDF, PnCDF, HxCDF, and HpCDF and higher levels of HxCDD, HpCDD, OCDD, and OCDF.

Age- and sex-matched pooled blood samples from sprayed and nonsprayed areas in a central mainland region were collected as was a pooled sample from individuals in direct contact with the pesticide. Results are presented in parts per trillion on a lipid basis. In each case, the control blood has a lower TEq than blood from exposed persons. The sample from those living in sprayed areas has a slightly higher TEq than those who handled the pesticide directly, possibly from direct contact or from ingestion of contaminated food. The pattern of congeners in exposed persons is consistent with intake of dioxins from PCP, but not of dibenzofurans.

PCDD/F levels in breast milk from exposed mothers are compared with those of control milk from the Chinese general population on a lipid basis in parts per trillion. Again, dioxin congeners, particularly 2,3,7,8-TCDD, 1,2,3,7,8-PnCDD, and OCDD, are noticeably higher in the exposed mothers and the TEq of the breast milk sample from sprayed areas is 5.4 versus 2.6 from the control group. The low PCDD/F and total TEq of all of these samples from mainland China are similar to levels reported in adipose tissue by us previously.[103] Milk lipid typically has lower levels of dioxins when compared with blood lipid levels. Since China is the world's most populous country, with over one billion inhabitants, the finding of relatively low dioxin and dibenzofuran tissue levels suggests that much of the earth's population does not yet have high dioxin body tissue levels.

Table 9
PCDD/PCDF in Human Milk and Blood from China from Pentachlorophenol (Na-PCP)-Sprayed Areas Compared with Controls

		Blood ppt, lipid					Breast milk ppt, lipid	
		15–19 years old		Men and women over 40 years old		Individuals in contact with Na-PCP		
	Na-PCP sample (ppb)	Control $N = 50$	Sprayed areas $N = 50$	Control $N = 50$	Sprayed areas $N = 50$	$N = 26$	Control $N = 50$	Sprayed areas $N = 50$
2,3,7,8-TCDD	1.1	<1.2	2.2	nd(1.2)	4.6	3.0	0.64	1.4
1,2,3,7,8-PnCDD	1.9	1.6	5.3	3.1	9.5	7.2	0.7	3.4
1,2,3,4,7,8-HxCDD	238.0	1.8	14.0	3.8	27.8	22.1	0.74	11.1
1,2,3,6,7,8-HxCDD	n.a.	4.3	4.2	4.9	8.9	9.0	1.5	2.2
1,2,3,7,8,9-HxCDD	1.8	1.7	1.7	2.6	2.3	2.9	0.66	0.79
Total HxCDD	272.7	7.8	20.0*	11.3	36.6*	33.9*	2.9	14.09
1,2,3,4,5,6,7-HpCDD	190.0	11.6	15.2	17.5	15.7	24.1	3.3	7.1
OCDD	420.0	104.1	568.0	117.0	748	1148	26.8	103.0
2,3,7,8-TCDF	0.2	<4.2	2.1	2.7	1.4	1.5	2.0	0.47

1,2,3,7,8-PnCDF	0.2	<1.6	nd(1.0)	nd(1.0)	nd(1.0)	nd(1.0)	0.37	0.23
2,3,4,7,8-PnCDF	0.2	2.7	1.1	2.7	1.9	2.4	1.6	0.88
Total PCDF	4.4*	3.5	1.1	2.7	1.9	2.4	1.97	1.11
1,2,3,4,7,8-HxCDF	12.4	3.4	3.1	4.7	4.9	16.4	1.0	1.5
1,2,3,6,7,8-HxCDF	na	2.1	1.5	3.0	2.1	2.3	0.77	0.52
1,2,3,7,8,9-HxCDF	0.4	nd(1.0)	nd(1.0)	nd(1.0)	nd(1.1)	nd(1.0)	nd	nd
2,3,4,6,7,8-HxCDF	0.4	1.9	1.0	2.7	2.0	1.2	0.33	0.28
Total HxCDF	16.9	7.4	5.6	10.4	9.0	20.0	2.1	2.3
1,2,3,4,6,7,8-HpCDF	2.5	5.1	3.6	7.7	4.1	4.9	0.71	0.55
1,2,3,4,7,8,9-HpCDF	0.1	nd(1.2)	nd(1.3)	<2.3	nd(2.4)	nd(1.6)	0.06	0.04
Total Hp-CDF	4.7*	5.1	3.6	8.9	4.1	4.9	0.77	0.59
OCDF	80.5	<5.0	5.9	<5.0	7.5	5.2	0.33	0.37
Total PCDDs	888*	125.7	610.8	148.9	814	1216	34	129
Total PCDFs	110	20.6	18.3	27.2	24	34	7	5
Total PCDD/Fs	998	146.3	629.1	176.1	838	1250	41	134
TEq (NATO–EPA)	29.9	4.8	9.0	5.7	16.5	14.8	2.6	5.4

Totals are rounded. For < values, 1/2 was used for calculation of totals. *These totals include non–2,3,7,8-substituted congeners. In addition, Na-PCP sample had other congeners not listed on the table for a total of 3.9 Total TCDD, 8.2 Total PCDD, 191.5 total HpCDD.

3.1.11. Yusho and Yu-cheng, Rice Oil Poisonings in Japan and Taiwan

Rice oil contaminated with PCBs, PCDFs, and a small amount of PCDDs was used in cooking and ingested by over 1850 individuals in Japan in 1968 and over 2000 persons in Taiwan in 1979 in remarkably similar incidents. These exposures caused illness similar to those documented in animal studies of dioxins and PCBs.[29,86,87,104] The toxic effects were believed to be primarily from the dibenzofurans, especially the 2,3,7,8-substituted penta-, hexa-, and hepta-chlorinated congeners. PCBs are believed to have contributed a lesser amount of toxicity and a small contribution is thought to have come from the dioxins.

Table 10 presents dioxins and dibenzofuran levels from four different tissues collected from two Yusho patients.[103] The first three tissue samples, fat, liver, and lung, were obtained from a 59-year-old male who died in March, 1977, approximately 8 years after exposure. The fourth sample, uterine tissue, was obtained from a female Yusho patient in 1985, 17 years after exposure. The mean results of six individual adipose tissue analyses from the Japanese general population are shown for comparison. The Yusho patients exhibit elevated levels of PnCDF, HxCDF, and HpCDF, the congeners contained in the rice oil, years after initial exposure.

In Taiwan, a series of studies were conducted on the victims and children born to the rice oil-poisoned mothers. A variety of ill health effects were noted, beginning at birth, indicating transplacental exposure and intrauterine damage.[105] Placentas obtained from five Taiwanese Yu-cheng patients 5 to $5^1/_2$ years after exposure were analyzed for PCDDs and PCDFs and are compared with results from U.S. general population placentas in Table 11. Elevations of the penta-, hexa-, and hepta-

Table 10
PCDD and PCDF Levels in Tissues from Two Yusho Patients
Compared with Japanese General Population (ppt, lipid)

	Patient #1			Patient #2	Japanese general population
	Fat	Liver	Lung	Uterus	Fat
2,3,7,8-TCDD	ND	ND	ND	ND	6.6
1,2,3,7,8-PnCDD	33.8	ND	37.5	ND	13
1,2,3,4,7,8/1,2,3,6,7,8-HxCDD	70.6	2733	365	ND	86
Total HpCDD	ND	733	245	ND	69
OCDD	273.5	29700	1915	ND	1360
2,3,7,8-TCDF	3.5	ND	10.5	ND	3.1
2,3,4,7,8-PnCDF	2132	49667	1825	2600	33
1,2,3,4,7,8/1,2,3,6,7,8-HxCDF	2927	177000	2050	3100	69
1,2,3,4,6,7,8-HpCDF	323.5	46333	250	ND	NA
1,2,3,4,6,8,9-HpCDF	ND	ND	38	ND	7.1
1,2,3,4,7,8,9-HpCDF	ND	1433	ND	ND	NA
OCDF	ND	ND	145	ND	NA
Total PCDDs	377.9	33166	2569.5	0	1534.6
Total PCDFs	5386	274433	4318.5	5700	429.1
Total PCDD/Fs	5763.9	307599	6888	5700	1963.7

Table 11
PCDDs and PCDFs in Taiwanese Placenta Samples Compared with 14 American Placentas (ppt, lipid)

	TEQ	American	Taiwan 12	Taiwan 13	Taiwan 14	Taiwan 15	Taiwan 16	Taiwan 17
2,3,7,8-TCDD	1	2.4	0.68	0.86	0.06	1.8	6.2	1.28
1,2,3,7,8-PeCDD	0.5	4.0	4.7	7.9	23.5	13.7	36.4	14.8
1,2,3,4,7,8-HxCDD	0.1	2.4	1.3	—	—	—	—	—
1,2,3,6,7,8-HxCDD	0.1	15.9	18.9	39.7	405	206	395	197.8
1,2,3,7,8,9-HxCDD	0.1	3.2	2.64	2.68	13.7	24.25	74	18.9
1,2,3,4,6,7,8-HpCDD	0.01	36.2	17	14.6	45.6	51	63.4	74.4
OCDD	0.001	282	189.6	119.2	294	1006	815.4	912.3
2,3,7,8-TCDF	0.1	1.9	1.64	0.82	4.2	4.1	9.4	1.61
2,3,4,7,8-PeCDF	0.5	3.6	744.8	984.9	5623.9	6.05	0.4	5.67
1,2,3,7,8-PeCDF	0.05	<1.0	1.64	1.18	10.33	5649.7	12558	2718.9
1,2,3,4,7,8-HxCDF	0.1	4.0	2131	163.92	29806.3	14595	26532	16902.78
1,2,3,6,7,8-HxCDF	0.1	2.0	40.5	55	624.3	272.4	—	—
1,2,3,7,8,9-HxCDF	0.1	1.7	<0.32	<0.23	<0.33	<0.4	<0.6	<0.28
2,3,4,6,7,8-HxCDF	0.1	2.0	<0.27	<0.09	<0.27	<0.35	<0.5	<0.22
1,2,3,4,6,7,8-HpCDF	0.01	6.3	48.9	520	762.3	387	430.6	463.1
1,2,3,4,7,8,9-HpCDF	0.01	<0.1	7.4	52	76.3	63.4	46.4	41.1
OCDF	0.001	<5.0	1.68	4.68	6.87	3.9	9.6	6.72
Total PCDDs		346.2	234.8	184.9	783.67	1303	1390.2	1220
Total PCDFs		23.5	2977.7	4288.3	36914.4	20981.35	39586.4	20139.9
Total PCDD/Fs		369.7	3212.5	4473.2	37698.1	22284.1	40976.6	21359.5
Total PCDD TEq		7.2	5.68	9	56.1	33.2	73	32.0
Total PCDF TEq		2.9	590.8	770.8	5864.3	1777.2	3287.0	1834.3
Total PCDD/F TEq		10.1	596.06	780	5920	1810	3360	1866

[a]Totals are rounded. Half of "<" values are used in totals. —, level included 123678-HxCDD/F. Samples obtained 5 to 5½ years after exposure.

Table 12

Total PCDD, PCDF, and TEq in Human Blood from General Populations, 1980–91 (ppt, lipid)[a]

	Vietnam (south) Dong Nai Pool = 33 (whole)	Vietnam (north) Hanoi Pool = 32 (whole)	Germany Mean N = 85 (whole)	America Pool = 100 (plasma)	Russia Baikal City Pool = 8 (whole)	Russia St. Petersburg Pool = 50 (whole)	Guam Mean N = 10 (whole)
Total PCDDs	1938	126	788	1499	88	130	1810
Total PCDFs	133	41	98	92	49	31	80
Total PCDD/Fs	2071	167	886	1591	137	161	1890
PCDD TEq	31	4	20	30	8	11	24
PCDF TEq	18	6	22	11	10	6	8
PCDD/F TEq	49	10	42	41	18	17	32

[a]Totals are rounded.

dibenzofuran congeners make a large contribution to the total toxic equivalents in these Taiwanese placentas, consistent with intake from the rice oil for the mothers and transplacental transfer to the fetuses. The placental dioxin toxic equivalent values, from 596 to 5920 ppt, reflect maternal body burden and compare with the U.S. value of 10 ppt, which is typical of blood or fat tissue TEq levels in industrial countries that characteristically vary between 15 and 40 ppt.

3.2. General Population PCDD and PCDF Levels

Dioxin human tissue levels and patterns vary in different geographical areas as can be seen from Tables 12, 13, and 14.[25,75,77–79,81,103,106–109] With some exceptions, dioxin levels are considerably higher than dibenzofuran levels and dioxin TEqs are higher than dibenzofuran TEqs. General population background human tissue dioxin levels can be valuable when used with matched controls for comparison when medically evaluating an exposed cohort. With the exception of Vietnam, where dioxin levels are high in the south reflecting Agent Orange and industrial contamination and low in the north reflecting their absence, there is usually little geographic variation within a given country, based on current data.

A somewhat different pattern and level of dioxin, dibenzofuran, and PCB congeners are present in various tissues of the human body, even after adjusting on a lipid basis to normalize to each tissue's lipid content.[35,50–52,110] This can be more striking for the higher chlorinated PCDD/Fs than for the lower chlorinated congeners as was noted previously for blood and adipose tissue. In estimation of actual exposure to and intake of PCDD/Fs, it is necessary to take into account exposure history, time after exposure, age, country of origin, the tissue sampled, and wet weight versus lipid-adjusted values. Methods of lipid quantification are not standard and may also play an important role in the interpretation of dioxin tissue levels.

General population dioxin tissue levels are considerably lower in less industrial countries, such as Thailand, China, India, Africa, the north of Vietnam, and

Table 13
Mean Adipose Tissue Levels for General Populations (1980s) (ppt, lipid)[a]

						Vietnam	
	USA $N = 15$	Germany $N = 4$	China $N = 7$	Japan $N = 6$	Canada $N = 46$	South, $N = 41$	North, $N = 26$
Total PCDDs	558	942	247	1535	1217	814	133
Total PCDFs	32	140	53	92	65	57	21
Total PCDD/Fs	590	1082	300	1627	1282	871	154
PCDD TEq	19	29	6	24	25	26	2
PCDF TEq	5	40	12	14	11	4	2
PCDD/F TEq	24	69	18	38	36	30	4

[a]Totals are rounded.

Table 14
Total PCDD and PCDF Levels in Pooled Milk Samples from Various Countries and Vietnamese Cities[a] (ppt, lipid)

	USA	Japan	Canada	Germany	South Africa		Pakistan	Vietnam			Thailand	Cambodia	Russia
					White	Black		Da Nang	HCM City	Hanoi			
	$N = 43$	$N = 6$	$N = 200$	$N = 185$	$N = 18$	$N = 6$	$N = 7$	$N = 11$	$N = 38$	$N = 30$	$N = 10$	$N = 8$	$N = 23$
Total PCDD	367	1085	493	289	356	280	249	406	302	104	82	77	73
Total PCDF	31	44	41	58	23	20	19	133	34	23	9	9	28
Total PCDD/Fs	398	1129	534	347	379	300	268	539	336	127	91	86	101
PCDD TEq	12	12	18	13	9	7	9	18	13	5	1	2	5
PCDF TEq	8	15	8	14	4	2	4	16	6	4	2	1	7
PCDD/F TEq	20	27	26	27	13	9	13	34	19	9	3	3	12

[a]Da Nang and HCM (Ho Chi Minh) City are located in the south of Vietnam, Hanoi is located in the north of Vietnam. Russian samples from Moscow, Baikalsk, Irkutsk, Novosibirsk, and Kachug.

Russia.[75,77–79,81,103,106–109] Higher levels of dioxins and dibenzofurans in human tissues are characteristic of more industrial countries at this time, because of higher chemical use and contamination. European human tissue specimens have a characteristic elevation of 2,3,4,7,8-PnCDF, relative to U.S. and Canadian samples, possibly from the more common use of leaded gasoline in Europe. German tissue generally has higher PCB content, presumably from greater contamination of the environment and food. These patterns are characteristic and could, in theory, be used in forensic medicine to help identify human tissue and establish country of origin. The persistence of dioxins, dibenzofurans, and PCBs is striking as seen following exposure in some human and environmental samples. Body burden may persist decades after environmental exposure, presenting long-term health risks.

3.3. PCBs in Human Tissue

An important new finding has been that coplanar, mono-, and di-*ortho*-substituted PCBs contribute a large share to the total dioxinlike toxicity in human tissue from industrial countries when calculated using proposed conservative dioxin toxic equivalency factors.[111] Table 15 shows the measured level of 34 dioxin and dioxinlike congeners and calculated dioxin toxic equivalents in blood from a pooled sample of six and the average of 50 individual analyses of male adults. The mono-*ortho*-substituted PCBs contribute 36.5 and 50.4 ppt TEq to the Missouri and Michigan samples, respectively. The coplanar PCBs contribute 6.8 and 13.5 ppt TEq to the Missouri and Michigan samples, respectively. Additively, the coplanar, mono-, and di-*ortho* PCBs markedly increase the total dioxinlike toxicity in these general population samples, adding 44.7 and 66.3 to total dioxin equivalents of 69 and 93 ppt.

These data, like those from Japan[86,112] and Atlanta, Georgia,[113,114] suggest that PCB human body burden and total current dose of dioxinlike chemicals may have to be reconsidered in general populations and in exposed groups. Because dioxins are now found in all humans, the term *nonexposed* can no longer be considered an accurate designation of the general population and the important tissue becomes the extent of the exposure.

Considering the large amount of PCBs currently found in human tissue relative to the smaller dioxin and dibenzofuran levels, the potential health effect of the dioxins may recently have been overestimated and that of PCBs underestimated. However, since for many toxic endpoints the effects of PCBs, PCDDs, and PCDFs are additive, the total dioxin and dioxinlike toxicity may be more important to consider. Recent work described in other chapters of this book suggests that the dioxinlike toxicity of PCBs may be less than first estimated,[118] but that nondioxinlike PCBs may be more toxic than previously thought to be the case.[115–117]

4. CONCLUSIONS AND DISCUSSION

For the most part, dioxins, dibenzofurans, and PCBs are historically new synthetic chemicals.[7,8,11,12] Because of the lack of information concerning kinetics in humans

Table 15
Mean Level of PCDDs, PCDFs, and PCBs in Individual Analyses from Michigan and Pooled Blood from Missouri (ppt, lipid)

	Missouri[a]	TEq	Michigan[b]	TEq
1) 2,3,7,8-TCDD	3.4	3.38	3.8	3.8
2) 1,2,3,7,8-PeCDD	7.1	3.35	9.3	4.6
3) 1,2,3,4,7,8-HxCDD	—[c]	—[c]	9.8	0.68
4) 1,2,3,6,7,8-HxCDD	67.5	6.75	72.1	7.2
5) 1,2,3,7,8,9-HxCDD	13.4	1.34	11.9	1.2
6) 1,2,3,4,6,7,8-HpCDD	155.0	1.55	118.6	1.2
7) OCDD	1208.0	1.21	793.9	0.8
8) 2,3,7,8-TCDF	3.19	0.3	2.3	0.2
9) 1,2,3,7,8-PeCDF	ND(2.1)	0.06	1.2	0.06
10) 2,3,4,7,8-PeCDF	7.0	3.50	8.8	4.4
11) 1,2,3,4,7,8-HxCDF	9.4	0.94	10.6	1.1
12) 1,2,3,6,7,8-HxCDF	6.04	0.6	6.9	0.6
13) 2,3,4,6,7,8-HxCDF	ND(5.7)	0.29	2.8	0.3
14) 1,2,3,7,8,9-HxCDF	ND(5.7)	0.29	2.8	0.3
15) 1,2,3,4,6,7,8-HpCDF	20.2	0.20	19.6	0.2
16) 1,2,3,4,7,8,9-HpCDF	ND(6.7)	0.34	3.1	0.03
17) OCDF	ND(16.7)	0.01	9.3	0.01
18) #77 Tetra PCB	34.2	0.34	78.6	0.8
19) #126 Penta PCB	49.8	4.98	104.4	10.4
20) #169 Hexa PCB	29.9	1.50	45.8	2.3
21) #28 2,4,4′-Tri PCB	10148	10.15	7170	7.0
22) #74 2,4,4′,5-Tetra PCB	7602	7.60	14330	14.3
23) #105 2,3,3′,4,4′-Penta PCB	3200	3.20	6928	6.9
24) #118 2,3′,4,4′,5-Penta PCB	11346	11.35	16213	16.2
25) #156 2,3,3′,4,4′,5-Hexa PCB	4202	4.20	5988	6.0
26) #99 2,2′,4,4′,5-Penta PCB	5328	0.11	11361	0.2
27) #128 2,2′,3,3′,4,4′-Hexa PCB	1200	0.02	2104	0.04
28) #138 2,2′,3,4,4′,5-Hexa PCB	14784	0.30	26297	0.5
29) #153 2,2′,4,4′,5,5′-Hexa PCB	23666	0.47	40055	0.8
30) #170 2,2′,3,3′,4,4′,5-Hepta PCB	4260	0.09	6620	0.1
31) #180 2,2′,3,4,4′,5,5′-Hepta PCB	12728	0.25	19034	0.4
32) #183 2,2′,3,4,4′,5′,6-Hepta PCB	1402	0.03	2534	0.05
33) #185 2,2′,3,4,5,5′,6-Hepta PCB	852	0.02	1284	0.03
34) #187 2,2′,3,4′,5,5′,6-Hepta PCB	3588	0.07	7378	0.2
Total PCDDs (1–7)	1454.4	17.6	1019	19.5
Total PCDFs (8–17)	64	6.5	67	7.2
Total coplanars (18–20)	114	6.8	229	13.5
Total mono-*ortho*s (21–25)	36498	36.5	50629	50.4
Total di-*ortho*s (26–34)	67808	1.4	116667	2.3
Grand total	105938	68.8	168611	92.9

[a]Pool, $N = 6$
[b]TCDD is mean of 44 individual analyses, all other congeners are mean of 50.
[c]1,2,3,4,7,8/1,2,3,6,7,8-HxCDD is sum of both.

and health effects of the dioxins, dibenzofurans, and PCBs, alone, in combination with similar chemicals, and in combination with chemically dissimilar chemicals, we are at an early learning stage in understanding how to use these measurements in medical evaluation and treatment. Metabolic, kinetic, clinical, and epidemiological studies, relying, as they must, on nondeliberate exposures of these toxic chemicals in humans, must be performed to add to the body of knowledge essential for more skillful use of these sensitive and specific biomarkers of exposure.[48]

Compared with the extremely crude and nonspecific biomarkers available to estimate dioxin exposure several decades ago, such as chloracne (a nonspecific skin rash or acne sometimes observed after exposure to chlorinated organic chemicals), chemists, physicians, and toxicologists have made considerable progress working together in this multidisciplinary area. However, there is still a long road before us in medicine and public health to learn to fully exploit the chemists' and others' newly developed and striking skills. Hopefully, with multidisciplinary scientific collaboration on exposure assessment and clinical consequences, it will be possible to move with more than deliberate speed to learn how better to help our patients and to understand the levels of dioxins and related chemicals which are associated with adverse health consequences in humans. It is clear, however, that these new tools for exposure assessment for dioxins and related chemicals place us far in advance of what is usual in the practice of occupational medicine today with respect to most chemical exposures.

ACKNOWLEDGMENTS. Some of the studies cited herein have been made possible by assistance from the Christopher Reynolds Foundation, the Samuel Rubin Foundation, the CS Fund, Church World Service, CIDSE, the Agent Orange Commission and Department of Public Health of the Commonwealth of Massachusetts, the State of Michigan, as well as the American Association for the Advancement of Sciences and the National Academy of Sciences, the latter for the work done originally in Vietnam and with Vietnamese milk, food, and wildlife specimens, ca. 1969–1974. The cooperation and expert dioxin chemistry of R. Baughman, T. Tiernan, J. J. Ryan, M. Gross, S. Raisanen, P. Fürst, O. Päpke, C. Rappe, K. Olie, J. Stanley, and K. Boggess are gratefully acknowledged. The expert technical assistance in preparation of this chapter is due to Mrs. Ruth Stento and Ms. Karen Charles. We especially thank all persons who participated in these studies by allowing tissue samples to be studied.

REFERENCES

1. H. Beck, K. Eckart, W. Mathar, and R. Wittkowski, PCDD and PCDF body burden from food intake in the Federal Republic of Germany, *Chemosphere* **18,** 417–424 (1989).
2. B. Birmingham, B. Thorpe, R. Frank, R. Clement, H. Tosine, G. Fleming, J. Ashman, J. Wheeler, B. D. Ripley, and J. J. Ryan, Dietary intake of PCDD and PCDF from food in Ontario, Canada, *Chemosphere* **19,** 507–512 (1989).
3. P. Fürst, C. Fürst, and W. Groebel, Levels of PCDDs and PCDFs in food-stuffs from the Federal Republic of Germany, *Chemosphere* **20**(7/9), 787–792 (1990).
4. A. J. Schecter, P. Fürst, C. Fürst, W. Groebel, J. D. Constable, S. Kolesnikov, A. Beim,

A. Boldonov, E. Trubitsun, B. Vlasov, Hoang Dinh Cau, Le Cau Dai, and Hoang Tri Quyng, Levels of chlorinated dioxins, dibenzofurans, and other chlorinated xenobiotics in food from the Soviet Union and the south of Vietnam, *Chemosphere* **20**(7/9), 799–806 (1990).
5. A. J. Schecter, P. Fürst, C. Fürst, M. Grachev, A. Beim, and V. Koptug, Levels of dioxins, dibenzofurans and selected other chlorinated organic compounds in food from Russia, *Chemosphere* **25,** 2009–2015 (1992).
5A. A. Schecter, J. Startin, C. Wright, M. B. Kelly, O. Päpke, A. Lis, M. Ball, and J. Dolsoy, Congener specific levels of dioxins and dibenzofurans in US food and estimated daily dioxin toxic equivalent intake, *Environ. Health Perspec.* **102,** 962–966 (1994).
6. J. M. Czuczwa and R. A. Hites, Environmental fate of combustion-generated polychlorinated dioxins and furans, *Environ. Sci. Technol.* **18,** 444–450 (1984).
7. J. M. Czuczwa, B. D. McVeety, and R. A. Hites, Polychlorinated dibenzo-p-dioxins and dibenzofurans in sediments from Siskiwit Lake, Isle Royale, *Science* **226,** 568–569 (1984).
8. J. M. Czuczwa and R. A. Hites, Sources and fate of PCDD and PCDF, *Chemosphere* **15,** 1417–1420 (1986).
9. J. M. Czuczwa and R. A. Hites, Airborne dioxins and dibenzofurans: Sources and fates, *Environ. Sci. Technol.* **20,** 195–200 (1986).
10. J. M. Czuczwa, F. Niessen, and R. A. Hites, Historical record of polychlorinated dibenzo-p-dioxins and dibenzofurans in Swiss lake sediments, *Chemosphere* **14**(9), 1175–1179 (1985).
11. A. J. Schecter, A. Dekin, N. C. A. Weerasinghe, S. Arghestani, and M. L. Gross, Sources of dioxins in the environment: A study of PCDDs and PCDFs in ancient, frozen Eskimo tissue, *Chemosphere* **17,** 627–631 (1988).
12. H. Y. Tong, M. L. Gross, A. J. Schecter, S. J. Monson, and A. Dekin, Sources of dioxins in the environment: Second stage study of PCDD/Fs in ancient human tissue and environmental samples, *Chemosphere* **20**(7/9), 987–992 (1990).
13. A. J. Schecter, B. D. Eitzer, and R. A. Hites, Chlorinated dioxin and dibenzofuran levels in sediments collected from rivers in Vietnam, 1984–6, *Chemosphere* **18,** 831–834 (1989).
14. A. J. Schecter and J. J. Ryan, Polychlorinated dibenzo-para-dioxin and dibenzofuran levels in human adipose tissues from workers 32 years after occupational exposure to 2,3,7,8-TCDD, *Chemosphere* **17**(5), 915–920 (1988).
15. A. J. Schecter and J. J. Ryan, Brominated and chlorinated dioxin blood levels in a chemist 34 years after exposure to 2,3,7,8-tetrachlorodibenzodioxin and 2,3,7,8-tetrabromodibenzodioxin, *Chemosphere* **23**(11/12), 1921–1924 (1991).
16. D. L. Phillips, A. B. Smith, V. W. Burse, G. K. Steele, L. L. Needham, and W. H. Hannon, Half-life of polychlorinated biphenyl in occupationally exposed workers, *Arch. Environ. Health* **44,** 351–354 (1989).
17. A. J. Schecter, Dioxins and dibenzofurans in exposed workers: Serial tissue levels in a worker exposed in a PCB transformer fire cleanup and blood levels in three exposed chemists, *Chemosphere* **25,** 1117–1122 (1992).
18. A. J. Schecter, J. J. Ryan, and P. J. Kostyniak, Decrease over a six year period of dioxin and dibenzofuran tissue levels in a single patient following exposure, *Chemosphere* **20,** 911–917 (1990).
19. M. Wolff and A. J. Schecter, Accidental exposure of children to polychlorinated biphenyls, *Arch. Environ. Contam. Toxicol.* **20,** 449–451 (1991).
20. A. J. Schecter, J. S. Stanley, K. Boggess, Y. Masuda, J. Mes, M. Wolff, P. Fürst, C. Fürst,

H. McGee, K. Wilson-Yang, and B. Chisholm, Polychlorinated biphenyl levels in the tissues of exposed and non-exposed humans, *Environ. Health Perspect.* **102,** 1 (1994).

21. A. J. Schecter, J. D. Constable, S. Arghestani, H. Tong, and M. L. Gross, Elevated levels of 2,3,7,8-tetrachlorodibenzodioxin in adipose tissue of certain U.S. veterans of the Vietnam war, *Chemosphere* **16**(8/9), 1997–2002 (1987).
22. A. J. Schecter, H. McGee, J. S. Stanley, and K. Boggess, Dioxin, dibenzofuran, and PCB, including coplanar PCB levels in the blood of Vietnam veterans in the Michigan Agent Orange Study, *Chemosphere* **25**(1/2), 205–208 (1992).
23. R. W. Baughman, *Tetrachlorodibenzo-p-dioxins in the environment: High resolution mass spectrometry at the picogram level,* Ph.D. thesis, Harvard University, Cambridge, MA (1974).
24. R. W. Baughman and M. Meselson, An analytical method for detecting TCDD (dioxin): Levels of TCDD in samples from Vietnam, *Environ. Health Perspect.* **September,** 27–35 (1973).
25. A. J. Schecter, Dioxins and related chemicals in humans and the environment, in: *Banbury Report 35: Biological Basis for Risk Assessment of Dioxins and Related Compounds* (M. Gallo, R. J. Scheuplein, and K. A. van der Heijden, eds.), pp. 169–213, Cold Spring Harbor Laboratory Press, Cold Spring Harbor, NY (1991).
26. J. S. Stanley, K. M. Bauer, K. Turman, K. Boggess, and P. Cramer, Final Report, in: *Determination of Body Burdens for Polychlorinated Dibenzo-p-dioxins (PCDDs) and Polychlorinated Dibenzofurans (PCDFs) in California Residents* (Midwest Research Institute, ed.), State of California Air Resources Board (1989).
27. A. J. Schecter, J. J. Ryan, and J. D. Constable, Chlorinated dibenzo-p-dioxin and dibenzofuran levels in human adipose tissue and milk samples from the north and south of Vietnam, *Chemosphere* **15**(9/12), 1613–1620 (1986).
28. C. Rappe, H. R. Buser, H. Kuroki, and Y. Masuda, Identification of polychlorinated dibenzofurans (PCDFs) retained in patients with Yusho, *Chemosphere* **7,** 259 (1978).
29. Y. Masuda, R. Kagawa, H. Kuroki, M. Kuratsune, T. Yoshimura, I. Taki, M. Kusuda, F. Yamashita, and M. Hayashi, Transfer of polychlorinated biphenyls from mothers to foetuses and infants, *Food Cosmet. Toxicol.* **16,** 543–546 (1978).
30. C. Rappe, M. Nygren, H. R. Buser, and T. Kauppinen, Occupational exposure to polychlorinated dioxins and dibenzofurans, in: *Chlorinated Dioxins and Related Compounds—Impact on the Environment* (O. Hutzinger, R. W. Frei, E. Merian, and F. Pocchiari, eds.), Pergamon Press, Elmsford, NY (1982).
31. M. L. Gross, J. O. Lay, Jr., P. A. Lyon, D. Lippstreu, N. Kangas, R. L. Harless, S. E. Taylor, and A. E. Dupuy, Jr., 2,3,7,8-tetrachlorodibenzo-p-dioxin levels in adipose tissue of Vietnam veterans, *Environ. Res.* **33,** 261 (1984).
32. A. J. Schecter, T. O. Tiernan, M. L. Taylor, G. F. vanNess, J. H. Garrett, D. J. Wagel, and G. Gitlitz, The use of fat biopsies to estimate patient exposure to polychlorinated dibenzofurans, polychlorinated dibenzoparadioxins, biphenylenes, and polychlorinated biphenyl isomers after an electrical transformer fire in Binghamton, New York, *American Chemical Society, 186th National Meeting, Washington, DC* **23,** 159–162 (1983).
33. A. J. Schecter, Contamination of an office building in Binghamton, New York by PCB's, dioxins, furans and biphenylenes after an electrical panel and electrical transformer incident, *Chemosphere* **12**(4/5), 669–680 (1983).
34. A. J. Schecter and T. O. Tiernan, Occupational exposure to polychlorinated dioxins, polychlorinated furans, polychlorinated biphenyls and biphenylenes after an electrical panel and transformer accident in an office building in Binghamton, NY, *Environ. Health Perspect.* **60,** 305–313 (1985).

35. A. J. Schecter, J. J. Ryan, and G. Gitlitz, Chlorinated dioxin and dibenzofuran levels in human adipose tissues from exposed and control populations, in: *Chlorinated Dioxins and Dibenzofurans in Perspective* (C. Rappe, G. Choudhary, and L. H. Keith, eds.), pp. 51–65, Lewis Publishers, Chelsea, MI (1986).
36. G. Eadon, L. Kaminsky, J. Silkworth, K. M. Aldous, D. Hilker, P. O'Keefe, R. Smith, J. Gierthy, J. Hawley, N. Kim, and A. Decaprio, Calculation of 2,3,7,8-TCDD equivalent concentrations of complex environmental contaminant mixtures, *Environ. Health Perspect.* **70,** 221–227 (1986).
37. Pilot Study on International Information Exchange on Dioxins and Related Compounds, *International Toxicity Equivalency Factor (I-TEF) Method of Risk Assessment for Complex Mixtures of Dioxins and Related Compounds,* Report No. 176, pp. 1–26, North Atlantic Treaty Organization, Committee on the Challenges of Modern Society (1988).
38. Pilot Study on International Information Exchange on Dioxins and Related Compounds, *Scientific Basis for the Development of the International Toxicity Equivalency Factor (I-TEF) Method of Risk Assessment for Complex Mixtures of Dioxins and Related Compounds,* Report No. 178, pp. 1–56, North Atlantic Treaty Organization, Committee on the Challenges of Modern Society (1988).
39. USEPA, *Interim procedures for estimating risks associated with exposures to mixtures of chlorinated dibenzo-p-dioxins and dibenzofurans (CDDs and CDFs) and 1989 update,* U.S. Department of Commerce, National Technical Information Service, Springfield, VA, PB90-145756 (1989).
40. A. J. Schecter and J. J. Ryan, Blood and adipose tissue levels of PCDDs/PCDFs over three years in a patient after exposure to polychlorinated dioxins and dibenzofurans, *Chemosphere* **18,** 635–642 (1989).
41. O. Päpke, M. Ball, Z. A. Lis, and K. Scheunert, PCDD/PCDF in whole blood samples of unexposed persons, *Chemosphere* **19,** 941–948 (1989).
42. P. W. Albro, W. B. Crummett, A. E. Dupuy, M. L. Gross, M. Hanson, R. L. Harless, F. D. Hileman, D. Hilker, C. Jason, J. L. Johnson, L. L. Lamparski, B. P. Y. Lau, D. D. McDaniel, J. L. Meehan, T. J. Nestrick, M. Nygren, P. O'Keefe, T. L. Peters, C. Rappe, J. J. Ryan, L. M. Smith, D. L. Stalling, N. C. A. Weerasinghe, and J. M. Wendling, Methods for the quantitative determination of multiple, specific polychlorinated dibenzo-p-dioxin and dibenzofuran isomers in human adipose tissue in the parts-per-trillion range. An interlaboratory study, *Am. J. Pathol.* **57**(13), 2717–2725 (1985).
43. World Health Organization, *Levels of PCBs, PCDDs and PCDFs in human milk and blood: Second round of quality control studies—Environment and Health in Europe #37,* pp. 1–76, FADL Publishers, Denmark (1991).
44. C. Rappe, S. Tarkowski, and E. Yrjanheikki, The WHO/EURO quality control study on PCDDs and PCDFs in human milk, *Chemosphere* **18,** 883–890 (1989).
45. S. Tarkowski and E. Yrjanheikki, WHO coordinated intercountry studies on levels of PCDDs and PCDFs in human milk, *Chemosphere* **19,** 995–1000 (1989).
46. A. J. Schecter, T. O. Tiernan, M. Taylor, G. F. vanNess, J. H. Garrett, D. J. Wagel, G. Gitlitz, and M. Bogdasarian, Biological markers after exposure to polychlorinated dibenzo-p-dioxins, dibenzofurans, biphenyls and biphenylenes. Part I: Findings using fat biopsies to estimate exposure, in: *Chlorinated Dioxins and Dibenzofurans in the Total Environment II* (L. Keith, C. Rappe, and G. Choudhary, eds.), pp. 215–245, Ann Arbor Science, Butterworths, Boston (1985).
47. H. Buser, Formation, occurrence and analysis of polychlorinated dibenzofurans, dioxins and related compounds, *Environ. Health Perspect.* **60,** 259–267 (1985).
48. Institute of Medicine, *Veterans and Agent Orange: Health Effects of Herbicides Used in Vietnam,* National Academy Press, Washington, DC (1993).

49. J. J. Ryan and Y. Masuda, Elimination of polychlorinated dibenzofurans (PCDFs) in humans from the Yusho and Yucheng rice oil poisonings, Presented at Dioxin '91, Research Triangle Park, NC, September 26 (Abstract) (1991).
50. D. G. Patterson, P. Fürst, L. O. Henderson, S. G. Isaacs, L. R. Alexander, W. E. Turner, L. L. Needham, and H. Hannon, Partitioning of in vivo bound PCDDs/PCDFs among various compartments in whole blood, *Chemosphere* **19,** 135–142 (1989).
51. A. J. Schecter, J. J. Ryan, J. D. Constable, R. Baughman, J. Bangert, P. Fürst, K. Wilmers, and R. P. Oates, Partitioning of 2,3,7,8-chlorinated dibenzo-p-dioxins and dibenzofurans between adipose tissue and plasma lipid of 20 Massachusetts Vietnam veterans, *Chemosphere* **20**(1/7), 951–958 (1990).
52. A. J. Schecter, O. Päpke, M. Ball, and J. J. Ryan, Partitioning of dioxins and dibenzofurans: Whole blood, blood plasma and adipose tissue, *Chemosphere* **23**(11/12), 1913–1919 (1991).
53. H. Poiger and C. Schlatter, Pharmacokinetics of 2,3,7,8-TCDD in man, *Chemosphere* **15,** 1489–1494 (1986).
54. J. L. Pirkle, W. H. Wolff, D. G. Patterson, L. L. Needham, J. E. Michalck, J. C. Miner, M. R. Peterson, and D. L. Phillips, Estimates of the half-life of 2,3,7,8-tetrachloro-dibenzo-p-dioxin in Vietnam veterans of Operation Ranch Hand, *J. Toxicol. Environ. Health* **270,** 165–171 (1989).
55. P. Mocarelli, L. L. Needham, A. Marocchi, D. G. Patterson, P. Brambilla, P. M. Gerthoux, L. Meazza, and V. Carreri, Serum concentrations of 2,3,7,8-tetrachlorodibenzo-p-dioxin and test results from selected residents of Seveso, Italy, *J. Toxicol. Environ. Health* **32,** 357–366 (1991).
56. A. J. Schecter, J. Constable, J. V. Bangert, J. Wiberg, M. Hansson, M. Nygren, and C. Rappe, Isomer specific measurement of polychlorinated dibenzodioxin and dibenzofuran isomers in human blood from American Vietnam veterans two decades after exposure to Agent Orange, *Chemosphere* **18**(1/6), 531–538 (1989).
57. A. J. Schecter, J. D. Constable, J. V. Bangert, H. Tong, S. Arghestani, S. Monson, and M. Gross, Elevated body burdens of 2,3,7,8-tetrachlorodibenzodioxin in adipose tissue of United States Vietnam veterans, *Chemosphere* **18**(1/6), 431–438 (1989).
58. A. H. Westing, Herbicides in war: Past and present, in: *Herbicides in War: The Long-Term Ecological and Human Consequences* (A. H. Westing, ed.), pp. 1–24, Taylor & Francis, London (1984).
59. H. Werner, *Epidemiology of Aging,* pp. 137–154, U.S. Department of Health, Education and Welfare, Washington, DC (1970).
60. J. S. Stanley, R. E. Ayling, P. H. Cramer, K. R. Thornburg, J. C. Remmers, J. J. Breen, J. Schwemberger, H. K. Kang, and K. Watanabe, Polychlorinated dibenzo-p-dioxin and dibenzofuran concentration levels in human adipose tissue samples from the continental United States collected from 1971 through 1987, *Chemosphere* **20**(7/9), 895–902 (1990).
61. J. S. Stanley, J. Onstot, and T. Sack, PCDDs and PCDFs in human adipose tissue from the EPA FY 82 NHATS repository, *Chemosphere* **15**(9/12), 1605–1612 (1986).
62. P. C. Kahn, M. Gochfeld, M. Nygren, M. Hansson, C. Rappe, H. Velez, T. Ghent-Guenther, and W. Wilson, Dioxins and dibenzofurans in blood and adipose tissue of Agent Orange-exposed Vietnam veterans and matched controls, *J. Am. Med. Assoc.* **259,** 1661–1667 (1988).
63. G. D. Lathrop, W. H. Wolfe, S. G. Machado, J. E. Michalek, T. G. Karrison, J. C. Miner, W. D. Grubbs, M. R. Peterson, and W. F. Thomas, *Epidemiologic investigation of health effects in Air Force personnel following exposure to herbicides: First followup examination, Results, Vol. 1,* National Technical Information Service, Springfield, VA (1987).
64. A. J. Schecter, H. McGee, J. S. Stanley, and K. Boggess, Chlorinated dioxin, di-

benzofuran, coplanar, mono-ortho, and di-ortho substituted PCB congener levels in blood and semen of Michigan Vietnam veterans compared with levels in Vietnamese exposed to Agent Orange, *Chemosphere* **27,** 241–252 (1993).

65. A. J. Schecter, H. McGee, J. S. Stanley, and K. Boggess, Chlorinated dioxin, dibenzofuran and coplanar PCB levels in blood and semen of Michigan Vietnam veterans, *Organohalogen Compounds Extended Abstracts from Dioxin '92* **9,** 231–234 (1992).
66. K. E. Boggess and J. S. Stanley, *Analysis of human blood and semen samples for PCDDs, PCDFs, and PCBs in support of the State of Michigan Vietnam Veteran Agent Orange Study, Report Nos. 9829-A and 9940-A,* Midwest Research Institute, Kansas City, MO (1993).
67. M. C. Hatch and Z. A. Stein, Agent Orange and risks to reproduction: The limits of epidemiology, *Teratogen. Carcinogen. Mutagen.* **6,** 185–202 (1986).
68. N. C. A. Weerasinghe, A. J. Schecter, J. C. Pan, R. L. Lapp, D. E. Giblin, J. L. Meehan, L. Hardell, and M. L. Gross, Levels of 2,3,7,8-tetrachlorodibenzo-p-dioxin (2,3,7,8-TCDD) in adipose tissue of U.S. Vietnam veterans seeking medical assistance, *Chemosphere* **15**(9/12), 1787–1794 (1986).
69. W. Wolfe, J. Michalek, J. C. Miner, A. Rahe, J. Silva, W. F. Thomas, W. D. Grubbs, M. B. Lustik, T. G. Karrison, R. H. Roegner, and D. E. Wiliams, Health status of Air Force veterans occupationally exposed to herbicides in Vietnam. I. Physical health, *J. Am. Med. Assoc.* **264,** 1824–1831 (1990).
70. J. E. Michalek, W. H. Wolfe, and J. C. Miner, Health status of Air Force veterans occupationally exposed to herbicides in Vietnam. II. Mortality, *J. Am. Med. Assoc.* **264,** 1832–1836 (1990).
71. H. K. Kang, K. K. Watanabe, J. Breen, J. Remmers, M. G. Conomos, J. S. Stanley, and M. Flicker, Dioxins and dibenzofurans in adipose tissue of U.S. Vietnam veterans and controls, *Am. J. Public Health* **81,** 344–349 (1991).
72. M. J. Fett, Australian veterans health studies: The mortality report, part I: A retrospective cohort study of mortality among Australian national servicemen of the Vietnam conflict era, and an executive summary of the mortality report, *Australian Government Public Service Publication* (1984).
73. Centers for Disease Control, Health status of Vietnam veterans. II. Physical health. The Centers for Disease Control Vietnam Experience Study, *J. Am. Med. Assoc.* **259,** 2708–2714 (1988).
74. A. J. Schecter, H. Y. Tong, S. J. Monson, and M. L. Gross, Levels of 2,3,7,8-TCDD in silt samples collected between 1985–86 from rivers in the north and south of Vietnam, *Chemosphere* **19,** 547–550 (1989).
75. A. J. Schecter, O. Päpke, M. Ball, M. Grachev, A. Beim, V. Koptug, Hoang Dinh Cau, Le Cao Dai, Hoang Tri Quynh, Nguyen Ngoc Thi Phuong, and Huynh Kim Chi, Dioxin and dibenzofuran levels in human blood samples from Guam, Russia, Germany, Vietnam and the USA, *Chemosphere* **25,** 1129–1134 (1992).
76. A. J. Schecter, O. Päpke, M. Ball, Hoang Dinh Cau, Le Cao Dai, Nguyen Quang Ming, Hoang Trong Quynh, Nguyen Ngoc Thi Phuong, Pham Hoang Phiet, Huynh Kim Chi, Dieu Thieu Vo, J. D. Constable, and J. Spencer, Dioxin and dibenzofuran levels in blood and adipose tissue of Vietnamese from various locations in Vietnam in proximity to Agent Orange spraying, *Chemosphere* **250,** 1123–1128 (1992).
77. A. J. Schecter, P. Fürst, C. Fürst, O. Päpke, M. Ball, Le Cao Dai, Hoang Tri Quynh, Nguyen Thi Ngoc Phuong, A. Beim, B. Vlasov, V. Chongchet, J. D. Constable, and K. Charles, Dioxins, dibenzofurans and selected chlorinated organic compounds in human milk and blood from Cambodia, Germany, Thailand, the U.S.A., the U.S.S.R., and Vietnam, *Chemosphere* **23**(11/12), 1903–1912 (1991).

78. A. J. Schecter, H. Y. Tong, S. J. Monson, M. L. Gross, S. Raisanen, T. Karhunen, E. K. Osterklund, J. D. Constable, Hoang, Dinh Cau, Le Cao Dai, Hoang Tri Quynh, Ton Duc Lang, Nguyen Thi Ngoc Phuong, Phan Hoang Phiet, and D. Vu, Human adipose tissue dioxin and dibenzofuran levels and "dioxin toxic equivalents" in patients from the north and south of Vietnam, *Chemosphere* **20**(7/9), 943–950 (1990).
79. A. J. Schecter, J. R. Startin, M. Rose, C. Wright, I. Parker, D. Woods, and H. Hansen, Chlorinated dioxin and dibenzofuran levels in human milk from Africa, Pakistan, southern Vietnam, the southern USA and England, *Chemosphere* **20**(7/9) 919–926 (1990).
80. A. J. Schecter, H. Y. Tong, S. J. Monson, M. L. Gross, and J. Constable, Adipose tissue levels of 2,3,7,8-TCDD in Vietnamese adults living in Vietnam, 1984–87, *Chemosphere*, **18,** 1057–1062 (1989).
81. A. J. Schecter, P. Fürst, J. J. Ryan, C. Fürst, H.-A. Meemken, W. Groebel, J. Constable, and D. Vu, Polychlorinated dioxin and dibenzofuran levels from human milk from several locations in the United States, Germany and Vietnam, *Chemosphere* **19,** 979–984 (1989).
82. A. J. Schecter, M. Gross, and J. J. Ryan, Biological markers of exposure to chlorinated dioxins and dibenzofurans in the United States and Vietnam, in: *Hazardous Materials Disposal: Siting and Management* (Ma. Chatterji, ed.), pp. 79–90, Gower Publishers, Aldershot, England (1987).
83. A. J. Schecter, J. J. Ryan, and J. D. Constable, Polychlorinated dibenzo-p-dioxin and polychlorinated dibenzofuran levels in human breast milk from Vietnam compared with cow's milk and human breast milk from the North American continent, *Chemosphere* **16**(8/9), 2003–2016 (1987).
84. A. J. Schecter, J. J. Ryan, M. Gross, N. C. A. Weerasinghe, and J. D. Constable, Chlorinated dioxins and dibenzofurans in human tissues from Vietnam, 1983–84, in: *Chlorinated Dioxins and Dibenzofurans in Perspective* (C. Rappe, G. Choudhary, and L. H. Keitheds.), pp. 35–50, Lewis Publishers, Chelsea, MI (1986).
85. W. J. Rogan, B. C. Gladen, K. L. Hung, S. L. Koong, L. Y. Shih, J. S. Taylor, Y. C. Wu, D. Yang, N. B. Ragan, and C. C. Hsu, Congenital poisoning by polychlorinated biphenyls and their contaminants in Taiwan, *Science* **241,** 334–336 (1988).
86. Y. Masuda and H. Yoshimura, Polychlorinated biphenyls and dibenzofurans in patients with Yusho and their toxicological significance: A review, *Am. J. Ind. Med.* **5,** 31–44 (1984).
87. M. Kuratsune and R. Shapiro (eds.), *PCB Poisoning in Japan and Taiwan,* Liss, New York (1984).
88. W. J. Rogan, B. C. Gladen, J. D. McKinney, N. Carreras, P. Hardy, J. Thullen, J. Tinglestad, and M. Tully, Neonatal effects of transplacental exposure to PCBs and DDE, *J. Pediatr.* **109,** 335–341 (1986).
89. W. J. Rogan, B. C. Gladen, J. D. McKinney, N. Carreras, P. Hardy, J. Thullen, J. Tingelstad, and M. Tully, Polychlorinated biphenyls (PCBs) and dichlorodiphenyl dichloroethene (DDE) in human milk: Effects of maternal factors and previous lactation, *Am. J. Public Health* **76,** 172–177 (1986).
90. T. A. Mably, D. L. Bjerke, R. W. Moore, A. Gendron-Fitzpatrick, and R. E. Peterson, In utero and lactational exposure of male rats to 2,3,7,8-tetrachlorodibenzo-p-dioxin: Effects on spermatogenesis and reproductive capability, *Toxicol. Appl. Pharmacol.* **114,** 118–126 (1992).
91. T. A. Mably, R. W. Moore, and R. E. Peterson, In utero and lactational exposure of male rats to 2,3,7,8-tetrachlorodibenzo-p-dioxin: Effects on androgenic status, *Toxicol. Appl. Pharmacol.* **114,** 97–107 (1992).
92. T. A. Mably, R. W. Moore, R. W. Goy, and R. E. Peterson, In utero and lactational exposure of male rats to 2,3,7,8-tetrachlorodibenzo-p-dioxin: Effects on sexual behavior

and the regulation of luteinizing hormone secretion in adulthood, *Toxicol. Appl. Pharmacol.* **114,** 108–117 (1992).

93. K. Olie, P. L. Vermeulen, and O. Hutzinger, Chlorodibenzo-p-dioxins and chlorodibenzofurans are trace components of fly ash and flue gas of some municipal incinerators in The Netherlands, *Chemosphere,* **6,** 455–459 (1977).
94. J. W. A. Lustenhouwer, K. Olie, and O. Hutzinger, Chlorinated dibenzo-p-dioxins and related compounds in incinerator effluents: A review of measurements and mechanisms of formation, *Chemosphere* **9,** 501–522 (1980).
95. L. C. Dickson, D. Lenoir, and O. Hutzinger, Surface-catalyzed formation of chlorinated dibenzodioxins and dibenzofurans during incineration, *Chemosphere* **19,** 277–282 (1989).
96. A. J. Schecter, R. Malkin, O. Päpke, M. Ball, and P. W. Brandt-Rauf, Dioxin levels in blood of municipal incinerator workers, *Med. Sci. Res.* **19,** 331–332 (1991).
97. A. J. Schecter and J. J. Ryan, Chlorinated and brominated dioxin levels in the blood of a chemist who became ill after synthesizing 2,3,7,8-TCDD and 2,3,7,8-TbrDD, in: *Organohalogen Compounds,* Vol. 4, pp. 141–144, Ecoinforma Press, Bayreuth (1991).
98. A. J. Schecter and J. J. Ryan, Persistent brominated and chlorinated dioxin blood levels in a chemist, *J. Occup. Med.* **34,** 702–707 (1992).
99. A. J. Schecter, B. J. Poiesz, P. W. Brandt-Rauf, O. Päpke, and M. Ball, Dioxin levels in blood of AIDS patients and controls, *Med. Sci. Res.* **19,** 273–275 (1991).
100. A. J. Schecter, J. J. Ryan, O. Päpke, M. Ball, and A. Lis, Elevated dioxin levels in the blood of male and female Russian workers with and without chloracne 25 years after phenoxyherbicide exposure: The Ufa "Khimprom" incident, *Chemosphere* **27,** 253–258 (1993).
101. A. J. Schecter and J. J. Ryan, Exposure of female production workers and their children in Ufa, Russia to PCDDs/PCDFs/planar PCBs, in: *Organohalogen Compounds: Short Papers from Dioxin '93* (H. Fiedler et al., eds.), pp. 55–58, Federal Environmental Agency, Vienna (1993).
102. A. Fischbein and M. S. Wolff, Conjugal exposure to polychlorinated biphenyls (PCBs), *Br. J. Ind. Med.* **44,** 284–286 (1987).
103. J. J. Ryan, A. J. Schecter, Y. Masuda, and M. Kikuchi, Comparison of PCDDs and PCDFs in the tissues of Yusho patients with those from the general population in Japan and China, *Chemosphere* **16**(8/9), 2017–2025 (1987).
104. J. E. Huff, Chemical toxicity and chemical carcinogenesis. Is there a causal connection? A comparative morphological evaluation of 1500 experiments, in: *Mechanisms of Carcinogenesis in Risk Identification* (H. Vainio, P. N. Magee, D. B. McGregor, and A. J. McMichael, eds.), pp. 428–466, International Agency for Research on Cancer, Lyon, France (1992).
105. W. J. Rogan, Yu-cheng, in: *Halogenated Biphenyls, Terphenyls, Naphthalenes, Dibenzodioxins and Related Products* (R. D. Kimbrough and A. A. Jensen, eds.), pp. 401–415, Elsevier, Amsterdam (1989).
106. A. J. Schecter, J. J. Ryan, and J. D. Constable, Chlorinated dioxins and dibenzofurans in human milk from Japan, India, and the United States of America, *Chemosphere* **18**(1/6), 975–980 (1989).
107. A. J. Schecter, P. Fürst, C. Fürst, W. Groebel, S. Kolesnikov, M. Savchenkov, A. Beim, A. Boldonov, E. Trubitsun, and B. Vlasov, Levels of dioxins, dibenzofurans, and other chlorinated xenobiotics in human milk from the Soviet Union, *Chemosphere* **20**(7/9), 927–934 (1990).
108. A. J. Schecter, A. Di Domenico, L. Turrio-Baldassarri, and J. J. Ryan, Dioxin and dibenzofuran levels in the milk of women from four geographical regions in Italy as

compared to levels in other countries, *Organohalogen Compounds Extended Abstracts from Dioxin '92* **9,** 227–230 (1992).

109. A. J. Schecter, O. Päpke, M. Ball, and Y. Masuda, Distribution of dioxins and dibenzofurans in blood from Japan, Israel, Russia, Guam, Vietnam, Germany, and the U.S.A., *Organohalogen Compounds Extended Abstracts from Dioxin '92* **9,** 239–242 (1992).
110. A. J. Schecter, J. Mes, and D. Davies, Polychlorinated biphenyl (PCB), DDT, DDE and hexachlorobenzene (HCB) and PCDD/F isomer levels in various organs in autopsy tissue from North American patients, *Chemosphere* **18,** 811–818 (1989).
111. S. Safe, Polychlorinated biphenyls (PCBs), dibenzo-p-dioxins (PCDDs), dibenzofurans (PCDFs), and related compounds: Environmental and mechanistic considerations which support the development of toxic equivalency factors (TEFs), *Crit. Rev. Toxicol.* **21,** 51–88 (1990).
112. J. J. Ryan, D. Levesque, L. G. Panopio, W. F. Sun, Y. Masuda, and H. Kuroki, Elimination of polychlorinated dibenzofurans (PCDFs) and polychlorinated biphenyls (PCBs) from human blood in the Yusho and Yu-cheng rice oil poisonings, *Arch. Environ. Contam. Toxicol.* **24,** 504–512 (1993).
113. D. G. Patterson, C. R. Lapeza, E. R. Barnhart, D. F. Groce, and V. W. Burse, Gas chromatographic/mass-spectrometric analysis of human serum for non-ortho (coplanar) and ortho substituted PCBs using isotope-dilution mass spectrometry, *Chemosphere* **19,** 127–134 (1989).
114. D. G. Patterson, G. D. Todd, W. E. Turner, S. G. Isaacs, and L. L. Needham, Levels of non-ortho-substituted polychlorinated biphenyls, dibenzo-p-dioxins, and dibenzofurans in human serum and adipose tissue, in: *Organohalogen Compounds,* Vol. 4, pp. 133–136, Ecoinforma Press, Bayreuth (1990).
115. R. F. Seegal and W. Shain, Neurotoxicity of polychlorinated biphenyls: The role of ortho-substituted congeners in altering neurochemical function, in: *The Vulnerable Brain and Environmental Risks,* Volume 2, *Toxins in Food* (R. L. Isaacson and K. F. Jensen, eds.), pp. 169–195, Plenum Press, New York (1992).
116. M. J. DeVito, W. Maier, J. Diliberto, and L. S. Birnbaum, Comparative ability of various PCBs, PCDFs and TCDD to induce hepatic cytochrome P-450 1A1 and 1A2 activity following 4 weeks of treatment, *Fundam. Appl. Toxicol.* **20,** 125–130 (1993).
117. C. D. Hebert, M. W. Harris, M. R. Elwell, and L. S. Birnbaum, Relative toxicity and tumor-producing ability of 2,3,7,8-tetra chlorodibenzo-p-dioxin (TCDD), 2,3,4,7,8-pentachlorodibenzofuran (PCDF), and 1,2,3,4,7,8-hexachlorodibenzofuran (HCDF) in hairless mice, *Toxicol. Appl. Pharmacol.* **102,** 362–377 (1990).
118. U. G. Ahlgorg, G. C. Beckina, L. S. Birnbaum, A. Brouwer, H. J. G. M. Derks, M. Feeley, G. Golor, A. Hanberg, J. C. Larsen, A. K. D. Liem, S. H. Safe, C. Schlatter, F. Waern, M. Younes, and E. Yrjanheikki, Toxic equivalency factors for dioxin-like PCBs, *Chemosphere* **2816,** 1049–1067 (1994).

Chapter 15

Human Health Effects of Polychlorinated Biphenyls

William J. Nicholson and Philip J. Landrigan

1. INTRODUCTION

Industrial use of PCBs began in 1929 and since then PCBs have found wide use as dielectric fluids in electrical transformers and capacitors, as heat exchange or hydraulic fluids, and in a variety of other commercial applications. PCBs are a group of 209 structurally different chlorobiphenyl congeners consisting of two linked phenyl rings on which can be located from 1 to 10 chlorine atoms. Commercially produced PCBs consist of mixtures of from 50 to 90 individual congeners. The principal product used in the United States was manufactured by the Monsanto Company and sold under the trade name "Aroclor." Table 1 lists the approximate percentages of PCBs in the most commonly used Aroclor mixtures. While many isomers exist for a specific number of chlorine atoms per molecule, nearly half the PCB congeners do not occur in the commercial mixtures.

PCBs are often used in conjunction with tri- or tetra-chlorobenzenes. In addition, variable amounts of contaminants may be present in some mixtures. Among contaminants of considerable concern are the polychlorinated dibenzofurans (PCDFs), which are formed during the pyrolysis of PCBs and can be formed during PCB production. During the mid-1970s Bowes and co-workers found contamination of Aroclor 1248, 1254, and 1260 by tetra-, penta-, and hexa-CDF at the μg/g level; no PCDFs were found in Aroclor 1016.[2] The data on Aroclor 1254 and 1260 were confirmed by Rappe and Buser.[3] Thus, studies of humans exposed to PCBs are complicated by the multiplicity of PCB congeners in the exposure and may also suffer from confounding by other potentially toxic compounds.

William J. Nicholson and Philip J. Landrigan • Department of Community Medicine, Mount Sinai School of Medicine, New York, New York 10029.

Dioxins and Health, edited by Arnold Schecter. Plenum Press, New York, 1994.

Table 1
The Composition and Degree of Chlorination of Commercial Mixtures of PCBs[a]

Commercial PCB product	Percent chlorine		Percentage of isomers with indicated number of chlorine atoms									
			1	2	3	4	5	6	7	8	9	10
		Number of isomers possible:	3	12	24	42	46	42	24	12	3	1
Aroclor 1232	32		26	29	24	14						
Aroclor 1016	41		2	19	57	22						
Aroclor 1242	42		3	13	28	30	22	4				
Aroclor 1248	48			2	18	40	36	4				
Aroclor 1254	54					11	49	34	6			
Aroclor 1260	60						12	38	41	8	1	

[a]From Ref. 1.

With the exception of Aroclor 1016, which contains 41% chlorine by weight, the last two digits of the numerical designation represents the weight percentage of chlorine in the mixture. From their first production until manufacturing ceased in 1977, 647,700 metric tonnes were produced in the United States.[1] Of this amount about 40% was used as dielectric and insulating fluids in electrical capacitors and transformers. Between 1970 and 1972, sales of PCBs for other than closed electrical equipment was voluntarily stopped by the Monsanto Company.[4]

Different Aroclors tended to be used in different calendar years and for different purposes. In U.S. electrical equipment manufacturing, Aroclor 1260 was more used prior to 1950, as was Aroclor 1254. Aroclor 1242 was the dominant mixture used in the 1950s and 1960s. It was phased out in 1971 and replaced with Aroclor 1016, which has about the same percentage of chlorine by weight, but very few PCBs with four or more chlorine atoms. This variation in usage leads to variation in the type of human exposure over time and also geographically. Descriptions of health effects in exposed populations may make reference to the type of PCB mixture, but the exposure information over time usually is uncertain.

1.1. Sources of Human Exposure

PCBs are extremely stable and resistant to acids, alkalies, hydrolysis, and heat. These properties were important for their industrial use, but they also contributed to their wide presence and persistence in the human environment. Mean PCB air concentrations prior to 1980 in rural and marine locations ranged from 0.1 to 1 ng/m^3 and up to 10 ng/m^3 in urban areas.[5] Decreases from these values may have occurred in recent years, but releases to the atmosphere still occur from contaminated disposal sites, releases from electrical equipment and fires. Moreover, even higher concentrations have been found in buildings, perhaps from PCBs released from electrical appliances and fluorescent lighting ballasts.[6] Nevertheless, inhalation is not the major source of exposure to members of the general population; the mean adult daily exposure is likely to be less than 1 ng/kg.

Air exposures and also dermal exposures are of substantial importance in the workplace. PCB concentrations as high as 10.5 mg/m^3 in air were reported by Elkins[7] in capacitor manufacturing. However, levels reported in such plants in more recent years were usually less than 1 mg/m^3.[8–10] At a concentration of 200 μg/m^3 and breathing 1 m^3/hr during moderate work, a 70-kg adult would have an exposure of about 30 μg/kg per day from inhalation. This exposure could be increased considerably, perhaps by severalfold, through skin exposure during work.

Past improper disposal of PCBs and PCB-containing equipment in landfills also can be a source of continuing exposure through skin contact by workers and others. Occasional releases of PCBs and other compounds to the environment can occur from fires or other accidents involving electrical transformers or capacitors. A well-described accident is that which occurred in the State Office Building in Binghamton, NY, during February, 1981.[11] Lightning strikes to power equipment are another source of human exposure.[12] While large amounts of PCBs and PCDFs can be released in such accidents, rapid responses by appropriately trained emergency and decontamination crews can limit population exposures.

PCBs have contaminated waterways, either from the discharge of waste containing PCBs into rivers and lakes or from the atmospheric deposition of PCBs into large water systems such as the Great Lakes. Once in a water system, PCBs absorb onto suspended particulates and settle to the bottom. Over time, PCBs from both the water and sediment can be taken up by the system biota and enter the food chain, where the PCBs concentrate. Levels in fish can exceed by 100,000 times the water levels in the system in which they live. For example, during the late 1970s, in a region of the Hudson River north of Troy, NY, which is contaminated with PCBs, river water usually had concentrations less than 1 ppb, whereas fish in the region could be contaminated to levels in excess of 100 ppm on a wet weight basis.[13]

Fish is one of the principal sources of PCB intake in the general population. Others are chicken and meat products. These are much less contaminated than fish, but eaten in greater quantity by the population. In recent years most market fish concentrations in the United States were less than 1 ppm; the geometric mean of samples collected in 1984 was 0.39 ppm.[14] Considering such fish concentrations as representative of eaten fish and assuming a consumption of 15 pounds/year, the estimated intake for a 70-kg adult is about 0.1 μg/kg per day. PCB concentrations tend to be higher in sport fish than in market fish; consumers of the same quantity of trout or salmon may be expected to have a greater intake. For example, Michigan residents who consumed 40 kg of fish per year, including sport fish, were found to have serum PCB concentrations of 366 ppb, compared with levels of 7 ppb in individuals who ate no fish.[15]

1.2. Evidence of Human Exposure

As mentioned above, the composition of PCBs to which individuals are exposed is highly variable. The composition is further altered by selective degradation of certain congeners in the ecosystem and enhancement of others in the food chain. Once in the human body, further alteration of the distribution of congeners occurs. Lower chlorinated PCBs are more readily and rapidly metabolized, yielding a greater percentage of the higher chlorinated congeners over time. Because PCBs are lipophilic, they concentrate in the adipose tissues of fish, other food products, and humans. With widespread contamination of the food chain, virtually all humans have measurable quantities of PCBs in their body tissues. Serum levels of PCBs in the general population are now generally less than 10 ng/ml, but can be higher depending on food consumption, especially sport fish.[16] Median adipose PCB concentrations in adults are now less than 1 mg/kg adipose tissue; prior to 1973 the median would have exceeded 1 mg/kg.[17] In the United States, median PCB levels in breast milk fat range from 1.5 to 2 mg/kg. A survey of 1057 nursing mothers in Michigan found a mean level of 1.5 ppm on a lipid basis.[18] The median concentration in a survey of 733 women in North Carolina was 1.8 ppm, with a maximum concentration of 17 ppm.[19] As human milk has from 2.5 to 4.5% fat, consumption of 150 ml/kg of milk per day by a baby could lead to average consumption levels of 6–12 μg/kg per day, with extreme levels as high as 100 μg/kg per day.

1.3. Rice Oil Exposures in Japan and Taiwan

PCBs were originally thought to be of low toxicity, in part because of their high chemical stability, but a poisoning incident in Japan during 1968 demonstrated that severe health consequences could arise from their ingestion. The poisoning incident resulted from the accidental contamination of rice oil by PCBs used as a heat exchange fluid during the rice oil production and other chemicals formed from PCBs. At the temperature of use, some of the PCBs had been converted to PCDFs. The clinical manifestations of those affected was termed *Yusho,* the Japanese term for rice oil disease. While the Yusho symptoms were initially ascribed to PCBs, it was soon established that the PCDFs were the predominant cause of the disease and of a later similar poisoning in Taiwan, the resulting disease being given the name *Yu-cheng*. Nevertheless, because of the severity of the Yusho symptoms and the uncertainty of its causation, numerous epidemiological studies were initiated to obtain data on the mortality and morbidity of workers exposed to high concentrations of PCBs. This review will focus largely on the results of the PCB studies. However, because of the similarity of effects and mode of action of PCBs and PCDFs in animals, it is worthwhile to consider briefly the Yusho incident to obtain guidance in evaluating human health effects from relatively pure PCB exposures.

Approximately 1800 individuals were identified as having been exposed during this accidental poisoning episode.[20] The principal manifestations of Yusho were a variety of lesions of the skin and mucous membranes. Black comedones, acneform eruptions, pigmentation of the face, eyelids, gingiva, and nails, and hypersecretion of the meibomian glands were commonly found. As time progressed, these symptoms diminished somewhat, but systemic disorders such as headache, gastric problems, joint swelling and pain, numbness of extremities and bronchitis-like symptoms developed in some of the individuals. Women experienced irregular menstrual cycles and weights of children born to exposed mothers were less than normal. Further, the children had retarded growth, abnormal tooth development, and some showed pigmentation characteristic of the exposure. Blood chemistries of exposed individuals revealed increased levels of triglycerides and decreased bilirubin. While jaundice was initially reported as a subjective symptom, no later reference to it has been made. These findings and the relative absence of alterations of the concentrations of enzymes reflecting liver function in the serum would suggest that hepatocellular injury (separate from an increased cancer risk) was not a major factor in patients with Yusho disease.[20–22] The above findings cannot be attributed to the Yusho patients' PCB exposure. Their PCB serum concentrations were much less than those individuals heavily exposed to PCBs occupationally, in whom most of the above symptoms were absent.[21]

1.4. Toxic Equivalent Factors

In addition to PCDFs, effects from exposure to the chlorinated dibenzodioxins (PCDDs) are also similar to those of the PCBs. Poland and co-workers[23,24] have demonstrated that the halogenated aromatic hydrocarbons, especially 2,3,7,8-tetrachlorodibenzodioxin (TCDD), are potent inducers of aryl hydrocarbon hydroxylase (AHH) and that the toxicity of the various isomers of the PCBs, PCDDs, and PCDFs strongly correlates with AHH inducibility. A binding receptor for TCDD and other chlorinated hydrocarbons, the Ah locus, in the liver cytosol has been identified, to which the binding affinity is directly proportional to AHH inducibility and congener toxicity. This suggests that many of the effects of the PCDDs, the PCDFs, and the PCBs are mediated through this receptor.

Safe[25] has reviewed the data on the toxicity of the different halogenated aromatic hydrocarbons relative to TCDD, the most toxic of the group. Binding to the Ah locus is found to be greatest for planar molecules of the size of TCDD (3×10 Å). Structure–binding relationships for the PCBs suggest that the most toxic compounds are those where the two phenyl rings can be coplanar. These PCBs have chlorines at both *para* positions and at at least two *meta* positions, but no chlorines at *ortho* positions (*ortho*-chlorines prevent coplanarity). These PCB congeners, 3,3′,4,4′-tetra-, 3,4,4′,5-tetra-, 3,3′,4,4′,5-penta-, and 3,3′,4,4′,5,5′-hexachlorobiphenyl, are approximately isosteric with TCDD and are more likely to have TCDD-like activity. Thus, the greater activity of those PCBs having a planar structure, and the similarity of action of the PCBs, the PCDFs, and the PCDDs can be understood on the basis of molecular structure. The relative toxicity of various compounds has been characterized by individual "toxic equivalent factors" (TEFs). Those estimated by Safe are listed in Table 2 and show the coplanar PCBs to have $^1/_{10}$th to $^1/_{100}$th the toxicity of 2,3,7,8-TCDD. The TEFs are estimated from a variety of toxicity data, which are not fully available for all congeners. Thus, they are only a "best estimate" for a given toxic effect.

Only recently are studies of PCB exposure or effect done with good information

Table 2
Relative Toxic Potencies and Proposed TEFs for PCB Congeners[a]

Congener	Potency range[b] (*in vivo* and *in vitro*)	TEF
2,2′,4,4′,5-Penta-CB	0.3–0.0006	0.1
3,3′,4,4′,5,5′-Hexa-CB	0.1–0012	0.05
3,3′,4,4′-Tetra-CB	0.009–0.00008	0.01
Mono-*ortho* coplanar PCBs	0.00045–0.0000014	0.001
Di-*ortho* coplanar PCBs	0.00002	0.00002

[a]From Safe.[25]
[b]The potencies of the individual congeners were determined relative to 2,3,7,8-TCDD for several different Ah receptor-mediated responses.

on the concentrations of individual congeners. Obviously, because of the substantial difference in TEFs for different congeners, congener- or congener-group-specific exposure information is useful. At present, however, virtually all health effects data that we have are related to total PCBs and, thus, this review will not attempt to relate human health effects to specific congener concentrations.

2. HUMAN MORBIDITY STUDIES OF INDIVIDUALS EXPOSED TO PCBs

2.1. PCB Effects on the Skin and Other Cutaneous Tissues

Skin lesions have long been associated with exposures to chlorinated aromatic hydrocarbons, especially the chlorinated naphthalenes. One of the first outbreaks of chloracne associated with PCB exposure was reported by Jones and Alden in 1936[28] who described the development of chloracne in 23 of 24 workers exposed to PCBs (and perhaps some PCDFs as well) during distillation processes. See Taylor[26,27] for a review of environmental chloracne.

In one of the first cross-sectional clinical studies of heavily exposed occupational groups, Fischbein *et al.*[29,30] identified skin lesions as the dominant abnormalities associated with PCB exposure. A history of skin rashes was reported by 39% of an examined population (326 male and female capacitor manufacturing employees) and burning sensations by 25%. Development of acne following employment was reported by 11% of workers. Hyperpigmentation, thickening of the skin, and nail discoloration were each reported by approximately 3% of the population. Thirty-seven percent of the examined individuals were found to have current skin abnormalities with the probability of occurrence correlating with serum PCB concentrations at a statistically significant level. The most commonly found abnormalities were erythema, swelling, dryness, and thickening. Sixteen individuals (5%) were found to have acneform eruptions and 48 (15%) showed conjunctival and palpebral abnormalities characteristic of PCB effects.

Other investigators have reported similar abnormalities among other groups exposed to PCBs. Ten cases of acne and/or folliculitis and five cases of dermatitis occurred among 80 capacitor manufacturing workers examined in Italy by Maroni *et al.*[31,32] They also noted two cases of bleeding hemangiomas; however, one had existed since birth but began to bleed following employment. The second was in an individual who also developed leukemia. The man had been employed for 16 years since the age of 30. All of the workers with chloracne were employed in high-exposure jobs. Their blood PCB concentrations ranged from 300 to 500 ppb.

In an examination of individuals having a somewhat lower exposure than those in the two previous groups, Smith *et al.*[33] reported correlations of mucous membrane and skin irritation with serum PCB levels. However, no clinical abnormalities attributable to PCB exposures were observed. Other reports of chloracne among capacitor manufacturing workers include those of Ouw *et al.*[34] and Baker *et al.*[35] A study of 55 transformer repairmen and 56 group-matched controls was conducted by Emmett *et al.*[36] Current exposures were generally to Aroclor 1260 with some to Aroclor 1242; recently measured air concentrations were mostly under 20 $\mu g/m^3$. Some neurobehavioral and irritant symptoms were found more frequently among the transformer employees, but were not related to PCB exposure. Comedones were also more frequent in the PCB-exposed group but no definite case of chloracne was identified. One notable finding of the study was the report by two PCB-exposed individuals of melanoma removal.

2.2. PCB Alterations of Blood Chemistry

Several studies have been conducted of groups of workers exposed to various concentrations of PCBs. The results of serum blood analyses in these various studies are summarized in Table 3, which indicates whether any statistically positive association of particular liver enzymes with serum PCB concentration was found. Most of the results are negative and, when positive, the effect is relatively minor. Nevertheless, in considering the overall results, SGOT and GGTP, particularly, are increased with exposure to PCBs.

Table 4 lists the association of serum lipid concentration with serum PCB level. Statistically significant increases in triglycerides and total cholesterol were associated in some studies with increased serum PCB concentrations. However, some studies did not fully control for body mass and the associated increase in serum lipids. Baker *et al.*[35] and Smith *et al.*[33] also noted an association of decreased high-density lipoproteins with PCB concentrations. The possible clinical significance of these findings is uncertain, but the association of increased triglycerides and decreased high-density lipids with an increased risk of coronary artery disease must be considered.

One of the dramatic features of animal studies is the increase in the induction of P-450 enzymes in the livers of rodents exposed to PCBs. P-450a and P-450c levels are

Table 3
Effect of PCB Exposure on Blood Enzyme Concentrations

Study	No. of subjects	Average PCB concn. (ppb serum)	Correlation with serum PCB[a]					
			SGOT	SGPT	GGTP	Alk P	LDH	Bili
Baker *et al.*[35]	148	17–75[b]	NS	NS	NS	NS	NS	NS
Maroni *et al.*[31]	80	41–1319[c]	NS	(+)	(+)			
Kreiss *et al.*[40]	458	3–158[c]		NS	+			NS
Chase *et al.*[37]	120	10–312	+	NS	NS			
Smith *et al.*[33]	317	1–3300L[c,d]	+	NS	+	NS	NS	NS
		1–250H[c,d]	+	NS	+	NS	NS	NS
Fischbein[39]	227	48L, 124H[e]	NS	NS	+male	NS	+fem	NS
Lawton *et al.*[41]	194	57–2270L[f]	NS	NS	+	NS	NS	–
		12–392H[f]	NS	NS	+	NS	NS	–
Emmett[38]	55	1–300[c]	NS	NS	+	NS	NS	NS

[a]NS, not significant; +, – indicates statistically significant correlation.
[b]Range of group means.
[c]Range of individual means.
[d]L, low chlorinated PCBs; H, high chlorinated PCBs.
[e]Study population means.
[f]95% range.

increased substantially in animals following administration of highly chlorinated PCBs. However, only limited human data exist. Alvares *et al.*[42] measured antipyrine clearance in five PCB-exposed capacitor manufacturing workers and found that the antipyrine half-life decreased significantly (from 15.6 to 10.5 hr). A slightly smaller effect was found among transformer repairers by Emmett.[38]

Table 4
Effect of PCB Exposure on Blood Lipid Concentrations

Study	No. of subjects	Average PCB concn. (ppb serum)	Correlation with PCB	
			Cholesterol	Triglycerides
Baker *et al.*[35]	148	17–75[a]	NS	+
Kreiss *et al.*[40]	458	3–158[b]	+	NS
Chase *et al.*[37]	120	10–312	NS	+
Smith *et al.*[33]	317	1–3300L[b,c]	+	NS
		1–250H[b,c]	+	+
Fischbein *et al.*[29]	321	48L, 124H[d]	(+)	NS
Lawton *et al.*[41]	194	57–2270L[e]	NS	NS
		12–392H[e]	NS	NS
Emmett[38]	55	1–300[b]	NS	NS

[a]Range of group means.
[b]Range of individual means.
[c]L, low chlorinated PCBs; H, high chlorinated PCBs.
[d]Study population means.
[e]95% range.

2.3. Liver Abnormalities

The above-mentioned increases in the serum concentrations of liver enzymes indicate a PCB effect on the liver. However, the clinical consequences of the findings are uncertain. Of greater concern is the report by Maroni *et al.*[32] of hepatomegaly in 14 of 80 examined individuals. None had a history of excess alcohol or drug intake. In addition to the examination findings, increased concentrations of GGTP, serum ornithine-carbamoyl transferase, and serum alanine aminotransferase (SGPT) were commonly noted among those with enlarged livers. Other clinical studies have not reported liver disease of such severity, if at all. For example, Fischbein *et al.*[29] found only four individuals with significantly enlarged livers among 326 examined capacitor manufacturing workers. Further, two were in individuals with a high alcohol intake.

2.4. PCB-Associated Cardiovascular Effects

The occasional finding of serum lipid abnormalities suggests the possibility of a relationship between PCB exposure and cardiovascular disease. Of the various studies of groups occupationally exposed to PCBs, only Smith *et al.*[33] mention the results of an analysis to determine the correlation between serum PCBs and increased blood pressure. There an apparent correlation was found to be related to the confounding factors of age and sex.

Analyses of health data from community groups exposed to relatively high concentrations of PCBs have shown associations of elevated blood pressure with increased serum PCB concentration. Kreiss *et al.*[40] considered data from a group that had high levels of serum PCBs from ingestion of fish. A correlation was found between serum PCB concentration and elevated blood pressure, particularly diastolic, increased GGTP concentration and serum cholesterol level. When the influence of possible confounding factors, including age, sex, and body mass, was removed, the statistically significant correlation with serum lipids no longer obtained, but the association of PCBs with elevated blood pressure remained. This finding is unusual, considering the relatively low exposure level (17.2 ppb, range 3.2–158) of the group. Stehr-Green *et al.*[43] found a statistically significant association of high blood pressure with serum PCB concentration in a study of residents near three chemical waste sites. The association decreased to a nonsignificant level ($p = 0.08$) when age and smoking were controlled for. The confounding factor of obesity was not considered.

2.5. PCB Respiratory Effects

Two studies have been reported of spirometric values measured in individuals exposed to PCBs during capacitor production in the same facility. One by Warshaw *et al.*[44] found a substantial prevalence of reduced forced vital capacity (FVC), both among smokers and non-smokers. Decreased FVCs were present in 14.9% of males and 13.1% of females tested using normal prediction equations developed by Morris *et al.*[45] Exposure to pneumoconiosis producing dust could not account for these findings; the prevalence of x-ray abnormalities among the workers was extremely low.

The above results were initially confirmed by a study of the same population conducted by the medical staff of the company. However, in a later study by Lawton *et al.*,[46] it was found that their reduced FVC results were probably the result of technician inexperience and inadequate expiratory efforts on the part of those studied. The later Lawton *et al.* analysis indicated a 2.9% incidence of restrictive pulmonary dysfunction compared with 16.2% in their earlier study. Prediction equations of Knudson *et al.*[47] were used. The reason for the difference in the two studies of roughly the same population could lie in the use of the different prediction equations. Those of Morris *et al.* predict 5 to 10% higher FVC values in adults than those of Knudson *et al.* As with the possible association of increased blood pressure with serum PCB concentration, the single study indicating decreased FVC among PCB-exposed workers must be confirmed in other populations.

2.6. Reproductive Effects

One of the significant results of PCB exposure to animals has been the pregnancy loss found in a study of monkeys by Allen *et al.*[48] In two groups of eight impregnated monkeys, five normal births occurred among those exposed to 2.5 ppm of Aroclor 1248 (in their diet for 7 months) and one in those exposed to 5 ppm. Spontaneous abortions accounted for three pregnancy losses in the lower exposed group and four in the higher. At 5 ppm, one stillbirth occurred and two animals failed to conceive. The significance of these monkey studies is enhanced by the finding of substantial reproductive problems among Yusho mothers.[20] Two stillbirths occurred among thirteen pregnancies and reduced birth weight was common. Of children born to Yusho and Yucheng parents, skin pigmentation and low birth weights were commonly found.[21,49]

A study evaluating birth outcomes of women exposed to pure PCBs was that of the New York State Department of Health.[50] An analysis of 51 births to women employed in high-exposure areas of two capacitor manufacturing facilities showed the infants had lower birth weights than 337 infants born to women who worked in low-exposure areas. The observed birth rate difference between the lower and higher exposure group was attributed to a shortened gestation period rather than to retardation of intrauterine growth.

Several studies have been done on the human reproductive effects from environmental PCB exposure. One by Fein *et al.*[51] compared 242 infants of mothers who consumed PCB-contaminated fish with 71 infants whose mothers did not. The results of the analysis indicated that exposed infants were 160 to 190 g lighter than controls and their head circumference, 0.6 to 0.7 cm smaller. Gestational ages were significantly shorter and maturity scores significantly lower. Analysis of covariance showed that none of these effects were attributable to any of 37 potential confounding variables. However, the analysis utilized only *t* tests or chi-square tests on dichotomized variables. Multiple regression analyses were not undertaken. In contrast, a study of Green Bay, Wisconsin, births found a small positive association of birth size with fish consumption.[52] However, the PCB levels from such consumption were lower than those of Fein *et al.* and New York State.

In a group of 858 children enrolled in a North Carolina breast milk study, PCB levels were not correlated with weight or with frequency of physician visits for common childhood illnesses.[53] In this study, the median level in the milk fat of 733 mothers at birth was 1.8 ppm, a level that is higher than typically found in the United States.[19]

2.7. Developmental Effects

The effects described above result from the mothers' exposures to PCBs and PCDFs and from the exposures of the fetuses *in utero* from transplacental transfer of the chlorinated compounds.[54,55] As described above, breast milk contains PCBs and nursing infants accumulate an additional burden. Both transplacental and postnatal exposures can contribute to altered development. No abnormalities at birth were attributed to PCBs in the large North Carolina study mentioned above.[53] However, in a later study,[56] using the Bayley Scales of Infant Development beginning at age 6 months, lower psychomotor development scores at both 6 and 12 months were associated with higher transplacental exposure to PCBs. No association was found with breast feeding exposures.

Positive developmental effects have also been reported in studies of a cohort, previously studied by Fein *et al.*,[51] of 236 children born to mothers who consumed Lake Michigan sport fish.[57,58] When evaluated at age 4 years, prenatal exposure as measured by cord blood was associated with lower weight. Prenatal exposure was also associated with poorer short-term memory on both verbal and quantitative tests in a dose-dependent manner. Reduced activity was associated with current serum PCB concentration and was strongest among children of mothers who nursed for at least 1 year. The data demonstrate the continued effect of prenatal exposure and additional effects from postnatal PCB ingestion via breast milk.

2.8. Other Effects

Immune dysfunction[59,60] and hepatic porphyria[61,62] have been associated with environmental exposure to polybrominated biphenyls and found among Yusho or Yucheng patients. However, porphyria was not present in the populations studied by Smith *et al.*[33]; no data have been reported on the immune status of populations exposed to commercial PCBs. Kilburn *et al.*[63] reported a significant impairment of memory and cognitive function among 14 firemen exposed to PCBs in a transformer fire compared with other firemen. Following a detoxification program, scores on memory tests improved. It was noted, however, that the initial neurological impairment was not related to serum or adipose PCB concentration.

2.9. Summary

The predominant finding among individuals exposed occupationally to PCBs is an increased prevalence of abnormal dermatological symptoms. Acneform eruptions, folliculitis, and possible skin thickening may result from high exposures to PCBs, particularly of the higher chlorinated isomers. Commonly reported symptoms from PCB exposure are burning sensations in the eyes or skin and rashes. The rashes and burning sensations disappear after cessation of exposure, but some skin lesions may persist for years. Other consistent findings include elevation of various liver enzymes, particularly GGTP and SGOT, and serum triglyceride and cholesterol concentrations. Liver enzyme induction has also been documented, which may account for the commonly found elevated GGTP concentrations. Some workers in one examined group were found to have varying degrees of hepatomegaly. However, this has not been a consistent finding among other populations exposed occupationally to high concentrations of PCBs. Studies among exposed individuals in the general population suggest the possible association of increased diastolic blood pressure with increased serum PCB level. One study of workers found an association between PCB exposure and reduced forced vital capacity. However, these last two findings require confirmation in other studies.

Two studies have suggested that infants born to mothers exposed to PCBs were smaller. A possible causal factor in both the lower birth weights and head sizes was a shortened gestation period associated with increased serum PCB level. Other studies during early childhood have associated lower psychomotor development and poorer short-term memory with *in utero* PCB exposure.

3. EXPERIMENTAL ANIMAL STUDIES

3.1. Introduction

A large number of studies have evaluated the potential carcinogenic effect of PCBs in a variety of animal species and exposure circumstances. Virtually all of these studies have focused on liver carcinogenesis and on the effect of PCBs either to induce cancer directly or to modulate the response of other liver carcinogens. The earliest study by Ito *et al.*[64] demonstrated that highly chlorinated PCBs induced hepatocellular carcinomas in mice and promoted the carcinogenic effect of benzene hexachloride in the same species. A variety of subsequent studies have confirmed these basic findings. The emerging data suggest that the role of PCBs in the development of hepatocellular carcinoma is a highly complex one. They may act as promoters of cells initiated spontaneously or by other previously administered carcinogens. They may also enhance or inhibit carcinogenesis by virtue of their stimulation of liver enzymes, which, in turn, can alter the metabolism of subsequently administered carcinogens. PCBs appear to have little or no genotoxic effect. They are negative in the Ames Salmonella test,[65,66] the micronucleus test,[65] the V79 Chinese hamster assay,[67] and do not produce chromosomal alterations[68] or dominant lethal mutations in rats.[69]

Inasmuch as PCBs may play a direct role in cancer promotion and an indirect one in cancer initiation, it is appropriate to review briefly the data on cancer initiation and promotion in general in order to appreciate the observed effects in animals and the potential effects in humans. Virtually all PCB studies have been concerned with hepatocellular carcinogenesis and, thus, the following brief review of initiation–promotion data will focus on liver cancer from chemical exposures. To the extent that the mechanism of action of PCBs is similar in human and animal species, the data on experimental carcinogenesis are important for the determination of the appropriate analysis of epidemiological studies of individuals exposed to PCBs. Further, in order to appreciate fully the role of PCBs as cancer risk modifiers in liver tumor development, we will also briefly review the existing information on the various liver changes associated with hepatocarcinogenesis and the effect of the various PCBs on the induction of liver enzymes that might alter the metabolism of externally administered carcinogenic agents.

3.2. Initiation and Promotion Effects in Hepatocarcinogenesis

A multistage model of carcinogenesis has long been proposed. Berenblum and Shubik[70] first demonstrated that chemically induced skin carcinogenesis could be separated into initiation and promotion stages. Their experiments with croton oil showed that the effect of an initiating carcinogen could be greatly enhanced by the later application of a promoting agent, which by itself was noncarcinogenic. Application of a promoter prior to an initiator had a substantially reduced effect. Subsequently, initiation and promotion effects have been demonstrated in the development of cancer in a wide variety of tissues, including the liver.[71] The initial two-stage model has been extended to include multiple proliferation and promotional stages.[72]

A multistage model of carcinogenesis has also been suggested for human cancer, based on the observed power law dependence of cancer incidence at a variety of sites.[73–76] The models have ranged from proposals of multiple (up to six or seven) mutations (or carcinogenic events) occurring in the same or adjacent cells[73,77] to models that involved preferential clonal development of altered cell lines.[75,78] Some or all of the stages may be effected by external carcinogens. For those stages susceptible to action by an external carcinogen, it would be expected that the probability of progression to the next stage would be proportional to the time that an agent, or its active metabolite, is present at a reaction site. A constant exposure to environmental carcinogens would then introduce a power of time for each stage that is effected by a particular external agent. It would also introduce a power of dose. Stages that depend on spontaneous changes occurring randomly in time also introduce a power of time for each such stage. While the model is a statistical one, based on the age dependence of cancer, it makes explicit predictions for the age, dose, and time dependence of cancer from external agents. Although direct evidence of initiation and promotion is unavailable in humans, the time courses of certain occupationally induced cancers are in accord with the model's predictions and suggest late stage or promotional action for some and early stage or initational action for others.[79]

A number of distinct pathological phenomena (or stages) have been identified in the development of hepatic carcinoma. (See Pitot and Sirica[80] for a good review.) These include the development of foci of cellular alteration or "liver islands," areas of colangiofibrosis, and neoplastic nodules.[81] The foci are areas devoid of glucose-6-phosphatase, canalicular adenosine triphosphatase, and β-glucuronidase activity. They also typically demonstrate the presence of GGTP. The cells in these regions are indistinguishable from normal hepatocytes except through special staining techniques. The areas of colangiofibrosis or "bile duct proliferation" are characterized by foci or areas of hyperbasophilic, atypical ducts in a fibrous stroma, usually with excess collagen formation. The neoplastic or "hyperplastic" nodules, sometimes referred to as hepatomas, are distinct spherical lesions containing substantially altered cells. There usually is a sharp demarcation of the lesion from the surrounding, unaffected liver. Each of the above appears well before the development of hepatocellular carcinoma induced by external agents.

It has been found that the number of enzyme-altered foci or neoplastic nodules may decrease following cessation of a carcinogenic exposure,[82] suggesting some regression of promotional effects. However, not all nodules totally regress and cells within those which have regressed continue to demonstrate the presence of a "preneoplastic" antigen,[83] indicating the presence of a permanent cellular alteration. Further, in some circumstances, it has been demonstrated that the enzyme-altered foci appear to be clonal developments of single altered cells.[84]

While the studies are not completely definitive, a useful model for the description of hepatocellular carcinoma considers the development of the altered foci and neoplastic nodules to be early stages in the carcinogenesis process. Exposure to an initiating carcinogen leads to altered cells, some of which may develop into the foci, subsequently into neoplastic nodules and, perhaps, ultimately into hepatocellular carcinoma. The development of foci, neoplastic nodules, and frank carcinoma is enhanced by the administration of various promoting agents. After administration of a promoting chemical ceases, but prior to the development of carcinoma, regression of the preneoplastic lesion can occur, with a concomitant reduction in cancer risk. However, cells within that lesion may still contain a memory of prior alteration that may again be stimulated to progress to hepatocellular carcinoma. Among the agents that have been especially effective in promoting liver cancer in rodents are phenobarbital (PB), dichlorodiphenyltrichloroethane (DDT), and PCBs.[85–87] Because of the strong relationship of preneoplastic lesions with hepatocellular carcinoma, many studies investigating initiation–promotion activity within the liver did so on the basis of these changes rather than on the results of the much longer studies of carcinoma development.

3.3. Biochemical Effects of PCBs and Similar Chemicals

Among other effects, PCBs are potent inducers and inhibitors of various liver enzymes.[88–91] In this activity there is a wide variability among the effects of different PCB isomers present in the commercial mixtures. It would appear that many PCB isomers exhibit a "mixed-type" induction pattern reflecting induction properties that simulate both PB and 3-methylcholanthrene (3-MC). It has been found that treatment of rats with PB-type inducers enhances the production of several cytochrome P-450-dependent monooxygenases and cytochromes P-450a, P-450b, and P-450e. In contrast, 3-MC induces different monooxygenases and cytochromes P-450a, P-450c, and P-450d. All of the above cytochromes, as well as the PB and 3-MC monooxygenases are induced by Aroclor 1254. This enzyme induction capability of the PCBs is of considerable significance in that many of the induced monooxygenases play a role in the metabolism of various other carcinogenic chemicals. The enzyme-inducing activity of PCBs is mimicked, but at much lower concentrations, by the PCDFs and PCDDs.

3.4. Experimental PCB Carcinogenesis Studies

Eight studies provide direct evidence of the carcinogenic potential of PCBs in rodents. However, the numbers of animals in most study groups was relatively small (less than 50) and several of the studies only followed animals for a limited portion of their life span (as short as 32 weeks). Nevertheless, the studies demonstrate reasonable consistency. Differences in tumor or focus incidence between studies are understandable in terms of the differences in dosage, durations of exposure and follow-up, species sensitivity, and degrees of PCB chlorination. As with general toxicity, the higher chlorinated PCBs (penta- or hexa-) are the more carcinogenic. Data on relative carcinogenicity of a particular PCB in different species are lacking. However, in terms of gross toxicity, guinea pigs and chickens appear to be the most sensitive species, with rats, monkeys, mice, rabbits, and hamsters increasingly more resistant to the toxic effects of PCBs.[91] Additionally, sex differences may be manifest. While the above generalities can be stated, the data on species or sex sensitivity for any specific toxic effect are extremely limited.

3.4.1. Mice

The first study on PCB-induced liver tumorogenesis is that of Ito *et al.*,[64] who fed dd male mice a diet to which 500, 250, or 100 ppm of Kanechlor 300, Kanechlor 400, or Kanechlor 500 had been added. Neoplastic nodules and hepatocellular carcinomas were seen after 32 weeks only in the group exposed to 500 ppm of Kanechlor 500 in their diet. There, 7 of 12 animals were found to have neoplastic nodules and 5 of 12, hepatocellular carcinomas.

In a second study of mice by Kimbrough and Linder,[92] administration of 300 ppm of Aroclor 1254 in the diet of mice for 11 months led to the development of hepatomas (neoplastic nodules) in 10 of 22 surviving mice. Only 1 of 24 mice fed the same diet for 6 months, followed by a 5-month recovery period, demonstrated hepatomas. No evidence of hepatocellular carcinoma was present.

3.4.2. Rats

Kimura and Baba[93] fed a varying diet of Kanechlor 400 to groups of ten male and female Donyru rats for a total of 32 weeks. Neoplastic nodules were found in all six

Table 5
Histopathological Findings in the Liver of Male Wistar Rats Given PCBs[a]

PCB in diet (ppm)	Cholangiofibrosis	Nodular hyperplasia
Kanechlor 500 (54% cholorine)		
1000	4/13	5/13
500	0/16	5/16
100	0/25	3/25
Kanechlor 400 (47% chlorine)		
1000	2/10	3/10
500	0/8	0/8
100	0/16	2/16
Kanechlor 300 (41% chlorine)		
1000	2/15	0/15
500	0/22	1/22
100	0/18	0/18

[a]From Ito *et al.*[64]

female rats fed a total of 1200 mg or more of PCB in their diet. No such changes were found in female rats fed 1100 mg or less or in any males fed up to 1800 mg of PCB. However, fatty degeneration was found in all study rats ingesting from 450 to 1800 mg of Kanechlor 400.

Ito *et al.*[64] also studied the effect of PCBs on rats. Table 5 lists the liver lesions found in rats fed varying amounts of Kanechlor 500, Kanechlor 400, and Kanechlor 300. All three Kanechlors produced liver changes related to the concentration in the diet. Again, the results were limited as the periods of exposure and observation were only for 52 weeks.

One of the more extensive studies of PCB carcinogenesis is that of Kimbrough *et al.*,[94] who exposed animals to 100 ppm Aroclor 1260 in their diet for 3 weeks to 22 months. The experimental group consisted of 184 female Sherman rats and tissues from a wide variety of organs were examined microscopically. The results of this study demonstrate a substantially increased incidence of hepatocellular carcinoma among the experimental animals (26/184 versus 1/173 in controls). Additionally, neoplastic nodules were found in 144 of 184 animals; none were present in controls. Altered foci were found in 182 of 184 rats versus 28 of 173 controls. Uterine cancers, including sarcomas of the endometrial stroma, were also elevated, but not to a statistically significant level.

Two government-sponsored chronic feeding studies have been reported, one in limited form by Calandra[95] and later in more detail by Levinskas.[96] In this study, Aroclor 1242, Aroclor 1254, or Aroclor 1260 at dietary levels of 1, 10, or 100 ppm was fed to groups of albino rats for 2 years. Changes attributable to the PCB exposure (focal hypertropy, nodular and ductal hyperplasia, and hepatoma) could be seen in groups exposed to each of the Aroclors at all exposure levels. However, no hepatocellular carcinomas were noted in the study.

A similar study was conducted by National Cancer Institute[97] in which Fischer 344 rats were fed 25, 50, or 100 ppm of Aroclor 1254 in their diet for 104–105 weeks. Again, altered foci were seen in all exposure groups. One and two hepatocellular carcinomas were seen, respectively, in the mid- and high-dose male groups and one in the low-dose female group. However, the group incidences of these malignancies were not statistically different from that of controls, and it was concluded that Aroclor 1254 was not deemed to be carcinogenic in the bioassay. It should be noted, however, that the two higher dose male groups had increased overall mortality rates compared with controls. Thus, fewer animals were at risk at older ages in the study. Based on a time course of hepatocellular carcinoma proportional to the fourth power of age, the expressed risk could have been reduced by as much as 40%. A second finding of note was an elevation of lymphomas and leukemias among male rats (controls, 3/24; low-dose, 2/24; mid-dose, 5/24; high-dose, 9/24). None of the group incidences are significantly different from that of the control. However, the dose-related trend is significant at the $p = 0.04$ level using the Cochran–Armitage test.[98] Most of the excess malignancies were lymphatic leukemias, although there were increases of granulocytic leukemia and various lymphomas in the two highest dose groups over that of controls. These were not significant, however. In the only other study to note the incidences of leukemias and lymphomas, Kimbrough *et al.*[94] did not observe a significant increase in lymphomas and leukemias (2 lymphomas and 0 leukemias in exposed animals versus 0 lymphomas and 1 leukemia in controls) among female Sherman rats.

The results of a long-term chronic feeding study by Norback and Weltman[99] are shown in Table 6. In this study, female rats fed 100 ppm of Aroclor 1260 in their diet for 16 months and 50 ppm for an additional 8 months showed a high incidence of

Table 6
Incidence of Liver Lesions and Carcinomas among Sprague–Dawley Rats Fed Aroclor 1260[a]

	Controls		Exposed[b]	
Histologic finding	Male ($N = 32$)	Female ($N = 49$)	Male ($N = 46$)	Female ($N = 47$)
Adenocarcinoma	0	0	0	24
Trabecular carcinoma[c]	0	0	2	19
Neoplastic nodule[d]	0	1	5	2
Cholangioma (simple)	2	2	14	21
Cholangioma (cystic)	0	1	2	5
Adenofibrosis	2	1	1	7

[a]From Norback and Weltman.[99]
[b]Exposure: 100 ppm in diet for 16 months following weaning, then 50 ppm for an additional 8 months.
[c]Animals with both trabecular carciomas and adenocarcinomas were placed only in the adenocarcinoma category.
[d]Animals with neoplastic nodules and carcinomas were placed only in the relevant carcinoma category.

adenocarcinoma and trabecular carcinoma in the liver. Ninety percent of the female animals had one or the other form of these hepatocellular carcinomas. In contrast, however, only 4% of the male animals fed the same diet demonstrated such malignancies. This finding of a sex-related risk is similar to that of Kimura and Baba,[93] noted above, and may be the result of the competitive interactions of hormones.

3.4.3. Dogs

Neither hepatocellular carcinoma nor neoplastic nodules were found in groups of four male or female dogs fed 1, 10, or 100 ppm of Aroclor 1242, 1254, or 1260.[95] The only exposure-related finding was a slight decrease in body weight gain in the high-exposed Aroclor 1254 and 1260 groups, as well as increases in liver weights and serum alkaline phosphatase levels in the groups exposed to 100 ppm of Aroclor 1260.

3.5. Promotion of Carcinogenesis by PCBs

The first study of the cancer-promoting effects of PCBs was that of Ito *et al.*,[64] who demonstrated that 250 ppm of Kanechlor 500, administered for 32 weeks with various concentrations of one of three isomers of benzene hexachloride, increased substantially the number of tumors over that produced by exposures to the benzene hydrochloride isomers alone. Hepatocellular carcinomas developed in 21 of 83 male dd mice fed 250 ppm PCB and benzene hexachloride versus 8 of 82 mice fed only benzene hexachloride.

Nishizumi[87] found that PCB, PB, and DDT administration after an exposure to diethylnitrosamine (DENA) in water at 50 ppm for 2 weeks enhanced the production of liver cancer. The greatest promoting activity was demonstrated by PCBs either alone or in combination with PB. A study by Preston *et al.*[100] confirmed the promotional effects of PCBs (Aroclor 1254) on DENA-induced tumorigenesis in Sprague–Dawley rats. Here, comparisons were made with PCBs especially purified of possible contamination by PCDFs (to less than 100 ppb) and PCBs contaminated with PCDFs (to 3 ppm). While the PCDF-contaminated PCBs were a slightly more effective promoting combination, the Aroclor 1254 alone quadrupled the number of hepatocellular carcinomas that developed.

Using an experimental cervical cancer model, Uchiyama and Chiba[101] found no effect of dietary ingestion of 100 ppm of PCBs following implantation of a 3-MC-impregnated thread within the cervix of rats. In contrast, an enhancement of the cervical cancer potential was demonstrated by the feeding of 100 ppm DDT for 8 weeks.

In addition to promoting hepatocellular carcinoma, PCBs have been found to enhance the production of enzyme-altered foci in rat livers.[102] Rats were initiated by a dose of 0.3 mmol/kg body wt of DENA 24 hr after two-thirds partial hepatectomy. Aroclor 1254 was administered 7, 28, and 49 days after the DENA and some rats were sacrificed 21 days after each dose of Aroclor. The livers of rats that received Aroclor

1254 either 7 or 7 and 28 days following initiation contained a significantly increased number of GGTP foci relative to solvent controls. The number of foci in rats administered Aroclor 1254 on days 7, 28, and 49 was identical to the number of rats dosed once or twice with Aroclor. It was hypothesized that the Aroclor 1254 (and PB) decreased the time required for the appearance of GGTP-positive foci without altering the final number of such foci that developed.

3.6. PCB-Altered Metabolism of Carcinogens

Separate from their role as a cancer promoter, PCBs can also affect carcinogenesis by altering the metabolism of subsequently applied carcinogenic agents, either by increasing the metabolism of a carcinogen requiring activation or by reducing a direct-acting carcinogen's effectiveness by enzyme-mediated deactivation. Deml *et al.*[103] have demonstrated enhancement of carcinogenesis by the prior treatment with PCBs of benzo [*a*] pyrene (BAP)-initiated, enzyme-altered foci. Here, prior administration of PCBs, which strongly induce AHH, increased the number of altered foci by two orders of magnitude compared with application of BAP alone. Additionally, subsequent applications of PCBs demonstrated a promotional effect by increasing altered foci by an additional factor of three over a regimen with discontinued PCB treatment.

Inhibition of carcinogenesis was demonstrated by Makiura *et al.*[104] who showed that coadministration of PCBs reduced the number of liver lesions produced by the carcinogens, 3′-methyl-4-dimethylaminoazobenzene, *N*-2-fluorenylacetamide, and/or DENA. The authors suggested that PCBs were strong inducers of the enzymes that metabolized the carcinogens to chemicals of lesser toxicity. Similar inhibition of carcinogenesis has been found in offspring exposed to PCBs *in utero* and via breast milk after birth.[105] Here, Wistar rats fed 50 ppm of DENA for 5 weeks had significantly fewer liver tumors than offspring of mothers unexposed during pregnancy. In a test of the role of PCBs in a two-stage mouse skin tumorigenesis assay, it was found that neither PCBs, TCDD, nor polybrominated biphenyls promoted skin tumors initiated by 7,12-dimethylbenz[*a*]anthracene (DMBA).[106] In contrast, PCBs and especially TCDD inhibited the formation of skin papillomas when applied for various periods prior to administration of the DMBA–phorbol ester regimen.

3.7. Summary of Animal Studies

All chronic animal studies, with the exception of those in dogs, have demonstrated that PCB exposure produces precancerous lesions (neoplastic nodules or altered foci). These lesions have been produced in some exposure circumstances by PCBs of the most commonly used chlorinations (40 to 60%). The incidence of the various lesions, however, is strongly correlated to the degree of chlorination, the higher chlorinated PCBs producing the greater effect. These findings are not the result of a small contamination of the PCBs by PCDFs; specially purified material showed virtually the same effect as material with a few parts per million of PCDF. While detailed studies have not been made of the relationship between the presence of PCB-induced lesions and hepatocellular carcinoma, studies of the development of hepatocellular carcinoma in other animal studies would suggest that they are precursor lesions. The probability that cancer will develop from such lesions, considering that there is a continued body burden of PCBs in exposed animals, has not been studied.

High incidences of hepatocellular carcinoma were found in two studies in which Aroclor 1260 was fed to rats for over 1 year. Kanechlor 500 was also shown to produce hepatocellular carcinomas in mice and a study in rats with Aroclor 1254 was marginally positive. All of the negative chronic carcinogenesis studies, with one exception, were of durations less than 1 year and utilized relatively few animals, limiting the relevance of the results. Further, limitations in terms of intensity of exposure may have precluded the development of hepatocellular carcinoma in studies of low chlorinated PCBs. Again, as with the development of neoplastic lesions, the higher chlorinated compounds were the most carcinogenic for a given exposure. In addition to hepatocellular carcinoma, PCBs were also found to produce a statistically significant increase in lymphomas and leukemias, based on an exposure-related trend. No significant increase was observed in the one other study that noted the incidences of leukemias and lymphomas. PCBs were also shown to greatly enhance the effect of some initiating carcinogens, when administered subsequently. The evidence that PCBs are animal carcinogens is strong and unequivocal. The primary mode of action would appear to be that of a promoting agent. However, their action need not involve the prior administration of initiating carcinogens; cells altered spontaneously or by endogenous factors are capable of transformation into carcinomas by PCBs.

In addition to promotional activity, particularly in the liver, PCBs can alter the metabolism of other carcinogens, if administered previously. Here, their action can either increase or decrease cancer incidence, depending on whether a carcinogen is activated or deactivated. Based on known metabolism of a particular carcinogen, one could make predictions as to the effect of prior administration of PCBs. However, no data exist as to the effect of PCB exposure to humans early in life when its effect would be greatest.

4. HUMAN MORTALITY STUDIES

4.1. Introduction

Substantial data have been published by four research groups in three countries on the mortality experience of worker cohorts exposed to PCBs during the manufacturing of electrical capacitors.[107–112] Other studies have provided information on unusual clusters of malignancies believed possibly to have arisen from exposure to PCBs.[113,114] The results of the cohort studies were varied. Those of Brown and Jones[107,108] suggested an excess of liver and biliary tract cancer and possibly rectal cancer, while those of Bertazzi *et al.*[109,110] suggested an excess of hematological neoplasms and gastrointestinal cancer, and those of Sinks *et al.*[112] found a significant increase in deaths from melanoma. The study of Gustavsson *et al.*[111] found too few deaths to enable any conclusions to be drawn. The results of no one study were sufficiently definitive to allow a firm conclusion to be drawn regarding the human carcinogenicity of PCBs.

It is possible to combine the results of the four research groups using the published data as well as additional data supplied by the researchers. Because it increases statistical power, this procedure, termed meta-analysis, has the potential to provide stronger evidence for any relationship that may exist between cancer at some site and PCB exposure. However, any overall combined mortality estimate would necessarily have to utilize crude estimates of expected deaths for some sites of interest. Further, no data were available that would allow combining of data according to duration or intensity of exposure.

4.2. Industrial Hygiene Measurements and Population Exposure Estimates

The study cohorts came from facilities of five different companies, some involving separate plants for heavy- and light-duty capacitor manufacturing. The facilities studied by Brown and Jones were those of two different electrical equipment manufacturers, one with a plant located in southeastern Massachusetts and one with two plant facilities located in upstate New York. The population of Sinks *et al.* came from a plant of a third company located in central Indiana. The plant studied by Bertazzi *et al.* was located north of Milan, Italy, and that of Gustavsson *et al.* in Stockholm, Sweden.

From published descriptions, the work activities in each facility appeared to be similar. Typical processes during the manufacturing of large industrial capacitors involved the forming of "capacitor packs" consisting of rolls of paper, film, and foil which were covered with cardboard and bound. These were tested for electrical quality, inserted into metal capacitor boxes and leads connected to the outside. Following assembly the capacitors were filled with PCBs. While this was done in later years utilizing automated equipment, flood filling chambers were utilized often in the process. In some cases the capacitors were heated in impregnation ovens, which released PCB fumes after opening. Once filled, the capacitors were placed in a vacuum to remove any moisture and the filling holes sealed by soldering. The finished capacitors were tested and, if satisfactory, cleaned with a solvent, usually trichloroethylene. Capacitors were then painted, dried, and packed for shipping. Capacitors that failed the electrical tests were sent to a salvage operation where the reusable components, including PCBs, were removed and the capacitors reconstructed as appropriate.

The smaller electrical capacitors, such as those used in household lighting fixtures or appliances, were also constructed by forming rolls of paper, film, and foil which were inserted into cases. Capacitors were filled with dielectric fluids in open baths, placed under vacuum, and sealed by either crimping or soldering. After washing with phosphates, trichloroethylene, or other solvent, the capacitors were tested and packed for shipping.

Obviously, during the open filling operations of either small or large capacitors, the air exposures above the PCB baths would be close to the vapor pressure of the PCBs, which depends strongly on temperature, but often would be expected to exceed 1 mg/m^3. In the bath filling operation of one facility, workers walked on slatted wooden floors to prevent slipping on pools of PCBs spilled onto flat concrete surfaces. This circumstance indicates that airborne exposure must have been substantial. In addition, the opportunity for skin contact was extremely high. Although workers usually wore rubber boots and rubberized clothing in high contact areas to protect themselves, protective clothing may not have been fully protective. Dispersal of the PCBs onto workplace surfaces was common. Surface contamination as high as 0.159 mg/cm^2 was measured in the Italian plant during 1977; levels as high as 0.006 mg/cm^2 were measured in 1982, two years after the cessation of PCB use. To the extent that air from the filling, sealing, or salvage operations was carried to other parts of a manufacturing facility, widespread exposure occurred. The degree of exposure, of course, depended on the circumstances of the manufacturing plant.

Table 7 lists exposure measurements available for the five manufacturing facilities reviewed here. As can be seen in the high-exposure areas, air concentrations approached 1 mg/m^3 and evidence for widespread dispersal into the case and component manufacturing areas was present. It should be emphasized that these measurements were made during the late 1970s, in some cases after installation of engineering controls and process alteration. For example, the National Institute for Occupational Safety and Health (NIOSH) measurements made at Brown plant 1 were taken after the quantity of PCBs used in capacitors was reduced to 25% of that used previously and after installation of ventilation equipment. The improvement in air quality from these measures is seen in the difference between the NIOSH and earlier company measurements. While the industrial hygiene data that are available are extremely limited, they suggest that the time-weighted average workplace air exposures of electrical capacitor manufacturing workers ranged from concentrations in excess of 1 mg/m^3 in the high-exposure areas to general plantwide concentrations of 0.05 to 0.1 mg/m^3. There is no evidence for substantially different airborne concentrations in the different plants here reviewed.

Dermal absorption also can be an important, and much more variable, route of body entry for PCBs. Observations on the potential for exposure of workmen during transformer maintenance by Lees *et al.*[115] indicated numerous instances of dermal contact, whether by contact with PCB-containing oil or PCB-contaminated surfaces. The amount of dermal absorption was not estimated, but the authors point out that the absorption of one drop of PCB fluid (0.05 ml) would result in an intake of 54 mg PCB. This is equivalent to the intake from an 8-hr air exposure of 7 mg/m^3.

In addition to possibly different exposure conditions in the five facilities under review, the average exposure durations of the individuals in the several cohorts differed substantially because of different cohort entry criteria for duration of employment used by the various investigators. Table 8 lists descriptive information about each of the four groups for which epidemiological data are to be reviewed. Individual exposure information was unavailable for the members of any group, although Brown and Jones selected for observation only workers deemed to have high exposures. For one cohort, even information on the jobs held by all cohort members was lacking.

Table 7
Exposure Measurements in Plants from Which PCB-Exposed Cohorts Were Established

Activity	NY (Brown 1) NIOSH (1977)[a]	NY (Brown 1) Company (1975)[b]	MA (Brown 2) NIOSH (1977)[a]
A. U.S. plants — Average air concentrations (μg/m³)			
Degrease			855 (4)[c]
Treat, impregnation, seal	175 (8)	792 (4)	791 (20)
Quality control, test, salvage	207 (16)	410 (1)	309 (8)
Assembly	99 (6)	251 (3)	
Case fabrication, foil production, winding, preassembly	43 (7)	40 (2)	92 (13)
Packing, shipping	109 (8)		90 (2)
Maintenance	150 (1)		280 (1)
Sinks *et al.* (1977)[112]			
Capacitor processing, baking	92 (42)		
Adjacent areas	59 (8)		
Office and paint room	16 (2)		
B. Italian plant (Bertazzi); Maroni (1977)[31]			
High-power	154 (4)		
Low-power capacitors	193 (3)		
Small household capacitors	59 (2)		
C. Swedish plant (1973)[111]			
Capacitor fill and heat treat	705 (1)		

[a]NIOSH industrial hygiene surveys of March and April, 1977, quoted in Brown and Jones.[107] Quantity of PCBs used in capacitors had recently been reduced to 25% of former amount.
[b]Company industrial hygiene information (personal communication).
[c]Number of samples in parentheses.

4.3. Review of Epidemiological Studies

Tables 9 and 10 summarize mortality results for each study, separately for males and females. The information listed in these tables is that provided by the individual researchers for this review. For cancer mortality it is more extensive than that provided in the published studies or those in preprint form. Details relative to each separate study are provided below. For the sites originally considered in a given study, the expected and observed deaths are identical with those published for the studies of Brown[108] and Gustavsson *et al.*[111] The expected deaths for the Bertazzi cohort are those calculated using the rates of the region (Lombardy) in which the plant is located, rather than those of the local town or of the nation. These are deemed to best represent

Table 8
Characteristics of Four PCB Mortality Studies

Characteristic	Brown[108]	Gustavasson *et al.*[111]	Bertazzi *et al.*[109]	Sinks *et al.*[112]
No. of subjects				
Male	1258	142	594	2742
Female	1309	0	1556	846
Minimum period of employment	3 months	6 months	1 week	1 day
First year of follow-up	1940	1965	1946	1957
Final year of follow-up	1982	1980	1982	1986
Minimum latency	3 months	1 year	1 week	1 day
Number of deaths				
Male	141	21	30	192 (male and female)
Female	154	NA	34	
Tracing completeness	96.2%	100.0%	99.5%	96.2%
Average employment duration	3.4 years[a]	6.5 years[b]	<5 years[c]	Unknown
Calendar years of plant PCB usage	38/46/51–77[d]	60–78	46–80	57–77[d]

[a]Employment duration is that in "PCB-exposed" jobs.
[b]Cohort members were those employed in work having direct PCB exposure.
[c]Cohort members included all plant employees.
[d]Three separate plants of two companies were included in the cohort of Brown.

the rates for the study population. In none of the individual cohorts was the number of deaths sufficient to allow meaningful analyses to be made with to respect to exposure or time categories that might relate to cancer risk. Thus, only data on overall cohort cause-specific mortality are presented in the tables.

4.3.1. Brown and Jones[107] and Brown[108]

The first mortality study of workers exposed to PCB was that of Brown and Jones,[107] who published data from a retrospective cohort study of 2567 workers in two production complexes using PCBs during the manufacture of electrical capacitors. The study cohort was defined as all workers who accumulated at least 3 months of employment at any time in areas of the plants where there was a potential exposure to PCBs, as designated by the company and confirmed by a NIOSH industrial hygiene survey. In the first follow-up of the group, the vital status of each individual was determined as of January 1, 1976. From 1940 through 1976, 160 individuals of both sexes had died compared with 182.35 expected. While the number of cancer deaths was less than expected (39 observed versus 43.79 expected), excess mortality occurred for rectal cancer (4 observed versus 1.19 expected) and cancer of the liver, gallbladder, and biliary passages (3 observed versus 1.07 expected). Most of the rectal cancers occurred among females employed in one of the plants, resulting in an excess for that plant that achieved statistical significance at the 95% level (3 observed versus 0.50 expected). In the case of the liver cancers, a follow-up study by Brown[108] found that one of the three

Table 9
Expected (E) and Observed (O) Cancer Mortality among PCB-Exposed Male Capacitor Manufacturing Workers in Four Cohorts

		Bertazzi		Brown 1		Brown 2		Gustavsson		Total	
Cause of death	ICD8	E	O	E	O	E	O	E	O	E	O
All causes	00–999	30.88	30	88.28	80	71.69	61	22.12	21	212.97	192
Cardiovascular disease	390–458	7.73	8	36.89	43	26.57	30	11.40	8	82.59	89
All malignant neoplasms	140–209	7.24	14	17.66	10	13.65	7	5.39	7	43.94	38
Pharynx, buccal cavity	146–149	0.25[a]	0	0.43	0	0.33	0	0.08[a]	0	1.09	0
Esophagus	150	0.19[b]	0	0.41	0	0.30	0	0.12[b]	0	1.02	0
Stomach	151	1.08[b]	2	0.80	0	0.56	1	0.66[b]	0	3.10	3
Intestine, except rectum	153	0.35[c]	0	1.55	1	1.15	0	0.41[c]	0	3.46	1
Rectum	154	0.21[c]	0	0.49	1	0.35	0	0.22[c]	0	1.27	1
Liver, biliary passages, and gall bladder	155–156	0.22	2	0.27	1	0.20	0	0.23	1	0.92	4
Pancreas	157	0.24[b]	2	0.94	0	0.70	1	0.36[b]	0	2.24	3
Larynx	161	0.27[a]	0	0.26	0	0.19	0	0.03[a]	0	0.75	0
Bronchus, trachea, lung, and other respiratory	162–163	1.75	3	6.05	5	4.63	0	1.05	2	13.48	10
Bone	170	0.09[a]	0	0.09	0	0.10	0	0.03[a]	0	0.31	0
Skin	172–173	0.06[a]	0	0.40	0	0.36	1	0.08[a]	0	0.90	1
Prostate	185	0.47[a]	1	0.92	0	0.56	1	0.90[a]	1	2.85	3
Urinary bladder	188	0.30[a]	0	0.45	1	0.31	1	0.18[a]	1	1.24	3
Kidney	189	0.10[a]	0	0.46	1	0.35	1	0.14[a]	1	1.05	3
Brain, nervous system	191–192	0.12[a]	1	0.66	0	0.59	0	0.19[a]	0	1.56	1
Lymphomas	200–203	0.46	1	1.08	0	0.95	1	0.31	1	2.80	3
Leukemias	204–207	0.41	2	0.74	0	0.65	0	0.22	0	2.02	2
All other malignancies		0.67	0	1.66	0	1.37	0	0.18	0	3.88	0

[a]Estimated from the ratio of national age-standardized site rates to all cancer rates (1975) times the study expected deaths for all cancer.
[b]Estimated from the ratio of national age-standardized site rates to all GI cancer rates (1975) times the study expected deaths for all GI cancer.
[c]Estimated from the ratio of national age-standardized site rates to combined colon–rectum cancer rates (1975) times the study expected deaths for all cancer.

Table 10
Expected (E) and Observed (O) Cancer Mortality among PCB-Exposed Female Capacitor Manufacturing Workers in Three Cohorts

Cause of death	ICD8	Bertazzi		Brown 1		Brown 2		Total	
		E	O	E	O	E	O	E	O
All causes	00–999	25.03	34	41.96	36	115.55	118	182.54	188
Cardiovascular disease	390–458	4.29	2	13.53	15	38.56	32	56.38	49
All malignant neoplasms	140–209	8.80	12	13.49	8	34.84	37	57.13	57
Pharynx, buccal cavity	146–149	0.08[a]	1	0.15	1	0.36	1	0.59	3
Esophagus	150	0.06[b]	0	0.11	0	0.27	0	0.44	0
Stomach	151	0.81[b]	1	0.37	0	1.01	0	2.19	1
Intestine, except rectum	153	0.49[c]	0	1.38	2	3.67	5	5.54	7
Rectum	154	0.20[c]	0	0.30	0	0.81	3	1.31	3
Liver, biliary passages, and gall bladder	155–156	0.18	0	0.24	0	0.65	3	1.07	3
Pancreas	157	0.19[b]	0	0.56	1	1.47	0	2.22	1
Larynx	161	0.03[a]	0	0.04	0	0.10	1	0.17	1
Bronchus, trachea, lung, and other respiratory	162–163	0.36	1	1.65	2	4.01	2	6.02	5
Bone	170	0.09[a]	0	0.05	0	0.14	0	0.28	0
Skin	172–173	0.08[a]	0	0.20	0	0.53	0	0.81	0
Breast	174	1.99	2	3.28	1	8.28	8	13.55	11
Uterine cervix	180	0.08[a]	0	0.67	1	1.78	2	2.53	3
Other uterus	181–182	0.77[a]	0	0.46	0	1.25	1	2.48	1
Ovary, fallopian tube	183	0.33[a]	1	1.07	0	2.73	3	4.13	4
Urinary bladder	188	0.11[a]	0	0.13	0	0.37	0	0.61	0
Kidney	189	0.10[a]	0	0.20	0	0.51	0	0.81	0
Brain, nervous system	191–192	0.16[a]	1	0.47	0	1.01	0	1.64	1
Lymphomas	200–203	0.90	4	0.66	0	1.75	4	3.31	8
Leukemias	204–207	0.84	0	0.41	0	1.14	0	2.39	0
All other malignancies		0.95	1	1.09	0	3.00	4[d]	5.04	5

[a]Estimated from the ratio of national age-standardized rates to all cancer rates (1975) times the study expected deaths for all cancer.
[b]Estimated from the ratio of national age-standardized rates to all GI cancer rates (1975) times the study expected deaths for all GI cancer.
[c]Estimated from the ratio of national age-standardized rates to combined colon–rectum cancer rates (1975) times the study expected deaths for combined colon–rectum cancer.
[d]Includes one malignancy certified as liver cancer (NOS), ICD8, 197.8. Pathology analysis showed the tumor to be metastatic from another site.

cancers was metastatic from another site, thus weakening the association based on death certification. Further, limited analyses according to time from onset of first exposure did not indicate any increasing risks with greater latencies.

The mortality experience of the above cohort was updated through 1982.[108] During the additional 7 years of follow-up, there were no additional deaths from cancer of the rectum and the original SMR of 346 decreased to 211. On the other hand, two additional deaths from cancer of the liver and biliary passages occurred, both among females. For the whole cohort, the SMR for liver and biliary tract cancer (ICD 155 and 156) was 210 if the above-mentioned liver cancer that was metastatic is not included. If this cancer is included, the resulting SMR achieves statistical significance at the $p < 0.05$ level using a one-sided test. (The argument for including it is that general population rates include some liver cancers that are improperly listed on the certificate of death and that one should compare like data.) Setting aside the argument for the moment, it must be recognized that only very high risks can be expected to achieve statistical significance if the cause of death is relatively rare. The finding of a more than twofold excess of cancer of the liver, gallbladder, and biliary passages, while of marginal statistical significance, is strengthened by the much stronger animal data which indicate liver cancer, and perhaps also biliary tract cancer, to be a possible outcome from PCB exposure.

Brown used U.S. mortality rates to calculate the expected number of deaths. The liver and biliary tract cancer rates for whites in the counties in and about which the plants were located were both lower and higher than the U.S. rates (34% lower for males and 12% lower for females in the counties about the New York plant; 23% higher for males and 7% higher for females in the county in which the Massachusetts plant is located).[116] Further, individuals whose job titles indicated an exposure to trichloroethylene, a liver carcinogen in animals, were eliminated from the study cohort of Brown and Jones. Thus, a consideration of possible epidemiological biases that might affect the results for liver and biliary tract cancer does not indicate any that would suggest a lower risk than that in Brown's paper and in Tables 9 and 10. On the other hand, the suggested elevated risk for cancer of the rectum is weakened by similar considerations. The local rectal cancer rates for both the New York and Massachusetts plants are significantly ($p < 0.05$) higher than U.S. rates (by 33% to 39%).

4.3.2. Bertazzi *et al.*[109,110]

In the initial study of Bertazzi *et al.*,[110] all production employees who accumulated at least 6 months of service in an electrical capacitor manufacturing facility between 1946 and 1970 were included in the study population. Mortality was observed from 1954 through 1978 and compared with that of the town in which the factory was located. Among male workers, 8 malignancies were observed versus 3.3 expected ($p = 0.040$). Cancer was also elevated among females (6 observed versus 2.3 expected). The greatest cancer excesses were of the lymphatic and hematopoietic system among both males and females and of digestive cancer among males.

The above cohort was expanded by adding nonproduction workers and including all individuals with at least 1 week of employment subsequent to 1946.[109] The rationale for inclusion of short-term workers was that chloracne cases were found among plant employees with fewer than 6 months of employment. While information on date of hire and residence was available, work histories were available only for workers hired subsequent to 1978. The updated mortality experience compared the observed number of deaths by cause to those expected, based on both national mortality rates and rates of the town in which the plant was located (population 150,000). Overall cancer and cancers of the lymphatic and hematopoietic system were elevated substantially for both men and women when compared with expected numbers of deaths calculated using national rates. Among the men, there was also a significant elevation of cancer of the gastrointestinal tract.

Expected numbers of deaths were also calculated using local rates. These were similar to those calculated using national rates for males, but the number of female deaths was about 40% lower. Investigation of the significant difference between these expected numbers led to the conclusion that the expected rates for individuals under the age of 45 were potentially inaccurate because of the few deaths that occurred among individuals of such ages in the town.[117] As most of the females employed in the plant had yet to reach age 45, these rates unduly influence the overall expected mortality. A comparison of age-standardized town rates with regional and national rates shows relatively close agreement. To avoid the inaccuracies associated with small numbers of deaths and to best reflect regional rates, expected numbers in Tables 9 and 10 were calculated using rates of the Lombardy region.

Among the various cancers, one leukemia among males and two lymphomas among women occurred within 2 years of first employment. It is unlikely that these three deaths can be associated with PCB exposure. Of note, however, is the finding of a hepatocellular carcinoma and a cancer of the biliary tract. As with all studies having relatively few deaths, the results are only suggestive of possible risks.

4.3.3. Gustavsson *et al.*[111]

A small group of 142 male Swedish capacitor manufacturing workers was followed from 1965 through 1982. Twenty-one deaths from all causes occurred, seven of which were from cancer. Among the cancer deaths, two were of the lung and one of the ampulla of Vater (ICD 156) In addition, one individual developed both a malignant lymphoma and a mesenchymal tumor. Analyses were conducted by job title and duration of exposure, but did not provide any information allowing an association to be made between the observed mortality and PCB exposure. The very few deaths limit the conclusions that can be drawn from this study. Swedish national rates were used for the calculation of expected deaths; these may underestimate the deaths to be expected in Stockholm. However, the use of these rates will have only a trivial effect on the overall results of this review.

4.3.4. Sinks *et al.*[112]

A capacitor manufacturing plant in Indiana has been studied by Sinks *et al.*[112] The cohort consisted of 3588 men and women who worked for at least 1 day between January 1, 1957, and March 31, 1977. Follow-up continued until June 30, 1986. Air measurements taken in 1977 revealed levels lower than those measured in other plants, but this is likely to be the result of precautions implemented in recent years because of concern for PCB exposures.

The mortality analysis (Table 11) does not indicate any unusual pattern, except for a substantial and significant increase in death from melanoma (8 observed versus 2.0

Table 11
Expected and Observed Cancer Mortality among 3588 PCB-Exposed Male and Female Capacitor Manufacturing Workers, 1957–1977[a]

Cause of death	Exp.	Obs.	SMR
All causes	283.3	192	0.7**
All malignant neoplasms	63.7	54	0.8
Pharynx, buccal cavity	1.7	0	—
Digestive organs	13.9	8	0.6
Liver, biliary passages, and gallbladder	0.8	1	1.1
Pancreas	2.8	2	0.7
Rectum	1.2	1	0.8
Respiratory system	20.2	15	0.7
Kidney	1.5	2	1.3
Skin[b]	2.0	8	4.1**
Brain and nervous system	2.8	5	1.8
Lymphatic and hematopoietic tissue	7.2	7	1.0
All other malignancies	8.5	5	0.6
Diseases of the heart	85.4	60	0.7**
Diseases of the respiratory system	12.3	10	0.8
Accidents	41.1	28	0.7*
Violence	21.5	24	0.6

[a]From Sinks *et al.*[112]
[b]The expected numbers of deaths included basal cell carcinoma, squamous cell carcinoma, and malignant melanoma combined. All observed skin cancer deaths were caused by malignant melanoma.
$^{*}p < 0.05$; $^{**}p < 0.01$.

expected) and a nonsignificant increase in brain tumors (5 observed versus 2.8 expected). All eight melanoma deaths occurred 5 or more years after first employment, but one individual was diagnosed with melanoma 2 months prior to employment. However, a ninth worker died of melanoma shortly after the conclusion of follow-up. Proportional hazards analyses investigated the relation of exposure to these two causes of death. In the case of the melanoma deaths, no relationship with estimated cumulative exposure was found, while the estimated cumulative exposure of those with brain malignancies was higher than that for other workers. There was only one melanoma death in the other cohort studies reviewed here, but a medical survey of 55 transformer repairmen by Emmett *et al.*[36] identified two previous melanomas in the group. Additionally, two deaths from melanoma were reported by Bahn *et al.*[113] in a group exposed to PCBs in a refinery. Two brain malignancies occurred in the cohort of Bertazzi (0.3 expected), but not in the others.

4.3.5. Bahn *et al.*[113]

Two cases of malignant melanoma were reported among 31 men exposed to Aroclor 1254 in a refinery. While the excess is statistically significant ($p < 0.001$), the individuals were also exposed to other chemicals within the refinery. Thus, the relationship to PCB exposure is more uncertain than in the above study of Sinks *et al.*

4.3.6. Shalet *et al.*[114]

Three kidney cancers were reported among a group of utility workers who maintained power equipment, including transformers. The workers were also exposed to other agents, including solvents, herbicides, and electromagnetic fields. The size of the work force from which the cases came was not specified. Kidney malignancy has not been notably elevated in other PCB mortality studies.

4.3.7. Kuratsune *et al.*[118]

Table 12 shows the mortality experience of Japanese Yusho patients. While any excess mortality cannot be attributed to PCBs because of the concurrent high PCDF exposure, the spectrum of mortality is of interest because of the similar toxicity of PCBs and PCDFs. Among 1761 patients followed through 1983, 79 deaths occurred among males and 41 among females, versus 66.13 and 48.90 expected. Cancer deaths were significantly increased among males (33 observed versus 15.51 expected), but not among females (8 observed versus 10.55 expected). Nine male and two female deaths from liver cancer occurred. The expected numbers were, respectively, 1.61 and 0.66, using national rates, and 2.34 and 0.79, using local rates. Additionally, eight lung cancer deaths occurred among males versus 2.45 expected.

Table 12
Expected[a] and Observed Deaths among Patients with Yusho[b]

	Deaths					
	Male			Female		
Cause of death	Obs.	Exp.	SMR	Obs.	Exp.	SMR
All causes	79	66.1	119	41	48.9	84
All malignant neoplasms	33	15.5	213	9	10.6	76
Stomach	8	5.7	140	0	3.3	0
Liver	9	1.6	559	2	0.7	304
		2.3[c]	385		0.8[c]	253
Lung, trachea, bronchus	8	2.5	326	0	0.9	0
Heart disease	10	9.5	106	9	7.7	118
Cerebrovascular disease	8	14.6	55	5	12.0	42
Liver disease	6	2.3	265	2	0.7	274

[a]Expected deaths calculated using national rates.
[b]From Kuratsune *et al.*[118]
[c]Expected deaths calculated using rates of Fukuoka and Nagasaki prefectures.

Table 13
Summary of Observed and Expected Cancer Deaths among Five Cohorts of PCB Capacitor Manufacturing Workers

	Obs.	Exp.	SMR	
All causes	572	678.9	84	
Cardiovascular disease	198	224.4	88	
All malignant neoplasms	149	164.8	90	
Digestive organs	35	40.4	87	
Rectum	5	3.8	132	
Liver, biliary passages, and gallbladder	8	2.8	285	$p = 0.008$[a]
Pancreas	6	7.3	82	
Respiratory organs	31	40.6	76	
Skin	9	3.7	243	$p = 0.014$[a]
Kidney	5	3.4	147	
Brain, nervous system	7	6.0	117	
Lymphatic and hematopoietic tissue	20	17.8	112	
All other malignancies	42	52.9	79[b]	

[a] p values are one-sided, values of 0.05 or less are indicated.
[b] Includes one cancer certified as liver cancer (NOS); ICD8, 197.8; pathology specimen showed cancer to be metastatic from another site.

4.4. Summary of Overall Epidemiological Study Results

A review of Tables 9, 10, and 11 shows that each study, by itself, does not provide definitive information on cancer risk from exposure to PCBs. Table 13 combines the cancer results from the various studies and serves to indicate the sites for which further analyses may be warranted. In considering the data of Table 13, it must be remembered that the cohorts from which the data came have different average durations and intensities of exposure. Further, in three of the cohorts, follow-up for some of the cohort members began within 1 year of first exposure to PCBs; in one a latency of 10 years was utilized. It is also important to recall the conclusions of the various experimental animal studies when reviewing Table 13. There, very strong evidence from several studies demonstrated that PCBs induced liver cancer, probably as a promoter acting in concert with other carcinogenic agents or processes.

A review of the data in Table 13 shows a statistically significant increase of cancer of the liver, biliary tract, and gallbladder. Of the eight cancers, all appear to be primary; two are of the liver, four are of the extrahepatic biliary tract, one is of the gallbladder, and the site of one is not given. The finding of a substantial number of preneoplastic lesions in the biliary tracts of rats fed Aroclor 1260 (Table 6) provides strong support for combining the results of the liver and biliary tract. If one considers the results for the biliary tract and gallbladder alone, the excess for those sites is also significant ($p < 0.05$). All of these cancers occurred 10 years after onset of employment in a PCB-using facility. The results of the analysis provide strong evidence that PCBs are associated with cancer of the liver, biliary tract, and gallbladder in humans.

While only eight such deaths were identified in all studies, the high and statistically significant SMR, 285, is strengthened by the concordance of animal data, results of human exposure to structurally similar PCDF compounds, and the observation of an appropriate time relationship for cancer risks.

The only other significant increase is for skin melanoma. In contrast to liver and biliary tract cancer, where cases were found in virtually all facilities, the melanomas basically come from one plant. This suggests the possibility of a cofactor unique to the Indiana plant in the etiology. On the other hand, the reports by Emmett and Bahn are supportive of a PCB role. While the overall results are suggestive, further data are required before a definitive conclusion can be drawn.

REFERENCES

1. *PCBs, PCDDs and PCDFs: Prevention and Control of Accidental and Environmental Exposures* (Environmental Health Series, No. 23), WHO Regional Office for Europe, Copenhagen (1987).
2. G. W. Bowes, M. J. Mulvihill, B. R. T. Simoneit, *et al.*, Identification of chlorinated dibenzofurans in American polychlorinated biphenyls, *Nature* **256,** 305–307 (1975).
3. R. Rappe and J. R. Buser, Chemical properties and analytical methods, in: *Halogenated Biphenyls, Terphenyls, Naphthalenes, Dibenzodioxins and Related Products* (R. D. Kimbrough, ed.), pp. 41–76, Elsevier/North-Holland, Amsterdam (1980).
4. D. Wood, Chlorinated biphenyl dielectrics—Their utility and potential substitutes, in: *Proceedings of the National Conference on Polychlorinated Biphenyls, November 19–21, 1975, Chicago,* pp. 317–324, EPA-560/6-75-004, Environmental Protection Agency, Office of Toxic Substances, Washington, DC (1976).
5. S. I. Eisenreich, B. B. Looney, and J. D. Thornton, Airborne organic contaminants in the Great Lakes ecosystem, *Environ. Sci. Technol.* **15,** 30–38 (1981).
6. K. E. MacLeod, Polychlorinated biphenyls in indoor air, *Environ. Sci. Technol.* **15,** 926–928 (1981).
7. H. B. Elkins, *The Chemistry of Industrial Toxicology,* pp. 149–150, 319–323, Wiley, New York (1959).
8. *Criteria for a Recommended Standard. . . . Occupational Exposure to Polychlorinated Biphenyls,* National Institute for Occupational Safety and Health, U.S. Government Printing Office, Washington, DC (1977).
9. D. P. Brown and M. Jones, Mortality and industrial hygiene study of workers exposed to polychlorinated biphenyls, *Arch. Environ. Health* **36,** 120–129 (1981).
10. T. Sinks, G. Steele, A. B. Smith, *et al.*, Mortality among workers exposed to polychlorinated biphenyls, *Am. J. Epidemiol.* **136,** 389–396 (1992).
11. A. Schecter and T. Tiernan, Occupational exposure to polychlorinated dioxins, polychlorinated furans, polychlorinated biphenyls, and biphenylenes after an electrical panel and transformer accident in an office building in Binghamton, NY, *Environ. Health Perspect.* **60,** 305–313 (1985).
12. M. S. Wolff and A. Schecter, Accidental exposure of children to polychlorinated biphenyls, *Arch. Environ. Contam. Toxicol.* **20,** 449–453 (1991).
13. E. G. Horn, L. J. Hetling, and T. J. Tofflemire, The problem of PCBs in the Hudson River system, *Ann. N.Y. Acad. Sci.* **320,** 591–609 (1979).

14. C. J. Schmitt, J. L. Zajicek, P. H. Peterman, *et al.*, National contaminant biomonitoring program: Residues of organochlorine chemicals in U.S. freshwater fish, 1976–1984, *Arch. Environ. Contam. Toxicol.* **19,** 748–781 (1990).
15. H. A. Anderson, General population exposure to environmental concentrations of halogenated biphenyls, in: *Halogenated Biphenyls, Terphenyls, Naphthalenes, Dibenzodioxins and Related Products* 2nd ed. (R. D. Kimbrough, ed.), pp. 325–344, Elsevier/North-Holland, Amsterdam (1989).
16. K. Kreiss, Studies on populations exposed to polychlorinated biphenyls, *Environ. Health Perspect.* **60,** 193–199 (1985).
17. F. W. Kutz, P. H. Wood, and D. P. Bottimore, Organochlorine pesticides and polychlorinated biphenyls in human adipose tissue, *Rev. Environ. Contam. Toxicol.* **120,** 1–82 (1991).
18. T. M. Wickizer, L. B. Brilliant, R. Copeland, and R. Tilden, Polychlorinated biphenyl contamination of nursing mothers' milk in Michigan, *Am. J. Public Health* **71,** 132–137 (1981).
19. W. J. Rogan, B. C. Gladen, J. D. McKinney, *et al.*, Polychlorinated biphenyls (PCBs) and dichlorodiphenyl dichloroethene (DDE) in human milk: Effects of maternal factors and previous lactation, *Am. J. Public Health* **76,** 172–177 (1986).
20. M. Kuratsune, T. Yoshimura, J. Matsuzaka, and A. Yamaguchi, Epidemiologic study on Yusho, a poisoning caused by ingestion of rice oil contaminated with a commercial brand of polychlorinated biphenyls, *Environ. Health Perspect.* **1,** 119–128 (1972).
21. M. Kuratsune, Y. Masuda, and J. Nagayama, Some of the recent findings concerning Yusho, in: *Proceedings of the National Conference on Polychlorinated Biphenyls, November 19–21, 1975, Chicago,* pp. 14–29, EPA-560/6-75-004, Environmental Protection Agency, Office of Toxic Substances, Washington, DC (1976).
22. Y. Masuda, Health status of Japanese and Taiwanese after exposure to contaminated rice oil, *Environ. Health Perspect.* **60,** 321–325 (1985).
23. A. Poland, W. E. Greenlee, and A. S. Kende, Studies on the mechanism of action of the chlorinated dibenzo-p-dioxins and related compounds, *Ann. N.Y. Acad. Sci.* **320,** 214–230 (1979).
24. A. Poland, J. Knutson, and E. A. Glover, A consideration of the mechanism of action of 2,3,7,8-tetrachlorodibenzo-p-dioxin and related halogenated aromatic hydrocarbons, in: *Human and Environmental Risks of Chlorinated Dioxins and Related Compounds* (R. E. Rucker, A. L. Young, and A. P. Gray, eds.), pp. 539–559, Plenum Press, New York (1983).
25. S. Safe, Toxicology, structure–function relationship and human and environmental health impacts of polychlorinated biphenyls: Progress and problems, *Environ. Health Perspect.* **100,** 259–268 (1992).
26. J. S. Taylor, Chloracne—A continuing problem, *Cutis* **13,** 585–591 (1974).
27. J. S. Taylor, Environmental chloracne: Update and review, *Ann. N.Y. Acad. Sci.* **320,** 295–307 (1979).
28. J. W. Jones and J. S. Alden, An acneiform dermatergosis, *Arch. Dermatol. Syphilol.* **33,** 1022–1034 (1936).
29. A. Fischbein, M. S. Wolff, R. Lilis, *et al.*, Clinical findings among PCB-exposed capacitor manufacturing workers, *Ann. N.Y. Acad. Sci.* **320,** 703–715 (1979).
30. A. Fischbein, M. S. Wolff, J. Bernstein, *et al.*, Dermatological findings in capacitor manufacturing workers exposed to dielectric fluids containing polychlorinated biphenyls (PCBs), *Arch. Environ. Health* **37,** 69–74 (1982).
31. M. Maroni, A. Colombi, S. Cantoni, *et al.*, Occupational exposure to polychlorinated

biphenyls in electrical workers. I. Environmental and blood polychlorinated biphenyls concentrations, *Br. J. Ind. Med.* **38,** 49–54 (1981).
32. M. Maroni, A. Colombi, G. Arbosti, *et al.*, Occupational exposure to polychlorinated biphenyls in electrical workers. II. Health effects, *Br. J. Ind. Med.* **38,** 55–60 (1981).
33. A. B. Smith, J. Schloemer, L. K. Lowry, *et al.*, Metabolic and health consequences of occupational exposure to polychlorinated biphenyls, *Br. J. Ind. Med.* **39,** 361–369 (1982).
34. H. K. Ouw, G. R. Simpson, and D. S. Siyale, The use and health effects of Aroclor 1242, a polychlorinated biphenyl in an electrical industry, *Arch. Environ. Health* **31,** 181–194 (1976).
35. E. L. Baker, Jr., P. J. Landrigan, C. J. Glueck, *et al.*, Metabolic consequences of exposure to polychlorinated biphenyls (PCB) in sewage sludge, *Am. J. Epidemiol.* **112,** 553–563 (1980).
36. E. A. Emmett, A. Maroni, J. M. Schmith, *et al.*, Studies of transformer repair workers exposed to PCBs. 1. Study design, PCB concentrations, questionnaire, and clinical examination results, *Am. J. Ind. Med.* **13,** 415–427 (1988).
37. K. Chase, O. Wong, D. Thomas, *et al.*, Clinical and metabolic abnormalities associated with occupational exposure to polychlorinated biphenyls (PCBs), *J. Occup. Med.* **24,** 109–114 (1982).
38. E. A. Emmett, Polychlorinated biphenyl exposure and effects in transformer repair workers, *Environ. Health Perspect.* **60,** 185–192 (1985).
39. A. Fischbein, Liver function tests in workers with occupational exposure to polychlorinated biphenyls (PCBs): Comparison with Yusho and Yu-cheng, *Environ. Health Perspect.* **60,** 145–150 (1985).
40. K. Kreiss, M. M. Zack, R. D. Kimbrough, *et al.*, Association of blood pressure and polychlorinated biphenyl levels, *J. Am. Med. Assoc.* **245,** 2505–2509 (1981).
41. R. W. Lawton, M. R. Ross, J. Feingold, and J. F. Brown, Jr., Effects of PCB exposure on biochemical and hematological findings in capacitor workers, *Environ. Health Perspect.* **60,** 165–184 (1985).
42. A. P. Alvares, A. Fischbein, H. E. Anderson, and A. Kappas, Alterations in drug metabolism in workers exposed to polychlorinated biphenyls, *Clin. Pharmacol. Ther.* **31,** 140–146 (1977).
43. P. A. Stehr-Green, E. Welty, G. Steele, and K. Steinberg, Evaluation of potential health effects associated with serum polychlorinated biphenyl levels, *Environ. Health Perspect.* **70,** 255–259 (1986).
44. R. Warshaw, A. Fischbein, J. Thornton, *et al.*, Decrease in vital capacity in PCB-exposed workers in a capacitor manufacturing facility, *Ann. N.Y. Acad. Sci.* **320,** 277–283 (1979).
45. J. F. Morris, A. Koski, and L. C. Johnson, Spirometric standards for healthy nonsmoking adults, *Am. Rev. Respir. Dis.* **103,** 57–67 (1971).
46. R. W. Lawton, M. R. Ross, and J. Feingold, Spirometric findings in capacitor workers occupationally exposed to polychlorinated biphenyls (PCBs), *J. Occup. Med.* **28,** 453–456 (1986).
47. R. J. Knudson, R. C. Slatin, M. D. Lebowitz, and B. Burrows, The maximal expiratory flow–volume curve. Normal standards, variability and effects of age, *Am. Rev. Respir. Dis.* **113,** 587–600 (1976).
48. J. R. Allen, D. A. Barsotti, L. K. Lambrecht, and J. P. Van Miller, Reproductive effects of halogenated aromatic hydrocarbons on nonhuman primates, *Ann. N.Y. Acad. Sci.* **320,** 419–425 (1979).
49. W. J. Rogan, B. C. Gladen, K. L. Hung, *et al.*, Congenital poisoning by polychlorinated biphenyls and their contaminants in Taiwan, *Science* **241,** 334–336 (1988).

50. P. R. Taylor, C. E. Lawrence, H. Hwang, and A. S. Paulson, Polychlorinated biphenyls: Influence on birthweight and gestation, *Am. J. Public Health* **74,** 1153–1154 (1984).
51. G. G. Fein, J. L. Jacobson, S. W. Jacobson, *et al.*, Prenatal exposure to polychlorinated biphenyls: Effects on birth size and gestational age, *J. Pediatr.* **105,** 315–320 (1984).
52. E. Dar, M. S. Kanerek, H. A. Anderson, and S. C. Sonzogni, Fish consumption and reproductive outcomes in Green Bay Wisconsin, *Environ. Res.* **59,** 189–201 (1992).
53. W. J. Rogan, B. C. Gladen, J. D. McKinney, *et al.*, Polychlorinated biphenyls (PCBs) and dichlorodiphenyl dichloroethene (DDE) in human milk: Effects on growth, morbidity and duration of lactation, *Am. J. Public Health* **77,** 1294–1297 (1987).
54. A. Schecter, O. Päpke, and M. Ball, Evidence for transplacental transfer of dioxins from mother to fetus: Chlorinated dioxin and dibenzofuran levels in the livers of stillborn infants, *Chemosphere* **21,** 1017–1022 (1990).
55. J. G. Koppe, K. Olie, and J. van Wijnen, Placental transport of dioxins from mother to fetus, *Dev. Pharmacol. Ther.* **18,** 9–13 (1992).
56. B. C. Gladen, W. J. Rogan, P. Hardy, *et al.*, Development after exposure to polychlorinated biphenyls and dichlorodiphenyl dichloroethene transplacentally and through human milk, *J. Pediatr.* **113,** 991–995 (1988).
57. J. L. Jacobson, S. W. Jacobson, and H. E. B. Humphrey, Effects of exposure to PCBs and related compounds on growth and activity in children, *Neurotoxicol. Teratol.* **12,** 319–326 (1990).
58. J. L. Jacobson, S. W. Jacobson, and H. E. B. Humphrey, Effects of in utero exposure to polychlorinated biphenyls and related contaminants on cognitive functioning in young children, *J. Pediatr.* **116,** 38–45 (1990).
59. J. G. Bekesi, J. P. Roboz, S. Solomon, *et al.*, Persistent immune dysfunction in Michigan residents exposed to polybrominated biphenyls, *Adv. Immunopharm. 2:* 33–39 (1983).
60. K. J. Chang, K. W. Hseih, T. P. Lee, *et al.*, Immunologic evaluation of patients with polychlorinated biphenyl poisoning: Determination of lymphocyte subpopulations, *Toxicol. Appl. Pharmacol.* **61,** 58–63 (1981).
61. J. J. T. W. A. Strik, M. Doss, G. Schraa, *et al.*, Coproporphyrinuria and chronic hepatic porphyria type A found in farm families from Michigan (U.S.A.) exposed to polybrominated biphenyls (PBB), in: *Chemical Porphyria in Man* (J. J. T. W. A. Strik and J. H. Koeman, eds.), pp. 29–53, Elsevier/North-Holland, Amsterdam (1979).
62. K. J. Chang, F. J. Lu, and T. C. Tung, Studies on patients with polychlorinated biphenyl poisoning. 2. Determination of urinary coproporphyrin, uroporphyrin, delta-aminolevulinic acid and porphobilinogen, *Res. Commun. Chem. Pathol. Pharmacol.* **30,** 547–553 (1980).
63. K. H. Kilburn, R. J. Warshaw, and M. G. Shields, Neurobehavioral dysfunction in firemen exposed to polychlorinated biphenyls (PCBs): Possible improvement after detoxification, *Arch. Environ. Health* **44,** 345–350 (1989).
64. N. Ito, H. Nagasaki, M. Arai, *et al.*, Histopathologic studies on liver tumorigenesis induced in mice by technical polychlorinated biphenyls and its promoting effect on liver tumors induced by benzene hexachloride, *J. Natl. Cancer Inst.* **51,** 1637–1646 (1973).
65. J. A. Heddle and W. R. Bruce, Comparison of tests for mutagenicity or carcinogenicity using assays for sperm abnormalities, formation of micronuclei, and mutation in Salmonella, in: *Origins of Human Cancer* (H. H. Hiatt *et al.*, eds.), pp. 1549–1557, Cold Spring Harbor Laboratory Press, Cold Spring Harbor, NY (1977).
66. R. E. McMahon, J. C. Cline, and C. Z. Thompson, Assay of 855 test chemicals in ten tester strains using a new modification of the Ames test for bacterial mutagens, *Cancer Res.* **39,** 682–693 (1979).

67. M. L. Hattula, Mutagenicity of PCBs and their pyrosynthetic derivatives in cell-mediated assay, *Environ. Health Perspect.* **60,** 255–257 (1985).
68. R. Hoopingarner, A. Samuel, and D. Krause, Polychlorinated biphenyl interactions with tissue culture cells, *Environ. Health Perspect.* **1,** 155–158 (1972).
69. S. Green, F. M. Sauro, and L. Freidman, Lack of dominant lethality in rats treated with polychlorinated biphenyls (Aroclors 1242 and 1254), *Food Cosmet. Toxicol.* **13,** 507–510 (1975).
70. I. Berenblum and P. Shubik, A new quantitative approach to the study of the stages of chemical carcinogenesis in the mouse skin, *Br. J. Cancer* **1,** 383–391 (1947).
71. E. Farber, Carcinogenesis—Cellular evolution as a unifying thread, *Cancer Res.* **33,** 2537–2550 (1973).
72. T. J. Slaga, Overview of tumor promotion in animals, *Environ. Health Perspect.* **50,** 3–14 (1983).
73. J. C. Fisher and J. H. Holloman, A hypothesis for the origin of cancer foci, *Cancer* **4,** 916–918 (1951).
74. P. Armitage and R. Doll, The age distribution of cancer and a multi-stage theory of carcinogenesis, *Br. J. Cancer* **3,** 1–12 (1954).
75. P. Armitage and R. Doll, A two-stage theory of carcinogenesis in relation to the age distribution of human cancer, *Br. J. Cancer* **11,** 161–169 (1957).
76. R. Doll, An epidemiological perspective of the biology of cancer, *Cancer Res.* **38,** 3573–3583 (1978).
77. C. O. Nordling, A new theory on the cancer inducing mechanism, *Br. J. Cancer* **1,** 68–72 (1953).
78. L. C. Fisher, Multiple mutation theory of carcinogenesis, *Nature* **181,** 651–652 (1958).
79. N. E. Day and C. C. Brown, Multistage models and primary prevention of cancer, *J. Natl. Cancer Inst.* **64,** 977–989 (1980).
80. H. Pitot and A. Sirica, The stages of initiation and promotion in hepatocarcinogenesis, *Biochim. Biophys. Acta* **605,** 191–215 (1980).
81. R. A. Squire and M. H. Levitt, Report of a workshop on classification of specific hepatocellular lesions in rats, *Cancer Res.* **35,** 3214–3223 (1975).
82. G. W. Teebor and F. F. Becker, Regression and persistence of hyperplastic hepatic nodules induced by N-2-fluorenylacetamide and the relationship to hepatocarcinogenesis, *Cancer Res.* **31,** 1–6 (1971).
83. E. Farber, On the pathogenesis of experimental hepatocellular carcinoma, in: *Hepatocellular Carcinoma* (K. Okuda and R. L. Peters, eds.), pp. 1–24, Wiley, New York (1976).
84. F. Scherer and M. Hoffman, Probable clonal genesis of cellular islands induced in rat liver by diethylnitrosamine, *Eur. J. Cancer* **7,** 369–371 (1971).
85. C. Peraino, R. J. M. Fry, E. Staffeldt, and J. P. Christopher, Comparative enhancing effects of phenobarbital, amobarbital, diphenylhydantoin, and dichlorodiphenyltrichloroethane in 2-acetylaminofluorene-induced hepatic tumorigenesis in the rat, *Cancer Res.* **35,** 2884–2890 (1975).
86. C. Peraino, R. J. M. Fry, E. Staffeldt, and W. E. Kisieleski, Effects of varying the exposure to phenobarbital on its enhancement of 2-acetylaminofluorene-induced hepatic tumorigenesis in the rat, *Cancer Res.* **33,** 2701–2705 (1973).
87. M. Nishizumi, Effect of phenobarbital, dichlorodiphenyltrichloroethane, and polychlorinated biphenyls on diethylnitrosamine-induced hepatocarcinogenesis, *Gann* **70,** 835–837 (1979).
88. C. L. Litterst, T. M. Farber, A. M. Baker, and E. J. Van Loon, Effect of polychlorinated biphenyls on hepatic microsomal enzymes in the rat, *Toxicol. Appl. Pharmacol.* **23,** 112–122 (1972).

89. T. S. Chen and K. P. DuBois, Studies on the enzyme inducing effect of polychlorinated biphenyls, *Toxicol. Appl. Pharmacol.* **26,** 504–512 (1973).
90. S. Safe, S. Bandiera, T. Sawyer, *et al.,* PCB's: Structure–function relationships and mechanism of action, *Environ. Health Perspect.* **60,** 47–56 (1985).
91. E. E. McConnell, Comparative toxicity of PCB's and related compounds in various species of animals, *Environ. Health Perspect.* **60,** 29–33 (1985).
92. R. D. Kimbrough and R. E. Linder, Induction of adenofibrosis and hepatomas of the liver in BALB/cJ mice by polychlorinated biphenyls (Aroclor 1254), *J. Natl. Cancer Inst.* **53,** 547–552 (1974).
93. N. T. Kimura and T. Baba, Neoplastic changes in the rat liver induced by polychlorinated biphenyl, *Gann* **64,** 105–108 (1973).
94. R. D. Kimbrough, R. A. Squire, R. E. Linder, *et al.,* Induction of liver tumors in Sherman strain female rats by polychlorinated biphenyl Aroclor 1260, *J. Natl. Cancer Inst.* **55,** 1453–1459 (1975).
95. J. C. Calandra, Summary of toxicological studies on commercial PCB's, in: *Proceedings of the National Conference on Polychlorinated Biphenyls, November 19–21, 1975, Chicago,* pp. 35–42, EPA-560/6-75-004, Environmental Protection Agency, Office of Toxic Substances, Washington, DC (1976).
96. G. J. Levinskas, A review and evaluation of carcinogenicity studies in mice and rats and mutagenicity studies with polychlorinated biphenyls, Monsanto Company (1981).
97. National Cancer Institute, *Bioassay of Aroclor 1254 for possible carcinogenicity,* Technical Report Series, No. 38, U.S. Department of Health, Education and Welfare, Washington, DC (1978).
98. P. Armitage, *Statistical Methods in Medical Research,* pp. 362–365, Wiley, New York (1971).
99. D. Norback and R. Weltman, Polychlorinated biphenyl induction of hepatocellular carcinoma in the Sprague–Dawley rat, *Environ. Health Perspect.* **60,** 97–105 (1985).
100. B. D. Preston, J. P. Van Miller, R. W. Moore, *et al.,* Promoting effects of polychlorinated biphenyls (Aroclor 1254) and polychlorinated dibenzofuran-free Aroclor 1254 on diethylnitrosamine-induced tumorigenesis in the rat, *J. Natl. Cancer Inst.* **66,** 509–515 (1981).
101. M. Uchiyama and T. Chiba, Co-carcinogenic effect of DDT and PCB feedings on methylcholanthrene-induced chemical carcinogenesis, *Bull. Environ. Contam. Toxicol.* **12,** 687–693 (1974).
102. M. A. Pereira, S. L. Hwerren, A. L. Britt, *et al.,* Promotion by polychlorinated biphenyls of enzyme-altered foci in rat liver, *Cancer Lett.* **15,** 185–190 (1982).
103. E. Deml, D. Oesterle, and F. J. Wiebel, Benzo(a)pyrene initiates enzyme-altered islands in the liver of adult rats following single pretreatment and promotion with polychlorinated biphenyls, *Cancer Lett.* **19,** 301–304 (1983).
104. S. Makiura, H. Aoe, S. Sugihara, *et al.,* Inhibitory effect of polychlorinated biphenyls on liver tumorigenesis in rats treated with 3′-methyl-4-dimethylaminoazobenzene, N-2-fluorenylacetamide, and diethylnitrosamine, *J. Natl. Cancer Inst.* **53,** 1253–1257 (1974).
105. M. Nishizumi, Reduction of diethylnitrosamine-induced hepatoma in rats exposed to polychlorinated biphenyls through their dams, *Gann* **71,** 910–912 (1980).
106. D. L. Berry, T. J. Slaga, J. DiGiovanni, *et al.,* Studies with chlorinated dibenzo-p-dioxins, polybrominated biphenyls, and polychlorinated biphenyls in a two-stage system of mouse skin tumorigenesis: Potent anticarcinogenic effects, *Ann. N.Y. Acad. Sci.* **320,** 405–414 (1979).
107. D. P. Brown and M. Jones, Mortality and industrial hygiene study of workers exposed to polychlorinated biphenyls, *Arch. Environ. Health* **36,** 120–129 (1981).

108. D. P. Brown, Mortality of workers exposed to polychlorinated biphenyls—An update, *Arch Environ. Health* **42,** 333–339 (1987).
109. P. A. Bertazzi, L. Riboldi, A. Pesatori, *et al.*, Cancer mortality of capacitor manufacturing workers, *Am. J. Ind. Med.* **11,** 165–176 (1987).
110. P. A. Bertazzi, C. Zocchetti, S. Guercilena, *et al.*, Mortality study of male and female workers exposed to PCBs, in: *Prevention of Occupational Cancer—International Symposium* (Occupational Safety and Health Series 46), pp. 242–248, International Labour Office, Geneva (1982).
111. P. Gustavsson, C. Hogstedt, and C. Rappe, Short-term mortality and cancer incidence in capacitor manufacturing workers exposed to polychlorinated biphenyls (PCBs), *Am. J. Ind. Med.* **10,** 341–344 (1986).
112. T. Sinks, G. Steele, A. B. Smith, *et al.*, Mortality among workers exposed to polychlorinated biphenyls, *Am. J. Epidemiol.* **136,** 389–398 (1992).
113. A. K. Bahn, I. Rosenwaike, N. Herrmann, *et al.*, Melanoma after exposure to PCB's, *N. Engl. J. Med.* **295,** 450 (1976).
114. S. L. Shalat, L. D. True, L. E. Fleming, and P. E. Pace, Kidney cancer in utility workers exposed to polychlorinated biphenyls (PCBs), *Br. J. Ind. Med.* **46,** 823–824 (1989).
115. P. S. J. Lees, M. Corn, and P. N. Breysse, Evidence for dermal absorption as the major route of body entry during exposure of transformer maintenance and repairmen to PCBs, *Am. Ind. Hyg. Assoc. J.* **48,** 257–264 (1987).
116. *U. S. Cancer Mortality Rates and Trends, 1950–1979* (3 Vols.), EPA-600/1-83-015, National Cancer Institute/Environmental Protection Agency, Government Printing Office, Washington DC (1983).
117. P. A. Bertazzi, University of Milan, Personal communication.
118. M. Kuratsune, K. Nakamura, M. Ikeda, and T. Hirohata, Analysis of deaths seen among patients with Yusho, Abstract: Dioxin 86 Symposium, Fukuoka, Japan (1986).

Chapter 16

Cancer Epidemiology

Lennart Hardell, Mikael Eriksson, Olav Axelson, and Shelia Hoar Zahm

1. THE SCOPE OF EPIDEMIOLOGY

The assessment of causal relationships between exposure and disease usually requires epidemiological studies. Randomized trials on humans are rarely achievable but can sometimes be accomplished in a secondary follow-up of the long-term side effects, e.g., of drugs. The association between smoking and lung cancer is a classical example of important information discovered by epidemiological studies. Other important epidemiological achievements include detecting the relationship between asbestos exposure and mesothelioma, arsenic exposure and lung cancer, and vinyl chloride and hemangiosarcoma of the liver. The epidemiological approach has been criticized because of its nonexperimental nature, but there is no other ethically acceptable way to confirm whether toxicologic observations of an adverse health effect are relevant for humans. However, epidemiological studies must be carefully conducted and evaluated since methodologic limitations may influence the studies as well as the interpretation of the findings.

Epidemiological investigations focusing on etiologic aspects are essentially of case–control or cohort design. Correlational studies regarding incidence or morbidity data in relation to some crude population exposure data are usually not persuasive. In case–control studies data on various exposures are considered among both persons with the disease and other individuals representing the source population for the cases,

Lennart Hardell • Department of Oncology, Örebro Medical Centre, S-701 85 Örebro, Sweden. Mikael Eriksson • Department of Oncology, University Hospital, S-901 85 Umeå, Sweden. Olav Axelson • Department of Occupational Medicine, University Hospital, S-581 85 Linköping, Sweden. Shelia Hoar Zahm • Occupational Studies Section, Environmental Epidemiology Branch, Division of Cancer Etiology, National Cancer Institute, Rockville, Maryland 20892.
Dioxins and Health, edited by Arnold Schecter. Plenum Press, New York, 1994.

i.e., controls. This type of study tends to be the only alternative for rare diseases. The cohort approach, on the other hand, follows a defined population exposed to a certain agent and compares the disease outcome to that experienced by a reference population without exposure to that agent. It is necessary that the diseases of interest are reasonably common; otherwise, very large cohorts are required to permit any conclusions.

2. EPIDEMIOLOGICAL STUDIES ON DIOXIN-EXPOSED POPULATIONS

Since the 1970s, many epidemiological studies have been published on persons exposed to polychlorinated dibenzodioxins (PCDDs), mainly as occurring in various types of pesticides, especially some of the chlorinated phenoxy herbicides. These individuals have been either producers or users of chemicals in which dioxins might have occurred as impurities. Several studies have been of the case–control design and have involved the quite rare group of malignant tumors known as soft-tissue sarcomas (STS) but also the more common malignant lymphomas. These studies were initiated because of clinical observations about a possible association. Also, cohorts of pesticide applicators or persons involved in the production of dioxin-contaminated chemicals have been studied.

There have also been several cohort studies of other populations with potential exposure to dioxins, such as workers in paper and pulp production, Vietnam veterans, and members of the general population, e.g., after the Seveso and Yusho accidents.

3. CLINICAL OBSERVATIONS

The first report linking human malignant disease with dioxin exposure occurred in 1976 when three male patients with STS treated at the Department of Oncology in Umeå in northern Sweden were found to have been occupationally exposed to phenoxyacetic acids. Subsequent review of the medical records of the Department identified another four cases with exposure to phenoxyacetic acids.[1] At that time there was also some indication that Swedish railroad workers who sprayed various herbicides suffered an increased risk of cancer.[2,3] Furthermore, in 1979 eleven patients with non-Hodgkin's lymphoma (NHL) and exposure to phenoxyacetic acids and/or chlorophenols were described from the same Department as the case report on STS.[4] These various observations instigated two Swedish case–control studies on STS,[5,6] and one on malignant lymphoma.[7,8] The results of these studies, described below, seem to have stimulated researchers in many countries to conduct similar investigations. Subsequently, some cohort studies on dioxin-exposed workers have also been published. The results from these different studies have in certain respects been conflicting. Not surprisingly, therefore, a debate on the results has followed, not only considering the scientific aspects but also with economic overtones.[9]

4. CASE–CONTROL STUDIES

4.1. The Swedish Studies

In addition to the above-mentioned studies performed in the late 1970s, two more case–control studies of STS were conducted by the same research group,[10,11] and additional studies have also been conducted in Sweden on STS[12,13] and NHL.[14] Also, some other types of malignant diseases have been studied focusing on exposure to dioxin-containing chemicals, i.e., with regard to nasopharyngeal cancer,[15] primary liver cancer,[16] and multiple myeloma.[17]

In Swedish forestry the predominant herbicide exposure has been a combination of 2,4-dichlorophenoxyacetic acid (2,4-D) and 2,4,5-trichlorophenoxyacetic acid (2,4,5-T) used to combat hardwoods. In agriculture, the predominant herbicide exposures have been the phenoxyacetic acids 4-chloro-2-methylphenoxyacetic acid (MCPA) and 2,4-D, which is particularly reflected in a study from southern Sweden.[6]

Regarding chlorophenols, exposure to pentachlorophenols were reported by workers in sawmills, carpentry, and certain other occupations, as well as in leisure time activities. On the other hand, no subject reported exposure to trichlorophenol, which has been used very little in Sweden, and which is known to be contaminated by 2,3,7,8-tetrachlorodibenzo-*p*-dioxin (TCDD).

4.1.1. Soft-Tissue Sarcoma Studies

In general, similar methods were used in several of the Swedish studies.[5,6,10–12] The cases were drawn from cancer registers and the controls were extracted from population registers. In two of the studies,[10,12] a second control group of malignant diseases other than STS was used. In one of these studies,[10] subjects with malignant lymphoma or nasopharyngeal cancer were excluded as controls since these types of tumors might relate to the exposure at issue.[7,8,15] Furthermore, for deceased cases deceased controls were used to equalize data assessment. They were identified from the National Register on Causes of Death. The next of kin were traced by contacting parishes where cases and controls were registered at the time of their deaths.

In the various studies referred to, an extensive self-administered questionnaire was used to obtain a complete working history for each case and control. Questions were asked as to specific job categories and exposures, smoking habits, leisure time exposure to chemicals, and so forth. These detailed questions elicited a broad exposure history and are therefore also likely to have obscured the hypotheses under investigation from the study subjects. If the answers were unclear or incomplete, the respondent was phoned for further information by an interviewer.

In some of the Swedish studies,[5,6,10,11] a minimum exposure time of 1 day was required for subjects to be classified as having been exposed to phenoxyacetic acids or chlorophenols. All exposure to these substances within 5 years before the diagnosis

Table 1
Odds Ratios and 95% Confidence Intervals (in Parentheses) in Swedish Case–Control Studies

Study	Chlorophenols[a]	Phenoxyacetic acids	Ref.
Soft-tissue sarcoma			
STS I	6.6 (2.4–18)	5.3 (2.4–12)	5
STS II	3.3 (1.3–8.1)	6.8 (2.6–17)	6
STS III	N.C.[b]	3.3 (1.4–8.1)	10
STS IV	5.3 (1.7–16)	1.3 (0.7–2.6)	11
STS V	1.6 (0.8–3.3)[d]		12
STS VI	N.A.[c]	2.4 (0.8–7.0)	13
Malignant lymphoma			
HD[e] + NHL	8.4 (4.2–17)	4.8 (2.9–8.1)	7
NHL	9.4 (3.6–25)	5.5 (2.7–11)	7,86
HD	6.5 (2.2–19)	5.0 (2.4–10)	8
NHL	N.C.	4.9 (1.3–18)[d]	14
HD	N.C.	3.8 (0.7–21)[d]	14
Other malignancies			
Colon I	1.8 (0.6–5.3)	1.3 (0.6–2.8)	22
Colon II	0.9 (0.2–3.5)	1.3 (0.7–2.4)	23
Nasopharynx, nasal	6.7 (2.8–16)	2.1 (0.9–4.7)	15
Primary liver cancer	2.2 (0.7–7.3)	1.7 (0.7–4.4)	16
Multiple myeloma	1.1 (0.6–1.9)[d]	2.2 (1.1–2.7)[d]	17

[a]Regarding chlorophenols, only high-grade exposure is presented.
[b]N.C., not calculated since few were exposed.
[c]N.A., not assessed.
[d]90% confidence interval.
[e]HD = Hodgkin's disease.

was excluded, because of the generally suggested latency time required for chemical carcinogenesis. Exposure to chlorophenols was classified into high grade, i.e., 1 week or more continuously or at least 1 month totally over the years, and low grade with less exposure than that.

All four most similarly designed studies on STS[5,6,10,11] demonstrated an association between exposure to phenoxyacetic acids or chlorophenols, as shown in Table 1, where the results of all Swedish studies are presented. The odds ratios in the two later studies were somewhat lower than in the earlier studies. This may partly reflect a later time period of exposure in the subjects, with a possibility of other exposure conditions, e.g., less dioxin-contaminated phenoxy herbicides.

Another above-mentioned case–control study from southeast Sweden enrolled 96 cases with STS, 450 randomly selected population controls, and 200 cancer controls.[12] Increased odds ratios (ORs) were obtained for gardeners (OR 4.1, 90% CI 1.0–14), railroad workers (OR 3.1, 90% CI 0.6–14), construction workers with exposure to impregnating agents (OR 2.3, 90% CI 0.5–8.9), and unspecified workers with potential exposure to phenoxy herbicides and/or chlorophenols (OR 1.7, 90% CI 0.3–7.3). Hence, the classification of exposure mainly relied only on occupational categories.[12]

Still another case–control study was performed as part of The Scandinavian Joint

Care Program on STS. It included 79 cases from Sweden and Finland and 226 controls, both males and females. Only cases with histopathologically high-grade STS were included. Exposure to phenoxyacetic acids yielded an OR of 2.4 (95% CI 0.8–7.0).[13] No information was given regarding different types of phenoxy herbicides. Chlorophenol exposure was not assessed.

4.1.2. Metaanalysis on STS and Potential Dioxin Exposure

The most similarly designed Swedish case–control studies on STS[5,6,10,11] have now been aggregated for a metaanalysis and the results are presented in Table 2.[18] The data were stratified on study. In one of the investigations, persons with other types of malignant diseases were used as controls in addition to population controls.[10] These cancer controls were now excluded, and only population controls were used in the analyses presented below. Although metaanalysis can be criticized on methodological grounds, it may be justified to use this method here in order to obtain a summary view of these studies along with some information on the dose–response relationships.

Exposure to dioxins was evaluated by using the knowledge on contamination of phenoxyacetic acids and chlorophenols by different isomers. An increased risk for STS was associated with exposure to all dioxins, i.e., both for TCDD and for dioxins other than TCDD. A dose–response effect for duration of exposure was shown with a significant trend ($p < 0.001$).[18]

Table 2

Mantel–Haenszel Odds Ratios (OR) and 90% Confidence Intervals (CI) Adjusted by Study for STS among Persons Exposed to All Dioxins, TCDD, and Dioxins Other Than TCDD in Four Case–Control Studies[5,6,10,11] Involving 434 Cases and 948 Controls[a]

	Unexposed	Exposed < 1 year	Exposed ≥ 1 year
All dioxins			
Cases	352	58	24
Controls	865	74	9
OR	1.0	2.4	6.4
CI	—	1.7–3.4	3.5–12
TCDD			
Cases	352	40	6
Controls	865	39	2
OR	1.0	3.0	7.2
CI	—	2.0–4.5	2.6–20
Other dioxins			
Cases	352	18	18
Controls	865	35	7
OR	1.0	1.7	6.2
CI	—	0.98–2.9	2.9–13

[a]All subjects were exposed for at least 1 day and a minimum latency period of 5 years was used.

4.1.3. Malignant Lymphoma Studies

The case–control study on malignant lymphoma, published in 1980, included both Hodgkin's disease (HD) and NHL.[7,8] The study included 60 HD cases, 105 NHL cases, 4 unclassifiable lymphoma cases, and 335 controls. As in the studies on STS, statistically significant ORs were found for exposure to phenoxyacetic acids or chlorophenols, as shown in Table 1. In further analyses this finding applied for both types of malignant lymphoma.[8] The study has now been analyzed regarding NHL only. The material was stratified for age and vital status. Exposure to phenoxyacetic acids yielded an OR of 5.5 (95% CI 2.9–10). High-grade exposure to chlorophenols resulted in OR = 9.4 (95% CI 3.9–23) and low-grade exposure OR = 3.3 (95% CI 1.7–6.5). Exposure to organic solvents yielded increased risks for both HD and NHL.

Another case–control study on 54 cases of HD and 106 NHL cases and 175 referents also indicated an excess risk from exposure to phenoxyacetic acids but also to other agents, especially solvents with regard to NHL.[14] By logistic regression analysis, increased ORs were obtained for phenoxyacetic acids and HD (OR 3.8, 90% CI 0.7–21) as well as NHL (OR 4.9, 90% CI 1.3–18). Still higher ORs were obtained for exposure to creosote, but the numbers were small. Interestingly, farming as such appeared to protect against NHL, implying negative confounding. This protective effect may help explain why no excess risk for NHL was seen on a rather crude cohort basis in a study on Swedish farmers; this is because of the lack of data on exposure to different agents in that study.[19]

4.1.4. Other Malignant Diseases

Multiple myeloma is a lymphoproliferative malignancy related to malignant lymphoma. Several reports have described an association with farming.[20] As presented in Table 1, a recently published Swedish case–control study on multiple myeloma[17] found an increased OR for exposure to phenoxyacetic acids, but not for chlorophenols.

Nasal cancer has long been associated with exposure to hardwood dust, and different hypotheses regarding the responsible mechanism have been tested in epidemiological studies. However, in one study of nasal and nasopharyngeal cancer, both malignancies showed a rather strong association with exposure to chlorophenols in sawmills, and a weaker relationship with exposure to phenoxy herbicides (see Table 1).[15] Chlorophenols have been used as wood preservatives, and exposure to these substances including contaminating dioxins has occurred in different occupational procedures where inhalation of sawdust from treated wood cannot be avoided.

An increase in the incidence of primary liver cancer has been reported from northern Vietnam, and an association with spraying with phenoxy herbicides has been postulated.[21] A case–control study in northern Sweden on primary liver cancer in men[16] yielded ORs of 1.7 (95% CI 0.7–4.4) and 2.2 (95% CI 0.7–7.3) for exposure to phenoxyacetic acids and chlorophenols, respectively, i.e., rather inconclusive results (Table 1).

4.1.5. Comments on Design Issues

Case–control studies using questionnaires or interviews for assessment of exposure may always be criticized by suggesting some kind of observational bias such as recall bias or interviewer bias. Hence, there is the possibility that the cases might remember more hazardous exposures than controls because of their cancer diagnosis. However, it is very unlikely that individuals with some particular kind of cancer would better remember a particular type of exposure than individuals with some other type of cancer. It is certainly worth noticing therefore that a case–control study on colon cancer,[22] performed in the same area with the same methodology as some of the other studies,[5,7,10] did not show any increased risk for exposure to phenoxyacetic acids or chlorophenols. As shown in Table 1, there was also no clear indication of risk in a later case–control study on colon cancer where more detailed questions specifically on exposure to dioxin-containing compounds were included.[23]

Interviewer bias might theoretically have been introduced in the initial three Swedish studies[5–7] by the fact that no special precautions were taken to prevent the telephone interviewer from knowing the case/control status during the supplementary telephone calls. To test the potential influence of such a bias, a comparison was made with ORs calculated only on the basis of the information in the questionnaires. The similar results clearly indicate that the results could not be explained by interviewer bias in this respect.[22]

Furthermore, in two of the later studies on STS,[10,11] as well as in the study on multiple myeloma,[17] all telephone interviews and all coding of collected data were done blinded with respect to the case/control status of the subjects although some individuals may have revealed their status during the interviews. When the exposure assessment was based mainly on occupation,[12] an effect was still observed, although weaker, as should be expected when exposure information is less specific.

Another question of concern is the matter of confounding. Spurious association may occur in an epidemiological study if another causative agent was associated with the exposure in the study population, and if this agent was not adequately controlled for in the design or analysis of the study. This other agent would then necessarily have to exert a stronger effect, or at least an equally strong effect as that appearing as associated with the exposure under study. No such strong confounding factor has ever been identified in connection with either STS or malignant lymphomas and exposure to phenoxy herbicides or chlorophenols, however, although the question of confounding has been echoing through the years.[24–26] The observed association with the exposure is therefore very unlikely to result from confounding.

4.2. The Studies from Kansas, Nebraska, Iowa, and Minnesota

Researchers from the U.S. National Cancer Institute have conducted three case–control studies in four U.S. midwestern agricultural states investigating the role of

Table 3
Number of White Male NHL Cases and Controls and Odds Ratios by Days per Year of Exposure to 2,4-D in Kansas[27] and Nebraska[29]

	Days/year	No. of NHL cases	No. of controls	Odds ratio
Kansas[a]	0	37	286	1.0
	1–2	6	17	2.7
	3–5	4	16	1.6
	6–10	4	16	1.9
	11–20	4	9	3.0
	21+	5	6	7.6
Nebraska[b]	0	54	184	1.0
	1–5	16	44	1.2
	6–20	12	25	1.6
	21+	3	4	3.3

[a]Kansas: p value for trend, 0.0001.
[b]Nebraska: p value for trend, 0.051.

herbicides in relation to malignant diseases (Table 3). The Kansas study, a population-based case–control study of NHL, HD, STS, and colon cancer, found a strong association between the phenoxyacetic acid herbicide 2,4-D and NHL, but no association between herbicide use and STS, HD, or colon cancer.[27,28] Among 2,4-D users, the risk of NHL increased with frequency of herbicide exposure to an OR of 7.6 (95% CI 1.8–32) for farmers exposed more than 20 days per year (Table 3). Only three cases and 18 controls reported use of 2,4,5-T and all but two of these controls has also used 2,4,-D.

In Nebraska, the population-based case–control study of NHL, HD, multiple myeloma, and chronic lymphocytic leukemia also found an association between NHL and 2,4-D.[29] Risks were lower than in Kansas, but the patterns were similar. Risk increased with frequency of exposure to an OR of 3.3 among farmers exposed more than 20 days per year. Among 2,4-D users, risk also increased the longer the farmers continued to work in potentially contaminated clothing after pesticide application. If the farmers changed clothing immediately, the OR was 1.1. Changing at the end of the day was associated with an OR of 1.5, while waiting to change until the following day or later was associated with an OR of 4.7 (95% CI 1.1–22). Detailed results for the other malignancies in this study have not been reported as yet.

The Iowa/Minnesota study investigated NHL[30] and leukemia.[31] For NHL modest nonsignificant increases in risk were observed for use of 2,4-D or 2,4,5-T. If handled at least 20 years prior to the interview, ORs increased to 1.3 (95% CI 0.9–1.8) and 1.7 (95% CI 0.8–3.6) for 2,4-D and 2,4,5-T, respectively. No information on frequency of use was presented. Leukemia, likewise, was associated with only modest nonsignificant increases of risk for 2,4-D and 2,4,5-T exposure. MCPA, however, was associated with an OR of 1.9 (95% CI 0.8–4.3). When the analyses were restricted to persons first exposed at least 20 years before the interview, risks increased for MCPA (OR 2.4, 95% CI 0.7–2.8) and 2,4,5-T (OR 1.8, 95% CI 0.8–4.0).

4.3. The Washington State Study

This case–control study included 128 cases of STS, 576 of NHL, and 694 controls.[32] The exposure assessment was based on categorizing job titles, activities, and chemical preparations reported during interviews by potential exposure. Therefore, there may be less certainty regarding exposure data compared with other case–control studies where subjects themselves named specific exposures. The relative prevalence of 2,4-D and 2,4,5-T, and thus TCDD, among the phenoxy acids used in the study region was also unclear in the published report.

No association with STS or NHL was found in the analyses based on the assessment of potential exposure to phenoxyacetic acids or chlorophenols. Of interest, however, is the elevated risk of NHL among men who had been farmers (OR 1.3, 95% CI 1.0–1.7) or forestry herbicide applicators (OR 4.8, 95% CI 1.2–19). Furthermore, those potentially exposed to phenoxy herbicides in any occupation for 15 years or more during the period prior to 15 years before diagnosis of NHL also had an increased risk (OR 1.7, 95% CI 1.0–2.8).

In the light of the results in the Swedish studies, it is also interesting that this study showed elevated risks for STS among exposed persons with Scandinavian surnames. Thus, high estimated potential for phenoxy acid exposure gave an OR of 2.8 (95% CI 0.5–15.6), and for chlorophenol exposure an OR of 7.2 (95% CI 2.1–24.7). Corresponding elevated risks were not found for NHL.

Another interesting finding was elevated relative risks associated with self-reported histories of chloracne, which gave an OR of 3.3 (95% CI 0.8–14.0) for STS, and 2.1 (95% CI 0.6–7.0) for NHL.

4.4. The New Zealand Studies

In New Zealand a research group has performed two studies on STS[33,34] and one on NHL.[35] Except for an additional control group in the NHL study, all controls were patients diagnosed with other cancers. This is essential to remember when interpreting the results, because recent cohort studies indicate a certain degree of general carcinogenic effects by dioxins.

Furthermore, the cases were not fully comparable with those in some of the Swedish studies,[5,6,10,11] because patients with STS in parenchymatous organs as the stomach were not included. In the Swedish studies showing the highest risks,[5,6,10,11] cases with these locations of disease comprised about 40% of the total.

The criteria for assessing exposure to phenoxyacetic acids and chlorophenols are not fully stated in the papers. Exposure of the subjects was categorized as "potential" or "probable or definite," e.g., based on job titles, activities, or information on specific chemicals, which should be expected to decrease any existing risks by dilution of exposure.

The first STS study[33] consisted of 82 cases and 92 controls. It showed ORs between 1.3 and 1.6 for phenoxyacetic acids and chlorophenols, with the higher ratios if only "probable or definite" exposure of more than 1 day and at least 5 years' latency was included. The risk increased to 3.0 (95% CI 1.1–8.3) if only exposure in farmers was considered. Moreover, work within tanneries or meat workers pelt departments with potential exposure to chlorophenols gave an OR of 7.2.

The second study on STS[34] encompassed 51 cases and 315 controls, which were derived from the study on NHL. In contrast to the first study, this study yielded an OR of 0.8 (95% CI 0.3–1.9) for exposure to phenoxyacetic acids. The report on this study was very brief, however, and no details were given on chlorophenol exposure, and no OR restricted to farmers can be calculated with the data available.

The study on NHL[35] consisted of 183 cases and 338 controls. No significantly increased OR for exposure to phenoxyacetic acids or chlorophenols was obtained, but the risk estimates varied from 1.1 to 1.5 between different subgroups.

NHL risk did not increase with duration of exposure to phenoxy herbicides but did increase to 2.2 (95% CI 0.4–12.6) in the 10–19 frequency of use per year category, then dropped off to 1.1 (95% CI 0.3–4.1) in the greater than 19 times per year category.[36]

A somewhat strange pattern in the control groups in the New Zealand studies is the high percentage of railway workers, i.e., 7.6–12%. This may raise the question whether the controls represent the source population for the cases.[37] There was an unexplained increased risk for railway workers of 3.2 with regard to STS in the first study.[33]

The authors of the New Zealand studies discuss the fact that herbicide spraying is a full-time occupation in that country, and that none of the STS or NHL cases had occurred among the approximately 1500 current and former commercial sprayers. However, only about 0.1 would be expected among the commercial sprayers in each of the two STS studies, and about 0.3 case among the commercial sprayers in the NHL study.[38] Therefore, the absence of commercial sprayers among the cases is not evidence against the association. In fact, a case of STS occurred in a commercial sprayer in 1984, i.e., 2 years after the end of recruitment to the second New Zealand STS study (letter dated January 15, 1984, Tasman Pulp and Paper Company Ltd, Medical Centre, Kawerau, N.Z.).

4.5. The Italian Study

In 1986 an Italian case–control study on STS was published.[39] It consisted of 37 male and 31 female cases, and 85 male and 73 female controls. The study was conducted in a region of northern Italy where the principal agricultural crop is rice. In this area rice weeding was a predominantly female occupation involving manual labor and mainly dermal contact with phenoxy herbicides. Rice weeding among women was associated with a relative risk of 2.3 (95% CI 0.7–7.7) in the study. No increased risk was seen in men.

5. COHORT STUDIES ON PRODUCERS AND USERS OF DIOXIN-CONTAMINATED CHEMICALS

5.1. Some Early Cohort Studies on Pesticide Users

Some early studies of workers exposed to various pesticides appeared in the early 1970s. One of these studies concerned a cohort of 348 railroad workers who had been spraying various herbicides along the Swedish railroads, especially amitrol, diurone, phenoxyacetic acids, both 2,4-D and 2,4,5-T, as well as potassium chlorate and monurone.[2] These workers were found to have an excess of malignant tumors, first thought to be related to amitrol exposure. In a further analysis, however, the greatest effect turned out to be among those with combined exposure to both amitrol and phenoxy herbicides with a significantly increased risk of 3.4 for all tumors.[40] For those with exposure mainly to phenoxy herbicides there were two stomach cancers with only 0.33 expected.

Similar observations appeared from East Germany, where pesticide sprayers, again exposed to a mixture of both phenoxyacetic acids and other herbicides as well as various insecticides, were found to suffer from an excess of lung cancers.[41] However, a Finnish cohort on 1926 men who had been spraying phenoxy herbicides along the roadsides had no absolute but some proportional increase in cancer mortality.[42] Considering the exposure pattern in the Nordic studies, it seems likely that the Swedish cohort might have had relatively more exposure to dioxins and also a combination with a variety of other agents.

5.2. The United States Cohort

An important contribution in the field of epidemiology on dioxin exposure was published in 1991 by a group representing the National Institute of Occupational Safety and Health (NIOSH).[43] A total of 5172 persons who had been involved in the production of TCDD-contaminated chemicals in any of 12 plants in the United States were included in the cohort. Of these persons, 172 had been incorporated in previously published small cohort studies from two companies, whereas 5000 were identified as "assigned to a production or maintenance job in a process involving TCDD contamination." The follow-up period ended in 1987, and the comparisons were made with the U.S. population.

A total of 265 cancer deaths were observed, whereas 229.9 were expected yielding a standardized mortality ratio (SMR) of 115 (95% CI 102–130). In a subcohort of 1520 persons with a latency of 20 years or more and exposure of at least 1 year the SMR for all cancer was 146 (95% CI 121–176). In the total cohort SMR for STS was 338 (95% CI 92–865), and in the subgroup defined above the SMR was 922 (95% CI 190–2695).

Four STS deaths were found in the total cohort. Review of tissue specimens of these cases had been previously performed.[44] Two of the cases were thereby reclassified as not STS. However, erronous information on death certificates must be considered to have been equally frequent in the death certificates of the reference group, i.e., the U.S. population. Thus, this type of directed reanalysis of death certificate diagnosis in only the cohort cases is problematic. It may be noted also that two more STS deaths were found in the NIOSH cohort but they were according to classification principles assigned to ICD codes for other sites of malignant diseases, and therefore not included in the SMR value for STS. Furthermore, another STS death was reported by the authors among a group of 139 workers with chloracne who did not meet the entry criteria for the cohort.

The SMR for NHL was 137 (95% CI 66–254), but there was no increased SMR in the subcohort with exposure. The subcohort had an increased risk for respiratory system cancers with an SMR of 142 (95% CI 103–192).

To verify TCDD exposure in the cohort, dioxin levels in serum were measured in a sample of 253 persons. The levels of TCDD, adjusted for lipids, correlated well with years of exposure.

5.3. The German BASF Cohort

Persons employed at a German chemical manufacturing facility where 2,4,5-trichlorophenol was produced were enrolled in a cohort study.[45] A total of 247 workers potentially exposed to TCDD were followed from 1953 until 1987. In this mortality study, 78 persons had died including 23 from malignant diseases, at the end of the follow-up period.

The SMR for all malignant diseases was 117 (90% CI 80–166). When workers with chloracne were examined separately, the SMR for all malignancies was 139 (90% CI 87–211). Considering a latency period of at least 20 years after the first exposure to TCDD, the SMR was increased to 201 (90% CI 122–315) for all malignancies and to 252 (90% CI 99–530) for lung cancer based on 5 cases. Nonsignificantly increased SMRs were also found for cancer in buccal cavity and pharynx, stomach, colon and rectum, although based on very few cases.

No deaths from STS or NHL were reported, but only 0.1 STS case and 0.6 NHL case would be expected based on the NIOSH study data.

5.4. The German Boehringer–Ingelheim Cohort

A mortality follow-up of 1583 workers (1184 men and 399 women) employed in a chemical plant in Germany that produced herbicides, including processes contaminated with TCDD, has been reported.[46] Vital status of workers hired between 1952 and 1984 was determined as of 1989. As reference, both the national mortality statistics of West Germany and deaths in a cohort of male gas workers were used. Since the results did not differ by reference group, figures based on the general population mortality statistics are presented here.

The SMR for total cancer mortality was increased to 124 in men (95% CI 100–152). Among men with 20 or more years the SMR was 187 (95% CI 111–295). In a subgroup with high exposure to TCDD, the SMR for all malignant diseases was 142 (95% CI 98–199) overall and 254 (95% CI 110–500) for persons who had been employed for at least 20 years. The accuracy in the assigning of TCDD exposure was evaluated by analyses of TCDD in adipose tissue from 48 members of the cohort.

Increased SMRs for men were found for malignant diseases in hypopharynx, esophagus, stomach, larynx, lung, prostate, kidney, and hematopoietic system. These results were based on comparatively small numbers, however, and not significant.

Among women, 20 deaths from malignant neoplasms were found corresponding to an SMR of 94 (95% CI 58–145). Only the risk for breast cancer was elevated with an SMR of 215 (95% CI 98–409) based on 9 deaths.

5.5. The IARC Study

In 1991 the International Agency for Research on Cancer (IARC) published a cohort mortality study which encompassed 18,910 production workers or sprayers from ten countries potentially exposed to phenoxyacetic herbicides or chlorophenols.[47] Exposure data were collected through questionnaires, job histories, and factory or spraying records. Workers were classified as exposed (n = 13,482), probably exposed (n = 416), unexposed (n = 3951), and there was also a group with unknown exposure (n = 541). Cause-specific national death rates were used as the reference.

In the exposed group, the SMR for all malignant neoplasms was 101 (95% CI 92–110). An excess risk based on four observed deaths was noted for STS with an SMR of 196 (95% CI 53–502). In a group with 10–19 years since first exposure the SMR was 606 (95% CI 165–1552) for STS in the whole cohort, and among sprayers the corresponding SMR was 882 (95% CI 182–2579).

Risks also appeared to be increased for cancers of the testicle, thyroid, other endocrine glands, and nose and nasal cavity, but were based on small numbers of deaths.

Five additional cases of STS were recorded for cohort members who were alive at the end of follow-up, or who had died with certified cause of death other than the ICD code 171 (used for STS).

Regarding NHL, 14 deaths were observed among men and 1 among women. These numbers did not represent significantly increased SMRs.

5.6. Studies on Vietnam Veterans

Both case–control and cohort studies have been performed to evaluate any health effects, particularly malignant diseases, among veterans who fought in the Vietnam war, where phenoxy herbicides were used in the warfare.[48–50] Common to all of these studies are considerable difficulties in assessment of exposure, and often indirect measures have been used, e.g., service in certain corps or areas with potential exposure to the sprayings.

TCDD levels in serum and adipose tissue clearly show that such indirect exposure criteria do not accurately identify or rank persons actually exposed to the dioxin-contaminated herbicides.[51] Interpretation of all of these studies must thus be done with caution. Despite this reservation, the cohort study on Massachusetts veterans[52] is interesting, since it showed a significantly increased risk for STS among veterans with potential exposure.

One study that seems to illustrate the problem of assessing exposure is the mortality cohort of 1261 Air Force veterans participating in Operation Ranch Hand and responsible for the aerial herbicide spraying missions in Vietnam.[53] Serum TCDD measurements on a small subset of this cohort, i.e., those who were believed to be most exposed, show that only very few had high TCDD levels, with only 5 subjects with more than 200 ppt. Thus, the absence of any increased cancer deaths in this cohort is not very informative.

A series of case–control studies on Vietnam veterans regarding NHL, STS, HD, nasal cancer, nasopharyngeal cancer, and primary liver cancer indicated approximately 50% excess for NHL but not for the other cancer forms among veterans compared with men who did not serve in Vietnam.[54–56] These studies involved rather large numbers, i.e., 1157 NHL, 342 STS, 310 HD, 48 nasal cancer, 80 nasopharyngeal cancer, and 130 liver cancer cases along with 1776 control subjects obtained through random digit dialing. Service in Vietnam was reported by 99 NHL, 26 STS, 28 HD, 2 nasal, 3 nasopharyngeal, and 8 liver cancer and 133 controls. Although rather detailed information was obtained about location in Vietnam, job duties, and self-perceived exposure to pesticides, there seems to be quite some uncertainty as to whether or not an individual was actually exposed. Moreover, except for NHL the results were based on low numbers of cases who had served in Vietnam. Regarding NHL and self-reported possible contact with Agent Orange, the referent group for these comparisons ("other Vietnam veterans") included blue water Navy servicemen who were found to be at significantly elevated risk of NHL. Including these sea-based men in the referent group would depress the risk estimates associated with self-reported exposure to Agent Orange. The data should be reanalyzed using only land-based Vietnam veterans without self-reported exposure as the referent. The nonpositive outcome of most of these studies is therefore not particularly persuasive of no effect in the light of all other studies regarding phenoxy herbicide and TCDD exposure.

6. COHORT STUDIES ON GENERAL PUBLIC AFTER ACCIDENTS

6.1. The Seveso Study

Through an accident in a chemical plant in Seveso, Italy, in 1976, the general public in the area was exposed to TCDD. Although the follow-up period is short, some interesting results have already been published in a ten-year mortality study.[57] In the second 5-year period of follow-up there was some increase in mortality both from NHL and from STS, though statistically rather imprecise. For all types of malignancies combined, no increase has been seen as yet, however.

Cancer incidence findings for the first postaccident decade have also been published.[58,59] In Zone B, i.e., the second closest area to the factory, an increase in hepatobiliary cancers, mainly primary liver cancer (OR 2.0), and extrahepatic bile ducts and gallbladder cancer in women (OR 5.4) (statistically significant), was found. For both males and females, statistically significantly increased risks for hematopoietic malignancies were observed, (ORs 2.3 and 2.0, respectively). In zone R, next to zone B, the risk of STS was elevated twofold whereas the risk of uterine cancer was significantly lowered. For more details, see Bertazzi and di Domenico (this volume).

6.2. The Yusho Study

In 1968, approximately 1900 persons accidentally consumed polychlorinated biphenyls (PCBs) and dibenzofurans (PCDFs) through poisoned rice oil ("Yusho," meaning oil disease) in Japan. In a follow-up study of this cohort,[60] a significantly increased risk of liver cancer in males (SMR 559, $p < 0.01$), and a nonsignificant increase of the same malignancy in females (SMR 304), was noted. This finding is well in accordance with the known hepatocarcinogenicity of PCBs in animals. Furthermore, a significantly increased risk of lung cancer was also seen in males (SMR 326, $p < 0.01$).

A similar accident ("Yu-cheng") occurred in 1979 in Taiwan, but no follow-up study of that cohort has been published.

7. COHORT STUDIES ON PULP AND PAPER MILL WORKERS

The bleaching of pulp and paper by chlorine may result in various chlorinated organic compounds, among them PCDDs. Three cohort studies of pulp and paper mill workers, who have a potential exposure to PCDDs in their work environment, have been published.[61–63] Two were cancer mortality stuies from the U.S.,[61,62] and the third one was a Finnish incidence study.[63] No increase in cancer in general or in certain specified tumor types, e.g., NHL, was noted in any of the studies. STS was not studied specifically, however.

There are other studies indicating some excess cancer risk in the pulp and paper industry, however. A case–control study encompassing 4070 men deceased during the period 1950–1970 revealed an increased mortality from secondary tumors of the liver and lung and there was also a slightly increased risk for malignant lymphomas, leukemias, and cancers of the pancreas and stomach.[64] An excess of stomach cancer has also appeared in another small study from a Swedish paper mill.[65] Others have found an increased risk of lung cancer among paper mill workers, probably related to asbestos exposure, however.[66] In the interpretation of these findings, it is important to consider that the actual exposure to dioxins is not known in the various studies.

8. ENVIRONMENTAL EXPOSURE THROUGH POLLUTION

Chlorophenols may contaminate areas surrounding sawmills where they are used. Thus, both the soil and water may be contaminated. The population may be exposed through the drinking water or by eating contaminated food, e.g., fish. In a recent Finnish study, an increased risk for both STS and NHL was reported in a municipality using drinking water and consuming fish from a lake contaminated with chlorophenols.[67] The OR for STS was 8.9 (95% CI 1.8–44) and for NHL 2.8 (95% CI 1.4–5.6). The risk for HD was 1.5 (95% CI 0.4–5.0).

A higher incidence of NHL has been reported in males living in a rice-growing area where 2,4-D and 2,4,5-TP, i.e., the propionic acid of 2,4,5-T, have been identified in the soil and water.[68] The rate in the most polluted area was two times higher than that in the rest of the territory. Regarding HD and STS, the absolute numbers were too small to allow meaningful inter-area comparisons.

A large number of roadways, arenas, yards, and other surface sites in Missouri were contaminated with dioxins including TCDD in 1971. Waste by-products from a hexachlorophene and 2,4,5-T production facility in southwestern Missouri were mixed with waste oils and sprayed for dust control throughout the state including the town of Times Beach. In TCDD-exposed subjects, depressed cell-mediated immunity and altered T-lymphocyte subsets have been reported.[69–71] Thymic hormone levels were examined in a group of 94 persons presumed to be TCDD-exposed from living in contaminated residential areas. Compared with a matched control group of 105 unexposed persons, the exposed group had significantly lower mean α_1-thymosin level.[72] This is in agreement with the finding that the thymus is a target organ in experimental animals exposed to TCDD.[73] There are no reports on the cancer incidence among persons with residential TCDD exposure.

In the early 1970s, three trucking terminals in St. Louis, Missouri, were sprayed with TCDD-contaminated waste oil to control dust. Approximately 600 workers were employed at these sites. Among these workers one self-reported case of porphyria cutanea tarda and STS has been published.[74] In the early 1970s, this individual had worked for several years as a truck driver at one of the TCDD-contaminated trucking terminals. He reported that his feet, legs, hands, and arms frequently became covered with oil from the terminal. He had no other history of TCDD exposure. No follow-up data on cancer incidence in the entire cohort of workers are available.

Animal Exposure

A study from New Zealand examined the prevalence of small-intestinal adenocarcinoma in 20,678 female sheep.[75] Exposure to phenoxyacetic acids, picolinic herbicides, or both was associated with increased tumor rates, which were significant for each herbicide. Exposure to recently sprayed feedstuffs was associated with a significantly larger increase in tumor rate than exposure to less recently sprayed food. No additional effect was noted for exposure to TCDD.

A significantly increased risk for seminomas was found in U.S. military working dogs who had served in Vietnam during the war (OR 1.9, 95% CI 1.2–3.0).[76] Exposures, or potential exposures, of interest were tetracycline, malathion, picloram (in Agent White), and 2,4-D and 2,4,5-T (Agent Orange) with contaminating TCDD.

Of interest is the association between canine malignant lymphoma and the use of 2,4-D on the lawns of the dog owners.[77] Risk increased with the area of herbicide lawn treatments per year.

9. GENERAL CONCLUSIONS

Our knowledge regarding the human carcinogenicity of TCDD has rapidly increased over the last decade. Although the carcinogenicity of dioxins and dioxinlike compounds, such as PCDFs and PCBs, must be investigated further, we believe that at present there are fairly conclusive data in certain respects. The evidence has generated considerable debate which has included economic and even political considerations.[9,78,79] Regarding specific types of malignant tumors, STS as well as NHL by now seem to be fairly clearly associated with exposure to phenoxy herbicides and related chlorinated phenols. Especially STS seems to be related to dioxins as judged by the results of studies that have appeared from different countries and research groups.

For NHL, there is epidemiological evidence for the association with phenoxy herbicides, but not TCDD specifically. A possibility, supported by the results in the U.S. case–control studies,[27,29,30] is that phenoxyacetic acids per se, or dioxin isomers other than TCDD are the causative agents for malignant lymphoma. In one of the Swedish studies there were five cases and no controls exposed to MCPA only, and seven cases and one control exposed to 2,4,-D only.[7] TCDD may play a role, however, through its immunotoxic capabilities. It is well established that immunosuppression increases the risk for malignant diseases, e.g., malignant lymphoma.[80]

The phenoxy herbicides used in Sweden included both 2,4,5-T and 2,4-D and also MCPA. Excesses of STS and NHL were observed. In the United States, the phenoxyacetic acid exposures were primarily to 2,4-D. Excesses of NHL, but not STS, were observed, In the industrial populations exposed to TCDD, STS excesses were observed. It is also worth noticing that HD has appeared in excess in Sweden but not in the United States or Canada. In a Canadian investigation, herbicide exposure was indicated as a risk factor for NHL.[81] In the 1960s, 90% and in the 1970s 75% of the herbicides used in Canada was 2,4-D.

A possible explanation might be that 2,4,5-T/TCDD exposure elevates STS risk while 2,4-D elevates NHL risk. The Swedish excess of NHL may have actually been related to 2,4-D, not the 2,4,5-T exposure. However, it must also be mentioned that commercial 2,4-D has been found to be contaminated with TCDD.[82]

There is also evidence of an increased risk for lung cancer and stomach cancer among TCDD-exposed subjects. TCDD has antiestrogenic effects in animals, which may explain the fact that different tumors are induced in male and female animals, e.g., liver cancer in female but not male rats. There are also epidemiological studies indicating an association between TCDD exposure and liver cancer in humans.[16,21,58,59] Women who are exposed to TCDD might develop different tumors than men. However, it is noteworthy that TCDD-treated female rats have decreased incidence of mammary gland tumors whereas an increased SMR for breast cancer was found in the Boehringer–Ingelheim cohort.[46] In contrast to the German findings, cancer incidence data from Zone B of Seveso showed a decreased incidence of breast and uterus cancer, i.e., estrogen-dependent malignancies.[58,59]

The mechanism by which TCDD causes cancer is not fully understood. It is a potent promoter in carcinogenesis probably through the aryl hydrocarbon (Ah) receptor. A receptor–ligand complex is formed which binds to specific promoter elements of the gene. The dioxin receptor may have a regulatory function which increases the expression of oncogenes and/or decreases the expression of tumor suppressor genes. TCDD also affects the regulation of growth factors and other steroid hormone receptors. These receptor-dependent mechanisms are not related to classic mutational carcinogenesis, however. Interactions with mutagens or with endogenous regulators of cell differentiation and proliferation may explain the hormonelike and carcinogenic effects of dioxins. This may also explain the variety of tumors induced in experimental animals and humans by TCDD exposure.[83–85]

REFERENCES

1. L. Hardell, Soft-tissue sarcomas and exposure to phenoxy acids: A clinical observation, *Läkartidningen* **74,** 2753–2754 (1977).
2. O. Axelson and L. Sundell, Herbicide exposure, mortality and tumor incidence. An epidemiological investigation on Swedish railroad workers, *Work Environ. Health* **11,** 21–28 (1974).
3. O. Axelson and L. Sundell, Phenoxy acids and cancer, *Läkartidningen* **74,** 2887–2888 (1977).
4. L. Hardell, Malignant lymphoma of histiocytic type and exposure to phenoxyacetic acids or chlorophenols, *Lancet* **1,** 55–56 (1979).
5. L. Hardell and A. Sandström, Case–control study: Soft-tissue sarcomas and exposure to phenoxyacetic acids or chlorophenols, *Br. J. Cancer* **39,** 711–717 (1979).
6. M. Eriksson, L. Hardell, N. O. Berg, T. Möller, and O. Axelson, Soft-tissue sarcomas and exposure to chemical substances: A case–referent study, *Br. J. Ind. Med.* **38,** 27–33 (1981).
7. L. Hardell, M. Eriksson, P. Lenner, and E. Lundgren, Malignant lymphoma and exposure to chemicals, especially organic solvents, chlorophenols and phenoxy acids: A case–control study, *Br. J. Cancer* **43,** 169–176 (1981).
8. L. Hardell and N. O. Bengtsson, Epidemiological study of socioeconomic factors and clinical findings in Hodgkin's disease, and reanalysis of previous data regarding chemical exposure, *Br. J. Cancer* **48,** 217–225 (1984).
9. O. Axelson and L. Hardell, Storm in a cup of 2,4,5-T, *Med. J. Aust.* **144,** 612–613 (1986).

10. L. Hardell and M. Eriksson, The association between soft-tissue sarcomas and exposure to phenoxyacetic acids. A new case–referent study, *Cancer* **62,** 652–656 (1988).
11. M. Eriksson, L. Hardell, and H. O. Adami, Exposure to dioxins as a risk factor for soft tissue sarcoma: A population based case–control study, *J. Natl. Cancer Inst.* **82,** 486–490 (1990).
12. G. Wingren, M. Fredriksson, H. Noorlind Brage, B. Nordenskjöld, and O. Axelson, Soft tissue sarcoma and occupational exposures, *Cancer* **66,** 806–811 (1990).
13. H. Olsson, T. A. Alvegård, H. Härkönen, L. Brandt, and T. Möller, Epidemiological studies on high-grade soft-tissue sarcoma within the framework of a randomised trial in Scandinavia, in: T. A. Alvegård, *Management and Prognosis of Patients with High-Grade Soft-Tissue Sarcoma,* Thesis, University of Lund, Sweden (1989).
14. B. Persson, A. M. Dahlander, M. Fredriksson, H. Noorlind Brage, C.-G. Ohlson, and O. Axelson, Malignant lymphomas and occupational exposures, *Br. J. Ind. Med.* **46,** 516–520 (1989).
15. L. Hardell, B. Johansson, and O. Axelson, Epidemiological study on nasal and nasopharyngeal cancer and their relation to phenoxy acid or chlorophenol exposure, *Am. J. Ind. Med.* **3,** 247 (1982).
16. L. Hardell, N. O. Bengtsson, U. Jonsson, S. Eriksson, and L.-G. Larsson, Aetiological aspects on primary liver cancer with special regard to alcohol, organic solvents and acute intermittent porphyria—An epidemiological investigation, *Br. J. Cancer* **50,** 389–397 (1984).
17. M. Eriksson and M. Karlsson, Occupational and other environmental factors and multiple myeloma: A population based case–control study, *Br. J. Ind. Med.* **49,** 95–103 (1992).
18. L. Hardell, M. Eriksson, M. Fredriksson, and O. Axelson, Dioxin and mortality from cancer, *N. Engl. J. Med.* **324,** 1810 (1991).
19. K. Wiklund, B. M. Lindefors, and L. E. Holm, Risk of malignant lymphoma in Swedish agricultural and forestry workers, *Br. J. Ind. Med.* **45,** 19–24 (1988).
20. A. Blair and S. Hoar Zahm, Cancer among farmers, *Occupational Medicine: State of the Art Reviews,* **6,** 335–354 (1991).
21. D. D. Van, Herbicides as a possible cause of liver cancer, in: *Herbicides in War, The Long-Term Ecological and Human Consequences* (A. H. Westing, ed.), pp. 119–121, Taylor & Francis, London (1984).
22. L. Hardell, Relation of soft-tissue sarcoma, malignant lymphoma and colon cancer to phenoxy acids, chlorophenols and other agents, *Scand. J. Work Environ. Health* **7,** 119–130 (1981).
23. M. Fredriksson, N. O. Bengtsson, L. Hardell, and O. Axelson, Colon cancer, physical activity, and occupational exposure. A case–control study, *Cancer* **63,** 1838–1842 (1989).
24. P. Cole, Direct testimony before the Environmental Protection Agency of the United States of America, Washington, DC, Exhibit 860, pp. 2–24 (1980).
25. Royal Commission on the Use and Effects of Chemical Agents on Australian Personnel in Vietnam, "Final Report," Australian Government Publishing Service, Canberra, Vol. 4, pp. 90–180 (1985).
26. T. Colton, Herbicide exposure and cancer (editorial), *J. Am. Med. Assoc.* **256,** 1176–1178 (1986).
27. S. K. Hoar, A. Blair, F. F. Holmes, C. D. Boysen, R. J. Robel, R. Hoover, and J. F. Fraumeni, Jr., Agricultural herbicide use and risk of lymphoma and soft-tissue sarcoma, *J. Am. Med. Assoc.* **256,** 1141–1147 (1986).
28. S. K. Hoar, A. Blair, F. F. Holmes, C. Boysen, R. J. Robel, Herbicides and colon cancer, *Lancet* **1,** 1277–1278 (1985).

29. S. Hoar Zahm, D. D. Weisenburger, P. A. Babbitt, R. C. Saal, J. B. Vaught, K. P. Cantor, and A. Blair, A case–control study of non-Hodgkin's lymphoma and the herbicide 2,4-dichlorophenoxyacetic acid (2,4-D) in eastern Nebraska, *Epidemiology* **1,** 349–356 (1990).
30. K. P. Cantor, A. Blair, G. Everett, R. Gibson, L. F. Burmeister, L. M. Brown, L. Schuman, and F. R. Dick, Pesticides and other agricultural risk factors for non-Hodgkin's lymphoma among men in Iowa and Minnesota, *Cancer Res.* **52,** 2447–2455 (1992).
31. L. M. Brown, A. Blair, R. Gibson, G. D. Everett, K. P. Cantor, L. M. Schuman, L. F. Burmeister, S. F. Van Lier, and F. Dick, Pesticide exposure and other agricultural risk factors for leukemia among men in Iowa and Minnesota, *Cancer Res.* **50,** 6585–6591 (1990).
32. J. S. Woods, L. Polissar, R. K. Severson, L. S. Heuser, and B. G. Kulander, Soft-tissue sarcoma and non-Hodgkin's lymphoma in relation to phenoxyherbicide and chlorinated phenol exposure in western Washington, *J. Natl. Cancer Inst.* **78,** 899–910 (1987).
33. A. H. Smith, N. E. Pearce, D. O. Fisher, H. J. Giles, C. A. Teague, and J. K. Howard, Soft tissue sarcoma and exposure to phenoxyherbicides and chlorophenols in New Zealand, *J. Natl. Cancer Inst.* **73,** 1111–1117 (1984).
34. A. H. Smith and N. E. Pearce, Update on soft tissue sarcoma and phenoxyherbicides in New Zealand, *Chemosphere* **15,** 1795–1798 (1986).
35. N. E. Pearce, A. H. Smith, J. K. Howard, R. A. Sheppard, H. J. Giles, and C. A. Teague, Non-Hodgkin's lymphoma and exposure to phenoxyherbicides, chlorophenols, fencing work, and meat works employment: A case–control study, *Br. J. Ind. Med.* **43,** 75–83 (1986).
36. N. Pearce, Phenoxy herbicides and non-Hodgkin's lymphoma in New Zealand: Frequency and duration of herbicide use, *Br. J. Ind. Med.* **46,** 143–144 (1989).
37. O. Axelson, Pesticides and cancer risks in agriculture, *Med. Oncol. Tumor Pharmacother.* **4,** 207–217 (1987).
38. J. Waterhouse, C. Muir, K. Shanmugaratnamet, J. Powell, D. Peacham, S. Whelan, and W. Davis (eds.), *Cancer Incidence in Five Continents,* Vol. IV, *International Agency for Research on Cancer,* Lyon (1982).
39. P. Vineis, B. Terracini, G. Ciccone, A. Cignetti, E. Colombo, A. Donna, L. Maffi, R. Pisa, P. Ricco, E. Zanini, and P. Comba, Phenoxy herbicides and soft-tissue sarcomas in female rice weeders. A population-based case-referent study, *Scand. J. Work Environ. Health* **13,** 9–17 (1986).
40. O. Axelson, L. Sundell, K. Andersson, C. Edling, C. Hogstedt, and H. Kling, Herbicide exposure and tumor mortality; an updated epidemiological investigation on Swedish railroad workers, *Scand. J. Work Environ. Health* **6,** 73–79 (1980).
41. E. Barthel, Increased risk of lung cancer in pesticide-exposed male agricultural workers, *J. Toxicol. Environ. Health* **8,** 1027–1040 (1981).
42. V. Riihimäki, S. Asp, and S. Hernberg, Mortality of 2,4-dichlorophenoxyacetic acid and 2,4,5-trichlorophenoxyacetic acid herbicide applicators in Finland. First report on an ongoing prospective study, *Scand. J. Work Environ. Health* **8,** 37–42 (1982).
43. M. A. Fingerhut, W. E. Halperin, D. A. Marlow, L. A. Piacitelli, P. A. Honchar, M. H. Sweeny, A. L. Greife, P. A. Dill, K. Steenland, and A. J. Suruda, Cancer mortality in workers exposed to 2,3,7,8-tetrachlorodibenzo-p-dioxin, *N. Engl. J. Med.* **324,** 212–218 (1991).
44. M. A. Fingerhut, W. E. Halperin, P. A. Honchar, A. B. Smith, D. M. Groth, and W. O. Russel, An evaluation of reports of dioxin exposure and soft tissue sarcoma pathology

among chemical workers in the United States, *Scand. J. Work Environ. Health* **10,** 299–303 (1984).
45. A. Zober, P. Messerer, and P. Huber, Thirty-four-year mortality follow up of BASF-employees exposed to 2,3,7,8-TCDD after the 1953 accident, *Int. Arch. Occup. Environ. Health* **62,** 139–157 (1990).
46. A. Manz, J. Berger, J. Dwyer, D. Flesch-Janys, S. Nagel, and H. Waltsgott, Cancer mortality among workers in chemical plant contaminated with dioxin, *Lancet* **338,** 959–964 (1991).
47. R. Saracci, M. Kogevinas, P. A. Bertazzi, B. H. Bueno de Mesquita, D. Coggon, L. M. Green, T. Kauppinen, K. A. L'Abbé, M. Littorin, E. Lynge, J. D. Mathews, M. Neuberger, J. Osman, N. Pearce, and R. Winkelman, Cancer mortality in workers exposed to chlorophenoxy herbicides and chlorophenols, *Lancet* **338,** 1027–1032 (1991).
48. H. K. Kang, L. Weatherbee, P. P. Breslin, Y. Lee, and B. Shepard, Soft-tissue sarcomas and military service in Vietnam: A case comparison group analysis of hospital patients, *J. Occup. Med.* **28,** 1215–1218 (1986).
49. N. A. Dalager, H. K. Kang, V. Burt, and L. Weatherbee, Non-Hodgkin's lymphoma among Vietnam veterans, *J. Occup. Med.* **33,** 774–779 (1991).
50. T. R. O'Brien, P. Decouflé, and C. A. Boyle, Non-Hodgkin's lymphoma in a cohort of Vietnam veterans, *Am. J. Public Health* **81,** 758–760 (1991).
51. S. D. Stellman and J. M. Stellman, Estimation of exposure to Agent Orange and other defoliants among American troops in Vietnam: A methodological approach, *Am. J. Ind. Med.* **9,** 305–321 (1986).
52. M. D. Kogan and R. W. Clapp, Soft-tissue sarcoma mortality among Vietnam veterans in Massachusetts, 1972 to 1983, *Int. J. Epidemiol.* **17,** 39–43 (1988).
53. J. E. Michalek, W. H. Wolfe, and J. C. Miner, Health status of Air Force veterans occupationally exposed to herbicides in Vietnam. II. Mortality, *J. Am. Med. Assoc.* **264,** 1832–1836 (1990).
54. E. A. Brann, The Selected Cancers Cooperative Study Group. The association of selected cancers with service in US military in Vietnam. I. Non-Hodgkin's lymphoma, *Arch. Intern. Med.* **150,** 2473–2483 (1990).
55. E. A. Brann, The Selected Cancers Cooperative Study Group. The association of selected cancers with service in US military in Vietnam. II. Soft-tissue and other sarcomas, *Arch. Intern. Med.* **150,** 2485–2492 (1990).
56. E. A. Brann, The Selected Cancers Cooperative Study Group. The association of selected cancers with service in US military in Vietnam. III. Hodgkin's disease, nasal cancer, nasopharyngeal cancer, and primary liver cancer, *Arch. Intern. Med.* **150,** 2495–2505 (1990).
57. P. A. Bertazzi, C. Zochetti, A. C. Pesatori, S. Guercilena, M. Sanarico, and L. Radice, Ten-year mortality study of the population involved in the Seveso incident in 1976, *Am. J. Epidemiol.* **129,** 1186–1200 (1989).
58. A. C. Pesatori, D. Consonni, A. Tironi, M. T. Landi, C. Zocchetti, and P. A. Bertazzi, Cancer morbidity in the Seveso area, 1976–1986, *Chemosphere* **25,** 209–212 (1992).
59. A. C. Pesatori, D. Consonni, A. Tironi, M. T. Landi, C. Zocchetti, and P. A. Bertazzi, Cancer morbidity (1977–1986) of a young population living in the Seveso area, in: *Dioxin '92, 12th International Symposium on Dioxins and Related Compounds,* Vol. 10, pp. 271–273, Finnish Institute of Occupational Health, Helsinki (1992).
60. M. Kuratsune, Y. Masuda, and J. Nagayama, Some of the recent findings concerning Yusho in: Proceedings of the National Conference on Polychlorinated Biphenyls, Chicago, Nov. 19–21 (1975).

61. C. F. Robinson, R. J. Waxweiler, and D. P. Fowler, Mortality among production workers in pulp and paper mills, *Scand. J. Work Environ. Health* **12,** 552–560 (1986).
62. P. K. Henneberger, B. G. Ferris, Jr., and R. R. Monson, Mortality among pulp and paper workers in Berlin, New Hampshire, *Br. J. Ind. Med.* **46,** 658–664 (1989).
63. J. Jäppinen, E. Pukkala, and S. Tola, Cancer incidence of workers in a Finnish sawmill, *Scand. J. Work Environ. Health* **15,** 18–23 (1989).
64. G. Wingren, B. Persson, K. Thorén, and O. Axelson, Mortality pattern among pulp and paper mill workers in Sweden: A case–referent study, *Am. J. Ind. Med.* **20,** 769–774 (1991).
65. G. Wingren, H. Kling, and O. Axelson, Gastric cancer among paper mill workers, *J. Occup. Med.* **27,** 715 (1985).
66. K. Thorén, G. Sällsten, and B. Järvholm, Mortality from asthma, chronic obstructive pulmonary disease, respiratory system cancer, and stomach cancer among paper mill workers: A case–referent study, *Am. J. Ind. Med.* **19,** 729–737 (1991).
67. P. Lampi, T. Hakulinen, and T. Luostarinen, Cancer incidence following chlorophenol exposure in a community in southern Finland, *Arch. Environ. Health* **47,** 167–175 (1992).
68. P. Vineis, F. Faggiano, M. Tedeschi, and G. Ciccone, Incidence rates of lymphomas and soft-tissue sarcomas and environmental measurements of phenoxy herbicides, *J. Natl. Cancer Inst.* **83,** 362–363 (1991).
69. A. P. Knutsen, Immunologic effects of TCDD exposure in humans, *Bull. Environ. Contam. Toxicol.* **33,** 763–781 (1984).
70. R. E. Hoffman, P. A. Stehr-Green, K. B. Webb, R. G. Evans, A. P. Knutsen, W. F. Schramm, J. L. Staake, B. B. Gibson and K. K. Steinberg, Health effects of long-term exposure to 2,3,7,8-tetrachlorodibenzo-p-dioxin, *J. Am. Med. Assoc.* **255,** 2031–2038 (1986).
71. R. G. Evans, K. B. Webb, A. P. Knutsen, S. T. Roodman, D. W. Roberts, J. R. Bagby, W. A. Garrett, and J. S. Andrews, A medical follow-up of the health effects of long-term exposure to 2,3,7,8-tetrachlorodibenzo-p-dioxin, *Arch. Environ. Health* **43,** 273–278 (1988).
72. P. A. Stehr-Green, P. H. Naylor, and R. E. Hoffman, Diminished thymosin$_{alpha\text{-}1}$ levels in persons exposed to 2,3,7,8-tetrachlorodibenzo-p-dioxin, *J. Toxicol. Environ. Health* **28,** 285–295 (1989).
73. J. G. Vos, J. A. Moore, and J. G. Zinkl, Toxicity of 2,3,7,8-tetrachlorodibenzo-p-dioxin (TCDD) in C57B1/6 mice, *Toxicol. Appl. Pharmacol.* **29,** 229–241 (1974).
74. W. Hope, D. Lischwe, W. Russell, and S. Weiss, Porphyria cutanea tarda and sarcoma in a worker exposed to 2,3,7,8-tetrachlorodibenzodioxin—Missouri, *J. Am. Med. Assoc.* **251,** 1536–1537 (1984).
75. K. W. Newell, A. D. Ross, and R. M. Renner, Phenoxy and picolinic acid herbicides and small-intestinal adenocarcinoma in sheep, *Lancet* **2,** 1301–1305 (1984).
76. H. M. Hayes, R. E. Tarone, H. W. Casey, and D. L. Huxsoll, Excess of seminomas observed in Vietnam service U.S. military working dogs, *J. Natl. Cancer Inst.* **82,** 1042–1046 (1990).
77. H. M. Hayes, R. E. Tarone, K. P. Cantor, C. R. Jessen, M. Dennis, M. McCurnin, and R. C. Richardson, Case–control study of canine malignant lymphoma: Positive association with dog owner's use of 2,4-dichlorophenoxyacetic acid herbicides, *J. Natl. Cancer Inst.* **83,** 1226–1231 (1991).
78. L. Hardell and O. Axelson, Storm in a cup of 2,4,5-T, *Med. J. Aust.* **145,** 299 (1986).
79. L. Hardell and O. Axelson, The boring story of Agent Orange and the Australian Royal Commission, *Med. J. Aust.* **150,** 602 (1989).

80. L. J. Kinlen, A. G. R. Sheil, J. Peto, and R. Doll, Collaborative United Kingdom–Australian study of cancer in patients treated with immunosuppressive drugs, *Br. Med. J.* **2,** 1461–1466 (1979).
81. D. T. Wigle, R. M. Semenciw, K. Wilkins, D. Riedel, L. Ritter, H. L. Morrison, and Y. Mao, Mortality study of Canadian male farm operators: Non-Hodgkin's lymphoma mortality and agricultural practices in Saskatchewan, *J. Natl. Cancer Inst.* **82,** 575–582 (1990).
82. H. Hagenmaier, Determination of 2,3,7,8-tetrachlorodibenzo-p-dioxin in commercial chlorophenols and related products, *Fresenius Z. Anal. Chem.* **325,** 603–606 (1986).
83. E. K. Silbergeld and T. A. Gasiewicz, Dioxins and Ah receptor, *Am. J. Ind. Med.* **16,** 455–474 (1989).
84. E. K. Silbergeld, Carcinogenicity of dioxins, *J. Natl. Cancer Inst.* **83,** 1198–1199 (1991).
85. G. W. Lucier, Receptor-mediated carcinogenesis, in: *Mechanisms of Carcinogenesis in Risk Identification* (H. Vainio, P. N. Magee, D. B. McGregor, and A. J. McMichael, eds.), IARC Scientific Publication No. 116, pp. 87–112, International Agency for Research on Cancer, Lyon (1992).
86. L. Hardell, M. Eriksson, A. Degerman, Exposure to phenoxyacetic acids, chlorophenols, or organic solvents in relation to histopathology, stage, and anatomical localization of non-Hodgkin's lymphoma. *Cancer Research* **54,** 2386–2389 (1994).

Chapter 17

Reproductive Epidemiology of Dioxins

Anne Sweeney

1. INTRODUCTION: BACKGROUND

Dioxin [for purposes of this chapter defined as 2,3,7,8-tetrachlorodibenzo-*para*-dioxin (TCDD)] has been dubiously hailed as "one of the most perplexing and potentially dangerous chemicals ever to pollute the environment."[1] It is a ubiquitous contaminant, produced as an unwanted by-product during the manufacture of many industrial and agricultural chemicals, as well as from incineration of municipal waste. It is believed that ingestion of contaminated food is the most likely source of dioxin exposure for the general population.[1,2]

The origin of concerns regarding a potential link between exposure to chlorinated dioxins and adverse reproductive events can be traced to early animal studies reporting increased incidence of developmental abnormalities in rats and mice exposed early in gestation to 2,4,5-trichlorophenoxyacetic acid (2,4,5-T).[3] This was of grave concern, as the U.S. military's most widely used herbicide during the Vietnam conflict at that time, Agent Orange, was composed of approximately equal proportions by weight of the *n*-butyl esters of 2,4-dichlorophenoxyacetic acid (2,4-D) and 2,4,5-T. The latter is contaminated during manufacture by TCDD.

Several TCDD-contamination episodes have occurred, resulting in the identification of human health effects attributed to high-dose exposure to TCDD, including the skin disorders of chloracne and hyperpigmentation, as well as liver damage and polyneuropathies.[4] Studies of workers (generally males) exposed to phenoxy acids and chlorophenols have attempted to ascertain health effects of TCDD exposure at greater than background levels. Two well-known contamination episodes occurred in Japan in

---

Anne Sweeney • School of Public Health, University of Texas Health Science Center at Houston, Houston, Texas 77025.

Dioxins and Health, edited by Arnold Schecter. Plenum Press, New York, 1994.

1968 (Yusho) and Taiwan in 1979 (Yu-cheng), where thousands of persons consumed cooking oil contaminated with polychlorinated biphenyls, polychlorinated dibenzofurans, and polychlorinated quaterphenyls. Adverse reproductive effects observed in these populations included increases in spontaneous abortions and low birth weight, increased bilirubin levels, conjunctivitis with enlarged sebaceous glands of the eyelid, pigmented gums, and hypoplastic deformed nails.[5–7] These incidents are described in detail in Hsu *et al.* (this volume). Perhaps the most widely known investigations of potential dioxin reproductive toxicity in humans are the studies of male Vietnam veterans and their offspring who were at risk of exposure to Agent Orange. The focus of this review is the examination of epidemiologic evidence regarding the relationship between dioxin exposure and adverse reproductive effects in humans.

The examination of reproductive and developmental disorders poses several unique challenges to the researcher as compared with other health outcomes. First, in order to understand both normal and pathologic reproduction, evaluation should include paternal and maternal, and sometimes fetal, contributions. The recent increased interest in male-mediated reproductive toxicity emphasizes the need to consider the couple as the unit of analysis in many reproductive study settings.

The second challenge to researchers is the interrelatedness of the spectrum of reproductive endpoints available for study, such as fertility and early pregnancy loss. Most studies have been restricted to the examination of recognized pregnancies only. These events were enumerated and classified as having terminated in a spontaneous abortion, or a stillbirth or live birth, with or without a congenital anomaly. It must be stressed that these individuals were able to produce a conception that survived long enough to be clinically recognized, and this may not be representative of all persons exposed to 2,3,7,8-TCDD. Fecundity (the capability to conceive), fertility (the capability to produce live children), and early pregnancy loss (those conceptions that do not survive to be recognized by usual diagnostic methods) as related to 2,3,7,8-TCDD exposure have not been evaluated. Clearly, these endpoints impact on the rates of reproductive outcomes occurring later in the reproductive spectrum.

Another unique feature of reproductive effects is the changing vulnerability of the developing organism throughout gestation. Exposure to a single teratogen throughout pregnancy may result in different effects at various stages of gestation. The window of susceptibility varies among different teratogens, and therefore knowledge of the timing of exposure is critical in these studies. Moreover, critical periods of exposure may differ depending on the parental mediator of exposure and the agent of concern.

Finally, although not restricted to studies of reproductive events, care must be given to the collection and analysis of confounding variables. These factors also may need to be obtained for both mother and father, with attention to the timing of specific characteristics, such as smoking or a change in occupation preceding or during pregnancy. These are some points to keep in mind as the reader reviews the studies of 2,3,7,8-TCDD and reproductive effects presented here.

One dilemma encountered when attempting to review the epidemiologic literature dealing with dioxins and reproductive effects is the categorization of studies of sufficient similarity to allow for comparative analysis. These studies vary greatly in the nature (occupational, environmental) and route (inhalation, digestion, absorption) of exposure; in the reproductive outcomes examined (often multiple endpoints were considered, and case definition differed across studies); in the assessment of parental exposure (maternal, paternal, or both); and in the timing of exposure relative to the pregnancy.

The reproductive effects of dioxins in humans were succinctly and elegantly reviewed by Hatch in 1984.[8] In her review, Hatch employed type of exposure—i.e., populations who were exposed occupationally, environmentally, or through industrial accident or military service—as the classification scheme for the research presented. The same approach has been followed in this review. The earlier investigations of dioxin and reproductive effects, i.e., those conducted prior to 1984, are presented separately from the more recent studies. The development of assays in the mid-1980s to quantitate 2,3,7,8-TCDD in serum and adipose tissue allowing individual measurements of exposure warrants this dichotomy of the research.

A brief description of the epidemiologic criteria for causation is warranted and is included in the section that evaluates the available body of research. The chapter concludes with a summary of the research to date and suggestions for future examinations of this issue.

When the efforts of multiple researchers trying to understand the relationship between dioxin and reproductive events are presented chronologically, the evolution of the more sophisticated techniques evidenced in the later studies can be appreciated. The history of research in this area provides a wonderful example of the progress of epidemiologic studies in eliminating the weaknesses of, and refining the strengths of, previous work to produce studies that at the very least can explain the troubling inconsistencies noted in earlier work. This chronology also emphasizes the need for interdisciplinary approaches to this multifaceted issue.

For those less historically minded, the summary alone will hopefully provide a brief history of the dioxin and reproductive events research, a description of what is known today as a result of this work, and some suggestions for future research based on what we have learned from the past.

2. REVIEW OF THE LITERATURE PRIOR TO 1984

2.1. Occupational Studies prior to 1984

The early studies of occupational dioxin exposure and reproductive effects focused on potential paternally-mediated effects with exposures occurring at varying intervals relative to conception. Townsend *et al.* interviewed 370 wives of employees exposed to dioxins at the Dow Michigan Division in Midland, Michigan (63% of those eligible for the study), and 345 control wives of Dow employees who were presumably not exposed to dioxin (62% of the eligible control pool).[9] Exposure classification was determined by an industrial hygienist familiar with the processes performed at the plant. Employees were considered exposed to dioxin if they had been assigned for at least 1 month to specific jobs associated with chlorophenol processes between 1939 and 1975. All outcomes were reported by the employee's spouse.

There was no systematic attempt to ascertain the reason(s) for the high refusal rate in both cohorts. "Unsolicited reasons" for refusal included divorce, death of spouse, or no pregnancies; no breakdown by cohort was provided. The possibility of differential rates of infertility or early pregnancy loss, as reflected by the reported absence of pregnancy, was not addressed in this study.

For the multiple endpoints of spontaneous abortion, stillbirth, birth defects, infant mortality, and childhood morbidity, no significant association between dioxin exposure and any adverse event was identified. Adjusted odds ratios for adverse outcomes and TCDD exposure were stillbirths (OR = 0.97, 95% CI 0.38–2.36), spontaneous abortion (OR = 0.96, 95% CI 0.65–1.42), infant deaths (OR = 0.82, 95% CI 0.30–2.09), congenital malformations (OR = 1.08, 95% CI 0.63–1.83), and all unfavorable outcomes (OR = 0.98, 95% CI 0.72–1.32).

Given the very long interval during which both exposure and event could have occurred, and the minimum requirement for paternal exposure being 1 month at any time during the interval, a dioxin effect, if it existed, would have been diluted by this approach unless the damage was "irreparable," as defined in this report. The authors do not discuss whether time since last exposure was considered in the analysis.

In another study of occupational exposure to 2,4,5-T, Smith *et al.* evaluated 548 professional chemical sprayers and their spouses in New Zealand.[10] Mailed surveys were used to ascertain spraying activities from the males and reproductive histories from their spouses. A group of 441 agricultural contractors and their spouses served as controls. This study had impressive response rates of 89% for the exposed and 83% for the nonexposed groups. The investigators noted that wives of the chemical sprayers anecdotally reported assisting their husbands in spraying activities, some performing this task during their pregnancy. No association between herbicide exposure and the outcomes of spontaneous abortion (OR = 0.89, 95% CI 0.6–2.4) or congenital malformations (OR = 1.2, 95% CI 0.6–1.3) were identified.

The questionnaires were completed in 1980, and elicited information on spraying activities and reproductive events that occurred between 1969 and 1980, another rather long period of time for recall of exposure and events. Reported pregnancy outcomes were categorized into three groups based on whether or not the fathers had sprayed any chemicals at any time during, or previous to, the calendar year in which the pregnancy occurred, and whether 2,4,5-T had been used. There were 1122 pregnancies reported among the 441 control spouses and 1172 among the 548 spouses of the exposed sprayers. If all other factors were considered equivalent in both cohorts (maternal age, socioeconomic status, and maternal smoking histories were shown to be similar), then there appear to be 220 fewer pregnancies observed in the exposed group than might be expected.

This study is limited by the lack of information on the total number of conceptuses, and the high probability of exposure to chemicals other than 2,4,5-T. While exposure levels were not quantitated in this study, a later study of a subset from this same population was conducted to estimate 2,3,7,8-TCDD exposure in this group.

In 1988, nine pesticide applicators with the greatest number of years and months per year of pesticide application were selected for a serum TCDD analysis (see Table 1). The mean serum TCDD level, adjusted for total lipids, among applicators (53.3 ppt) was ten times that of matched control subjects (5.6 ppt). However, exposure in the reproductive study was based on self-reports of pesticide application, and research has demonstrated repeatedly that self-reports do not correlate with documented serum TCDD levels.[11] Therefore, it would be helpful if serum TCDD levels could be obtained in a subset of those applicators with lower self-reported exposures.

A clinical epidemiology study conducted in 1979 examined workers involved with the manufacture of 2,4,5-T between 1948 and 1969 in Nitro, West Virginia.[12] All active and retired plant employees exposed to the 2,4,5-T process during that 22-year interval comprised the eligible pool of "exposed" subjects. The control group was comprised of current and former plant employees who were never associated with the 2,4,5-T process, according to company records. The response rates for these cohorts were 61% (n = 204) and 46% (n = 163), respectively.

A reproductive history was obtained from these male employees during an interview and clinical examination. No attempt was made to verify these reports of live births, infant deaths, miscarriages, birth defects, and stillbirths with either spouse or through medical records. There were no significant differences in rates of any adverse outcome by exposure status. Given the poor response rates, crude measure of exposure, and lack of verification for paternally reported reproductive histories, this study was not likely to detect an association between dioxin and reproductive events if one existed.

Two ecologic studies of the relationship between paternal dioxin exposure as determined by job title and birth defects yielded conflicting results. Balarajan and McDowell reported an excess of cleft lip and/or palate among children born to fathers potentially exposed to 2,4,5-T in agricultural, gardening, and forestry occupations during 1974–1979 in Great Britain.[13] Malformations were identified through a birth defects surveillance program. Using birth certificate data, which results in underascertainment of birth defects, for the Oxford and West Berkshire areas that encompassed a slightly earlier period from 1965 to 1977, Golding and Sladden were unable to confirm this finding.[14]

2.2. Environmental Studies prior to 1984

The problem of documentation of exposure is perhaps of even greater concern in earlier studies of subjects *environmentally* exposed to dioxin that lack individual TCDD measures. The route of exposure (inhalation, ingestion, absorption); length and intensity of exposure; and the timing of the exposure are more difficult to estimate in free-living populations compared with workers in an occupational setting.

Selection bias is also a critical concern in these early environmental investigations because there was no equivalent to company records in the occupational setting that defined the population at risk. Proximity of residence to a contaminated site was generally the best option available for identification of the population at risk of exposure. Issues such as the length of time at that residence, the amount of time the subject spent in the home, and the occurrence of the contamination episode relative to the time spent at home, all based on subject self-report, may have impacted greatly on the degree of exposure.

Volunteer bias is an additional problem, as evidenced in the Times Beach, Missouri, investigations that relied on subjects responding to public health authorities' requests for participation.[15,16] Moreover, it is extremely difficult to conduct epidemiologic investigations under crisis situations such as the industrial accident that occurred in Seveso, Italy. Given these limitations, the efforts of these investigators have provided valuable impetus into the refinement of study designs and the development of more sophisticated techniques to explore this issue.

In 1979, a report was issued by the Environmental Protection Agency describing the results of a study conducted in Alsea, Oregon, that assessed the relationship between 2,4,5-T used in forest management and spontaneous abortion.[17] Exposure was determined ecologically by reviewing herbicide application records for a quarter of the study area and extrapolating this measure to the entire study area. While this investigation reported an increased spontaneous abortion rate in areas where forests were sprayed compared with the unsprayed areas, the study is widely acknowledged to have deficiencies that seriously limit interpretation of the results.[18] Use of 2,4,5-T in other areas, including pastures, orchards, and rice-growing regions, was not assessed. Examination of spontaneous abortion was restricted to those events identified in hospital records. Serious underascertainment of events resulted since most women in the study area sought medical attention from private physicians.

Two additional reports concerning 2,4,5-T and birth defects appeared that same year, with conflicting findings. Nelson *et al.* examined sex- and race-specific rates of facial clefts during the period 1948–1974 in Arkansas, according to areas with estimated 2,4,5-T application based on rice acreage.[19] Birth certificates were screened and checked against the records of the Crippled Children's Services, an agency involved with active statewide case-detection and management of facial clefts. A total of 1201 cases of cleft lip and/or palate were identified over this interval.

The investigators found increasing trends over time in facial clefts for both the high- and low-exposure groups. This was attributed to improved case ascertainment although the "crudity of the index of exposure" and the limitations of ecologic studies were acknowledged.

Field and Kerr[20] evaluated the relationship between neural tube defects and 2,4, 5-T usage in the New South Wales area of Australia for the years 1965–1976. Rates of neural tube defects were examined against the amounts of 2,4,5-T applied throughout the whole of Australia during the previous calendar year. A linear correlation (no coefficient was reported) between previous year's usage and combined birth rate of anencephaly and meningomyelocele was found. This relationship "disappeared" in the final 2 years of the study. It was during this 2-year period that monitoring of 2,4,5-T was initiated in Australia to ensure a TCDD concentration of less than 0.1 ppm.

Seasonal variation in rates of neural tube defects in New South Wales has been documented, with the peak rate occurring among summer conceptions. Summer was also the period of heaviest spraying of 2,4,5-T in New South Wales, according to a survey of local authorities. These investigators proposed the view that "neural tube defects originate from the interaction of a 'predisposed' embryonic genotype with environmental factors." Again, the ecologic nature of this study precludes any inference of causation.

In a similarly designed study, Thomas evaluated neural tube defects, cystic kidney disease, and stillbirths in Hungary during the period 1969–1976.[21] Utilizing the Hungarian birth defects registry for case ascertainment, rates of these birth defects were compared against the previous year's usage of 2,4,5-T. Despite a marked increase in 2,4,5-T usage, no increased incidence in any defect was identified. In a follow-up report through 1980, Thomas and Czeizel noted the same results, as well as a decrease in recognized spontaneous abortion rates from 16.3% to 11.8% between 1970 and 1980.[22]

The examination of stillbirths in reproductive epidemiology studies has been described as a "very poor proxy for spontaneous abortion."[22] Likewise, the assessment of recognized spontaneous abortions is now generally acknowledged to be a poor approximation of early pregnancy loss, as discussed below. Thus, the decrease noted in recognized spontaneous abortion rates in the later Hungarian report may also be construed as evidence that additional, more sensitive research is required in the area of events surrounding conception and early pregnancy.

Hanify *et al.* plotted the rates of all diagnosed congenital malformations (excluding miscarriages at less than 28 weeks' gestation) identified by hospital records against average exposure concentrations of 2,4,5-T in the Northland region of New Zealand.[23] Exposure data were obtained from company records detailing site, date, and amount of 2,4,5-T application. As no spraying was performed in Northland in the early half of the 1960s, comparisons were made between the "unexposed" years of 1959–1965 and the "exposed" years 1972–1976.

A significant association between spray activity and talipes only was observed. This defect was the most common malformation identified overall, with 49 cases in the unexposed group and 52 among the exposed, after exclusion of stillbirths, neonatal deaths, and hip dislocation defects. The total number of malformations, and thus the frequency of any individual defect, were quite small in this ecologic study.

2.3. The Seveso, Italy, Dioxin Accident of 1976

The 1976 accident at a chemical plant near Meda, Italy, in which a toxic cloud containing TCDD, among other contaminants, was released has been well described.[24,25] Following the explosion, an extensive surveillance system was initiated to monitor the health of the exposed population. Rates of reproductive events were examined by "zones" of exposure. Three zones were defined, depending on the concentration of TCDD found in the soil.

Zone A was the area of highest TCDD concentration, and housed 733 people. Zone B was considered a low-level contamination area with 4800 people, and Zone R, with the largest population of 32,000, was determined to be a low- or no-exposure area. Because of the embryotoxic and teratological effects of TCDD exposure noted in laboratory animals, there was great concern for pregnant women residing in the contaminated areas. A registry system for all births in the study area was instituted and attempts were made to measure fetal losses, congenital malformations, and postnatal development for a majority of the infants born in the three zones.

The early investigations of spontaneous abortion following the accident were complicated by several factors. In order to document an increase in the rate of an event, good baseline information is required. Reliable background data for spontaneous abortion in the study area were not available. The reported rates of miscarriage in the 6 months after the accident (July through December, 1976), 9.6% in the contaminated area and 12.6% in the contamination-free area, do not exceed the worldwide frequency for background rates of spontaneous abortion of 15–20%. During the first 6 months of 1977, the spontaneous abortion rate in the contaminated area rose to 12.2%, while remaining virtually unchanged in the contamination-free area during this time.[25]

In addition, the impact of induced abortions sought after the accident is difficult to estimate, particularly since abortion was illegal in Italy in 1976. The birth rate for the entire study area declined between 1976 and 1980, and the small number of conceptions available for study limits the power of these studies to detect an association between dioxin exposure and spontaneous abortion.

These same caveats should also be applied when considering the rates of congenital malformations reported following the accident. No reliable information regarding rates of birth defects in this area was available until the establishment of the Seveso Congenital Malformations Registry, 6 months after the exposure incident. Reggiani reported that the congenital malformation rate in Seveso increased from 0.13% in 1976, when no systematic surveillance for birth defects was performed, to 0.87% in the first half of 1977.[25] For the 12 months of 1977, systematic identification of malformations in the study area documented 38 birth defects among the 2774 newborns (1.4%); these defects occurred mainly in the less contaminated areas. Reasons for these very low rates as compared with background levels of 2–3% are not clear. The potential contribution of TCDD to the congenital malformation rate cannot be separated out from improved case ascertainment.

The influence of both spontaneous and induced abortions on the birth defect rate is likewise unknown. Pocchiari *et al.* described one study in which 30 therapeutic and 4 spontaneous abortions from the Seveso area were examined.[24] There were no indications of mutagenic, teratogenic, or embryotoxic effects that could be attributed to TCDD exposure. However, it was difficult to determine maternal exposure status for any of the cases. A more recent study examined the association between TCDD and cytogenetic abnormalities in fetuses from abortions induced shortly after the Seveso accident.[26] The frequency of aberrant cells and the mean number of aberrations per damaged cell were significantly higher in the exposed infants. This study was also hampered by a poor definition of exposure, and no measure of TCDD in the fetal tissues was reported. This approach, however, offers important and exciting possibilities for future research as genetic techniques in the area of DNA-adduct analyses become more widely available.

2.4. Studies of Exposure to Agent Orange by Military Veterans and Vietnamese Civilians prior to 1984

The problem of exposure documentation has also been a highly controversial issue in studies of potential exposure to Agent Orange in Vietnam and adverse health effects among Vietnam veterans and residents of Vietnam and their offspring. An early study conducted among the Vietnamese population examined the period from 1960 to 1969.[27] Exposure was dichotomized into pre- or light-spraying years from 1960 to 1965 and heavy-spraying years from 1966 to 1969. A total of 480,087 births, 16,166 stillbirths, 2866 hydatidiform moles, and 2355 congenital malformations of all types were examined in this study. Pregnancy outcome data were collected from a total of 22 hospitals.

Increases in the rates of stillbirths, molar pregnancies, and congenital malformations were noted in the coastal plain and delta areas following heavy spraying, although the authors emphasized the slight downward trend observed for all outcomes in the countrywide rates. Several biases in this sampling approach have been identified, which severely limit the interpretation of the study's findings. The births examined were not representative of the births in the country during that period. In addition, the hospital records were incomplete, and transport of the mothers to the selected hospitals resulted in uncertainty regarding maternal residence during the pregnancy.

A second investigation conducted in Vietnam utilized HERBS tapes (military records detailing Agent Orange spraying missions) covering the period from 1965 to 1971 to determine maternal exposure status according to area of residence.[28] HERBS data were matched to hospital records indicating the date of birth to imply "estimated date of conception" and to maternal residence at birth to imply "potential for maternal exposure." Birth outcome data were collected from hospital records, which, as in the above study, were subject to inaccuracies and incompleteness.

No association between spraying of Agent Orange and any type of birth defect or perinatal mortality was noted. Although cleft lip defects increased in proportion to other malformations during the period of heavy spraying, the total number of birth defects declined and continued to decrease after spraying activities ceased.

Finally, Australian investigators examined the relationship between *service* in Vietnam during 1962–1972 and birth defects.[29] Cases were infants born with any of a defined set of congenital malformations in any of 34 hospitals in New South Wales, Victoria, and the Australian Capital Territory from 1966 to 1979. Control infants were matched to cases on maternal age, and hospital and time of birth, yielding 8500 matched pairs for analysis.

Fathers of case and control infants were matched against a list of every member who had served in the Australian Army during the specified time period. No associations were detected for Vietnam service and total birth defects (OR = 1.02, 95% CI 0.8–1.3), or for any of the approximately 100 birth defects examined. Length of service in Vietnam, time between deplanement and conception, and Vietnam service prior to and following conception were considered in the analysis.

A critical point that should be emphasized regarding the Australian study is the assessment of service in Vietnam as the exposure of interest. The author clearly stated that investigations had indicated that exposure to herbicides was "infrequent and probably very low in Australian troops in Vietnam; the study does not exclude possible effects of herbicides in situations of substantial exposure."[29]

A series of unpublished studies conducted by Vietnamese investigators was reviewed by Hatch[30] and should be mentioned in this review. While there are also limitations to these reports, including incomplete background data for rates of reproductive events and sparsity of the epidemiologic details provided in the studies, the assessment of Vietnamese populations offered the opportunity to evaluate pregnancies with various patterns of parental exposure.

Investigations conducted in northern Vietnam assessed pregnancies with no maternal exposure to spraying activities; paternal exposure was presumed to have occurred only when the father had performed military service in the south. Studies of couples in southern Vietnam compared reproductive outcomes observed in sprayed versus nonsprayed areas, and represent potential associations between either maternal and/or paternal herbicide exposure and risk of adverse pregnancy outcome.

Three studies examined presumed paternal herbicide exposure and birth defects among pregnancies in northern Vietnam. Lang and Van compared the frequency of birth defects during the period 1975–1978 as a function of the father's having served in the south of Vietnam.[30] Among 2547 offspring whose fathers had never served in the south, the birth defect rate was 6 per thousand ($n = 15$). Among 511 offspring whose fathers had served in the south, the rate was 29 per thousand ($n = 15$). Similar results were obtained in a study by Lang *et al.* at unspecified agricultural and handicraft cooperatives in northern Vietnam.[30] The congenital malformation rate among "exposed" pregnancies (paternal service in southern Vietnam) was 23–26 per thousand (71 or 82 of 3147). In comparison, the rate among the unexposed pregnancies was 5 per thousand (10 of 2172).

In what is perhaps the most stringent of the Vietnamese studies by Can *et al.*, 40,064 women from three rice-growing districts in northern Vietnam were assessed (although few details are provided on the method of selection).[30] All of the women were married and pregnant at least once during the war, had no history of tuberculosis, syphilis, or malaria, or of using antibiotics or hormones during pregnancy. "Detailed" interviews were conducted by physicians and midwives, and district health records were consulted in an attempt to validate reported pregnancies and outcomes. Only pregnancies conceived during the conflict were considered in this study.

There was a total of 121,993 pregnancies among the 29,041 women whose spouses were "nonexposed" and 32,069 pregnancies among the 11,023 women whose spouses had served in the south. Service in south Vietnam was associated with an increased rate of spontaneous abortion ($p = 0.05$). The increased rate of congenital malformations among the exposed pregnancies (0.6%) compared with the unexposed (0.4%) was marginally significant ($p = 0.10$).

In an attempt to confirm this finding, these investigators conducted a case–control study in which they examined a random sample of 61 families of children who had survived with a birth defect. A control group of 183 families of normal children matched on maternal age, number of deliveries, living environment, and age was selected. Forty-nine percent of children with a birth defect had a father who had served in south Vietnam compared with 21% of the children without malformations, yielding an odds ratio of 3.6.

Additional unpublished studies conducted in Vietnam by Vietnamese investigators were reviewed by Constable and Hatch in 1985.[31] The reader is referred to this source for details of the studies. It was concluded that studies of presumptive paternal herbicide exposure prior to or at conception were suggestive of a relationship with congenital malformations. The evidence for an association with spontaneous abortion was "less convincing," and for molar pregnancies no relationship appeared to exist. Those studies that examined *both* maternal and paternal herbicide exposure were also suggestive of a relation to birth defects, as well as spontaneous abortion, stillbirths, and molar pregnancies. Two follow-up studies by these same investigators supported these earlier findings.[32,33]

In the next section, the studies of dioxin and reproductive effects that have been published since 1984 will be presented in detail.

3. REVIEW OF THE LITERATURE FROM 1984 TO 1992

During the interval since 1984, assays to measure TCDD in adipose tissue, milk, and serum were being tested and refined.[34–36] Several investigators have since utilized these assays in (usually small) subsets of their study populations to describe exposure to 2,3,7,8-TCDD in their total sample and also in attempts to validate their assumptions regarding magnitude of exposure for their study subjects. Subsets were generally selected to represent subjects designated as "high" versus "low" exposure by the study investigators, utilizing various information sources including self-reports, company or military records, and so forth. Table 1 presents the results of the exposure analyses.[37–45]

These data illustrate wide variability in groups presumed to have been exposed to 2,3,7,8-TCDD at levels above background ($\leq$10 ppt). For example, the mean and median serum levels of Vietnam ground combat troops with service in areas heavily sprayed with Agent Orange did not exceed the levels found in the U.S. general population. There is evidence for significantly higher exposure to 2,3,7,8-TCDD among *certain subgroups* of Vietnam veterans[41,44] as well as residents of Vietnam,[45] Seveso,[37] and Missouri, USA.[38]

An important finding resulting from these investigations was the inability of the exposure indices to classify exposure status in individuals.[11] Thus, 2,3,7,8-TCDD analyses in small subsets of the total study sample, selected on the basis of presumed exposure as defined by an "exposure index" (either military or government records, 2,3,7,8-TCDD levels in soil samples, or self-reports), have demonstrated that the exposure classification schemes utilized in these studies were not valid.

It is also important to note that with the exception of the Ranch Hand study, these subsets were selected to describe 2,3,7,8-TCDD exposure in the total study sample and

Table 1
Individual TCDD Levels (ppt/lipids) in Selected Populations

Study population	Specimen	Range[a]	Mean	Median	Ref.
Seveso, Italy residents	Serum				37
10 Zone A		828–56,000	19,144	16,600	
10 former Zone A		1770–10,400	5,240	4,540	
10 non-ABR zone		nd–137	—	—	
39 Missouri residents with history of TCDD exposure	Adipose tissue	2.8–750	79.7	17.0	38
57 Missouri residents with no known TCDD exposure		1.4–20.2	7.4	6.4	
9 New Zealand pesticide applicators	Serum	3.0–131.0	53.3	37.6	39
9 controls		2.4–11.3	5.6	9.3	
646 Vietnam ground combat troops with service in heavily sprayed areas	Serum	nd–45	4.2	3.8	40
97 non-Vietnam veterans		nd–15	4.1	3.8	
10 "regularly" exposed Vietnam veterans	Blood	—	46.3	25.1	41
	Adipose tissue	—	41.7	15.4	
10 Vietnam veterans with "little or no" exposure	Blood	—	6.6	5.3	
	Adipose tissue	—	5.1	5.4	
7 non-Vietnam veterans	Blood	—	4.3	3.9	
	Adipose tissue	—	3.2	3.5	
26 Vietnam veterans	Adipose tissue	nd–11	5.8	—	42
36 Vietnam veterans	Adipose tissue	—	13.4	10.0	43
79 non-Vietnam veterans		—	12.5	11.4	
80 civilians		—	15.8	11.8	
791 Ranch Hands	Serum	0–617.8	—	12.8	44
942 controls		0–54.8	—	4.2	
Vietnamese populations: 16 OB/GYN patients from a South Vietnam hospital	Adipose tissue	nd–103	20	13.5	45

[a]nd, not detectable.

not for an examination of the relationship between 2,3,7,8-TCDD and reproductive events. In addition, the data from the Ranch Hand population indicated a serum 2,3,7,8-TCDD half-life of 7.1 years,[46] but serum samples in this group were collected and analyzed at 11- and 15-year intervals following exposure. Questions regarding the impact of initial dose, age, gender, and pregnancy itself on half-life in humans remain unanswered at this time.

With the exception of the Seveso population, specimens were obtained and analyzed several years following exposure. Seveso residents, therefore, offer a unique population for the evaluation of the half-life of TCDD in men, women, and children over a wide age spectrum. Approximately 30,000 samples were collected and frozen in the weeks following the accident; these are the only results reported to date.

3.1. Environmental Studies

In the most recent report concerning reproductive effects among the Seveso cohort, the frequency of birth defects by zone of residence was examined for the period 1977–1982.[47] Outcomes were documented in the Seveso Congenital Malformations Registry, which included all live births and stillbirths (n = 15,291) that occurred during this 6-year period to women residing in the study area. Exposure was defined by residence in the contaminated zones (as described in the previous section on the early Seveso studies). A total of 742 malformations were identified, yielding a rate of 4.8%.

No major malformations occurred among 26 births in Zone A. There were 25 cases among 435 births in Zone B (5.7%), and in Zone R the rate was 4.5% (110 of 2439 births). Relative risks for zones AB versus the non-ABR area were calculated for the following categories of birth defects: total (OR = 1.2, 90% CI 0.88–1.64, total isolated (OR = 0.97, 90% CI 0.58–1.64), major (OR = 1.02, 90% CI 0.64–1.61), mild (OR = 1.44, 90% CI 0.92–2.24), multiple (OR = 0.71, 90% CI 0.14–3.70), and syndromes (OR = 1.85, 90% CI 0.57–6.05) of birth defects.

The authors acknowledged the lack of power to detect associations for specific defects (see Table 2). In addition, the limitations of defining exposure by area of residence and the possibility of underascertainment of birth defects because of spontaneous or induced abortions were discussed.

3.2. The Times Beach, Missouri, TCDD Episode

In 1971, a waste oil dealer in Missouri disposed of waste sludge, containing approximately 29 kg TCDD, by mixing it with waste oils as a dust control spray, which was subsequently used throughout the state. Public health officials began receiving reports of toxicity among animals in the contaminated areas. After a child who had been playing in one of the areas developed acute hemorrhagic cystitis, an investigation to determine the etiology of these events was launched. It was not until 3 years later that TCDD was identified in the soil. The Environmental Protection Agency then analyzed soil samples and produced a list of the TCDD levels in the soil of contaminated areas. Forty-four areas with TCDD soil levels of $\geq$ 1 ppb were identified; nine of these areas were primarily residential areas.

General media announcements from health officials were made, urging persons potentially exposed to TCDD to participate in a survey and health screening process. People were warned of their potential exposure by virtue of their residence, employment, or engagement in recreational activities in the contaminated sites. A pilot study was initiated in 1983, in which a subset of those persons who had responded to the media announcements was assessed.[16] Approximately 800 completed questionnaires

Table 2

Results of Studies Examining tahe Effect of Dioxin on Reproductive Outcomes in Humans, 1984–1992

Exposed group	Control group	Type of exposure	Data source: exposure/outcome	Outcome	Outcome in exposed (*n*)	Outcome in unexposed (*n*)	OR	95% CI
2900 births in Zones A, B, and R, Seveso, Italy, 1977–1981 (Mastroiacova *et al.*[47])	12,391 births in study area outside Zones A, B, and R	TCDD cloud released from chemical plant accident	TCDD soil analysis/ Seveso Congenital Malformations Registry	Total birth defects	137	605	0.97	0.83–1.13[a]
				Multiple birth defects	10	38	1.12	0.63–2.02
				Syndromes	5	29	0.74	0.33–1.63
				Major birth defects	67	343	0.83	0.67–1.04
				Minor birth defects	70	262	1.14	0.92–1.42
68 persons residing in areas of high TCDD concentration (Stehr *et al.*[16])	36 persons with no known contact with contaminated soil	Contact with TCDD-contaminated soil	EPA soil analyses for TCDD/interview	Infertility (males)	—	—	—	NS
				Impotence	—	—	—	NS
				Infertility (females)	—	—	—	NS
154 residents of Quail Run Mobile Home Park (Hoffman *et al.*[48])	155 residents non-exposed mobile park home	Contact with soil sprayed with TCDD for dust control	EPA soil analyses for TCDD/interview	Fetal deaths	—	—	—	NS
				Spontaneous abortions	—	—	—	NS
				Congenital malformations	—	—	—	NS
402 pregnancies to exposed mothers (Stockbauer *et al.*[49])	804 pregnancies to unexposed mothers (matched on maternal age and race, hospital and year of birth, plurality)	Contact with soil sprayed with TCDD for dust control	EPA soil analyses for TCDD/vital statistics and hospital records	Birth defects (all)	17	42	0.78	0.40–1.47
				Major birth defects	15	35	0.84	0.40–1.66
				Multiple birth defects	2	11	0.34	0.03–1.65
				Fetal deaths	4	5	1.60	0.32–7.43
				Infant deaths	5	5	2.00	0.46–8.69
				Perinatal deaths	6	9	1.33	0.39–4.20
				Low birth weight	27	36	1.59	0.89–2.81
				Very low birth weight	1	4	0.50	0.01–5.05
				IUGR	14	26	1.09	0.50–2.28
191 births to women resident near a herbicide plant, Sweden (Forsberg and Nordstrom[50])	2569 births among women not resident near a herbicide plant, Sweden	Dioxins, dibenzofurans, phenoxy acids, chlorophenols	Residence listings near plant/hospital records	Spontaneous abortion	28	248	1.6	1.03–2.49
4926 infants from the Metropolitan Atlanta Congenital Defects Program (Erickson *et al.*[52])	3029 infants from Georgia Vital Statistics Records	Vietnam military service	Self-reported, and Exposure Opportunity Index/ birth defects registry and vital statistics	Total birth defects (96 subcategories also examined)	428	268	0.97	0.83–1.14

7924 Vietnam veterans (CDC[56,57])	7364 non-Vietnam veterans	Vietnam military service	Military records/self-reports	Total birth defects	826	590	1.3	1.2–1.4
				Spontaneous abortion			1.3	1.2–1.4
				Stillbirth			0.9	0.7–1.1
				Low birth weight	100	87	1.1	0.8–1.4
				Childhood cancer	25	17	1.5	0.8–2.8
				Sperm abnormalities				
				Concentration	51	20	2.3	1.2–4.3
				Motility	91	58	1.2	0.8–1.8
				Morphology	51	29	1.6	0.9–2.8
1791 offspring of Vietnam veterans (Sub-study #1, CDC[56,57])	1575 offspring of non-Vietnam veterans	Vietnam military service	Military records/self-report and hospital record verification	Total birth defects	130	112	1.0	0.8–1.4
				Low birth weight	51	37	1.1	0.7–1.8
				Perinatal mortality	58	54	1.0	0.7–1.5
				Suspected birth defects	21	21	0.9	0.5–1.7
127 offspring of Vietnam veterans (Sub-study #2, CDC[56,57])	94 offspring of non-Vietnam veterans	Vietnam military service	Military records/self-report and hospital record verification	Cerebrospinal malformations	26	12	1.8	0.8–4.0
2858 Vietnam veterans (Stellman *et al.*[58])	3933 non-Vietnam veterans	Vietnam military service	Survey/survey	Diffculty conceiving	349	363	1.2	NS
				Time to conception	4.4 years	4.4 years	—	NS
				Birth weight	—	—	—	NS
				Spontaneous aboraticn	231	195	1.7	1.4–2.0
201 spontaneous abortion cases at Boston Hospital for Women (Aschengrau and Monson[59])	1119 full-term births at Boston Hospital for Women	Vietnam military service	Military records/hospital records	Spontaneous abortion	8	44	0.9	0.42–1.9
966 infants with late adverse pregnancy outcomes at Boston Hospital for Women (Aschengrau and Monson[60])	998 normal term infants at Boston Hospital for Women	Vietnam military service	Military records/hospital records	Birth defects	55	656	1.3	0.9–1.9
				Stillbirth	18	151	1.8	0.4–3.9
				Neonatal deaths	11	189	0.9	0.2–4.2
					5	52	1.5	1.0–3.1
					3	36	1.2	0.5–1.7
2533 conceptions among 791 Ranch Hand personnel (Wolfe *et al.*[44])	2074 conceptions among 942 non-Ranch Hand personnel	Spraying/handling of Agent Orange	Serum TCDD levels/hospital and medical records	Spontaneous abortion	28	72	0.9	0.52–1.4
				Total birth defects	36	84	0.9	0.54–0.86
				Sperm abnormalities	39	172	0.9	0.57–1.4

[a]90% confidence interval.

were screened for selection of participants; it was not clear if this number represents all of the questionnaires that had been returned up to that time.

This phase of the study was intended to identify potential problems for future investigations. Persons determined to be at "high" versus "low" risk for dioxin exposure, based on their completed surveys, were selected. "High risk" was defined as either (1) reported residence or occupation in areas with TCDD levels between 20 and 100 ppb for at least 2 years or in areas with TCDD levels > 100 ppb for at least 6 months, or (2) participation in activities requiring close contact with soil in areas with TCDD concentrations as described for similar periods of time. "Low-risk" persons were determined to have had no access to, or "regular high-soil-contact activities" in any known contaminated area. Controls were frequency matched with the high exposed group on type of exposure site, age, sex, race, and socioeconomic status.

Sixty-eight "high-risk" persons (83% of those eligible) and 36 "low-risk" persons (90% response) were evaluated through physical, neurological, and dermatological examinations, laboratory tests, and interviews. Information on reproductive outcomes was obtained during the interview administered to the subjects "or the nearest relative." None of the outcomes observed, including reported sexual dysfunction and "female reproductive health problems," differed significantly by exposure status although high-risk women had a later mean age at menarche ($p = 0.06$). No sample sizes or other statistical test results were reported. A total of 30 births were available for assessment in this sample. This study is severely limited by the volunteer nature of the sample, high exposure misclassification, and lack of verification of reported outcomes.

The following year, a more intensive study was undertaken to test the results found in the pilot study.[48] The exposed group consisted of residents of the Quail Run Mobile Home Park in Gray Summit, Missouri, where TCDD levels in the soil were measured at up to 2200 ppb. Data on 95 of the approximately 207 households in the park were available; 154 persons (74%) who were both "eligible and interested" agreed to participate. The unexposed group was comprised of 155 residents of a nonexposed trailer park, representing 77% of those "both eligible and interested" in participation.

The protocol was similar to that employed in the pilot study described above. The authors reported that "no differences were found between the exposed and unexposed groups in the frequency of reproductive disorders or adverse pregnancy outcomes, such as fetal deaths, spontaneous abortions, and children with congenital malformations." No sample sizes or statistical test results for any of the outcomes were reported. Clearly, this study was not designed to investigate the relationship between dioxin exposure and reproductive outcomes, and very little can be learned about this association from these results.

In a retrospective cohort study by Stockbauer *et al.*, the association between TCDD and reproductive outcomes was examined among residents of contaminated areas in Missouri.[49] All live births and stillbirths that occurred in the nine residential areas identified as contaminated with TCDD during the period 1972–1982 were identified through Missouri vital statistics records. A total of 402 births were examined from six of the residential areas. TCDD level in the soil from these six areas ranged from 241 to 2200 ppb. No TCDD levels were reported for the three areas in which no births occurred during this interval, which would have been of interest to note.

A reference group of 804 unexposed births, matched for maternal age and race, hospital and year of birth, and plurality were selected from the vital statistics records. Medical records were abstracted for ascertainment of birth defects in the matched sets. In addition, the births were linked to a statewide birth defect register which had recently been initiated. Data on several potentially confounding variables were obtained from birth certificates.

The exposed mothers tended to be less educated, had more children, were more likely to be in the extremes of the prepregnant weight distribution, and were more likely to smoke cigarettes. Statistical testing for these differences was not reported. Moreover, statistical adjustment for these potential confounders was performed only in the birth weight analyses, although there was little change in the birth weight risk ratios after adjustment.

Increased risk ratios were reported in the exposed group for infant death (OR = 2.0), fetal death (OR = 1.6), perinatal death (OR = 1.3), low birth weight (AOR = 1.5), and several subcategories of malformations, although none of these findings achieved statistical significance (Table 2).

Two approaches were utilized to determine a dose-response relationship. In the first, the study data set was matched against the Missouri central listing of dioxin-exposed persons, which yielded 98 of the exposed mothers and none of the control mothers. These 98 women were then dichotomized into "high" (n = 20) or "low" (n = 78) exposure groups. High exposure was defined as residence for at least 6 months in areas with $\geq$ 100 ppb TCDD in the soil or $\geq$ 2 years in areas with TCDD level from 20 to 100 ppb. Low exposure was defined as residence in areas with similar TCDD levels as described above but for less than the required time period, or residence at sites with 1–19 ppb TCDD. No evidence for a dose–response relationship was observed with this analysis.

The second approach involved categorizing the births into two different intervals relative to the spraying of the soil with the TCDD-contaminated sludge in 1971–1973. When births in the 1972–1974 period were compared with births from 1975 to 1982, the authors reported that "the only birth outcomes with higher risk ratios in the earlier time period were very low birth weight, birth defects, and major birth defects." None of these observations were statistically significant, but sample sizes and testing results were not provided in the paper.

The authors acknowledged the possibility of exposure misclassification, and also the "modest" power of the study to detect associations because of the small sample size.

In 1985, a report was published from Sweden that described the release of toxic contaminants by a herbicide company into the soil and groundwater surrounding the village of Teckomatorp.[50] The wastes contained phenoxy acids and chlorophenols, as well as dioxins and dibenzofurans. Following the closure of the factory, a medical task force was established to assess possible health effects among the residents of Teckomatorp. The authors reported that two previous studies, one which conducted a field survey to ascertain miscarriages, and a second which utilized the Swedish Register of Congenital Malformations and the Medical Birth Record system, found no evidence for increased spontaneous abortion rates related to contamination.

The company had been involved with the manufacture of herbicides from 1965 until it was closed down in 1977. "All obtainable" hospital records from 1965 to 1979 for women less than 35 years of age were reviewed. "Exposed" births were those from Teckomatorp ($n = 191$). Three surrounding areas described as the "most suitable comparison group for our study" supplied the nonexposed births ($n = 2569$). Maternal age and smoking status did not differ between the exposed and unexposed groups.

The spontaneous abortion rate was 15% in the exposed area compared with 9% in the unexposed area ($p = <0.01$). These figures are at or below the general background rate of 15–20%. Because increasing complaints of four smells emanating from the herbicide company were reported during 1974, the rate of miscarriages between 1974 and 1979 was examined separately. The rate in Teckomatorp increased to 20%, while the rate in the control area remained stable.

The conclusions that can be drawn from this study are very limited because of the small sample size and lack of detail provided concerning sample selection.

3.3. Studies of Military Veterans Published from 1984 to 1992

3.3.1. Evaluation of Exposure

Evidence from earlier studies to determine if Agent Orange exposure increased the risk of adverse pregnancy outcomes among Vietnam veterans had been described as "sparse, sometimes off the point, sometimes conflicting. . . ."[51] It was also evident that the general dissatisfaction with these studies had a common factor: the lack of a valid measure of dioxin exposure. Once the assays to document individual TCDD exposure became available and served as the gold standard for assessing the exposure assumptions made by study investigators, this concern regarding exposure misclassification was shown to be justified.

There are still unanswered questions regarding the metabolism and elimination of TCDD in humans, and which cofactors are related to these functions. Hopefully, research in this area, examining both identified populations at high probability of exposure, e.g., Seveso, and future cohorts currently being assembled, will provide these answers.

Until 1992, when the first study to examine reproductive outcomes among Vietnam veterans based on individual exposure measurements of 2,3,7,8-TCDD was published, this remained the major criticism of the research. Even in the Ranch Hand study, however, the lack of data on dioxin level at the time of conception continues to instill uncertainty about the findings.

3.3.2. The Validity of "Exposure Indices"

3.3.2a. The CDC Case–Control Study's Exposure Opportunity Index. The CDC conducted a large case–control study that examined the relationship between service in Vietnam and risk of congenital malformations.[52] Although service in Vietnam was the major independent variable, two additional measures of exposure were assessed. Vietnam veterans were asked if they believed they had been exposed to Agent Orange. In addition, an exposure opportunity index (EOI), developed by the Army Agent Orange Task Force, assigned a score estimating he likelihood of exposure based on places and times of Vietnam service. These scores ranged from 1 (minimal) to 5 (high) opportunity for Agent Orange exposure.

In a later separate study, Kahn *et al.* measured individual serum 2,3,7,8-TCDD levels in 10 Vietnam veterans who had handled Agent Orange "regularly" while in Vietnam, 10 Vietnam veterans with little or no exposure to Agent Orange, and 27 Vietnam-era veterans.[41] The lipid-adjusted levels of TCDD among those men who had handled Agent Orange were significantly elevated (median = 25.1 ppt) compared with the Vietnam control veterans (median = 5.3 pg/g) and the Vietnam-era veterans (median = 3.9 pg/g) ($p < 0.01$).

The CDC then conducted a study, using the serum assay as the standard, to evaluate the validity of both self-reported exposure to Agent Orange and the EOI used in the earlier case–control study of military service in Vietnam and birth defects.[40] The results revealed a poor correlation (no correlation coefficient was provided) between both of these exposure estimates and serum TCDD levels. In addition, the distributions of serum 2,3,7,8-TCDD levels were nearly identical (median = 3.8 ppt) among 646 ground combat troops who had served in heavily sprayed areas compared with 97 veterans who had never served in Vietnam.

3.3.2b. The Ranch Hand Study's Exposure Index. In 1984, the United States Air Force released preliminary results in a 20-year study designed to examine the personnel responsible for conducting the aerial spraying of herbicides in Vietnam.[53] The analyses in this baseline study were based on cohort status, i.e., Ranch Hands ("exposed") versus controls ("nonexposed"). In an attempt to determine a link between exposure and clinical endpoints, an exposure index was developed to estimate individual 2,3,7,8-TCDD exposure. The exposure index developed for this study was defined as the product of a TCDD weighting factor and the gallons of TCDD herbicides sprayed during the veteran's tour divided by the number of Ranch Hands sharing his duties during his tour.

In 1988, a report describing a United States Air Force and CDC collaborative pilot study utilizing the serum 2,3,7,8-TCDD assay among 200 Air Force Ranch Hand personnel was published. It was noted that the Ranch Hand personnel had significantly higher serum 2,3,7,8-TCDD levels than controls.[54] These data also indicated that the Ranch Hands as a whole were not as highly exposed to 2,3,7,8-TCDD as compared with the occupationally exposed NIOSH cohort[55] and the Seveso population.[37]

In 1992, the report describing reproductive outcomes among Ranch Hand personnel became available.[44] In addition to outcome verification, serum 2,3,7,8-TCDD levels were measured in a sample of Ranch Hands (n = 791) and the comparison population (n = 942). A comparison of the exposure index used in the baseline Ranch Hand study with individual 2,3,7,8-TCDD levels also revealed "considerable misclassification" among the study subjects.

As a result of these investigations, it became clear that the likelihood of exposure misclassification in studies of the relationship between 2,3,7,8-TCDD and reproductive events without direct measures of individual exposure casts considerable doubt on the validity of the findings. Studies that defined exposure as paternal military service in Vietnam should be evaluated without inference to 2,3,7,8-TCDD exposure.

3.3.2c. Individual Studies of Military Veterans 1984–1992. In 1984, the CDC released a report describing the results of the large case–control study that examined the relationship between service in Vietnam (not dioxin exposure *per se*) and risk of congenital malformations.[52] Case-group babies were infants with serious structural malformations born between 1968 and 1980 and registered with the Metropolitan Atlanta Congenital Defects Program (MACDP). Of the 7133 eligible cases, maternal interviews were obtained for 4929 (69%). Control babies were selected from Georgia vital statistics records, and were frequency matched to cases on race, and year and hospital of birth. Of the 4246 eligible controls, maternal interviews were obtained for 3029 (71%). Paternal interview rates were 56 and 57%, respectively.

While response rates were similar by case status overall, among nonwhites significantly more cases were not interviewed. The major reason for nonparticipation was the inability to locate subjects rather than subjects refusing to enroll.

Among the children with congenital malformations, 428 (9%) were fathered by Vietnam veterans and 268 (9%) were fathered by non-Vietnam veterans. The OR for service in Vietnam and birth defects of any type was 0.97 (95% CI 0.83–1.14). ORs were also calculated for 96 separate categories of birth defects with no significant associations observed.

The EOI, developed for this study as described above, also was not associated with total birth defects. However, significant associations were observed for spina bifida, cleft lip ± cleft palate, neoplasms, and coloboma. With the exception of the last defect, all three also showed evidence of a dose–response relationship.

For the analysis that examined potential associations between self-reported Agent Orange exposure and birth defects, the frequency of exposure among fathers of each type of malformation was compared with the frequency among fathers with all other defects in an attempt to reduce recall bias. Twenty-five percent (n = 74) of the Vietnam veterans reported that they believed they were exposed to Agent Orange during their service in Vietnam. The analysis of self-reported exposure and birth defects was negative on all counts, in contrast to the EOI analysis which found significant associations as described above. The small numbers of cases of many individual defects resulted in a virtual lack of power to detect any associations for these specific defects. In addition, the documented poor correlation between both self-reported exposure and the EOI with serum TCDD levels allows no inferences to be drawn from this study regarding the relationship between dioxin exposure and adverse reproductive effects.

Finally, as part of a large multifaceted study that was mandated by Congress, the CDC evaluated the risk of service in Vietnam and adverse reproductive outcomes.[56,57] The Vietnam Experience Study (VES) protocol involved two phases. In the first phase, a random sample of male veterans who met eligibility criteria related to military service was selected.

Of the eligible "exposed" group, i.e., those veterans who had served in Vietnam, 84% (n = 7924) agreed to participate; and 84% (n = 7364) of the nonexposed (those who had not served in Vietnam) were enrolled. A random sample of these 15,288 veterans who completed the telephone interview was then selected for the second phase, which consisted of a comprehensive medical examination. For this phase, the response rates were 75% (n = 2490) for the Vietnam veterans and only 63% (n = 1972) for the non-Vietnam veterans group.

During the interview, veterans were questioned about the following reproductive events: miscarriage, induced abortion, ectopic pregnancy, live births, stillbirths, birth defects, leukemia and other childhood cancers, and major health problems or impairments during the first 5 years of life.

A preliminary analysis of the interview data revealed that the Vietnam veterans had reported 40–50% more birth defects than the non-Vietnam veterans. In addition, a difference between the cohorts for certain subcategories of malformations, including spina bifida, cleft lip ± cleft palate, and hydrocephalus, was noted. Therefore, a substudy was conducted to compare the rates of total birth defects in Vietnam and non-Vietnam veterans as these events were recorded on hospital records.

The eligible sample for this study consisted of 2282 veterans who had not yet received their medical examinations, as obtaining the additional information required for the substudy as well as permission to obtain hospital records for their offspring would be easier in this group. Hospital records were obtained for 92% (n = 1791) of the offspring of Vietnam veterans and 91% (n = 1575) of the offspring of non-Vietnam veterans.

When birth defects were identified from hospital records, there was no association of Vietnam service with total, major, minor, or suspected birth defects. From the telephone interview data, the OR for Vietnam service and total birth defects was 1.3 (95% CI 1.2–1.4); in the hospital records substudy, the OR was 1.1 (95% CI 0.7–1.8). It was concluded that this finding supported the explanation of differential reporting in the telephone interview.

The ORs for selected categories of birth defects calculated from both the telephone interview and hospital records study are presented in Table 3. The rate of birth defects increased for both cohorts when malformations were identified in the medical records. The main report on reproductive outcomes, which stated that the second objective of the substudy was to assess the extent of differential reporting between the two cohorts, provided a description of the analysis of misclassification. Overall, the agreement between veterans reports and hospital records for the presence of a birth defect was "relatively poor" *for both cohorts*. Positive predictive value, sensitivity, and the kappa statistic were slightly lower among Vietnam veterans (Table 4).

It was further stated that there was no evidence of selection bias or participation bias in this substudy, because there were no differences between cohorts in health histories, demographic or military covariates, and the participation of both groups of veterans was high. However, the subjects in this substudy were selected from the group of veterans who completed the physical examination. The response rate of this phase of the study was not high, as noted above, with only 63% of the non-Vietnam veterans participating in the medical examination. While characteristics of the subset who

Table 3
Odds Ratios for Selected Categories of Birth Defects for the Telephone Interview and Hospital Records Study in the Vietnam Experience Study, 1989[a]

	Rate (per 1000)			
Defect category	Vietnam veterans	Controls	OR	95% CI
	Telephone interview			
Total	64.6	49.5	1.3	1.2–1.4
	Hospital records study			
Total	72.6	71.1	1.0	0.8–1.4
Major	28.5	23.5	1.1	0.7–1.8
Minor	32.4	34.3	1.0	0.7–1.5

[a]Adapted from the Centers for Disease Control Vietnam Experience Study, Health status of Vietnam veterans III: Reproductive outcomes and child health, *J. Am. Med. Assoc.* **259**, 2715–2719 (1988).

agreed to undergo physical examination did not differ from the telephone interview sample, no data on those who refused the exam, and no reasons for refusing to participate, were provided in this paper.

Adjusted odds ratios (AOR) for two additional outcomes verified in the hospital records substudy, low birth weight (AOR = 1.1, 95% CI 0.8–1.4) and perinatal mortality (AOR = 1.6, 95% CI 0.8–3.1), did not differ by Vietnam service status. A race-specific analysis of total, major, and minor defects did show on increased risk for total and minor birth defects among black veterans as compared with white veterans.

Table 4
Results of the Misclassification Analyses for Birth Defects in the Hospital Records Substudy, Vietnam Experience Study, 1989[a]

Vietnam veterans		Non-Vietnam veterans	
PPV[b]	24.8%	PPV	32.9%
NPV	95.2%	NPV	95.8%
Sensitivity	27.1%	Sensitivity	30.3%
Specificity	94.7%	Specificity	96.2%
% agreement	90.6%	% agreement	92.4%
Kappa index	20.9%	Kappa index	27.6%

[a]Adapted from the Centers for Disease Control Vietnam Experience Study, Health status of Vietnam veterans Volume V: Reproductive outcomes and child health, CDC, Atlanta (1989).
[b]PPV, positive predictive value; NPV, negative predictive value.

The AOR was 3.3 (95% CI 1.5–7.5) for total defects and 2.9 (95% CI 1.1–8.0) for minor malformations. These findings are based on very small numbers, however, and on multiple occurrences of two minor defects in two families.

From data obtained during the telephone interview, the AOR for Vietnam service and spontaneous abortion was 1.3 (95% CI 1.2–1.4). Although an excess among Vietnam veterans was noted across all three trimesters, only the association in the first trimester was significant. There was no attempt to confirm this endpoint in the hospital records. No significant differences were observed for the reproductive outcomes of induced abortion (AOR = 1.0, 95% CI 0.9–1.2), stillbirths (AOR = 0.9, 95% CI 0.7–1.1), ectopic pregnancies (AOR = 1.0, 95% CI 0.7–1.2), or childhood cancers (OR = 1.5, 95% CI 0.8–2.8).

In addition to the reported excess of total birth defects in the telephone interview, Vietnam veterans also reported more neural tube defects and hydrocephalus than the non-Vietnam veterans. Therefore, a second substudy was undertaken to examine the increase in cerebrospinal malformations (CSMs) reported by Vietnam veterans in the telephone interview. In this substudy, an attempt was made to obtain hospital records for all of the offspring identified in the telephone interview who met one of the following criteria: (1) offspring with a CSM as reported by a veteran, (2) those with a reported condition that suggested a CSM, or (3) all reported stillbirths.

Of the 403 offspring reported to have a CSM, 109 were ineligible, 58 because of conception prior to military service, and 51 that were classified as miscarriage. Again, the issue of the impact of spontaneous abortion on rates of birth defects is introduced.

The CSM substudy was limited by a poor response rate among the non-Vietnam veterans. The sample thus consisted of 127 offspring of Vietnam veterans (82.5%) and 94 children of the non-Vietnam veterans (67%). Compared with fathers who participated in the CSM study, nonparticipants were more likely to be nonwhite, less educated, unmarried, younger at the time of their child's birth, and have lower general technical test scores. There were also differences in paternal covariates between the participating cohorts that could confound the findings. Compared with Vietnam veterans, non-Vietnam veterans were better educated, had higher general technical test scores, were more likely to be married, were older when their child was born, were less likely to consume alcohol, and were more likely to have had a nontactical primary military occupational specialty (MOS) in the Army, and to have served in both the later and earlier time periods.

The number of CSMs reported by the veterans that were verified by hospital records was examined separately by stillbirth and live birth status. Among the reported stillbirths, five CSMs were documented among children of Vietnam veterans and six among the non-Vietnam veterans. Ten of these eleven CSMs had not been reported by the veterans. Positive predictive values derived from this analysis were 6.5% for Vietnam veterans and 8.1% for non-Vietnam veterans.

Among the live births, 21 of 49 reported CSM cases were noted on hospital records for the Vietnam veterans; 6 of the reported 20 cases were observed among the non-Vietnam veterans group, yielding positive predictive values of 43 and 30%, respectively.

Tables 35 and 36 in the main report listed the fathers' descriptions of the birth defects in their offspring obtained through the telephone interview, by cohort. It was intriguing to note that among Vietnam veterans, for 22 of the 55 reported cases of birth defects the hospital record finding was "none." In 5 additional cases, the hospital record finding was "none of the nervous system." Of these 27 nondocumented reports of birth defects, 3 cases (11%) were reported as having died within the first year of life. Death during the first year of life was also reported for 1 of the 8 (12%) unverified CSMs in the non-Vietnam veterans cohort. These findings are sufficiently intriguing to warrant follow-up of these infants through death certificates.

In summary, the CDC investigators concluded that for "most reproductive and child health outcomes studied, Vietnam veterans were more likely to report an adverse event than were non-Vietnam veterans." In the two substudies conducted to compare rates that were identified through hospital records, no significant differences in adverse outcomes between the two cohorts were determined. However, the ability of this study to address the issue of dioxin exposure and reproductive outcome is limited. The question of bias still remains in the two substudies. In addition, exposure was defined as service in Vietnam and adverse outcomes of pregnancy, which does not provide insight into the question of the reproductive toxicity of Agent Orange.

The relation of self-reported exposure to Agent Orange and reproductive outcomes was part of a study conducted among 6810 American Legionnaires who had served during the Vietnam war.[58] Information was obtained through a questionnaire mailed to 2860 veterans (42%) who had served in Southeast Asia and 3933 veterans (58%) who served elsewhere. No association between Agent Orange exposure and difficulty with conception, time to conception of the first child, or infant birth weight was observed. However, the proportion of spontaneous abortion was significantly higher among the spouses of veterans who served in Vietnam (7.6%) compared with controls (5.5%) ($p < 0.001$). These figures were well below the background rate for recognized spontaneous abortion in the general population.

The significance of these findings are limited by the inaccuracy of self-reported exposure, the lack of outcome verification through medical records, and the selection of veterans from the American Legion organization, as they may not be representative of veterans who served during the Vietnam conflict.

A case–control study to investigate the relationship between paternal military service in Vietnam and risk of spontaneous abortion was conducted at Boston Hospital for Women.[59] Cases, identified through hospital records, were spontaneous abortions $\leq$ 27 weeks' gestation that occurred between July of 1976 and August of 1978. Paternal identifying information from the hospital birth records was linked with national and state military records to identify those fathers who had served in Vietnam. Frequency of service in Vietnam was compared for cases and all full-term live births at the hospital during this time period. No association was detected (OR = 0.88, 95% CI 0.42–1.86).

These same investigators conducted a subsequent study that expanded the outcomes examined to include late adverse pregnancy outcomes.[60] Case infants identified through hospital records were comprised of 857 congenital malformations, 61 stillbirths, and 48 neonatal deaths that occurred at the hospital during August, 1977, and

March, 1980. "Exposure" was defined using the same method as the previous study. Frequency of paternal service in Vietnam was compared for cases and 998 controls. Again, no association with any of these later adverse outcomes was detected.

3.3.2d. The Baseline Ranch Hand Study 1984. This initial report of the health of Ranch Hand personnel involved with the dissemination of herbicides during the Vietnam conflict utilized cohort status (Ranch Hands versus comparisons) as the basis of the analysis.[53] This group of veterans included those who served in Vietnam during 1962–1965, when Herbicides Purple, Pink, and Green were sprayed. These herbicides had higher TCDD concentrations (13, 66, and 66 ppm, respectively) than Herbicide Orange with 2 ppm TCDD.

The study is a matched cohort design with the comparison group selected from cargo mission personnel who also flew in Southeast Asia (SEA) but who were not directly exposed to Agent Orange. Controls were matched to the Ranch Hands for age, race, and occupational category. In 1981, it was discovered that 18% of the controls selected had *not* served in SEA and replacements were chosen. Subsequent statistical analysis indicated that the replacement controls differed from the originals in several covariates. It was therefore decided to test the main hypotheses of this study utilizing the original control group subjects. Most analyses were presented separately for the "original" and the "replacement" control groups, which makes interpretation of the results quite complex.

The protocol consisted of a comprehensive personal and family health questionnaire and a physical examination, including an in-depth laboratory analysis. The response rates for each phase of the protocol were quite different both within and between cohorts. Compliance with the questionnaire phase was 97% (n = 11740 for the Ranch Hands and 93% (n = 956) for controls. In the physical examination phase, compliance dropped to 87% (n = 1045) for the Ranch Hands and 76% (n = 773) for controls.

Nonresponders were "on the average" younger than participants; more detail and statistical testing results would have been helpful here. Ranch Hand enlisted personnel had higher participation rates than officers, and black Ranch Hand officers had lower participation rates than nonblack officers. The difference in the response rates for the physical examination phase of the protocol was ascribed partially to the active encouragement of the Ranch Hand Association for participation and the intense media coverage that the study received. The authors stated that the "majority" of reasons given for nonparticipation were "no time–no interest" and passive refusal. Again, a detailed description of reasons for refusal by cohort status would have been useful.

The reproductive outcomes evaluated in this phase of the study were ascertained through questionnaires obtained from both the veterans and their spouses/partners. A total of 7399 conceptions were analyzed in this preliminary report. There were 3293 conceptions among 1174 Ranch Hands, and 4106 among the 1531 controls.

No significant differences were reported for four "measures of fertility": (1) the number of childless marriages, (2) the number of couples having achieved their desired family size, (3) the number of childless marriages per total number of marriages, and

(4) the number of conceptions per years spent together, which included nonmarital relationships. The fertility analysis was performed on the total number of conceptions reported and was not adjusted for any confounding variables. Moreover, exposure in this analysis was defined by a simple dichotomy of Ranch Hands versus controls.

In order to examine the relationship of TCDD exposure and spontaneous and induced abortion, stillbirths, and live births, exposure was stratified by pre- and post-SEA service. The unadjusted analysis indicated that Ranch Hands had increased spontaneous abortion rates in both pre-SEA duty ($p = 0.06$) and post-SEA duty ($p = 0.13$). The report qualified this by stating that inferences based on these analyses which were not adjusted for key factors affecting pregnancy outcome "are of questionable value," although no similar qualification was given for the fertility analysis. After adjustment for maternal age, smoking and alcohol use, and paternal age, no significant difference was observed for spontaneous abortion.

Among the live births with complete data obtained to allow for adjustment of cofactors, no difference in risk of prematurity was noted. However, estimate of gestational age was based on parental report, which is not a sensitive measure of gestational length, and it was not clear whether prematurity was analyzed as only a dichotomous variable (< 37 weeks, ≥ 37 weeks), or as a continuous variable. No analyses for birth weight differences by exposure status were performed, which was unfortunate given the controversy regarding the finding of lower birth weight among infants exposed to PCDDs and PCDFs.[5–7]

Unadjusted analyses were conducted to examine the relationship between exposure and neonatal death, infant death, physical handicaps, birth defects, and learning disabilities. These analyses were stratified by pre- and post-SEA service periods. The results indicated that Ranch Hands were more likely to report physical handicaps ($p = 0.07$), birth defects ($p = 0.08$) and neonatal death ($p = 0.02$) in the post-SEA analysis. After adjustment for the maternal and paternal covariates described above, the relationship with birth defects achieved statistical significance ($p = 0.04$); the other relationships were not statistically significant.

Twelve of the seventy-six birth defects reported to have occurred among the Ranch Hands after post-SEA service were skin anomalies (ICD Code 757). Because these 12 anomalies represented a very broad range of conditions from mild birthmarks to serious conditions, they were excluded from subsequent analysis. When these anomalies are excluded, this relationship is no longer statistically significant ($p = 0.14$) although "still of interest."

Finally, semen samples from Ranch Hands ($n = 560$) and controls ($n = 409$) were analyzed for sperm count and morphology. The response rates for this parameter were 72.5 and 76.5%, respectively, although some of the samples submitted were ineligible for analysis because of prior vasectomies and orchiectomies. Linear regression techniques examined sperm count (as a continuous variable) and percentages of sperm with abnormal morphology as the dependent variables. Independent variables were age and exposure to industrial chemicals. No differences between groups in either parameter were identified.

This finding contrasts with the semen analysis results performed among 324 Vietnam veterans and 247 non-Vietnam veterans in the VES.[61] That analysis indicated that Vietnam veterans had significantly lower mean sperm concentrations ($p = 0.05$) and twice the proportion with values below the clinical reference value (20 million cells/ml) than the non-Vietnam veterans. In addition, the mean percentage of morphologically normal cells was significantly decreased in the Vietnam veterans. These analyses were adjusted for 12 covariates, although exposure to industrial chemicals was not among them.

3.3.2e. The Ranch Hand Study–Reproductive Outcomes. The significant association between Ranch Hand status and birth defects found in the previous study was sufficiently troubling to launch a massive project to verify *all* reported conceptions and pregnancy outcomes through medical record abstraction. In addition to outcome verification, individual TCDD exposure documentation via serum assay was undertaken in 1988 for Ranch Hands and controls. In 1992, the Air Force released the results of the first study that examined the relationship between direct measure of individual serum TCDD and verified reproductive outcomes.[44] A total of 4607 conceptions were examined in this study; 2533 were contributed by 791 Ranch Hands, and 2074 were contributed by 942 controls.

Ranch Hand personnel were shown to have significantly higher serum TCDD levels compared with the controls (Table 1). The median values were 12.8 and 4.2 ppt, respectively. The 98th percentile for Ranch Hands was 166.4 ppt; for controls, 10.4 ppt. Dioxin levels were determined in 1987. These results were used to estimate initial doses received during the veterans tour in SEA. However, no attempt was made to estimate dioxin level at the time of conception.

Several analyses of reproductive effects by serum TCDD level were trichotomized into: low ($\leq$ 10 ppt), medium (15–$\leq$33.3 ppt), and high ($>$ 33.3 ppt) dioxin groups.

The fertility analysis that was performed in the earlier study was not repeated according to level of serum dioxin, which was a disappointing omission. There was a significant variation in the association between dioxin and miscarriage with time since SEA tour among Ranch Hands with current serum dioxin levels $>$10 ppt ($p = 0.014$). This was attributed to the low miscarriage rate among the pre-SEA Ranch Hands with current dioxin levels $>$33.3 ppt (Table 5). Among the post-SEA conceptions, a linear trend can be seen in spontaneous abortions with increasing serum TCDD levels among Ranch Hands who had "late tours" in SEA, i.e., less than or equal to 18.6 years had elapsed between their tour of duty and the serum TCDD analysis. The opposite trend is noted in Ranch Hands with "early tours," i.e., greater than 18.6 years had elapsed between the end of duty and the 1987 blood draw. It was concluded that dioxin did not affect the rates of miscarriage because it seemed "implausible that dioxin would act differently in the two groups."

An alternative explanation might be that there *is* a relationship, but that it cannot be detected by this type of analysis. The data in Table 5 are misleading in that the dioxin level at the time of conception is not considered. Assuming a half-life of 7 years in humans, it would seem reasonable, for example, that the two groups of Ranch Hands in the current "low" dioxin category with post-SEA conceptions may have had very different dioxin levels at the time their children were conceived. This is possible

Table 5

Rates of Miscarriage (per 1000) by Pre- and Post-Vietnam Tour Status and Time since Tour of Duty, among 1475 Ranch Hands with >10 ppt Serum Dioxin, Ranch Hand Study, 1992[a]

		Miscarriage rate (No./n) Current dioxin level:			
Time of conception	Time since tour (years)	10–14.9 ppt	15–33.3 ppt	>33.3 ppt	p
Pretour	≤18.6	142.0 (23/162)	146.8 (32/218)	48.8 (2/41)	0.014
	>18.6	123.9 (14/113)	159.4 (33/207)	166.7 (16/96)	
Posttour	≤18.6	92.1 (7/76)	136.6 (22/161)	168.5 (15/89)	
	>18.6	237.3 (14/59)	198.6 (29/146)	121.5 (13/107)	

[a]Adapted from Wolfe *et al.*, An epidemiologic investigation of health effects in Air Force personnel following exposure to herbicides: Reproductive outcomes, U.S. Air Force, Brooks Air Force Base, Texas (1992).

because the early tour veterans had more time to decrease their body burden of TCDD before their blood samples were drawn in 1987 than did their late tour counterparts.

Table 2 includes the ORs for spontaneous abortion, total birth defects, and low sperm count in the Ranch Hand study. These measures were calculated for veterans with >33.3 versus >10–33.3 ppt serum TCDD levels for spontaneous abortion and birth defects; and >33.3 versus <10 ppt TCDD for low sperm count.

The main report also stated that the "expected dose-pattern" for dioxin and total adverse reproduction outcomes (miscarriage, tubal pregnancy, other noninduced abortive pregnancy, or stillbirth) is the "linear one in which the highest anomaly rate occurs at the highest levels of dioxin." If a linear response is assumed, might this imply that very early pregnancy losses occur at the highest dioxin levels, so that the conceptus would not survive long enough to be clinically recognized? Or, are very early pregnancy losses and clinically recognized spontaneous abortions two separate entities with different dioxin thresholds? Such a scenario has been suggested to explain changes in spontaneous abortions observed after exposure to radiation in Hiroshima.[62]

These questions are of interest because the rate of each of these endpoints may directly affect the rates of all subsequent reproductive outcomes that are available for examination. The miscarriages assessed in this study are most likely late spontaneous abortions, as 99.6% of reported miscarriages were verified through medical records.

An analysis in which the 1987 dioxin levels are used to estimate dioxin level at time of conception would be a worthwhile effort. If a relationship between paternal dioxin level and adverse reproductive outcome does exist, this may help to determine the dose–response pattern of the relationship.

No evidence was found to support an association between dioxin and "total adverse outcomes," defined as miscarriage, tubal pregnancy, other (noninduced) abortive pregnancy, or stillbirth, or total conceptions. These findings should be viewed with caution in view of the above comments concerning the unexplored area of events early in gestation.

Overall, there was little convincing evidence to support an association between birth weight, examined as both a continuous variable and dichotomized (<2500 g or ≥2500 g) and paternal dioxin level. Analyses that adjusted for covariates included maternal and paternal age, maternal alcohol and smoking, and race of the father. No assessment of dioxin and prematurity was reported.

The potential association between cohort status and birth defects was examined for all defects combined, and an additional 12 categories of malformations. The only categories with sufficient numbers of verified post-SEA cases to detect a relative risk of 2 were total birth defects (229 cases among 1045 Ranch Hands and 289 cases among 1602 controls); and musculoskeletal deformities (132 cases among Ranch Hands and 180 among controls).

A significant variation was observed in the association between total birth defects ($p = 0.03$), defects of the respiratory system ($p = 0.03$), and urinary system abnormalities ($p = 0.04$), by Ranch Hand versus control status with time of conception (pre- or post-SEA). All of these findings were related to a lower rate among Ranch Hands in the pre-SEA conceptions and a higher rate among the post-SEA conceptions for the Ranch Hands.

Analyses of birth defects by dioxin level did not find any "consistent patterns" to support an association. For example, among children of flying enlisted and enlisted ground personnel, children of Ranch Hands in the low dioxin category had higher rates (433 per 1000 and 317 per 1000) than children of controls with background dioxin levels (229 per 1000). However, rates of children of enlisted ground personnel in the high dioxin group were not significantly elevated. Again, these analyses were not based on dioxin level at time of conception. Moreover, if high dioxin levels were related to early pregnancy loss, these results would make more biological sense.

There was also a significant association between high dioxin levels and neonatal death (OR = 5.5, 95% CI 1.5–20.7). Insufficient numbers ($n = 13$) precluded the calculation of an AOR for this finding.

Finally, no association was detected between dioxin level and either sperm count or percentage of abnormal sperm. These analyses were based on semen samples that had been collected in 1982.

4. SUMMARY

While the evidence on dioxin and reproductive effects that has appeared since 1984 may still be described as "sometimes off the point, sometimes conflicting," this painstaking research process has been successful in elucidating some of the reasons for this confusion. Although it would be premature to commit to a position regarding a "cause and effect" relationship between dioxin exposure and reproductive effects, the direction for future research efforts has been refined. In this section, the major epidemiologic research will be summarized and evaluated according to epidemiologic criteria for causation.

Epidemiologic investigations are not conducted under the tightly controlled conditions that govern the laboratory sciences. There are guidelines, however, that provide structure to the process of examining epidemiologic evidence and deciding whether that evidence is sufficient to result in some form of remedial action. While the terminology may differ depending on the reference, there are six criteria that are consistently employed by epidemiologists engaged in such an evaluation process: consistency; temporality; magnitude of the association; specificity; dose–response relationship; and biological plausibility. Each of these criteria will be defined and discussed as it pertains to the dioxin–reproductive effects debate.

1. Consistency of the association. If an association between a factor and a disease is demonstrated across a variety of studies utilizing different designs and different populations, the argument for causation is strengthened. The replication of an association under different conditions decreases the likelihood that confounding is responsible for the observed association. Consistency is a powerful criterion for causation but *only when* "the variables under test (exposure, outcomes) are similar enough" to justify the comparison of the various studies' findings.[51]

It should also be determined *a priori* that each study included in the critical evaluation process is in adherence to basic epidemiologic principles governing study design and analysis. Deficient studies with suspect results should be excluded. While this is not meant to imply that such studies have no worth, as invaluable information has often been derived from these studies which improve on prior examinations of the issue, they have no place in the evaluation process. Unfortunately, in studies of dioxin–reproductive effects in humans, the probability of exposure misclassification forces exclusion of much of the research to date.

2. Temporality of the relationship. If factor X is a cause of disease Y, then X must precede Y in time. It is obvious that if this criterion is *not* satisfied, no valid argument for causation can be made. However, restricting the examination of reproductive events to those that occurred after the exposure does not in and of itself satisfy this time order criterion. Several factors must also be considered, such as the half-life of the contaminant in the body and the concentration present at the time of conception.

3. Magnitude of the association. This criterion refers to the degree to which the measure of association (e.g., OR or relative risk) exceeds the null value of 1. The stronger the association between exposure and effect, the more convincing is the argument for causation. There is no definitive cutpoint to numerically define a meaningful measure of association. Other factors, such as the prevalence of the exposure in the population, impact on the significance of the measure. Because so many adverse health conditions are multifactorial in etiology, a general rule of thumb is that a relative risk less than 2 renders a cause–effect relationship less likely.

With the exception of the finding in the VES that Vietnam veterans were more than twice as likely as non-Vietnam veterans to have low sperm concentrations (OR = 2.3, 95% CI 1.2–4.3), no effect measure greater than 2 was noted in any of these investigations. This is not surprising given the limitations of the studies described above, particularly with regard to exposure misclassification. Therefore, the trends across studies carry more important than "statistically significant" results.

A critical element that should always accompany the effect measure is a confidence interval. Placement of an interval around the measure enables us to quantitate the result for a more meaningful interpretation. An OR of 30 is quite impressive, but if the 95% CI is 0.9–200, we are less impressed with the magnitude of the association!

4. Specificity of the association. Specificity refers to the uniqueness of the association between a factor and an outcome. If the relationship were absolute, then *only* factor X would be related to *only* disease Y. It is indeed rare to encounter this type of association, which renders this criterion generally less useful in the evaluation process.

5. Dose–response relationship. When the risk of disease increases with the dose or gradient of exposure, the evidence for causation is strengthened. It should be emphasized that there are many possible dose–response patterns, which may result in different threshold levels for different endpoints. Because of the exposure misclassification bias present in most of the dioxin research, with the exception of the Ranch Hand study, it is not valid to attempt to determine dose–response relationships. Elucidation of a dose–response pattern may be derived utilizing data from the Ranch Hand study to calculate TCDD levels at the time of the reproductive event. However, with regard to early spontaneous abortion, this analysis would not be able to address those losses occurring early in gestation.

6. Biological plausibility. According to this criterion, the observed association between exposure and effect should be consistent with existing theory and information from other scientific disciplines. Certainly one would feel more secure in the causation debate if the biological basis for an observed association can be explained. However, biological *implausibility* may simply reflect gaps in existing scientific knowledge that could explain the relationship.

A growing body of animal research described elsewhere in this volume lends biological plausibility to the association between dioxin and most of the reproductive endpoints evaluated in these studies, with the notable exception of molar pregnancies. There is evidence that dioxin affects testis and accessory gland weight, testicular morphology, spermatogenesis and fertility in males. A model for a paternally mediated dioxin effect on congenital malformations has not been reported. Among female animals, the primary reproductive endpoints that have been examined include decreased fertility and pregnancy loss.

The mechanism by which TCDD causes reproductive and developmental effects has not been well described, although considerable insight has been gained from research focusing on the Ah receptor. While the Ah receptor has been linked with birth defects in several mouse strains, it appears that the mechanism of effect may be dependent on the outcome evaluated, as well as other dioxin congeners to which the population is exposed. Clearly, these relationships in humans have not been adequately investigated.

A variety of study designs, including case–control, ecologic, cross-sectional, and historical cohort designs, have addressed the issue of dioxin and reproductive effects. Unfortunately, the different criteria for case definition across studies make it difficult to compare the results. In addition, the method of case ascertainment for certain endpoints influences the rate of events observed. The VES substudies of veteran-reported birth defects compared with those identified through hospital records demonstrated that rates of self-reported outcomes differed by exposure status. Moreover, predictive value of self-reported events was poor in both cohorts. In contrast, rates of birth defects in the Ranch Hand study were similarly reported by the Ranch Hands and controls. Both groups underreported 7% of birth defects in children conceived prior to their SEA tour, and 14% occurring after their tour of duty.

Equally disappointing has been the discovery in the Times Beach, Missouri, CDC, and Ranch Hand studies that self-reported dioxin exposure and exposure indices developed from TCDD soil analyses and military records to estimate individual dioxin exposure are poorly correlated with serum TCDD levels. Thus, because of the likelihood of exposure misclassification in those studies lacking direct measures of exposure, the findings from studies of Vietnam veterans should be restricted to describing associations with military service in Vietnam and reproductive effects. Those studies conducted among Vietnamese populations, while subject to exposure misclassification as well, may offer a greater probability of both previous exposure as well as continued exposure through the contaminated food chain.

Specific Reproductive Endpoints

1. Spontaneous abortions. Miscarriages were investigated in several studies with different designs and varied patterns of parental exposure. Events were generally ascertained by self- or spousal-report. When case ascertainment was through medical records, such as in the Ranch Hand study or the Vietnamese investigations, the events are by definition restricted to those miscarriages that were clinically recognized.

Research in the area of early pregnancy loss indicates that 30–50% of all conceptions are lost prior to or during implantation.[63] The rate of loss between implantation and expected first menstrual period ranges from 22 to 30%.[64,65] Thus, it is clear that restriction of the examination of pregnancy loss to those events that are ascertained through self-reports or medical records results in a large proportion of the outcome of interest being missed. In studies of environmental factors and spontaneous abortion in which there is lack of information concerning the conditions surrounding conception, "the conflation of different doses with different effects can mislead."[66] Because of these discrepancies, it would not be meaningful to pool the results of the research on the association between dioxin exposure and miscarriage to judge the "consistency" of the association.

Overall, it must be acknowledged that the data compiled to date are inadequate to address this issue. To simply enumerate and compare the number of "positive" versus "negative" studies to ascertain consistency in the research would be inappropriate. The reasons for this have been described above in detail, with emphasis on high (40–50%) exposure misclassification, small sample sizes, lack of data on dioxin levels at the time of conception, and the unknown impact of early pregnancy loss on rates of subsequent adverse outcomes.

The animal and human evidence for a dioxin–pregnancy loss relationship is sufficiently suggestive to warrant further investigation. Several studies of various designs and populations have demonstrated weak but consistent associations,[7,25,30–33,50,58] whereas others have not.[9,10,22,24,27–29,49,52,56] (These studies include those which should be restricted to the assessment of military service in Vietnam and reproductive events.) The Ranch Hand study leaves several questions unanswered, including the determination of a dose–response level *at the time of conception* and adverse reproductive outcomes, as well as the impact of early pregnancy losses on rates of spontaneous abortion and birth defects that survive long enough to be "counted."

2. Congenital malformations. The confusing evidence regarding the relationship between dioxin exposure and birth defects results not only from the same limitations described above for the studies of miscarriage, but also from the lack of power to evaluate specific types of malformations. In order to increase the power to detect a potential relationship, the studies have combined all birth defects together and calculated an OR for total birth defects. Given emerging evidence for etiologic heterogeneity among the subgroups of birth defects,[67] it is possible that this approach might dilute the effect measure.

These studies should also be stratified by type of parental exposure, i.e., paternal, maternal, or both. Biologically plausible mechanisms for birth defects resulting from paternal exposure to toxic substances have not been well researched. It has been known for many years that temporary infertility may occur after exposure of human males to certain toxic substances. However, animal research suggests that spermatogenesis is "particularly resilient" after exposure to these chemicals.[68] If dioxin were related to malformations among the offspring conceived after paternal service in Vietnam, it is implied that the effect must occur premeiotically. Some animal studies have found that spermatogonia and spermatocytes (premeiotic spermatogenic cells) were able to repair DNA after exposure to toxic agents whereas spermatids and spermatozoa did not have this capability.[69]

Again, it must be acknowledged that the data compiled to address this issue are also inadequate. A few studies[13,20,30–33] have suggested an association, including those investigations of the Yusho and Yu-Cheng incidents.[5–7] Many studies have failed to find a relationship between dioxin and birth defects.[9,10,14,19,21,22,27–29,49,52,56] In view of the serious limitations imposed by these studies, it must be concluded that the relationship between paternal dioxin exposure and congenital malformations remains unknown. However, if *military service in Vietnam* is the exposure of interest, there is little evidence to support an association with birth defects.

3. "Miscellaneous" endpoints. Additional reproductive outcomes that were evaluated in a subset of the studies include molar pregnancies (in the Vietnamese studies), infant birth weight, neonatal and infant death, and child cancer and mortality. Mainly because of small sample sizes, it is difficult to reach conclusions regarding neonatal, infant, and child mortality and childhood cancers. However, the increased risk for neonatal death observed in the Ranch Hand study, the only study with individual TCDD levels, should be further investigated. Available evidence does not support an association between paternal dioxin level and low birth weight[44,56]; a maternally mediated effect with documentation of TCDD exposure has not been examined.

The VES found a significant relationship between service in Vietnam and sperm abnormalities, while the Ranch Hand study was not able to confirm these findings when exposure was defined both by cohort status and by serum TCDD levels. However, the data on alterations in male reproductive hormone levels associated with occupational exposure to TCDD[70] warrant further research in these areas.

In conclusion, the research to date has been successful in resolving some of the confusion surrounding the conflicting evidence for an association of dioxin exposure and various reproductive endpoints in humans. High occurrence of exposure misclassification, differences in case definitions across studies, and small sample sizes have severely limited the power of these studies to address these questions. Additional research that includes a measure of dioxin level at the time of conception for both mother and father is necessary if the effect of dioxins on the spectrum of reproductive outcomes is to be understood.

ACKNOWLEDGMENTS. The author gratefully acknowledges the valuable comments and suggestions on this chapter contributed by Drs. Brenda Eskanazi, Gary Kimmel, and Beatrice Selwyn.

REFERENCES

1. Commentary on 2,3,7,8-tetrachlorodibenzo-para-dioxin (TCDD), Prepared by the Scientific Review Committee of the American Academy of Clinical Toxicology, *Clin. Toxicol.* **23,** 191–204 (1985).
2. W. Rottluff, K. Teschke, C. Hertzman, *et al.,* Sources of dioxins and furans in British Columbia, *Can. J. Public Health* **81,** 94–100 (1990).
3. K. D. Courtney and J. A. Moore, Teratogenicity studies with 2,4,5-trichlorophenoxyacetic acids and 2,3,7,8-TCDD, *Toxicol. Appl. Pharmacol.* **20,** 396–403 (1971).
4. A. P. Poland, D. Smith, G. Metter, and P. Possick, A health survey of workers in a 2,4-D and 2,4,5,-T plant, *Arch. Environ. Health* **22,** 316–327 (1971).
5. W. J. Rogan, PCBs and cola-colored babies: Japan 1968 and Taiwan 1979, *Teratology* **26,** 259–261 (1982).
6. Y. Y. Yen, S. J. Lan, Y. C. Ko, *et al.,* Follow-up study of reproductive hazards of multiparous women consuming PCB-contaminated rice oil in Taiwan, *Bull. Environ. Contam. Toxicol.* **43,** 647–655 (1989).
7. W. J. Rogan, B. C. Gladen, K.-L. Hung, *et al.,* Congenital poisoning by polychlorinated biphenyls and their contaminants in Taiwan, *Science* **241,** 334–336 (1988).
8. M. Hatch, Reproductive effects of the dioxins, in *Public Health Risks of the Dioxins* (W. W. Lowrance, ed.), pp. 255–275, William Kaufmann, Los Altos, CA (1984).
9. J. C. Townsend, K. M. Bodner, P. F. D. VanPeenen, *et al.,* Survey of reproductive events of wives of employees exposed to chlorinated dioxins, *Am. J. Epidemiol.* **115,** 695–713 (1982).
10. A. H. Smith, D. O. Fisher, N. Pearce, and C. J. Chapman, Congenital defects and miscarriages among New Zealand 2,4,5-T sprayers, *Arch. Environ. Health* **37,** 197–200, (1982).
11. L. L. Needham, D. G. Patterson, and V. N. Houk, Levels of TCDD in selected human populations and their relevance to human risk assessment, in *Banbury Report 35. Biological Basis for Risk Assessment of Dioxins and Related Compounds* (Gallo, Scheuplein, and Heijden, eds.), pp. 229–257, Cold Spring Harbor Laboratory Press, Cold Spring Harbor, NY (1991).
12. R. R. Suskind and V. S. Hertzberg, Human health effects of 2,4,5-T and its toxic contaminants, *J. Am. Med. Assoc.* **251,** 2372–2380 (1984).
13. R. Balarajan and M. McDowell, Congenital malformations and agricultural workers (letter), *Lancet* **1,** 1112–1113 (1983).

14. J. Golding and T. Sladden, Congenital malformations and agricultural workers (letter), *Lancet* **1,** 1393 (1983).
15. K. B. Webb, The pilot Missouri health effect study, *Bull. Environ. Contam. Toxicol.* **33,** 662–672 (1984).
16. P. A. Stehr, G. Stein, H. Falk, *et al.*, A pilot epidemiologic study of possible health effects associated with 2,3,7,8-tetrachlorodibenzo-p-dioxin contaminations in Missouri, *Arch. Environ. Health* **41,** 16–22 (1986).
17. U.S. Environmental Protection Agency, Six years' spontaneous abortion rates in Oregon areas exposed to forest 2,4,5-T spray practices, EPA, Washington, DC (1979).
18. M. Hatch and J. Kline, Spontaneous abortion and exposure during pregnancy to the herbicide 2,4,5-T, U.S. Environmental Protection Agency, Washington, DC (1981).
19. C. J. Nelson, J. F. Holson, H. G. Green, and D. W. Gaylor, Retrospective study of the relationship between agricultural use of 2,4,5-T and cleft palate occurrence in Arkansas, *Teratology* **19,** 377–384 (1979).
20. B. Field and C. Kerr, Herbicide use and incidence of neural-tube defects, *Lancet* **1,** 1341–1342 (1979).
21. H. F. Thomas, 2,4,5-T use and congenital malformation rates in Hungary (letter), *Lancet* **2,** 214–215 (1980).
22. H. F. Thomas and A. Czeizel, Safe as 2,4,5-T? (letter), *Nature* **295,** 276 (1982).
23. J. A. Hanify, P. Metcalf, C. L. Nobbs, and K. J. Worsley, Aerial spraying of 2,4,5-T and human birth malformations: An epidemiological investigation, *Science* **212,** 349–351 (1981).
24. F. Pocchiari, V. Silvano, and A. Zampieri, Human health effects from accidental release of tetrachlorodibenzo-p-dioxin (TCDD) at Seveso, Italy, *Ann. N.Y. Acad. Sci.* **320,** 311–320 (1977).
25. G. Reggiani, Medical problems raised by the TCDD contamination in Seveso, Italy, *Arch. Toxicol.* **40,** 161–188 (1978).
26. M. L. Tenchini, C. Cimaudo, G. Puchetti, *et al.*, A comparative cytogenetic study on cases of induced abortions in TCDD-exposed and nonexposed women, *Environ. Mutagen.* **5,** 73–85 (1983).
27. R. T. Cutting, T. H. Phuoc, J. Ballo, *et al.*, Congenital malformations, hydatidiform moles, and stillbirths in the Republic of Vietnam 1960–1969, U.S. Government Printing Office, Washington, DC (1970).
28. P. Kunstadter, *A Study of Herbicides and Birth Defects in the Republic of Vietnam,* National Academy Press, Honolulu (1982).
29. Case–control study of congenital anomalies and Vietnam service. Report to the Minister for Veterans' Affairs, Australian Government Printing Service, Canberra (1983).
30. M. Hatch, in: *Herbicides and War: The Long-Term Ecological and Human Consequences* (A. H. Westing, ed.), Taylor & Francis, London (1984).
31. J. D. Constable and M. C. Hatch, Reproductive effects of herbicide exposure in Vietnam: Recent studies by the Vietnamese and others, *Teratogen. Carcinogen. Mutagen.* **5,** 231–250 (1985).
32. L. D. Huong, N. T. N. Phuong, T. T. Thuy, and N. T. K. Hoan, An estimate of the incidence of birth defects, hydatidiform mole and fetal death in utero between 1952 and 1985 at the obstetrical and gynecological hospital of Ho Chi Minh City, Republic of Vietnam, *Chemosphere* **18,** 805–810 (1989).
33. N. T. N. Phuong, T. T. Thuy, and P. K. Phuong, An estimate of differences among women giving birth to deformed babies and among those with hydatidiform mole seen at the OB-GYN hospital of Ho Chi Minh City in the south of Vietnam, *Chemosphere* **18,** 801–803 (1989).

34. C. Rappe, P.-A. Berqvist, M. Hansson, *et al.*, Chemistry and analysis of polychlorinated dioxins and dibenzofurans in biological samples, in: *Banbury Report 18. Biological Mechanisms of Dioxin Action* (A. Poland and R. D. Kimbrough, eds.), pp. 17–25, Cold Spring Harbor Laboratory Press, Cold Spring Harbor, NY (1984).
35. D. G. Patterson, J. S. Holler, C. R. Lapeza, *et al.*, High resolution gas chromatography/high resolution mass spectrometric analysis of human adipose tissue for 2,3,7,8-tetrachlorodibenzo-p-dioxin, *Anal. Chem.* **58,** 705–713 (1986).
36. A. Schecter, J. J. Ryan, and G. Gitlitz, Chlorinated dioxin and dibenzofuran levels in human adipose tissues from exposed and control populations, in: *Chlorinated Dioxins and Dibenzofurans in Perspective* (C. Rappe, J. Choudhary, and L. Keith, eds.), pp. 51–65, Lewis Publishers, Chelsea, MI (1986).
37. P. Mocarelli, L. Needham, A. Marocchi, *et al.*, Serum concentrations of 2,3,7,8-tetrachlorodibenzo-p-dioxin and test results from selected residents of Seveso, Italy, *J. Toxicol. Environ. Health* **32,** 357–366 (1991).
38. D. G. Patterson, R. E. Hoffman, L. Needham, *et al.*, 2,3,7,8-tetrachlorodibenzo-p-dioxin levels in adipose tissue of exposed and control persons in Missouri, *J. Am. Med. Assoc.* **256,** 2683–2686 (1986).
39. A. H. Smith, D. G. Patterson, M. L. Warner, *et al.*, Serum 2,3,7,8-tetrachlorodibenzo-p-dioxin levels of New Zealand pesticide applicators and their implications for cancer hypothesis, *J. Natl. Cancer Inst.* **84,** 104–108 (1992).
40. The Centers for Disease Control Veterans Health Studies. Serum 2,3,7,8-tetrachlorodibenzo-p-dioxin levels in US Army Vietnam-era veterans, *J. Am. Med. Assoc.* **260:**1249–1254 (1988).
41. P. C. Kahn, M. Gochfeld, M. Nygren, *et al.*, Dioxins and dibenzofurans in blood and adipose tissue of Agent Orange-exposed Vietnam veterans and matched controls, *J. Am. Med. Assoc.* **259,** 1661–1667 (1988).
42. A. Schecter, J. D. Constable, J. V. Bangerf, *et al.*, Elevated body burdens of 2,3,7,8-tetrachlorodibenzo-p-dioxin in adipose tissue of US Vietnam veterans, *Chemosphere* **18,** 431–438 (1989).
43. H. K. Kang, K. K. Watanabe, J. Breen, *et al.*, Dioxins and dibenzofurans in adipose tissue of US Vietnam veterans and controls, *Am. J. Public Health* **81,** 344–349 (1991).
44. W. H. Wolfe, J. E. Michalek, J. C. Miner, and A. J. Rahe, Air Force Health Study. An epidemiologic investigation of heath effects in Air Force personnel following exposure to herbicides: Reproductive outcomes, U.S. Air Force, Brooks Air Force Base, Texas (1992).
45. N. T. N. Phuong, B. S. Hung, D. Q. Vu, and A. Schecter, Dioxin levels in adipose tissue of hospitalized women living in the south of Vietnam 1984–85 with a brief review of their clinical histories, *Chemosphere* **19,** 933–936 (1989).
46. J. L. Pirkle, W. H. Wolfe, D. G. Patterson, *et al.*, Estimates of the half-life of 2,3,7,8-tetrachloridibenzo-p-dioxin in Vietnam veterans of Operation Ranch Hand. *J. Toxicol. Environ. Health* **27,** 165–171 (1989).
47. P. Mastroiacovo, A. Spagnolo, E. Marni, *et al.*, Birth defects in the Seveso area after TCDD contamination, *J. Am. Med. Assoc.* **259,** 1668–1672 (1988).
48. R. E. Hoffman, P. A. Stehr-Green, K. B. Webb, *et al.*, Health effects of long-term exposure to 2,3,7,8-tetrachlorodibenzo-p-dioxin, *J. Am. Med. Assoc.* **255,** 2031–2038 (1986).
49. J. W. Stockbauer, R. E. Hoffman, W. F. Schramm, and L. D. Edmonds, Reproductive outcomes of mothers with potential exposure to 2,3,7,8-tetrachlorodibenzo-p-dioxin, *Am. J. Epidemiol.* **128,** 410–419 (1988).
50. B. Forsberg and S. Nordstrom, Miscarriages around a herbicide manufacturing company in Sweden, *Ambio* **14,** 110–111 (1985).

51. M. C. Hatch and Z. A. Stein, Agent Orange and risks to reproduction: The limits of epidemiology, *Teratogen. Carcinogen. Mutagen.* **6,** 185–202 (1986).
52. J. D. Erickson, J. Mullinare, P. W. McClain, *et al.*, Vietnam veterans risks for fathering babies with birth defects, *J. Am. Med. Assoc.* **252,** 903–912 (1984).
53. G. D. Lathrop, W. H. Wolfe, R. A. Albanese, and P. M. Moynihan, Air Force Health Study (Project Ranch Hand II). An epidemiologic investigation of health effects in Air Force personnel following exposure to herbicides, USAF School of Aerospace Medicine, Brooks Air Force Base, Texas (1984).
54. W. H. Wolfe, J. E. Michalek, J. C. Miner, and M. R. Peterson, Serum 2,3,7,8-tetrachlorodibenzo-p-dioxin levels in Air Force Health Study personnel, *Morbid. Mortal. Weekly Rep.* **37,** 309–311 (1988).
55. L. A. Piacitelli, M. H. Sweeney, D. D. G. Patterson, *et al.*, Serum levels of 2,3,7,8-substituted PCDDs and PCDFs among workers exposed to 2,3,7,8-TCDD contaminated chemicals, *Chemosphere* **25,** 251–254 (1992).
56. The Centers for Disease Control Vietnam Experience Study. Health status of Vietnam veterans. Volume V: Reproductive outcomes and child health, CDC, Atlanta (1989).
57. The Centers for Disease Control Vietnam Experience Study. Health status of Vietnam veterans III. Reproductive outcomes and child health, *J. Am. Med. Assoc.* **259,** 2715–2719 (1988).
58. S. D. Stellman, J. M. Stellman, and J. F. Sommer, Health and reproductive outcomes among American Legionnaires in relation to combat and herbicide exposure in Vietnam, *Environ. Res.* **2,** 150–174 (1988).
59. A. Aschengrau and R. R. Monson, Paternal military service in Vietnam and risk of spontaneous abortion, *J. Occup. Med.* **7,** 618–623 (1989).
60. A. Aschengrau and R. R. Monson, Paternal military service in Vietnam and risk of late adverse pregnancy outcomes, *Am. J. Public Health* **10,** 1218–1224 (1990).
61. The Centers for Disease Control Vietnam Experience Study. Health status of Vietnam veterans II. Physical health, *J. Am. Med. Assoc.* **259,** 2708–2714 (1988).
62. R. W. Miller and W. J. Blot, Small head size after in utero exposure to atomic radiation, *Lancet* **2,** 784–787 (1972).
63. A. T. Hertig, J. Rock, and E. C. Adams, Thirty-four fertilized human ova, good, bad, and indifferent, recovered from 210 women of known fertility: A study of biologic wastage in early human pregnancy, *Pediatrics* **23,** 202–211 (1959).
64. A. J. Wilcox, C. R. Weinberg, J. F. O'Connor, *et al.*, Incidence of early loss of pregnancy, *N. Engl. J. Med.* **319,** 189–194.
65. A. M. Sweeney, M. R. Meyer, J. H. Aarons, *et al.*, Evaluation of methods for the prospective identification of early fetal losses in environmental epidemiology, *Am. J. Epidemiol.* **127,** 843–850 (1988).
66. J. Kline, Z. Stein, and M. Susser, Criteria for teratogenesis: Specificity, in: *Conception to Birth: Epidemiology of Prenatal Development*, pp. 18–30, Oxford University Press, London (1989).
67. M. Khoury, Epidemiology of birth defects, *Epidemiol. Rev.* **11,** 244–248 (1989).
68. J. H. Pearn, Teratogens and the male, *Med. J. Aust.* **2,** 16–20 (1983).
69. I. P. Lee and R. L. Dixon, Factors influencing reproducion and genetic toxic effects on male gonads, *Environ. Health Perspective* **24,** 117–127 (1978).

Chapter 18

Chemical, Environmental, and Health Aspects of the Seveso, Italy, Accident

Pier Alberto Bertazzi and Alessandro di Domenico

1. INTRODUCTION

Today *dioxin* is a familiar word, but very few people in Italy had heard of it before the Seveso accident of July 10, 1976. Ever since, TCDD and related chemicals[1,2] have been dealt with profusely by the domestic and international scientific communities. Because of the sensitizing effect it had, the Seveso case should be considered a turning point in environmental risk management. For instance, it is recalled here that, in February of 1988, the Commissione Consultiva Tossicologica Nazionale (CCTN)—Italy's National Toxicology Commission—established a set of maximum tolerable limits (MTLs) for polychlorinated dibenzodioxins (PCDDs) and dibenzofurans (PCDFs) in several environmental matrices.[3,4] These limits were based on the 1987 array of U.S. EPA toxicity equivalence factors (TEFs) to convert PCDD and PCDF analytical data into "toxicity equivalents of TCDD" (TE units),[1] and also on much of the technical experience gained by dealing with the Seveso event.

2. THE ACCIDENT OF JULY 10, 1976

2,4,5-Trichlorophenol (TCP) production at the Givaudan–Hoffmann–LaRoche ICMESA plant at Meda (Milan, Italy) was started in 1969–1970 and brought up to full-

Pier Alberto Bertazzi • Institute of Occupational Health, Epidemiology Section, University of Milan, 20122 Milan, Italy. Alessandro di Domenico • Laboratory of Comparative Toxicology and Ecotoxicology, Italian National Institute of Health, 00161 Rome, Italy.
Dioxins and Health, edited by Arnold Schecter. Plenum Press, New York, 1994.

scale levels in the years that followed with a big production increase ($>10^5$ kg/year) over the 1974–June 1976 period. TCP was obtained by a discontinuous process based on hydrolyzing 1,2,4,5-tetrachlorobenzene to sodium trichlorophenate with sodium hydroxide in the presence of xylene and ethylene glycol, and transforming the trichlorophenate to TCP by acidifying it with hydrochloric acid. The hydrolysis reaction was carried out inside ICMESA's Department B, by utilizing a 10-m^3 stainless steel reactor, in 7000-kg batches of chemicals.[5–8]

At 12:37 p.m. on July 10, 1976, an exothermic reaction raised the temperature and pressure inside the reactor beyond limits, thereby causing a safety device, consisting of a rupture disk set at ~3.8 atm (380 kPa), to blow out.[5–9] TCDD production was also increased to an unknown extent. The safety device was mounted on an exhaust pipe that was directly connected to the reactor cover and, passing through the roof of Department B, ended up in the open. When the rupture disk collapsed, the overheated fluid mixture burst through the pipe out into the open air propelled by the thrust of the built-up pressure. The chemical cloud that left the reactor entrained nearly 2900 kg of organic matter, including at least 600 kg of sodium trichlorophenate and an amount of TCDD which is still being evaluated. The visible part of the cloud rose up to about 50 m; it subsequently subsided and fell back to earth, but was wind-driven over a wide area of territory (Fig. 1).[5,10–13]

Emission gradually decreased until it ceased altogether. Within less than 2 hr of the accident, chemicals settled on the ground as far as 6 km south of ICMESA, or were dispersed by wind streams.[6,13,14] Serious environmental contamination followed: the leaves of plants near ICMESA, courtyard animals, and birds were seriously affected, many dying within a few days of the accident. At the same time, dermal lesions among humans who had been exposed to the toxic alkaline cloud began to appear. About 10 days after the accident, TCDD was found in various types of samples collected near the plant.

Toward the end of the runaway reaction, reactor temperatures are thought to have increased well above 300°C, thus causing extensive mineralization of residual organic substances.[5,7,15,16] After blowout, ~2300 kg of residual chemicals was present in the reactor; their approximate composition was sodium chloride (72%), decomposed organic matter, and some 250–300 g TCDD.[7,8,17] TCDD amount was later reestimated at approximately 600 g.

3. RISK MANAGEMENT MEASURES AND ASSESSMENT

3.1. Definition of Areas at Different Risk Levels

As a first step, the information available on the location of toxic and pathological events—regarding vegetation, animals, and humans[18]—and on the airstream pattern at blowout time was used to draw an approximate diagram of the contaminated area. This was further confirmed by chemical monitoring of TCDD in the soil carried out under emergency conditions.[6,13,19–21] Within 5 weeks of the accident, the area hit was

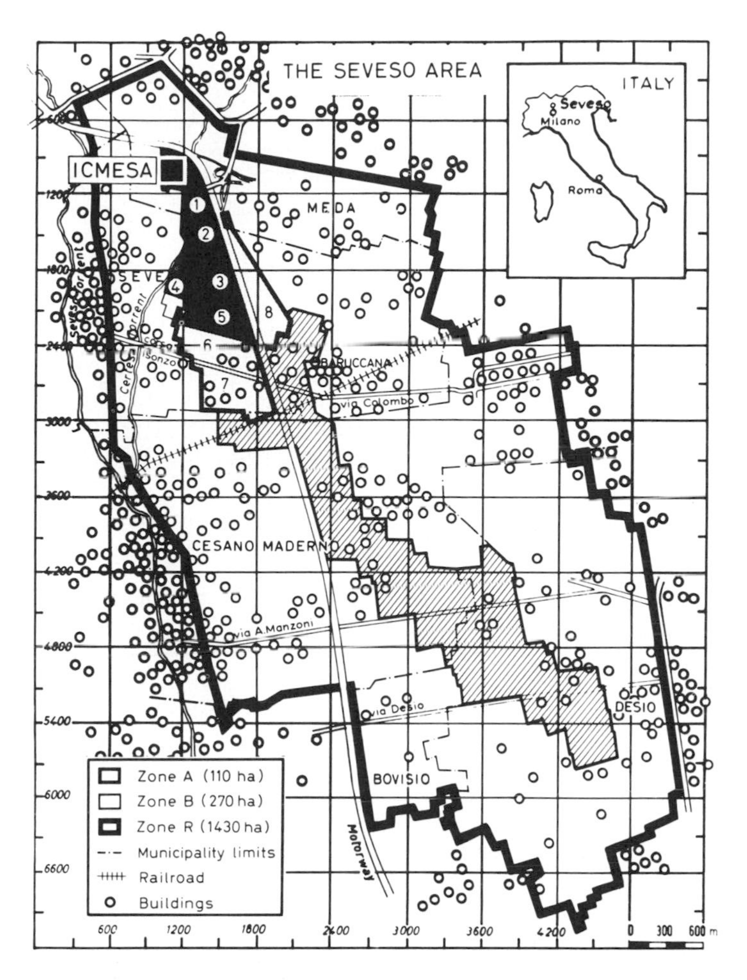

Figure 1. The contaminated area at its maximum extension. Zone A fanned out from ICMESA to the south and contained most of the TCDD that had escaped. Zone B was the natural extension of Zone A along the main diffusion pathway of the TCDD-containing cloud. Both Zones A and B were enclosed by a larger territory coded as Zone R. The grid unit cell was a 50-m square for Zone A and a 150-m square for Zones B and R. Numerals identify Subzones A_1–A_8. (Adapted from di Domenico *et al.*[13])

subdivided into Zones A, B, and R, in descending TCDD and toxicological risk levels (Fig. 1).

The borderline between Zone A and Zone B was set at average TCDD concentrations in the soil of ≤ 50 μg/m²; the boundaries of Zones A and B in Zone R were fixed where average contamination was ≤ 5 μg/m². Zone R included the remaining territory where detectable levels of TCDD (formally, ≥ 0.75 μg/m²) were found. In all cases, borderlines were established following preexisting natural or artificial divisions, in general compliance with the contamination pattern made out.[5,6,8,13,19,22] Since TCDD concentration detected in Zone A ranged over more than four orders of magnitude

Table 1
Summary of Some Analytical Features of Zones A, B, and R, and Breakdown of Zone A into Subzones A_1–A_8[a]

	TCDD levels (μg/m²) according to different mappings[b]				
		January 1977[c]			
Zone A subzone	September 1976 (means)	Means	Ranges	March 1978[c] (means)	1979–1980 (means)
A_1	1.6 E+3[d]	4.5 E+2	<0.75–5.5 E+3	6.2 E+2	9.9 E+2
A_2	2.5 E+3	4.3 E+2	6.1–1.7 E+3	4.0 E+2	1.2 E+3
A_3	1.3 E+2	3.5 E+2	<0.75–2.0 E+3	1.6 E+2	1.9 E+2
A_4	2.6 E+2	1.7 E+2	<0.75–9.0 E+2		1.1 E+2
A_5	1.2 E+2	6.3 E+1	<0.75–4.3 E+2		7.4 E+1
A_6	9.1 E+1	3.1 E+1	<0.75–2.7 E+2		
A_7	1.2 E+1	1.7 E+1	<0.75–9.2 E+1		
A_8[e]	<5.0				
Zone B	1976–1977 (mean)		1978 (mean)	1980–1981 (mean)	1983 (mean)
	3.4		4.3	3.2	2.8
Zone R	~0.5[f]				

[a]From di Domenico *et al.*,[43] La Porta *et al.*,[30] and Reggiani.[8]
[b]The configuration of Subzones A_1–A_7 changed with time and the data reported provide only broad indications of area-specific contamination levels. Mappings identified according to di Domenico *et al.*[13] and La Porta *et al.*[30]
[c]These mappings are now considered to be biased. Yet they are very important as examples of early Zone A investigations based on a 50-m-square grid and because of the many soil site checks carried out.
[d]1.6 E+3 = 1600.
[e]Because of the low TCDD concentrations, Subzone A_8 was soon reclassified as part of Zone B.
[f]Fading into background with time elapsing and agronomical works.

(from <0.75 μg/m^2 to >20 mg/m^2),[8,13,23] the zone was broken down into Subzones A_1–A_8, each characterized by a somewhat lower range of TCDD levels (Table 1).

On July 26, 1976, approximately 200 people were evacuated from a 15-ha area immediately southeast of ICMESA[5,6,8,13,22,24–26]—the first part of what shortly afterward would be defined as Zone A. Following further analytical findings concerning the TCDD contents of soil and vegetation samples, a few days later the entire Zone A (over 730 inhabitants altogether) was evacuated. Zones B and R were subjected to area-specific hygiene regulations including prohibition of farming, consuming local agricultural products, and keeping poultry and other animals (Table 2). Indeed, many major risk management measures and policies were already established during the emergency period (July 10–August 15, 1976).

3.2. TCDD Contamination Maps

Beginning with the emergency period, soil monitoring was performed repeatedly during the first decade following the Seveso incident (1976–1986) for at least three

Table 2
Measures Designed to Limit Human Exposure in the Seveso Area[a]

- Measures affecting people's behavior
  - Evacuation of Zone A
    - To avoid any contact with highly contaminated soil, some 160 houses in Zone A were evacuated
    - Hunting prohibited for approximately 8 years
  - Precautions for Zone B residents
    - Intensification of personal hygiene
    - No animal breeding or vegetable planting
    - Daily relocation of children up to 12 years old and pregnant women
    - Abstention from procreation
    - Minimization of air dust level (vehicle speed limit, 30 km/hr)
    - Careful emptying of vacuum cleaners
    - Hunting prohibited for approximately 8 years
  - Precautions for Zone R residents
    - Intensification of personal hygiene
    - No animal breeding or vegetable planting
    - Pets to be fed with food from areas other than Zones A, B, and R
    - Hunting prohibited for approximately 8 years
- Measures acting on the environment
  - Measures common to all zones
    - All locally bred animals (mostly chickens and rabbits) slaughtered and their carcasses transferred to Zone A
    - Honey collected and disposed of in Zone A
  - Defoliation and agronomic activities
    - Defoliation aimed at removing TCDD deposited on grass and leaves; before cutting, green parts sprayed with vinyl acetate glue to fix TCDD
    - Ploughing performed repeatedly to dilute TCDD in topsoil layer and facilitate its dissipation
  - Reclamation of Subzones A_6 and A_7
    - Cleanup carried out by four teams, each working on a 4-hr/day shift; workers wore individual protective equipment
  - Schools
    - Thorough interior cleanup and floor washing daily
    - Monthly analytical checking on indoor contamination status for over a year after the accident. Procedure discontinued after TCDD levels consistenly found $<0.01\ \mu g/m^2$ (tolerable limit)

[a] Adapted from Fortunati and La Porta.[26]

reasons: for reassessment of TCDD distribution patterns and levels with time, reassessment and updating of risk estimates and risk management measures, and as a backup tool to determine the effectiveness of remedial actions and reclaiming operations. Thousands of soil samples were collected and assayed according to criteria, techniques, and procedures often developed ad hoc to meet the requirements of a unique and unexpected case for which no reference to a former experience in Italy could be made. In general, the analytical tools and setups grew more sophisticated and reliable as time went by[13,14,23,24,27]; however, studies have proven that the sets of soil data up to 1979 may have seriously underestimated TCDD levels.[23,28,29]

Aside from the Zone A map of September, 1976, for which sampling was carried out on a polar coordinate reference system, all other mappings were obtained by establishing sampling sites on a north–south grid.[5,13,20,30,31] Generally, sampling frequency provided one datum per cell, although cases of higher frequency are often encountered. The usual way of building and representing a contamination map was to asssociate the magnitude of the analytical outcomes to graphic symbols (for instance, a larger dot for a higher TCDD level) classified by discrete ranges, and thereby make a graphic report by using the pertinent reference system. However, data sets were also constructed to exhibit TCDD levels as bidimensional isoconcentration curves or histograms (Figs. 2 and 3),[19,32–35] or tridimensional histograms or curves.[28,31,34,36] In

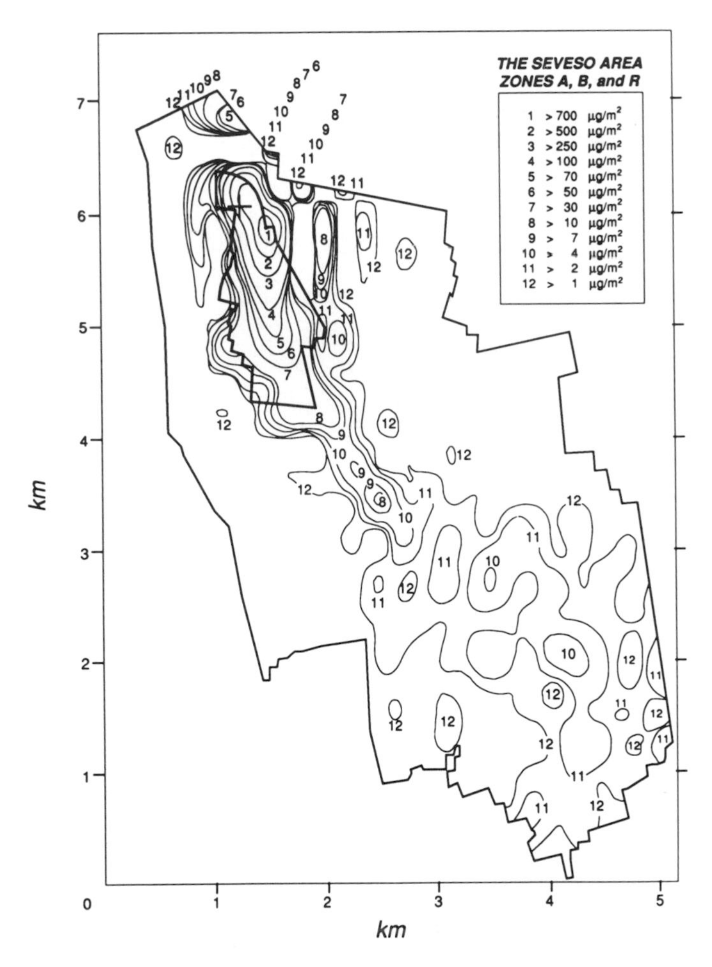

Figure 2. TCDD isoconcentration curves drawn on the basis of approximately 1100 soil measurements carried out in 1976–1977; large areas void of curves are associated with lack of data. Zone A profile is visible within the larger Zone R boundaries. The picture visualizes to some extent the ups and downs of TCDD distribution, greater in Zone A than elsewhere. (Adapted from Belli *et al.*[32])

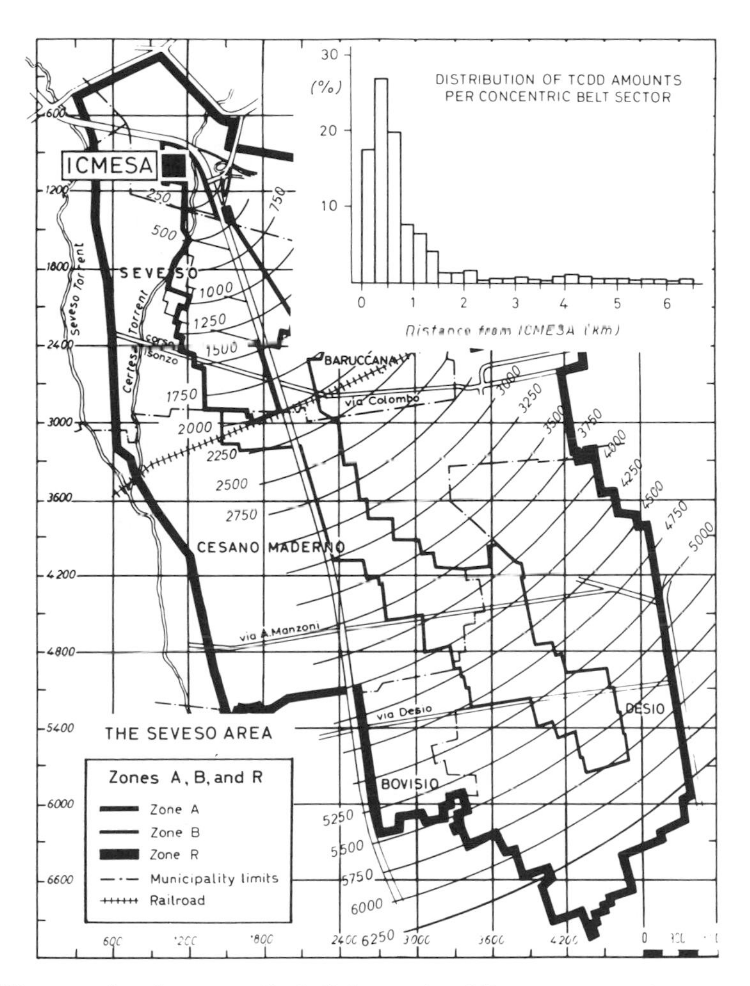

Figure 3. The contaminated area was sliced all the way into 250-m concentric belt sectors centered on ICMESA. TCDD measurements were summed up to yield sector-specific total amounts whose normalized (100%) sum provided the distribution graph in the upper right corner. The histogram shows the exponentiallike decrease of TCDD levels in soil with increasing distance from source.

particular, Belli *et al.*[37,38] have recently provided a preliminary tridimensional representation by means of fractal models, currently under development.

As mentioned, contamination maps served different purposes. By integrating concentration levels in soil, 0.1- to 3-kg estimates of total TCDD present within Zone R boundaries were obtained[8,13,19,21,33]; however, values < 0.5 kg were prefered. It should be emphasized that, because of the biases on soil data sets as previously stated, all such figures may well be underestimated by one to two orders of magnitude. Lastly, of the total TCDD amount analytically estimated, 40 to 50% was found to be present within 0.5 km from ICMESA, $> 85\%$ occurred within Zone A boundaries, and $> 99\%$ was recovered within Zone A and Zone B boundaries.

3.3. Early Assessment and Management of Risk

The first risk assessment for the Seveso area was carried out in order to establish a set of tolerable environmental limits assuming that TCDD intake from ingestion of contaminated soil was the most important factor of hazard. At this stage, several toxicological features of TCDD were already known[8]; however, the risk assessment and management were substantially based on the following issues[39]:

- A no-observed-effect level (NOEL) from subchronic and carcinogenicity studies in rats of 1 ng/day per kg bw[40,41]
- A safety factor of 1000
- Of all inhabitants, children most liable to be at risk
- A tolerable daily intake (TDI) estimated at 1 pg/kg bw

Since reclamation works in farmable soil were aimed at restoring average TCDD concentrations to < 0.75 μg/m^2, it was estimated that the adopted TDI would allow a 20-kg child a daily ingestion of up to ~3 g (!) of soil. In the absence of sound data, the bioavailability of soilborne TCDD was conservatively[42] set at 100%.

In light of the above, tolerable limits for land (soil), housing interiors, equipment, and other matrices and items were set by regional law. Definition of risk areas—namely, Zones A, B, and R—was carried out by taking into account the TCDD levels detected predominantly in the 7-cm topsoil layer, and therefore extensive use of the "surface density" unit μg/m^2 was made, *the unit surface being defined as a 1-m square with a 7-cm thickness.* TCDD surface densities were converted to the commoner ng/kg concentration units multiplying by an average factor of 8. The following was obtained for the different risk areas and matrices[5,13,22,39,43]:

Farmable land	<0.75 μg/m^2	>6 ng/kg
Not farmable land	≤ 5 μg/m^2	≤ 40 ng/kg
Limit for evacuation	>50 μg/m^2	>400 ng/kg
Building outdoor surfaces	≤ 0.75 μg/m^2	
Building indoor surfaces	≤ 0.01 μg/m^2	

In 1988 these limit values were taken as a reference to set the current maximum tolerable limits. The early risk assessment and management described was focused primarily on defining the area(s) to be evacuated and those subject to hygiene and sanitary restrictions.

3.4. Assessment of Exposure and Risk for Residents of Zones B and R

In Zone B—a typical agricultural–industrial setting with a number of houses that had small private gardens—the original contamination levels were between 5 and 50 μg/m^2. Since this area was next to the evacuated area and along the main line of contamination, risk assessment for the nearly 5000 people of the zone was regarded as a critical issue by health authorities and local communities. This was partly a result of risk management policies adopted during the emergency period and based on an analytical layout of TCDD in soil which was incomplete or not fully reliable: indeed, several soil monitorings taken up to 1985 confirmed that the TCDD contamination pattern was very uneven[39] (e.g., see Fig. 2). Because of this, setting the Zone A–Zone B borderline must be considered as somewhat arbitrary.

Therefore, in 1984 a new risk assessment was carried out for Zone B inhabitants (Table 3, upper secton). Compared with that of the emergency period, the new assessment made use of better, more reliable, analytical information and, in particular, of recent data on cancer response in rodents.[41,44–47] A TCDD lifetime daily intake of

Table 3

Examples of Risk and Exposure Assessments Carried out in 1984 for Families of Zone B Having Vegetable Gardens and Courtyard Animals[a]

Risk assessment (per million people)[b]		
Exposure route	TCDD intake[c]	Extra cancer risk
Soil,[d] ingestion[e]	4.6 E−3[f]	1.6 E−1
dermal contact	1.7 E−3	6.1 E−2
inhalation[g]	6.0 E−4	2.1 E−2
Drinkable water[h]	≤2.9 E−2	≤1.0
Vegetables	5.7	2.0 E+2
Meat and dairy products	2.4	8.6 E+1
Fish consumption[i]	~0	~0
Totals	8.1	2.9 E+2

Exposure assessment (TCDD intake)[c,j]					
Exposure group[k]	Vegetables	Courtyard animals	Other animals	Total[l]	Total[m]
Group 1				3.5 E−2	~0.007
Group 2	3.5			3.5	~0.7
Group 3	3.5	6.7		1.0 E+1	2.0
Group 4	3.5	6.7	2.3 E+1	3.3 E+1	6.7

[a]Model and assumptions: mean TCDD level in most Zone B soil, 1.5 μg/m² (12 ng/kg); average individual weight, 70 kg; length of lifetime exposure, 70 years.
[b]From Fortunati.[39]
[c]In pg/day per kg bw.
[d]Intake and risk figure breakdown obtained by the authors on the basis of original references[39,48] and the later work by di Domenico.[4] Zone B inhabitants not consuming local food were expected to have only a TCDD intake as per Group 1 below.
[e]The only risk evaluated in 1976.
[f]4.6 E−3 = 0.0046.
[g]Dust level in air, ≤0.14 mg/m³ (see text).
[h]TCDD level in drinkable water, ≤1 pg/liter (see text).
[i]In Zone B, no surface waters to collect fish.
[j]From Pocchiari *et al.*[25] TCDD intakes from drinkable water, vapors, and local cow and human milk considered negligible or not evaluated for lack of data.
[k]Group 1, people not consuming local food; Group 2, people consuming local vegetables; Group 3, people consuming local vegetables and meat from small courtyard animals; Group 4, people consuming local vegetables and animal food. For all groups, background intake from contaminated soil and dust as per Group 1.
[l]Dissipation of TCDD from soil not considered.
[m]Dissipation of TCDD considered (over a 70-year exposure).

0.28 pg/kg bw was estimated to determine an extra cancer risk of 1×10^{-5} and a simplified linear relationship was assumed to exist between cancer response and low TCDD exposures. According to the model employed, most (80%) of Zone B was characterized by a mean TCDD level of 1.5 $\mu g/m^2$ (12 ng/kg), whereas the remaining 20% portion had a mean contamination level that was 10 times as high; the following exposure routes were considered[25,39,48]:

- Ingestion of soil
- Absorption from dermal contact with soil
- Inhalation of contaminated dust
- Contribution of drinkable water
- Ingestion of vegetables grown in home gardens
- Consumption of animal products (primarily chickens and rabbits) from the area

As the resident population was under 5000 people who were only partly exposed to ingesting contaminated food, the conclusion was reached that the conservatively estimated extra cancer incidence of $\sim 3 \times 10^{-4}$ (Table 3, upper section) could not be observed or that it would not be statistically significant. However, owing to the relatively high value of the risk estimate, a more detailed analysis of various parameters (e.g., the rate of TCDD transmigration from soil to vegetation and vegetables, residents' eating habits, including egg and milk consumption not considered in former estimates) was deemed to be advisable, especially where lack of experimental data had required large extrapolations. Part of the analysis required experimental work such as the use of controlled cultivated plots and animal farms in Zone B.[39] In view of the incremental risk magnitude, Lombardy's Regional Government also made arrangements for operations to dilute and possibly replace the contaminated topsoil layer in vegetable gardens and agricultural areas in order to further reduce the risk from consuming vegetables and farm animals from contaminated areas. At the same time, the agricultural procedures, such as plowing, were expected to facilitate TCDD dissipation from topsoil.

In 1984 a specific study was carried out to estimate Zone B population exposure to TCDD, should hygienic restrictions be removed[25]; the study provided a general backup for the risk assessment described above. The following points were used for assessment:

- Average TCDD level in soil was assumed to be 12 ng/kg.
- The entire population (<5000 people) was broken down into four groups with different exposure potentials.
- Information on local dietary habits and food production was based on an ad hoc survey.
- Food consumption estimates were corrected for fraction of food consumed by people not living in Zone B.
- TCDD levels in different food items estimated from analytical data or from soil–vegetation, soil–animal, and soil–vegetation–animal translocation factors were derived from experimental data.
- TCDD half-life in soil was estimated to be 10 years.

The lower section of Table 3 summarizes the outcomes obtained; figures based on TCDD constant levels in soil appear to be fivefold larger than those taking into account TCDD dissipation. In their work, the authors acknowledged the lack of sound data to describe TCDD distribution in Zone B soil and the relationship between the chemical levels in soil and those in plants and animals, so that a "reasonably conservative" approach in estimating risk had been adopted in order to protect human health. On the whole, the estimates indicated that, as potential TCDD intakes might reach values that could not be disregarded, comprehensive analytical investigation on TCDD levels in vegetable and animal foods from Zone B was considered to be essential before removing the ban on farming and consuming local products. In retrospect, it is interesting to compare the results of Table 3, lower section, with the TDI for TCDD presently adopted in Italy and, for instance, recommended at a consultation of the World Health Organisation (TDI = 10 pg/day per kg bw) which is considerably higher than current U.S. EPA guidelines.[3,4,49]

Assessments of exposure and risk for the inhabitants of Zones B and R were carried out by di Domenico and Zapponi.[50] The study was based on the environmental analytical data available and assumed that TCDD intake through food produced locally was zero because of the existing hygienic restrictions; the exposure period considered was from 6 months after the accident onward. Under such conditions, the extra lifetime cancer risk for the cohort at the highest exposure did not exceed $\sim 1 \times 10^{-5}$.

4. REHABILITATION OPERATIONS

After the emergency period, Lombardy's Regional Government issued a special law and set up the Special Office for Seveso to handle sanitary, social, economic, and organizational problems.[22,39,51] By appointing five task groups acting within the Special Office, five rehabilitative programs were implemented and completed. They covered:

- Contamination control and reclamation operations
- Sanitary control for the population and veterinary assistance
- Social assistance (including schools and replacement of dwellings for the evacuated population)
- Reconstruction of homes and public buildings when not reclaimable
- Economic aid to productive groups and facilities (e.g., artisans, industries) damaged by the incident

The five programs were aimed at normalizing life in the contaminated areas on the basis of benefit-to-cost evaluations taking into account the technical, economical, and social aspects as well as public opinion. As an example, in the following an outline is provided of the procedures adopted to reclaim Subzones A_6 and A_7—characterized by relatively low levels of contamination as compared wtih other Zone A subzones (Table 1)—and Zones B and R.

Subzones A_6 and A_7 had much social, economical, and political importance in

Table 4

Outline of Cleanup Methods Employed for Rehabilitation of Subzones A_6 and A_7[a]

Building interiors

Items not reclaimed

Textiles (curtains, carpets, clothing, etc.); furniture with textile upholstery; wooden floors; food products; household appliances; low-valued miscellaneous indoor material. All this material was classified, listed, removed (starting from attics down to living quarters and basements), and transferred as such to Subzone A_5 acting as a temporary storage place.

Items reclaimed

Building interiors were vacuum-cleaned and washed thoroughly with water and detergent (wallpaper was removed). Cleansing liquids were sucked, collected in steel containers, and transferred to Subzone A_5. Walls were repainted and floors, if possible, refinished.

Furniture was vacuum-cleaned and washed with solvent-impregnated cloths (later disposed of in steel containers to be transferred to Subzone A_5). Reclaiming started from higher floors and, in flats, from rooms farthest from entrance. Entrance halls were left until the end. Furniture was refinished.

External surfaces

External surfaces were washed with pressurized (60 bar) water. Vegetation was cut, ground, and transferred to Subzone A_5. Topsoil was manually or mechanically removed to a depth of 30 cm or more; scarification was halted when residual TCDD levels at excavation bttom matched the pertinent reference limit. Trucks used in reclaiming were not allowed to run over reclaimed areas and had special routes. The soil removed was carefully loaded on trucks to avoid spilling during loading and transfers. Operations in reclaimed areas were carried out while keeping dust level in air as low as possible by spraying water.

[a]From La Porta and Fortumati.[51]

that they had accommodated about 67% of the evacuated population.[5,8,22,43,51] Together they covered some 32 ha; their minimum distance from ICMESA was 1200 m. TCDD levels in soil had been found up to 270 $\mu g/m^2$; however, estimated mean values were 31 $\mu g/m^2$ in Subzone A_6 and 17 $\mu g/m^2$ in Subzone A_7 (Table 1). The methodology adopted for reclaiming involved mechanical removal of soil with TCDD from surfaces to reach concentrations within the tolerable limits: for instance, removal of the top 25 cm of soil also removed an average of $> 90\%$ of total TCDD present.[52,53] As summarized in Table 4, several cleanup methods were employed.

During reclamation operations, building interiors and exteriors were repeatedly checked by means of wipe and/or scrape testing; topsoil cores were taken in gardens.[13,24,27,43] Approximately 700 assays were carried out on the 87 buildings and surrounding gardens. At the end of reclamation operations, all TCDD levels were below the established tolerable limits.[30,51] The same holds for agricultural and farming areas which were checked by sampling topsoil cores in 56 sites selected on the reference grid. Buildings, gardens, and agricultural and farming areas were finally reconditioned and analytically rechecked at random. When these checks gave favorable outcomes, cleanup operations were considered teminated; at that point, health authorities granted clearance for reentry of evacuated people. The very uneven distribution of TCDD in the soil and on internal and external building surfaces (from <0.01 up to some $\mu g/m^2$) led to the adoption of a specific statistical approach when assessing the effectiveness of reclamation operations.[43]

Rehabilitation of Zones B and R was started in 1977.[5,22] It eventually entailed soil removal in public and private gardens, agricultural treatments of topsoil layers to dilute surface TCDD and help its dissipation, and covering with fresh uncontaminated soil. Indeed, simply by ploughing (and mixing), the TCDD levels in the upper 7-cm soil layer could drop by as much as a factor of 3 to 6.[52,54] The effectiveness of agricultural treatments—e.g., ploughing, harrowing, sowing—was increased by repeating treatments during the same year and over subsequent years.[5,8,22] The rapid effect of dilution was accompanied by the slower dissipation processes which were facilitated by bringing TCDD from deeper to surface topsoil layers. Simple, rather inexpensive, and efficacious ploughing alone was largely employed to rehabilitate areas of agricultural and farming interest.[54]

5. ENVIRONMENTAL IMPACT OF THE ACCIDENT

5.1. TCDD Vertical Distribution in Soil

Many in-field investigations were carried out to study the vertical gradient of TCDD in soil and its permeation capacity under the action of meteoric water.[19,52] In fact, one major concern was that the chemical, by permeating the soil, might endanger the local groundwater system. As indicated, TCDD vertical distribution and mobility also had relevant implications on soil reclamation processes and, in a broad sense, on risk assessment and management. Most investigations were performed in Zone A, as the higher TCDD levels facilitated measurements at greater depths and, by intensifying the eventual permeation process, were expected to make it more liable to be detected. Statistical analysis and mathematical modeling were used by different authors to describe the distribution of TCDD in Seveso soil as a function of depth and time.[29,52,55]

In general, TCDD levels dropped steeply between the depths of 4 to 12 cm, and at a much lesser rate from a depth of ~8 cm downward. Until the fall of 1976, most (>90%) TCDD was generally contained in the first 7–8 cm of topsoil, which is why it became customary to monitor soil by coring an ~7-cm topsoil layer. In 1977, > 75% of TCDD was still on average contained in the top 12-cm layer of soil. At Sites A11 and A12 (Table 5)—as in other cases—the vertical gradient in soil was studied at different times. By comparing the available data in the upper six 4-cm-thick soil layers, it may be seen that distribution patterns obtained from investigations carried out on October 15, 1976, and on April 1, 1977, exhibited significant differences in that some TCDD appeared to have moved from the upper layers downward; no further significant changes were observed in the next 8 months. It may be pointed out that the period of greater downward movement was also characterized by a remarkable amount of local atmospheric precipitation. Out of all of the data available, Sites A11 and A12 of Zone A have been used here as a paradigm for the entire set of data; however, occasional significant deviations were observed from site to site even when close to each other.

Table 5 also summarizes the results of an investigation in five Zone B sites; they appear to match the general distribution pattern described above.[52]

Table 5
Vertical Distribution of TCDD in Soil at Sites A11 and A12 in Zone A Investigated in 1976 and 1977, and at Sites B1–B5 in Zone B Sampled on November 11, 1976[a]

Depth (cm)	TCDD level (μg/m²)[b]					
	A11[c]	A11[d]	A11[e]	A12[c]	A12[d]	A12[e]
0–4	1.2 E+3[f]	3.1 E+1	1.6 E+1	1.4 E+3	9.2 E+1	3.4 E+2
4–8	2.0 E+2	3.0	9.1 E−1	1.4 E+2	5.6 E+1	1.5 E+2
8–12	3.3 E+1	1.1	1.2	2.5 E+1	3.7 E+1	5.1 E+1
12–16	2.1 E+1	2.3 E−1	4.7 E−1	<2.0 E−2	4.2 E+1	5.7 E+1
16–20	1.6 E+1	1.4	1.9 E−1	6.4	4.1	4.2 E+1
20–24	1.7 E+1	1.9 E−1	2.5 E−1	3.6	5.7	2.6 E+1
24–28	2.0	2.5	9.0 E−2	<2.0 E−2	3.0	9.7
28–32	<2.0 E−2	8.8 E−1	1.3 E−1	2.5	2.5	7.4
32–36	<2.0 E−2	1.0 E+1	4.0 E−2	6.0 E−1	2.6	1.1
36–40	<2.0 E−2		5.0 E−2	1.1	2.5	7.3 E−1
40–44	<2.0 E−2			<2.0 E−2		
	B1	B2	B3	B4	B5	
0–4	5.3	6.2	9.7	6.4	1.2 E+1	
4–8	1.0 E−1	<5.0 E−2	5.0 E−1	1.0 E−1	3.0 E−1	
8–12	<5.0 E−2	<5.0 E−2	<5.0 E−2	3.0 E−1	4.0 E−1	
12–16	<5.0 E−2	<5.0 E−2	2.0	<5.0 E−2	1.0 E−1	
16–20	2.0 E−1	<5.0 E−2	<5.0 E−2	<2.0 E−2	<5.0 E−2	

[a]From di Domenico *et al.*[52]
[b]Estimated amount of TCDD in a parallelepiped of soil with a 1-m² base and height equal to the layer thickness.
[c]Sampling on October 15, 1976.
[d]Sampling on April 1, 1977.
[e]Sampling on December 1, 1977.
[f]1.2 E+3 = 1200.

5.2. TCDD Levels in Sediments and in Ground and Surface Waters

From shortly after the accident, sampling and analyses were periodically carried out on surface water streams of the area hit as far south as the River Lambro. This river and the Torrent Seveso are both part of the Milan surface water system. Outcomes were found to be consistently negative for TCDD at a detection threshold <10 pg/liter.[20,24] Monthly determinations conducted on pipeline and ground waters also provided consistent negative results, even when detection sensitivity was ~1 pg/liter.

During the same period, sediment samples from the Torrents Certesa and Seveso were also assayed; the former flowed into the latter after passing through a stretch of the most contaminated sector of Zone A (Fig. 1). Results were positive (on the order of 1 ng/kg) within the first few kilometers downstream from their confluence, but negative in samplings performed farther downstream. The intense rainfalls of the fall of 1976 and the following winter caused Torrent Seveso to repeatedly overflow its embankments at the point of entry into Milan, thus depositing silt on adjacent areas. In silt samples from the first four floods, no TCDD was detected; however, TCDD just above the detection threshold (~1 ng/kg) was found in silt from the fifth flood.[20,24]

5.3. TCDD Levels in Atmospheric Particulates

TCDD transported by suspended air particulates (<100 μm) was monitored by high volume samplers in June, 1977; however, on a regular basis, dustfall jars were extensively utilized to collect settling dust (>10 μm) (Fig. 4).[20,24,56] Monitoring was carried out to determine whether fine air dust from topsoil of the more contaminated areas was hazardous for the less contaminated or uncontaminated neighboring areas. As a preventive and risk management measure, atmospheric monitoring was implemented and tested in particular during the Zone A reclamation works of 1977–1980, since the latter involved mostly mechanical removal of topsoil layers (scarification). Of specific interest in this context, and for the amount of data available, are the results from settling air dust samples (July 1977–June 1980); the few findings from high-volume samplers—including measures of dust level (up to 0.14 mg/m^3)—showed general agreement with those of settling dust samples.

TCDD was detected sporadically and at different times by the dustfall jar network. Apart from Sampling Site (SS) 2 in the highly contaminated Subzone A_1, most individual findings were below the detection threshold (~0.1 ng/m^2 per day) and only seldom were detectable levels present; maximum values of 0.39 and 0.35 ng/m^2 per day were measured, respectively, at SS 1 in July, 1978, and at SS 6 in June, 1978. On the contrary, TCDD was detected in most samples from SS 2; peak deposition values of 0.75, 0.79, and 0.49 ng/m^2 per day were observed in September, 1977, July, 1978, and in the July 5–September 15 period of 1979, respectively. SS 2 showed seasonal variations in both dust and TCDD deposition rates, these normally reaching a peak during the summer.

In dust, TCDD levels were found to range between ~0.06 (detection threshold)[24] and 2.1 ng/g—the latter determined in the summer of 1977—with the higher values detected at SS 2 during the summer periods.[20,56] The TCDD amount in the dust decreased with increasing distance from areas with higher levels of contamination, so that airborne TCDD could not really significantly affect TCDD levels in the top layers of soil. The data from the dustfall jar at SS 6 did not fit into the above picture very well, probably because of a nearby municipal solid waste incineration plant.

In order to increase the sensitivity of detection, the individual sediment extracts from a sampling site were pooled and reanalyzed. Six or seven extracts of sequential samplings were employed for each pool. In most cases, the time-averaged TCDD levels resulted in values being below or near detection threshold (~0.01 ng/m^2 per day), although the 1978 March-through-October pools from SSs 1 and 6 provided values approximately six times as high. The highest value of 0.23 ng/m^2 per day was measured at SS 2 during the same period. A slowly declining trend in TCDD fallout was observed after 1978.

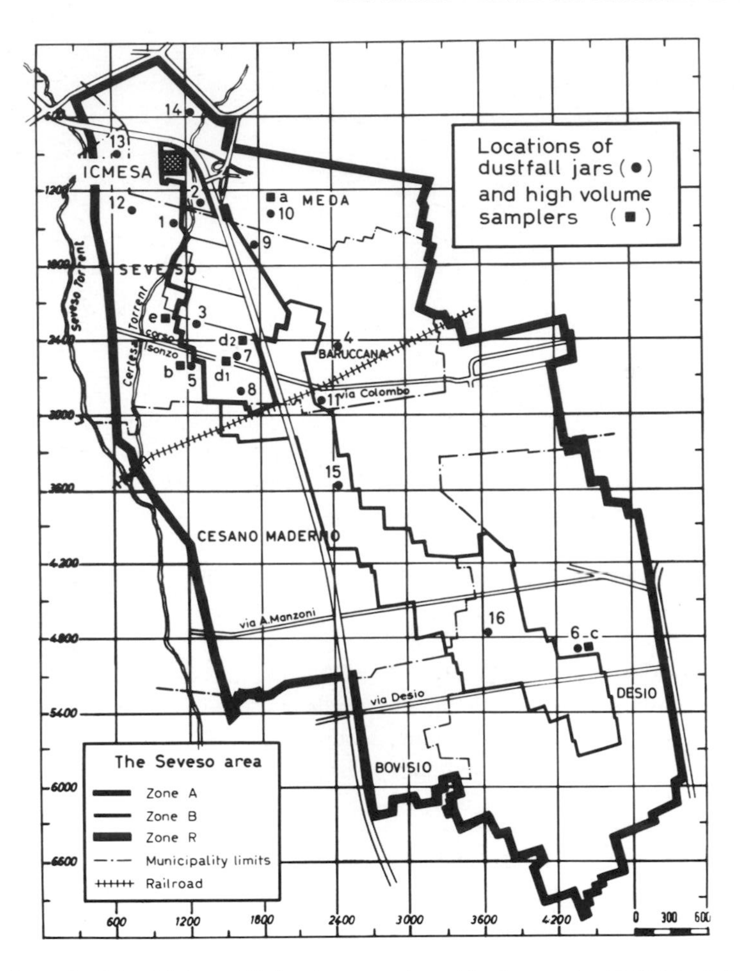

Figure 4. Network of dustfall jar and high-volume samplers utilized to monitor TCDD in air in the Seveso area. A dustfall jar is a passive collector of atmospheric precipitation and settling dust. Sampling periods normally lasted between 3 and 5 weeks. TCDD was assessed through the assay of sediment, whose amount was also determined after drying. No TCDD was ever detected in the aqueous matrix at a detection threshold of ~10 pg/liter. (Adapted from di Domenico *et al.*[56])

5.4. TCDD Levels in Plant Tissues

Shortly after the accident, samples of leaves, grass, and vegetables were collected in the area (especially in Zone A along the main diffusion pathway) and assayed. Many of these samples exhibited the effects of a severe chemical action, such as withering, burns, and yellow spots—effects that were less and less evident with increasing distance from ICMESA. As the south–southeasterly distance from the plant increased, TCDD levels in the vegetable samples decreased rapidly, following a negative exponential trend similar to that observed for topsoil (Fig. 3). For instance, in samples collected in July, 1976, concentrations ranged between 50 and 0.01 mg/kg fresh tissue at distances of approximately 175 to 2000 m, respectively, from the source.[20]

Between 1978 and 1981, over 150 epigeal and hypogeal samples of several edible vegetable species (carrot, lettuce, onion, potato, and small radishes) from Zone R were assayed for contamination by TCDD and other TCDD isomers at a detection threshold of 0.2 ng/kg fresh vegetable tissue.[20] At levels close to 1 ng/kg, TCDD was identified only in two epigeal samples (relative frequency, 1.30%). In several cases (8.44%), the compound was detected at between 0.2 and 0.8 ng/kg, but this turned out to be unreliable because of uncertain identification. In nine cases (5.84%), analytical assessment could not be carried out because of interferences. In all of the other samples, TCDD was not detected. 1,3,6,8- and 1,3,7,9-TCDDs were quantitated in the 1.2–2.4 ng/kg range in epigeal samples (22.9%; 18.2% of the total set); however, they were not detected in hypogeal parts. Analysis of both types of samples provided negative outcomes in most cases (56.8%); uncertain results between 0.2 and 0.8 ng/kg accounted for the remaining determinations.

The levels of TCDD in wheat and oat grown in 12 experimental plots in Zone R were checked in 1980 by sampling ears with 10-cm stems during the maturation period.[20] Of 103 whole ear samples, not one contained detectable levels (>0.5 ng/kg) of TCDD, but 1,3,6,8- and 1,3,7,9-TCDD were detected with a frequency of 16.3% in the concentration ranges 0.76–2.2 and ~0.3–1.7 ng/kg, respectively. TCDD and its isomers were not found in 97 samples of wheat and rye kernels collected after harvesting.

In general terms, the above observations agree with a study by Wipf *et al.*[57] According to these authors, immediately after the accident, TCDD levels in samples of vegetable material from Zone A reached values up to some mg/kg; however, in the newly grown vegetation of the following years, TCDD levels were seen to drop by several orders of magnitude. In 1977, TCDD was not detected (<1 ng/kg) in the flesh of apples, pears, and peaches, nor in corncobs and kernels, all samples obtained from plants grown in a highly contaminated area; however, levels of the compound up to ~0.1 μg/kg were detected in the peel of the fruit and in the sheaths of the corncobs, this suggesting that contamination of vegetables was primarily by aerial route. Studies with carrots grown in a highly contaminated medium provided further evidence of the absence of a significant TCDD uptake from soil. No measurable amount of TCDD was detected in vegetable samples—including carrots—collected in Zone R in 1977–1980.

With respect to the Seveso accident, other studies performed on plants have been reported by Cocucci *et al.*,[58] Facchetti *et al.*,[59,60] and Sacchi *et al.*[61]

5.5. TCDD Levels in Wildlife and Domestic Animals

In the first few years after the accident, TCDD was detected in field mice and other wildlife specimens captured in areas characterized by considerable contamination (Table 6, upper section). In particular, chemical levels in the bodies of field mice were comparable to those in the soil, thereby providing some evidence for bioaccumulation. Earthworms were also tested as indicators of soil contamination. In 1980, samples of

Table 6
Impact of TCDD on Wildlife and Domestic Animals in the First Few Years after the Accident

TCDD levels in wildlife[a]

Animal	N	Tissue	Finding frequency	TCDD level (μg/kg) Mean	TCDD level (μg/kg) Range
Field mouse	14	Whole body	14/14	4.5	~0.07–49
Hare	5	Liver	3/5	7.7	2.7–13
Toad	1	Whole body	1/1	0.2	
Snake	1	Liver	1/1	2.7	
		Adipose tissue	1/1	16	
Earthworm[b]	2	Whole body	1/2	12	

Rabbit mortality[c]

Zone	Rabbits (N)	Deaths	Mortality (%)
A	1,089	348	31.9
B	4,814	426	8.8
R	18,982	1,288	6.8
Total	24,885	2,062	8.3

TCDD levels in rabbit liver[d]

Zone	Rabbits (N)	Finding frequency (%)	TCDD level (μg/kg) Mean ± S.E.[e]	TCDD level (μg/kg) Range
A	67	97	85 ± 12	3.7–630
B	19	84	90 ± 25	7.0–380
R	137	81	26 ± 6	0.27–460
Background[f]	86	13	13 ± 6	0.32–55

[a]From Fanelli *et al.*[62]
[b]Each sample was a 5-g pool of earthworms.
[c]July 10–August 31, 1976. From Fanelli *et al.*[64]
[d]July 1976–July 1978. From Fanelli *et al.*[64]
[e]Standard error.
[f]Surrounding areas.

two earthworm species were collected in areas characterized by TCDD levels in soil of between ~0.06 and 9.2 μg/kg. It was observed that TCDD concentrations in the earthworm bodies were correlated with those in the soil; the earthworm bioaccumulation factor for TCDD from soil was estimated at ~10.[62,63]

In 1976, the entire Seveso area had been 81,000 domestic animals, many of which died within a short time following the accident. In Zones A, B, and R, animal breeding before the accident was mostly in family-size farms, where more than 80,000 poultry, rabbits, and other small animals were raised primarily for household consumption. At the time of the accident, there were also 349 cattle, 233 pigs, 49 horses, 49 goats, and 21 sheep. Cattle fed mostly on fodder made of grass, hay, and sliced corn collected near the farms; during winter, the fodder was mixed with commercial feed. Pigs and poultry fed on commercial products from distant areas. Rabbits were given mainly grass harvested randomly, often from nearby farms but sometimes far away. After the accident, many rabbits died; to a lesser extent, mortality was observed in poultry and other small animals. These deaths began some days after the release of the cloud containing TCDD and the other chemicals; deaths increased remarkably in the following 2 weeks, and then decreased for months afterward.[63,64]

The total number of spontaneous deaths registered by the end of August, 1976, was 3281, of which 2062 were accounted for by rabbits (Table 6, middle section).[63,64] In particular, a mortality rate up to 100% was noted in animals fed on green fodder from contaminated areas, whereas a much lower mortality incidence was observed in animals on commercial feed or fodder collected before the accident or far away from ICMESA. During the first weeks, 2294 autopsies were carried out on animals (usually on rabbits or poultry) from nearly 1200 farms of the Seveso area. In addition to the normal incidence of known pathological signs, in ~15% of the farms on Zones A, B, and R many subjects showed pathological conditions never previously seen.

As the animal studies were being carried out, analytical detection of TCDD in the tissues of dead animals was also implemented to ascertain whether exposure to the chemical had occurred, establish a possible dose–response relationship, and determine if the rabbit could be used for biomonitoring owing to its capability to bioaccumulate TCDD in the liver (μg/kg range) before showing pathological signs (Table 6, lower section).[18,63,64] It should be mentioned that a comparative study on the uptake in the rabbit of TCDD in different formulations—including accident-contaminated Seveso soil—indicated that Seveso soilborne TCDD had a bioavailability on average 68% lower than solvent-borne (free) TCDD.[42]

Many of the farms kept dairy cattle, generally from five to ten cows. On average, daily milk production was ~10 liters/cow and destined for local consumption. Approximately 2 weeks after the accident, cows from Zone A were transferred to a special cowshed under sanitary surveillance. Cattle breeding in Zones B and R continued under veterinary control, but the animals were fed safe fodder from elsewhere. All of these cows were slaughtered in 1977 and 1978. In a first sampling campaign (July 27–August 28, 1976), milk was collected from cows living in areas at different distances from ICMESA, each farm usually providing a single sample obtained by pooling the individual milk samples from each cow of the farm.[63,65] The outcomes of this investigation proved that higher TCDD levels (from 76 up to 7900 ng/liter) were present in milk samples from farms close to the chemical plant; milk from outside Zone R and from the south end of Zone B did not contain detectable amounts of TCDD (<40 ng/liter) or contained TCDD at levels <200 ng/liter.[18,63,65]

5.6. Reassessment of TCDD Persistence Model in Soil

Several efforts have been made to investigate the TCDD temporal trend in Seveso soil. Early studies were based essentially on Zone A data obtained between the accident and 1979.[5,20,55,66,67] More recent studies have utilized data from Zone A as well as from Zones B and R.[29,68] The subject has gained even more interest as the early TCDD figures for soil may have been affected by a large bias,[23,28,29] and yet they are critical for a correct interpretation of the chemical environmental dynamics. In particular, recovery yields were shown to never exceed 75% even when analytical procedures were at their best (since late 1979). On the other hand, Zone A findings before 1979 appeared to be affected by TCDD losses which, depending on the specific area and TCDD level in soil, might have reached values on the order of 70–90%.[23] However, accurate estimates of the analytical losses could not be provided because of the functional (but false) assumption used for model computing that no TCDD had disappeared from soil with time: this assumption has the effect of yielding optimistic loss figures.

In light of the above, the ongoing studies on the TCDD temporal trend in Seveso soil have the purposes of providing a more reliable description of TCDD persistence pattern, estimating analytical losses prior to late 1979, estimating the total amount of the compound released or, at least, deposited in the area hit, and possibly yielding new exposure and risk assessment figures for the people exposed. In fact, the assessment of TCDD environmental dynamics and levels in soil through time is critical to evaluating the long-term exposure of the once-exposed populations and the associated lifetime cancer risk.[35] Results obtained so far, relative to the first three points, are summarized below.

The data to serve as a basis for the current analysis were selected from the general data base according to stringent selection rules, and cover a time span of 5.5 years.[14,23] Figure 5 exhibits the TCDD time-trend regression function obtained by using a two-exponential model; analysis outcome details are summarized in Table 7.[14,69] According to the present model, the TCDD that remained after the period of rapid vanishing settled into the soil with little interaction with those environmental factors responsible for its initial disappearance. It is suggested that the initial (<0.5 year) rapid reduction

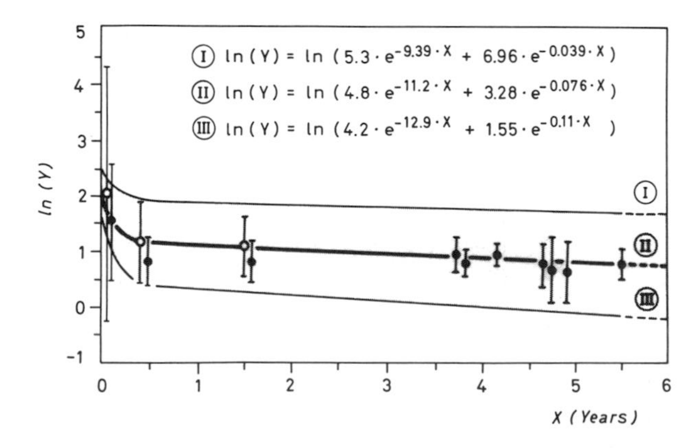

Figure 5. The two-exponential model developed to describe TCDD persistence in Seveso soil. The function has a steep downslope within < 0.5 year from the origin and then almost flattens out for longer times; in other words, in Seveso soil TCDD levels diminished rapidly shortly after blowout, but did so at a much slower rate over the following 5 years. (Adapted from di Domenico *et al.*[69])

Table 7
Summary of Some Indicative Results of TCDD Persistence Studies in the Soil of the Seveso Area

Model function

$\ln(Y) = \ln(q1 \cdot e^{-k3 \cdot X} + q2 \cdot e^{-k4 \cdot X})$ $\quad q1 = Y^{o} \cdot k1$, $\quad q2 = Y^{o} \cdot k2$

Function parameters

Quantity	Mean estimate	Standard deviation	95% confidence limits	Units/comments
Fast dynamics[a]				
$q1$	4.8	2.5	4.2–5.3	Normalized[b]
$k3$	11.2	7.6	9.39–12.9	years^{-1}
$t^{1/2}[k3]$	0.0619[c]		0.0537–0.0738[c]	years
Slow dynamics[b,d]				
$q2$	3.28	0.33	1.55–6.96	Normalized[b]
$k4$	0.076	0.022	0.039–0.11	years^{-1}
$t^{1/2}[k4]$	9.1		6.2–17	years
Other estimates				
y^{o}	8.0		5.7–12	Normalized[b]
$k1$	0.59		0.43–0.73	
$k2$	0.41		0.27–0.57	

Regression equations[e]

$\ln(Y) = \ln(5.3 \cdot e^{-9.39 \cdot X} + 6.96 \cdot e^{-0.039 \cdot X})$ Upper boundary

$\ln(Y) = \ln(4.8 \cdot e^{-11.2 \cdot X} + 3.28 \cdot e^{-0.076 \cdot X})$ Intermediate

$\ln(Y) = \ln(4.2 \cdot e^{-12.9 \cdot X} + 1.55 \cdot e^{-0.11 \cdot X})$ Lower boundary

[a] Arithmetic means and their standard deviations and confidence limits obtained from a specific simulation treatment.[69]
[b] Reference to Cerlesi *et al.*[14]
[c] $\langle t^{1/2}[k3] \rangle$ = 23 days; 95% confidence limits = 20–27 days.
[d] Means and standard deviations from regression analysis.
[e] Compare with curves of Fig. 5.

was mainly related to the action of UV sunlight and heat-promoted volatilization on the TCDD which was then distributed very superficially on both vegetation and soil. Furthermore, the heavy rainfalls of the fall and winter of 1976 washed down the contaminated outer surfaces and disturbed the top layer of soil, moving TCDD deeper into soil and away from direct solar radiation, air streams, and precipitation. In the process, a dilution or loss of that fraction of TCDD loosely bound to the earth matrix occurred so that, after about half a year, only TCDD that was strongly bound to soil remained.

On these grounds, it has been concluded that two different half lives describe the TCDD persistence trend at Seveso: on average, they have been estimated to be 23 days (fast-vanishing dynamics) and 9.1 years (slow-vanishing dynamics). The equation representing the lower boundary of regression curve 95% confidence region provides an estimate of the total TCDD released (30–40 kg) which is fully compatible with what is known of the chemistry and physics of the accident. Based on such an equation, these half lives assume the values of the corresponding 95% lower confidence limits (20 days and 6.2 years, respectively); further, the fast-vanishing dynamics appears to have been associated with the disappearance of most of the TCDD (~70%). Unfortunately, not much may be said as yet on how TCDD was released and distributed over the territory and, therefore, how biased the early contamination maps are.

Biodegradation was likely to be of negligible importance in TCDD reduction over the period explored (first 5.5 years).[70,71]

6. BIOLOGICAL DATA ON HUMAN EXPOSURE

Seven months after the ICMESA accident, a 55-year-old woman died of pancreatic adenocarcinoma, which had spread to the liver and to the extrahepatic bile ducts and had metastasized in the lungs. The subject had lived in an area of Zone A characterized by a mean TCDD level of ~200 $\mu g/m^2$. At the moment of the accident, the woman was eating inside her home with the windows and doors open; she also consumed vegetables from her garden in the 4 days following the event. The woman remained in her home until July 26. The first symptoms related to the tumor became clinically evident at the end of October 1976, and laboratory confirmation after exploratory laparotomy arrived a month later. As the subject had been exposed to the chemicals released from ICMESA by inhalation, ingestion, and dermal contact, TCDD presence in her tissues was investigated after death occurred. The following results were obtained and reported on a whole or wet weight basis, rather than the currently traditional lipid basis[72]:

Fat	1840 pg/g
Pancreas	1040
Liver	150
Thyroid	85
Brain	60
Lung	60
Kidney	40
Blood	~6

These data provided a preliminary picture of TCDD distribution in human tissues following what could be defined as an "acute environmental exposure." Based on their findings, the authors suggested classifying tissues into four groups with, respectively, high (fat and pancreas), medium (liver), low (thyroid, brain, lung, and kidney), and very low (blood) levels of the chemical—the specific levels being eventually in association with preferential accumulation routes.

Approximately 10 years after the Seveso accident, analytical methods were improved and became available to measure TCDD levels in small blood samples.[73–76]

From the 1976–1985 laboratory medical examinations following the Seveso accident, there were some 30,000 1- to 3-ml serum samples which had been kept stored at −30°C since the time they were collected.[77–79] In 1988, a set of thirty 1976 serum and plasma samples was selected for TCDD assessment. The analytical results of this first set of determinations concern a selected group of ten Zone A children with chloracne, a group of ten Zone A adults without chloracne (however, a sample o lost), and ten subjects from the surrounding noncontaminated area. It sh out that these are the first subjects whose results of laboratory medical and individual clinical histories (from 1976 through 1985) have been ass quantitative measurements of TCDD levels in samples of blood taken near ume of acute exposure.[79]

The lipid-adjusted levels of serum TCDD for the three groups were determined. Detected TCDD ranges in children with chloracne and in adults without chloracne were 830–56,000 and 1800–10,000 ppt, respectively. Although any comparison may be biased by age-related factors, based on groups, chloracne appears to be associated primarily with higher TCDD levels. However, it is also remarkable that blood lipid TCDD levels as high as 10,000 ppt following acute exposure are compatible with the absence of such a skin disorder. Among the ten nonexposed subjects, only one had detectable levels of TCDD (~140 ppt), which possibly resulted from a <1% carryover from a preceding assay of a sample with a high level of TCDD.

7. EARLY HEALTH FINDINGS

Medical examination programs were initiated with the following aims: to ascertain early adverse health consequences in the exposed population, to give guidance for the allocation of service and resources, to identify needs and suggest areas for further surveillance and research.

The earliest sign of adverse effects in humans became apparent when, on the sixth day, 19 children were admitted to local hospitals with skin lesions caused by contact of uncovered parts of the body with caustic chemicals contained in the cloud.

7.1. Chloracne

By the end of 1976, 34 cases of chloracne were diagnosed in persons under 15 years of age. It was then decided to examine all children attending nurseries, infant and primary school of the contaminated areas. Nearly 90% of them had skin examinations. By April, 1977, 187 cases of overt chloracne were diagnosed by an expert panel, and 164 (88%) were in children. Chloracne distribution closely resembled the TCDD contamination pattern (Table 8). In Zone A, 61 cases were observed, of whom 19 were adults. The 42 cases among children corresponded to 19.6% of all children living in this Zone, indeed a large proportion, which was even higher when only the most contaminated subzones were considered. The prevalence of chloracne was 0.5% in Zone B, and 0.7% in Zone R. This distribution of chloracne cases provided also the first hint of a possibly inaccurate definition of the boundaries of contamination zones. As a matter of fact, in certain parts of Zone R in the municipalities of Seveso and Meda, the proportion of chloracne in children was higher than in Zone B. Thus, for example, in the Zone R suburb of Polo, located in the top right-hand corner of the accident scenario, southeast of the plant (Fig. 1), out of 750 children, 19 (2.5%) were diagnosed as having chloracne. Fifty-one cases occurred in children residing outside

Table 8
Distribution of 164 Chloracne Cases Diagnosed up to April, 1977, in Children Having Their Residence in Different Areas of the Contaminated Territory or Outside

Area	Total population aged 3–14 years	Chloracne cases	Percent
Zone A total	214	42	19.6
Zone A max.[a]	54	26	48.1
Zone B	1,468	8	0.5
Zone R	8,680	63	0.7
Zone R Polo[b]	750	19	2.5
Other	48,263	51	0.1

[a]Includes only the most contaminated part of Zone A.
[b]Subzone located near the plant.

the designated contamination zones. Where they were at the time of the accident could not be established with certainty. By mid-1978, six additional cases were detected, bringing the total number to 193. No further cases of chloracne were discovered or reported after this time.[80–82]

The higher frequency of chloracne in children than in adults is possibly explained by the fact that the former had more opportunities to come in contact with the toxic cloud components and to ingest or inhale them through outdoor activities, contact with soil, vegetation, dirt outside their homes, and so forth. Other relevant explanatory factors might be the absence of a systematic screening in adults, which possibly left unnoticed other existing cases, or a difference in susceptibility to dioxin effects at different ages.

7.2. Subjective Symptoms and Laboratory Tests

The comparison of 146 cases and 182 controls (nonchloracne children, age and sex matched to the cases, selected from the area) revealed a significantly higher frequency of nausea, lack of appetite, vomiting, abdominal pain, headache, and eye irritation among the former. In addition, they exhibited a higher frequency of abnormal values for the liver enzymes gamma-glutamyltranspeptidase (GGT) and alanine aminotransferase (AAT), and for urinary aminolevulinic acid (ALA-U), a porphyrin precursor. Chloracne subjects from Zone A had biochemical abnormalities more frequently than those from other zones.[80]

Health examination results of 18 subjects over 14 years of age affected by chloracne were also reported (no control group was concurrently examined). There was a high frequency (around 25%) of self-reported symptoms, and of signs of liver enlargement. In 1977, biochemical tests showed elevated serum cholesterol values of higher than 230 mg/100 ml in eight subjects, and ALA urinary concentration outside the reference range in five subjects.[83]

7.3. Peripheral Neuropathy

Peripheral neurological changes were among the signs of TCDD toxicity. Persons evacuated from Zone A were invited to undergo neurological examination in 1977 and 1978. A nonexposed population served as reference. Electrophysiological and clinical signs of peripheral neuropathy were nearly three times as frequent among Seveso residents having either raised serum liver enzyme levels or chloracne (12/55 or 22%) than among controls (13/168 or 8%). When only subjects younger than 20 years with chloracne were considered, the relative frequency rose up to nearly five times.[84]

7.4. Enzyme Induction

In the early months after the accident, urinary D-glucaric acid excretion, an indirect index of enzyme induction, was measured in 14 children with and 17 children without chloracne, all from Zone A. The former had a significantly higher level of D-glucaric acid in urine.[85]

It is understandable how in the hectic, early postaccident period these observations were often lacking formal design and proper conduct. For example, it was not easy to control biases, such as those linked to interview and reporting, or assure standardization of diagnostic procedures, proper selection of controls, adequate size of the sample, and so forth. Notwithstanding these limitations, the reported early surveys, which mainly concerned chloracne cases, showed beyond any doubt that at least a portion of the population had been exposed to the powerful toxin 2,3,7,8-TCDD.

8. SURVEILLANCE PROGRAMS

Surveillance programs were then designed in order to continue over time the health monitoring of those subjects exhibiting immediate, acute effects (e.g., chloracne cases), and identify in the affected population at large the possible occurrence of health effects in the short- to mid-term period. One of the major problems was the large size of the population in the active surveillance. This difficulty was augmented by the lack of a preexisting validated information system and by the time constraints which led many different teams to be called into the area, with diminished possibilities for quality control and procedure standardization.

8.1. Spontaneous Abortion

According to animal data, an increase of spontaneous abortions in pregnant exposed women was to be expected. Ascertainment of spontaneous abortion is, in general, a difficult surveillance task. Specific problems further complicated the picture in our case. One was the absence of a valid, on-going data collection system; another involved the moral and political issues related to legalization of abortion; and, finally, a very active birth control campaign was carried out which may have decreased conception rates. In such a context, the completeness, accuracy, and quality of data remained questionable. Nevertheless, several attempts were made to report and interpret the occurrence of spontaneous abortions in the area.

Analysis by trimester (from the accident to early 1978) and by municipality showed some time-related variability with the highest abortion rate seemingly occurring in the earliest trimester in the contaminated areas. However, it was difficult to determine the possible contribution of TCDD exposure, or of exposure to a "chemical disaster" as such, and to exclude a major role of biases related to information recording and data collection, etc. Results were considered inconclusive, if at all valid.[18]

In another report,[82] crude estimates drawn from vital statistics sources were provided. A decrease in the birth rate was observed in 1977–1980 in the entire Lombardy region, and not just in the accident area. The proportion of abortions compared to live births per year was not considered to depart from "the generally accepted abortion rate of 10–20%."

In a third analysis, notifications of spontaneous abortions to county medical officers were used. Spontaneous abortion rate in 1977 was higher than in 1976, but not departing from historical rates estimated from 1973 onwards. An improved physician's care in notification was considered a probable explanation for the change in the postaccident period. These data were then supplemented with information directly provided by hospitals in the region (Table 9). An increased pregnancy loss percent was seen in late 1976, with a subsequent fall (again, a change in physicians' notification attitude?). Zone B rates showed a further increase in mid-1977, and were consistently higher than those of Zone R and the noncontaminated area; the excess was statistically significant only in the third trimester of 1977. Data quality was considered questionable, and results lacked consistency.[86]

8.2. Cytogenetics

The first chromosome analysis was performed in 1976 at the request of hospitals where eight children aged 2 to 10 years and four pregnant women with skin lesions presumably caused by TCDD exposure had been admitted. For comparison, the earliest available results on ten unmatched subjects were adopted. The proportion of aberrant cells was higher in the exposed children and women, but only when gaps were included.[87]

A more extensive cytogenetics study included 301 subjects, as indicated in Table

Table 9

Pregnancy Loss Rate [Abortions/(Births + Stillbirths + Abortions) × 100] by Exposure Zone and Trimester[a,b]

Zone		Jul–Sep 1976	Oct–Dec 1976	Jan–Mar 1977	Apr–Jun 1977	Jul–Sep 1977	Oct–Dec 1977
B	No. of abortions	3	4	5	8	10	4
	Pregnancy loss %	11.1	22.2	17.2	28.5	31.2	13.7
R	No. of abortions	19	17	15	17	16	20
	Pregnancy loss %	13.7	16.3	12.7	12.5	11.4	13.8
Noncontaminated	No. of abortions	74	94	119	81	67	99
	Pregnancy loss %	11.0	14.8	16.6	13.0	10.5	14.3

[a]Source: Special Office for Seveso, Lombardy Region.
[b]Zone A is not shown because very few pregnancies occurred.

Table 10
Cytogenetic Study in Seveso, 1977: Frequency of Chromosomal Aberrations in Lymphocytes

			% aberrant cells	
Exposure	No. of subjects	No. of mitoses	Including gaps	Excluding gaps
Acute	145	6470	2.49	0.99
Chronic	69	3040	2.53	0.92
Controls	87	3958	1.64	0.48

10. Those with acute exposure were subjects living in the most contaminated area near the accident plant; workers employed in that plant were considered as having had long-term ("chronic") exposure; age- and sex-matched controls were people living in the surrounding noncontaminated area. Proportion of aberrant cells was higher among the exposed, but statistical analysis showed that the only significant difference was among the scorers of the five laboratories involved in the survey. A further analysis of a larger number of mitoses on selected samples from the three exposure categories was then carried out. Differences between exposure categories did not become significant even after correction for interobserver variability. No consistent evidence of chromosomal effects associated with TCDD exposure was thus provided by this study.[87]

A third set of data was obtained after examining maternal peripheral blood, placenta and umbilical cord, and fetal tissues of induced abortions in 19 women from the Seveso area, and 16 women not known to be exposed to environmental mutagens and teratogens who had abortions for nonmedical reasons. Within a pattern of marked variability, no significant differences were observed in the frequency of individual types of aberration, average number of aberrations per aberrant cell, or frequency of polyploids in maternal blood and placenta between the two groups. Instead, fetal tissues of exposed pregnancies exhibited aberrant cell frequency significantly higher and a greater number of aberrations per damaged cell than control pregnancies. Several factors might explain these findings, including those related to growth in culture. In addition, the possibility of preexisting chromosomal damage in fetal cells could not be ruled out. No differences were seen regarding those pregnancies started in Seveso before versus those after the accident. Thus, the extent to which the increased frequency of chromosomal aberrations in fetal tissue reflected maternal exposure to TCDD could not be established.[88]

8.3. Birth Defects

The first set of data available consisted of 30 cases of induced abortion (of which 3 were from Zone A, and 5 from Zone B), and four spontaneous abortions (from Zone R), all of which occurred in 1976 after the accident. Embryological and histo-morphological investigations were conducted on this material, and no indications of

Table 11
Relative Risk and 90% Confidence Interval
for Selected Groups of Malformations
in the TCDD-Contaminated Area (Zones A, B, and R)
versus the Surrounding Noncontaminated Territory

	1977–1982	First quarter 1977
Total defects	0.97 (0.83–1.13)	1.49 (0.64–3.45)
Major defects	0.83 (0.67–1.04)	0.93 (0.26–3.32)
Mild defects	1.14 (0.92–1.42)	2.50 (0.79–7.94)

mutagenic, teratogenic, or fetotoxic effects attributable to TCDD were detected. In 23 induced abortions, no anomalies or organic alterations could be found. In six other cases, various morphological alterations were visible; some were probably artifacts, and some were of borderline pathological significance. The four spontaneous and the one remaining induced abortions were probably related to dioxin exposure, but this link could not be proved. Investigations were limited by the fact that in the majority of cases fetal tissues were incomplete.[89]

At the beginning of 1977 a congenital malformation registry was established, which included all live births and stillbirths to women residing in the accident area in July, 1976. Data were collected for the period 1977–1982; 742 malformed infants were registered out of a total of 15,291 births (live and still). Out of 26 births in Zone A, no cases of major malformations were found. In none of the three exposure zones (A, B, or R) was the frequency of mild, major, or combined defects significantly higher than in the reference population. Table 11 shows relative risks for the entire surveillance period and for the first quarter of 1977, when children had probably been exposed to TCDD during the first week of gestation (140 births in all). None of the relative risk values were statistically significant. Major information biases were excluded, whereas the small number of exposed pregnancies, especially in Zones A and B, might have precluded the identification of low-risk and/or very rare defects.[90]

A piece of related information came from the mortality and cancer incidence studies (dealt with in more detail below). Mortality related to congenital malformations was not increased among those born after the accident.[91] In the same period, for none of the cancer types examined was the incidence in the exposed young population significantly increased.[92]

8.4. Follow-Up of Special Groups

Long-term effects on the *peripheral nervous system* (PNS) were explored in 152 chloracne cases who agreed to participate in a survey conducted between October, 1982, and May, 1983, and in 123 subjects without chloracne, frequency-matched by sex and age, who volunteered to serve as the reference group. None of the subjects had a clear-cut peripheral neuropathy, but clinical and electrophysiological signs of PNS involvement were, significantly, nearly twice as frequent in the chloracne group than in controls. In particular, there were 11 cases versus 2 controls presenting at least two bilateral clinical signs, and 25 cases versus 9 controls exhibiting at least one abnormal electrophysiological function.[93]

Forty-eight children aged 3 to 8 years from Zone A underwent repeated examinations from November, 1976, to May, 1979, for the study of *immunologic effects*. Control subjects were selected from the school population of a nearby noncontaminated town. Total serum complement hemolytic activity (CH50) had significantly higher values among the exposed subjects at each examination. Exposed children also exhibited higher values for lymphocyte responses to phytohemagglutinin (PHA) and to pokeweed mitogen (PWM), and in the absolute number of lymphocytes of peripheral blood. Results for other tests failed to show clearly diverging values between the exposed and control subjects. Consistently increased values were more evident in children with chloracne.[94]

Induction of microsomal enzymes in the liver was one of the best documented TCDD effects in laboratory animals. An indirect test of enzyme induction, urinary D-glucaric acid, was evaluated between 1976 and 1979 in different groups of the exposed population and controls. Children from Zone A with chloracne had, in 1976, significantly higher levels of D-glucaric acid in urine than children without skin lesions. In 1979, children who had left Zone A had levels similar to those of controls, whereas in children still living in Zone B the urinary excretion was significantly higher. In 1980, however, urine samples of the Zone B children showed significantly lower levels. In 1981, 34 children evacuated in 1976 from Zone A had normal values, whereas 61 children from Zone B and 59 children from a Zone R sector very close to the plant (i.e., Polo) had urinary D-glucaric acid levels almost significantly elevated. Adults living in Zone B ($N = 117$) had significantly higher levels of D-glucaric acid excretion in 1978 than controls ($N = 127$) from a noncontaminated area.[85]

Children from the three contaminated zones were followed from 1976 to 1982 to examine whether *liver function* and *lipid metabolism* showed alterations as possible consequences of TCDD exposure. In all, nearly 400 children aged 6 to 10 years at the time of the accident were examined on a yearly basis. The only clear-cut difference in test values between exposed and reference children was seen in 1976 and 1977, when boys living in the most contaminated Zone A exhibited consistently higher levels of GGT and AAT. The increase was slight, perhaps attributable to TCDD exposure, and disappeared with time. No alterations of blood concentrations of cholesterol and triglycerides were found.[95]

Repeated surveys on the group of 193 *chloracne cases* and on unexposed control subjects matched for age and gender were conducted until 1985, with a participation rate between 70 and 80%. No significant differences or temporal trends in mean values of liver enzymes and lipids were detected. A decrease in cholesterol and triglyceride values was apparent in the chloracne group between 1976 and 1982. At the end of the follow-up in 1984, one subject had persistent chloracne, and five had chloracne scars on face and forearms. Motor and sensory nerve conduction velocity was measured, and neither significant differences between groups nor temporal trends were observed. Apart from skin signs, no clinical or systemic sequelae of chloracne were thus detected 8 years after first exposure.[96]

Another special group surveilled was comprised of *workers* employed in *decontamination operations* in the area. These people entered the most contaminated part of Zone A under strict personal protection and environmental measures. They underwent preemployment medical examination for eligibility. The values of a set of preemployment tests (e.g., liver function, lipid and heme metabolism) were compared with the same values after 9 months, and with those of an unexposed group. No significant changes were detected.[97] Later analysis of cleanup workers' experience confirmed that, on average, the safety measures taken had been effective. No TCDD-related clinical health impairment was found (as, for example, chloracne, liver impairment, peripheral neuropathy, porphyria cutanea tarda), and no significant differences in biochemical outcomes compared with unexposed subjects were detected. Nine subjects left for non-health-related reasons, and five for negative job fitness evaluation; for two of them a transient effect of exposure to TCDD could not be completely excluded.[98]

Between 1976 and 1985, laboratory tests were periodically performed on the 20 subjects whose serum was assayed for TCDD, to detect possible alterations of the liver, kidney, bone marrow, lipid metabolism, and immune system function. However, only modest, transient, small departures from the normal ranges were observed in four children with chloracne, four Zone A adults, and one referent subject. None of the observed alterations had pathological significance either with respect to the number of tests involved or with respect to the extent of the alteration.[79]

9. LONG-TERM MORTALITY AND CANCER INCIDENCE STUDIES

All surveillance programs were supervised, and their results periodically evaluated by an International Steering Committee which ended its work in 1984. Their conclusion was that chloracne represented the only health outcome clearly attributable to the accidental exposure to TCDD. No conclusion could be reached at that time regarding long-term effects. Surveillance programs were discontinued, but long-term investigations were designed in order to examine mortality and cancer incidence.

As time passed, it appeared that there might be a migration away from the area by those people who had most suffered physically, emotionally, or economically because of the accident; and they may well have been the most relevant to the determination of late health effects of accident exposure. In addition, a dilution phenomenon related to the moving out of exposed and moving in of nonexposed subjects was to be expected. In order to avoid these sources of bias, a cohort approach was adopted. The study population was thus comprised of all persons ever resident in one of the 11 towns within the accident scenario, at any time from the date of the accident onwards (including newborns and immigrants), irrespective of their current residence. The information about towns and street addresses allowed attributing subjects to one of the three exposure zones or to the surrounding noncontaminated area. Admission into the study cohort was discontinued as of December 31, 1986; no potential for exposure was deemed to exist any more for newcomers into the area after that date.

The follow-up was based on individual information recorded on vital statistics registries which are maintained by every municipality in Italy. When a person moved outside the study area, the towns concerned were contacted successively, until the person was located. The tracing turned out to be successful for over 99% of study subjects, and for them vital status and, when deceased, cause-of-death information became available.

Because of the absence of a national registration system of cancer cases, the cancer incidence study had, instead, to be limited to people residing within the Lombardy region (nearly 9,000,000 inhabitants). The linkage of the information on all hospitalizations in the region with the records of cohort members allowed the identification of the study subjects admitted/discharged with a diagnosis mentioning cancer. Original medical records were reexamined to ascertain true diagnosis and date of occurrence of cancer. The ascertainment rate for cancer morbidity was close to 95%.

People living in the territory surrounding Zone R, not contaminated by TCDD, were the source of reference data. They shared with the index population the main characteristics related to living and occupational environment, personal habits, and social and educational background.

Results for mortality in the $10^1/_2$-year period following the accident have been reported.[99] A statistically significant increase in mortality from chronic ischemic heart disease was noted in males in the early postaccident period. The increase was highest [relative risk (RR) = 3.2] in Zone A. Among females, an increased pattern of cardiovascular mortality was noted (RR in Zone A = 1.9). Two mechanisms have been hypothesized to explain the increased pattern, i.e., TCDD toxicity and postdisaster stress which might have precipitated preexisting conditions. Both may have contributed.[100] The relative risk for all cancer deaths was slightly below 1. Among males, suggestive increases were seen for soft tissue sarcomas, melanoma, and myeloid leukemia.

The mortality study was descriptive in nature, hence results did not permit conclusively associating any of the unusual cancer mortality findings with exposure to TCDD in 1976. Other limiting factors were the short time period elapsed since first exposure, the small number of deaths from certain causes, and exposure definition based on soil TCDD levels rather than on individual biological indicators.

Cancer incidence findings for the first postaccident decade are available and are summarized in Table 12. Cancer cases in Zone A were too few (seven cases among males and seven among females) to elicit any meaningful conclusion. In Zone B, where the relative risk for all cancers was exactly 1.0, four specific relative risks attracted attention: two were increases (hepatobiliary tract and hematopoietic tissue) and two decreases (breast and uterus). Liver is certainly one of the target organs of TCDD toxicity. The noted hepatobiliary increase was mainly sustained by primary liver cancer in men (four cases, RR = 2.1; confidence interval 95%, CI_{95} = 0.8–5.7), and by extrahepatic bile ducts and gallbladder cancer in women (four cases, RR = 4.8, CI_{95} = 1.7–13.5, statistically significant). Another suggested site of TCDD action is

Table 12
Cancer Incidence, 1977–1986, in the Population Aged 20–74 Years Living in the TCDD-Contaminated Area: Results for Selected Cancers among Males and Females Combined

	Zone B			Zone R		
Cancer site	Cases observed	RR[a]	CI_{95}[b]	Cases observed	RR	CI_{95}
All cancers	115	1.0	0.8–1.2	790	0.9	0.9–1.0
Digestive system	30	1.0	0.7–1.4	211	0.9	0.8–1.0
Hepatobiliary tract	10	2.3	1.2–4.4	23	0.7	0.5–1.1
Respiratory system	24	1.0	0.7–1.5	163	1.0	0.8–1.1
Soft tissue sarcoma	0	—		8	2.3	1.0–5.1
Skin	5	0.9	0.4–2.1	41	1.0	0.7–1.4
Breast[c]	10	0.7	0.4–1.3	113	1.1	0.9–1.3
Genitourinary	18	0.9	0.5–1.4	133	0.9	0.7–1.1
Uterus	2	0.4	0.1–1.5	23	0.6	0.4–0.9
Hematopoietic tissue	15	2.1	1.2–3.5	45	0.9	0.6–1.2
Hodgkin's disease	3	2.6	0.9–9.0	7	0.9	0.4–2.0
Non-Hodgkin's lymphoma	4	1.6	0.6–4.3	23	1.3	0.8–2.0
Multiple myeloma	4	3.9	1.4–1.7	4	0.5	0.2–1.4
Myeloid leukemia	3	2.8	0.9–9.0	7	0.9	0.4–2.1

[a]RR, relative risk.
[b]CI_{95}, 95% confidence interval.
[c]Females only.

hematopoietic tissue. The increase in Zone B was statistically significant, and consistently affected males and females (RR = 2.3 and 1.9, respectively). In particular, lymphoreticulosarcoma among males (three cases, RR = 5.3, CI_{95} = 1.6–17.5), and multiple myeloma among females (two cases, RR = 5.1, CI_{95} = 1.2–21.6) showed statistically significant increases.[101] Thus, the population of Zone B exhibited the clearest suggestions of a possibly increased cancer occurrence, a finding that might be consistent with their postaccident experience (they remained in the contaminated area, and their compliance to restrictive regulations was never evaluated). The low incidence of estrogen-dependent cancers (breast and uterus) was also of great interest since TCDD is known to exert a powerful antiestrogenic action.[102]

The remarkable result in Zone R was, instead, the elevated risk of soft tissue sarcomas, along with the significantly lowered risk of uterine cancer. Soft tissue sarcoma is another tumor associated by several investigations with TCDD exposure.[103] The increase noted in Zone R was twofold, and of borderline statistical significance.

Relevant to the hypothesis of an association of the noted cancer occurrence pattern with TCDD exposure was the fact that the largest increases for hepatobiliary cancer, hematologic neoplasms, and soft tissue sarcomas, as well as the most striking decreases for uterus (mainly corpus uteri) and breast cancer, were estimated among people residing for the longest period in the contaminated area.[101]

Table 13
Cancer Incidence, 1977–1986, in the Population Aged 0–19 Years Living in the TCDD-Contaminated Area: Results for Selected Cancers

Cancer sites	Cases observed	Cases expected	Relative risk	95% confidence intervals
All	23	18.2	1.26	0.8–2.0
Ovary and uterine adnexa	2	0.0	—	—
Nervous system	5	3.4	1.45	0.5–3.9
Brain	4	3.0	1.32	0.4–4.0
Thyroid[a]	2	0.4	4.66	0.7–33.1
Non-Hodgkin's lymphoma	2	1.3	1.54	0.3–7.6
Hodgkin's lymphoma	3	1.9	1.54	0.4–5.7
Lymphatic leukemia	3	2.8	1.07	0.3–3.7
Myeloid leukemia	2	0.9	2.30	0.6–9.2

[a]Cases are restricted to females.

The above results concern people 20–74 years of age. Mortality and cancer incidence of the young members of the cohort (1–20 years of age) were analyzed separately. Mortality data showed an increase of leukemia deaths above expectations, although statistically nonsignificant, in both males and females. A suggestive increase of congenital anomalies was also noted; however, five out of the seven observed anomalies in the contaminated area turned out to have occurred in children born before the accident.[91] Cancer incidence was only slightly above expectations. Given the small number of events, results are presented in Table 13 for the entire contaminated area (A + B + R). Of some interest were the increased risks for thyroid cancer and myeloid leukemia, even though statistically nonsignificant. The suggestion of a possible association between exposure to TCDD and thyroid cancer finds support in experimental and human data. The increased risk of myeloid leukemia is quite consistent with the findings in the adult population.[92]

Once again, results of this analysis should be viewed with caution because of the limitations of the study, in particular the very limited number of events involved and the absence of information on individual exposure.

To interpret these interim results of the long-term investigations, one can note that in Zone A, where inhabitants certainly had a heavy, short-term (few weeks) exposure, the population size was small, and the study had no power to detect rare events. No cases of cancer that could be considered as "sentinel" events occurred. In Zone B the population was larger, and hence there was greater power of the study, while exposure was on the average lower, but potentially much longer than in Zone A. Here, two types of cancer known to probably be associated with TCDD exposure were in excess (hepatobiliary and lymphopoietic), and one can hypothesize that a balance of exposure intensity and sample size was probably achieved in this population subgroup. Another very rare tumor possibly associated with TCDD (i.e., soft tissue sarcoma) was, instead, increased in the largest but least contaminated area (Zone R). Taken together, these results support the hypothesis of an increased, albeit modestly so, cancer risk, and certainly motivate further investigations. However, they do not prove conclusively that an association with TCDD exists. In fact, results only refer to the first postaccident decade. Second, exposure assessment was based on environmental measurements only, and exposure categories were based on official residence within zones whose boundaries seemed, at least in part, debatable. Finally, because of the rarity of the health outcomes of interest, this relatively large study suffered from sample size problems.

10. OVERALL INTERPRETATION AND PERSPECTIVES

Overall, three sets of health results are available, which are summarized in the Appendix. Their interpretation is limited by the weaknesses inherent in several studies.

Results of short- and mid-term surveillance programs clearly documented that the accident caused toxic damage to the population. Chloracne was the only health effect consistently established. Other health outcomes known to be possibly associated with dioxin exposure were investigated. For none of them could an unusual pattern, either for the frequency or for the type of outcome concerned, attributable to TCDD be firmly established. These investigations, however, suffered from all of the constraints related to a postdisaster scenario, which often precluded proper design (e.g., lack of referent subjects), efficient conduct (e.g., compliance to the scheduled examinations), and valid conclusion (e.g., information and selection bias).

The second set of data came from the follow-up of small, selected groups, mainly chloracne children. Results failed to show health impairment in these heavily exposed subjects up to 8–9 years after the accident. In addition, they showed that chloracne is reversible and not necessarily associated, at least in the time span considered, with other systemic effects. Longer follow-up and larger size are necessary to corroborate these conclusions. Moreover, results in these few subjects cannot be easily generalized until susceptibility factors are taken into consideration.

The last set of data are provided by early results of long-term investigations. After 10 years, these investigations uncovered an increase of cardiovascular mortality early after the accident possibly explicable in terms of precipitation of preexisting diseases caused by the stressful experience following the chemical disaster. In addition, a slight departure from the background occurrence of certain tumors was observed. All cancer departures from expectations are consistent with previous experimental and human data and are plausible from a biological perspective. However, the small number of events, and the still short observation period prevent sound conclusions. An additional limitation is exposure classification based on soil contamination data which do not actually indicate exposure and body burden.

The Seveso cohort provides unique opportunities for the investigation of TCDD effects in humans. In fact, it is the largest observable group of persons with documented special exposure to TCDD with the possible exception of Agent Orange in Vietnam; the blood levels measured are the highest so far reported, and exposure seemed relatively free of other known or persistent contaminants.

A great deal of work is still to be done, and research plans are being adjusted to reflect the rapidly increasing knowledge of TCDD toxicity, mode and mechanisms of action, environmental movement, targets, and so forth. Three research paths may be particularly rewarding and are presently being pursued.

The first is related to the fact that time since initial exposure was too short to reveal a clear pattern of possible TCDD cancer effects. Accordingly, the study will continue so as to examine the health experience of the population for at least two full decades after the accident.

The second path stems from the fact that exposure to TCDD probably varied among people even within the same contamination zone. Thus, zone categorization did not reflect either in absolute or relative terms the actual level and difference of TCDD exposure. A better definition of individual exposure would greatly improve interpretation of cancer findings. To this end, in a sample of the population, the association with TCDD of the cancer types which exhibited an increased incidence will be studied in more depth with a case–control approach, aimed particularly at better qualifying and quantifying exposure (also with the use of biological markers) such as blood dioxin measurements, and controlling sources of confounding or modification of the effect (e.g., occupation, diet, personal habits).

Meanwhile, TCDD levels in the blood of Seveso subjects are being measured from previous specimens and additional samples are expected to be measured.[78] Recently, Needham *et al.*[104] presented some preliminary findings concerning Zone B residents: since detected TCDD levels appeared to be approximately 5–20 times as high as background, the conclusion was drawn that in 1976 some of these residents were heavily exposed. In addition, subjects who immigrated into Zone B after July of 1976 were tested for TCDD in their blood after some years of residence: the levels detected were comparable with background.

A third promising study concerns the existence of a genetically determined interindividual variability modulating TCDD effects in human cells. In particular, to evaluate the role of susceptibility factors in determining the cancer risk, cancer cases and controls with the same potential for exposure will be studied and examined with a variety of laboratory assays (including Ah receptor, gene markers of polymorphism and inducibility, blood TCDD.)

These studies are foreseen to last for a long period of time, and are expensive and complex. A sensible question is: are they worth conducting? Our answer is yes, if one considers that TCDD is now known to be a widespread contaminant of our environment and represents one of the major environmental health problems in industrial countries. An improved knowledge of its effects on humans will constitute an important contribution to the protection of human health and quality of our environment.

ACKNOWLEDGMENTS. We gratefully acknowledge the help of Professor P. Mocarelli, University Department of Clinical Pathology, Hospital of Desio (Milan), and Dr. G. Belli, Department of Nuclear and Theoretical Physics, University of Pavia (Pavia), for having provided some reference material for the pertinent sections of this chapter. We are also indebted to G. Briancesco of the Istituto Superiore di Sanità for preparing the drawings, and to S. Holt for language assistance in preparing the manuscript.

APPENDIX: SYNOPSIS OF THE HEALTH INVESTIGATIONS PERFORMED ON THE POPULATION POSSIBLY EXPOSED TO TCDD FOLLOWING THE SEVESO, ITALY, ACCIDENT IN 1976

Population	Outcome	Results (references)	Period	Remarks
Dermatologic screening of some 16,000 children aged 0–14 years. Adults spontaneously reporting for examination	Chloracne	A total of 187 cases asertained (164 aged <14 years). Strong positive association of frequency and severity of cases with soil TCDD level. 20% of Zone A children affected.[80,82]	1976–77	Seven additional cases diagnosed during 1977–78. No further cases after 1978.
Children with chloracne (n = 146), controls (n = 182)	Symptoms and lab tests	Increased frequency of GI symptoms, headache, and eye irritation. Higher frequency of abnormal GGT, GPT, and ALA-U.[80]	1977	Probable interview/reporting bias for symptoms. Controls not properly selected.
Some 300 residents evacuated from Zone A and some 300 referents	Peripheral neuropathy	Two- to threefold increased frequency of abnormalities among subjects with signs of TCDD exposure (chloracne or increased liver enzyme levels).[84]	1977–78	No acute polyneuropathy found. All nonsevere cases.
Women admitted to obstetrics departments in Lombardy and residing in the polluted area	Spontaneous abortions	No clear changes in pregnancy loss rate by trimester or towns of residence.[18]	1976–78	No historical data for comparison. Poor data quality.
Census and vital statistics data	Spontaneous abortions	No apparent increase.[82]	1977–80	Indirect estimate. Possible reporting bias.
Pregnancies in the contaminated area	Spontaneous abortions	Slight increase in 1977 compared with 1976. Abortion rates lower than external reference. Internal comparison: highest rates in Zone B.[86]	1973–77	Possible reporting bias. Lack of consistency.
Eight children and four pregnant women with skin lesions; 10 adult controls	Chromosome aberrations	No significant increase in peripheral lymphocytes.[87]	1976	No suitable controls available. Small size.
145 Zone A residents, 69 plant workers, and 87 controls	Chromosome aberrations	Suggestive, nonsignificant increase in Zone A residents and plant workers.[87]	1976–77	Large variation mainly accounted for by interobserver differences.
Induced abortions in TCDD-exposed (n = 19) and nonexposed (n = 16) women	Chromosome aberrations	No increase in exposed maternal blood and placenta. Possible increase in exposed fetal tissues.[88]	1976–79	Technical factors other than exposure relevant to interpretation.

(*continued*)

APPENDIX (*Continued*)

Population	Outcome	Results (references)	Period	Remarks
Induced (n = 30) and spontaneous (n = 4) abortions in exposed women	Birth defects	No indications of mutagenic, teratogenic, and fetotoxic effects.[89]	1976	Incomplete fetal tissues for examination.
Some 15,000 births to TCDD-exposed and nonexposed women	Birth defects	Frequency in exposed not significantly higher. No association of soil contamination levels with type/frequency of defects.[90]	1977–82	Small size to detect very rare defects.
48 Zone A children and 48 controls	Immune function	Increased complement activity, lymphocyte response to lectins, and number of peripheral lymphocytes in exposed.[94]	1976–79	Examination repeated every 5–6 months. Poor compliance of controls. Increased immune reactivity followed by immunodepression?
Adults (n = 117) and children from Zones A (n = 81), B (n = 112), and R (n = 121)	Enzyme induction	D-Glucaric acid excretion increased in exposed at all ages. Highest values in chloracne cases. After 1980 return to normal values.[85]	1976–81	In 1979 Zone B children showed highest excretion levels. High exposure in parts of Zone R suggested.
Children 6–10 years from Zone A (n = 69), B (n = 83), and R (n = 241)	Liver enzyme Lipid metabolism	Slight abnormalities in GGT and ALT activity in boys with highest exposure. Abnormalities disappeared with time.[95]	1976–82	Examination repeated every year.
152 chloracne cases and 123 controls	Peripheral neuropathy	Increased frequency of clinical and electrophysiological signs of PNS involvement.[93]	1982–83	No evidence of clinical peripheral neuropathy according to WHO criteria.
Some 150 chloracne cases and a comparison group from an unexposed area	Skin lesion Neurophysiology Liver enzyme Lipid metabolism	No significant differences between groups. No temporal trends. Absence of abnormalities 10 years after exposure.[96]	1982–85	In 1984, one subject still had chloracne and five had persisting scars.

36 decontamination workers and 36 unexposed controls	Urine Porphyrin Liver enzyme Lipid metabolism Blood count	No significant differences. No changes over time. No association of tests mean values with length of work.[97,98]	1980–84	Environmental measures and personal protection to minimize exposure adopted. Preemployment health examination.
20 Zone A residents and 10 unexposed subjects	Serum TCDD Liver function Lipid metabolism Immune function Enzyme induction	In chloracne cases highest TCDD concentration ever reported in humans. Very high concentration also in Zone A residents without chloracne. TCDD not detected in unexposed. No significant alterations in any of the tests.[79]	1976–85	Small, highly selected sample. Generalization of results premature.
All persons aged 20–74 years living in contaminated zones and some 180,000 reference subjects	Mortality	Increased cardiovascular mortality early after the accident in the most affected zone. Suggested departure from expectations for several cancers.[99]	1976–86	Probable "disaster" (stress) rather than TCDD effect. Short latency. Ecological definition of exposure.
All persons aged 0–19 years living in contaminated zones and some 60,000 reference subjects	Mortality	Suggestive, nonsignificant increase of leukemia.[91]	1976–86	Same limitations as above, plus small size.
All persons aged 20–74 years living in contaminated zones and some 180,000 reference subjects	Cancer incidence	Increased occurrence of hepatobiliary cancer, soft tissue sarcoma, and hematologic neoplasms. Decrease of estrogen-dependent tumors.[101]	1977–88	Consistency with existing experimental and human evidence. Short latency. Partial control of confounding.
All persons aged 0–19 years living in contaminated zones and some 60,000 reference subjects	Cancer incidence	Suggestive, nonsignificant increase of myeloid leukemia and thyroid cancer.[92]	1977–86	Same limitations as above, plus small size.

REFERENCES

1. U.S. EPA, Interim Procedures for Estimating Risks Associated with Exposures to Mixtures of Chlorinated Dibenzo-*p*-dioxins and Dibenzofurans (CDDs and CDFs) and 1989 Update, EPA/625/3-89/016, Risk Assessment Forum, Environmental Protection Agency, Washington, DC (1989).
2. WHO, Polychlorinated Dibenzo-*para*-dioxins and Dibenzofurans, *Environmental Health Criteria 88,* International Programme on Chemical Safety, World Health Organization, Geneva (1989).
3. CCTN, Parere della Commissione Consultiva Tossicologica Nazionale sui PCDD e PCDF—Valutazione Tossicologica delle Policlorodibenzodiossine (PCDD) e dei Policlorodibenzofurani (PCDF) in Riferimento alla Loro Presenza nell'Ambiente, *Serie Relazioni* **89**(3), pp. 5–55, Instituto Superiore di Sanità, Rome (1989).
4. A. di Domenico, Guidelines for the definition of environmental action alert thresholds for polychlorodibenzodioxins and polychlorodibenzofurans, *Regul. Toxicol. Pharmacol.* **11,** 8–23 (1990).
5. V. Silano, Case study: Accidental release of 2,3,7,8-tetrachlorodibenzo-*p*-dioxin (TCDD) at Seveso, Italy, in: *Emergency Response to Chemical Accidents—Interim Document 1* (P. H. Jones and A. Gilad, eds.), pp. 167–203, International Programme on Chemical Safety, Regional Office for Europe, World Health Organization, Copenhagen (1981).
6. G. U. Fortunati, The Seveso accident, *Chemosphere* **14,** 729–737 (1985).
7. A. di Domenico, V. Silano, G. Viviano, and G. Zapponi, in: *A Report of NATO/CCMS Working Group on Management of Accidents Involving the Release of Dioxins and Related Compounds* (A. di Domenico and A. E. Radwan, eds.), ISTISAN 88/8, pp. 125–134, Istituto Superiore di Sanità, Rome (1988).
8. G. Reggiani, in: *Agent Orange and Its Associated Dioxin: Assessment of a Controversy* (A. L. Young and G. M. Reggiani, eds.), pp. 227–269, Elsevier, Amsterdam (1988).
9. J. Sambeth, What really happened at Seveso, *Chem. Eng.* **90,** 44–47 (1983).
10. R. L. Rawls and D. A. O'Sullivan, Italy seeks answers following toxic release, *Chem. Eng. News* **54,** 27–35 (1976).
11. A. W. M. Hay, Tetrachlorodibenzo-*p*-dioxin release at Seveso, *Disasters* **1,** 289–308 (1977).
12. J. Peterson, Seveso: The event, *Ambio* **7,** 232–233 (1978).
13. A. di Domenico, V. Silano, G. Viviano, and G. Zapponi, Accidental release of 2,3,7,8-tetrachlorodibenzo-*p*-dioxin (TCDD) at Seveso, Italy. II. TCDD distribution in the soil surface layer, *Ecotoxicol. Environ. Saf.* **4,** 298–320 (1980).
14. S. Cerlesi, A. di Domenico, and S. P. Ratti, 2,3,7,8-Tetrachlorodibenzo-*p*-dioxin (TCDD) persistence in the Seveso (Milan, Italy) soil, *Ecotoxicol. Environ. Saf.* **18,** 149–164 (1989).
15. M. H. Milnes, Formation of 2,3,7,8-tetrachlorodibenzodioxin by thermal decomposition of sodium 2,4,5-trichlorophenate, *Nature* **232,** 395—396 (1971).
16. L. Canonica, Seveso: Considerazioni e commenti, *Chim. Ind.* (Milan) **59,** 87–89 (1977).
17. G. U. Fortunati, in: *Technological Response to Chemical Pollutions* (P. Bonizzoni and S. Meroni, eds.), pp. 47–60, Ufficio Speciale di Seveso, Regione Lombardia, Milan (1985).
18. A. Bonaccorsi, R. Fanelli, and G. Tognoni, In the wake of Seveso, *Ambio* **7,** 234–239 (1978).
19. F. Pocchiari, A. di Domenico, V. Silano, and G. Zapponi, Accidental release of 2,3,7,8-

tetrachlorodibenzo-*p*-dioxin (TCDD) at Seveso: Assessment of environmental contamination and of effectiveness of decontamination treatments, in: *The Proceedings of the Sixth International CODATA Conference* (B. Dreyfus, ed.), pp. 31–37, Pergamon Press, Elmsford, NY (1979).

20. F. Pocchiari, A. di Domenico, V. Silano, and G. Zapponi, in: *Accidental Exposure to Dioxins—Human Health Aspects* (F. Coulston and F. Pocchiari, eds.), pp. 5–35, Academic Press, New York (1983).
21. F. P. Foraboschi and G. U. Fortunati, La decontaminazione di estese superfici e lo smaltimento di grandi volumi di materiali a basso inquinamento di diossina, in: *Technological Response to Chemical Pollutions* (P. Bonizzoni and S. Meroni, eds.), pp. 35–45, Ufficio Speciale di Seveso, Regione Lombardia, Milan (1985).
22. F. Pocchiari, V. Silano, and G. Zapponi, The chemical risk management process in Italy. A case study: The Seveso accident, *Sci. Total Environ.* **51,** 227–235 (1986).
23. S. Cerlesi, A. di Domenico, and S. P. Ratti, Recovery yields of early analytical procedures to detect 2,3,7,8-tetrachlorodibenzo-*p*-dioxin (TCDD) in soil samples at Seveso, Italy, *Chemosphere* **18,** 989–1003 (1989).
24. F. Cattabeni, A. di Domenico, and F. Merli, Analytical procedures to detect 2,3,7,8-TCDD at Seveso after the industrial accident of July 10, 1976, *Ecotoxicol. Environ. Saf.* **12,** 35–52 (1986).
25. F. Pocchiari, F. Cattabeni, G. Della Porta, G. U. Fortunati, V. Silano, and G. Zapponi, Assessment of exposure to 2,3,7,8-tetrachlorodibenzo-*p*-dioxin (TCDD) in the Seveso area, *Chemosphere* **15,** 1851–1865 (1986).
26. G. U. Fortunati and V. La Porta, in: *A Report of NATO/CCMS Working Group on Management of Accidents Involving the Release of Dioxins and Related Compounds* (A. di Domenico and A. E. Radwan, eds.), pp. 49–52, ISTISAN 88/8, Instituto Superiore di Sanità, Rome (1988).
27. A. di Domenico, V. Silano, G. Viviano, and G. Zapponi, Accidental release of 2,3,7,8-tetrachlorodibenzo-*p*-dioxin (TCDD) at Seveso, Italy. I. Sensitivity and specificity on analytical procedures adopted for TCDD assay, *Ecotoxicol. Environ. Saf.* **4,** 283–297 (1980).
28. G. Belli, S. Cerlesi, and S. P. Ratti, in: *Technological Response to Chemical Pollutions* (P. Bonizzoni and S. Meroni, eds.), pp. 129–133, Ufficio Speciale di Seveso, Regione Lombardia, Milan (1985).
29. S. P. Ratti, G. Belli, A. Lanza, S. Cerlesi, and G. U. Fortunati, The Seveso dioxin episode: Time evolution properties and conversion factors between different analytical methods, *Chemosphere* **15,** 1549–1556 (1986).
30. V. La Porta, M. Occa, and M. Matteo, in: *Technological Response to Chemical Pollutions* (P. Bonizzoni and S. Meroni, eds.), pp. 135–153, Ufficio Speciale di Seveso, Regione Lombardia, Milan (1985).
31. G. Belli, S. Cerlesi, E. Milani, and S. P. Ratti, Statistical interpolation model for the description of ground pollution due to TCDD produced in the 1976 chemical accident at Seveso in the heavily contaminated Zone A, *Toxicol. Environ. Chem.* **22,** 101–130 (1989).
32. G. Belli, G. Bressi, E. Calligarich, S. Cerlesi, and S. P. Ratti, in: *Chlorinated Dioxins and Related Compounds—Impact on the Environment* (O. Hutzinger, R. W. Frei, E. Merian, and F. Pocchiari, eds.), pp. 155–172, Pergamon Press, Elmsford, NY (1982).
33. A. Cavallaro, G. Tebaldi, and R. Gualdi, Analysis of transport and ground deposition of the TCDD emitted on 10 July 1976 from the ICMESA factory (Seveso, Italy), *Atmos. Environ.* **16,** 731–740 (1982).
34. S. P. Ratti, G. Belli, A. Lanza, and S. Cerlesi, in: *Chlorinated Dioxins and Dibenzofurans*

in Perspective (C. Rappe, G. Choudhary, and L. H. Keith, eds.), pp. 467–476, Lewis Publishers, Chelsea, MI (1986).
35. S. P. Ratti, G. Belli, P. A. Bertazzi, G. Bressi, S. Cerlesi, and F. Panetsos, TCDD distribution on all the territory around Seveso: Its use in epidemiology and a hint into dynamical models, *Chemosphere* **16,** 1765–1773 (1987).
36. S. Cerlesi, G. Belli, and S. P. Ratti, in: *Technological Response to Chemical Pollutions* (P. Bonizzoni and S. Meroni, eds.), pp. 121–128, Ufficio Speciale di Seveso, Regione Lombardia, Milan (1985).
37. G. Belli, G. Bressi, L. Carrioli, S. Cerlesi, M. Diani, S. P. Ratti, and G. Salvadori, An attempt to provide a fractal model for the description of the TCDD distribution in all the territory around Seveso (Milan, Italy), *Chemosphere* **20,** 1567–1573 (1990).
38. G. Belli, S. P. Ratti, and G. Salvadori, An empirical fractal model for the TCDD distribution on the Seveso (Milan, Italy) territory, *Toxicol. Environ. Chem.* **33,** 201–218 (1991).
39. G. U. Fortunati, in: *A Report of NATO/CCMS Working Group on Management of Accidents Involving the Release of Dioxins and Related Compounds* (A. di Domenico and A. E. Radwan, eds.), pp. 113–120, ISTISAN 88/8, Instituto Superiore di Sanità, Rome (1988).
40. R. J. Kociba, P. A. Keeler, C. N. Park, and P. J. Gehring, 2,3,7,8-Tetrachlorodibenzo-*p*-dioxin (TCDD): Results of a 13-week oral toxicity study in rats, *Toxicol. Appl. Pharmacol.* **35,** 553–574 (1976).
41. R. J. Kociba, D. G. Keyes, J. E. Beyer, R. M. Carreon, C. E. Wade, D. A. Dittenber, R. P. Kalnins, L. E. Frauson, C. N. Park, S. D. Barnard, R. A. Hummel, and C. G. Humiston, Results of a two-year chronic toxicity and oncogenicity study of 2,3,7,8-tetrachlorodibenzo-*p*-dioxin in rats, *Toxicol. Appl. Pharmacol.* **46,** 279–303 (1978).
42. A. Bonaccorsi, A. di Domenico, R. Fanelli, F. Merli, R. Motta, R. Vanzati, and G. A. Zapponi, The influence of soil particle adsorption on 2,3,7,8-tetrachlorodibenzo-*p*-dioxin biological uptake in the rabbit, *Arch. Toxicol. Suppl.* **7,** 431–434 (1984).
43. A. di Domenico, V. Silano, G. Viviano, and G. Zapponi, Accidental release of 2,3,7,8-tetrachlorodibenzo-*p*-dioxin (TCDD) at Seveso, Italy. III. Monitoring of residual TCDD levels in reclaimed buildings, *Ecotoxicol. Environ. Saf.* **4,** 321–326 (1980).
44. R. J. Kociba, D. G. Keyes, J. E. Beyer, R. M. Carreon, and P. J. Gehring, Long-term toxicologic studies of 2,3,7,8-tetrachlorodibenzo-*p*-dioxin (TCDD) in laboratory animals, *Ann. N.Y. Acad. Sci.* **320,** 397–404 (1979).
45. R. J. Kociba and B. A. Schwetz, Toxicity of 2,3,7,8-tetrachlorodibenzo-*p*-dioxin (TCDD), *Drug Metab. Rev.* **13,** 387–406 (1982).
46. NTP, Carcinogenesis Bioassay of 2,3,7,8-Tetrachlorodibenzo-*p*-dioxin (CAS No. 1746-01-6) in Swiss–Webster Mice (Dermal Study), National Toxicology Program, Technical Report Series No. 201, NIH Publication No. 82-1757, National Institutes of Health, Research Triangle Park, NC (1982).
47. NTP, Carcinogenesis Bioassay of 2,3,7,8-Tetrachlorodibenzo-*p*-dioxin (CAS No. 1746-01-6) in Osborne–Mendel Rats and B6C3F1 Mice (Gavage Study), National Toxicology Program, Technical Report Series No. 209, NIH Publication No. 82-1765, National Institutes of Health, Research Triangle Park, NC (1982).
48. R. D. Kimbrough, H. Falk, P. Stehr, and G. Fries, Health implications of 2,3,7,8-tetrachlorodibenzodioxin (TCDD) contamination of residential soil, *J. Toxicol. Environ. Health* **14,** 47–93 (1984).
49. WHO, Consultation on Tolerable Daily Intake from Food of PCDDs and PCDFs, Regional Office for Europe, World Health Organization (Copenhagen): Summary report of the meeting held in Bilthoven, Netherlands, December 4–7, 1990 (1991).

50. A. di Domenico and G. A. Zapponi, 2,3,7,8-Tetrachlorodibenzo-*p*-dioxin (TCDD) in the environment: Human health risk estimation and its application to the Seveso case as an example, *Regul. Toxicol. Pharmacol.* **6,** 248–260 (1986).
51. V. La Porta and G. U. Fortunati, in: *A Report of NATO/CCMS Working Group on Management of Accidents Involving the Release of Dioxins and Related Compounds* (A. di Domenico and A. E. Radwan, eds.), ISTISAN 88/8, pp. 74–77, Istituto Superiore di Sanità, Rome (1988).
52. A. di Domenico, V. Silano, G. Viviano, and G. Zapponi, Accidental release of 2,3,7,8-tetrachlorodibenzo-*p*-dioxin (TCDD) at Seveso, Italy. IV. Vertical distribution of TCDD in soil, *Ecotoxicol. Environ. Saf.* **4,** 327–338 (1980).
53. G. Tarelli, R. Pezzano, and F. Pinelli, in: *Technological Response to Chemical Pollutions* (P. Bonizzoni and S. Meroni, eds.), pp. 73–79, Ufficio Speciale di Seveso, Regione Lombardia, Milan (1985).
54. A. Piepoli and P. Federico, Aspetti ingegneristici degli interventi di bonifica, in: *Technological Response to Chemical Pollutions* (P. Bonizzoni and S. Meroni, eds.), pp. 61–72, Ufficio Speciale di Seveso, Regione Lombardia, Milan (1985).
55. G. Belli, G. Bressi, E. Calligarich, S. Cerlesi, and S. P. Ratti, in: *Chlorinated Dioxins and Related Compounds—Impact on the Environment* (O. Hutzinger, R. W. Frei, E. Merian, and F. Pocchiari, eds.), pp. 137–153. Pergamon Press, Elmsford, NY (1982).
56. A. di Domenico, V. Silano, G. Viviano, and G. Zapponi, Accidental release of 2,3,7,8-tetrachlorodibenzo-*p*-dioxin (TCDD) at Seveso, Italy. VI. TCDD levels in atmospheric particles, *Ecotoxicol. Environ. Saf.* **4,** 346–356 (1980).
57. H. K. Wipf, E. Homberger, N. Neuner, U. B. Ranalder, W. Vetter, and J. P. Vuilleumier, in: *Chlorinated Dioxins and Related Compounds—Impact on the Environment* (O. Hutzinger, R. W. Frei, E. Merian, and F. Pocchiari, eds.), pp. 115–126, Pergamon Press, Elmsford, NY (1982).
58. S. Cocucci, F. Di Gerolamo, A. Verderio, A. Cavallaro, G. Colli, A. Gorni, G. Invernizzi, and L. Luciani, Absorption and translocation of tetrachlorodibenzo-*p*-dioxin by plants from polluted soil, *Experientia* **35,** 482–484 (1979).
59. S. Facchetti, A. Balasso, C. Fichtner, G. Frare, A. Leoni, and C. Mauri, in: *Technological Response to Chemical Pollutions* (P. Bonizzoni and S. Meroni, eds.), pp. 231–242, Ufficio Speciale di Seveso, Regione Lombardia, Milan (1985).
60. S. Facchetti, A. Balasso, C. Fichtner, G. Frare, A. Leoni, C. Mauri, and M. Vasconi, Studies on the absorption of TCDD by some plants species, *Chemosphere* **15,** 1387–1388 (1986).
61. G. A. Sacchi, P. Viganò, G. Fortunati, and S. M. Cocucci, Accumulation of 2,3,7,8-tetrachlorodibenzo-*p*-dioxin from soil and nutrient solution by bean and maize plants, *Experientia* **42,** 586–588 (1986).
62. R. Fanelli, M. G. Castelli, G. P. Martelli, A. Noseda, and S. Garattini, Presence of 2,3,7,8-tetrachlorodibenzo-*p*-dioxin in wildlife living near Seveso, Italy: A preliminary study, *Bull. Environ. Contam. Toxicol.* **24,** 460–462 (1980).
63. R. Fanelli, C. Chiabrando, and A. Bonaccorsi, TCDD contamination in the Seveso incident, *Drug. Metab. Rev.* **13,** 407–422 (1982).
64. R. Fanelli, M. P. Bertoni, M. G. Castelli, C. Chiabrando, G. P. Martelli, A. Noseda, S. Garattini, C. Binaghi, V. Marazza, and F. Pezza, 2,3,7,8-Tetrachlorodibenzo-*p*-dioxin toxic effects and tissue levels in animals from the contaminated area of Seveso, Italy, *Arch. Environ. Contam. Toxicol.* **9,** 569–577 (1980).
65. R. Fanelli, M. P . Bertoni, M. Bonfanti, M. G. Castelli, C. Chiabrando, G. P. Martelli,

M. A. Noè, A. Noseda, S. Garattini, C. Binaghi, V. Marazza, F. Pezza, D. Pozzoli, and G. Cicognetti, 2,3,7,8-Tetrachlorodibenzo-*p*-dioxin levels in cow's milk from the contaminated area of Seveso, Italy, *Bull. Environ. Contam. Toxicol.* **24,** 634–639 (1980).
66. A. di Domenico, V. Silano, G. Viviano, and G. Zapponi, Accidental release of 2,3,7,8-tetrachlorodibenzo-*p*-dioxin (TCDD) at Seveso, Italy. V. Environmental persistence of TCDD in soil, *Ecotoxicol. Environ. Saf.* **4,** 339–345 (1980).
67. A. di Domenico, G. Viviano, and G. Zapponi, in: *Chlorinated Dioxins and Related Compounds—Impact on the Environment* (O. Hutzinger, R. W. Frei, E. Merian, and F. Pocchiari, eds.), pp. 105–114, Pergamon Press, Elmsford, NY (1982).
68. S. P. Ratti, G. Belli, G. Bressi, S. Cerlesi, and C. Zocchetti, An empirical model to describe the TCDD distribution on all the territory around Seveso and the time dependence of its parameters, *Chemosphere* **18,** 921–924 (1989).
69. A. di Domenico, S. Cerlesi, and S. P. Ratti, A two-exponential model to describe the vanishing trend of 2,3,7,8-tetrachlorodibenzodioxin (TCDD) in the soil at Seveso, northern Italy, *Chemosphere* **20,** 1559–1566 (1990).
70. R. Hütter and M. Philippi, in: *Chlorinated Dioxins and Related Compounds—Impact on the Environment* (O. Hutzinger, R. W. Frei, E. Merian, and F. Pocchiari, eds.), pp. 87–93, Pergamon Press, Elmsford, NY (1982).
71. M. Philippi, J. Schmid, H. K. Wipf, and R. Hütter, A microbial metabolite of TCDD, *Experientia* **38,** 659–661 (1982).
72. S. Facchetti, A. Fornari, and M. Montagna, Distribution of 2,3,7,8-tetrachlorodibenzo-*p*-dioxin in the tissues of a person exposed to the toxic cloud at Seveso, *Adv. Mass Spectrom.* **8B,** 1405–1414 (1980).
73. D. J. Patterson, Jr., L. Hampton, C. R. Lapeza, Jr., W. T. Belser, V. Green, L. Alexander, and L. L. Needham, High-resolution gas chromatographic/high-resolution mass spectrometric analysis of human serum on a whole-weight and lipid basis for 2,3,7,8-tetrachlorodibenzo-*p*-dioxin, *Anal. Chem.* **59,** 2000–2005 (1987).
74. D. G. Patterson, Jr., W. E. Turner, L. R. Alexander, S. Isaacs, and L. L. Needham, The analytical methodology and method performance for the determination of 2,3,7,8-TCDD in serum for the Vietnam veteran Agent Orange validation study, the Ranch Hand validation and half-life studies, and selected NIOSH worker studies, *Chemosphere* **18,** 875–882 (1989).
75. D. G. Patterson, Jr., P. Fürst, L. R. Alexander, S. G. Isaacs, W. E. Turner, and L. L. Needham, Analysis of human serum for PCDDs/PCDFs: A comparison of three extraction procedures, *Chemosphere* **19,** 89–96 (1989).
76. D. G. Patterson, Jr., W. E. Turner, S. G. Isaacs, and L. R. Alexander, A method performance evaluation and lessons learned after analyzing more than 5,000 human adipose tissue, serum, and breast milk samples for polychlorinated dibenzo-*p*-dioxins (PCDDs) and dibenzofurans (PCDFs), *Chemosphere* **20,** 829–836 (1990).
77. P. Mocarelli, F. Pocchiari, and N. Nelson, Preliminary report: 2,3,7,8-Tetrachlorodibenzo-*p*-dioxin exposure to humans–Seveso, Italy, *Morbid. Mortal. Weekly Rep.* **37,** 733–736 (1988).
78. P. Mocarelli, D. G. Patterson, Jr., A. Marocchi, and L. L. Needham, Pilot study (Phase II) for determining polychlorinated dibenzo-*p*-dioxin (PCDD) and polychlorinated dibenzofuran (PCDF) levels in serum of Seveso, Italy, residents collected at the time of exposure: Future plans, *Chemosphere* **20,** 967–974 (1990).
79. P. Mocarelli, L. L. Needham, A. Marocchi, D. G. Patterson, P. Brambilla, P. M. Gerthoux, L. Meazza, and V. Carreri, Serum concentrations of 2,3,7,8-Tetrachlorodibenzo-*p*-

dioxin and test results from selected residents of Seveso, Italy, *J. Toxicol. Environ. Health* **32,** 357–366 (1991).
80. F. Caramaschi, G. Del Corno, C. Favaretti, S. E. Giambelluca, E. Montesarchio, and G. M. Fara, Chloracne following environmental contamination by TCDD in Seveso, Italy, *Int. J. Epidemiol.* **10,** 135–143 (1981).
81. G. Del Corno, C. Favaretti, F. Caramaschi, S. E. Giambelluca, E. Montesarchio, F. Bonetti, and C. Volpato, Distribution of chloracne cases in the Seveso area following contamination by TCDD, *Ig. Mod.* **77,** 635–658 (1982).
82. G. M. Reggiani, in: *Halogenated Biphenyls, Terphenyls, Naphthalenes, Dibenzodioxins and Related Products* (R. D. Kimbrough and A. A. Jensen, eds.), pp. 445–470, Elsevier, Amsterdam (1989).
83. S. E. Giambelluca, C. Favaretti, G. Del Corno, F. Caramaschi, E. Montesarchio, F. Bonetti, and C. Volpato, Chloracne and clinical impairment in a group of subjects over 14 years of age, exposed to TCDD in the Seveso area, *Ig. Mod.* **77,** 675–680 (1982).
84. G. Filippini, B. Bordo, P. Crenna, N. Massetto, M. Musicco, and R. Boeri, Relationship between clinical and electrophysiological findings and indicators of heavy exposure to 2,3,7,8-tetrachlorodibenzo-dioxin, *Scand. J. Work Environ. Health* **7,** 257–262 (1981).
85. G. Ideo, G. Bellati, A. Bellobuono, and L. Bisanti, Urinary D-glucaric acid excretion in the Seveso area, polluted by tetrachlorodibenzo-p-dioxin (TCDD): Five years of experience, *Environ. Health Perspect.* **60,** 151–157 (1985).
86. L. Bisanti, F. Bonetti, F. Caramaschi, G. Del Corno, C. Favaretti, S. Giambelluca, E. Marni, E. Montesarchio, V. Puccinelli, G. Remotti, C. Volpato, and E. Zambrelli, Experience of the accident of Seveso, *Acta Morphol. Acad. Sci. Hung.* **28,** 139–157 (1980).
87. L. De Carli, A. Mottura, F. Nuzzo, G. Zei, M. Tenchini, M. Fraccaro, B. Nicoletti, G. Simoni, and P. Mocarelli, Cytogenetic investigation of the Seveso population exposed to TCDD, in: *Plans for clinical and epidemiologic follow-up after area-wide chemical contamination,* Proceedings of an International Workshop, pp. 292–317, National Academy Press, Washington, DC (1982).
88. M. L. Tenchini, C. Grimaudo, G. Pacchetti, A. Mottura, S. Agosti, and L. De Carli, A comparative cytogenetic study on cases of induced abortions in TCDD-exposed and nonexposed women, *Environ. Mutagen.* **5,** 73–85 (1983).
89. H. Rehder, L. Sanchioni, F. Cefis, and A. Gropp, Pathologisch-embryologische untersuchungen an abortusfallen im Zusammenhang mit dem Seveso-Ungluck, *Schweiz. Med. Wochenschr.* **108,** 1617–1625 (1978).
90. P. P. Mastroiacovo, A. Spagnolo, E. Marni, L. Meazza, R. Bertollini, and G. Segni, Birth defects in the Seveso area after TCDD contamination, *Am. Med. Assoc.* **259,** 1668–1672 (1988).
91. P. A. Bertazzi, C. Zocchetti, A. C. Pesatori, S. Guercilena, D. Consonni, A. Tironi, and M. T. Landi, Mortality of a young population after accidental exposure to 2,3,7,8-tetrachlorodibenzodioxin, *Int. J. Epidemiol.* **21,** 118–123 (1992).
92. A. C. Pesatori, D. Consonni, A. Tironi, M. T. Landi, C. Zocchetti, and P. A. Bertazzi, Cancer morbidity (1977–1986) of a young population living in the Seveso area, in: *Dioxin '92: Toxicology, Epidemiology, Risk Assessment and Management, Organohalogen Compounds,* Vol. 10, pp. 271–273, Finnish Institute of Occupational Health, Helsinki (1992).
93. S. Barbieri, C. Pirovano, G. Scarlato, P. Tarchini, A. Zappa, and M. Maranzana, Long-term effects of 2,3,7,8-tetrachlorodibenzo-p-dioxin on the peripheral nervous system:

Clinical and neurophysiological controlled study on subjects with chloracne from the Seveso area, *Neuroepidemiology* **7,** 29–37 (1988).

94. G. G. Sirchia, in: *Plans for clinical and epidemiologic follow-up after area-wide chemical contamination,* Proceedings of an International Workshop, pp. 234–266, National Academy Press, Washington, DC (1982).
95. P. Mocarelli, A. Marocchi, P. Brambilla, P. M. Gerthoux, D. S. Young, and N. Mantel, Clinical laboratory manifestations of exposure to dioxin in children. A six-year study of the effects of an environmental disaster near Seveso, Italy, *J. Am. Med. Assoc.* **256,** 2687–2695 (1986).
96. G. Assennato, D. Cervino, E. A. Emmett, G. Longo, and P. Merlo, Follow-up of subjects who developed chloracne following TCDD exposure at Seveso, *Am. J. Ind. Med.* **16,** 119–125 (1989).
97. I. Ghezzi, P. Cannatelli, G. Assennato, F. Merlo, P. Mocarelli, P. Brambilla, and F. Sicurello, Potential 2,3,7,8-tetrachlorobenzo-p-dioxin exposure of Seveso decontamination workers. A controlled prospective study, *Scand. J. Work Environ. Health* **8,** (Suppl. 1), 176–179 (1982).
98. G. Assennato, P. Cannatelli, P. Emmett, I. Ghezzi, and P. Merlo, Medical monitoring of dioxin clean-up workers, *Am. Ind. Hyg. Assoc. J.* **11,** 586–692 (1989).
99. P. A. Bertazzi, C. Zocchetti, A. C. Pesatori, S. Guercilena, M. Sanarico, and L. Radice, Ten-year mortality study of the population involved in the Seveso incident in 1976, *Am. J. Epidemiol.* **129,** 1187–1200 (1989).
100. P. A. Bertazzi, Industrial disasters and epidemiology. A review of recent experiences, *Scand. J. Work Environ. Health* **15,** 85–100 (1989).
101. P. A. Bertazzi, A. C. Pesatori, D. Consonni, A. Tironi, M. T. Landi, and C. Zocchetti, Cancer incidence in a population accidentally exposed to 2,3,7,8-tetrachlorodibenzo-para-dioxin, *Epidemiology* **4,** 398–406 (1993).
102. S. Safe, B. Astroff, M. Harris, T. Zacharewski, R. Dickerson, M. Romkes, and L. Biegel, 2,3,7,8-Tetrachlorodibenzo-p-dioxin (TCDD) and related compounds as antiestrogens: Characterization and mechanism of action, *Pharmacol. Toxicol.* **69,** 400–409 (1991).
103. M. A. Fingerhut, W. E. Halperin, D. A. Marlow, L. A. Piacitelli, P. A. Honchar, M. H. Sweeney, A. L. Greife, P. A. Dill, K. Steenland, and A. Suruda, Cancer mortality in workers exposed to 2,3,7,8-tetrachlorodibenzo-p-dioxin, *N. Engl. J. Med.* **324,** 212–218 (1991).
104. L. L. Needham, P. Mocarelli, D. G. Patterson, Jr., A. Marocchi, P. Brambilla, P. M. Gerthoux, L. Meazza, and V. Carreri, Findings from our Seveso study and comparisons with other exposed populations, Presented at the 12th International Symposium on Dioxins and Related Compounds "DIOXIN '92," August 24–28, Tampere, Finland (1992).

Chapter 19

The Yusho Rice Oil Poisoning Incident

Yoshito Masuda

1. INTRODUCTION

A mass poisoning, called Yusho, occurred in western Japan, mainly in Fukuoka and Nagasaki prefectures, in 1968. Yusho was caused by ingestion of rice oil that was contaminated with Kanechlor-400, a commercial brand of Japanese polychlorinated biphenyls (PCBs).[1] It was later found that the rice oil had been contaminated not only with PCBs but also with polychlorinated dibenzofurans (PCDFs),[2] polychlorinated quaterphenyls (PCQs),[3] and others. Consequently, Yusho was a poisoning by a mixture of PCBs, PCDFs, PCQs, and others. A very similar mass poisoning, called Yu-cheng, occurred in central Taiwan in 1979, 11 years after the Japanese Yusho incident.[4,5] These two incidents of food poisonings are very valuable as a source of information concerning the toxic effects of these chemicals on humans. Several books and reviews on the broad fields of poisonings have been published.[6–10] This chapter updates the rice oil poisonings, primarily focusing on Yusho.

2. EPIDEMIOLOGICAL STUDY

An epidemic of a strange skin disease similar to chloracne was announced to the public in Fukuoka, Japan, in October, 1968. Kuratsune *et al.*[11] examined outbreak conditions of this epidemic disease called "Yusho," namely oil disease. The most common initial symptoms were increased eye discharge and swelling of eyelids, acneform eruption and follicular accentuation, and pigmentation. Most of the patients

Yoshito Masuda • Daiichi College of Pharmaceutical Sciences, Fukuoka 815, Japan.
Dioxins and Health, edited by Arnold Schecter. Plenum Press, New York, 1994.

(99%) were affected during 1968, 55% of the occurrences being concentrated in the three months from June to August. All of the patients had used Kanemi brand rice oil and the oil was produced or shipped by the Kanemi company on February 5 and 6, 1968, or soon thereafter. Gas chromatographic and X-ray fluorescence analyses of the rice oil revealed that only the sample produced or shipped in the beginning of February was contaminated by a large amount of chlorine (maximum 462 ppm); none of the oils shipped in the other months were contaminated with more than a trace amount of chlorine. Kanechlor had been used at the company in its equipment for heating the processed oil at a reduced pressure in order to remove odors from the rice oil. It is believed that it must have leaked from the heating pipe and contaminated the oil, because small holes were discovered in the old pipe.

3. TOXIC AGENTS IN RICE OIL

When the Yusho incident was disclosed in 1968, the rice oils were analyzed for PCBs by gas chromatography and X-ray fluorometry, estimated 2000 to 3000 ppm of Kanechlor-400 in the rice oil based on the organic chlorine content,[1] since no specific method for PCBs was available at that time. Samples of the rice oil produced on February 5 or 6 were analyzed for PCBs using the standard analytical method, yielding approximately 1000 ppm of PCBs in the rice oil.[12] PCBs in the rice oil showed a somewhat different gas chromatographic pattern from that of Kanechlor-400, indicating that Kanechlor-400-contaminated rice oil was heated under reduced pressure and eliminated some amounts of lower chlorinated PCBs which have shorter retention times on the gas chromatogram.[13]

The oil was found to contain 5 ppm of PCDFs, about 250 times the concentration (0.02 ppm) expected from the concentration of PCDFs in other unused Kanechlor-400.[2] This finding was confirmed by Miyata *et al.*[14] The marked increase of PCDFs in the oil could have occurred in the following way. The Kanechlor-400 used as a heat transfer medium for deodorizing rice oil was heated to higher than 200°C for a long time and PCBs were gradually converted to PCDFs. The PCBs with increased PCDF concentrations leaked to the rice oil through small holes in the heating pipe. The conversion of PCBs to PCDFs by heating at higher temperatures was confirmed by Miyata and Kashimoto[15] and Nagayama *et al.*[16] Concentrations of PCBs, PCDFs, and PCQs in the rice oil and Kanechlor-400 and their ratios were reported as shown in Table 1 by Kashimoto and Miyata.[17] PCDFs in the rice oil were composed of more than 40 congeners including toxic congeners of 2,3,7,8-tetra-, 1,2,3,7,8-penta-, 2,3,4,7,8-penta-, 1,2,3,4,7,8-hexa-, and 1,2,3,6,7,8-hexa-CDFs.[18,19] A total of 74 PCBs and 47 congeners of tetra- through octa-PCDFs were quantitatively determined in the rice oil.[20] The concentrations of major PCDF congeners are given in Table 2. It was recently discovered that high chlorinated dibenzofurans decompose during the alkaline treatment of samples, a common analytic cleanup process for PCB and PCDF analyses. By an improved analytical process, hepta-CDFs were found to be major components in the rice oil, which contained 160 ppm of PCBs.[21]

Table 1
Concentrations of PCBs, PCDFs, and PCQs in Yusho Oil and Kanechlor-400[a]

	Concentration (ppm)			Ratio (%)	
	PCBs	PCDFs	PCQs	PCDFs/PCBs	PCQs/PCBs
Yusho oil produced					
Feb. 5, 1968	968	7.4	866	0.76	89
Feb. 9, 1968	151	1.9	490	1.3	320
Feb. 10, 1968	155	2.3	536	1.4	350
Feb. 11, 1968	43.7	0.48	—	1.1	—
Feb. 15, 1968	12.3	0.085	—	0.69	—
Feb. 18, 1968	1.8	0.012	—	0.67	—
Unused KC-400	999,800	33	209	0.003	0.021
Used KC-400	968,400	510	31,000	0.052	3.2
	999,000	277	690	0.028	0.069
	971,900	20	28,000	0.002	2.9

[a]Kashimoto and Miyata.[17]

At an early stage after the Yusho incident, 2000–3000 ppm of PCBs was thought to be contained in the rice oil by X-ray fluorometry. Actual levels of PCBs were about 1000 ppm, one-half to one-third of the above PCB levels. The rice oil was expected to contain a large amount of chlorinated compounds other than PCBs and PCDFs. Miyata *et al.*[3,22] detected 866 ppm of PCQs consisting of penta- through deca-chlorinated congeners in the rice oil as shown in Table 1. Presence of PCQs in the rice oil was confirmed by Kamps *et al.*[23] and Yamaguchi and Masuda.[24]

Polychlorinated quaterphenyl ethers (PCQEs), namely dipolychlorobiphenyl ethers, and polychlorinated terphenyls were also identified as minor components in the fraction of PCQs from the rice oil.[3] Formation of PCQEs from PCBs was accompanied by the formation of PCQs in the heating process of Kanechlor-400 at high temperatures.[22,25] Polychlorinated sexiphenyls, namely trimers of PCBs, 70 ppm, were separated from the PCQ fraction from the rice oil by gel permeation chromatography.[24] In the PCDF fraction from the rice oil, polychlorinated naphthalene[18] and polychlorinated phenyldibenzofurans[26] were identified in small quantities. Kashimoto *et al.*[21] found 0.13 ppm of PCDDs and 1.41 ppm of three coplanar PCBs in a sample of the rice oil which contained 169 ppm of PCBs. Tanabe *et al.*[20] also quantified the congeners of PCDDs and coplanar PCBs. These concentrations are shown in Table 2.

Toxicities of individual congeners of PCDDs, PCDFs, and PCBs were evaluated relative to 2,3,7,8-tetra-CDD (TCDD) toxicity.[27–29] Using the TCDD toxic equivalent factors, the International Organization for PCDDs and PCDFs[27] and the Nordic Council for PCBs,[29] concentrations of TCDD equivalents in the rice oil were calculated as shown in Table 2. Total TCDD equivalents in the rice oil was calculated to be 0.98 ppm, of which the amount contributed from PCDFs was 91%, from PCBs 8%, and from PCDDs 1%.

Table 2
Concentrations of PCDD, PCDF, and PCB Congeners and TCDD Toxic Equivalents (TEQ) in the Rice Oil[a]

	Concentration (ppb)	TEQ factor used	TCDD TEQ (ppb)
2,3,7,8-Tetra-CDD	nd	1	0
Other Tetra-CDDs	3	—	0
1,2,3,7,8-Penta-CDD	7	0.5	3.5
Other Penta-CDDs	77	—	0
2,3,7,8-Hexa-CDDs	71	0.1	7.1
Other Hexa-CDDs	203	—	0
1,2,3,4,6,7,8-Hepta-CDD	185	0.01	1.9
Other Hepta-CDDs	160	—	0
Octa-CDD	120	0.001	0.1
Total PCDDs	826		12.6
2,3,7,8-Tetra-CDF	660	0.1	66
Other Tetra-CDFs	2,570	—	0
1,2,3,7,8-Penta-CDF	525	0.05	26
2,3,4,7,8-Penta-CDF	1,350	0.5	675
Other Penta-CDFs	3,580	—	0
2,3,7,8-Hexa-CDFs	1,225	0.1	123
Other Hexa-CDFs	1,259	—	0
2,3,7,8-Hepta-CDFs	267	0.01	3
Other Hepta-CDFs	42	—	0
Octa-CDF	76	0.001	0
Total PCDFs	11,600		893
3,3′,4,4′-Tetra-CB	11,500	0.0005	5.75
3,3′,4,4′,5-Penta-CB	630	0.1	63
3,3′,4,4′,5,5′-Hexa-CB	27	0.01	0.27
2,3′,4,4′,5-Penta-CB	32,000	0.0001	3.2
2,3,3′4,4′-Penta-CB	28,000	0.0001	2.8
2,3,3′,4,4′,5-Hexa-CB	2,950	0.001	3
Other mono-*ortho* PCBs	91,800	—	0
Di-*ortho* PCBs	135,100	—	0
Tri-*ortho* PCBs	7,580	—	0
Other PCBs	71,000	—	0
Total PCBs	380,000		78
Totals	392,400		984

[a]Values are average of two samples. Totals are rounded. Data from Tanabe *et al.*[20] TEQ factors used are from Kutz *et al.*[27] for PCDDs and PCDFs and from Ahlborg *et al.*[29] for PCBs.

4. INTAKE OF THE CONTAMINATED RICE OIL

A survey of 141 Yusho patients who consumed the rice oil containing 920, 866, and 5 ppm of PCBs, PCQs, and PCDFs, respectively, showed that the average consumption of the rice oil was 688 ml in total and 506 ml during the latent period before

Table 3
Estimated Intakes of Rice Oil and TCDD Toxic Equivalents (TEQ) by Yusho Patients, Mean and Range in Parentheses[a]

	Rice oil	TCDD TEQ
Average total intake per capita	688 ml	0.62 mg
	(195–3375)	(0.18–3.04)
Average intake during latent period	506 ml	0.456 mg
	(121–1934)	(0.11–1.74)
Average daily intake	0.171 ml/kg	154 ng/kg
	(0.031–0.923)	(28–832)
Smallest intake during latent period	121 ml	0.11 mg
Smallest daily intake during latent period	0.031 ml/kg	28 ng/kg

[a]TCDD TEQ are calculated by 0.98 ppm in Yusho oil and 0.92 of oil density. Data from Hayabuchi *et al.*[30]

illness was apparent. Therefore, the total amounts of PCBs, PCQs, and PCDFs ingested by a patient were 633, 596, and 3.4 mg, respectively, on average and the amount ingested during the latent period were 466, 439 and 2.5 mg, respectively.[30] The smallest amounts ingested by a patient during the latent period were estimated to be 111, 105, and 0.6 mg, respectively. The average concentration of TCDD toxic equivalents in the rice oil was determined to be 0.98 ppm (Table 2), and the intakes of TCDD equivalents by patients were calculated. Table 3 lists the intake of rice oil and TCDD equivalents by Yusho patients. The clinical severity of illness and the blood PCB levels showed a close positive correlation with the total amount of oil consumed but not with the amount of oil consumed per kilogram body weight per day.[30,31] This may indicate that during exposure to these highly persistent toxic substances, the level of toxic substances in the body increased to the level needed for development of the toxic symptoms of Yusho.

5. TOXIC AGENTS IN TISSUES AND BLOOD OF YUSHO PATIENTS

5.1. PCBs

Table 4 summarizes the concentrations of PCBs in adipose tissue and liver.[32–34] Soon after the onset of Yusho in 1968, PCB concentrations in the adipose tissue were very high compared with the later levels of Yusho patients and controls. The level of PCBs in the adipose tissue rapidly decreased to several parts per million in the next year, 1969. However, these levels, which were only slightly higher than the controls, were maintained until recent years. The level of PCBs in the liver was considerably lower than that of the adipose tissue in the same patient. In 1973, five years after the onset of Yusho, average blood levels of PCBs in Yusho patients ($N = 41$) and controls ($N = 37$) were determined to be 7 and 3 ppb, respectively.[35] After that time, average

Table 4
Concentrations of PCBs in Tissues of Yusho Patients and Controls

Case	Sex	Age	Date of death or surgical operation	PCB concentration (ppm, wet basis) Adipose tissue	Liver
1[a]	F	Stillborn	October 1968	0.02	0.07
2[b]	M	About 17	November 1968	76 (face)	—
				13 (abdomen)	—
3[b]	?	Adult	November 1968	46	—
4[a]	M	17	July 1969	1.3	0.14
5[a]	M	25	July 1969	2.8	0.2
6[a]	M	46	May 1972	4.3	0.08
7[a]	M	59	March 1977	1.2	0.006
8–14[c]	M,F	43–55	February 1986	1.0–5.7	—
Controls					
A(N = 31)[a]	M,F	0–61	1981	1.39[d] (0.09–13)	0.05[d] (0.01–0.2)
B(N = 11)[c]	M,F	29–61	February 1986	0.44–1.3	—

[a]Masuda and Kuratsune.[32]
[b]Goto and Higuchi.[33]
[c]Iida *et al.*[34]
[d]Mean and range in parentheses.

blood PCB levels of Yusho patients were found to be 6.1 ppb in 1979 (N = 64),[36] 3.08 ppb (seven PCB congeners) in 1983 (N = 18),[37] and 7.9 ppb in 1991 (N = 9). These levels were always only two to three times higher than the control levels. The average concentrations of PCBs in the blood of Yusho patients were found to gradually decrease after 1975.[38] The same PCB congeners were identified in the blood of Yusho patients and persons from the general population. However, the PCB congeners in Yusho patients differed quantitatively from those found in nonexposed persons. Table 5 shows that the composition of PCB congeners present in blood recently obtained from Yusho patients is still quite different from those of nonexposed persons, being characterized by low concentrations of 2,3′,4,4′,5-penta-CB and much higher concentrations of 2,3,3′,4,4′,5-hexa-CB.[37,39] This characteristic difference has been adopted as one of the criteria for diagnosis of Yusho.[40] The characteristic type is classified into three, as follows. Type A: Peculiar to Yusho, its gas chromatographic pattern is quite different from that observed in normal persons; type B: an intermediate pattern between types A and C; and type C: commonly observed in the blood PCBs of the general population. Biological half lives of these PCB congeners were determined in three Yucheng patients who had very high blood PCB levels of 156–397 ppb.[41] The half-life of 2,3′,4,4′,5-penta-CB, 1.16 years, was much shorter than those of 2,2′,4,4′,5,5′-hexa-CB, 4.28 years, and 2,3,3′,4,4′,5-hexa-CB, 4.21 years. The shorter half-life of the penta congener may partly cause the peculiar pattern in Yusho patients. Among the

Table 5
Concentrations (Means ± S.D.) of PCB Congeners in 1983 Blood Samples of Yusho Patients[a]

	Concentration (ppt)		
	Yusho patients (N = 18)	Controls (N = 27)	Ratio Yusho/controls
2,3′,4,4′,5-Penta-CB	55 ± 44	71 ± 95	0.77
2,2′,4,4′,5,5′-Hexa-CB	1086 ± 678	348 ± 430	3.12
2,3,3′,4,4′-Penta-CB	38 ± 26	27 ± 35	1.41
2,2′,3,4,4′,5′-Hexa-CB	522 ± 294	139 ± 160	4.02
2,3,3′,4,4′,5-Hexa-CB	836 ± 343	50 ± 33	16.7
2,2′,3,4,4′,5,5′-Hepta-CB	311 ± 239	91 ± 124	3.42
2,2′,3,3′,4,4′,5-Hepta-CB	198 ± 142	45 ± 55	4.40
Total	3080 ± 1440	763 ± 922	4.04

[a]Masuda *et al.*[37]

PCB congeners identified in Yusho patients, 2,3,3′,4,4′,5-hexa-CB showed strong enzyme-inducing activity in the liver and marked atrophy of the thymus in rats.[42] Therefore, 2,3,3′,4,4′,5-hexa-CB was considered to be one of the PCB congeners most causally related to the symptoms of Yusho. Recently, Kashimoto *et al.*[21] and Tanabe *et al.*[20] reported the presence of highly toxic coplanar PCBs in the tissues of Yusho patients, 3,3′,4,4′,5-penta-CB, a coplanar PCB, being measured at 330 and 410 ppt in intestines and 720 ppt in adipose tissue, respectively. The levels of three coplanar PCBs were very low in the tissues relative to other PCBs, being 0.06–0.6% in Yusho patients and 0.03–0.08% in controls.[21] However, dioxinlike toxicity of 3,3′,4,4′,5-penta-CB is the highest among PCBs as shown in Table 6, since its TCDD toxic equivalent factor is much higher than those of other PCBs.

Only several selected congeners of PCBs in the rice oil were retained in the body of patients as described above and most of the PCB congeners had disappeared from the body within a year by excretion or by being metabolized to hydroxy and methylsulfone PCBs. The methylsulfone PCBs which were probably derived from the PCBs ingested with the rice oil were identified in the tissues of Yusho patients.[43] The concentration (fat basis) of methylsulfone PCBs was higher in the lung (0.67 ppm) than in the adipose tissue (0.07 ppm), contrasting with the concentrations of PCBs in the tissues, 0.8 and 1.3 ppm, respectively.[44]

5.2. PCDFs

Although Yusho patients ingested more than 40 different PCDF congeners with rice oil,[18] only several particular PCDF congeners have been retained in the tissues of patients.[19] Most of the retained PCDF congeners had all lateral (2, 3, 7, and 8)

Table 6
Concentrations of TCDD Toxic Equivalents (TEQ)
in the Tissues of a Yusho Patient[a]

	TCDD TEQ (wet basis, ppt)		
	Yusho patient[b]		
	Adipose	Liver	Control[c] (adipose)
2,3,7,8-Tetra-CDD	<1.9	<0.7	3.7
1,2,3,7,8-Penta-CDD	9	<1.0	3.2
1,2,3,4,7,8-Hexa-CDD	<0.15	<0.03	0
1,2,3,6,7,8-Hexa-CDD	16	2.2	3.4
1,2,3,7,8,9-Hexa-CDD	<0.15	<0.03	0.57
1,2,3,4,6,7,8-Hepta-CDD	<0.11	<0.05	0.33
Octa-CDD	0.23	1.3	0.51
Total PCDDs	25.2	3.5	11.7
2,3,7,8-Tetra-CDF	4.4	4.7	0.31
1,2,3,7,8-Penta-CDF	1.45	4.95	0.025
2,3,4,7,8-Penta-CDF	850	1150	5.5
1,2,3,4,7,8-Hexa-CDF	130	840	0.56
1,2,3,6,7,8-Hexa-CDF	14	223	0.53
2,3,4,6,7,8-Hexa-CDF	<0.15	<0.45	0.14
1,2,3,7,8,9-Hexa-CDF	0	0	0
1,2,3,4,6,7,8-Hepta-CDF	0.95	15	0.03
1,2,3,4,7,8,9-Hepta-CDF	0	0	0
Octa-CDF	0	0	0
Total PCDFs	1001	2238	7.1
3,3′,4,4′-Tetra-CB	0.35	0.065	0.175
3,3′,4,4′,5-Penta-CB	72	5.4	33
3,3′,4,4′,5,5′-Hexa-CB	3.8	0.5	0.9
2,3′,4,4′,5-Penta-CB	0.35	0.03	—
2,3,3′,4,4′-Penta-CB	0.21	0.02	—
2,3,3′,4,4′,5-Hexa-CB	24.7	0.93	—
Total PCBs	101.4	6.94	34
Totals	1128	2248	53

[a]Data from Tanabe *et al.*[20] TEQ factors used are from Kutz *et al.*[27] for PCDD/Fs and Ahlborg *et al.*[29] for PCBs.
[b]Male, died in 1977 at age 59.
[c]Mean of 12 persons.

positions chlorinated and all of the congeners apparently excreted had two vicinal hydrogenated C atoms in at least one of the two rings. Concentrations of major PCDF congeners identified in the tissues and blood of Yusho patients are shown in Table 7.[30,45–48] High concentrations of 2,3,4,7,8-penta-CDF up to 7 ppb were observed in the liver and adipose tissues in 1969, a year after the incident. Subsequently, such high concentrations of PCDF congeners were not found in the tissues of Yusho patients, except 5 ppb of hexa-CDF in the liver in 1977. Higher than control levels of PCDF

Table 7
Concentrations of PCDF Congeners in the Tissues of Yusho Patients (wet weight basis)

Tissue[a]	Year of sampling	PCBs (ppm)	PCDFs (ppb) 2378-	23478-	123478/123678-	1234678-
1						
Liver	1969	1.4	0.3	6.9	2.6	
2						
Liver	1969	0.2	0.02	1.2	0.3	
Adipose		2.8	0.3	5.7	1.7	
3						
Liver	1972	0.03	<0.01	0.3	0.03	
Adipose		4.3	nd	0.8	0.2	
4						
Adipose	1975	0.2	nd	0.1	0.5	
5						
Liver	1977	0.06	nd	1.49	5.31	1.39
Adipose		3.0	0.002	1.45	1.99	0.22
Lung		0.016	0.002	0.365	0.41	0.05
6						
Uterus	1985	0.005	nd	0.026	0.031	nd
7						
Comedo	1982	0.2	nd	0.36	0.39	0.1
8						
Adipose[b]	1986	2.2	0.003	1.4	0.51	
9						
Adipose[c]	1986	2.3	0.028	0.77	0.66	0.036
Blood[c]		0.0085	0.0025	0.0025	0.0004	
Control						
Adipose[d]	1986	1.1	0.007	0.02	0.02	0.009
Blood[d]		0.0033	0.00007	0.00006	0.00009	

[a]1–4: Kuroki and Masuda[45]; 5–6: Ryan *et al.*[46]; 7: Kuroki *et al.*[47]; 8: Iida *et al.*[34]; 9–control: Iida *et al.*[48]
[b]Average of seven patients.
[c]Average of six patients.
[d]Average of three controls.

congeners continued up to 1986, when the levels of PCDF congeners were up to 40 times higher than control levels, while PCB levels in the patients were only 2 to 3 times higher than the controls. It is noteworthy that the PCDF concentrations in the liver were almost as high as those in adipose tissue, while PCB concentrations were much lower in the former than in the latter. This relative abundance of PCDFs to PCBs in the liver and adipose tissue was also seen in normal persons.[14]

In Yu-cheng patients, the same PCDF congeners were identified in various tissues and blood,[49,50] with the concentrations of 2,3,4,7,8-penta-CDF in some patients being comparable in adipose tissue and liver with Yusho patients and being 10–40 times higher in blood relative to Yusho patients. Biological half lives of 2,3,4,7,8-penta-CDF and 1,2,3,4,7,8-hexa-CDF were determined to be 2.1 and 2.6 years, respectively, in the blood of three Yu-cheng patients, whereas in the blood of Yusho patients, their half lives were estimated to be about 10 years.[41,51] This discrepancy may be caused by the difference in PCDF concentrations in blood at the time of first samplings, being 13–42 ppt in Yu-cheng early after exposure and 1–5 ppt in Yusho later after exposure.

Among the PCDF congeners retained in the tissues of Yusho patients, 2,3,4,7,8-penta-CDF showed the highest enzyme inducing activities and acute toxicity in rats[42,52] and was the most highly accumulative in the livers of rats and monkeys.[53] Therefore, 2,3,4,7,8-penta-CDF is considered to be the most important etiologic agent for the Yusho symptoms. Relative toxicities of PCDD and PCDF congeners to 2,3,7,8-tetra-CDD (TEQ factor) have been estimated at many research organizations throughout the world.[27,54,55] Using international TEQ factors for PCDDs and PCDFs[27] and Nordic TEQ factors for PCBs,[29] toxic contributions of the PCDDs, PCDFs, and PCBs retained in a Yusho patient have been calculated and are shown in Table 6. 2,3,4,7,8-Penta-CDF showed highest toxicity in the liver and adipose tissue of patients. However, in the adipose tissue of a control subject, 3,3′,4,4′,5-penta-CB contributed the highest dioxinlike toxicity (62%) among PCDDs, PCDFs, and PCBs.

Iida *et al.*[56] recently found that combined administration of rice bran fiber and cholestyramine for 2 weeks increased (30–50%) fecal excretion of PCDFs in four Yusho patients.

5.3. PCQs

Table 8 shows the concentrations of PCQs and PCBs in adipose tissue, liver, blood, buccal mucosa, and hair of Yusho patients and normal controls.[36,57–60] In typical Yusho patients who have type A PCBs, PCQ concentrations in the tissues and blood were approximately the same or 2 to 4 times lower than the PCB levels. The concentrations of PCQs seemed to decrease with time, as the high concentration of PCQs (2400 ppb) in adipose tissue in 1969 was not observed in the same tissue in 1972 or 1984 when the level of PCQs in adipose tissue was on average 207 ppb. However, the PCQ concentrations in the tissues and blood of patients were always much higher than the corresponding concentrations in normal controls. In 1984, 16 years after the onset, PCQ levels in the adipose tissue and blood of patients were still more than 100 times higher than the corresponding levels in controls. In contrast, the PCB levels in Yusho patients were only 2 to 3 times higher than those of controls. In the government's officially certified Yusho patients, as the type of PCBs alters from A (typical Yusho) to B and C (control type), the PCQ concentrations in blood decrease in parallel with the PCB concentrations. Japanese workers who had been occupationally exposed to PCBs did not show any detectable levels of PCQs in blood although their PCB levels were as high as 33 ppb.[61] Since the PCQ concentrations were reflective of the amount of rice oil intake, blood PCQ levels have been adopted as one of the criteria for diagnosis of Yusho,[40] together with the type of PCBs.

Although PCQs were actually ingested by the patients with the rice oil and retained in the body for years, the major toxicities for Yusho were considered to be

Table 8
Concentrations of PCBs and PCQs in the Tissues and Whole Blood of Yusho Patients

Tissue	Year of sampling	Type of PCBs	Concn. (whole basis, ppb) PCBs	PCQs
1				
Adipose	1969	A	5091	2400
Liver		A	226	218
2				
Adipose	1972	A	6091	1444
Liver		A	69	144
3				
Intestine	1975	A	3472	1770
Liver		A	114	52
4				
Intestine	1977	A	3630	1125
Liver		A	68	27
5				
Blood (n = 29)	1979	A	7.3 ± 4.5	3.04 ± 2.11
Blood (n = 15)		B	5.4 ± 3.6	1.39 ± 1.34
Blood (n = 8)		C	2.7 ± 1.2	0.28 ± 0.19
6				
Blood (n = 11)	1979	A	6.2 ± 4.9	0.09–5.85
Blood (n = 20)		B/C	2.7 ± 1.4	<0.02–0.42
7				
Blood (n = 31)	1979	A	9.6 ± 6.4	2.9 ± 2.3
Blood (n = 4)		B	4.7 ± 5.2	2.0 ± 3.4
Blood (n = 29)		C	2.6 ± 1.1	0.02 ± 0.03
8				
Buccal mucosa (n = 27)	1983		279 ± 41	66 ± 13
Blood (n = 25)			7.20 ± 0.82	0.79 ± 0.13
9				
Adipose (n = 11)	1984		1579 ± 657	207 ± 112
Blood (n = 22)			5.36 ± 2.51	1.34 ± 1.11
Hair (n = 13)	1986		28.9 ± 18.1	0.35 ± 0.36
Control				
1				
Adipose (n = 3)	1978		762	2.2
Liver (n = 3)			37	0.7
5				
Blood (n = 29)	1979		2.3 ± 1.5	<0.02
6				
Blood (n = 18)	1979		3.3 ± 1.2	<0.02
7				
Blood (n = 23)	1979		2.9 ± 1.0	0.02 ± 0.03
8				
Buccal mucosa (n = 7)	1983		64.9 ± 16	<4
Blood (n = 7)			1.86 ± 0.13	<0.02
9				
Adipose (n = 40)	1984		778 ± 670	1.40 ± 0.96
Blood (n = 32)			2.43 ± 1.74	<0.02
Hair (n = 19)	1986		8.06 ± 5.60	<0.1

[a] 1–5: Kashimoto *et al.*[57]; 6: Takamatsu *et al.*[58]; 7: Iida *et al.*[36]; 8: Okumura *et al.*[59]; 9: Ohgami *et al.*[60]

caused by PCDFs and not by PCQs, as PCQs have been found to be much less toxic in rats and monkeys.[62]

6. CLINICAL FEATURES

Most patients were affected within the 9 months from February, 1968, when the contaminated rice oil was shipped to the market from Kanemi company, to October, 1968, when the epidemic of Yusho was reported to the public. The subjective symptoms of Yusho as reported by patients are summarized in Table 9. Pigmentation of nails, skin, and mucous membranes, distinctive hair follicles, acneform eruptions, increased eye discharge, increased sweating of the palms, and a feeling of weakness were the most notable symptoms.[11] Yu-cheng patients in Taiwan who ingested the rice oil contaminated with PCBs, PCDFs, and PCQs in almost the same manner as Yusho[63] showed symptoms very similar to those of Yusho patients.[64]

Table 9
Percent Distribution of Symptoms of Yusho Patients Examined before October 31, 1968[a]

Symptoms	Males (N = 89)	Females (N = 100)
Dark brown pigmentation of nails	83.1	75.0
Distinctive hair follicles	64.0	56.0
Increased sweating of palms	50.6	55.0
Acnelike skin eruptions	87.6	82.0
Red plaques on limbs	20.2	16.0
Itching	42.7	52.0
Pigmentation of skin	75.3	72.0
Swelling of limbs	20.2	41.0
Stiffened soles in feet and palms of hands	24.7	29.0
Pigmented mucous membrane	56.2	47.0
Increased eye discharge	88.8	83.0
Hyperemia of conjunctiva	70.8	71.0
Transient visual disturbance	56.2	55.0
Jaundice	11.2	11.0
Swelling of upper eyelids	71.9	74.0
Feeling of weakness	58.4	52.0
Numbness in limbs	32.6	39.0
Fever	16.9	19.0
Hearing difficulties	18.0	19.0
Spasm of limbs	7.9	8.0
Headache	30.3	39.0
Vomiting	23.6	28.0
Diarrhea	19.1	17.0

[a]Kuratsune *et al.*[11]

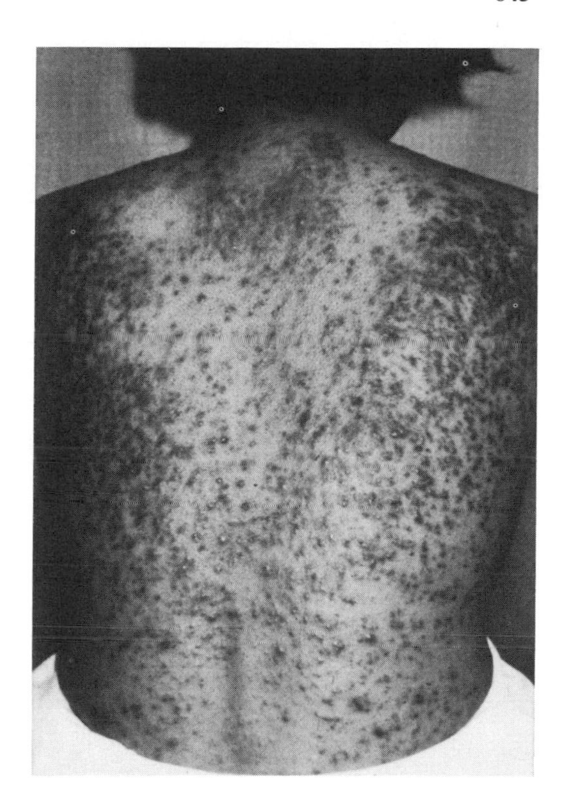

Figure 1. Acneform eruptions on the back of a Yusho patient (female, age 33, photographed in December, 1968).

6.1. Dermal Signs

The most notable symptoms of Yusho are dermal lesions such as follicular keratosis, dry skin, marked enlargement and elevation of the follicular orifice, comedo formation, and acneform eruption (Figs. 1 and 2).[65,66] The acneform eruptions develop in the face, cheek, jaw, back, axilla, trunk, external genitalia, and elsewhere. Dark-colored pigmentation of the corneal limbus, conjunctivae, gingivae, lips, oral mucosa, and nails is a specific finding of Yusho. Clinical symptoms of 72 patients were examined in relation to the types and concentrations of blood PCBs in 1973 and 1974, about 5 years after the onset. As shown in Table 10, dermatological symptoms were observed mostly for the type A group of blood PCBs. The type C group showed few such symptoms. General signs such as fatigue and headache were complaints of patients with all Yusho types of blood PCBs.[67] The skin symptoms have diminished gradually in the 10 years since the onset, while continual subcutaneous cyst formation with secondary infection was still occurring in a relatively small number of the severely affected patients.[68] According to the annual physical examination of 109 Yusho patients in 1986, the proportion of patients with dermal signs decreased to 59% from 72% in 1976.[69]

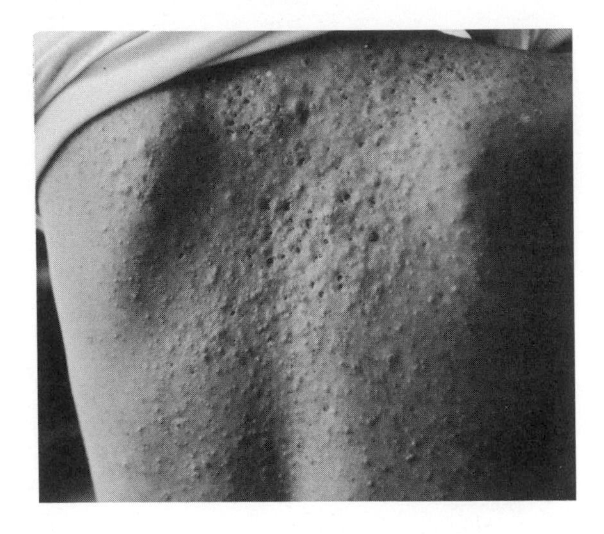

Figure 2. Acneform eruptions on the back of a Yu-cheng patient (male, age 18, photographed in May, 1980).

Table 10

Types and Concentrations of Blood PCBs and Incidence of Clinical Symptoms among 72 Yusho Patients from April, 1973, to March, 1974[a]

Clinical symptoms	Percent of incidence			
	Type A (N = 43)	Type B (N = 26)	Type C (N = 3)	Total (N = 72)
Pigmentation				
Skin	51	0	0	31
Palpebra	72	19	0	50
Gingiva	95	58	67	81
Nail	76	35	0	57
Acneform eruption	35	0	0	21
Comedo	35	23	0	29
Infection of skin	33	12	0	24
Deformation of nails	56	38	0	53
Alopecia	0	4	0	1
Tooth disorders	19	8	0	14
Meibomian gland hypersecretion	93	81	100	89
Fatigue	49	58	33	51
Fever	2	4	0	3
Phymata in articular region	9	8	0	8
Cough and sputum	65	46	33	57
Digestive disorder	35	46	67	40
Headache	47	35	33	42
Numbness of extremities	52	31	33	33
Menstrual disturbance	(5/17)29	(2/9)22		(7/26)27
Blood PCBs (ppb)	7.2 ± 4.9	4.3 ± 3.1	1.7 ± 0.2	5.9 ± 4.8

[a]Koda and Masuda.[67]

6.2. Ocular Signs

The main ocular signs right after onset were hypersecretion of the meibomian glands and abnormal pigmentation of the conjunctiva. Cystic swelling of the meibomian glands filled with yellow infarctlike contents was observed in the typical cases.[70] These signs have markedly subsided in the 10 years after the onset of Yusho, but 84% of the 75 patients showed somewhat abnormal changes of the tarsal glands, such as irregularity or disappearance of the gland pattern, infarction, and lithiasis, and 43% showed pigmentation of the eyelids and conjunctivae in 1978. Eye discharge was still a complaint of 64% of the patients in 1978.[71] The contents of the tarsal glands were found to contain PCBs in 1984. Type A PCBs were found in the contents of tarsal glands from some severely ill Yusho patients.[72] The ocular signs of both abnormal pigmentation of conjunctivae and hypersecretion of the meibomian glands were closely related to PCB concentrations and patterns in blood.[73] In an examination of the meibomian glands by transillumination and infrared photography in 1986, more than half of the examined eyes of 52 patients demonstrated loss of the meibomian glands, suggesting meibomian cyst formation, while only 10% of unaffected eyes in controls showed such abnormalities.[74] It is notable that the cheeselike material can be squeezed out from the meibomian glands in some patients even today.

6.3. Neurological Signs

Most Yusho patients complained of various neurological symptoms such as headache, numbness of the limbs, hypoesthesia, and neuralgic limbs. However, reduced sensory nerve conduction velocities of radial and/or sural nerve were observed in 9 of 23 cases examined soon after the poisoning, while reduced motor nerve conduction velocities of the ulnar and tibial nerves were seen only in 2 cases.[75] Electroencephalographic examination did not reveal any significant changes. It was thought, therefore, that central nervous damage was not clinically prominent or significant in Yusho patients.[76] Frequency of headache was 59.6% in 208 patients in 1973, being comparable to the frequency in controls. But no relationship was observed between blood PCB concentrations and patients with or without headache. Headache was reported by 12 of 41 patients having A-type blood PCBs and 19 of 46 patients having C-type blood PCBs. Frequency of headache was not related to the type of PCBs, in contrast to the dermal signs.[77]

6.4. Respiratory Signs

About 40% of 203 patients complained of persistent cough with expectoration, suffering from chronic bronchitis at an early stage after onset. Secondary bacterial infections were often observed in the respiratory system of patients. Immunity was examined in relation to the infection. The serum IgA and IgM levels of the patients were significantly lower than those of controls, while IgG was higher.[78] PCBs were identified in the sputa of patients at concentrations several times lower than those in blood.[79,80] Frequency and severity of the respiratory symptoms correlated well with the PCB concentrations in blood but not well with the type of PCBs.[81] Thereafter, the respiratory distress occurring in these patients improved gradually in the 10 years following onset. However, over the next 5 years little or no improvement of respiratory symptoms was observed in most cases.[82]

6.5. Endocrine Signs

Rapid adrenocorticotropic hormone (ACTH) tests, for detecting a disturbance in adrenocortical functions by examining serum 17-hydroxycorticosteroids after injections of ACTH, performed in 86 patients in the second year after onset showed no evidence of severe abnormalities of adrenocortical functions.[83] About 40% of 95 patients exhibited an elevated urinary excretion of 17-ketosteroids and 17-hydroxycorticosteroids. Androsterone, etiocholanolone, and dehydroepiandrosterone tended to increase in male patients and to decrease in females. The elevated excretion of these steroids was possibly related to the poisoning.[84] Irregular menstrual cycles were observed in 58% of 81 female patients in 1970. Frequency of the irregular cycles was not related to the severity of Yusho. Urinary excretion of estrogens, pregnanediol, and pregnantriol tended to be low in Yusho patients.[85] Thyroid function was investigated in 123 patients in 1984, 16 years after onset. The serum triiodothyronine and thyroxine levels were significantly higher than those of normal controls, while thyroid stimulating hormone levels were normal. There was no correlation between PCB levels and levels of these thyroid hormones.[86]

6.6. Liver Signs and Functional Changes

Abnormalities of the liver in gross appearance were rarely observed in Yusho patients. However, an electron microscopic study on the liver biopsy specimens from one male Yusho patient revealed a reduction of the rough-surfaced endoplasmic reticulum and hypertrophy of the smooth-surfaced endoplasmic reticulum,[87] suggesting the drug-metabolizing enzymes were induced in the liver of Yusho patients. The mitochondria showed morphological heterogeneity and inclusion bodies and giant mitochondria were frequently observed in the hepatic cells.[88] Basic liver functional tests such as glutamate oxaloacetic transaminase and glutamine pyruvic transaminase were within normal levels. However, accelerated erythrocyte sedimentation rate, high titer in thymol turbidity, increased M fraction of lactate dehydrogenase, and elevated alkaline phosphatase were observed in severe cases suggesting possible liver damage.[89,90] Serum ribonuclease levels in 101 patients were significantly higher than those in healthy controls in 1974.[91] The increased serum ribonuclease activity decreased gradually with time by 1978.[92] The serum bilirubin concentration, 0.48 mg/100 ml mean of

121 patients, was lower than the level of healthy adults and correlated inversely with the blood PCB levels and serum triglyceride concentrations.[93] The levels of urinary porphyrin in 16 and 71 Yusho patients were determined to be within the range of control levels except for one patient, who was suspected of porphyria cutanea tarda.[94,95] Marked elevated serum triglyceride was one of the abnormal laboratory findings peculiar to Yusho in its early stages. High serum triglyceride levels ranging from 200 to 600 mg/100 ml (normal 60–107 mg/100 ml) were observed in 12 of 24 cases, while total serum cholesterol remained unchanged and phospholipids tended to be somewhat lowered ranging from 94 to 172 mg/100 ml (normal 156–219 mg/100 ml).[96] Significant positive correlation was observed between serum triglyceride levels and blood PCB concentrations of 42 patients in 1973. The high level of serum triglyceride was also observed in typical patients with A-type blood PCBs.[97] For the 10 years following onset, follow-up study on serum triglyceride levels in 24 patients indicated that persistent hyperglyceridemia decreased to the normal range since 1973 for females and 1975 for males.[98] Subsequent chemical analysis of the sera of 110 patients in 1979 revealed that elevated serum triglyceride levels persisted in 26 of the patients.[99] Aryl hydrocarbon hydroxylase (AHH) activity in lymphocytes from 42 patients in 1985 was compared with the corresponding activity in 128 healthy nonsmokers. Both 3-methylcholanthrene-induced and noninduced AHH activities in the patients were significantly higher than those of controls, indicating that minute amounts of remaining PCDFs and PCBs are influencing the enzymatic effects in Yusho patients.[100]

6.7. Oral and Dental Signs

Pigmentation of the oral mucosa was one of the characteristic signs of Yusho. A high incidence (62.9%) of pigmentation of the gingivae and lips was observed in 70 Yusho patients in 1968 and 1969.[101] This pigmentation persisted for a long time and was still observed in 74 of 99 patients examined in 1982.[102] When the affected gingiva of one patient was surgically removed, the pigmentation recurred within a year.[103] The concentrations and type (A, B, C) of blood PCBs corresponded to the degree of oral pigmentation in 1973.[104] However, Yusho patients with type C blood PCBs also showed a somewhat high incidence of oral pigmentation.[103] Radiographic examination of the mouth of Yusho patients demonstrated anomalies in the number of teeth and in the shape of the roots and marginal bone resorption at the roots.[103,105] In an examination of 118 patients in 1984, 24.5% of the examined teeth had periodontal pockets on the mesial surface. This frequency was much higher than that (16%) in teeth of 54 control persons in Fukuoka.[105] These changes were considered to have been caused by the remaining PCBs and PCDFs which had been influencing endocrine factors in the body.

6.8. Babies Born to Patients and Infants

Eleven women who were officially certified Yusho patients delivered nine living and two stillborn babies in Fukuoka prefecture in 1968. Ten of them had shown the characteristic grayish dark brown skin at birth. Most of the babies also had dark-colored gingivae and nails and increased eye discharge.[106–108] The majority were small-for-date and their postnatal growth curves for body weight were similar in shape to the national standard curves, but evidently lower for some of the babies. The skin pigmentation faded 2 or 3 months after birth, but the increased eye discharge and nail pigmentation remained for more than a year in some of the babies. Even babies born to Yusho mothers 1 to 3 years after onset showed some pigmentation in the skin and gingivae at birth.[109] One Yusho mother gave birth to three babies with dark brown skin, the pigmentation being most serious in the first baby and diminishing steadily in the second and third babies. Two of seven babies fed the milk of Yusho mothers were diagnosed as Yusho; they are termed trans-milk Yusho babies, as they had been born before the mothers ingested the contaminated rice oil. The gains in height and body weight of 42 school children with Yusho were compared with those of controls before and after the poisoning. Both height and body weight gains of the boys with Yusho significantly decreased after the poisoning. The same tendencies were also observed in some of the girls with Yusho.[110] These temporal growth suppressions were reversed and thereafter tended to be close to the average values for the controls.[111]

7. DEATHS AMONG YUSHO PATIENTS

The total number of individuals officially registered as Yusho patients was 1862, of whom 149 had died by 1990. Kuratsune *et al.*[112] compared the number of deaths among 1761 patients registered as Yusho by the end of 1983 with the expected number of deaths calculated on the basis of the national death rates. The results are shown in Table 11. A significant excess mortality was observed for malignant neoplasms at all sites, cancer of the liver, and cancer of the lung, trachea, and bronchus in males. Excess mortality for liver cancer was not significant in females. The excess deaths from liver cancer were seen mainly in Fukuoka prefecture, while no such excess was seen in Nagasaki prefecture where 550 patients were registered. These findings suggest that the poisoning might have caused liver cancer in males in Fukuoka, although it is still too early to draw any firm conclusions from this cancer mortality study. In a related study of 5172 workers exposed to TCDD, excess mortality from all cancers combined, cancer of the respiratory tract, and soft tissue sarcoma appear to be related to TCDD exposure.[113] Excess mortality from cancer of the respiratory tract was found both in studies of TCDD and in the Yusho study involving the closely related PCDFs and PCBs.

8. RISK ASSESSMENT OF PCDDS, PCDFS, AND PCBS IN HUMANS

As shown in Table 3, the patients began to show manifestations of Yusho when their body burden of PCDFs and PCBs reached 456 μg of TCDD equivalents on average with a range of 110 to 1740 μg. The latency period for these symptoms was 71

Table 11
Observed and Expected Number of Deaths and SMR (O/E) by Cause of Death[a]

Cause of death	Males (887)			Females (874)		
	Observed	Expected	O/E	Observed	Expected	O/E
Total	79	66.13	1.19	41	48.90	0.84
Tuberculosis	1	1.26	0.79	0	0.50	0.00
Malignant neoplasms	33	15.51	2.13**	8	10.55	0.76
Esophagus	1	0.77	1.30	1	0.18	5.45
Stomach	8	5.69	1.40	0	3.26	0.00
Rectum, sigmoid colon, and anus	1	0.63	1.60	0	0.46	0.00
Liver	9	1.61	5.59**	2	0.66	3.04
Pancreas	1	0.71	1.41	1	0.46	2.18
Lung, trachea, and bronchus	8	2.45	3.26**	0	0.85	0.00
Breast	0	0.01	0.00	1	0.66	1.46
Uterus				1	0.58	1.71
Leukemia	1	0.45	2.23	0	0.32	0.00
Diabetes	1	0.75	1.34	0	0.69	0.00
Heart disease	10	9.46	1.06	9	7.65	1.18
Hypertensive disease	0	1.20	0.00	0	1.39	0.00
Cerebrovascular disease	8	14.61	0.55	5	12.03	0.42*
Pneumonia and bronchitis	3	3.17	0.95	0	2.33	0.00
Gastric and duodenal ulcer	0	0.73	0.00	1	0.34	2.96
Chronic liver disease and cirrhosis	6	2.26	2.65	2	0.73	2.74
Nephritis, nephrose syndrome, and nephrose	0	0.79	0.00	2	0.71	2.81
Accidents	5	4.66	1.07	2	1.32	1.52

[a]Kuratsune *et al.*[112]
$*p < 0.05$, $**p < 0.01$.

days on average with a range of 20 to 190 days and average daily intake during this period was calculated to be 107 ng/kg of TCDD equivalents, assuming a body weight of 60 kg. One patient showed some manifestations of Yusho when his body burden reached 210 μg of TCDD equivalents during the 135-day latency period. Daily intake during the latency period was calculated to be 28 ng/kg of TCDD equivalents, which was the smallest daily intake among the Yusho patients. Ryan *et al.*[114] calculated the body burden of TCDD equivalents associated with nausea and anorexia to be 2.2 μg/kg and that associated with chloracne to be 3 μg/kg, using the data for rice oil ingestion by Yusho patients. These values are equivalent to 132 and 180 μg for an adult patient and quite close to the smallest body burden of 210 μg TCDD equivalents for a patient.

The general population ingests small amounts of PCDDs, PCDFs, and PCBs from contaminated meat, fish, and dairy products; daily intakes were calculated to be 1 to 3 pg/kg of TCDD equivalents in Canada,[115] Germany,[116] Japan,[117] and other countries. PCDDs, PCDFs, and PCBs accumulate in the body and their equilibrated concentrations were estimated to be 10–50 ppt (fat basis) of TCDD equivalents in human tissues.[118] Breast milk was contaminated with PCDDs, PCDFs, and PCBs at about the same level of TCDD equivalents as for human tissues.[119] Babies fed breast milk were estimated to ingest 20–160 pg/kg per day of TCDD equivalents for several months, assuming that the baby consumes about 150 ml/kg body wt of breast milk which contains about 3% fatty materials.

When the daily intakes of TCDD equivalents by Yusho patients (28 and 107 ng/kg) are compared with those of the general population (1–3 pg/kg), there is a difference of more than four orders of magnitude. However, as the periods of ingestion differ greatly for the two groups, 71 and 135 days for Yusho patients and lifelong for the general population, the toxicity levels of PCDDs, PCDFs, and PCBs remaining in Yusho patients were only about 100 times higher than those of controls, about 2000 ppt of TCDD equivalents in adipose tissue of Yusho patients versus 20–50 ppt for the general population. When the intake of these chemicals by nursing babies is compared in the two populations, the intake of TCDD equivalents by breast-fed babies of the general population, a maximum of 160 pg/kg per day, is only two orders of magnitude lower than that of Yusho patients, a minimum of 28 ng/kg per day. Moreover, feeding periods for the toxic chemicals are very similar in the two groups, several months for babies of the general population and from 1 to 5 months for Yusho patients. Two babies were actually certified as Yusho patients after ingesting PCDFs and PCBs through breast-feeding from mothers with Yusho for 4 months in Nagasaki prefecture.[109] The Yusho research group at Kyushu University advised a Yusho mother not to breast-feed her baby in 1990 as her breast milk contained 2 ng/kg per day of TCDD equivalents. In the general population, Pluim *et al.*[120] found that exposure to high levels of PCDDs and PCDFs, both intrauterine and via breast milk, modulates the hypothalamic–pituitary–thyroid regulatory system in human newborns.

REFERENCES

1. H. Tsukamoto, S. Makisumi, H. Hirose, T. Kojima, H. Fukumoto, K. Fukumoto, M. Kuratsune, M. Nishizumi, M. Shibata, J. Nagai, Y. Yae, K. Sawada, M. Furukawa, H. Yoshimura, K. Tatsumi, K. Oguri, H. Shimeno, K. Ueno, H. Kobayashi, T. Yano, A. Ito, T. Okada, K. Inagami, T. Koga, Y. Tomita, T. Koga, Y. Yamada, M. Miyaguchi, M. Sugano, K. Hori, K. Takeshita, K. Manako, Y. Nakamura and N. Shigemori, The chemical studies on detection of toxic compounds in the rice bran oils used by the patients of Yusho, *Fukuoka Acta Med.* **60,** 496–512 (1969).
2. J. Nagayama, M. Kuratsune, and Y. Masuda, Determination of chlorinated dibenzofurans in Kanechlors and "Yusho oil," *Bull. Environ. Contam. Toxicol.* **15,** 9–13 (1976).
3. H. Miyata, T. Kashimoto, and N. Kunita, Studies on the compounds related to PCB (V). Detection and determination of unknown organochlorinated compounds in Kanemi rice oil caused Yusho, *J. Food Hyg. Soc.* **19,** 364–371 (1978).
4. S.-T. Hsu, C.-I. Ma, S. K.-H. Hsu, S.-S. Wu, N. H.-M. Hsu, and C.-C. Yeh, Discovery and epidemiology of PCB poisoning in Taiwan, *Am. J. Ind. Med.* **5,** 71–79 (1984).
5. S.-T. Hsu, C.-I. Ma, S. K.-H. Hsu, S.-S. Wu, N. H.-M. Hsu, C.-C. Yeh, and S.-B. Wu, Discovery and epidemiology of PCB poisoning in Taiwan: A four-year followup, *Environ. Health Perspect.* **59,** 5–10 (1985).

6. M. Kuratsune, An abstract of results of laboratory examination of patients with Yusho and of animal experiments, *Environ. Health Perspect.* **1,** 129–136 (1972).
7. K. Higuchi (ed.), *PCB Poisoning and Pollution*, pp. 1–179, Kodansha and Academic Press, Tokyo and New York (1976).
8. M. Kuratsune, Yusho, in: *Halogenated Biphenyls, Terphenyls, Naphthalenes, Dibenzodioxins and Related Products* (R. D. Kimbrough, ed.), pp. 287–302, Elsevier/North-Holland, Amsterdam (1980).
9. M. Kuratsune and R. E. Shapiro (eds.), PCB poisoning in Japan and Taiwan, *Am. J. Ind. Med.* **5,** 1–153 (1984).
10. M. Kuratsune, Yusho, with reference to Yu-cheng, in: *Halogenated Biphenyls, Terphenyls, Naphthalenes, Dibenzodioxins and Related Products* (R. D. Kimbrough and A. A. Jensen, eds.), pp. 381–400, Elsevier, Amsterdam (1989).
11. M. Kuratsune, Y. Yoshimura, J. Matsuzaka, and A. Yamaguchi, Epidemiological study of Yusho, a poisoning caused by ingestion of rice oil contaminated with commercial brand of polychlorinated biphenyls, *Environ. Health Perspect.* **1,** 119–128 (1972).
12. J. Nagayama, Y. Masuda, and M. Kuratsune, Chlorinated dibenzofurans in Kanechlors and rice oil used by patients with Yusho, *Fukuoka Acta Med.* **66,** 593–599 (1975).
13. Y. Masuda, R. Kagawa, and M. Kuratsune, Polychlorinated biphenyls in Yusho patients and ordinary persons, *Fukuoka Acta Med.* **65,** 17–24 (1974).
14. H. Miyata, T. Kashimoto, and N. Kunita, Detection and determination of polychlorodibenzofurans in normal human tissues and Kanemi rice oil caused "Kanemi Yusho," *J. Food Hyg. Soc.* **19,** 260–265 (1977).
15. H. Miyata and T. Kashimoto, Studies on the compounds related to PCBs (IV) Investigation on polychlorodibenzofuran formation, *J. Food Hyg. Soc.* **19,** 78–84 (1978).
16. J. Nagayama, M. Kuratsune, and Y. Masuda, Formation of polychlorinated dibenzofurans by heating polychlorinated biphenyls, *Fukuoka Acta Med.* **72,** 136–141 (1981).
17. T. Kashimoto and H. Miyata, Difference between Yusho and other kind of poisoning involving only PCBs, in: *PCBs and the Environment* (J. S. Wade, ed.), Vol. 3, pp. 1–26, CRC Press, Boca Raton, FL (1987).
18. H. R. Buser, C. Rappe, and A. Gara, Polychlorinated dibenzofurans (PCDFs) found in Yusho oil and used Japanese PCB, *Chemosphere* **7,** 439–449 (1978).
19. C. Rappe, H. R. Buser, H. Kuroki, and Y. Masuda, Identification of polychlorinated dibenzofurans (PCDFs) retained in patients with Yusho, *Chemosphere* **8,** 259–266 (1979).
20. S. Tanabe, N. Kannan, T. Wakimoto, R. Tatsukawa, T. Okamoto, and Y. Masuda, Isomer-specific determination and toxic evaluation of potentially hazardous coplanar PCBs, di benzofurans and dioxins in the tissues of "Yusho" and PCB poisoning victim and in the causal oil, *Toxicol. Environ. Chem.* **34,** 215–231 (1989).
21. Y. Kashimoto, H. Miyata, K. Takayama, and J. Ogaki, Levels of PCDDs, coplanar PCBs and PCDFs in patients with Yusho and the causal oil by HR-GC HR-MS, *Fukuoka Acta Med.* **78,** 325–336 (1987).
22. H. Miyata, Y. Murakami, and T. Kashimoto, Studies on the compounds related to PCB (VI). Determination of polychlorinated quaterphenyl (PCQ) in Kanemi rice oil caused "Yusho" and investigation on the PCQ formation, *J. Food Hyg. Soc.* **19,** 417–425 (1978).
23. L. R. Kamps, W. J. Trotter, S. J. Young, L. J. Carson, J. A. G. Roach, J. A. Sphon, J. T. Tanner, and B. McMahon, Polychlorinated quaterphenyls identified in rice oil associated with Japanese "Yusho" poisoning, *Bull. Environ. Contam. Toxicol.* **20,** 589–591 (1978).
24. S. Yamaguchi and Y. Masuda, Quantitative analysis of polychlorinated quaterphenyls in Yusho oil by high performance liquid chromatography, *Fukuoka Acta Med.* **76,** 132–136 (1985).

25. T. Yamaryo, T. Miyazaki, Y. Masuda, and J. Nagayama, Formation of polychlorinated quaterphenyls by heating polychlorinated biphenyls, *Fukuoka Acta Med.* **70,** 88–92 (1979).
26. H. Kuroki, Y. Ohmura, K. Haraguchi, and Y. Masuda, Identification of polychlorinated phenyldibenzofurans (PCPDFs) in the causal rice oil associated with Yusho, *Fukuoka Acta Med.* **80,** 190–195 (1989).
27. F. W. Kutz, D. P. Bottimore, E. W. Bretthauer, and D. N. McNelis, History and achievements of the NATO/CCMS pilot study on international information exchange on dioxins, *Chemosphere* **17**(11)**,** N2–N7 (1988).
28. S. Safe and D. Phil, Polychlorinated biphenyls (PCBs), dibenzo-p-dioxins (PCDDs), dibenzofurans (PCDFs), and related compounds: Environmental and mechanistic considerations which support the development of toxic equivalency factors (TEFs), *Crit. Rev. Toxicol.* **21,** 51–88 (1990).
29. U. Ahlborg, A. Hanberg, and K. Kenne, *Risk Assessment of Polychlorinated Biphenyls (PCBs),* pp. 1–99, Nordic Council of Ministers, Copenhagen (1992).
30. H. Hayabuchi, T. Yoshimura, and M. Kuratsune, Consumption of toxic oil by 'Yusho' patients and its relation to the clinical response and latent period, *Food Cosmet. Toxicol.* **17,** 455–461 (1979).
31. H. Hayabuchi, M. Ikeda, T. Yoshimura, and Y. Masuda, Relationship between the consumption of toxic rice oil and long-term concentration of polychlorinated biphenyls in the blood of Yusho patients, *Food Cosmet. Toxicol.* **19,** 53–55 (1981).
32. Y. Masuda and M. Kuratsune, Toxic compounds in the rice oil which caused Yusho, *Fukuoka Acta Med.* **70,** 229–237 (1979).
33. M. Goto and K. Higuchi, The symptomatology of Yusho (chlorobiphenyls poisoning) in dermatology, *Fukuoka Acta Med.* **60,** 409–431 (1969).
34. T. Iida, R. Nakagawa, S. Takenaka, K. Fukamachi, and K. Takahashi, Polychlorinated dibenzofurans (PCDFs) in the subcutaneous adipose tissue of Yusho patients and normal controls, *Fukuoka Acta Med.* **80,** 296–301 (1989).
35. Y. Masuda, R. Kagawa, K. Shimamura, M. Takada, and M. Kuratsune, Polychlorinated biphenyls in the blood of Yusho patients and ordinary persons, *Fukuoka Acta Med.* **65,** 25–27 (1974).
36. T. Iida, M. Keshino, S. Takata, S. Nakamura, K. Takahashi, and Y. Masuda, Polychlorinated biphenyls and polychlorinated quaterphenyls in human blood, *Fukuoka Acta Med.* **72,** 185-191 (1981).
37. Y. Masuda, S. Yamaguchi, H. Kuroki, and K. Haraguchi, Polychlorinated biphenyl isomers in the blood of recent Yusho patients, *Fukuoka Acta Med.* **76,** 150–152 (1985).
38. T. Iida, K. Fukamachi, K. Takahashi, and Y. Masuda, Time course variation of PCB levels and peak pattern on gas chromatogram in the blood of normal persons, *Fukuoka Acta Med.* **76,** 137–144 (1985).
39. H. Kuroki and Y. Masuda, Structures and concentrations of the main components of polychlorinated biphenyls retained in patients with Yusho, *Chemosphere* **6,** 469–474 (1977).
40. M. Kuratsune, M. Aono, and H. Yoshida, Foreword, *Fukuoka Acta Med.* **78,** 181–192 (1987).
41. Y. Masuda, H. Kuroki, K. Haraguchi, J. J. Ryan, and S.-T. Hsu, Elimination of PCDF and PCB congeners in the blood of patients with PCB poisoning in Taiwan, *Fukuoka Acta Med.* **82,** 262–268 (1991).
42. Y. Masuda and H. Yoshimura, Polychlorinated biphenyls and dibenzofurans in patients with Yusho and their toxicological significance: A review, *Am. J. Ind. Med.* **5,** 31–44 (1984).

43. K. Haraguchi, H. Kuroki, and Y. Masuda, Capillary gas chromatographic analysis of methylsulphone metabolites of polychlorinated biphenyls retained in human tissues, *J. Chromatogr.* **361,** 239–252 (1986).
44. K. Haraguchi, Y. Masuda, A. Bergman, and M. Olsson, PCB methylsulphone: Comparison of tissue levels in Baltic grey seals and a Yusho patient, *Fukuoka Acta Med.* **82,** 269–273 (1981).
45. H. Kuroki and Y. Masuda, Determination of polychlorinated dibenzofuran isomers retained in patients with Yusho, *Chemosphere* **7,** 771–777 (1978).
46. J. J. Ryan, A. Schecter, Y. Masuda, and M. Kikuchi, Comparison of PCDDs and PCDFs in the tissues of Yusho patients with those from the general population in Japan and China, *Chemosphere* **16,** 2017–2025 (1987).
47. H. Kuroki, M. Ohma, K. Haraguchi, Y. Masuda, and T. Saruta, Quantitative analysis of polychlorobiphenyl (PCB) and polychlorodibenzofuran (PCDF) isomers in the comedo and subcutaneous abscess of Yusho patients, *Fukuoka Acta Med.* **78,** 320–324 (1987).
48. T. Iida, H. Hirakawa, T. Matsueda, R. Nakagawa, S. Takenaka, K. Morita, Y. Narazaki, K. Fukamachi, K. Takahashi, and H. Yoshimura, Levels of polychlorinated biphenyls and polychlorinated dibenzofurans in the blood of Yusho patients and normal subjects, *Toxicol. Environ. Chem.* **35,** 17–24 (1992).
49. P. H. Chen, C.-K. Wong, C. Rappe, and M. Nygren, Polychlorinated biphenyls, dibenzofurans and quaterphenyls in toxic rice-bran oil and in the blood and tissues of patients with PCB poisoning (Yu-cheng) in Taiwan, *Environ. Health Perspect.* **59,** 59–65 (1985).
50. C. Rappe, M. Nygren, H. Buser, Y. Masuda, H. Kuroki, and P. H. Chen, Identification of polychlorinated dioxins (PCDDs) and dibenzofurans (PCDFs) in human samples, occupational exposure and Yusho patients, in: *Human and Environmental Risks of Chlorinated Dioxins and Related Compounds* (R. E. Tucker, A. L. Young, and A. P. Gray, eds.), pp. 241–253, Plenum Press, New York (1983).
51. J. J. Ryan, D. Levesque, L. G. Panopio, W. F. Sun, Y. Masuda, and H. Kuroki, Elimination of polychlorinated dibenzofurans (PCDFs) and polychlorinated biphenyls (PCBs) from human blood in the Yusho and Yu-cheng rice oil poisonings, *Arch. Environ. Contam. Toxicol.* **24,** 504–512 (1993).
52. S. Yoshihara, K. Nagata, H. Yoshimura, H. Kuroki, and Y. Masuda, Inductive effect on hepatic enzymes and acute toxicity of individual polychlorinated dibenzofuran congeners in rats, *Toxicol. Appl. Pharmacol.* **59,** 580–588 (1981).
53. H. Kuroki, Y. Masuda, S. Yoshihara, and H. Yoshimura, Accumulation of polychlorinated dibenzofurans in the livers of monkeys and rats, *Food Cosmet. Toxicol.* **18,** 387–392 (1980).
54. J. S. Bellin and D. G. Barnes, Interim procedures for estimating risks associated with exposures to mixtures of chlorinated dibenzo-p-dioxins and dibenzofurans (CDDs and CDFs), in: *Risk Assessment Forum,* pp. 1–27, EPA, Washington, DC (1987).
55. U. G. Ahlborg, H. Håkansson, F. Wærn, and A. Hanberg, *Nordisk Dioxinrisk Bedomining,* pp. 1–129, Nordisk Ministerrad, Copenhagen (1988).
56. T. Iida, H. Hirakawa, T. Matsueda, K. Fukamachi, H. Tokiwa, H. Tuji, and Y. Hori, Clinical trials of cholestyramine and a combination of rice bran fiber and cholestyramine for promotion of fecal excretion of retained polychlorinated dibenzofurans in Yusho patients, *Organohalogen Compounds* **10,** 105–108 (1992).
57. T. Kashimoto, H. Miyata, and N. Kunita, The presence of polychlorinated quaterphenyls in the tissues of Yusho patients, *Food Cosmet. Toxicol.* **19,** 335–340 (1981).
58. M. Takamatsu, M. Oki, K. Maeda, and T. Kashimoto, Relations between PCQ level and PCB pattern in the blood of Yusho patients, *Fukuoka Acta Med.* **72,** 192–197 (1981).
59. H. Okumura, N. Masuda, A. Akamine, and M. Aono, Concentration levels of the PCB

and PCQ, pattern of the PCB and ratio of CB% in buccal mucosa of patients with the PCB poisoning (Kanemi-Yusho), *Fukuoka Acta Med.* **78,** 358–364 (1987).
60. T. Ohgami, S. Nakano, F. Murayama, K. Yamashita, H. Irifune, M. Watanabe, N. Tsukazaki, K. Tanaka, H. Yoshida and Y. Rikioka, A comparative study on polychlorinated biphenyls (PCB) and polychlorinated quaterphenyls (PCQ) concentrations in subcutaneous fat tissue, blood and hair of patients with Yusho and normal control in Nagasaki Prefecture, *Fukuoka Acta Med.* **80,** 307–312 (1989).
61. T. Kashimoto, H. Miyata, S. Kunita, T.-C. Tung, S.-T. Hsu, K.-J. Chang, S.-Y. Tang, G. Ohi, J. Nakagawa, and S. Yamamoto, Role of polychlorinated dibenzofuran in Yusho (PCB poisoning), *Arch. Environ. Health* **36,** 321–326 (1981).
62. N. Kunita, S. Hori, H. Obana, T. Otake, H. Nishimura, T. Kashimoto, and N. Ikegami, Biological effect of PCBs, PCQs and PCDFs present in the oil causing Yusho and Yucheng, *Environ. Health Perspect.* **59,** 79–84 (1985).
63. C.-F. Lan, P. H.-S. Chen, L.-L. Shich, and Y.-H. Chen, An epidemiological study on polychlorinated biphenyls poisoning in Taichung area, *Clin. Med. (Taipei)* **7,** 96–100 (1981).
64. Y.-C. Lü and Y.-C. Wu, Clinical findings and immunological abnormalities in Yu-cheng patients, *Environ. Health Perspect.* **59,** 17–29 (1985).
65. H. Urabe and H. Koda, The dermal symptomatology of Yusho, in: *PCB Poisoning and Pollution* (K. Higuchi, ed.), pp. 105–123, Kodansha and Academic Press, Tokyo and New York (1976).
66. H. Urabe and M. Asahi, Past and current dermatological status of Yusho patients, *Environ. Health Perspect.* **59,** 11–15 (1985).
67. H. Koda and Y. Masuda, Relation between PCB level in the blood and clinical symptoms of Yusho patients, *Fukuoka Acta Med.* **66,** 624–628 (1975).
68. M. Asahi, H. Koda, H. Urabe, and S. Toshitani, Dermatological symptoms of Yusho alterations in this decade, *Fukuoka Acta Med.* **70,** 172–180 (1979).
69. S. Toshitani, M. Asahi, and H. Urabe, Dermatological findings in the general examination of Yusho in 1985–1986, *Fukuoka Acta Med.* **78,** 349–354 (1987).
70. H. Ikui, K. Sugi, and S. Uga, Ocular signs of chronic chlorobiphenyls poisoning "Yusho," *Fukuoka Acta Med.* **60,** 432–439 (1969).
71. T. Kohno and Y. Yamana, Ten-year follow-up on ocular manifestations of "Yusho" (accidental polychlorinated biphenyls poisoning), *Fukuoka Acta Med.* **70,** 181–186 (1979).
72. T. Kohno, T. Ohnishi, and H. Hironaka, Ocular manifestations and polychlorinated biphenyls in the tarsal gland contents of Yusho patients, *Fukuoka Acta Med.* **76,** 244–247 (1985).
73. Y. Ohnishi and T. Yoshimura, Relationship between PCB concentrations or patterns in blood and ocular signs among people examined for "Yusho," *Fukuoka Acta Med.* **68,** 123–127 (1977).
74. T. Kohno and Y. Ohnishi, In vivo transillumination of meibomian glands in Yusho patients, *Fukuoka Acta Med.* **78,** 355–357 (1987).
75. Y. Kuroiwa, Y. Murai, and T. Santa, Neurological and nerve conduction velocity studies on 23 patients with chlorobiphenyls poisoning, *Fukuoka Acta Med.* **60,** 462–463 (1969).
76. K. Nagamatsu and Y. Kuroiwa, Electroencephalographical studies on 20 patients with chlorobiphenyls poisoning, *Fukuoka Acta Med.* **62,** 157–158 (1971).
77. H. Iwashita, K. Shida, and Y. Masuda, Headache, paresthesia of the limbs and blood polychlorinated biphenyls (PCB) concentration in chronic PCB poisoning, *Fukuoka Acta Med.* **68,** 139–144 (1977).
78. N. Shigematsu, Y. Norimatsu, T. Ishibashi, M. Yoshida, S. Suetsugu, T. Kawatsu,

T. Ikeda, R. Saito, S. Ishimaru, T. Shirakusa, M. Kido, K. Emori, and H. Toshimitsu, Clinical and experimental studies on respiratory involvement in chlorobiphenyls poisoning, *Fukuoka Acta Med.* **62,** 150–156 (1971).

79. T. Kojima, Chlorobiphenyls in the sputa and tissues, *Fukuoka Acta Med.* **62,** 25–29 (1971).
80. N. Shigematsu, S. Ishimaru, T. Ikeda, and Y. Masuda, Further studies on respiratory disorders in polychlorinated biphenyls (PCB) poisoning. Relationship between respiratory disorders and PCB concentrations in blood and sputum, *Fukuoka Acta Med.* **68,** 133–138 (1977).
81. N. Shigematsu, S. Ishimaru, R. Saito, T. Ikeda, K. Matsuba, K. Sugiyama, and Y. Masuda, Respiratory involvement in polychlorinated biphenyls poisoning, *Environ. Res.* **16,** 92–100 (1978).
82. Y. Nakanishi, Y. Kurita, H. Kanegae, and N. Shigematsu, Respiratory involvement and immune studies in polychlorinated biphenyls and polychlorinated dibenzofurans poisoning, *Fukuoka Acta Med.* **76,** 196–203 (1985).
83. A. Watanabe, S. Irie, T. Nakashima, and S. Katsuki, Endocrinological studies on chlorobiphenyls poisoning, *Fukuoka Acta Med.* **62,** 159–162 (1971).
84. J. Nagai, M. Furukawa, A. Tojo, and T. Fujimoto, Colorimetric and gas-chromatographic determinations of urinary 17-ketosteroids. Survey of chlorobiphenyls poisoning patients by these methods, *Fukuoka Acta Med.* **62,** 51–65 (1971).
85. M. Kusuda, Yusho and female. Studies on sexual functions in female patients with rice oil poisoning, *Sanka to Fujinka (Obstet. Gynecol.)* **38,** 1063–1072 (1971).
86. K. Murai, K. Okamura, H. Tsuji, E. Kajiwara, H. Watanabe, K. Akagi, and M. Fujishima, Thyroid function in "Yusho" patients exposed to polychlorinated biphenyls (PCB), *Environ. Res.* **44,** 179–187 (1987).
87. C. Hirayama, T. Irisa, and T. Yamamoto, Fine structural changes of the liver in a patient with chlorobiphenyls intoxication, *Fukuoka Acta Med.* **60,** 455–461 (1969).
88. T. Yamamoto, C. Hirayama, and T. Irisa, Some observations on the fine structure of mitochondria in hepatic cells from a patient with chlorobiphenyls intoxication, *Fukuoka Acta Med.* **62,** 85–88 (1971).
89. M. Okumura and S. Katsuki, Clinical observation on Yusho (chlorobiphenyls poisoning), *Fukuoka Acta Med.* **60,** 440–446 (1969).
90. M. Okumura, Course of serum enzyme change in PCB poisoning, *Fukuoka Acta Med.* **63,** 396–400 (1972).
91. M. Yamanaka, K. Akagi, N. Hirao, and K. Murai, Abnormality of serum enzyme in PCB poisoning patients with special reference to ribonuclease, *Fukuoka Acta Med.* **66,** 617–619 (1975).
92. K. Akagi, K. Murai, T. Shikata, M. Yamanaka, and T. Omae, Serum ribonuclease in patients with PCBs poisoning, *Fukuoka Acta Med.* **70,** 211–214 (1979).
93. C. Hirayama, M. Okumura, J. Nagai, and Y. Masuda, Hypobilirubinemia in patients with polychlorinated biphenyls poisoning, *Clin. Chim. Acta* **55,** 97–100 (1974).
94. J.J.T.W.A. Strik, H. Kip. T. Yoshimura, Y. Masuda, and E. G. M. Harmsen, Porphyrins in urine of Yusho patients, in: *Chemical Porphyria in Man* (J. J. T. W. A. Strik and J. H. Koeman, eds.), pp. 63–68, Elsevier/North-Holland, Amsterdam (1979).
95. S. Nonaka, T. Shimoyama, T. Honda, and H. Yoshida, Analysis of urinary porphyrins in polychlorinated biphenyl poisoning (Yusho) patients, in: *Chemical Porphyria in Man* (J. J. T. W. A. Strik and J. H. Koeman, eds.), pp. 69–73, Elsevier/North-Holland, Amsterdam (1979).
96. H. Uzawa, Y. Ito, A. Notomi, and S. Katsuki, Hyperglyceridemia resulting from intake of rice oil contaminated with chlorinated biphenyls, *Fukuoka Acta Med.* **60,** 449–454 (1969).

97. M. Okumura, Y. Masuda, and S. Nakamuta, Correlation between blood PCB and serum triglyceride levels in patients with PCB poisoning, *Fukuoka Acta Med.* **65,** 84–87 (1974).
98. M. Okumura, M. Yamanaka, and S. Nakamuta, Ten year follow-up study on serum triglyceride levels in 24 patients with PCB poisoning, *Fukuoka Acta Med.* **70,** 208–210 (1979).
99. K. Akagi, K. Murai, and T. Shikata, Laboratory examination in PCBs poisoning patients with special reference to lipoprotein, *Fukuoka Acta Med.* **72,** 245–248 (1981).
100. J. Nagayama, C. Kiyohara, A. Fukuda, Y. Nakamura, T. Hirohata, M. Asahi, and T. Yoshimura, A study of aryl hydrocarbon hydroxylase activity in Yusho patients, *Fukuoka Acta Med.* **78,** 301–304 (1987).
101. M. Aono and H. Okada, Oral findings in Yusho, *Fukuoka Acta Med.* **60,** 468–470 (1969).
102. A. Akamine, I. Hashiguchi, T. Kishi, T. Furukawa, and M. Aono, Alteration in oral pigmentations of patients with Yusho, *Fukuoka Acta Med.* **74,** 284–288 (1983).
103. H. Fukuyama, Y. Anan, A. Akamine, and M. Aono, Alteration in stomatological findings of patients with Yusho (PCB poisoning) in the general examination, *Fukuoka Acta Med.* **70,** 187–198 (1979).
104. H. Fukuyama, Y. Hidaka, S. Sano, and M. Aono, Relation between blood PCB level and oral pigmentation in Yusho patients, *Fukuoka Acta Med.* **68,** 128–132 (1977).
105. A. Akamine, I. Hashiguchi, K. Maeda, Y. Hara, N. Chinjyu, Y. Iwamoto, and M. Aono, Prevalence of periodontal disease in patients with Yusho, *Fukuoka Acta Med.* **76,** 248–252 (1985).
106. I. Taki, S. Hisanaga, and Y. Amagase, Report on Yusho (chlorobiphenyls poisoning) pregnant women and their fetuses, *Fukuoka Acta Med.* **60,** 471–474 (1969).
107. A. Yamaguchi, T. Yoshimura, and M. Kuratsune, A survey on pregnant women having consumed rice oil contaminated with chlorobiphenyls and their babies, *Fukuoka Acta Med.* **62,** 117–122 (1971).
108. I. Funatsu, F. Yamashita, T. Yoshikane, T. Funatsu, Y. Ito, S. Tsugawa, M. Hayashi, T. Kato, M. Yakushiji, G. Okamoto, A. Arima, N. Adachi, K. Takahashi, M. Miyahara, Y. Tashiro, M. Shimomura, S. Yamasaki, T. Arima, T. Kuno, H. Ide, and T. Arima, A chlorobiphenyl induced fetopathy, *Fukuoka Acta Med.* **62,** 139–149 (1971).
109. T. Yoshimura, Epidemiological study on Yusho babies to mothers who had consumed oil contaminated by PCB, *Fukuoka Acta Med.* **65,** 74–80 (1974).
110. T. Yoshimura, A case control study on growth of school children with "Yusho," *Fukuoka Acta Med.* **62,** 109–116 (1971).
111. T. Yoshimura and M. Ikeda, Growth of school children with polychlorinated biphenyl poisoning or Yusho, *Environ. Res.* **17,** 416–425 (1978).
112. M. Kuratsune, Y. Nakamura, M. Ikeda, and T. Hirohata, Analysis of deaths seen among patients with Yusho. A preliminary report, *Chemosphere* **16,** 2085–2088 (1987).
113. M. A. Fingerhut, W. E. Halperin, D. A. Marlow, L. A. Piacitelli, P. A. Honchar, M. H. Sweeney, A. L. Greife, P. A. Dill, K. Steenland, and A. J. Suruda, Cancer mortality exposed to 2,3,7,8-tetrachlorodibenzo-p-dioxin, *N. Engl. J. Med.* **324,** 212–218 (1991).
114. J. J. Ryan, T. A. Gasiewicz, and J. F. Brown, Jr., Human body burden of polychlorinated dibenzofurans associated with toxicity based on the Yusho and Yucheng incidents, *Fundam. Appl. Toxicol.* **15,** 722–731 (1990).
115. B. Birmingham, A. Gilman, D. Grant, J. Salminen, M. Boddington, B. Thorpe, I. Wile, P. Toft, and V. Armstrong, PCDD/PCDF multimedia exposure analysis for the Canadian population: Detailed exposure estimation, *Chemosphere* **19,** 637–642 (1989).
116. H. Beck, K. Eckert, W. Mather, and R. Wittkowski, PCDD and PCDF body burden from food intake in the Federal Republic of Germany, *Chemosphere* **18,** 417–424 (1989).

117. K. Takayama, H. Miyata, O. Aozasa, M. Mimura, and T. Kashimoto, Dietary intake of dioxin-related compounds through food in Japan, *J. Food Hyg. Soc. Jpn.* **32,** 525–532 (1991).
118. Y. Masuda, Dioxin pollution in human body and its evaluation, in: *Problems of Dioxin Pollution and Prospects of Their Settlements,* pp. 105–120, Kogyo Gijyutsu Kai, Tokyo (1992).
119. E. Yrjänheikki, Levels of PCBs, PCDDs and PCDFs in breast milk, *Environ. Health* **34,** 1–90 (1989).
120. H. J. Pluim, J. G. Koppe, K. Olie, J. W. van der Slikke, J. H. Kok, T. Vulsma, D. van Tijn, and J. J. M. de Vijlder, Effects of dioxins on thyroid function in newborn babies, *Lancet* **339,** 1303 (1992).

Chapter 20

The Yu-cheng Rice Oil Poisoning Incident

Chen-Chin Hsu, Mei-Lin M. Yu, Yung-Cheng Joseph Chen, Yue-Liang Leon Guo, and Walter J. Rogan

1. INTRODUCTION

As human beings, we hope to learn from past experiences and not repeat our mistakes. Unfortunately, 11 years after the Japanese Yusho incident (see Chapter 19), a similar tragedy happened in Taiwan in 1979. A Japanese-produced PCB mixture (Kanechlor 400, 500) was used as the heat-transfer medium in the process of deodorization and decolorization of rice oil by a rice oil company in central Taiwan. PCBs and their heat-degraded by-products, polychlorinated dibenzofurans (PCDFs), ter- and quaterphenyls (PCTs and PCQs), leaked into the rice oil and intoxicated 2000 people who had consumed the oil. The initial clinical symptoms consisted of acne, pigmentation of the nails and skin, and hypersecretion of the meibomian glands. Because the disease was caused by ingestion of rice oil, the syndrome was referred to as Yu-cheng (pronounced Yo-Jun), which translates to oil disease, and the exposed subjects were referred to as the Yu-cheng cohort.

Since the incident occurred at a time when PCBs were recognized as serious

---

Chen-Chin Hsu and Yung-Cheng Joseph Chen • Department of Psychiatry, National Cheng Kung University Medical College, Tainan 70428, Taiwan, Republic of China. Mei-Lin M. Yu • Department of Public Health, National Cheng Kung University Medical College, Tainan 70428, Taiwan, Republic of China. Yue-Liang Leon Guo • Department of Environmental and Occupational Health, National Cheng Kung University Medical College, Tainan 70428, Taiwan, Republic of China. Walter J. Rogan • Intramural Research Program, National Institute of Environmental Health Sciences, Research Triangle Park, North Carolina 27709.

Dioxins and Health, edited by Arnold Schecter. Plenum Press, New York, 1994.

environmental pollutants, it immediately caught national and international attention. The Taiwan Provincial Health Department designated several hospitals as Yu-cheng treatment centers. The levels of PCBs and heat-degraded by-products were measured, cases with clinical symptoms were reported, and children born to exposed women during or after the outbreak were identified and have been intensively followed for physical and developmental outcomes.

This chapter describes the discovery and epidemiologic findings of the outbreak, chemical and toxicological data, clinical findings of the directly exposed Yu-cheng cohort and their children exposed to heat-degraded PCBs prenatally and by lactation, and includes updated data not available when the experience of this cohort was last reviewed.[1]

2. DISCOVERY AND EPIDEMIOLOGIC FINDINGS

The outbreak and the discovery of the cause of Yu-cheng have been reviewed by Hsu *et al.*[2–4] On May 21, 1979, a local health bureau in Taichung county, central Taiwan, was notified of an outbreak of skin disease among staff and students boarded at the Hwei-Ming School for Blind Children. The initial clinical symptoms consisted of acne, pigmentation of the nails and skin, and hypersecretion of sebaceous glands. Routine microbiological and toxicologic tests were done on various foodstuffs and water, but failed to yield any helpful result. When similar cases involving 85 of 150 workers from a nearby plastic shoe factory were reported in early September, 1979, an epidemiologic investigation was conducted and a common exposure, a rice bran cooking oil (C-rice bran oil) manufactured by a rice oil company in Changhua and purchased from an edible-oil store in Taichung, was identified. More cases from other companies and local households in both Changhua and Taichung counties were reported, and all were proved to have consumed the C-rice bran oil.

Local clinicians did not recognize the acne shown on the victim's faces and bodies as chloracne, and so analyses directed specifically toward acnegenic halogenated hydrocarbons were not done at first. After the cause of the outbreak, consumption of C-rice bran oil, was identified, and since both the etiology and the symptoms resembled the Japanese Yusho incident (see Chapter 19), the Taiwan Department of Health consulted Japanese scientists who had been involved in the 1968 Yusho investigation. Samples of C-rice bran oil from the Hwei-Ming School and the oil store in Taichung and blood from victims were analyzed and PCBs resembling Kanechlor 400, 500, were detected on October 6, 1979.

The etiology of the outbreak was announced by the government on October 8, 1979, and distribution of all C-rice bran oil was then prohibited. Local residents who had consumed the oil and physicians who had treated the victims started reporting cases to local health bureaus. As of February, 1983, 2022 Yu-cheng subjects were reported and included in the Yu-cheng registry that was maintained by the Taiwan Provincial Health Department. Figure 1 shows the distribution and the morbidity rate

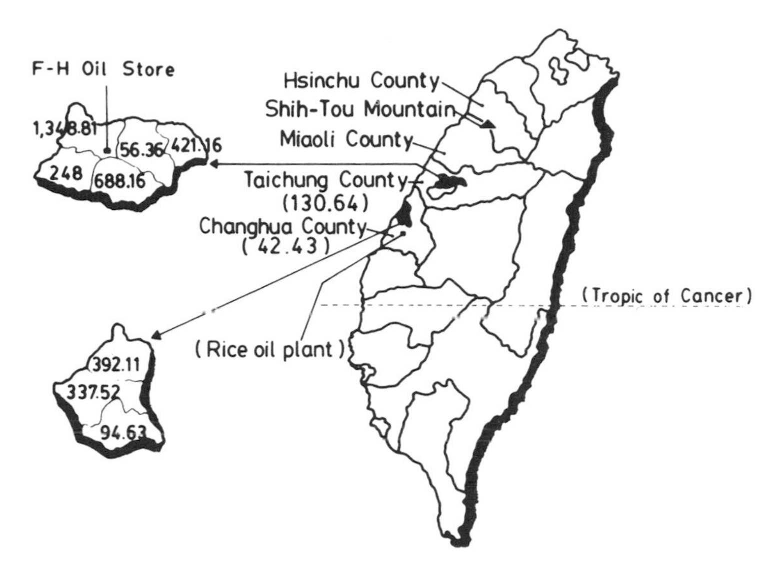

Figure 1. Geographic distribution of PCB-poisoned patients.

per 100,000 persons of Yu-cheng in different counties and villages. Sixty-eight percent of the victims were from Taichung, and 26% from Changhua county. Victims from other counties were mainly from two sources, monks and nuns in Shih-Tou Mountain temples who had purchased the contaminated oil in Taichung, and students from Miaoli county who had come to Taichung for technique-training courses. There were similar numbers of female and male Yu-cheng subjects, and more than 50% of the victims were less than 25 years of age. In Taichung, 60% of the victims were factory workers and 15% were students; in Changhua, the majority of the victims were students (39%), and 19, 16, and 14% were factory workers, housewives, and farmers, respectively. Generally speaking, the Yu-cheng cohort is a young cohort of low socioeconomic status.

3. ANALYTIC CHEMISTRY, EXPOSURE, AND LEVELS

3.1. Oil Contamination

The only oil samples that were positive for PCBs/PCDFs were those from Taichung oil store, the School for Blind Children, and from victims' homes. No PCBs/PCDFs were detected in oil samples from Changhua oil stores and the oil company; perhaps the owners had removed all contaminated oil after the etiology of the illness had been publicly released. Even though no machine containing PCBs was found at the oil company, the high PCB levels from both soil samples at the site and blood samples from plant workers suggested that PCBs had been used in the plant recently. In addition, the detection of PCQs and PCDFs in the toxic oil further suggested that PCBs had been subjected to high temperatures. It was suggested that pipes filled with PCBs, presumably Japanese-manufactured Kanechlor 400, 500, were used as the heat-transfer medium in the process of deodorization and decolorization of the C-rice bran oil. If the pipes developed leaks, the PCBs could then have mixed with the oil. Since PCB mixtures at room temperature are clear, colorless, and tasteless, a large amount of mixing could have gone unnoticed.

The chemicals from eight oil samples were measured using gas chromatography/mass spectrometry (GC/MS) by Chen *et al.*,[2,5–8] Kashimoto *et al.*,[9] and Masuda *et al.*[10] Seven samples contained about 53–100 ppm total PCBs, one-tenth of the level in the Japanese Yusho incident, while one sample contained 405 ppm. The PCB peaks present in the oil were those with a higher number of chlorine atoms and with relatively longer retention times; the percentages of tetra-, penta-, hexa-, and heptachlorobiphenyls were 33.8, 47.1, 12.4, and 4.5, respectively, a pattern closely resembling Kanechlor 400, 500; and the most prominent congeners were 2,4,5,3′,4′-(12.4%) and 2,3,4,3′,4′-penta-CB (11.5%).[2,11] There were about 0.01–1.68 ppm total PCDFs in the oil samples, with a constant PCDF/PCB ratio of about 0.1–0.3%. Thus, it appears that the PCBs were contaminated with a relatively constant amount of PCDFs, but the amount of PCBs/PCDFs that got into the rice oil varied. PCDFs with four, five, and six chlorines were present in the oil, and 2,3,4,8-tetra- and 2,3,4,7,8-penta-CDF were the major congeners.[2,11] The PCQs to PCBs ratio varies in the reports on the Taiwan oil from about 39%[9] to 180%.[10] This may represent analytic differences rather than real ones, since the PCQs are quantitated by a perchlorination procedure rather than by summing the values of individual isomers.

3.2. Exposure

Ninety-eight patients were interviewed by Lan *et al.*[12] on food and oil consumption patterns prior to and during the outbreak. It was estimated that patients consumed the contaminated oil on average at a rate of 1.4 kg/month (range 1.0–1.6 kg/month) for 2 to 3 (average 2.7) months before they became symptomatic, and then for another 6 months before the oil was withdrawn. The PCB and PCDF concentrations in the oil samples were about 67–99 and 0.21–0.40 ppm, respectively. Thus, the patients consumed on average 302 mg (range 196–457) of PCBs and 1.3 mg (range 0.5–1.9) of PCDFs before they developed symptoms, and about 1 g (range 0.7–1.4) of PCBs and 3.8 mg (range 1.8–5.6) of PCDFs total. Because the rice oil in Taiwan was withdrawn at a later time, the Taiwanese patients consumed about ten times as much contaminated oil as the Japanese patients; however, since the PCB/PCDF concentration in the Japanese oil was ten times that of the Taiwanese oil, patients from both countries consumed about the same amount of PCBs and PCDFs.

3.3. Levels

3.3.1. PCBs in Blood

Total blood PCB levels have been measured by several research groups at different times. Kashimoto *et al.*[9] and Masuda *et al.*[10] analyzed blood samples from two small groups of patients within the first year of exposure. P. H. Chen *et al.*[2,7,8] analyzed 165 blood samples from Yu-cheng patients attending a dermatology clinic within 18 months of exposure. Between 1979 and 1983, the Taiwan Provincial Department of Health collected 2378 serial blood samples from 1246 Yu-cheng subjects, with 1–8 samples per subject. Except 11 persons in the Department of Health collection, all persons tested in these studies had detectable PCB levels, and the mean levels ranged from 38 to 99 ppb (Table 1). These levels were higher than those of 92 Taiwanese blood donors, who had a mean PCB level of 9.8 ppb (range 0–25.3 ppb).[13] The PCB levels in the Yu-cheng subjects were also found to be higher than those of the Japanese Yusho subjects (mean PCB value of 72 Yusho patients was 5.9 ppb),[14] probably because the Japanese samples were drawn much longer after the Yusho incident. P. H. Chen *et al.*[15] studied the comparative elimination rates of individual PCB congeners from the blood of patients. The results indicate that the concentrations of those congeners with adjacent unsubstituted carbons at the *meta* and *para* positions, like 2,5,3′4′-tetra-, 2,3,2′,4′,5′-, 2,5,2′,3′,4′-, and 2,3,6,3′,4′-penta-CBs, declined faster than those of other congeners. Mean total serum PCB concentration of 32 Yu-cheng samples drawn in 1985 was found to have declined to about 15 ppb (range 0.6–86.8), and the major persistent congeners detected then were the hexachlorinated biphenyls, such as 2,2′,3,4,4′,5- and 2,2′,4,4′,5,5′-[16] (Table 1). In 1985, Rogan and colleagues collected blood samples from 21 children born to Yu-cheng women during or after the outbreak and 15 age, sex, neighborhood-matched controlled children; 14 of the 21 exposed children and 6 of the 15 controls had detectable PCBs in blood, the mean PCB levels being 8.45 ppb (range 0.12–77.8 ppb) in the *in utero*-exposed children and 3.49 ppb (range 1.06–6.99 ppb) in the controls.[17] In February–March, 1991, 12 years after the Yu-cheng incident, blood samples were collected from 113 *in utero*-exposed children and their age, sex, neighborhood, maternal age, and socioeconomic status-matched controls. Ryan *et al.*[18] reported the results from samples of 14 exposed (age range from 6.5 to 12.5 years) and 10 control children. Six of the fourteen exposed children had detectable PCBs in blood, with a mean of 2.99 ppb (range 0.94–6.8 ppb); the PCB level of the pooled control sample was 0.51 ppb.

3.3.2. PCDFs and PCQs in Blood

Serum PCDF and PCQ levels were measured by Kashimoto *et al.*[9] and Chen *et al.*[2,11] (Table 1). Chen *et al.* also identified 2,3,4,7,8-penta- and 1,2,3,4,7,8-hexa-CDF as the major detectable congeners in 10 samples. Lundgren *et al.*[16] identified the same congeners in 12 samples collected from different subjects in 1985, but at a lower

Table 1
Concentrations of PCBs and Heat-Degraded By-products in Blood of Directly Exposed Yu-cheng Subjects

	Year blood samples drawn	*n*	PCBs	PCDFs	PCQs	PCDF congeners					
						2,3,7,8 tetra	1,2,4,7,8 penta	2,3,4,7,8 penta	2,3,4,6,7 penta	1,2,3,4,7,8 hexa	1,2,3,4,6,7,8 hepta
Kashimoto *et al.* (1981)	1979–80	15	mean 60 ± 39 ppb median 54 range (4–188)	mean 0.14 ± 0.07 ppb median 0.09 range (<0.005–0.27)	mean 19.3 ± 13.0 ppb median 16.9 range (0.9–63.8)						
Masuda *et al.*	1979–80	23	mean 99 ± 163 ppb								
Chen *et al.* (1984), Chen and Hsu (1987)	1980–81	165	mean 38 ppb median 28 range (10–720)	range (0.02–0.2 ppb) (*n* = 10)			trace	det.[a]		det.	trace
Yu-cheng Registry	1979–83	2378[b]	mean 53.5 ppb median 60 range (0–853)								
Lundgren *et al.* (1988)	1985	32	mean 15.4 ± 19.0 ppb range (0.6–86.8)			<0.3 ppt range:	<0.3 ppt	2.7 ± 1.8 ppt (0.4–5.5) (*n* = 12)	<0.3 ppt	10.8 ± 4.9 ppt (0.7–18.6)	

[a]det., detectable.
[b]2378 serial measurements of 1246 subjects, 1–8 samples per subject.

concentration, suggesting some metabolism or excretion over time (Table 1). In 1991, the two congeners were still detectable in 9 of the 14 blood samples drawn from the *in utero*-exposed children by Ryan *et al.*[18]; the 2,3,4,7,8-penta-CDFs ranged from 89 to 570 ppt, lipid, in the exposed children, and 22 ppt, lipid, in the pooled control sample; the 1,2,3,4,7,8-hexa-CDF ranged from 120 to 1590 ppt, lipid, in the exposed children, and 18 ppt, lipid, in the control sample.

3.3.3. Tissue Levels

One patient had his blood and adipose tissue sampled 10 months after the incident and the total PCB levels in his blood and adipose tissue were 39 ppb and 12.8 ppm (wet weight basis).[8] Three autopsy studies have been reported, two of adults and one of a 22-month-old child.[2,4,8,19,20] In general, highest PCB levels were in fat or fatty tissues, and there was some evidence for elevated PCDF concentration in liver. The concentrations of total PCB in intestinal fat and total PCDF in the liver of a patient who died 2 years after the outbreak were 10.8 ppm and 35.1 ppb (wet weight basis), respectively.[2,19] The major PCDFs retained in the liver and other tissues were 1,2,3,4,7,8-hexa-, 2,3,4,7,8-penta-, and 1,2,4,7,8-penta-CDF. In the adult reported by Hsu *et al.*,[4] the PCB concentrations in bladder, colon, blood vessels, and heart were 10,208, 5889, 5320, and 3317 ppm (wet weight basis), but an adipose sample was not mentioned separately. The PCB concentrations in blood, liver, and adipose tissue of the 22-month-old child, who was exposed *in utero,* were 4, 17, and 814 ppb, respectively.[20] These levels were relatively low, but it is not clear that the analytical methods were comparable.

Schecter *et al.*[21] measured dioxin and dibenzofuran levels in six placentas of Yu-cheng women obtained in 1984 and 1985 and one general population placenta from a U.S. woman. The Yu-cheng placentas were found to contain elevated levels of two congeners; 2,3,4,7,8-penta-CDF ranged from 820 to 12,560 ppt, lipid, compared with 6.8 ppt, lipid, in the nonexposed placenta; and 1,2,3,4,7,8-hexa-CDF ranged from 2345 to 26,540 ppt, lipid, compared with 8.7 ppt, lipid, in the nonexposed placenta.

4. CLINICAL FINDINGS OF THE YU-CHENG COHORT

4.1. General Symptoms

Subjective complaints of the Yu-cheng subjects varied over the course of the illness. Ocular symptoms, such as increased eye discharge, swelling of eyelids, and disturbance of vision, were the major complaints at early stages, and as time went by, constitutional symptoms like general malaise, numbness of limbs, pruritus, and headache and dizziness became more obvious and serious.[22–24] Ten percent of the female victims were found to have abnormal menstruation. S.-J. Lan and Yen[25] reported decreased growth in both height and weight of 30 elementary school students in Changhua who had ingested the contaminated oil.

4.2. Dermatology

Skin symptoms of the Yu-cheng subjects have been studied in detail by two groups of dermatologists (Table 2). Both groups reported that mucocutaneous pigmentation was the most common symptom, occurring in at least 90% of the patients; it usually occurred at conjunctiva, gingiva, and buccal mucosa, nasal apex and ala, and finger- and toenails, and the hue varied from brown to brownish gray to gray.[22–24,26,27] The next most common symptom was acneform eruptions, predominantly open comedones, papules, and pustules, occurring in 75% of Wong's and 51% of Lü's patients. It differs from acne vulgaris in that it is found not only on the classical sites for acne but also on extremities, axillae, and external genital areas. Deformities of finger- and toenails were found in 68% of Wong's patients, and in 38% of Lü's patients. Lü also found follicular accentuation and horny plugs (accentuated and elevated hair follicles with blackish keratinous material plugs in the enlarged orifice) to be prominent (at least 21%) in their patients, especially in the age range 11–20. Other findings included keratotic plaques, horny growth like wart or callosity, of palms and soles, dry skin, and itching. When grouping the patients according to the Japanese Goto and Higuchi grading system (ocular sign alone for grade 0, pigmentation of nails and skin for grade I, comedo formation and follicular accentuation for grade II, localized acneform lesions and cysts for grade III, widespread and extensive distribution of the above lesions for grade IV[28]), 79.5% of Wong's patients and 61.6% of Lü's patients were in grades 0–II.[23,24,26,27]

Both Lü *et al.* and Wong *et al.* tried to relate the total blood PCB level to the severity of skin symptoms, but neither group showed any consistent association. These negative associations plus the fact that neither report had any control or background rates for the skin lesions made the interpretation difficult. However, the very high rates of mucocutaneous pigmentation, the unique distribution pattern of acneform eruptions, and the unusual picture of hyperkeratotic plaques make a strong case that these symptoms are related to the ingestion of PCB- and PCDF-contaminated rice bran oil. Histologic examination of 21 skin biopsies showed hyperkeratosis, increased pigmentation of epidermis, and cystic dilatation of hair follicles.[29] Except for the highly pigmented epidermis, the eruptions were indistinguishable from acne vulgaris. Fu studied the ocular manifestations of 117 patients and found a positive correlation between total blood PCB level and severity of conjunctival pigmentation.[30,31]

4.3. Neurology

Two groups of exposed subjects have been evaluated for neurologic dysfunction (Table 2). One hundred and fifty-five Yu-cheng subjects from the Hwei-Ming School for Blind Children, including students, staff, and their families, were seen by neurologists at the National Taiwan University Hospital in early 1980[32–34]; all of them had neurologic examinations and nerve conduction velocity (NCV) tests. The most common neurologic complaints were numbness and paresthesia of extremities (36.1%) and

Table 2
Selective Physical and Laboratory Findings among Directly Exposed Yu-cheng Cohort Subjects

Findings	Wong *et al.* (1982)	Lü *et al.* (1984, 1985)	Chang *et al.* (1980)	Chen *et al.* (1981, 1983, 1985)	Chia *et al.* (1981, 1984)	Lu *et al.* (1980)
$n =$	122	358	143	155	39	69
Hyperpigmentation	92%	90%				
Acne	75%	51%				
Nail dysplasia	68%	38%				
Enlarged meibomian glands		17%				
Peripheral sensory neuropathy[a]				36%	64%	
Headache and/or dizziness				35%	38%	
Slowed nerve conduction				52%	46% ($n = 35$)	
Abnormal EEG					22% ($n = 27$)	
		$n = 133$				
Hematocrit		$\downarrow$	$\downarrow$			
SGOT		$\uparrow$	$\uparrow$			
SGPT		$\uparrow$	$\uparrow$			
Triglycerides		$\uparrow$	$\uparrow$			
Bilirubin			$\downarrow$			
White blood cells		$\uparrow$	$\uparrow$			
T cells			$\downarrow$			
Active T cells			$\downarrow$			
Helper T cells			$\downarrow$			
Suppressor T cells			$\uparrow$			
γ-Globulin			$\downarrow$			
α_2-Globulin			$\uparrow$			
Increase urinary porphyrins						$\uparrow$ mean
Uro/coproporphyrin > 1						52%

[a]Paresthesia, numbness in the distal part of the extremities.

headache and/or dizziness (34.8%). Neurologic examination showed that 7.7% of the patients had decreased vibration sensation in lower limbs, and 5.8% had hearing impairment. A few patients had hyperalgesia, and absent or decreased ankle jerks. On NCV testing, 28.4% had sensory nerve slowing, 7.8% had motor nerve slowing, and an additional 16.1% had both deficits. However, the incidences of the above-mentioned symptoms/signs in the general population were not reported. When compared with values for 63–150 nonexposed normal subjects, the exposed subjects were on average 4 m/sec slower than the controls in both sensory and motor NCV. Patients with blood PCB levels of 24 ppb or greater had significantly slower peroneal nerve motor NCV than those with blood PCB levels below 24 ppb. For 65 patients with known blood PCQ levels, blood PCQ level was negatively associated with median nerve sensory NCV. Chen *et al.*[32] suggested that these deficits might be related to a PCB-induced porphyria, and they did porphyrin analyses on 24-hr urine samples from 48 patients. Although a higher percentage of patients with abnormal NCV had abnormal urinary porphyrins (63 versus 50%), the difference was not statistically significant. Ogawa[35] found motor paresis of the hind limbs and reduced motor NCV on experimentally induced PCB-poisoned rats, and histologic examination revealed segmental demyelination with loss of large nerve fibers in peripheral nerves; this may explain how PCB exposure affects NCV.

Thirty-nine patients admitted to the Veterans General Hospital for treatment of skin lesions were examined by Chia and colleagues for neurologic function in 1980.[36–38] Thirty-five of them (with 44 age-matched controls) had NCV tests, 27 had electroencephalograms (EEG), and 4 had cerebrospinal fluid (CSF) sample drawn. Paresthesia and numbness of extremities (64%), pain over the back, limbs, or orbits (41%), and headache (38%) were the most common neurologic symptoms. Thirty-one percent of the patients had slower sensory nerve conduction, and 29% had slower motor nerve conduction. The patients had both sensory and motor NCV 5–6 m/sec slower than 44 nonexposed healthy subjects. Six of the twenty-seven patients receiving EEG had a mildly abnormal EEG pattern of paroxysmal bilateral slow waves, occasionally mixed with spikes or sharp waves in the frontotemporal region, a pattern not specific to any defect. PCB levels in CSF were "normal" for all four patients. Twenty-eight of the thirty-nine patients were reexamined 2 years later, and it was found that all symptoms had diminished except for dizziness and absent or sluggish deep tendon reflex. Although all NCV had improved, the values were still 4 to 10% slower than for controls.

4.4. Immunology

There is a strong impression among physicians who care for the Yu-cheng victims that they suffer more frequent and more severe skin and respiratory infections. There is also a general belief among many of the victims that they have lowered resistance to infectious diseases. In 1979, Chang *et al.*[39] studied immune function in 143 patients from the Hwei-Ming School. When compared with the normal Taiwan laboratory values, total leukocyte count of the patients was elevated (9650 ± 2800 versus $7053/mm^3$ ± 1205) but with a normal differential, α_2-globulin was slightly increased, and γ-globulin was slightly decreased (Table 2). Delayed-type skin hypersensitive test using streptokinase/streptodornase solution was positive in 36% of patients compared with 79% of the Taiwan general population; this result suggests suppression of cellular immunity in the Yu-cheng patients.

Serum immunoglobulin tests on a subset of 30 patients who had blood PCB levels above 15 ppb and 23 age, sex-matched controls showed significant decreases in IgA and IgM in the exposed (185 ± 88 versus 245 ± 70 for IgA, and 105 ± 58 versus 173 mg/100 ml ± 48 for IgM), suggesting suppression of humoral immunity.[40] The 30 patients had two-thirds the percentage of T cells (42 versus 63%) of controls, and the percentages of active T cells (11 versus 22%) and "helper" T cells (22 versus 37%) were also decreased, while the percentages of B cells and "suppressor" T cells were not affected. This suggests that different types of lymphocytes may have different sensitivity toward PCB toxicity. Since "helper" T cells help the proliferation, differentiation, and immunoglobulin secretion of B cells, when the percentage of "helper" T cells decreases, serum immunoglobulin, like IgA and IgM, decreases too.[40] There were rough correlations among serum PCB concentration, clinical grade of the severity of the skin lesions, decreasing percentages of "active" T cells, and decreased size of induration on skin hypersensitive testing.[41]

A significantly lower percentage of phagocytes, monocytes, and polymorphonuclear leukocytes of the same patients bear immunoglobulin and complement receptors.[42] Phagocytes are responsible for the elimination of the infectious microorganisms, and the decrease of phagocyte complement and immunoglobulin receptors may be associated with the lowered resistance to infectious disease of the Yu-cheng patients. Three years later, using a newly developed monoclonal antibody technique, the percentage of T cells and "active" T cells of the same patients recovered to normal, yet the percentage of "helper" T cells was still low.[43] Wu *et al.*[44] observed enhanced lymphocyte proliferation *in vitro* in response to phytohemagglutinin, pokeweed mitogen, and purified protein derivative (PPD) in 83 patients in the first year after the outbreak, and they found further increases in both spontaneous and mitogen-induced *in vitro* lymphocyte proliferation in a subset of 30 patients studied 3 years after the exposure.[43] This suggests that the effect of heat-degraded PCBs on lymphocyte function still existed 3 years after the exposure.

4.5. Liver Function, Urinary Porphyrin, and Laboratory Findings

Blood chemistry was studied in 143 patients from Hwei-Ming School for Blind Children in late 1979 by Chang *et al.*[39] and 133 patients in a special clinic by Lü and Wong.[23] The profiles of the two groups were relatively similar (Table 2): mild anemia of a normocytic type, leukocytosis, elevations of the liver enzymes, SGOT, SGPT, and alkaline phosphatase but not BUN or LDH. Serum triglyceride and conjugated bilirubin were elevated but total bilirubin was reduced. The low bilirubin is probably related to induction of bilirubin UDP-glucuronyl transferase in liver by PCBs and enhancement of conjugation of bilirubin to glucuronic acid which results in rapid biliary excretion of bilirubin.[39] Chen and Shen[45] studied the association between blood PCB level and serum triglyceride in 104 patients admitted to VGH for skin lesions, and no association was found. The serum triglycerides pose an unsolved analytical problem, since they might lead to an artifactual increase in serum PCBs, which are largely present in fat. There is no consensus on whether or how to adjust for the fat content of serum when reporting PCB concentrations; with adipose tissue measurement this problem would not exist.

Fifteen percent of body's heme production occurs in the liver. When there are enzyme deficiencies along the biosynthesis pathway, abnormal and excessive porphyrins, by-products of the biosynthesis, and other heme precursors appear in body fluid. Hepatic porphyria is caused by damage to the cell membrane of the hepatocytes and is characterized by abnormal urinary porphyrin excretion.[46] Type B hepatic porphyria, greater excretion of uroporphyrin than coproporphyrin, is the first or mildest form of porphyria induced by PCBs in laboratory animals,[47] and is relatively specific to exposures to PCBs and hexachlorobenzene. F.-J. Lu and colleagues studied urinary excretion of porphyrins and heme precursors in 69 blind students from the Hwei-Ming School and 20 healthy volunteers.[48,49] The mean 24-hr excretion of uroporphyrin for the exposed students was 41.2 ± 24.6 mg/24 hr, compared with 13.6 ± 11.8 for the controls; levels of excretion of δ-aminolevulinic acid, a heme precursor, for the exposed and the controls were 1.0 ± 0.6 and 0.7 ± 0.3, respectively, and red cell δ-aminolevulinic acid dehydratase (δ-ALAD) activities were depressed. Excretion of coproporphyrin and porphobilinogen was not affected. The mean uro/copro ratio was 1.4 ± 1.3 in the exposed, and 0.5 ± 0.3 in the controls. Thirty-six of the sixty-nine exposed students had ratios above 1 while none of the 20 controls had a ratio that high. δ-ALAD activities of 23 exposed subjects from Changhua county were depressed in buffered solutions of different pH when compared with those of 20 healthy volunteers.[50]

4.6. Chromosomes

PCBs and PCDFs are not considered to be mutagens. However, their substantial metabolic effects on systems responsible for metabolism of potentially mutagenic xenobiotics make their long-term *in vivo* action somewhat unpredictable. Wuu and Wong[51] studied chromosomes in lymphocytes of 36 exposed patients and 10 unexposed healthy controls in 1982. Nineteen of the thirty-six patients, compared with none of the controls, had chromosome or chromatid aberrations; the aberrations included gaps, breaks, exchanges, and acentric fragments. Lundgren *et al.*[16] assessed sister chromatid exchange (SCE) levels in 35 Yu-cheng women and 24 sex, neighborhood-matched controls in 1985, and found little difference using the conventional assay; the mean frequencies were 7.3 in exposed and 7.6 in controls. However, after adding α-naphthoflavone (ANF), which may be metabolized by PCB/PCDF-induced lymphocytes into more reactive SCE-causing genotoxic metabolites, into the assay, the SCE frequency in the exposed became greater than that of the controls (mean frequencies 10.8 and 8.9). Significant dose–response relationships were observed between ANF-induced SCEs and serum concentrations of total PCBs and of several PCB congeners, 2,2′4,4′,5′-penta-, 2,2′,3,4,4′,5-, 2,2′,4,4′,5,5′-, 2,3,3′,4,4′,5-hexa-, 2,2′,3,3′,4,4′,5-, 2,2′,3,4,4′,5,5′-hepta-PCBs.

4.7. Placenta Studies

One of the most consistent effects of PCBs and related chemicals is induction of activities of monooxygenase enzymes which are involved in the metabolism of chemical carcinogens, such as benzo[a]pyrene. Wong and colleagues[52] studied placental tissues from 4 Yu-cheng women who delivered their babies in 1983, 12 Taiwanese hospital controls, and smoking and nonsmoking volunteers from the obstetric service at the University of North Carolina Hospital. Placental homogenates of the exposed women showed large increases in monooxygenase enzymes, including aryl hydrocarbon hydroxylase, 7-ethoxycoumarin *O*-deethylase, and diol, quinone, and phenolic metabolites of benzo[*a*]pyrene.

5. CLINICAL FINDINGS OF THE *IN UTERO*-EXPOSED YU-CHENG CHILDREN

5.1. Teratology

Infants born to Yu-cheng women have been studied in some depth. Nine infants born between October of 1979 and December of 1980 were evaluated for birth defects at birth by three groups of researchers.[20,53,54] All were noted to have hyperpigmentation of the skin, especially on the lips, gingivae, and nails, and hypersecretion of the meibomian gland (Fig. 2). K.-C. Wong and Hwang[53] also reported skin desquamation, black color of the nose, and deformed nails. The twins examined by Law *et al.*[54] and another infant studied by Lan *et al.*[20] had swelling of the upper eyelid, respiratory distress, and pneumonia. Six of these nine infants were small for gestational age. Eight of these infants have been reviewed in Miller's[55] and Rogan's[56] papers.

5.2. General Symptoms and Porphyrins

Lan *et al.*[57] reported the birth weight of 49 children born to Yu-cheng women between 1979 and 1985. The gestational age-adjusted birth weights for female and male exposed babies were respectively 83 and 87% of those of the "normal" babies, and the deficits were significant in the first and second child born after the outbreak, but not in the third.

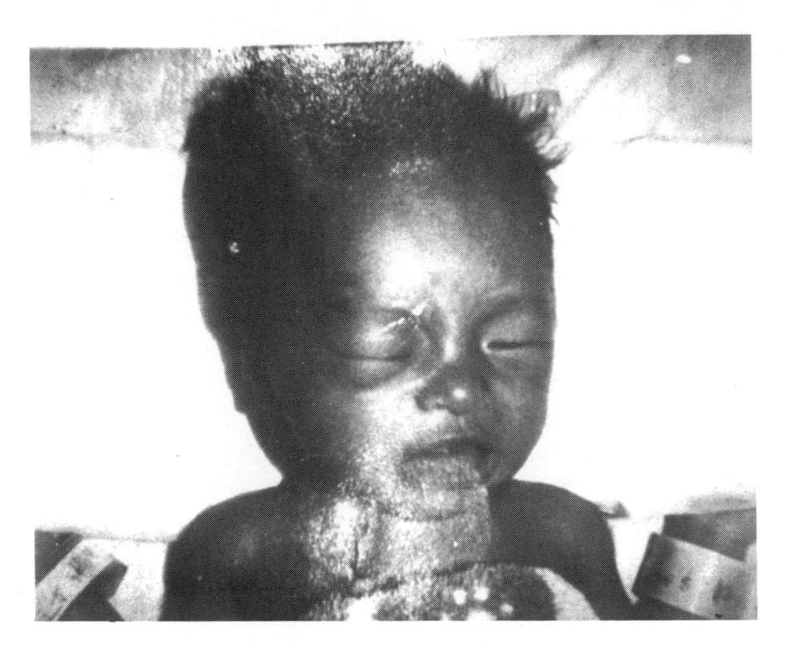

Figure 2. Hyperpigmentation of the skin of a newborn baby.

In April of 1985, Rogan *et al.*[58] identified 132 *in utero*-exposed children born between June of 1978 and March of 1985 and 190 age, sex, neighborhood-matched control children. Thirty of the exposed children also were breast-fed and thus had postnatal chemical exposure. In most cases this number is too small to allow separation of effects related specifically to chemical exposure through breast milk. Parents of 128 exposed and 115 control children were interviewed, and 117 *in utero*-exposed and 108 control children attended physical examinations. The exposed children had lower reported birth weight (mean $\pm$ S.E.; 2749 $\pm$ 46 g in exposed versus 3228 $\pm$ 40 g in control babies) and were 7% lighter and 3% shorter at examination. Reported medical history showed that 24% of the exposed children versus 4% of the control children had bronchitis, and the exposed children were more likely to have abnormalities on pulmonary auscultation at examination. One hundred and seven pairs of children were reexamined in February of 1992, seven years after the initial examination; the Yu-cheng children were still 2% shorter than the controls; and the exposed parents reported a 1.3-fold excess of upper respiratory infection and a 3-fold excess of otitis media during the 6 months prior to the examination.[59]

Spot urines, one-time voided urine specimen, were collected in 1985 for 75 exposed children, 74 controls, and 12 siblings of the exposed children.[60] Four of the *in utero*-exposed children (5%), two controls, and one sibling had a type B hepatic porphyria (i.e., a uroporphyrin/coproporphyrin ratio of greater than 1). Total porphyrin excretion was elevated in the exposed children as a group (95 versus 81 μg/liter), and 11% of the exposed children compared with 3% of the control children had total

Table 3
Selective Physical and Laboratory Findings among Directly Exposed Yu-cheng Cohort Subjects and *in Utero*-Exposed Children

Finding	Directly exposed adults and children[a] $n = 480$	*In utero*-exposed children (at 1985 (exam.), $n = 117$
Hyperpigmentation	90–92%	43%
Acne	51–75%	17%
Nail dysplasia	38–68%	32%
Enlarged meibomian glands	17%	10%
Urinary porphyrins	↑ mean ($n = 69$)	↑ ($n = 75$)
Uro/coproporphyrin > 1	52% ($n = 69$)	5% ($n = 75$)

[a]Note: the groups of directly exposed patients examined do not represent a random sample of the entire exposed cohort.

urinary porphyrin concentrations greater than 200 μg/liter. This suggests that *in utero* exposure to PCBs and their heat-degraded by-products causes a mild disturbance in porphyrin metabolism; however, the frequency in the *in utero*-exposed group is lower than that in the directly exposed cohort (Table 3).

5.3. Dermatology

Information on birth and medical history of the 128 *in utero*-exposed and 115 control children identified in 1985 by Rogan *et al.*[58] was derived from questionnaires filled out by parents. A significantly higher percentage of exposed children were reported to have hyperpigmentation, conjunctivitis, swelling of the eyelid, eye discharge, deformed or small nails, natal teeth, and swollen gums at birth; they also had more deformed finger- and toenails and acne scars in their lifetime history.[58,61] Children born later after the exposure did not differ from those born earlier.[61] On examination, the exposed children had a much higher rate of dystrophic fingernails (Fig. 3) and dystrophic or pigmented toenails than controls; dystrophic fingernails were more specific, occurring only in the exposed children. They also had increased rates of hyperpigmentation (Fig. 4), acne (Fig. 5), and scaly or keratotic disorders than the controls. The dermatologic symptoms of these children were similar but milder than those of the directly exposed cohort subjects (Table 3). In addition, the exposed children had more generalized itching, localized skin infections, and hair loss.[61] The effects in the children are more apparent in nails, hair, teeth, gums, skin hyperpigmentation, and growth and development; the characteristic defect was described by Rogan *et al.* as a type of acquired ectodermal dysplasia, which means the defects were formed on the ectodermal layer during the embryonic stage.[58]

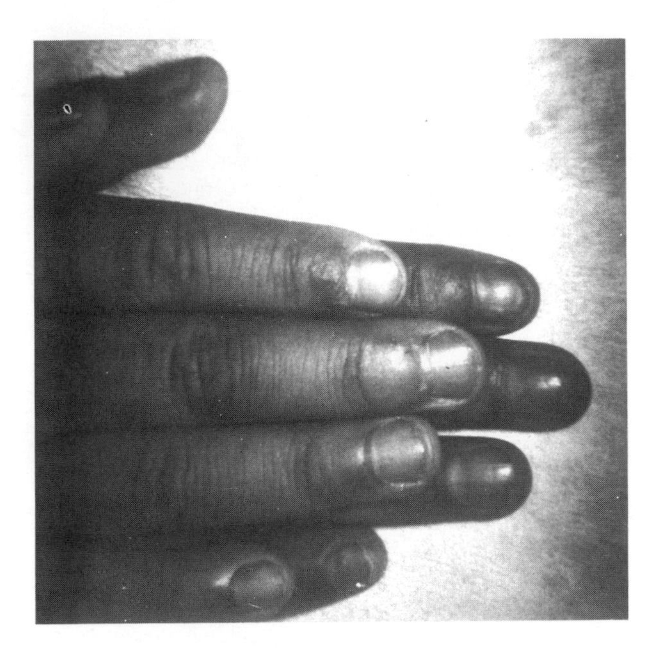

Figure 3. Pigmented and slightly dystrophic fingernails of a 6-year-old boy on bottom. Control child's nails on top.

5.4. Neurology, Cognitive and Behavioral Development

In 1985, the 117 *in utero*-exposed children and their controls examined by Rogan *et al.*[58] were also given neurologic tests. Ten percent of the exposed versus three percent of the control children were considered by the neurologists to have developmental or psychomotor delay. This group of controls was originally selected to provide background rates for physical findings, and they were thought to be too loosely

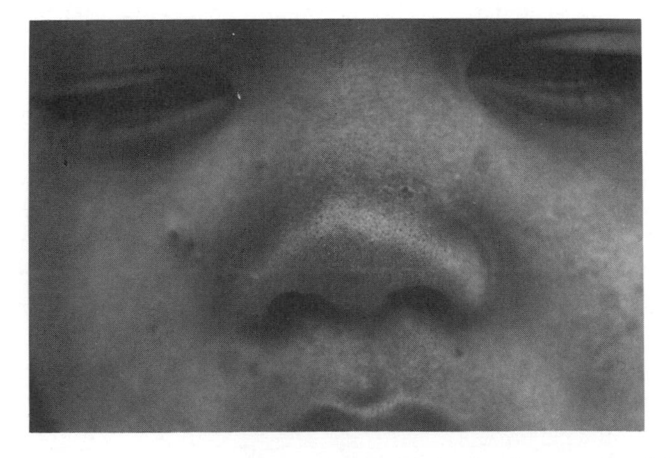

Figure 4. Hyperpigmentation over nasal apex and ala of an *in utero*-exposed child.

Figure 5. Acneform eruptions of an *in utero*-exposed child.

matched for developmental assessment, so a new set of controls who matched 118 *in utero*-exposed children on age, sex, maternal age, parents' combined educational level and occupation, and neighborhood were selected 6 months after the initial survey. These 118 pairs of children were given age-appropriate cognitive and behavioral assessments. The exposed children scored 4 to 7 points lower than the control children on all cognitive tests except for verbal IQ, on which they scored the same, on the Chinese version of the Wechsler Intelligence Scale for Children, revised version (WISC-R); and they had worse scores on Rutter's Behavioral Scale. The same group of children have been followed by Hsu and his research team with cognitive and behavioral tests biannually since the fall of 1985. The yearly reports published by this team[62–68] continued for 6 years to show a mild yet consistent deficit of the *in utero* PCB-exposed children on all scales of all cognitive tests. On behavioral assessment, the exposed children were more intense in reaction and more negative in quality of mood on temperament assessment, had more psychosomatic, habit, and behavioral problems than the controls on Rutter's Behavioral Scale, and were hyperactive on a modified Werry–Weiss–Peters Activity Scale. Chen *et al.*[69] merged data from 12 rounds of assessment and examined the hypothesis that the time elapsed between the exposure and the child's birth does affect the degree of development delay, and that the effect diminishes as the children grow. Their results showed that the exposed children scored approximately 5 points lower on all scales of all IQ tests for the ages of 4 through 7 years. Children born up to 5 or 6 years after their mothers' exposure continued to be affected, and the effect of *in utero* PCB and PCDF exposure on children's cognitive development persisted in the children up to the age of 7 years. Cognitive function was also evaluated in the youngest elder siblings of 15 *in utero*-exposed children, and WISC-R IQ scores of the *in utero*-exposed children were 9 to 19 points lower than those of their youngest elder siblings at each year from 1985 to 1990.[70]

Yu *et al.*[17] examined the relationship between developmental findings and physical and historical findings for the *in utero*-exposed children. They found that the exposed children who were shorter, or who were reported to have exhibited neonatal symptoms of intoxication, such as eye discharge, eyelid swelling, and hyperpigmentation, or who ever had deformed nails had greater developmental delay. However, there was little relationship between other physical findings of the 1985 examination or measures of maternal exposure based on either serum PCB level or degree of clinical symptoms and developmental scores.

6. MORTALITY

As of February, 1983, 24 deaths among 2022 Yu-cheng cohort subjects had been reported. Half of the deaths were caused by hepatoma, liver cirrhosis, or liver diseases with hepatomegaly.[4] It should be noted that Taiwan has an extraordinary incidence of hepatitis B (15%), and liver cirrhosis and liver cancer are common. Eight of the first thirty-nine babies born to Yu-cheng women died of pneumonia, bronchitis, sepsis, and prematurity, causing a very high infant mortality rate (20.5 versus 1% in the general population).[4] There were three more deaths among the Yu-cheng offspring noted by 1985,[58] and one of the 118 *in utero*-exposed children followed by Hsu's team died in a car accident during the 6-year developmental follow-up.[71] No comprehensive mortality data are available after the 1985 report.

7. THE MANAGEMENT AND TREATMENT OF THE VICTIMS

Cholestyramine, a basic anion-exchange resin shown in an animal study[72] to bind chlordecone in rat intestine, increases the excretion of chlordecone into the feces, and decreases its content in the tissues by 30–52% after 2 weeks of treatment, was suggested as a promising therapy for treating chronic poisoning with chlordecone and possibly with other lipophilic toxins. Cholestyramine was used in a clinical trial at the Veterans General Hospital (VGH) in Taipei.[73] After 2 to 8 months of treatment, the blood PCB levels of 20 patients receiving cholestyramine did not decline more than those of the placebo group; however, 45% of the treatment group, compared with 3% of the control group, were reported to have greater relief of their clinical symptoms. Sixteen patients with severe clinical symptoms voluntarily participated in a fasting trial in mid-1981 and/or early 1982.[74] After fasting for 7 or 10 days, all patients experienced improvement of symptoms such as headache, lumbago, arthralgia, cough, sputa, and acneform eruptions. Chinese medicine and acupuncture have been used by China Medical College to treat the victims; however, no document or publication has been released on the treatment effect.

8. SUMMARY AND FUTURE RESEARCH PLAN

It has been 14 years since the Yu-cheng outbreak; and although the painful memory of the outbreak may have faded somewhat, among those involved some of the chemicals persist in the body and continue to affect the directly exposed Yu-cheng cohort and their offspring. Most information regarding the outbreak and the effects of direct exposure to the chemicals came from case reports and clinical observations of subgroups of the cohort during the first 6 years after the incident. Most of the observations were done on small populations with no or poorly defined controls. Since April of 1985, more carefully designed epidemiologic studies of children born to Yu-cheng mothers during or after the outbreak and proper controls have provided thorough information of physical, cognitive, and behavioral effects of *in utero* heat-degraded PCB exposure.

To understand the effect of PCBs and heat-degraded by-products on various organs or systems of the body, several research projects are ongoing or in the planning stage. To investigate the long-term health effects of direct exposure to these chemicals, the directly exposed Yu-cheng cohort who had ingested the contaminated oil and proper controls have been identified and the mortality rate, cancer and other chronic disease morbidity rates, and reproductive functions of the two groups will be studied and compared. Blood samples from all directly and *in utero*-exposed subjects and pooled control samples will be collected, PCB and PCDF congener levels will be measured, and the relations of blood chemical levels to physical findings will be evaluated. For the children identified in 1985, who will go through puberty in the next few years, the caffeine breath test[75] will be given to test the PCB/PCDF effect on cytochrome P450 IA2 activity in liver, and the effect of the altered enzyme activity on the metabolism of sexual hormones and sexual maturation of Yu-cheng children. A second *in utero* PCB-exposed child cohort, children born to either exposed mothers or exposed fathers between June of 1985 and July of 1992, has been identified, and the research questions of whether PCBs continue to affect offspring through transplacental transfer and nursing 6 to 13 years after the exposure, and whether PCB/PCDF affects the offspring through paternal reproductive pathways will be studied.

The findings of these studies will provide dose-related etiologic information not usually available in most occupational or environmental exposure settings. There may be a unique opportunity to test a specific hypothesis about virus–chemical interaction with respect to hepatocarcinogenesis. Taiwan has a high incidence of hepatitis B infection (15%), and a consequent high rate of hepatoma. In Japan, preliminary data from the Yusho cohort with almost identical exposure suggested a sixfold excess of liver cancer mortality in men and a threefold excess in women[76]; hepatoma has been shown in animal studies to be caused by PCB exposure.[77–81] Also, as immunosuppresants, the PCBs and PCDFs may contribute to a decreased ability to fight against cancers or infectious diseases. Thus, it is plausible to study the possible interaction between two factors.

It is very unlikely that there will be another similar incident, but the potential for exposure to PCBs that have not been so heat-degraded or to dibenzofurans and dioxins is much greater, through accidents, improper disposal practices and transport of PCB-containing transformers and capacitors, and contaminated food. The findings from the investigations of the Yu-cheng incident surely provide clues about the toxicities that background PCB exposure may cause, and because dibenzofurans and dioxins are thought to have a similar mechanism of action, the information found regarding this incident may contribute to the understanding of human health response to these compounds.

REFERENCES

1. W. J. Rogan, Yu-cheng, in: *Halogenated Biphenyls, Terphenyls, Naphthalenes, Dibenzodioxins and Related Products* (R. D. Kimbrough and A. A. Jensen, eds.), 2nd fully revised edition, pp. 401–415, Elsevier, Amsterdam (1989).
2. P. H. Chen and S.-T. Hsu, PCB poisoning from toxic rice-bran oil in Taiwan, in: *PCBs and the Environment* (J. S. Waid, ed.), pp. 27–38, CRC Press, Boca Raton, FL (1987).
3. S.-T. Hsu, C.-I. Ma, S. K.-H. Hsu, S.-S. Wu, N. H.-M. Hsu, and C.-C. Yeh, Discovery and epidemiology of PCB poisoning in Taiwan, *Am. J. Ind. Med.* **5,** 71–79 (1984).
4. S.-T. Hsu, C.-I. Ma, S. K.-H. Hsu, S.-S. Wu, N. H.-M. Hsu, C.-C. Yeh, and S.-B. Wu, Discovery and epidemiology of PCB poisoning in Taiwan: A four-year followup, *Environ. Health Perspect.* **59,** 5–10 (1985).
5. P. H. Chen, K.-T. Chang, and Y.-D. Lu, Polychlorinated biphenyls and polychlorinated dibenzofurans in the toxic rice-bran oil that caused PCB poisoning in Taichung, *Bull. Environ. Contam. Toxicol.* (now *Arch. Environ. Contam. Toxicol.*) **26,** 489–495 (1981).
6. P. H. Chen, K.-T. Chang, and Y.-D. Lu, Toxic compounds in the cooking oil which caused PCB poisoning in Taiwan. I. Levels of polychlorinated biphenyls and polychlorinated dibenzofurans [in Chinese; English summary], *Clin. Med. (Taipei)* **7,** 71–76 (1981).
7. P. H. Chen, Y.-D. Lu, M.-H. Yang, and J.-S. Chen, Toxic compounds in the cooking oil which caused PCB poisoning in Taiwan. II. The presence of polychlorinated quaterphenyls and polychlorinated terphenyls [in Chinese; English summary], *Clin. Med. (Taipei)* **7,** 77–82 (1981).
8. P. H. Chen, M.-L. Luo, C.-K. Wong, and C.-J. Chen, Polychlorinated biphenyls, dibenzofurans, and quaterphenyls in the toxic rice-bran oil and PCBs in the blood of patients with PCB poisoning in Taiwan, *Am. J. Ind. Med.* **5,** 133–145 (1984).
9. T. Kashimoto, H. Miyata, S. Kunita, T.-C. Tung, S.-T. Hsu, K.-J. Chang, S.-Y. Tang, G. Ohi, J. Nakagawa, and S. Yamamoto, Role of polychlorinated dibenzofuran in Yusho (PCB poisoning), *Arch. Environ. Health* **36,** 321–326 (1981).
10. Y. Masuda, H. Kuroki, T. Yamaryo, K. Haraguchi, M. Kuratsune, and S.-T. Hsu, Comparison of causal agents in Taiwan and Fukuoka PCB poisonings, *Chemosphere* **11,** 199–206 (1982).
11. P. H . Chen, C.-K. Wong, C. Rappe, and M. Nygren, Polychlorinated biphenyls, dibenzofurans and quaterphenyls in toxic rice-bran oil and in the blood and tissue of patients with PCB poisoning (Yu-cheng) in Taiwan, *Environ. Health Perspect,* **59,** 59–65 (1985).
12. C.-F. Lan, P. H. Chen, L.-L. Shieh, and Y.-H. Chen, An epidemiological study on polychlorinated biphenyls poisoning in Taichung area [in Chinese; English summary], *Clin. Med. (Taipei)* **7,** 96–100 (1981).
13. S.-H. Liu, Y.-C. Ko, T.-L. Huang, *et al.,* Residues of PCBs in blood of ordinary persons in central Taiwan [in Chinese], *Annu. Rep. Bureau Health Taiwan* **1,** 269–275 (1985).

14. Y. Masuda, R. Kagawa, K. Shimamura, M. Takada, and M. Kuratsune, Polychlorinated biphenyls in the blood of Yusho patients and ordinary persons, *Fukuoka Acta Med.* **65,** 25–27 (1974).
15. P. H. Chen, M.-L. Luo, C.-K. Wong, and C.-J. Chen, Comparative rates of elimination of some individual polychlorinated biphenyls from the blood of PCB-poisoned patients in Taiwan, *Food Chem. Toxicol.* **20,** 417–425 (1982).
16. K. Lundgren, G. W. Collman, S. Wang-Wuu, T. Tiernan, M. Taylor, C. L. Thompson, and G. W. Lucier, Cytogenetic and chemical detection of human exposure to polyhalogenated aromatic hydrocarbons, *Environ. Mol. Mutagen.* **11,** 1–11 (1988).
17. M.-L. Yu, C.-C. Hsu, B. C. Gladen, and W. J. Rogan, *In utero* PCB/PCDF exposure: Relation of developmental delay to dysmorphology and dose, *Neurotoxicol. Teratol.* **13,** 195–202 (1991).
18. J. J. Ryan, B. P. Y. Lau, Y. M. Masuda, and Y.-L. Guo, Mass spectrometry as a tool for the measurement of PCDFs/COPCBs/CAPCBs in human blood and application to the two rice oil poisonings, A keynote address presented at the International Conference on Biological Mass Spectrometry, Kyoto, September 20–24 (1992).
19. P. H. Chen and R. A. Hites, Polychlorinated biphenyls and dibenzofurans retained in the tissues of a deceased patient with Yucheng in Taiwan, *Chemosphere* **12,** 1507 –1516 (1983).
20. S.-J. Lan, S.-Y. Tang, and Y.-C. Ko, The effects of PCB poisoning; a study of a transplacental Yu-cheng baby: report of a case [in Chinese; English summary], *Kaohsiung J. Med. Sci.* **3,** 64–68 (1987).
21. A. Schecter, J. R. Startin, C. Wright, M. Kelly, G. Lucier, and K. Charles, Dioxin and dibenzofuran levels in Yucheng placentas and control placentas comparing dioxin/dibenzofuran levels with receptor binding and enzyme induction, Presented at Dioxin '92, Tampere, Finland, August 24–28 (1992).
22. Y.-Y. Lee, P.-N. Wong, Y.-C. Lü, C.-C. Sun, Y.-C. Wu, R.-Y. Lin, S.-H. Jee, K.-Y. Ng, and H.-P. Yeh, An outbreak of PCB poisoning, *J. Dermatol. (Tokyo)* **7,** 435–441 (1980).
23. Y.-C. Lü and P.-N. Wong, Dermatological, medical, and laboratory findings of patients in Taiwan and their treatments, *Am. J. Ind. Med.* **5,** 81-115 (1984).
24. Y.-C. Lü and Y.-C. Wu, Clinical findings and immunological abnormalities in Yu-cheng patients, *Environ. Health Perspect.* **59,** 17–29 (1985).
25. S.-J. Lan and Y.-Y. Yen, Study of the effects of PCBs poisoning on the growth of primary school children [in Chinese; English summary], *Kaohsiung J. Med. Sci.* **2,** 682–687 (1986).
26. P.-C. Cheng, C.-J. Chen, C.-K. Wong, and P. H. Chen, Dermatological survey of 122 PCB poisoning patients in comparison with blood PCB levels [in Chinese; English summary], *Clin Med. (Taipei)* **7,** 15–22 (1981).
27. C.-K. Wong, C.-J. Chen, P.-C. Cheng, and P. H.-S. Chen, Mucocutaneous manifestations of polychlorinated biphenyls (PCB) poisoning: A study of 122 cases in Taiwan, *Br. J. Dermatol.* **107,** 317–323 (1982).
28. M. Goto and K. Higuchi, The symptomatology of Yusho (PCB poisoning) in dermatology, *Fukuoka Acta Med.* **60,** 409–431 (1969).
29. P.-C. Cheng and K.-Y. Liu, Dermatopathological findings of PCB poisoning patients [in Chinese; English summary], *Clin. Med. (Taipei)* **7,** 41–44 (1981).
30. Y.-A. Fu, Ocular manifestations of PCB poisoning and its relationships between blood PCB levels and ocular findings [in Chinese; English summary], *Clin. Med. (Taipei)* **7,** 28–34 (1981).
31. Y.-A. Fu, Ocular manifestation of polychlorinated biphenyls intoxication, *Am. J. Ind. Med.* **5,** 127–132 (1984).
32. R.-C. Chen, Y.-C. Chang, K.-J. Chang, F.-J. Lu, and T.-C. Tung, Peripheral neuropathy

caused by chronic polychlorinated biphenyls poisoning [in English; Chinese summary], *J. Formosan Med. Assoc.* **80,** 47–54 (1981).

33. R.-C. Chen, Y.-C. Chang, T.-C. Tung, and K.-J. Chang, Neurological manifestations of chronic polychlorinated biphenyls poisoning [in English; Chinese summary], *Proc. Natl. Sci. Counc. A ROC* **7,** 87–91 (1983).
34. R.-C. Chen, S.-Y. Tang, H. Miyata, T. Kashimoto, Y.-C. Chang, K.-J. Chang, and T.-C. Tung, Polychlorinated biphenyl poisoning: Correlation of sensory and motor nerve conduction, neurologic symptoms, and blood levels of polychlorinated biphenyls, quaterphenyls, and dibenzofurans, *Environ. Res.* **37,** 340–348 (1985).
35. M. Ogawa, Electrophysiological and histological studies of experimental chlorobiphenyls poisoning, *Fukuoka Acta Med.* **62,** 74–78 (1971).
36. L.-G. Chia and F.-L. Chu, Neurological studies on polychlorinated biphenyl (PCB)-poisoned patients, *Am. J. Ind. Med.* **5,** 117–126 (1984).
37. L.-G. Chia and F.-L. Chu, A clinical and electrophysiological study of patients with polychlorinated biphenyl poisoning, *J. Neurol. Neurosurg. Psychiatry* **48,** 894–901 (1985).
38. L.-G. Chia, M.-S. Su, R.-C. Chen, Z.-A. Wu, and F.-L. Chu, Neurological manifestations in polychlorinated biphenyls (PCB) poisoning [in Chinese; English summary], *Clin. Med. (Taipei)* **7,** 45–61 (1981).
39. K.-J. Chang, J.-S. Chen, P.-C. Huang, and T.-C. Tung, Study of patients with polychlorinated biphenyls poisoning. I. Blood analyses of patients [in Chinese; English summary], *J. Formosan Med. Assoc.* **79,** 304–313 (1980).
40. K.-J. Chang, K.-H. Hsieh, T.-P. Lee, S.-Y. Tang, and T.-C. Tung, Immunologic evaluation of patients with polychlorinated biphenyl poisoning: Determination of lymphocyte subpopulations, *Toxicol. Appl. Pharmacol.* **61,** 58–63 (1981).
41. K.-J. Chang, K.-H. Hsieh, S.-Y. Tang, T.-C. Tung, and T.-P. Lee, Immunologic evaluation of patients with polychlorinated biphenyl poisoning: Evaluation of delayed-type skin hypersensitive response and its relation to clinical studies, *J. Toxicol. Environ. Health* **9,** 217–223 (1982).
42. K.-J. Chang, K.-H. Hsieh, T.-P. Lee, and T.-C. Tung, Immunologic evaluation of patients with polychlorinated biphenyl poisoning: Determination of phagocyte Fc and complement receptors, *Environ. Res.* **28,** 329–334 (1982).
43. Y.-C. Wu, R.-P. Hsieh, and Y.-C. Lü, Altered distribution of lymphocyte subpopulations and augmentation of lymphocyte proliferation in chronic PCB poisoned patients [in English; Chinese summary], *Chin. J. Microbiol. Immunol.* **17,** 177–187 (1984).
44. Y.-C. Wu, Y.-C. Lü, H.-Y. Kao, C.-C. Pan, and R.-Y. Lin, Cell-mediated immunity in patients with polychlorinated biphenyl poisoning [in English; Chinese summary], *J. Formosan Med. Assoc.* **83,** 419–429 (1984).
45. C.-J. Chen and R.-L. Shen, Blood PCB level and serum triglyceride in PCB poisoning patients [in Chinese; English summary], *Clin Med. (Taipei)* **7,** 66–70 (1981).
46. M. Doss, Pathobiochemical transition of secondary coproporphyrinuria to chronic hepatic porphyria in humans, *Klin. Wochenschr.* **58,** 141–148 (1980).
47. J. A. Goldstein and S. Safe, Mechanism of action and structure–activity relationships for the chlorinated dibenzo-*p*-dioxins and related compounds, in: *Halogenated Biphenyls, Terphenyls, Naphthalenes, Dibenzodioxins and Related Products* (R. D. Kimbrough and A. A. Jensen, eds.), 2nd fully revised edition, pp. 239–293, Elsevier, Amsterdam (1989).
48. F.-J. Lu, K.-J. Chang, S.-C. Lin, and T.-C. Tung, Studies on patients with polychlorinated biphenyls poisoning: Determination of urinary coproporphyrin, uroporphyrin, δ-aminolevulinic acid, and porphobilinogen [in Chinese; English summary], *J. Formosan Med. Assoc.* **79,**990–995 (1980).

49. K.-J. Chang, F.-J. Lu, F.-C. Tung, and T.-P. Lee, Studies on patients with polychlorinated biphenyl poisoning. 2. Determination of urinary coproporphyrin, uroporphyrin, δ-aminolevulinic acid and porphobilinogen, *Res. Commun. Chem. Pathol. Pharmacol.* **30,** 547–554 (1980).
50. F.-J. Lu, S.-H. Wang, Y.-C. Wu, and R.-Y. Lin, δ-aminolevulinic acid dehydratase test for polychlorinated biphenyls poisoning [in Chinese; English summary], *J. Formosan Med. Assoc.* **83,** 27–33 (1984).
51. K.-D. Wuu and C.-K. Wong, A chromosomal study on blood lymphocytes of patients poisoned by polychlorinated biphenyls [in English; Chinese summary], *Proc. Natl. Sci. Counc. B ROC* **9,** 67–69 (1985).
52. T. K. Wong, R. B. Everson, and S.-T. Hsu, Potent induction of human placental mono oxygenase activity by previous dietary exposure to polychlorinated biphenyls and their thermal degradation products, *Lancet* **1,** 721–724 (1985).
53. K.-C. Wong and M.-Y. Hwang, Children born to PCB poisoning mothers [in Chinese; English summary], *Clin. Med. (Taipei)* **7,** 83–87 (1981).
54. K.-L. Law, B.-T. Hwang, and I.-S. Shaio, PCB poisoning in newborn twins [in Chinese; English summary], *Clin. Med. (Taipei)* **7,** 88–91 (1981).
55. R. W. Miller, Teratogenesis, *Environ. Health Perspect.* **60,** 211–214 (1985).
56. W. J. Rogan, PCBs and cola-colored babies: Japan, 1968, and Taiwan, 1979, *Teratology* **26,** 259–261 (1982).
57. S.-J. Lan, Y.-Y. Yen, C.-H. Yang, C.-Y. Yang, and E.-R. Chen, A study on the birth weight of transplacental Yu-cheng babies [in Chinese; English summary], *Kaohsiung J. Med. Sci.* **3,** 273–282 (1987).
58. W. J. Rogan, B. C. Gladen, K.-L. Hung, S.-L. Koong, L.-Y. Shih, J. S. Taylor, Y.-C. Wu, D. Yang, N. B. Ragan, and C.-C. Hsu, Congenital poisoning by polychlorinated biphenyls and their contaminants in Taiwan, *Science* **241,** 334–336 (1988).
59. S.-H. Ju, Y.-J. Chen, Y.-C. Chen, and C.-C. Hsu, Follow-up study of growth and health of children born to mothers intoxicated by polychlorinated biphenyls [abstract], *Pediatr. Res.* **28,** 93A (1992).
60. B. C. Gladen, W. J. Rogan, N. B. Ragan, and F. W. Spiert, Urinary porphyrins in children exposed transplacentally to polyhalogenated aromatics in Taiwan, *Arch. Environ. Health* **43,** 54–58 (errata 348) (1988).
61. B. C. Gladen, J. S. Taylor, Y.-C. Wu, N. B. Ragan, W. J. Rogan, and C.-C. Hsu, Dermatological findings in children exposed transplacentally to heat-degraded polychlorinated biphenyls in Taiwan, *Br. J. Dermatol.* **122,** 799–808 (1990).
62. C.-C. Hsu, C.-C. Chen, W.-T. Soong, S.-J. Sue, C.-Y. Liu, C.-C. Tsung, S.-C. Lin, S.-H. Chang, and S.-L. Liao, A six-year follow-up study of intellectual and behavioral development of Yu-cheng children: Cross-sectional findings of the first fieldwork [in Chinese; English summary], *Chin. Psychiatry* **2,** 26–40 (1988).
63. C.-C. Chen, C.-C. Hsu, T.-L. Yeh, S.-C. Lin, and Y.-H. Duann, A six-year follow-up study of intellectual and behavioral development of Yu-cheng children: Cross-sectional findings of the second fieldwork study [in Chinese; English summary], *Chin. Psychiatry* **2,** 257–266 (1988).
64. T.-L. Yeh, C.-C. Hsu, C.-C. Chen, Y.-H. Duann, S.-C. Lin, M.-C. Wen, and M.-J. Su, A six-year follow-up study of intellectual and behavioral development of Yu-cheng children: Findings during the second year of fieldwork [in Chinese: English summary], *Chin. Psychiatry* **2,** 172–185 (1988).
65. Y.-C. Chen, C.-C. Hsu, W.-T. Soong, H.-C. Ko, C.-C. Chen, T.-L. Yeh, S.-C. Lin, M.-C. Wen, and M.-J. Su, A six-year follow-up study of intellectual and behavioral development

of Yu-cheng children: Findings during the third year of fieldwork [in Chinese; English summary]. *Chin. Psychiatry* **3,** 89–98 (1989).

66. C.-C. Hsu, Y.-C. Chen, W.-T. Soong, and H.-C. Ko, A six-year follow-up study of intellectual and behavioral development of Yu-cheng (oil disease) children: Cross-sectional findings of the fourth year fieldwork [in Chinese; English summary], *Chin. Psychiatry* **3**(Suppl. 1)**,** 101–111 (1989).
67. Y.-C. Chen, T.-L. Yeh, and C.-C. Hsu, A six year follow-up study on the intellectual and behavioral development of Yu-cheng (oil disease) children: Findings for the fifth year of fieldwork [in Chinese; English summary], *Chin. Psychiatry* **4,** 40–51 (1990).
68. Y.-C. Chen, Y.-L. Guo, and C.-C. Hsu, The cognitive and behavioral development of children prenatally exposed to polychlorinated biphenyls and contaminants: Sixth-year fieldwork report, *Chin. Psychiatry* **6,** 116–125 (1992).
69. Y.-C. Chen, Y.-L. Guo, C.-C. Hsu, and W. J. Rogan, Cognitive development of Yu-cheng ('oil disease') children prenatally exposed to heat-degraded PCBs, *J. Am. Med. Assoc.* **268,** 3213–3218 (1992).
70. Y.-C. Chen, Y.-L. Guo, and C.-C. Hsu, Cognitive development of children prenatally exposed to polychlorinated biphenyls (Yu-cheng children) and their siblings, *J. Formosan Med. Assoc.* **91,** 704–707 (1992).
71. C.-C. Hsu, Annual report to the National Science Council [in Chinese] (1992).
72. J. J. Boylan, J. L. Egle, and P. S. Guzelian, Cholestyramine: Use as a new therapeutic approach for chlordecone (kepone) poisoning, *Science* **199,** 893–895 (1978).
73. K.-M. Hung and C.-K. Wong, A preliminary report of the treatment results of PCB poisoning patients [in Chinese; English summary], *Clin. Med. (Taipei)* **7,** 92–95 (1981).
74. M. Imamura and T.-C. Tung, A trial of fasting cure for PCB-poisoned patients in Taiwan, *Am. J. Ind. Med.* **5,** 147–153 (1984).
75. G. H. Lambert, D. A. Schoeller, A. N. Kotake, C. Flores, and D. Hay, The effect of age, gender, and sexual maturation on the caffeine breath test, *Dev. Pharmacol. Ther.* **9,** 375–388 (1986).
76. M. Kuratsune, Yusho, with reference to Yu-cheng, in: *Halogenated Biphenyls, Terphenyls, Naphthalenes, Dibenzodioxins and Related Products* (R. D. Kimbrough and A. A. Jensen, eds.), 2nd fully revised edition, pp. 381–400, Elsevier, Amsterdam (1989).
77. N. T. Kimura and T. Baba, Neoplastic changes in the rat liver induced by polychlorinated biphenyl, *Jpn. J. Cancer Res. [Gann]* **64,** 105–108 (1973).
78. R. D. Kimbrough, R. A. Squire, R. E. Linder, J. D. Strandberg, R. J. Montali, and V. W. Burse, Induction of liver tumors in Sherman strain female rats by polychlorinated biphenyl Aroclor 1260. *J. Natl. Cancer Inst.* **55,** 1453–1459 (1975).
79. N. Ito, H. Nagasaki, S. Makiura, and M. Arai, Histopathological studies on liver tumorigenesis in rats treated with polychlorinated biphenyls, *Jpn. J. Cancer Res. [Gann]* **65,** 545–549 (1974).
80. R. D. Kimbrough and R. E. Linder, Induction of adenofibrosis and hepatomas of the liver in BALB/cJ mice by polychlorinated biphenyls (Aroclor 1254). *J. Natl. Cancer Inst.* **53,** 547–552 (1974).
81. H. Nagasaki, S. Tomii, T. Mega, M. Marugami, and N. Ito, Hepatocarcinogenicity of polychlorinated biphenyls in mice, *Jpn. J. Cancer Res. [Gann]* **63,** 805 (1972).

Index